TWO-PHASE FLOW, BOILING, AND CONDENSATION IN CONVENTIONAL AND MINIATURE SYSTEMS

Providing a comprehensive introduction to the fundamentals and applications of flow and heat transfer in conventional and miniature systems, this fully enhanced and updated edition covers all the topics essential for graduate courses on two-phase flow, boiling, and condensation.

Beginning with a concise review of single-phase flow fundamentals and interfacial phenomena, detailed and clear discussion is provided on a range of topics including two-phase hydrodynamics and flow regimes, mathematical modeling of gas–liquid two-phase flows, pool and flow boiling, flow and boiling in mini- and microchannels, external- and internal-flow condensation with and without noncondensables, condensation in small flow passages, and two-phase choked flow.

Numerous solved examples and end-of-chapter problems that include many common design problems likely to be encountered by students make this an essential text for graduate students. With up-to-date detail on the most recent research trends and practical applications, it is also an ideal reference for professionals and researchers in mechanical, nuclear, and chemical engineering.

S. Mostafa Ghiaasiaan is a Professor in the George W. Woodruff School of Mechanical Engineering at Georgia Institute of Technology. Before joining the faculty in 1991, Professor Ghiaasiaan worked in the Aerospace and Nuclear Power industry for eight years, conducting research and development activity on modeling and simulation of transport processes, multiphase flow, and nuclear reactor thermal-hydraulics and safety. Professor Ghiaasiaan has more than 200 publications on transport phenomena and multiphase flow, is a Fellow of the American Society of Mechanical Engineers (ASME), and has been an Executive Editor for the journal *Annals of Nuclear Energy* since 2006. He is also the author of the widely used graduate text *Convective Heat and Mass Transfer* (Cambridge University Press, 2011).

Two-Phase Flow, Boiling, and Condensation

IN CONVENTIONAL AND MINIATURE SYSTEMS

S. Mostafa Ghiaasiaan

Georgia Institute of Technology

CAMBRIDGE
UNIVERSITY PRESS

CAMBRIDGE
UNIVERSITY PRESS

University Printing House, Cambridge CB2 8BS, United Kingdom

One Liberty Plaza, 20th Floor, New York, NY 10006, USA

477 Williamstown Road, Port Melbourne, VIC 3207, Australia

4843/24, 2nd Floor, Ansari Road, Daryaganj, Delhi – 110002, India

79 Anson Road, #06–04/06, Singapore 079906

Cambridge University Press is part of the University of Cambridge.

It furthers the University's mission by disseminating knowledge in the pursuit of education, learning and research at the highest international levels of excellence.

www.cambridge.org
Information on this title: www.cambridge.org/9781107153301

First published 2008
First paperback edition 2014
Second edition 2017

Printed in the United States of America by Sheridan Books, Inc.

A catalog record for this publication is available from the British Library

Library of Congress Cataloging in Publication data
Ghiaasiaan, Seyed Mostafa, 1953–
Two-phase flow: boiling and condensation in conventional and miniature systems /
S. Mostafa Ghiaasiaan.
Includes bibliographical references and index.
ISBN 978-1-107-15330-1 (hardback)
1. Fluid dynamics.   2. Two-phase flow.   I. Title.
TA357.G4625   2007
620.1′064 – dc22   2007016309

ISBN 978-1-107-15330-1 Hardback

Additional resources for this publication at www.cambridge.org/ghiaasiaan

To my wife Pari Fatemeh Shafiei,
and my son Saam

Contents

PART TWO. BOILING AND CONDENSATION

Preface to the Second Edition

Since the publication of the first edition of this book significant advances have been made in the art and science of gas–liquid two-phase flow and phase-change phenomena, in particular with respect to flow and heat transfer in miniature and micro-systems. Furthermore, more emphasis is now placed on flow of pure and mixed refrigerants and cryogens owing to their expanding applications in industry. This edition is meant to reflect these changes, as well as address numerous helpful comments and suggestions that I have received from the users of the first edition.

The objectives, methods of presentation, and overall structure of this edition are the same as those for the first edition. The chapters and their general topics of discussion have thus remained unchanged. However, all chapters have been revised, although to different extents. Two-phase flow, boiling, and condensation in binary fluid mixtures; and flow and heat transfer in helically coiled flow passages, are among the new topics that have been added and discussed in some detail in several chapters due to their growing significance. Chapters 8, 10, and 14 have been revised most extensively in response to the rapidly evolving arena of flow and heat transfer in mini- and microchannels. A large number of new solved examples and end-of-chapter problems have also been added, most of which deal with refrigerants and cryogens.

I am indebted to numerous colleagues and former students for the completion of this book. Most recently, in fall of 2014, I had the pleasure of teaching Transport Phenomena in Multiphase Flow, a graduate-level course out of which this book actually originated. This was the most enjoyable two-phase flow and boiling class I had ever taught. Many of the newly added end-of-chapter problems have been solved and in some cases modified/corrected by the students of that class. I thank them all!

Preface to the First Edition

This book is the outcome of more than fifteen years of teaching graduate courses on nuclear reactor thermal-hydraulics and two-phase flow, boiling, and condensation to mechanical and nuclear engineering students. It is targeted to be the basis of a semester-level graduate course for nuclear, mechanical, and possibly chemical engineering students. It will also be a useful reference for practicing engineers.

The art and science of multiphase flow are indeed vast, and it is virtually impossible to provide a comprehensive coverage of all of their major disciplines in a graduate textbook, even at an introductory level. This textbook is therefore focused on gas–liquid two-phase flow, with and without phase change. Even there, the arena is too vast for comprehensive and in-depth coverage of all major topics, and compromise is needed to limit the number of topics as well as their depth and breadth of coverage. The topics that have been covered in this textbook are meant to familiarize the reader with a reasonably wide range of subjects, including well-established theory and technique, as well as some rapidly developing areas of current interest.

Gas–liquid two-phase flow and flows involving change-of-phase heat transfer apparently did not receive much attention from researchers until around the middle of the twentieth century, and predictive models and correlations prior to that time were primarily empirical. The advent of nuclear reactors around the middle of the twentieth century, and the recognition of the importance of two-phase flow and boiling in relation to the safety of water-cooled reactors, attracted serious attention to the field and led to much innovation, including the practice of first-principle modeling, in which two-phase conservation equations are derived based on first principles and are numerically solved. Today, the area of multiphase flow is undergoing accelerating expansion in a multitude of areas, including direct numerical simulation, flow and transport phenomena at mini- and microscales, and flow and transport phenomena in reacting and biological systems, to name a few. Despite the rapid advances in theory and computation, however, the area of gas–liquid two-phase flow remains highly empirical owing to the extreme complexity of the processes involved.

In this book I have attempted to come up with a balanced coverage of fundamentals, well-established as well as recent empirical methods, and rapidly developing topics. Wherever possible and appropriate, derivations have been presented at least at a heuristic level.

The book is divided into seventeen chapters. The first chapter gives a concise review of the fundamentals of single-phase flow and heat and mass transfer. Chapter 2 discusses two-phase interfacial phenomena. The hydrodynamics and mathematical modeling aspects of gas–liquid two-phase flow are then discussed in

Chapters 3 through 9. Chapter 10 rounds out Part One of the book and is devoted to the hydrodynamic aspects of two-phase flow in mini- and microchannels.

Part Two focuses on boiling and condensation. Chapters 11 through 14 are devoted to boiling. The fundamentals of boiling and pool boiling predictive methods are discussed in Chapter 11, followed by the discussion of flow boiling and critical and postcritical heat flux in Chapters 12 and 13, respectively. Chapter 14 is devoted to the discussion of boiling in mini- and microchannels. External and flow condensation, with and without noncondensables, and condensation in small flow passages are then discussed in Chapters 15 and 16. The last chapter is devoted to two-phase choked flow. Various property tables are provided in several appendices.

Frequently Used Notation

A	Flow area (m^2); atomic number
A_C	Flow area in the vena-contracta location (m^2)
A_d	Frontal area of a dispersed phase particle (m^2)
a	Speed of sound (m/s)
a_I''	Interfacial surface area concentration (surface area per unit mixture volume; m^{-1})
Bd	Bond number $= l^2 / \left(\frac{\sigma}{g\Delta\rho} \right)$
B_h	Mass-flux-based heat transfer driving force
$\tilde{B}_h$	Molar-flux-based heat transfer driving force
B_m	Mass-flux-based mass transfer driving force
$\tilde{B}_m$	Molar-flux-based mass transfer driving force
Bi	Biot number $= hl/k$
Bo	Boiling number $= q_w''/(Gh_{fg})$
C	Concentration (kmol/m^3)
C	Constant in Wallis's flooding correlation; various constants
c	Wave propagation velocity (m/s)
Ca	Capillary number $= \mu_L U/\sigma$
Cr	Crispation number $= \frac{\mu}{\sigma l} \left(\frac{k}{\rho C_P} \right)$
C_2	Constant in Tien–Kutateladze flooding correlation
C_C	Contraction ratio
C_D	Drag coefficient
C_{He}	Henry's coefficient (Pa; bar)
Co	Convection number $= (\rho_g/\rho_f)^{0.5}[(1-x)/x]^{0.8}$
C_P	Constant-pressure specific heat (J/kg·K)
$\tilde{C}_P$	Molar-based constant-pressure specific heat (J/kmol·K)
C_{sf}	Constant in the nucleate pool boiling correlation of Rohsenow
C_v	Constant-volume specific heat (J/kmol·K)
$\tilde{C}_v$	Molar-based constant-volume specific heat (J/kg·K)
C_0	Two-phase distribution coefficient in the drift flux model
D	Tube or jet diameter (m)
D_H	Hydraulic diameter (m)
Dn	Dean number $\left[= \mathrm{Re}_{D_H} (R_i/R_{cl})^{1/2} \right]$
Dn$_{eq}$	Equivalent Dean number
D	Mass diffusivity (m^2/s)
D_{ij}	Binary mass diffusivity for species i and j (m^2/s)
D $_{iG}$, **D**$_{iL}$	Mass diffusivity of species i in gas and liquid phases (m^2/s)

d	Bubble or droplet diameter (m)
d_{cr}	Critical diameter for spherical bubbles (m)
d_{Sm}	Sauter mean diameter of bubbles or droplets (m)
E	Eddy diffusivity (m^2/s)
$\mathbf{E_1}, \mathbf{E}$	One-dimensional and three-dimensional turbulence energy spectrum functions based on wave number (m^3/s^2)
$\mathbf{E_1^*}, \mathbf{E^*}$	One-dimensional and three-dimensional turbulence energy spectrum functions based on frequency (m^2/s)
$\mathbf{E}_B$	Bulk modulus of elasticity (N/m^2)
E_H	Eddy diffusivity for heat transfer (m^2/s)
Eo	Eötvös number $= g\Delta\rho l^2/\sigma$
e	Total specific convected energy (J/kg)
$\mathbf{e}$	Unit vector
F	Degrees of freedom; force (N); Helmholtz free energy (J); correction factor
F^I	Interfacial Helmholtz free energy (J)
F_i	Interfacial force, per unit mixture volume (N/m^3)
Fo	Fourier number $= \left(\dfrac{k}{\rho C_P}\right)\dfrac{t}{l^2}$
Fr	Froude number $= U^2/(gD)$
F_{vm}	Virtual mass force, per unit mixture volume (N/m^3)
F_w	Wall force, per unit mixture volume (N/m^3)
F_{wG}, F_{wL}	Wall force, per unit mixture volume, exerted on the liquid and gas phases (N/m^3)
F_σ	Surface tension force (N)
f	Fanning friction factor; frequency (Hz); distribution function (m^{-1} or m^{-3}); specific Helmholtz free energy (J/kg)
f'	Darcy friction factor
f^I	Specific interfacial Helmholtz free energy (J/m^2)
$\hat{f}$	Fugacity (Pa)
f_{cond}	Condensation efficiency
G	Mass flux (kg/m^2·s); Gibbs free energy (J)
G^I	Interfacial Gibbs free energy (J)
Ga	Galileo number $= \dfrac{\rho_L \Delta\rho g l^3}{\mu_L^2}$
Gr	Grashof number $= \left(\dfrac{gl^3}{v_L^2}\right)\left(\dfrac{\rho_L - \rho_g}{\rho_L}\right)$
Gz	Graetz number $= \dfrac{4Ul^2}{z}\left(\dfrac{\rho C_P}{k}\right)$
$\vec{g}$	Gravitational acceleration vector (m/s^2)
g	Specific Gibbs free energy (J/kg); gravitational constant ($= 9.807$ m/s^2 at sea level); breakup frequency (s^{-1})
g^I	Specific interfacial Gibbs free energy (J/m^2)
H	Heat transfer coefficient (W/m^2·K); height (m); enthalpy (J)
Hn	Helical coil number $\left[= \mathrm{Re}_{D_H}\left(R_i/R_c\right)^{1/2}\right]$
H_r	Radiative heat transfer coefficient (W m^2·K)
He	Henry number
h	Specific enthalpy (J/kg); mixed-cup specific enthalpy (J/kg); collision frequency function (m^3·s)

h_L	Liquid level height in stratified flow regime (m); specific enthalpy of liquid (J/kg)
h_{fg}, h_{sf}, h_{sg}	Latent heats of vaporization, fusion, and sublimation (J/kg)
$\tilde{h}_{fg}, \tilde{h}_{sf}, \tilde{h}_{sg}$	Molar-based latent heats of vaporization, fusion, and sublimation (J/kmol)
I_m	Modified Bessels function of the first kind and mth order
J	Diffusive molar flux (kmol/m^2·s)
Ja	Jakob number $= (\rho C_P)_L \Delta T / \rho_g h_{fg}$ or $C_{PL} \Delta T / h_{fg}$
J^{**}	Flux of a transported property in the generic conservation equations (Chapters 1 and 5)
J^*	Dimensionless superficial velocity in Wallis's flooding correlation
Ja*	Modified Jacob number $= \sqrt{\frac{\rho_L}{\rho_G} \frac{C_{PL} \Delta T}{h_{fg}}}$
j	Diffusive mass flux (kg/m^2·s); molecular flux (m^{-2}·s^{-1}); superficial velocity (m/s)
k	Thermal conductivity (W/m·K); wave number (m^{-1})
K	Loss coefficient; Armand's flow parameter; mass transfer coefficient (kg/m^2·s)
K	Parameter in Katto's DNB correlation (Chapter 13)
$\tilde{K}$	Molar-based mass transfer coefficient (kmol/m^2·s)
K*	Kutateladze number; dimensionless superficial velocity in Tien–Kutateladze flooding correlation
Ka	Kapitza number $= v_L^4 \rho_L^3 g / \sigma^3$
K_{hor}	Correction factor for critical heat flux in horizontal channels
Le	Lewis number $= \alpha / D$
L_B	Boiling length (m); bubble (vapor clot) length (m)
L_{heat}	Heated length (m)
L_{slug}	Liquid slug length (m)
l	Length (m); characteristic length (m)
l_D	Kolmogorov's microscale (m)
l_E	Churn flow entrance length before slug flow is established (m)
l_F	Length scale applied to liquid films (m)
M	Molar mass (kg/kmol); component of the generalized drag force (per unit mixture volume) (N/m^3)
Ma	Marangoni number $= \left(\frac{\partial \sigma}{\partial T} \right) \nabla T \frac{l^2}{\mu} \left(\frac{\rho C_P}{k} \right)$
Mo	Morton number $= g \mu_L^4 \Delta \rho / \left(\rho_L^2 \sigma^3 \right)$
$\vec{M}_{IK}$	Generalized interfacial drag force (N/m^3) exerted on phase k
$\vec{M}_{ID}$	Interfacial drag force term (N/m^3)
$\vec{M}_{IV}$	Virtual mass force term (N/m^3)
M_K	Signal associated with phase k
M_2	Constant in Tien–Kutateladze flooding correlation
m	Mass fraction; mass of a single molecule (kg); dimensionless constant
m	Mass (kg)
m''	Mass flux (kg/m^2·s)
N''	Molar flux (kmol/m^2·s)
$\vec{N}$	Unit normal vector
N_{Av}	Avogadro's number ($= 6.022 \times 10^{26}$ molecules/kmol)
N_{con}	Confinement number $= \sqrt{\sigma / g \Delta \rho} / l$
Nu	Nusselt number Hl/k

N_μ	Viscosity number $= \mu_L/[\rho_L\sigma\sqrt{\sigma/(g\Delta\rho)}]^{1/2}$
n	Number density (m^{-3}); number of chemical species in a mixture; dimensionless constant; polytropic exponent
p	Perimeter (m)
P	Pressure (N/m^2); Legendre polynomial
ΔP_P	Pump (supply) pressure drop (N/m^2)
ΔP_C	Channel (demand) pressure drop (N/m^2)
Pe	Péclet number $= Ul(\rho C_P/k)$
Pr	Prandtl number $= \mu C_P/k$
P_r	Reduced pressure $= P/P_{cr}$
Pr$_{turb}$	Turbulent Prandtl number
p_f	Wetted perimeter (m)
p_{heat}	Heated perimeter (m)
Q	Volumetric flow rate (m^3/s); dimensionless wall heat flux
q'	Heat generation rate per unit length (W/m)
q''	Heat flux (W/m^2)
$\dot{q}_v$	Volumetric energy generation rate (W/m^3)
R	Radius (m); gas constant (N·m/kg·K)
R_c	Radius of curvature (m)
R_C	Wall cavity radius (m)
R_{cl}	Coil radius in helically coiled tube (m)
R_t	Radius of torsion (m)
Re	Reynolds number $(\rho Ul/\mu)$
Re$_F$	Liquid film Reynolds number $= 4\Gamma_F/\mu_L$
R_j	Equilibrium radius of a jet (m)
$\dot{R}_l$	Volumetric generation rate of species l (kmol/m^3·s)
R_u	Universal gas constant ($= 8314$ N·m/kmol·K)
r	Distance between two molecules (Å) (Chapter 1); radial coordinate (m)
$\dot{r}_l$	Volumetric generation rate of species l (kg/m^3·s)
S	Sheltering coefficient; entropy (J/K); source and sink terms in interfacial area transport equations (s^{-1}·m^{-6}); distance defining intermittency (m)
Sc	Schmidt number $= \nu/\mathbf{D}$
Sh	Sherwood number $= Kl/\rho\mathbf{D}$ or $\tilde{K}l/C\mathbf{D}$
So	Soflata number $= [(3\sigma^3)/(\rho^3 g\nu^4)]^{1/5}$
Su	Suratman number $= \rho l\sigma/\mu^2$
S_r	Slip ratio
s	Specific entropy (J/kg·K)
$\vec{T}$	Unit tangent vector
T	Temperature (K)
T_r	Reduced temperature $= T/T_{cr}$
t	Time (s); thickness (m)
t_c	Characteristic time (s)
$t_{c,D}$	Kolmogorov's time scale (s)
t_{gr}	Growth period in bubble ebullition cycle (s)
t_{res}	Residence time (s)
t_{wt}	Waiting period in bubble ebullition cycle (s)

U	Internal energy (J)
U	Velocity (m/s); overall heat transfer coefficient (W/m^2·K)
U_B	Bubble velocity (m/s)
$U_{B,\infty}$	Rise velocity of Taylor bubbles in stagnant liquid (m/s)
U_r	Slip velocity (m/s)
U_τ	Friction velocity (m/s)
u	Specific internal energy (J/kg)
u	Velocity (m/s)
u_D	Kolmogorov's velocity scale (m/s)
V	Volume (m^3)
V_1	Volatility parameter (Section 12.16)
V_d	Volume of an average dispersed phase particle (m^3)
V_{gj}	Gas drift velocity (m/s)
V'_{gj}	Parameter defined as $V_{gj} + (C_0 - 1)\langle j \rangle$ (m/s)
v	Specific volume (m^3/kg)
We	Weber number $= \rho U^2 l/\sigma$
W	Width (m)
w	Interpolation length in some flooding correlations (m)
x	Quality
x_{eq}	Equilibrium quality
X	Mole fraction; liquid-side mole fraction (in gas–liquid two-phase systems); Martinelli's factor
Y	Gas-side mole fraction (in gas–liquid two-phase systems)
Z	Compressibility factor

Greek characters

α	Void fraction; wave growth parameter (s^{-1}); phase index
$\boldsymbol{\alpha}$	Thermal diffusivity (m^2/s)
α_k	In situ volume fraction occupied by phase k
β	Volumetric quality; phase index; parameter defined in Eq. (1.75); coefficient of volumetric thermal expansion (K^{-1}); dimensionless parameter
$\beta(V, V')$	Probability of breakup events of particles with volume V' that result in the generation of a particle with volume V (m^{-1})
β_{ma}	Rate factor for mass transfer
β_{th}	Dimensionless transpiration rate for heat transfer
Δ	Plate thickness (m)
δ	Kronecker delta; gap distance (m); thermal boundary layer thickness (m)
γ	Activity coefficient
δ_F	Film thickness (m)
δ_m	Thickness of the microlayer (m)
ε	Porosity; radiative emissivity; Bowring's pumping factor (Chapter 12); turbulent dissipation rate (W/kg); perturbation
ε_D	Surface roughness (m)
$\tilde{\varepsilon}$	Energy representing maximum attraction between two molecules (J)
Ψ	Parameter in Baker's flow regime map (Chapter 4)

ψ	Cavity side angle (rad or degrees); transported property (Chapters 1 and 5); stream function (m^2/s)
Φ^2	Two-phase multiplier for pressure drop
Φ	Two-phase multiplier for minor pressure drops; dissipation function (s^{-2})
φ	Velocity potential (m^2/s); pair potential energy (J)
φ	Transported property (Chapters 1 and 5); relative humidity
$\hat{\phi}$	Fugacity coefficient
χ	Correction factor in CHF correlations for binary mixtures
Γ	Volumetric phase change rate (per unit mixture volume) (kg/m^3·s); correction factor for the kinetic model for liquid–vapor interfacial mass flux; dimensionless coefficient; surface concentration of surfactants (kmol/m^2)
Γ_F	Film mass flow rate per unit width (kg/m·s)
γ	Specific heat ration (C_P/C_v); perforation ratio
η	Local pressure divided by stagnation pressure
η_c	Convective enhancement factor
η_{ch}	Choking point pressure divided by stagnation pressure
K	Curvature (m^{-1})
κ	von Kármán's constant
κ_B	Boltzmann's constant ($= 1.38 \times 10^{-23}$ J/K)
Π	Interfacial pressure (N/m)
λ	Molecular mean free path (m); wavelength (m); coalescence efficiency; parameter in Baker's flow regime map (Chapter 4)
λ_d	Fastest growing wavelength (m)
λ_H	Critical Rayleigh unstable wavelength (m)
λ_L	Laplace length scale (capillary length) $= \sqrt{\sigma/g\Delta\rho}$
μ	Viscosity (kg/m·s); chemical potential (J/kg)
ν	Kinematic viscosity (m^2/s)
π	Number of phases in a mixture; 3.1416
θ	Azimuthal angle (rad); angle of inclination with respect to the horizontal plane (rad or degrees); contact angle (rad or degrees)
θ'	Angle of inclination with respect to vertical (rad or degrees)
$\theta_0, \theta_a, \theta_r$	Equilibrium (static), advancing, and receding contact angles (rad or degrees)
ρ	Density (kg/m^3)
ρ'	Momentum density (kg/m^3)
σ	Surface tension (N/m)
σ_A, σ_A'	Smaller-to-lager flow area ratios in a flow-area change
$\tilde{\sigma}$	Molecular collision diameter (Å)
σ_A	Molecular scattering cross section (m^2)
σ_c, σ_e	Condensation and evaporation coefficients
T	Torsion (m^{-1})
τ	Molecular mean free time (s); shear stress (N/m^2)
$\bar{\bar{\tau}}$	Viscous stress tensor (N/m^2)
Ω	Azimuthal angle for film flow over horizontal cylinders (rad)
Ω_k, Ω_D	Collision integrals for thermal conductivity and mass diffusivity

ω	Angular frequency (rad/s); humidity ratio; dimensionless parameter (Chapter 17)
ξ	Chemical potential (J/kg); noncondensable volume fraction
ζ	Interphase displacement from equilibrium (m)
$\Im$	Tangential coordinate on the liquid–gas interphase

Superscripts

r	Relative
+	In wall units
•	In the presence of mass transfer
–	Average
–t	Time averaged
–tk	Time averaged for phase k
=	Tensor
*	Dimensionless
~	Molar based; dimensionless

Subscripts

avg	Average
B	Bubble
Bd	Bubble departure
b	Boiling; bulk
bp	Boiling point
c	Continuous phase; curved flow passage
ch	Choked (critical) flow
cond	Condensation
cont	Contraction
cr	Critical
d	Dispersed phase
dp	Dew point
eq	Equilibrium
ev	Evaporation
ex	Expansion
exit	Exit
f	Saturated liquid
f0	All vapor–liquid mixture assumed to be saturated liquid
fr	Frictional
FC	Forced convection
F	Liquid or vapor film
G	Gas phase
g	Saturated vapor; gravitational
g0	All liquid–vapor mixture assumed to be saturated vapor
GI	At interphase on the gas side
G0	All mixture assumed to be gas
h	Homogeneous
heat	Heated

I	Gas–liquid interface; irreversible
ideal	Ideal
in	Inlet
inc	Inception of waviness
L	Liquid phase
L0	All mixture assumed to be liquid
LI	At interphase on the liquid side
m	Mixture, mixture-average
ma	Mass transfer
n	Sparingly soluble (noncondensable) inert species
out	Outlet
R	Reversible
rad	Radiation
ref	Reference
res	Associated with residence time
s	"s" surface (gas-side interphase); isentropic; solid at melting or sublimation temperature; straight flow passage
sat	Saturation
SB	Subcooled boiling
slug	Liquid or gas slug
spin	Spinodal
TB	Transition boiling
TP	Two-phase
th	Thermal
tot	Total
turb	Turbulent
UC	Unit cell
u	"u" surface (liquid-side interphase)
V	Virtual mass force
v	Vapor when it is not at saturation; vapor in a multicomponent mixture; volumetric
W	Water
w	Wall
wG	Wall–gas interface
wL	Wall–liquid interface
z	Local quantity corresponding to location z
0	Equilibrium state

Abbreviations

BWR	Boiling water reactor
CFD	Computational fluid dynamics
CHF	Critical heat flux
DC	Direct-contact
DFM	Drift flux model
DNB	Departure from nucleate boiling
DNBR	Departure from nucleate boiling ratio
HEM	Homogeneous-equilibrium mixture

HM	Homogeneous mixture
MFB	Minimum film boiling
LOCA	Loss of coolant accident
NVG	Net vapor generation
OFI	Onset of flow instability
ONB	Onset of nuclear boiling
OSV	Onset of significant void
PWR	Pressurized water reactor

PART ONE

TWO-PHASE FLOW

1 Thermodynamic and Single-Phase Flow Fundamentals

1.1 States of Matter and Phase Diagrams for Pure Substances

1.1.1 Equilibrium States

Recall from thermodynamics that for a system containing a pure and isotropic substance that is at equilibrium, without any chemical reaction, and not affected by any external force field (also referred to as a P–v–T system), an equation of state of the following form exists:

$$f(P, v, T) = 0. \tag{1.1}$$

This equation, plotted in the appropriate Cartesian coordinate system, leads to a surface similar to Fig. 1.1, the segments of which define the parameter ranges for the solid, liquid, and gas *phases*. The substance can exist in a stable equilibrium state only on points located on this surface. Using the three-dimensional plot is awkward, and we often use the *phase diagrams* that are the projections of the aforementioned surface on P–v (Fig. 1.2) and T–v (Fig. 1.3) planes. Figures 1.2 and 1.3 also show where *vapor* and *gas* occur. The projection of the aforementioned surface on the P–T diagram (Fig. 1.4) indicates that P and T are interdependent when two phases coexist under equilibrium conditions. All three phases can coexist at the *triple point.*

To derive the relation between P and T when two phases coexist at equilibrium, we note that equilibrium between any two phases α and β requires that

$$g_\alpha = g_\beta, \tag{1.2}$$

where $g = u + Pv - Ts$ is the specific Gibbs free energy. For small changes simultaneously in both P and T while the mixture remains at equilibrium, this equation gives

$$dg_\alpha = dg_\beta. \tag{1.3}$$

From the definition of g one can write

$$dg = d\boldsymbol{u} + Pdv + vdP - Tds - sdT. \tag{1.4}$$

However, from the Gibbs' equation (also referred to as the first Tds relation) we have

$$Tds = d\boldsymbol{u} + Pdv. \tag{1.5}$$

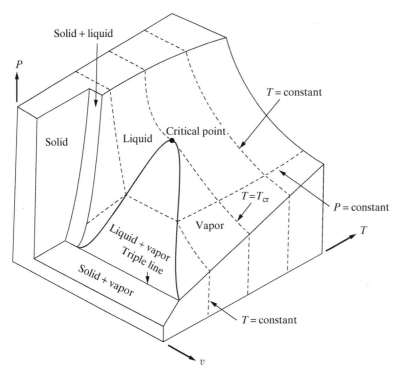

Figure 1.1. The P–v–T surface for a substance that contracts upon freezing.

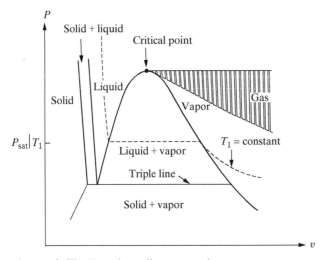

Figure 1.2. The P–v phase diagram.

We can now combine Eqs. (1.3) and (1.4) and write for the two phases

$$dg_\alpha = -s_\alpha dT + v_\alpha dP \tag{1.6}$$

and

$$dg_\beta = -s_\beta dT + v_\beta dP. \tag{1.7}$$

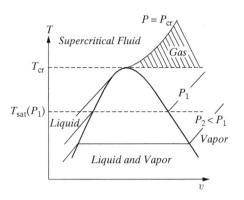

Figure 1.3. The T–v phase diagram.

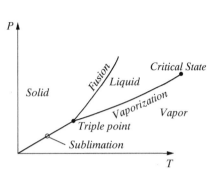

Figure 1.4. The P–T phase diagram.

Substitution from Eqs. (1.6) and (1.7) into Eq. (1.3) gives

$$\frac{dP}{dT} = \frac{s_\beta - s_\alpha}{v_\beta - v_\alpha}.$$ (1.8)

Now, for the reversible process of phase change of a unit mass at constant temperature, one has $q = T(s_\beta - s_\alpha) = (h_\beta - h_\alpha)$, where q is the heat needed for the process. Combining this with Eq. (1.8), the well-known *Clapeyron's relations* are obtained:

evaporation:

$$\frac{dP}{dT} = \left(\frac{dP}{dT}\right)_{\text{sat}} = \frac{h_{\text{fg}}}{T_{\text{sat}}(v_{\text{g}} - v_{\text{f}})},$$ (1.9)

sublimation:

$$\left(\frac{dP}{dT}\right)_{\text{sublim}} = \frac{h_{\text{sg}}}{T_{\text{sublim}}(v_{\text{g}} - v_{\text{s}})},$$ (1.10)

melting:

$$\left(\frac{dP}{dT}\right)_{\text{melt}} = \frac{h_{\text{sf}}}{T_{\text{melt}}(v_{\text{f}} - v_{\text{s}})}.$$ (1.11)

1.1.2 Metastable States

The surface in Fig. 1.1 defines the stable equilibrium conditions for a pure substance. Experience shows, however, that it is possible for a pure and unagitated substance to remain at equilibrium in superheated liquid ($T_{\text{L}} > T_{\text{sat}}$) or subcooled (supercooled) vapor ($T_{\text{G}} < T_{\text{sat}}$) states. Very slight deviations from the stable equilibrium diagrams are in fact common during some phase-change processes. Any significant deviation from the equilibrium states renders the system highly unstable and can lead to rapid and violent phase change in response to a minor agitation.

In the absence of agitation or impurity, spontaneous phase change in a metastable fluid (homogeneous nucleation) must occur because of the random molecular fluctuations. Statistical thermodynamics predicts that in a superheated liquid, for example, pockets of vapor covering a range of sizes are generated continuously while surface tension attempts to bring about their collapse. The probability of the formation of

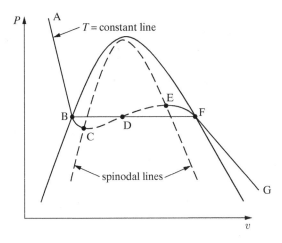

Figure 1.5. Metastable states and the spinodal lines.

vapor embryos increases with increasing temperature and decreases with increasing embryo size. Spontaneous phase change (homogeneous boiling) will occur only when vapor microbubbles that are large enough to resist surface tension and would become energetically more stable upon growth are generated at sufficiently high rates.

One can also argue based on classical thermodynamics that a metastable state is in principle possible only as long as (Lienhard and Karimi, 1981)

$$\left(\frac{\partial v}{\partial P}\right)_T \leq 0. \tag{1.12}$$

This condition implies that fluctuations in pressure are not followed by positive feedback, where a slight increase in pressure would cause volumetric expansion of the fluid, itself causing a further increase in pressure. When the constant-T lines on the P–v diagram are modified to permit unstable states, a figure similar to Fig. 1.5 results. The *spinodal lines* represent the loci of points where Eq. (1.12) with equal sign is satisfied. Lines AB and FG are constant-temperature lines for stable equilibrium states. Line BC represents metastable, superheated liquid. Metastable subcooled vapor occurs on line EF, and line CDE represents impossible (unstable) states.

Using the spinodal line as a criterion for nucleation does not appear to agree well with experimental data for homogeneous boiling. For pure water, the required liquid temperature for spontaneous boiling can be found from the following empirical correlation (Lienhard, 1976):

$$\frac{T_L}{T_{cr}} = 0.905 + 0.095 \left(\frac{T_{sat}}{T_{cr}}\right)^8, \tag{1.13}$$

where

    T_{cr} = critical temperature of water (647.15 K),
    T_L = local liquid temperature (K),
    T_{sat} = $T_{sat}(P_\infty)$ (K), and
    P_∞ = pressure of water.

EXAMPLE 1.1. Calculate the superheat needed for spontaneous boiling in pure atmospheric water.

Table 1.1. *Summary of parameters for Eq. (1.14).*

Conservation/transport law	ψ	ϕ	J^{**}
Mass	1	0	0
Momentum	$\vec{U}$	$\vec{g}$	$P\bar{\bar{I}} - \bar{\bar{\tau}}$
Energy	$\mathbf{u} + \frac{U^2}{2}$	$\vec{g} \cdot \vec{U} + \dot{q}_v/\rho$	$\vec{q}' + (P\bar{\bar{I}} - \bar{\bar{\tau}}) \cdot \vec{U}$
Thermal energy in terms of enthalpy	h	$\frac{1}{\rho}\left(\dot{q}_v + \frac{DP}{Dt} + \bar{\bar{\tau}} : \nabla\vec{U}\right)$	$\vec{q}'$
Thermal energy in terms of internal energy	$\mathbf{u}$	$\frac{1}{\rho}\left(\dot{q}_v + P\nabla \cdot \vec{U} + \bar{\bar{\tau}} : \nabla\vec{U}\right)$	$\vec{q}'$
Species l, mass-flux-based	m_l	$\frac{\dot{r}_l}{\rho}$	$\vec{j}_l$
Species l, molar-flux-based*	X_l	$\frac{\dot{R}_l}{\mathbf{C}}$	$\vec{J}_l$

* In Eq. (1.14), ρ must be replaced with $\mathbf{C}$.

SOLUTION. We have $P_\infty = 1.013$ bar; therefore $T_{\text{sat}} = 373.15$ K. The solution of Eq. (1.13) then leads to $T_L = 586.4$ K. The superheat needed is thus $T_L - T_{\text{sat}} = 213.3$ K.

Example 1.1 shows that extremely large superheats are needed for homogeneous nucleation to occur in pure and unagitated water. The same is true for other common liquids. Much lower superheats are typically needed in practice, owing to heterogeneous nucleation. Subcooled (supercooled) vapors in particular undergo fast nucleation (fogging) with a supersaturation (defined as $\frac{P_{\text{sat}}(T_G)-P}{P_{\text{sat}}(T_G)}$) of 1 % or so (Friedlander, 2000).

1.2 Transport Equations and Closure Relations

The local instantaneous conservation equations for a fluid can be presented in the following shorthand form (Delhaye, 1969):

$$\frac{\partial \rho\psi}{\partial t} + \nabla \cdot (\vec{U}\rho\psi) = -\nabla \cdot J^{**} + \rho\varphi, \tag{1.14}$$

where ρ is the fluid density, U is the local instantaneous velocity, f represents the transported property, y is the source term for f, and J^{**} is the flux of f. Table 1.1 summarizes the definitions of these parameters for various conservation laws. All these parameters represent the mass-averaged mixture properties when the fluid is multicomponent. Other parameters used in the table are defined as follows:

$\vec{g}$ = acceleration due to all external body forces,
$\dot{q}_v$ = volumetric energy generation rate,
$\dot{r}_l$ = mass generation rate of species l in unit volume,
$\dot{R}_l$ = mole generation rate of species l in unit volume,
$\vec{q}''$ = heat flux,
$\mathbf{u}$ = specific internal energy,
h = specific enthalpy, and
m_l, X_l = mass and mole fractions of species l.

Angular momentum conservation only requires that the tensor $(P\bar{\bar{I}} - \bar{\bar{\tau}})$ remain unchanged when it is transposed. The thermal energy equation, represented by either the fourth or fifth row in Table 1.1, is derived simply by first applying $\vec{U}$ (i.e., the dot product of the velocity vector) on both sides of the momentum conservation equation, and then subtracting the resulting equation from the energy conservation equation represented by the third row of Table 1.1. The energy conservation represented by the third row and the thermal energy equation are thus not independent from one another.

The equation set that is obtained by substituting from Table 1.1 into Eq. (1.14) of course contains too many unknowns and is not solvable without *closure relations*. The closure relations for single-phase fluids are either *constitutive relations*, meaning that they deal with constitutive laws such as the equation of state and thermophysical properties, or *transfer relations*, meaning that they represent some transfer rate law. The most obvious constitutive relations are, for a pure substance,

$$\rho = \rho(\mathbf{u}, P) \tag{1.15}$$

or

$$\rho = \rho(h, P). \tag{1.16}$$

For a multicomponent mixture these equations should be recast as

$$\rho = \rho(\mathbf{u}, P, m_1, m_2, \ldots, m_{n-1}) \tag{1.17}$$

or

$$\rho = \rho(h, P, m_1, m_2, \ldots, m_{n-1}), \tag{1.18}$$

where n is the total number of species. For a single-phase fluid, the constitutive relations providing for fluid temperature can be

$$T = T(\mathbf{u}, P) \tag{1.19}$$

or

$$T = T(h, P); \tag{1.20}$$

for a multicomponent mixture,

$$T = T(\mathbf{u}, P, m_1, m_2, \ldots, m_{n-1}) \tag{1.21}$$

or

$$T = T(h, P, m_1, m_2, \ldots, m_{n-1}). \tag{1.22}$$

In Eqs. (1.17) through (1.22), the mass fractions $m_1, m_2, \ldots, m_{n-1}$ can be replaced with mole fractions $X_1, X_2, \ldots, X_{n-1}$.

Let us assume that the fluid is Newtonian, and it obeys Fourier's law for heat diffusion and Fick's law for the diffusion of mass. The transfer relations for the fluid will then be

$$\bar{\bar{\tau}} = \tau_{ij}\mathbf{e}_i\mathbf{e}_j; \; \tau_{i,j} = \mu\left(\frac{\partial u_i}{\partial x_j} + \frac{\partial u_j}{\partial x_i}\right) - \frac{2}{3}\mu\delta_{ij}\nabla \cdot \vec{U}, \tag{1.23}$$

$$\vec{q}'' = -k\nabla\mathrm{T} + \sum_l \vec{j}_l h_l, \tag{1.24}$$

$$\vec{j}_l = -\rho \boldsymbol{D}_{lm}\nabla m_l, \tag{1.25}$$

$$\vec{J}_l = -\boldsymbol{CD}_{lm}\nabla X_l, \tag{1.26}$$

where $\mathbf{e}_i$ and $\mathbf{e}_j$ are unit vectors for i and j coordinates, respectively, and $\boldsymbol{D}_{lm}$ represents the *mass diffusivity* of species l with respect to the mixture.

Mass-flux- and molar-flux-based diffusion will be briefly discussed in the next section. The second term on the right side of Eq. (1.24) accounts for energy transport from the diffusion of all of the species in the mixture. For a binary mixture, one can use subscripts 1 and 2 for the two species, and the mass diffusivity will be $\boldsymbol{D}_{12}$. The diffusive mass transfer is typically a slow process in comparison with the diffusion of heat, and certainly in comparison with even relatively slow convective transport rates. As a result, in most nonreacting flows the second term on the right side of Eq. (1.24) is negligibly small.

The last two rows of Table 1.1 are equivalent and represent the transport of species l. The difference between them is that the sixth row is in terms of mass flux and its rate equation is Eq. (1.25), whereas the last row is in terms of molar flux and its rate equation is Eq. (1.26). A brief discussion of the relationships among mass-fraction-based and mole-fraction-based parameters will be given in the next section. A detailed and precise discussion can be found in Mills (2001). The choice between the two formulations is primarily a matter of convenience. The precise definition of the average mixture velocity in the mass-flux-based formulation is consistent with the way the mixture momentum conservation is formulated, however. The mass-flux-based formulation is therefore more convenient for problems where the momentum conservation equation is also solved. However, when constant-pressure or constant-temperature processes are dealt with, the molar-flux-based formulation is more convenient.

In this formulation, and everywhere in this book, we consider only one type of mass diffusion, namely the *ordinary diffusion* that is caused by a concentration gradient. We do this because in problems of interest to us concentration gradient-induced diffusion overwhelms other types of diffusion. Strictly speaking, however, diffusion of a species in a mixture can be caused by the cumulative effects of at least four different mechanisms, whereby (Bird *et al.*, 2002)

$$\vec{j}_l = \vec{j}_{l,d} + \vec{j}_{l,p} + \vec{j}_{l,g} + \vec{j}_{l,T}. \tag{1.27}$$

The first term on the right side is the concentration gradient-induced diffusion flux, the second term is caused by the pressure gradient in the flow field, the third term is caused by the external body forces that may act unequally on various chemical species, and the last term represents the diffusion caused by a temperature gradient, also called the *Soret effect*. A useful discussion of these diffusion terms and their rate laws can be found in Bird *et al.* (2002).

The conservation equations for a Newtonian fluid, after implementing these transfer rate laws in them, can be written as follows.

Mass conservation:

$$\frac{\partial \rho}{\partial t} + \nabla \cdot (\rho \vec{U}) = 0 \tag{1.28}$$

or

$$\frac{D\rho}{Dt} + \rho \nabla \cdot \vec{U} = 0. \tag{1.29}$$

Momentum conservation, when the fluid is incompressible and viscosity is constant:

$$\rho \frac{D\vec{U}}{Dt} = \frac{D(\rho\vec{U})}{Dt} = -\nabla P + \rho\vec{g} + \mu\nabla^2\vec{U}. \tag{1.30}$$

Thermal energy equation for a pure substance, in terms of specific internal energy:

$$\rho \frac{D\mathbf{u}}{Dt} = \nabla \cdot k\nabla T - P\nabla \cdot \vec{U} + \mu\Phi. \tag{1.31}$$

Thermal energy equation for a pure substance, in terms of specific enthalpy:

$$\rho \frac{Dh}{Dt} = \nabla \cdot (k\nabla T) + \frac{DP}{Dt} + \mu\Phi, \tag{1.32}$$

where the parameter Φ is the dissipation function (and where $\mu\Phi$ represents the viscous dissipation per unit volume). For a multicomponent mixture, the energy transport caused by diffusion is sometimes significant and needs to be accounted for in the mixture energy conservation. In terms of specific enthalpy, the thermal energy equation can be written as

$$\rho \frac{Dh}{Dt} = \nabla \cdot (k\nabla T) + \frac{DP}{Dt} + \mu\Phi - \nabla \cdot \sum_{l=1}^{n} \vec{j_l}h_l. \tag{1.33}$$

Chemical species mass conservation, in terms of partial density and mass flux:

$$\frac{\partial \rho_l}{\partial t} + \nabla \cdot (\rho_l\vec{U}) = \nabla \cdot (\rho\mathbf{D}_{12}\nabla m_l) + \dot{r}_l. \tag{1.34}$$

Chemical species mass conservation in terms of mass fraction and mass flux:

$$\rho \left[\frac{\partial m_l}{\partial t} + \nabla \cdot (m_l\vec{U}) \right] = \nabla \cdot (\rho\mathbf{D}_{12}\nabla m_l) + \dot{r}_l. \tag{1.35}$$

Chemical species mass conservation, in terms of concentration and molar flux:

$$\frac{\partial \mathbf{C}_l}{\partial t} + \nabla \cdot (\mathbf{C}_l\vec{U}) = \nabla \cdot (\mathbf{C}\mathbf{D}_{12}\nabla X_l) + \dot{R}_l. \tag{1.36}$$

Chemical species mass conservation, in terms of mole fraction and molar flux:

$$\mathbf{C} \left[\frac{\partial X_l}{\partial t} + \vec{U} \cdot \nabla X_l \right] = \nabla \cdot (\mathbf{C}\mathbf{D}_{12}\nabla X_l) + \dot{R}_l - X_l \sum_{j=1}^{n} \dot{R}_j. \tag{1.37}$$

1.3 Single-Phase Multicomponent Mixtures

By mixture in this chapter we mean a mixture of two or more chemical species in the same phase. Ordinary dry air, for example, is a mixture of O_2, N_2, and several noble gases in small concentrations. Water vapor and CO_2 are also present in air most of the time.

The *partial density* of species l, ρ_l, is simply the in situ mass of that species in a unit mixture volume. The mixture density ρ is related to the partial densities according to

$$\rho = \sum_{l=1}^{n} \rho_l, \tag{1.38}$$

with the summation here and elsewhere performed on all the chemical species in the mixture. The *mass fraction* of species l is defined as

$$m_l = \frac{\rho_l}{\rho}. \tag{1.39}$$

The *molar concentration* of chemical species i, ρ_l, is defined as the number of moles of that species in a unit mixture volume. The forthcoming definitions for the mixture molar concentration and the *mole fraction* of species l will then follow:

$$\mathbf{C} = \sum_{l=1}^{n} \mathbf{C}_l \tag{1.40}$$

and

$$X_l = \frac{\mathbf{C}_l}{\mathbf{C}}. \tag{1.41}$$

Clearly,

$$\sum_{l=1}^{n} m_l = \sum_{l=1}^{n} X_l = 1. \tag{1.42}$$

The following relations among mass-fraction-based and mole-fraction-based parameters can be easily shown:

$$\rho_l = M_l \mathbf{C}_l, \tag{1.43}$$

$$m_l = \frac{X_l M_l}{\sum_{j=1}^{n} X_j M_j} = \frac{X_l M_l}{M}, \tag{1.44}$$

$$X_l = \frac{m_l/M_l}{\sum_{j=1}^{n} \frac{m_j}{M_j}} = \frac{m_l M}{M_l}, \tag{1.45}$$

where M and M_i represent the mixture and chemical specific i molar masses, respectively, with M defined according to

$$M = \sum_{j=1}^{n} X_j M_j \tag{1.46a}$$

or

$$\frac{1}{M} = \sum_{j=1}^{n} \frac{m_j}{M_j}. \tag{1.46b}$$

When one component, say component j, constitutes the bulk of a mixture, then

$$M \approx M_j \tag{1.47}$$

and

$$m_l \approx \frac{X_l}{M_l} M_j. \tag{1.48}$$

In a gas mixture, *Dalton's law* requires that

$$P = \sum_{l=1}^{n} P_l, \tag{1.49}$$

where P_l is the *partial pressure* of species l. In a gas mixture the components of the mixture are at thermal equilibrium (the same temperature) at any location and any time and conform to the forthcoming constitutive relation:

$$\rho_l = \rho_l(P_l, T), \tag{1.50}$$

where T is the mixture temperature and P_l is the partial pressure of species l. Some or all of the components may be assumed ideal gases, in which case for the ideal gas component j, one has

$$\rho_j = \frac{P_j}{\frac{R_u}{M_j} T}, \tag{1.51}$$

where R_u is the universal gas constant. When all the components of a gas mixture are ideal gases, then

$$X_l = P_l/P. \tag{1.52}$$

EXAMPLE 1.2. The atmosphere of a laboratory during an experiment is at $T = 25\,°C$ and $P = 1.013$ bar. Measurement shows that the relative humidity in the lab is 77%. Calculate the air and water partial densities, mass fractions, and mole fractions.

SOLUTION. Let us start from the definition of relative humidity, φ:

$$\varphi = P_v/P_{sat}(T).$$

Thus,

$$P_v = (0.77)(3.14 \text{ kPa}) = 2.42 \text{ kPa}.$$

The partial density of air can be calculated by assuming air is an ideal gas at $25\,°C$ and pressure of $P_a = P - P_v = 98.91$ kPa to be $\rho_a = 1.156$ kg/m^3.

The water vapor is at $25\,°C$ and 2.42 kPa and is therefore superheated. Its density can be found from steam property tables to be $\rho_v = 0.0176$ kg/m^3. Using Eqs. (1.38) and (1.39), one gets $m_v = 0.015$. Equation (1.45) gives $X_v = 0.0183$.

EXAMPLE 1.3. A sample of pure water is brought into equilibrium with a large mixture of O_2 and N_2 gases at 1 bar pressure and 300 K temperature. The volume fractions of O_2 and N_2 in the gas mixture before it was brought into contact with the water sample were 22% and 78%, respectively. Solubility data indicate that the mole fractions of O_2 and N_2 in water for the given conditions are approximately 5.58×10^{-6} and 9.9×10^{-6}, respectively. Find the mass fractions of O_2 and N_2 in both liquid and gas phases. Also, calculate the molar concentrations of all the involved species in the liquid phase.

SOLUTION. Before the $O_2 + N_2$ mixture is brought in contact with water, we have

$$P_{O_2,\text{initial}}/P_{\text{tot}} = X_{O_2,\text{G,initial}} = 0.22,$$
$$P_{N_2,\text{initial}}/P_{\text{tot}} = X_{N_2,\text{G,initial}} = 0.78,$$

where $P_{\text{tot}} = 1$ bar. The gas phase after it reaches equilibrium with water will be a mixture of O_2, N_2, and water vapor. Since the original gas mixture volume was large, and given that the solubilities of oxygen and nitrogen in water are very low, we can write for the equilibrium conditions

$$P_{O_2,\text{final}}/(P_{\text{tot}} - P_v) = X_{O_2,\text{G,initial}} = 0.22, \qquad (\text{a-1})$$
$$P_{N_2,\text{final}}/(P_{\text{tot}} - P_v) = X_{N_2,\text{G,initial}} = 0.78. \qquad (\text{a-2})$$

Now, under equilibrium, we have

$$X_{O_2,\text{G,final}} \approx P_{O_2,\text{final}}/P_{\text{tot}}, \qquad (\text{b-1})$$
$$X_{N_2,\text{G,final}} \approx P_{N_2,\text{final}}/P_{\text{tot}}. \qquad (\text{b-2})$$

We have used the approximately equal signs in these equations because it was assumed that water vapor acts as an ideal gas. The vapor partial pressure will be equal to the vapor saturation pressure at 300 K, namely, $P_v = 0.0354$ bar. Equations (a-1) and (a-2) can then be solved to get $P_{O_2,\text{final}} = 0.2122$ bar and $P_{N_2,\text{final}} = 0.7524$ bar. Equations (b-1) and (b-2) then give $X_{O_2,\text{G,final}} \approx 0.2122$ and $X_{N_2,\text{G,final}} \approx 0.7524$, and the mole fraction of water vapor will be

$$X_{\text{G,V}} = 1 - (X_{O_2,\text{G,final}} + X_{N_2,\text{G,final}}) \approx 0.0354.$$

To find the gas-side mass fractions, first apply Eq. (1.46a), and then Eq. (1.44):

$$M_G = 0.2122 \times 32 + 0.7524 \times 28 + 0.0354 \times 18 \Rightarrow M_G = 28.49,$$

$$m_{O_2,\text{G,final}} = \frac{X_{O_2,\text{G,final}} M_{O_2}}{M_G} = \frac{(0.2122)(32)}{28.49} \approx 0.238,$$

$$m_{N_2,\text{G,final}} = \frac{(0.7524)(28)}{28.49} \approx 0.739.$$

For the liquid side, first get M_L, the mixture molecular mass number from Eq. (1.46a):

$$M_L = 5.58 \times 10^{-6} \times 32 + 9.9 \times 10^{-6} \times 28$$
$$+ [1 - (5.58 \times 10^{-6} + 9.9 \times 10^{-6})] \times 18 \approx 18.$$

Therefore, from Eq. (1.44),

$$m_{O_2,\text{L,final}} = \frac{5.58 \times 10^{-6}}{18}(32) = 9.92 \times 10^{-6},$$

$$m_{N_2,\text{L,final}} = \frac{9.9 \times 10^{-6}}{18}(28) = 15.4 \times 10^{-6}.$$

To calculate the concentrations, we note that the liquid side is now made up of three species, all with unknown concentrations. Equation (1.41) should be written out for every species, while Eq. (1.40) is also satisfied. These give four equations in terms of the four unknowns C_L, $C_{O_2,L,final}$, $C_{N_2,L,final}$, and $C_{L,W}$, where C_L and $C_{L,W}$ stand for the total molar concentrations of the liquid mixture and the molar concentration of water substance, respectively. This calculation, however, will clearly show that, owing to the very small mole fractions (and hence small concentrations) of O_2 and N_2,

$$C_L \approx C_{L,W} = \rho_L/M_L = \frac{996.6 \text{ kg/m}^3}{18 \text{ kg/kmol}} = 55.36 \text{ kmol/m}^3.$$

The concentrations of O_2 and N_2 could therefore be found from Eq. (1.41) to be

$$C_{O_2,L,final} \approx 3.09 \times 10^{-4} \text{ kmol/m}^3,$$
$$C_{N_2,L,final} \approx 5.48 \times 10^{-4} \text{ kmol/m}^3.$$

The extensive thermodynamic properties of an ideal single-phase mixture, except for entropy, when represented as *per unit mass* (in which case they actually become intensive properties) can all be calculated from

$$\xi = \frac{1}{\rho} \sum_{l=1}^{n} \rho_l \xi_l = \sum_{l=1}^{n} m_l \xi_l, \tag{1.53}$$

with

$$\xi_l = \xi_l(P_l, T), \tag{1.54}$$

where ξ can be any mixture property such as $\rho, \mathbf{u}, h,$ or s, and ξ_l is the same property for pure substance l. Similarly, the following expression can be used when specific properties (except entropy) are defined *per unit mole* for an ideal mixture

$$\tilde{\xi} = \frac{1}{C} \sum_{l=1}^{n} C_l \tilde{\xi}_l = \sum_{l=1}^{n} X_l \xi_l. \tag{1.55}$$

More details about multicomponent mixtures will be provided in Section 2.8.

With respect to diffusion, Fick's law for a binary mixture can be formulated as follows. First, consider the mass-flux-based formulation. The total mass flux of species l is

$$\vec{m}_l'' = \rho_l \vec{U} + \vec{j}_l = m_l(\rho \vec{U}) + \vec{j}_l, \tag{1.56}$$

where Fick's law for the diffusive mass flux is represented by Eq. (1.25), and the mixture velocity is defined as

$$\vec{U} = \sum_{i=1}^{I} m_l \vec{U}_l = \vec{G}/\rho. \tag{1.57}$$

Consider now the molar-flux-based formulation. The total molar flux of species l can be written as

$$\vec{N}_l'' = C_l \vec{\tilde{U}} + \vec{j}_l. \tag{1.58}$$

Fick's law is represented by Eq. (1.26), and the molar-average mixture velocity is defined as

$$\vec{U} = \sum_{l=1}^{I} X_l \vec{U}_l = \vec{G}/(\mathbf{C}\mathbf{M}). \qquad (1.59)$$

1.4 Phase Diagrams for Binary Systems

The phase diagrams discussed in Section 1.1 dealt with systems containing a single chemical species. In some applications, however, we deal with phase-change phenomena of mixtures of two or more chemical species. Examples include air liquefaction and separation and refrigerant mixtures such as water–ammonia and R-410A.

For a nonreacting P–v–T system composed of n chemical species, Gibbs' phase rule states that

$$F = 2 + n - \pi, \qquad (1.60)$$

where π is the number of phases and F is the number of degrees of freedom. For a single-phase binary mixture, $n = 2, \pi = 1$, and therefore $F = 3$, meaning that the number of independent and intensive thermodynamic properties needed for specifying the state of the system is three. All the equilibrium states of the system can then be represented in a three-dimensional coordinate system with P, T, and composition. We can use the mole fraction of one of the species (e.g., X_1) to specify the composition, in which case (P, T, X_1) will be the coordinate system. When two phases are considered in the binary system (say, liquid and vapor), then $n = 2, \pi = 2$, and therefore $F = 2$. The number of independent and intensive thermodynamic properties needed for specifying the state of the system will then be two, meaning that only two of the three coordinates in the (P, T, X_1) space can be independent. The two-phase equilibrium state will then form a two-dimensional surface in the (P, T, X_1) space. When all three phases at equilibrium are considered, $F = 1$, and the equilibrium states will be represented by a space curve.

Let us now focus on the equilibrium vapor–liquid system, and define species 1 as the more volatile species, i.e., the species with a lower boiling temperature, in the mixture. We are interested in the two-dimensional surface in the (P, T, X_1) space representing this equilibrium. For convenience, for binary two-phase mixtures we will use X and Y to represent the mole fractions in the liquid and vapor phases, respectively. Thus, X_1 will be the same as $X_{1,L}$, and Y_1 will be the same as $X_{1,G}$. Rather than working with the three-dimensional space, it is easier to work with the projection of the two-dimensional surface on (T, X_1) or (P, X_1) planes, and this leads to the "TX" and "PX" diagrams, displayed qualitatively in Figs. 1.6 and 1.7, respectively, for a *zeotropic* (also referred to as *non-azeotropic*) mixture. A binary mixture is called *zeotropic* when the concentration makeup of the liquid and vapor phases are never equal. A mixture of water and ammonia is a good example of a zeotropic binary system.

The behavior of a zeotropic binary system during evaporation can be better understood by following what happens to a mixture that is initially at state Z (subcooled liquid) that is heated at constant pressure. The process is displayed in Fig. 1.6. The mixture remains at the original concentration represented by $X_{1,m}$, the

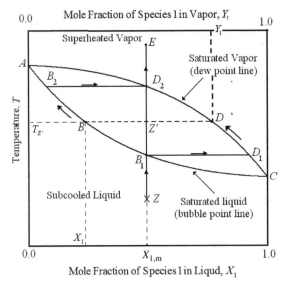

Figure 1.6. Constant-pressure phase diagram for a zeotropic (non-azeotropic) binary mixture.

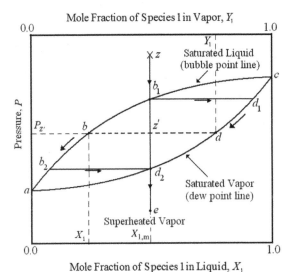

Figure 1.7. Constant-temperature phase diagram for a zeotropic (non-azeotropic) binary mixture.

mixture-average mole fraction of species 1, as long as it is in the subcooled liquid state, until it reaches the state B_1. With further heating of the mixture, the liquid and vapor phases will have different concentrations. The concentration of the liquid phase moves along the $B_1 B_2$ curve, whereas the concentration of the vapor phase follows the $D_1 D_2$ curve. When the mixture temperature is at $T_{Z'}$, for example, the mole fraction of species 1 is equal to Y_1 in the vapour phase, and it is equal to X_1 in the liquid phase. When evaporation is complete, the liquid will have the state corresponding to point B_2, and the vapor phase will correspond to point D_2. The line ABC is often referred to as the *bubble point line*, and the line ADC is called the *dew point line*. For refrigerants, the difference between the dew and bubble temperatures is called the *temperature glide*.

In Fig. 1.7, a process is displayed where an initially subcooled mixture with conditions corresponding to the point z is slowly depressurized while its temperature is maintained constant. Here as well, the concentration remains unchanged until point

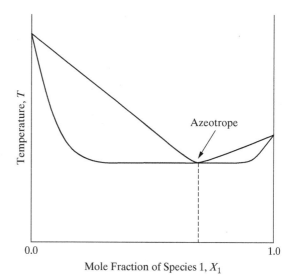

Figure 1.8. Constant-pressure phase diagram for a binary mixture that forms a single azeotrope.

b_1 is reached. With further depressurization the liquid phase will move on the $b_1 b_2$ curve, while the vapor moves along the $d_1 d_2$ curve. Complete evaporation of the mixture ends at point d_2, where the mole fraction of species 1 will remain constant with further depressurization.

For cooling and condensation of a binary system, the processes are similar to those displayed in Figs. 1.6 and 1.7, only in reverse. The straight lines such as BD in Fig. 1.6 and bd in Fig. 1.7 are referred to as *tie lines*. Tie lines have a useful geometric interpretation. It can be proved that

$$\frac{N_f}{N_g} = \frac{\overline{Z'D}}{\overline{Z'B}} = \frac{\overline{z'd}}{\overline{z'b}}, \tag{1.61}$$

where N_f and N_g are the total numbers of liquid and vapor moles in the mixture.

An *azeotrope* is a point at which the concentrations of the liquid and the vapor phases are identical. Some binary mixtures form one or more azeotropes at intermediate concentrations. A single azeotrope is more common and leads to TX and PX diagrams similar to Figs. 1.8 and 1.9. A mixture that is at an azeotrope behaves like a saturated single-component species and has no temperature glide. Azeotropic mixtures suitable for use as refrigerants are uncommon, however, because it is difficult to find one that satisfies other necessary properties for application as a refrigerant.

A mixture is called *near azeotropic* if during evaporation or condensation the liquid and vapor concentrations differ only slightly. In other words, the temperature glide during phase-change processes is very small for near-azeotropic mixtures. A good example is the refrigerant R-410A, which is a fifty–fifty per cent mass mixture of refrigerants R-32 and R-125, and its temperature glide for standard compressor pressures and temperatures is less than about 0.1 °C.

1.5 Thermodynamic Properties of Vapor–Noncondensable Gas Mixtures

Vapor–noncondensable mixtures are often encountered in evaporation and condensation systems. Properties of vapor–noncondensable mixtures are discussed in this

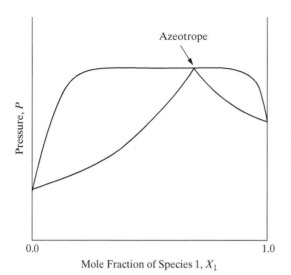

Figure 1.9. Constant-temperature phase diagram for a binary mixture that forms a single azeotrope.

section by treating the noncondensable as a single species. Although the noncondensable may be composed of a number of different gaseous constituents, average properties can be defined such that the noncondensables can be treated as a single species, as is commonly done for air. Subscripts v and n in the following discussion will represent the vapor and noncondensable species, respectively.

Air–water vapor mixture properties are discussed in standard thermodynamic textbooks. For a mixture with pressure P_G, temperature T_G, and vapor mass fraction m_v, the relative humidity φ and humidity ratio ω are defined as

$$\varphi = \frac{P_v}{P_{\text{sat}}(T_G)} \approx \frac{X_v}{X_{v,\text{sat}}} \tag{1.62}$$

and

$$\omega = \frac{m_v}{m_n} = \frac{m_v}{1 - m_v}, \tag{1.63}$$

where $X_{v,\text{sat}}$ is the vapor mole fraction when the mixture is saturated. In the last part of Eq. (1.62) it is evidently assumed that the noncondensable as well as the vapor are ideal gases. A mixture is saturated when $P_v = P_{\text{sat}}(T_G)$. When $\varphi < 1$, the vapor is in a superheated state, because $P_v < P_{\text{sat}}(T_G)$. In this case the thermodynamic properties and their derivatives follow the gas mixture rules.

EXAMPLE 1.4. Find $(\partial h_G / \partial P_G)_{T_G, m_v}$ for a binary vapor–noncondensable mixture assuming that the mixture does not reach saturation.

SOLUTION. The mixture specific enthalpy is defined according to Eq. (1.53):

$$h_G = m_v h_v + (1 - m_v) h_n.$$

From Eq. (1.60), the number of degrees of freedom for the system is three; therefore the three properties P_G, T_G, and m_v uniquely specify the state of the mixture. With T_G and m_v kept constant, one can write

$$\left(\frac{\partial h_G}{\partial P_G} \right)_{T_G, m_v} = m_v \left\{ \left(\frac{\partial h_v}{\partial P_v} \right) \left(\frac{\partial P_v}{\partial P_G} \right) \right\}_{T_G} + (1 - m_v) \left\{ \left(\frac{\partial h_n}{\partial P_n} \right) \left(\frac{\partial P_n}{\partial P_G} \right) \right\}_{T_G}. \tag{a}$$

Since $P_v \approx P_G X_v$ and $P_n \approx P_G(1 - X_v)$, and using the relation between m_v and X_v, one then has

$$\left(\frac{\partial P_v}{\partial P_G}\right)_{m_v} = X_v = \frac{m_v/M_v}{\frac{m_v}{M_v} + \frac{1-m_v}{M_n}},$$

$$\left(\frac{\partial P_n}{\partial P_G}\right)_{m_v} = (1 - X_v) = \frac{(1 - m_v)/M_n}{\frac{m_v}{M_v} + \frac{1-m_v}{M_n}}.$$

The specific enthalpy of an ideal gas is a function of temperature only. The noncondensable is assumed to be an ideal gas, therefore the second term on the right of Eq. (a) will be zero. The term $(\partial h_v/\partial P_v)_{T_G}$ on the right side of Eq. (a) can be calculated using vapor property tables.

The vapor–noncondensable mixtures encountered in evaporators and condensers are often saturated. For saturated mixtures, the following must be added to the other mixture rules:

$$T_G = T_{sat}(P_v), \tag{1.64}$$

$$\rho_v = \rho_g(T_G) = \rho_g(P_v), \tag{1.65}$$

$$h_v = h_g(T_G) = h_g(P_v). \tag{1.66}$$

Using the identity $m_v = \frac{\rho_v}{\rho_n+\rho_v}$ and assuming that the noncondensable is an ideal gas, one can show that

$$\frac{P_G - P_v}{\frac{R_u}{M_n}T_{sat}(P_v)}(1 - m_n) - \rho_g(P_v)m_n = 0. \tag{1.67}$$

Equation (1.67) indicates that P_G, T_G, and m_v are not independent. This is of course expected, because now the mixture has only two degrees of freedom. By knowing two parameters (e.g., T_G and m_v), Eq. (1.67) can be iteratively solved for the third unknown parameter (e.g., the vapor partial pressure when T_G and m_v are known). The variations of the mixture temperature and the vapor pressure are related by the Clapeyron relation, Eq. (1.9):

$$\frac{\partial T_G}{\partial P_v} = \frac{\partial T_{sat}(P_v)}{\partial P_v} = \frac{T_G v_{fg}}{h_{fg}}. \tag{1.68}$$

EXAMPLE 1.5. For a saturated vapor–noncondensable binary mixture, derive expressions of the forms

$$\left(\frac{\partial \rho_G}{\partial P_G}\right)_{X_n} = f(P_G, X_n)$$

and

$$\left(\frac{\partial \rho_G}{\partial X_n}\right)_{P_G} = f(P_G, X_n).$$

SOLUTION. Let us approximately write

$$\rho_G = \frac{P_G}{\frac{R_u}{M}T_G} = \frac{MP_G}{R_u T_{sat}(P_v)},$$

where $M = X_n M_n + (1 - X_n)M_v, T_G = T_{sat}(P_v),$ and $P_v = (1 - X_n)P_G.$ The argument of $T_{sat}(P_v)$ is meant to remind us that T_{sat} corresponds to $P_v = P_G(1 - X_n).$ Then

$$\left(\frac{\partial \rho_G}{\partial P_G}\right)_{X_n} = \frac{M}{R_u T_{sat}} - \frac{P_G M}{R_u T^2_{sat}}\left(\frac{\partial T_{sat}}{\partial P_G}\right).$$

Also, using the Clapeyron relation

$$\frac{\partial T_{sat}}{\partial P_G} = \frac{\partial T_{sat}}{\partial P_v}\frac{\partial P_v}{\partial P_G} = \frac{v_{fg} T_{sat}}{h_{fg}}(1 - X_n)$$

gives the result

$$\left(\frac{\partial \rho_G}{\partial P_G}\right)_{X_n} = \frac{M}{R_u T_{sat}} - \frac{P_v v_{fg} M}{R_u T_G h_{fg}}.$$

It can also be proved that

$$\left(\frac{\partial \rho_G}{\partial X_n}\right)_{P_G} = \frac{P_G}{R_u T_G}(M_n - M_v) + \frac{P_G^2 v_{fg} M}{R_u T_G h_{fg}}, \tag{1.69}$$

where v_{fg} and h_{fg} correspond to $T_{sat} = T_G.$

EXAMPLE 1.6. For a saturated vapor–noncondensable mixture, derive an expression of the form

$$\left(\frac{\partial h_G}{\partial m_n}\right)_{P_G} = f(P_G, m_n).$$

SOLUTION. Let us start with

$$h_G = (1 - m_n)h_g + m_n h_n, \tag{a}$$

where h_g is the saturated vapor enthalpy at $P_v = X_v P_G,$ with $X_v = (m_v M)/M_v,$ and with M defined as in Eq. (1.46). Treating the noncondensable gas as ideal, one can write

$$h_n = h_{n,ref} + \int_{T_{ref}}^{T_G} C_{P,n} dT,$$

where subscript ref represents a reference temperature for the noncondensable enthalpy. Noting that $h_g = h_g(P_v)$ and $P_v = P_G(1 - X_n),$ we have

$$\left(\frac{\partial h_G}{\partial m_n}\right)_{P_G} = -h_g + (1 - m_n)\frac{\partial h_g}{\partial P_v}\frac{\partial P_v}{\partial X_n}\frac{\partial X_n}{\partial m_n} + h_n + m_n\frac{\partial h_n}{\partial T_G}\frac{\partial T_G}{\partial m_n}. \tag{b}$$

By manipulation of this equation, one can derive

$$\left(\frac{\partial h_G}{\partial m_n}\right)_{P_G} = -h_g - P_G(1 - m_n)\frac{\partial X_n}{\partial m_n}\left(\frac{\partial h_g}{\partial P_v}\right) + h_n - m_n C_{P,n}\left(\frac{T_G v_{fg}}{h_{fg}}\right)P_G\left(\frac{\partial X_n}{\partial m_n}\right), \tag{c}$$

where, again, v_{fg} and h_{fg} correspond to $T_{sat} = T_G.$ If, for simplicity, it is assumed that $C_{P,n} = $ const. (a good assumption when temperature variations in the problem of interest are relatively small), then $h_n - h_{n,ref} = C_{P,n}(T_G - T_{ref}).$ The problem is solved by substituting from Eq. (1.45) for $\frac{\partial X_n}{\partial m_n}$. Note that Clapeyron's relation has been used for the derivation of the last term on the right side of this expression.

1.6 Transport Properties

1.6.1 Mixture Rules

The viscosity and thermal conductivity of a gas mixture can be calculated from the following expressions (Wilke, 1950):

$$\mu = \sum_{j=1}^{n} \frac{X_j \mu_j}{\sum_{i=1}^{n} X_i \phi_{ji}}, \tag{1.70}$$

$$k = \sum_{j=1}^{n} \frac{X_j k_j}{\sum_{i=1}^{n} X_i \phi_{ji}}, \tag{1.71}$$

$$\phi_{ji} = \frac{\left[1 + (\mu_j/\mu_i)^{1/2}(M_i/M_j)^{1/4}\right]^2}{\sqrt{8}[1 + (M_j/M_i)]^{1/2}}. \tag{1.72}$$

These rules have been deduced from gaskinetic theory and have proven to be quite adequate (Mills, 2001).

For liquid mixtures the property calculation rules are complicated and are not well established. However, for most dilute solutions of inert gases, which are the main subject of interest in this book, the viscosity and thermal conductivity of the liquid are similar to the properties of pure liquid.

With respect to mass diffusivity, everywhere in this book, unless stated otherwise, we will assume that the mixture is binary; namely, only two different species are present. For example, in dealing with an air–water vapor mixture (as it pertains to evaporation and condensation processes in air), we follow the common practice of treating dry air as a single species. Furthermore, we assume that the liquid only contains dissolved species at very low concentrations.

For the thermophysical and transport properties, including mass diffusivity, we rely primarily on experimental data. Mass diffusivities of gaseous pairs are approximately independent of their concentrations in normal pressures but are sensitive to temperature. The mass diffusion coefficients are sensitive to both concentration and temperature in liquids, however.

1.6.2 Gaskinetic Theory

Gaskinetic theory (GKT) provides for the estimation of the thermophysical and transport properties in gases. These methods become particularly useful when empirical data are not available. Simple GKT models the gas molecules as rigid and elastic spheres (hard spheres) that influence one another only by impact (Gombosi, 1994). When two molecules impact, furthermore, their directions of motion after collision are isotropic, and following a large number of intermolecular collisions the orthogonal components of the molecular velocities are independent of each other. It is also assumed that the distribution function of molecules under equilibrium is isotropic.

These assumptions, along with the ideal gas law, lead to the well-known Maxwell–Boltzmann distribution, whereby the fraction of molecules with speeds in the $|\vec{U}|$ to $|\vec{U} + d\vec{U}|$ range is given by $f(U)dU$, and

$$f(U) = \left(\frac{M}{2\pi R_u T}\right)^{3/2} e^{-\frac{MU^2}{2R_u T}}. \tag{1.73}$$

If the magnitude (absolute value) of velocity is of interest, the number fraction of molecules with speeds in the $|U|$ to $|U + dU|$ range will be equal to $F(U)dU$, where

$$F(U) = 4\pi U^2 f(U). \tag{1.74}$$

Let us define, for convenience,

$$\beta = \frac{m}{2\kappa_B T} = \frac{M}{2R_u T}, \tag{1.75}$$

where m is the mass of a single molecule and κ_B is Boltzmann's constant. (Note that $\frac{\kappa_B}{m} = \frac{R_u}{M}$.) In Cartesian coordinates, we will have for each coordinate i

$$\sqrt{\frac{\beta}{\pi}} \int_{-\infty}^{\infty} e^{-\beta U_i^2} dU_i = 1. \tag{1.76}$$

Various moments of the Maxwell–Boltzmann distribution can be found. For example, using Eq. (1.74), we get the mean molecular speed by writing

$$\langle |U| \rangle = 4\pi \left(\frac{\beta}{\pi}\right)^{3/2} \int_0^{\infty} e^{-\beta U^2} U^3 dU = \sqrt{\frac{8\kappa_B T}{\pi m}}. \tag{1.77}$$

Likewise, the average molecular kinetic energy can be found as

$$\langle E_{kin} \rangle = \frac{1}{2} m \langle U^2 \rangle = 2\pi \left(\frac{\beta}{\pi}\right)^{3/2} m \int_0^{\infty} e^{-\beta U^2} U^4 dU = \frac{3}{2} \kappa_B T. \tag{1.78}$$

The average speed of molecules in a particular direction (e.g., in the positive x direction in a Cartesian coordinate system) can be found by first noting that according to Eq. (1.73) the number fraction of molecules that have velocities along the x coordinate in the range U_x and $U_x + dU_x$ is

$$\left(\frac{M}{2\pi R_u T}\right)^{3/2} \int_{-\infty}^{+\infty} dU_y \int_{\infty}^{+\infty} dU_z e^{-\frac{M(U_x^2 + U_y^2 + U_z^2)}{2R_u T}} dU_x. \tag{1.79}$$

The average velocity in the positive direction will then follow:

$$\langle U_{x+} \rangle = \left(\frac{M}{2\pi R_u T}\right)^{3/2} \int_{-\infty}^{+\infty} dU_y \int_{\infty}^{+\infty} dU_z \int_0^{+\infty} e^{-\frac{M(U_x^2 + U_y^2 + U_z^2)}{2R_u T}} U_x dU_x. \tag{1.80}$$

Using Eq. (1.76), one can then easily show that

$$\langle U_{x+} \rangle = \sqrt{\frac{\beta}{\pi}} \int_0^{\infty} e^{-\beta U^2_{x+}} U_{x+} dU_{x+} = \sqrt{\frac{\kappa_B T}{2\pi m}}. \tag{1.81}$$

For an ideal gas, furthermore, the number density of gas molecules is

$$n = \rho N_{Av}/M = \frac{P}{k_B T},\qquad(1.82)$$

where N_{Av} is Avagadro's number. The flux of gas molecules passing, per unit time, in any particular direction (e.g., in the positive x direction in a Cartesian coordinate system), through a surface element oriented perpendicularly to the direction of interest, will be

$$j_{molec,x+} = n\langle U_{x+}\rangle = \frac{P}{\sqrt{2\pi k_B m T}} = \left(\frac{M}{2\pi R_u}\right)^{1/2}\frac{P}{m\sqrt{T}}.\qquad(1.83)$$

This expression, when multiplied by δA, the surface area of a very small opening in the wall of a vessel containing an ideal gas, will provide the rate of molecules leaking out of the vessel (*molecular effusion*) and is valid as long as the characteristic dimension of δA is smaller than the mean free path of the gas molecules. This expression is also used in the simplest interpretation of the molecular processes associated with evaporation and condensation, as will be seen in Chapter 2.

According to simple GKT, the gas molecules have a mean free path of (see Gombosi (1994) for detailed derivations):

$$\lambda = \frac{1}{\sqrt{2}n\sigma_A},\qquad(1.84)$$

where σ_A is the molecular scattering cross section. The molecular mean free time can then be found from

$$\tau = \frac{\lambda}{\langle|U|\rangle} = \frac{1}{\sqrt{2}\,n\sigma_A\langle|U|\rangle}.\qquad(1.85)$$

Given that random molecular motions and intermolecular collisions are responsible for diffusion in fluids, expressions for μ, k, and D can be found based on the molecular mean free path and free time. The simplest formulas derived in this way are based on the Maxwell–Boltzmann distribution, which assumes equilibrium. More accurate formulas can be derived by taking into consideration that all diffusion phenomena actually occur as a result of nonequilibrium. The transport of the molecular energy distribution under nonequilibrium conditions is described by an integrodifferential equation, known as the Boltzmann transport equation. The aforementioned Maxwell–Boltzmann distribution (Eq. (1.73) or (1.74)) is in fact the solution of the Boltzmann transport equation under equilibrium conditions. Boltzmann's equation cannot be analytically solved in its original form, but approximate solutions representing relatively slight deviations from equilibrium have been derived, and these nonequilibrium solutions lead to useful formulas for the gas transport properties. One of the most well-known approximate solutions to the Boltzmann equation for near-equilibrium conditions was derived by Chapman, in 1916, and Enskog, in 1917 (Chapman and Cowling, 1970). The solution leads to widely used expressions for gas transport properties that are only briefly presented and discussed in the following. More detailed discussions about these expressions can be found in Bird *et al.* (2002), Skelland (1974), and Mills (2001).

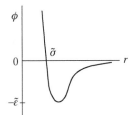

Figure 1.10. The pair potential energy distribution according to the Lennard-Jones 6–12 intermolecular potential model.

The interaction between two molecules as they approach one another can be modeled only when intermolecular forces are known. The force between two identical molecules, $\vec{F}$, defined to be positive when repulsive, can be represented in terms of a pair potential energy, ϕ, where

$$\vec{F} = -\nabla\phi(r), \qquad (1.86)$$

with r being the distance separating the two molecules. Several models have been proposed for ϕ [see Rowley (1994) for a concise review]; the most widely used among them is the empirical Lennard-Jones 6–12 model (Rowley, 1994):

$$\phi(r) = 4\tilde{\varepsilon}\left[\left(\frac{\tilde{\sigma}}{r}\right)^{12} - \left(\frac{\tilde{\sigma}}{r}\right)^{6}\right]. \qquad (1.87)$$

Figure 1.10 depicts Eq. (1.87). The Lennard-Jones model, like all similar models, accounts for the fact that intermolecular forces are attractive at large distances and become repulsive when the molecules are very close to one another. The function $\phi(r)$ in Lennard-Jones's model is fully characterized by two parameters: $\tilde{\sigma}$, the collision diameter, and $\tilde{\varepsilon}$, the energy representing the maximum attraction. Values of $\tilde{\sigma}$ and $\tilde{\varepsilon}$ for some selected molecules are listed in Appendix H. The force constants for a large number of molecules can be found in Svehla (1962). When tabulated values are not known, they can be estimated by using empirical correlations based on the molecule's properties at its critical point, liquid at normal boiling point, or the solid state at melting point (Bird *et al.*, 2002). In terms of the substance's critical state, for example,

$$\tilde{\sigma} \approx 2.44(T_{cr}/P_{cr})^{1/3} \qquad (1.88)$$

and

$$\tilde{\varepsilon}/\kappa_{B} \approx 0.77T_{cr}, \qquad (1.89)$$

where T_{cr} and $\tilde{\varepsilon}/\kappa_{B}$ are in degrees kelvin, P_{cr} is in atmospheres, and $\tilde{\sigma}$ calculated in this way is in ångströms. The Lennard-Jones model is used extensively in molecular dynamic simulations.

According to the Chapman–Enskog model, the gas viscosity can be found from

$$\mu = 2.669 \times 10^{-6}\frac{\sqrt{MT}}{\tilde{\sigma}^2\Omega_k}\,(\text{kg/ms}), \qquad (1.90)$$

where T is in kelvins, $\tilde{\sigma}$ is in ångströms, and Ω_k is a collision integral for thermal conductivity or viscosity. (Collision integrals for viscosity and thermal conductivity are equal, i.e., $\Omega_k = \Omega_\mu$.) For monatomic gases the Chapman–Enskog model predicts

$$k = k_{\text{trans}} = 2.5 C_v \mu = \frac{15}{4} \left(\frac{R_u}{M} \right) \mu. \tag{1.91}$$

For a polyatomic gas, the molecule's internal degrees of freedom contribute to the gas thermal conductivity, and

$$k = k_{\text{trans}} + 1.32 \left(C_P - \frac{5}{2} \frac{R_u}{M} \right) \mu. \tag{1.92}$$

The binary mass diffusivity of species 1 and 2 can be found from

$$D_{12} = D_{21} = 1.858 \times 10^{-7} \frac{\sqrt{T^3 \left(\frac{1}{M_1} + \frac{1}{M_2} \right)}}{\tilde{\sigma}_{12}^2 \Omega_D P} \, (\text{m}^2/\text{s}), \tag{1.93}$$

where P is in atmospheres, Ω_D represents the collision integral for the two molecules for mass diffusively, and

$$\tilde{\sigma}_{12} = \frac{1}{2} (\tilde{\sigma}_1 + \tilde{\sigma}_2), \tag{1.94}$$

$$\tilde{\varepsilon}_{12} = \sqrt{\tilde{\varepsilon}_1 \tilde{\varepsilon}_2}. \tag{1.95a}$$

Appendix I can be used for the calculation of collision integrals for a number of selected species. The collision integrals can also be calculated using the following curvefits (Neufeld *et al.*, 1972; Bird *et al.*, 2002)

$$\Omega_k = \Omega_\mu = \frac{1.16145}{\xi^{0.14874}} + \frac{0.52487}{\exp(0.77320\xi)} + \frac{2.16178}{\exp(2.43787\xi)} \tag{1.95b}$$

$$\Omega_D = \frac{1.06036}{\xi^{0.1561}} + \frac{0.193}{\exp(0.47635\xi)} + \frac{1.03587}{\exp(1.52996\xi)} + \frac{1.76474}{\exp(3.89411\xi)}, \tag{1.95c}$$

where $\xi = \left(\frac{\kappa_B T}{\tilde{\varepsilon}} \right)$.

1.6.3 Diffusion in Liquids

The binary diffusivities of solutions of several nondissociated chemical species in water are given in Appendix G. The diffusion of a dilute species 1 (solute) in a liquid 2 (solvent) follows Fick's law with a diffusion coefficient that is approximately equal to the binary diffusivity D_{12}, even when other diffusing species are also present in the liquid, provided that all diffusing species are present in very small concentrations.

Theories dealing with molecular structure and kinetics of liquids are not sufficiently advanced to provide reasonably accurate predictions of liquid transport properties. A simple method for the estimation of the diffusivity of a dilute solution is the Stokes–Einstein expression

$$D_{12} = \frac{\kappa_B T}{3 \pi \mu_2 d_1}, \tag{1.96}$$

Table 1.2. *Specific molar volume at boiling point for selected substances.*

Substance	$\tilde{V}_{b1} \times 10^3 \, (\text{m}^3/\text{kmol})$	T_b (K)
Air	29.9	79
Hydrogen	14.3	21
Oxygen	25.6	90
Nitrogen	31.2	77
Ammonia	25.8	240
Hydrogen sulfide	32.9	212
Carbon monoxide	30.7	82
Carbon dioxide	34.0	195
Chlorine	48.4	239
Hydrochloric acid	30.6	188
Benzene	96.5	353
Water	18.9	373
Acetone	77.5	329
Methane	37.7	112
Propane	74.5	229
Heptane	162	372

Note: After Mills (2001).

where subscripts 1 and 2 refer to the solute and solvent, respectively, and d_1 is the diameter of a single solute molecule, and can be estimated from $d_1 \approx \tilde{\sigma}$, namely, the Lennard-Jones collision diameter. Alternatively, it can be estimated from

$$d_1 \approx \left(\frac{6}{\pi} \frac{M_1}{\rho_1 N_{\text{Av}}} \right). \tag{1.97}$$

The Stokes–Einstein expression in fact represents the Brownian motion of spherical particles (solute molecules in this case) in a fluid, under the assumption of creep flow without slip around the particles. It is accurate when the spherical particle is much larger than intermolecular distances. It is good for estimation of the diffusivity when the solute molecule is approximately spherical and is at least five times larger than the solvent molecule (Cussler, 2009).

A widely used empirical correlation for binary diffusivity of a dilute and non-dissociating chemical species (species 1) in a liquid (solvent, species 2) is (Wilke and Chang, 1954)

$$\boldsymbol{D}_{12} = 1.17 \times 10^{-16} \frac{(\Phi_2 M_2)^{1/2} T}{\mu \tilde{V}_{b1}^{0.6}} \, (\text{m}^2/\text{s}), \tag{1.98}$$

where $\boldsymbol{D}_{12}$ is in square meters per second; $\tilde{V}_{b1}$ is the specific molar volume, in cubic meters per kilomole, of species 1 as liquid at its normal boiling point; μ is the mixture liquid viscosity in kg/m s; T is the temperature in kelvins; and Φ_2 is an association parameter for the solvent: $\Phi_2 = 2.26$ for water, 1.5 for ethanol, 1.9 for methanol and ethylene glycol, and 1 for benzene, methane, and other unassociated refrigerants and solvents (Mills, 2001; Bird *et al.*, 2002). Values of $\tilde{V}_{b1}$ for several species are given in Table 1.2.

1.7 Turbulent Boundary Layer Velocity and Temperature Profiles

Near-wall hydrodynamic and heat-transfer phenomena are crucial to many boiling and condensation processes. Examples include bubble nucleation, growth, and release during flow boiling, and flow condensation.

The universal velocity profile in a two-dimensional, incompressible turbulent boundary layer can be represented as (Schlichting, 1968)

a viscous sublayer:

$$u^+ = y^+ \quad y^+ < 5, \tag{1.99}$$

a buffer sublayer:

$$u^+ = 5 \ln y^+ - 3.05 \quad 5 < y^+ < 30, \tag{1.100}$$

and an inertial sublayer:

$$u^+ = \frac{1}{\kappa} \ln y^+ + B \quad 30 < y^+ \lesssim 400, \tag{1.101}$$

where $\kappa = 0.40$, $B = 5.5$, y is the distance from the wall, u is the velocity parallel to the wall, and

$$y^+ = \frac{y U_\tau}{v}, \tag{1.102}$$

$$u^+ = u/U_\tau, \tag{1.103}$$

$$U_\tau = \sqrt{\tau_w/\rho}. \tag{1.104}$$

This universal velocity profile can be utilized for determining the turbulent properties in the boundary layer. For example, according to the definition of the turbulent mixing length, l_m, one can write

$$\tau_w = \tau_{lam} + \tau_{turb} = \rho \left[v + l_m^2 \left| \frac{\partial u}{\partial y} \right| \right] \frac{\partial u}{\partial y}. \tag{1.105}$$

The mixing length is related to the turbulent eddy diffusivity by noting that

$$\tau_w = \rho (E + v) \frac{du}{dy}. \tag{1.106}$$

In a turbulent boundary layer near the wall, $\tau \approx \tau_w = $ const., and as a result Eq. (1.106) can be manipulated to derive the following two useful relations:

$$\frac{E}{v} = \left(\frac{du^+}{dy^+} \right)^{-1} - 1, \tag{1.107}$$

$$\left(1 + l_m^{+2} \left| \frac{du^+}{dy^+} \right| \right) \frac{du^+}{dy^+} = 1, \tag{1.108}$$

where im $l_m^+ = \frac{l_m U_\tau}{v}$. Equation (1.108) can be rewritten as

$$\left(\frac{dy^+}{du^+} \right)^2 - \frac{dy^+}{du^+} - l_m^{+2} = 0. \tag{1.109}$$

Equations (1.107) and (1.109), along with Eqs. (1.99)–(1.101) can evidently be used for calculating the eddy diffusivity distribution in the boundary layer.

Turbulent boundary layers support a near-wall temperature distribution when heat transfer takes place, which has a peculiar form when it is presented in appropriate dimensionless form. This "temperature law of the wall" is very useful and has been applied in many phenomenological models, as well as to the development of heat transfer correlations. The temperature law of the wall can be derived by noting that in a steady and incompressible two-dimensional boundary layer, when the heat transfer boundary condition at the wall is a constant heat flux, one can write

$$q''_y \approx q''_w = -\rho C_p(\alpha + E_H)\frac{\partial T}{\partial y} = -\rho C_P \left(\frac{\nu}{\text{Pr}} + \frac{E}{\text{Pr}_{\text{turb}}}\right)\frac{\partial T}{\partial y}, \tag{1.110}$$

where y is the distance from the wall, E_H is the eddy diffusivity for heat transfer, q''_y is the heat flux in the y direction, Pr_{turb} is the turbulent Prandtl number (which is typically ≈ 1 for common fluids), and T is the local time-averaged fluid temperature. Equation (1.110) can now be manipulated to get

$$T^+ = \frac{T_w - T(y)}{\frac{q''_w}{\rho C_P U_\tau}} = \int_0^{y+} \frac{dy^+}{1/\text{Pr} + E/(\nu\,\text{Pr}_{\text{turb}})}. \tag{1.111}$$

One can now proceed as follows. Assume $\text{Pr}_{\text{turb}} = 1$, which is a good approximation for common fluids. Using Eqs. (1.106), (1.107) and (1.99)–(1.101), find E/ν in each of the sublayers of the turbulent boundary layer. Substitute the latter into Eq. (1.111) and perform the integrations, noting that $E \approx 0$ in the viscous sublayer, and $\nu + E \approx E$ in the fully turbulent sublayer. The result will be the well-known temperature law of the wall (Martinelli, 1947):

$$T^+ = \begin{cases} \text{Pr}\,y^+ & \text{for } y^+ \leq 5, \tag{1.112} \\ 5\left\{\text{Pr} + \ln\left[1 + \text{Pr}\left(\frac{y^+}{5} - 1\right)\right]\right\} & \text{for } 5 < y^+ < 30, \tag{1.113} \\ 5\left\{\text{Pr} + \ln[1 + 5\,\text{Pr}] + \frac{1}{5K}\ln\left(\frac{y^+}{30}\right)\right\} & \text{for } y^+ \geq 30. \tag{1.114} \end{cases}$$

EXAMPLE 1.7. Subcooled water flows through a heated pipe that has an inner diameter of 2.5 cm. The mean velocity of water is 2.1 m/s. The pipe receives a wall heat flux of 2×10^5 W/m². Assuming that the pipe is hydraulically smooth, and using properties of water at 370 K, calculate and plot the profiles of velocity and temperature as a function of y, the distance from the wall.

SOLUTION. For water at the state given, $\rho = 960.6$ kg/m³, $\mu = 2.915 \times 10^{-4}$ kg/m s, $k = 0.664$ W/m·K, and $\text{Pr} = 1.85$. The Reynolds number will be $\text{Re} = 1.73 \times 10^5$. The wall Fanning friction factor can be found from Blasius's correlation, $f = 0.079\,\text{Re}^{-0.25} = 0.00387$. From there, we obtain $\tau_w = 0.5 f \rho \overline{U}^2 = 8.205$ N/m², and $U_\tau = \sqrt{\tau_w/\rho} = 0.0924$ m/s. We also need to calculate the wall temperature. Let us use the correlation of Dittus and Boelter, whereby

$$H = \frac{k}{D}(0.023\text{Re}^{0.8}\text{Pr}^{0.4}) = 12{,}113 \text{ W/m}^2\cdot\text{K},$$

$$T_w = \overline{T} + q''_w/H = 386.5 \text{ K}.$$

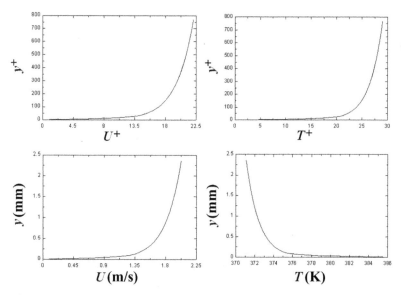

Figure E1.7.

One can now parametrically vary y^+, get y from Eq. (1.102), and then calculate u^+ from Eqs. (1.99)–(1.101). Knowing u^+, one can then calculate u from Eq. (1.103). Next, one should calculate T^+ from Eqs. (1.112)–(1.114), and from there calculate T from the left side of Eq. (1.111). The following table contains some typical calculated numbers. The calculations lead to the figures displayed in Fig. E1.7 .

y^+	y (mm)	u^+	u (m/s)	T^+	T(K)
1	0.0032 83	1	0.0924	1.849	385.5
10.1	0.03316	8.513	0.7868	14.55	378.7
101.1	0.332	17.04	1.575	23.92	373.7
201.2	0.6607	18.76	1.734	25.64	372.8
301.3	0.9893	19.77	1.827	26.65	372.3
401.5	1.318	20.49	1.893	27.37	371.9

Example 1.7 is a reminder that the velocity and temperature laws of the wall can be applied to internal flows as well.

The velocity law of the wall can alternatively be represented by recasting Eq. (1.107) as

$$u^+(y^+) = \int\limits_0^{y^+} \frac{dy^+}{\frac{E}{\nu}+1}. \tag{1.115}$$

This equation is of course for flow over a flat surface, but it can be applied to the flow field near a curved wall as long as the wall radius of curvature is much larger than the boundary layer thickness. For steady, incompressible flow inside tubes, with negligible body force effect, one can easily show that

$$\tau(y) = \tau_{w}\frac{(R-y)}{R}. \tag{1.116}$$

Using this expression, one can derive

$$u^{+} = \frac{1}{R^{+}}\int\limits_{0}^{y^{+}}\frac{(R^{+}-y^{+})dy^{+}}{\frac{E}{\nu}+1}, \tag{1.117}$$

where $R^{+} = RU_{\tau}/\nu$.

The temperature law of the wall can likewise be represented by Eq. (1.111) for a flat surface. These equations can be directly integrated to derive the velocity and temperature profiles, and from there one can obtain expressions for friction factors and heat transfer coefficients, when an appropriate eddy diffusivity model (or, equivalently, a mixing length model) is available. Several models that well represent the inner zones of the boundary layer (viscous sublayer and the buffer layer) have long been available. A widely used model is due to van Driest (1956):

$$l_{m} = \kappa y[1 - \exp(-y^{+}/A)], \tag{1.118}$$

where $A = 26$ for flat plates and $\kappa = 0.4$ is von Kármán's constant. It can be shown that Eqs. (1.115) and (1.118) lead to

$$u^{+} = \int\limits_{0}^{y+}\frac{2dy^{+}}{1 + \{1 + 0.64y^{+2}[1 - \exp(-y^{+}/26)]^{2}\}^{1/2}}. \tag{1.119}$$

Reichardt (1951) has proposed

$$\frac{E}{\nu} = k[y^{+} - y_{n}^{+}\tanh(y^{+}/y_{n}^{+})], \, y_{n}^{+} = 11. \tag{1.120}$$

This expression is for a flat surface. When applied to flow in a circular pipe, it can be used for $y^{+} < 50$, and for $y^{+} > 50$ one should use

$$\frac{E}{\nu} = \left(\frac{k}{3}\right)y^{+}\left[0.5 + \left(\frac{r^{+}}{R^{+}}\right)^{2}\right]\left(1 + \frac{r^{+}}{R^{+}}\right), \tag{1.121}$$

where $r^{+} = rU_{\tau}/\nu$.

The correlation of Deissler (1954) is

$$\frac{E}{\nu} = n^{2}u^{+}y^{+}[1 - \exp(-n^{2}u^{+}y^{+})], \tag{1.122}$$

where $n = 0.124$. When applied to flow in a circular pipe, this expression should be used for $y^{+} < 26$, and for $y^{+} > 26$, one should use

$$\frac{E'}{\nu} = \left\{\frac{y^{+}[1 - (y^{+}/R^{+})]}{2.5} - 1\right\}. \tag{1.123}$$

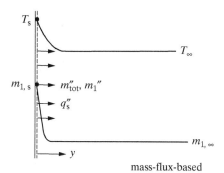

Figure 1.11. Heat and mass transfer between a surface and a fluid.

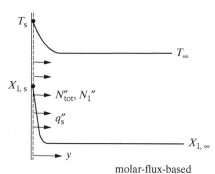

1.8 Convective Heat and Mass Transfer

When heat transfer alone takes place between a surface and a moving fluid, as shown in Fig. 1.11, then

$$q_s'' = -k \frac{\partial T}{\partial y}\bigg|_{y=0} = H(T_s - T_\infty), \tag{1.124}$$

where k is the thermal conductivity of the fluid. If very slow mass transfer takes place in a binary mixture (for example owing to the sublimation of the surface, when air flows over a naphthalene block), the mass flux of the transferred species (the vaporizing naphthalene in the aforementioned example) follows:

$$m_1'' = -\rho \boldsymbol{D}_{12} \frac{\partial m_1}{\partial y}\bigg|_{y=0} = K(m_{1,s} - m_{1,\infty}), \tag{1.125}$$

where ρ is the density of the fluid mixture and subscript 1 represents the transferred species (naphthalene vapor in the example).

Consider now the case where a finite mass flux m_{tot}'' passes through the surface. In this case, Fourier's and Fick's laws still hold. The transfer of mass, though, distorts the temperature and chemical species concentration profiles in the vicinity of the interphase. Equations (1.124) and (1.125) should then be replaced with

$$q_s'' = -k \frac{\partial T}{\partial y}\bigg|_{y=0} = \dot{H}(T_s - T_\infty), \tag{1.126}$$

$$m_1'' = m_{tot}'' m_{1,s} - \rho \boldsymbol{D}_{12} \frac{\partial m_1}{\partial y}\bigg|_{y=0} = m_{tot}'' m_{1,s} + \dot{K}(m_{1,s} - m_{1,\infty}). \tag{1.127a}$$

The modified heat and mass transfer coefficients $\dot{H}$ and $\dot{K}$ account for the blowing or suction effect caused by mass transfer. The first term on the right side of Eq. (1.127a) represents the convective transfer of species 1. When species 1 is the only transferred species (e.g., during evaporation or condensation in the presence of a noncondensable gas), this term becomes $m''_1 m_{1,\mathrm{s}}$. The mass flux will then be

$$m''_1 (1 - m_{1,\mathrm{s}}) = -\rho \boldsymbol{D}_{12} \left. \frac{\partial m_1}{\partial y} \right|_{y=0} = \dot{K}(m_{1,\mathrm{s}} - m_{1,\infty}). \tag{1.127b}$$

The effect of mass transfer on convection can be estimated by the Couette flow film model. This engineering model assumes that the interfacial heat and mass transfer resistances occur in a fluid film that can be modeled as a Couette flow (Mills, 2001; Kays *et al.*, 2005). The same results can be derived by using a stagnant film model (Ackerman, 1937; Bird *et al.*, 2002). Accordingly,

$$\frac{\dot{H}}{H} = \frac{m''_{\mathrm{tot}} C_{\mathrm{P}}}{\exp\left(\frac{m''_{\mathrm{tot}} C_{\mathrm{P}}}{H}\right) - 1} \tag{1.128}$$

and

$$\frac{\dot{K}}{K} = \frac{m''_{\mathrm{tot}}}{\exp\left(\frac{m''_{\mathrm{tot}}}{K}\right) - 1}. \tag{1.129}$$

The mass transfer affects the friction between the fluid and the surface as well, which can also be estimated by using the Couette flow film model and writing

$$\tau_{\mathrm{s}} = \dot{f}(1/2)\rho U_\infty^2, \tag{1.130}$$

$$\frac{\dot{f}}{f} = \frac{\beta}{e^\beta - 1}, \tag{1.131}$$

$$\beta = \frac{2m''_{\mathrm{tot}}}{\rho U_\infty f}, \tag{1.132}$$

where f is the skin friction coefficient when there is no mass transfer. Equations (1.128), (1.129), and (1.131) are convenient to use when mass fluxes are known. The Couette flow film model predictions can also be cast in the following forms, which are more convenient when mass fractions are known:

$$\frac{\dot{H}}{H} = \ln(1 + B_{\mathrm{h}})/B_{\mathrm{h}}, \tag{1.133}$$

$$\frac{\dot{K}}{K} = \ln(1 + B_{\mathrm{m}})/B_{\mathrm{m}}, \tag{1.134}$$

$$\frac{\dot{f}}{f} = \ln(1 + B_{\mathrm{f}})/B_{\mathrm{f}}, \tag{1.135}$$

$$B_{\mathrm{h}} = \frac{m''_{\mathrm{tot}} C_{\mathrm{P}}}{\dot{H}}, \tag{1.136}$$

$$B_{\mathrm{m}} = \frac{m_{1,\infty} - m_{1,\mathrm{s}}}{m_{1,\mathrm{s}} - m''_1/m''_{\mathrm{tot}}}, \tag{1.137}$$

$$B_{\mathrm{f}} = \frac{2m''_{\mathrm{tot}}}{\rho U_\infty f}, \tag{1.138}$$

The total mass flux and B_{m} can now be found by combining Eqs. (1.127a), (1.134), and (1.137), and that leads to

$$m''_{\mathrm{tot}} = \dot{K} B_{\mathrm{m}} = K \ln(1 + B_{\mathrm{m}}). \tag{1.139}$$

Molar-Flux-Based Formulation

The formulation of mass transfer and its effect on heat transfer and friction were thus far mass-flux-based. They can be put in molar-flux-based form, which is sometimes more convenient. In the molar-flux-based formulation, Eq. (1.125) will be replaced with

$$N_1'' = -\mathbf{C}D_{12}\frac{\partial X_1}{\partial y}\bigg|_{y=0} = \tilde{K}(X_{1,s} - X_{1,\infty}),\tag{1.140}$$

where $\mathbf{C}$ is the total molar concentration in the vicinity of the surface and $\tilde{K}$ is the molar-based mass transfer coefficient (in kmol/m²·s, for example). Equation (1.126) remains unchanged, and Eq. (1.127a) will be replaced with

$$N_1'' = N_{\text{tot}}''X_{1,s} - \mathbf{C}D_{12}\frac{\partial X_1}{\partial y}\bigg|_{y=0} = N_{\text{tot}}''X_{1,s} + \dot{\tilde{K}}(X_{1,s} - X_{1,\infty}).\tag{1.141a}$$

When species 1 is the only transferred species (e.g., during evaporation or condensation when a noncondensable is present), then $N_1'' = N_{\text{tot}}''$. Equation (1.141a) then can be recast as

$$N_1''(1 - X_{1,s}) = -\mathbf{C}D_{12}\frac{\partial X_1}{\partial y}\bigg|_{y=0} = \dot{\tilde{K}}(X_{1,s} - X_{1,\infty}).\tag{1.141b}$$

The predictions of the Couette flow film theory in molar-flux-based formulation will then be

$$\frac{\dot{H}}{H} = \frac{N_{\text{tot}}''\tilde{C}_{\text{P}}}{\exp\left(\frac{N_{\text{tot}}''\tilde{C}_{\text{P}}}{H}\right) - 1}\tag{1.142}$$

and

$$\frac{\dot{\tilde{K}}}{\tilde{K}} = \frac{N_{\text{tot}}''}{\exp\left(\frac{N_{\text{tot}}''}{\tilde{K}}\right) - 1},\tag{1.143}$$

where $\tilde{C}_{\text{P}}$ is also molar based (in kJ/kmol·K, for example). Equations (1.130), and (1.131) remain unchanged, and Eq. (1.132) is replaced with

$$\beta = \frac{2N_{\text{tot}}''}{CU_\infty f}.\tag{1.144}$$

Equations (1.131), (1.142), and (1.143) are convenient to use when the molar fluxes are known. When mole fractions are known and we need to calculate the molar fluxes, the following equations can be applied instead:

$$\frac{\dot{H}}{H} = \frac{\ln(1 + \tilde{B}_{\text{h}})}{\tilde{B}_{\text{h}}},\tag{1.145}$$

$$\frac{\dot{\tilde{K}}}{\tilde{K}} = \frac{\ln(1 + \tilde{B}_{\text{m}})}{\tilde{B}_{\text{m}}},\tag{1.146}$$

$$\frac{\dot{f}}{f} = \frac{\ln(1 + \tilde{B}_{\text{f}})}{\tilde{B}_{\text{f}}},\tag{1.147}$$

where

$$\tilde{B}_h = \frac{N''_{tot}\tilde{C}_P}{\dot{H}}, \tag{1.148}$$

$$\tilde{B}_m = \frac{X_{1,\infty} - X_{1,s}}{X_{1,s} - (N''_1/N''_{tot})}, \tag{1.149}$$

$$\tilde{B}_f = \frac{2N''_{tot}}{CU_\infty \dot{f}}. \tag{1.150}$$

The total molar flux will then follow:

$$N''_{tot} = \dot{\tilde{K}}\tilde{B}_m = \tilde{K}\ln(1 + \tilde{B}_m). \tag{1.151}$$

Heat and Mass Transfer Analogy

Parameters H, f, and K are in general obtained from empirical or analytical correlations. For low-velocity forced flows (where the compressibility effect is small), for example, we usually deal with correlations of the forms

$$f = f(\text{Re}), \tag{1.152}$$

$$\text{Nu} = \frac{Hl}{k} = \text{Nu}(\text{Re}, \text{Pr}), \tag{1.153}$$

$$\text{Sh} = \frac{\tilde{K}l}{\mathbf{C}\mathbf{D}_{12}} = \frac{Kl}{\rho\mathbf{D}_{12}} = \text{Sh}(\text{Re}, \text{Sc}). \tag{1.154}$$

The functions on the right-hand sides of these equations depend on the system geometry and configuration. Such correlations are in fact solutions to the conservation equations that govern the transport of momentum, heat, and mass. Equation (1.154), as noted, can be written in mass-flux form when $Kl/\rho\mathbf{D}_{12}$ is the left-hand side of the equation. In molar-flux form, the left side is $\tilde{K}l/\mathbf{C}\mathbf{D}_{12}$. The right side of the equation is the same for both cases, however.

The reader is probably familiar with the important analogy that exists between heat and momentum transfer processes, and this analogy has been applied in the past for the derivation of some of the widely used heat transfer correlations. It holds because of the similarity between dimensionless boundary layer momentum and thermal energy conservation equations. This similarity indicates that the solution of one system (momentum transfer) should provide the solution of the other (heat transfer). Thus, empirically correlated friction factors, which are generally simpler to measure, are used for the derivation of heat transfer correlations.

Since the physical laws that govern the diffusion of heat and mass (namely, Fourier's and Fick's laws) are mathematically identical, there is an analogy between heat and mass transfer processes as well. For many systems the dimensionless conservation equations governing heat and mass transfer processes are mathematically identical when the mass transfer rate is vanishing small, implying that the solution of one can be directly used for the derivation of the solution for the other. Accordingly, for any particular system, when a correlation similar to Eq. (1.153) for heat transfer is available, one can simply replace Pr with Sc and Nu with Sh, thereby deriving a correlation for mass transfer. The correlation obtained in this way is of course valid for the same flow conditions (i.e., the same ranges of Re or Gr). Furthermore, the procedure will be valid only when Sc and Pr have similar orders of magnitudes.

EXAMPLE 1.8. For flow across a sphere, the correlation of Ranz and Marshall (1952) for heat transfer at the surface of the sphere gives

$$\mathrm{Nu} = Hd/k = 2 + 0.3\,\mathrm{Re}_d^{0.6}\,\mathrm{Pr}^{0.33}.$$

Using this correlation, find the sublimation rate at the surface of a naphthalene sphere that is 2 mm in diameter and is moving at a velocity of 3 m/s with respect to atmospheric air. The naphthalene particle and air are both at 27 °C. For naphthalene vapor in air under atmospheric pressure, Sc = 2.35 at 300 K (Cho *et al.*, 1992; Mills, 2001). Furthermore, the vapor pressure of naphthalene can be estimated from Mills (2001):

$$P_v(T) = 3.631 \times 10^{13}\exp(-8586/T), \text{ where } T \text{ is in kelvin and } P_v \text{ is in pascals.}$$

SOLUTION. Let us use subscripts 1 and 2 to refer to air and naphthalene, respectively. At 300 K, for air $\nu_1 = 15.8 \times 10^{-6}\,\mathrm{m^2/s}$. Since the naphthalene partial pressure in air will be quite small, the air–naphthalene mixture viscosity will be approximately equal to the viscosity of air. This leads to $\boldsymbol{D}_{12} = \nu_1/\mathrm{Sc} = 6.7 \times 10^{-6}\,\mathrm{m^2/s}$. Also, it is reasonable to assume that the particle is isothermal. With $T = 300$ K, the naphthalene vapor pressure at the surface of the particle will be only $P_{v,s} = 13.5$ Pa. The mole fraction of naphthalene at the surface of the particle can then be found from

$$X_{2,s} = P_{v,s}/P_{tot} = 1.33 \times 10^{-4}.$$

for naphthalene, $M_2 \approx 128$. Using Eq. (1.44), we get $m_{2,s} \approx 5.9 \times 10^{-4}$.

By using the analogy between heat and mass transfer, the Ranz–Marshal correlation can be cast as

$$\mathrm{Sh} = \frac{Kd}{\rho_1 \boldsymbol{D}_{12}} = 2 + 0.3\,\mathrm{Re}_d^{0.6}\,\mathrm{Sc}^{0.33},$$

where, in view of the extremely low concentration of naphthalene vapor, we have used the density of air, ρ_1, to represent the density of the naphthalene–air mixture at the surface of the particle. For the numbers given, one gets

$$\mathrm{Re} = Ud/\nu_1 = 380$$

and

$$\mathrm{Sh} \approx 16.1 \Rightarrow K \approx 0.0636\,\mathrm{kg/m^2s}.$$

Given the very low mass fraction of naphthalene at the particle surface, one expects that the sublimation rate will be very small. Therefore, let us solve the problem assuming a vanishingly small mass transfer rate. We can then calculate the sublimation mass flux:

$$m_{2,s}'' = K(m_{2,s} - m_{2,\infty}) = 3.74 \times 10^{-5}\,\mathrm{kg/m^2 \cdot s},$$

where, for naphthalene mass fraction in the ambient air, $m_{2,\infty} = 0$ has been assumed. The very low mass flux confirms that the assumption of vanishingly small mass flux was fine. In other words, there is no need to correct the solution for the Stefan flow effect.

EXAMPLE 1.9. In the previous example, assume that the 2-mm-diameter sphere is a liquid water droplet and that the droplet and air are both at 25 °C. Calculate the evaporation rate at the droplet surface.

SOLUTION. Assuming that the droplet is isothermal, $T_s = 298$ K. Let us use subscripts 1 and 2 for water vapor and air, respectively. Water property tables then indicate that $P_{1,s} = 3141$ Pa. Therefore

$$X_{1,s} = P_{1,s}/P = 3141/1.013 \times 10^5 = 0.031.$$

The vapor mass fraction at the surface can now be found by using this value of $X_{1,s}$ in Eq. (1.44), and that gives $m_{1,s} = 0.0195$. Given the small concentration of water vapor in air, we can assume that the properties of the vapor–air mixture are the same as the properties of pure air, and from there we obtain $\nu = 15.6 \times 10^{-6}$ m²/s. Also, from the information in Appendix E, $\boldsymbol{D}_{12} = 2.6 \times 10^{-5}$ m²/s. Using the definition of the Schmidt number, we get $\text{Sc} = \nu/\boldsymbol{D}_{12} = 0.61$. As shown in the previous example, $\text{Re} = 384.6$. We can use the correlation of Ranz and Marshall, along with the analogy between heat and mass transfer, to get $\text{Sh} = \frac{Kd}{\rho_1 \boldsymbol{D}_{12}} = 2 + 0.3\text{Re}_d^{0.6} \text{Sc}^{0.33}$, and from there we obtain $K = 0.169$ kg/m²s. The mass fraction of water vapor far away from the droplet is zero; that is, $m_{1,\infty} = 0$ (because the air is assumed to be dry), and the mass transfer driving force can now be found by writing

$$B_m = (m_{1,\infty} - m_{1,s})/(m_{1,s} - 1) = 0.01985.$$

Since water vapor is the only transferred species, the evaporation mass flux can then be found from Eq. (1.139), and that leads to $m_1'' = m_{tot}'' = 3.31 \times 10^{-3}$ kg/m²s.

COMMENT. The droplet and air cannot remain at the same temperature, because evaporation at the surface cools the droplet.

PROBLEMS

1.1 The typical concentration of CO_2 in atmospheric air is 377 parts per million (PPM) by volume. Calculate the mass fractions of CO_2 and water vapor in air at room temperature when it is in equilibrium with water.

1.2 Prove Eq. (1.69) in the Solution to Example 1.5.

1.3 Prove Eq. (c) in the Solution to Example 1.6.

1.4 Calculate the viscosity and thermal conductivity of saturated air–water vapor mixtures under atmospheric pressure, for temperatures in the range 35–85 °C. Discuss the results, in particular with respect to the adequacy of neglecting the effect of water vapor on air properties.

1.5 Using the results of the Chapman–Enskog model, find the binary mass diffusivities of mixtures of the following species in air at 300 and 400 K temperatures and 1 bar pressure: H_2, He, and NO. Compare the results with data extracted from Appendix E.

1.6 Using the Chapman–Enskog model, estimate the binary mass diffusivities for the following pairs: H_2–water vapor, NO–water vapor, N_2–NH_3, and UF_6–Ar, for the conditions of Problem 1.5.

1.7 A long, 5-mm-diameter cylinder made of naphthalene is exposed to a cross-flow of pure air. The air is at 300 K temperature and flows with a velocity of 5 m/s. Estimate the time it takes for the diameter of the naphthalene cylinder to be reduced by 40 μm.

1.8 (a) Prove the temperature law of the wall of Martinelli (1947).

(b) An alternative to the expression for the buffer zone velocity profile is (Levich, 1962)

$$u^+ = 10\tan^{-1}(0.1y^+) + 1.2 \quad \text{for} \quad 5 < Y^+ < 30.$$

Derive equations similar to Martinelli's temperature law of the wall using Eqs. (1.99)–(1.101), along with this expression, for the dimensionless velocity distribution in the buffer zone.

1.9 Repeat the solution of Example 1.9, this time using the molar-flux-based formulation of mass transfer.

1.10 Consider the roof of a bus that is moving in still, atmospheric air with a speed of 95 km/h. The air temperature is 300 K.

(a) Assume that the car's roof surface is at 320 K because of solar radiation. Using an applicable correlation, calculate the convective heat flux between the roof's surface and the air at 4.5 m behind the leading edge of the roof. Neglect the surface roughness of the roof of the bus.

(b) Assume that a bug, that can be idealized as a sphere with 1.0 mm diameter, is trapped in the boundary layer at the location described in part (a) so that its center is 3.5 mm away from the wall. Estimate the drag force experienced by the bug. Also, estimate the time-average velocity difference across the bug's body.

You can find the drag coefficient for the bug from

$$C_D = \left[\sqrt{25/\text{Re}_d} + 0.5407\right]^2,$$

where d is the diameter of the bug.

(c) Estimate the air temperature that the bug experiences, and the temperature difference across the bug's body.

2 Gas–Liquid Interfacial Phenomena

2.1 Surface Tension and Contact Angle

2.1.1 Surface Tension

Liquids behave as if they are separated from their surroundings by an elastic skin that is always under tension and has the tendency to contract. Intermolecular forces are the cause of this tendency. For the molecules inside the liquid bulk, forces from all directions cancel each other out, and the molecules remain at near equilibrium. The molecules that are at the surface are pulled into the liquid bulk, however. According to gas and liquid kinetic theories, the surface of a liquid is in fact in a state of violent agitation, and the molecules at the surface are continuously replaced either through their motion into the liquid bulk or by evaporation and condensation at the interphase.

The interface between immiscible fluids can be modeled as an infinitely thin membrane that resists stretching and has a tendency to contract. Surface tension σ characterizes the interface's resistance to stretching.

The thermodynamic definition of surface tension is as follows. For a system at equilibrium that contains interfacial area,

$$dU = T\,dS - P\,dV + \sum_i \xi_i\,dm_i + \sigma\,dA_I, \tag{2.1}$$

where U is the system's internal energy, S is the entropy, σ is the surface tension, ξ_i is the chemical potential of species i, m_i is the total mass of species i, and A_I is the total interfacial area in the system. It is often easier to discuss surface tension in terms of Helmholtz and Gibbs free energies, which are defined, respectively, as

$$F = U - TS \tag{2.2}$$

and

$$G = U + PV - TS. \tag{2.3}$$

For a system at equilibrium, then

$$dF = -S\,dT - P\,dV + \sum_i \xi_i\,dm_i + \sigma\,dA_I \tag{2.4}$$

and

$$dG = -S\,dT + V\,dP + \sum_i \xi_i\,dm_i + \sigma\,dA_I. \tag{2.5}$$

The surface tension is thus related to the Hemholtz and Gibbs free energies according to

$$\sigma = \left(\frac{\partial F}{\partial A_{\mathrm{I}}}\right)_{T,V,N} = \left(\frac{\partial G}{\partial A_{\mathrm{I}}}\right)_{T,P,N} \tag{2.6}$$

where subscript N implies that the chemical makeup of the system remains unchanged.

A classical interpretation of surface tension is as follows. The work needed to increase the interfacial area in a system is

$$dW = \sigma\, dA_{\mathrm{I}}. \tag{2.7}$$

let us define f^{I} as the specific Helmholtz free energy for the interfacial area (i.e., Helmholtz free energy per unit interfacial area). Then, for a process without chemical reaction in which $T = $ const. and $V = $ const.,

$$dF = d\left(A_{\mathrm{I}} f^{\mathrm{I}}\right). \tag{2.8}$$

From Eqs. (2.6) and (2.8), one can write

$$\sigma = \left(\frac{\partial F}{\partial A_{\mathrm{I}}}\right)_{T,V,N} = f^{\mathrm{I}} + A\left(\frac{\partial f^{\mathrm{I}}}{\partial A_{\mathrm{I}}}\right)_{T,V,N}, \tag{2.9}$$

The second term on the right side is zero, because f^{I} does not depend on the magnitude of the interfacial area. Thus, for a process without chemical reaction in which $T = $ const. and $V = $ const., one has

$$\sigma = f^{\mathrm{I}}. \tag{2.10}$$

It can be similarly shown that, for a process without chemical reaction in which $T = $ const. and $V = $ const.,

$$\sigma = g^{\mathrm{I}}, \tag{2.11}$$

where g^{I} is the Gibbs free energy per unit interfacial area.

Consider now the interface between two pure, isothermal, and immiscible fluids (1) and (2), at equilibrium. For a segment of the interface defined by the orthogonal and infinitesimally short line segments δs_1 and δs_2 (see Fig. 2.1), the surface tension forces and the force resulting from an imbalance between phasic pressures need to be at equilibrium. For equilibrium in the $\vec{N}$ direction (the direction perpendicular to the interphase),

$$(P_1 - P_2) ds_1 ds_2 = 2\sigma\, ds_1 \sin\frac{d\theta_2}{2} + 2\sigma\, ds_2 \sin\frac{d\theta_1}{2} \approx \sigma\,(ds_1 d\theta_2 + ds_2 d\theta_1), \tag{2.12}$$

where subscripts 1 and 2 refer, respectively, to the fluids beneath and above the interphase in Fig. 2.1. This expression simplifies to

$$P_1 - P_2 = \sigma\left[\left(\frac{ds_1}{d\theta_1}\right)^{-1} + \left(\frac{ds_2}{d\theta_2}\right)^{-1}\right]. \tag{2.13}$$

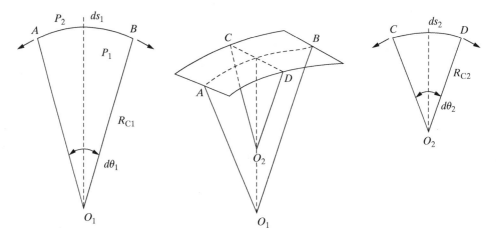

Figure 2.1. Surface tension forces.

By noting that $ds_1/d\theta_1 = R_{C1}$ and $ds_2/d\theta_2 = R_{C2}$, where R_{C1} and R_{C2} are the principal radii of curvature, Eq. (2.13) leads to

$$P_1 - P_2 = \sigma \left(\frac{1}{R_{C1}} + \frac{1}{R_{C2}} \right) = 2\sigma K_{12}, \qquad (2.14)$$

where K_{12} is the mean surface curvature. Equation (2.14) is the *Young–Laplace* equation. An important property of any surface is that at any point the mean curvature is a constant. Thus, for a sphere,

$$P_1 - P_2 = \frac{2\sigma}{R}. \qquad (2.15)$$

Extensive surface tension data for various liquids are available. For a liquid in contact with its own vapor the surface tension is a function of temperature and must satisfy the following obvious limit:

$$\sigma \to 0 \quad \text{as } P \to P_{cr}. \qquad (2.16)$$

Here P_{cr} represents the critical pressure. Empirical correlations for surface tension must account for this condition. For pure water, an accurate correlation is (International Association for the Properties of Water and Steam, 1994)

$$\sigma = 0.238(1 - T/T_{cr})^{1.25} \left[1 - 0.639 \left(1 - \frac{T}{T_{cr}} \right) \right], \qquad (2.17)$$

where T is in kelvins, σ is in newtons per meter, and $T_{cr} = 647.15$ K.

A useful and reasonably accurate empirical correlation for many liquids is

$$\sigma = a - bT. \qquad (2.18)$$

Table 2.1 contains surface tension data for several liquids and values of coefficients a and b for some (Jasper, 1972; Lienhard and Lienhard, 2005).

The preceding discussion dealt with surface tension of a pure liquid, in which case σ can be assumed to depend on temperature, and not on interphase curvature or any

Table 2.1. *Surface tensions of some pure liquids.*

Substance	Temperature, T (°C)	Surface tension, σ (N/m) $\times 10^3$	Substance	Temperature, T (°C)	Surface tension, σ (N/m) $\times 10^3$
Carbon dioxide	−40	13.14	Hydrogen	−258	2.8
	−30	10.82		−255	2.3
	−20	8.6		−253	1.95
	−10	6.5		−248	1.1
	0.0	4.55	Oxygen	−213	20.7
	10	2.77		−193	16.0
	20	1.21		−173	11.1
	30	0.06	Sodium	500	175
Ammonia	−70	59.1		700	160
	−50	51.1		900	140
	−30	43		1100	120
	−10	36.3	Potassium	500	105
	10	29.6		700	90
	20	26.4		900	76
	30	23.3		600	400
	40	20.3	Mercury	300	470
Benzene	10	30.2		400	450
	30	27.6		500	430
	50	24.9		600	400
	70	22.4			

Substance	Temperature range (°C)	a(N/m $\times 10^3$)	b(N/m °C $\times 10^3$)
Nitrogen	−195 to −183	26.42	0.2265
Oxygen	−202 to −184	−33.72	−0.2561
Carbon tetrachloride	15 to 105	29.49	0.1224
Mercury	5 to 200	490.6	0.2049
Methyl alcohol	10 to 60	24.00	0.0773
Ethyl alcohol	10 to 100	24.05	0.0832
Butyl alcohol	10 to 100	27.18	0.08983

external force field. In practice, some parameters can affect the surface tension, and one can write

$$\sigma = \sigma_0 - \sum \Pi_j, \qquad (2.19)$$

where σ_0 is the surface tension of pure liquid and Π_j is the *interfacial pressure* (in force per unit length newtons per meter in SI units) associated with mechanism j and is positive when the interfacial pressure is repulsive.

2.1.2 Contact Angle

When a liquid droplet is placed on a solid surface, the condition similar to that depicted in Fig. 2.2 is noticed. Under equilibrium, on a plane that is perpendicular to the three-phase contact line, a line tangent to the gas–liquid interphase and passing

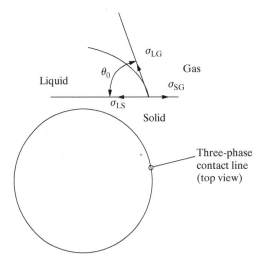

Figure 2.2. The liquid–gas–solid interface.

through the point where all three phases meet forms an angle θ_0, called the *contact angle* with the solid surface. When the three phases (solid, liquid, and gas) are at equilibrium, the net force acting on the point where the three phases meet must be zero. This requires that

$$\sigma_{SG} = \sigma_{LS} + \sigma_{LG} \cos \theta_0, \tag{2.20}$$

where σ_{SG} is the interfacial tension between solid and gas, σ_{LG} is the interfacial tension between liquid and gas (the surface tension of the liquid), and σ_{LS} is the interfacial tension between liquid and solid.

The work of adhesion between the solid surface and liquid, W_{SL}, can be defined as the amount of energy needed to separate the liquid from a unit solid surface area and thereby expose the separated solid and liquid unit surfaces to gas. It can easily be shown that

$$W_{SL} = \sigma_{SG} + \sigma_{LG} - \sigma_{LS}. \tag{2.21}$$

Combing Eqs. (2.20) and (2.21) leads to the *Young–Dupré equation*

$$W_{SL} = \sigma_{LG}(1 + \cos \theta_0). \tag{2.22}$$

The equilibrium contact angle θ_0 also characterizes the *surface wettability*. Complete wetting occurs when $\theta_0 \approx 0$, whereby the liquid attempts to spread over the entire solid surface. In contrast, complete nonwetting occurs when $\theta_0 \approx 180°$. Partial wetting occurs when $\theta_0 < 90°$; and partial nonwetting is encountered when $\theta_0 > 90°$. Surface wettability has an important effect on boiling incipience and nucleate boiling (Tong *et al.*, 1990; You *et al.*, 1990).

2.1.3 Dynamic Contact Angle and Contact Angle Hysteresis

Experiments show that the magnitude of the contact angle for a liquid–solid pair is not a constant; it depends on the relative motion between the solid–liquid–gas contact line and the solid surface (Schwartz and Tejada, 1972). In the absence of

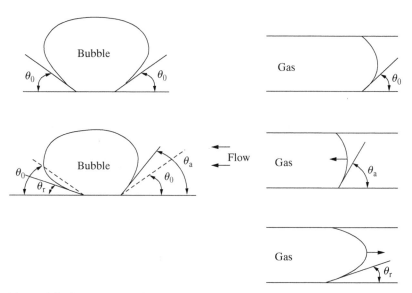

Figure 2.3. Contact angle hysteresis: θ_0 = equilibrium contact angle; θ_r = receding contact angle; θ_a = advancing contact angle.

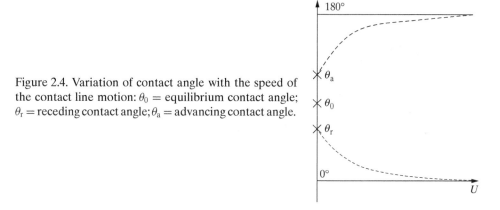

Figure 2.4. Variation of contact angle with the speed of the contact line motion: θ_0 = equilibrium contact angle; θ_r = receding contact angle; θ_a = advancing contact angle.

any motion, the static contact angle θ_0 is established. When the gas–liquid interphase moves toward the gas phase (i.e., when liquid spreads on the surface) we deal with advancing contact angle θ_a. Receding contact angle θ_r is observed when the gas–liquid interphase moves toward the liquid phase. In general $\theta_r < \theta_0 < \theta_a$ (see Fig. 2.3), and for inhomogeneous surfaces the receding and the advancing contact angles may depend on the speed of the gas–liquid–solid contact line with respect to the solid surface (Schwartz and Tejada, 1972; see Fig. 2.4). The difference between dynamic and static contact angles can be large. For example, for a Teflon–octane system where $\theta_0 = 26°$, $\theta_a = 48°$ for an advancing velocity of 9.7 cm/s (Schwartz and Tejada, 1972).

2.1.4 Surface Tension Nonuniformity

Nonuniformity in surface tension distribution over a gas–liquid interphase can lead to a net interfacial shear stress and cause flow in an otherwise quiescent fluid

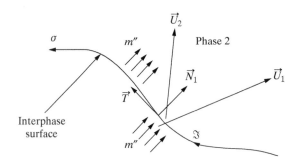

Figure 2.5. Schematic of a two-dimensional interphase.

field. The fluid motion caused by the surface tension gradient is referred to as the *Marangoni effect.*

Consider the two-dimensional flow field in Fig. 2.5, where the interphase separates the two phases 1 and 2. It can be shown that conservation of linear momentum for the interphase leads to

$$m''(\vec{U}_1 - \vec{U}_2) + (P_1\bar{\bar{I}} - \bar{\bar{\tau}}_1 - P_2\bar{\bar{I}} + \bar{\bar{\tau}}_2) \cdot \vec{N}_1 - \frac{\sigma}{R_C}\vec{N}_1 + \frac{d\sigma}{d\Im}\vec{T} = 0, \qquad (2.23)$$

where $\vec{T}$ is the unit tangent vector. The momentum balance in the direction perpendicular to the interphase can be derived by obtaining the scalar (dot) product of Eq. (2.23) with $\vec{N}_1$ to get

$$m''(u_{1,n} - u_{2,n}) + P_1 - P_2 + 2\mu_2\left(\frac{\partial u_{2,n}}{\partial n}\right) - 2\mu_1\left(\frac{\partial u_{1,n}}{\partial n}\right) - \frac{\sigma}{R_c} = 0, \quad (2.24)$$

where $u_{1,n}$ and $u_{2,n}$ represent components of the velocity vectors $\vec{U}_1$ and $\vec{U}_2$ in the direction perpendicular to the interphase and defined positive in the direction of $\vec{N}_1$. Likewise, for the tangential coordinate $\Im$, the dot product of Eq. (2.23) with $\vec{T}$ gives

$$\mu_1\left(\frac{\partial u_{1,n}}{\partial \Im} + \frac{\partial u_{1,\Im}}{\partial n}\right) - \mu_2\left(\frac{\partial u_{2,n}}{\partial \Im} + \frac{\partial u_{2,\Im}}{\partial n}\right) = \frac{\partial \sigma}{\partial \Im}. \qquad (2.25)$$

Clearly, the presence of nonuniformity in σ can affect the shear stresses on both sides of the interphase. Surface tension nonuniformities can result from the nonuniform distribution of the concentration of surface-active materials (leading to diffusocapillary flows), spatial variations of electric charges or surface potential (leading to electrocapillary flows), or the nonuniform interface temperature distribution (resulting in the thermocapillary flows).

2.2 Effect of Surface-Active Impurities on Surface Tension

Surfactants are typically polar molecules with one end having affinity with the liquid (hydrophilic when the liquid is water) and the other end of the molecule being repulsed by the liquid (hydrophobic for water). They tend to spread over the interphase, and when they are present in small quantities they tend to form a monolayer. The molecules in the monolayer impose a repulsive force on one another that is opposite to the compressive surface tension force. The result is a reduction in the surface tension by Π_Γ, the repulsive pressure of the adsorbed layer. The reduction of surface tension can be by up to five orders of magnitude. Usually $\Pi_\Gamma < \sigma_0$, and a stable interphase is maintained. If $\Pi_\Gamma > \sigma_0$, however, $\sigma < 0$ results, and the interphase tends to expand indefinitely. In liquid–liquid mixtures this would lead to emulsification. The repulsive pressure caused by a surfactant is constant only when the surfactant concentration at the interphase (i.e., the number of moles per unit interfacial surface area) is uniform. This can be the case only when the flow field is stagnant. However, fluid motion at the vicinity of the interphase generally causes the surfactant concentration to become nonuniform.

For a two-component dilute solution of a surfactant in a liquid, it can be proved that (Davies and Rideal, 1963)

$$\frac{\partial \Pi_\Gamma}{\partial C_\Gamma} = \frac{\Gamma R_u T}{C_\Gamma}. \tag{2.26}$$

Here Γ is the surface concentration of the surfactant (in kilomoles per meter squared in SI units) and C_Γ is the bulk molar concentration of the surfactant in the liquid (in kilomoles per meter cubed). This expression is called Gibbs' equation. The concentration of the surfactant on the interphase itself can be represented by the following transport equation (Levich, 1962; Probstein, 2003):

$$\frac{\partial \Gamma}{\partial t} + \nabla_I \cdot (\Gamma \vec{U}_I) = D_{\Gamma,I} \nabla_I^2 \Gamma - [D_\Gamma \nabla C_\Gamma] \cdot \vec{N}, \tag{2.27}$$

where $D_{\Gamma,I}$ is the binary surface diffusivity of the surfactant, D_Γ is the binary diffusivity of the surfactant with respect to the liquid bulk, and $\vec{U}_I$ is the velocity of the interphase. The unit normal vector $\vec{N}$ is oriented toward the gas phase. The last term on the right side of Eq. (2.27) accounts for the diffusion of the surfactant in the liquid bulk, and it can be neglected when the surfactant has a negligibly small solubility in the liquid. The operator ∇_I is the gradient on the interfacial surface. On the surface of a sphere with radius R, for example,

$$\nabla_I^2 \Gamma = \frac{1}{R^2 \sin\theta} \frac{\partial}{\partial \theta} \left(\sin\theta \frac{\partial \Gamma}{\partial \theta} \right) + \frac{1}{R^2 \sin^2\theta} \frac{\partial^2 \Gamma}{\partial \phi^2}, \tag{2.28}$$

$$\nabla_I \cdot (\Gamma \vec{U}_I) = \frac{1}{R \sin\theta} \frac{\partial}{\partial \theta} (\sin\theta \Gamma U_\theta) + \frac{1}{R \sin\theta} \frac{\partial}{\partial \phi} (\Gamma U_\phi). \tag{2.29}$$

Few data are available regarding the magnitude of $D_{\Gamma,I}$. Sakata (1969) has reported values of 10^{-9} to 10^{-8} m^2/s for myristic acid monolayers on water.

Surface-active impurities can have an important effect on gas–liquid interfacial hydrodynamics (Huang and Kintner, 1968; Springer and Pigford, 1970; Chang and Chung, 1985; Daiguji et al., 1977; Dey et al., 1997; Kordyban and Okleh, 1995). The interfacial waves can be significantly suppressed by surfactants, for example. The nonuniform stretching of the interphase during wave growth results in a net interfacial force that opposes the wave's further growth (Emmert and Pigford, 1954). The interfacial velocity can also be significantly reduced or even completely suppressed by surfactants during the motion of bubbles in liquids or the motion of droplets in gas. This in turn slows, or even completely stops, the internal circulation in the bubble or droplet. Using a constitutive relation of the form $\partial \Pi_\Gamma / \partial \Gamma = R_u T$, and using a surface diffusion coefficient range of $D_{\Gamma,I} = 10^{-9}$–10^{-3} m^2/s, with the higher

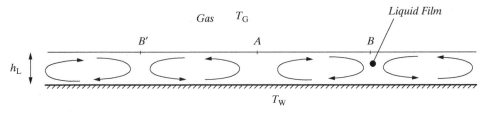

Figure 2.6. Bénard's circulation cells in a heated liquid film.

limit representing the diffusion of gaseous-type surfactants, Chang and Chung (1985) showed that the strength of the internal circulation of a spherical liquid droplet can be reduced by an order of magnitude owing to a small surfactant concentration on its surface. The internal circulation could be completely shut down with a high enough surfactant concentration at the interphase.

By spreading thin surfactant films on the surface of stagnant liquid pools, one can reduce the liquid evaporation rate. This is an application of surfactants when reduced evaporation is important. An example is the storage of highly radioactive spent nuclear fuel rods in water pools (Pauken and Abdel-Khalik, 1995). The spreading of thin liquid films on stagnant liquid surfaces has been investigated rather extensively in the past (Joos and Pinters, 1977; Foda and Cox, 1980; Camp and Berg, 1987; Dagan, 1984).

2.3 Thermocapillary Effect

The thermocapillary effect refers to the spatial variation of surface tension resulting from the nonuniformity of temperature on the gas–liquid interphase. The nonuniformity of surface tension leads to a net tangential force that can result in net force acting on a dispersed fluid particle or cause fluid motion in an otherwise quiescent flow field. As mentioned earlier, such fluid motion is referred to as the Marangoni effect.

One of the best-known surface tension-driven flows is the Bénard cellular flow that can occur in a thin liquid film (e.g., 1-mm-deep water film) heated from below (see Fig. 2.6). Warm liquid flows upward under point A, and from there flows toward points B and B'. While the liquid flows toward the latter points, its temperature diminishes owing to heat loss to the gas. Underneath points B and B' the cooled liquid flows downward toward the base of the liquid film. The temperature gradient that develops at the liquid–gas interphase, and its resulting surface tension gradient, drive the circulatory flow. The conditions necessary for the onset of the cellular motion can be modeled by using linear stability analysis (Scriven and Sterling, 1964). Such an analysis indicates that instability leading to the establishment of the recirculation cells depends on the following dimensionless parameters (Carey, 2008):

$$\text{Ma} = \frac{|T_w - T_l|}{h_L} \frac{|\frac{\partial \sigma}{\partial T}| h_L^2}{\alpha_L |\mu_L} \quad \text{(Marangoni number)}, \tag{2.30}$$

$$\text{Bi} = H_l h_L / k_L \quad \text{(Biot number)}, \tag{2.31}$$

$$\text{Bd} = \frac{\Delta \rho g h_L^2}{\sigma} \quad \text{(Bond number)}, \tag{2.32}$$

and

$$\text{Cr} = \frac{\mu_L \alpha_L}{\sigma h_L} \quad \text{(Crispation number)}. \tag{2.33}$$

The more general definition of the Marangoni number is

$$\text{Ma} = \left| \left(\frac{\partial \sigma}{\partial T} \right) \nabla T_I \right| \frac{l^2}{\alpha_L \mu_L}, \tag{2.34}$$

where l is a characteristic length and α_L is the thermal diffusivity of the liquid. The Marangoni number represents the ratio between the force arising from surface tension nonuniformity and viscous forces. The Marangoni number can also be interpreted as a thermal Peclet number by noting that a velocity scale can be defined by

$$U \sim \left| \left(\frac{\partial \sigma}{\partial T} \right) \nabla T_I \right| \frac{l}{\mu_L}$$

For a shallow liquid in a pan underlying a quiescent gas, for example, the maximum velocity in the liquid film (which occurs at the liquid layer surface) that results from the thermocapillary effect will be (Probstein, 2003)

$$U_{\text{max}} \approx \frac{\Delta T}{l} \frac{\left| \frac{\partial \sigma}{\partial T} \right| h_L}{\mu_L},$$

where l is the length of the pan, and ΔT is the temperature difference between the two ends of the liquid layer. (The two ends are separated from each other by a length l.) With the above representation of the capillary velocity scale, we then have

$$\text{Ma} = \text{Pe} = \frac{Ul}{\alpha_L}.$$

When the thermocapillary effect is the only mechanism causing nonuniformity in the surface tension, the right side of Eq. (2.25), which represents the interfacial force (force per unit width) in the tangential direction $\Im$, can be written as

$$\frac{\partial \sigma}{\partial \Im} = \frac{\partial T_I}{\partial \Im} \left(\frac{\partial \sigma}{\partial T} \right). \tag{2.35}$$

The term $(\partial \sigma / \partial T)$ for common liquids is often approximated as a constant, negative number.

EXAMPLE 2.1. The stationary, hemispherical micro vapor bubble shown in Fig. 2.7 is submerged in a stagnant thermal boundary layer. The vapor–liquid interfacial temperature is assumed to vary linearly with y, and the surface tension is a linear function of the interfacial temperature T_I. Derive an expression for the net thermocapillary force that acts on the bubble.

SOLUTION. The net surface tension force δF_σ that acts on the liquid phase can be written as

$$\delta \vec{F}_\sigma = (2\pi R_B \sin \theta) \left(\frac{\partial \sigma}{\partial \theta} \right) (-\vec{T}) d\theta, \tag{a}$$

$$\frac{\partial \sigma}{\partial \theta} = \left(\frac{\partial T_I}{\partial y} \right) \left(\frac{\partial y}{\partial \theta} \right) \left(\frac{\partial \sigma}{\partial T} \right) = -R_B \sin \theta \left(\frac{\partial T_I}{\partial y} \right) \left(\frac{\partial \sigma}{\partial T} \right), \tag{b}$$

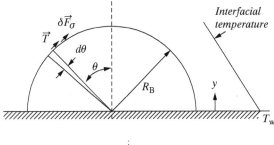

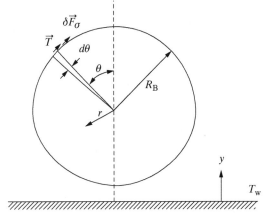

Figure 2.7. Thermocapillary effect in a hemispherical and spherical microbubble (Example 2.1).

where $\vec{T}$ is the unit tangent vector; the negative sign in front of it is because it is oriented against $d\theta$. We are interested only in the y component of $\delta\vec{F}_\sigma$, namely $\delta F_{\sigma,y} = \delta F_\sigma \sin\theta$. Therefore,

$$\delta F_{\sigma,y} = 2\pi R_B^2 \left(\frac{\partial T_I}{\partial y}\right)\left(\frac{\partial \sigma}{\partial T}\right)\sin^3\theta\, d\theta,$$

$$F_{\sigma,y} = \int_0^{\frac{\pi}{2}} \delta F_{\sigma,y} = 2\pi R_B^2 \left(\frac{\partial T_I}{\partial y}\right)\left(\frac{\partial \sigma}{\partial T}\right)\int_0^{\frac{\pi}{2}} \sin^3\theta\, d\theta$$

$$= \frac{4\pi}{3} R_B^2 \left(\frac{\partial T_I}{\partial y}\right)\left(\frac{\partial \sigma}{\partial T}\right). \tag{c}$$

The terms $\partial T_I/\partial y$ and $\partial\sigma/\partial T$ are both negative, meaning that $F_{\sigma,y} > 0$. A similar force, only in the opposite direction, will be imposed on the bubble. The thermocapillary force thus presses the bubble against the heated surface.

A similar analysis can be carried out when the interfacial temperature is an arbitrary function of y, giving

$$F_{\sigma,y} = 2\pi R_B^2 \int_0^{\frac{\pi}{2}} \left[\left(\frac{\partial \sigma}{\partial T}\right)\left(\frac{\partial T_I}{\partial y}\right)\right]\sin^3\theta\, d\theta, \tag{d}$$

where the integrand should be calculated at $y = R_B \cos\theta$. For a complete sphere (Problem 2.2), the thermocapillary force will be twice as large.

The analysis in Example 2.1 assumes no internal flow in the bubble. When the internal motion of the bubble is considered, Eq. (2.25) leads to the following boundary condition for the bubble:

$$\mu_{\text{G}} \left(\frac{\partial u_\theta}{\partial r} - \frac{u_\theta}{r} \right)_{r=R_{\text{B}}} - \mu_{\text{L}} \left(\frac{\partial u_\theta}{\partial r} - \frac{u_\theta}{r} \right)_{r=R_{\text{B}}} = \frac{1}{R_{\text{B}}} \left(\frac{\partial \sigma}{\partial \theta} \right)_{r=R_{\text{B}}}. \quad (2.36)$$

The hydrodynamic problem representing the motion of the bubble for $\text{Re}_{\text{B}} < 1$ (which justifies the neglection of inertial effects) can now be solved, provided that the temperature distribution over the bubble surface is known. If it is assumed that the bubble surface temperature is equal to the surrounding liquid temperature, and that the temperature distribution in the liquid is linear along coordinate y, then

$$\frac{1}{R_{\text{B}}} \left(\frac{\partial \sigma}{\partial \theta} \right)_{r=R_{\text{B}}} = -\sin \theta \left(\frac{\partial \sigma}{\partial T} \right) \left(\frac{\partial T_{\text{L}}}{\partial y} \right), \quad (2.37)$$

which is of course identical to the right-hand side of Eq. (b) in Example 2.1. A more realistic solution can be obtained, however, by noticing that the presence of the bubble will distort the temperature profile in its surrounding liquid. Young *et al.* (1959) reported on experiments in which small bubbles were kept stationary in a liquid pool by proper adjustment of the vertical temperature profile in the liquid. They also solved for the temperature profile for the interior and exterior of a spherical fluid particle suspended in another liquid by assuming steady state, and using the Hadamard–Rybczyński creep flow solution (Hadamard, 1911) for the hydrodynamics. (A useful and detailed derivation of the Hadamard–Rybczyński solution can be found in Chapter 8 of Levich (1962).) Their solution led to the following expression for the rise velocity of a spherical fluid particle (represented by subscript d) suspended in another stagnant liquid (represented by subscript c):

$$U_{\text{B}} = \frac{2g(\rho_{\text{c}} - \rho_{\text{d}})(\mu_{\text{d}} + \mu_{\text{c}})}{3\mu_{\text{c}}(2\mu_{\text{c}} + 3\mu_{\text{d}})} R_{\text{B}}^2 - \left(\frac{\partial \sigma}{\partial T} \right) \left(\frac{\partial T_{\text{L}}}{\partial y} \right) \frac{2k_{\text{c}}}{(2\mu_{\text{c}} + 3\mu_{\text{d}})(2k_{\text{c}} + k_{\text{d}})} R_{\text{B}}, \quad (2.38)$$

where y is the vertical upward (with respect to gravity) coordinate and $(\partial T_{\text{L}}/\partial y)$ represents the temperature gradient away from and undisturbed by the droplet. The solution of Young *et al.* for a small bubble suspended in liquid, when the approximations $\mu_{\text{d}}/\mu_{\text{c}} \approx 0$ and $k_{\text{d}}/k_{\text{c}} \approx 0$ are used, then leads to

$$\frac{4}{3} \pi R_{\text{B}}^3 (\rho_{\text{L}} - \rho_{\text{G}}) \vec{g} + 4\pi \mu_{\text{L}} R_{\text{B}} \vec{U} + 2\pi R_{\text{B}}^2 \left(\frac{\partial \sigma}{\partial T} \right) (\nabla T_{\text{L}}) = 0, \quad (2.39)$$

where $\vec{U}$ is the steady velocity of the bubble. The first term is the buoyancy force if it is multiplied by -1, and the second term is the drag force according to the classical Hadamard–Rybczyński solution for creep flow around an inviscid bubble (Hadamard, 1911). The third term represents the thermocapillary force, which tends to move the bubble in the direction of increasing temperature. The model of Young *et al.* has been compared with microgravity droplet migration data and has been found to do well for very small droplets with diameters of about 11 μm (Braum *et al.*, 1993). For larger droplets the qualitative dependence of the migration velocity on droplet size and the liquid temperature gradient appears to be correctly predicted by Eq. (2.38). It overpredicts the migration velocity for drops that have diameters of the order of 1 mm and larger, however (Wozniak, 1991; Xie *et al.*, 1998).

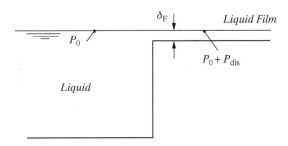

Figure 2.8. The disjoining pressure in a thin liquid film.

The thermocapillary effect is often unimportant in common thermal processes because of the dominance of hydrodynamic effects and buoyancy forces. It is, however, important in microscale phase-change processes and, in particular, in microgravity. It plays a role during the growth of vapor bubbles in subcooled boiling (Kao and Kenning, 1972; Marek and Straub, 2001). It has also been argued that in microgravity conditions the Marangoni effect can be an effective replacement for the buoyancy-dominated convection in normal gravity (Straub et al., 1994).

2.4 Disjoining Pressure in Thin Films

For ultrathin liquid films (less than about 100 μm in thickness) on solid surfaces, the proximity of the solid molecules to the liquid molecules in the vicinity of the liquid–gas interphase affects the pressure in the liquid film. The long-range intermolecular forces are responsible for this effect. This phenomenon can be modeled by defining a *disjoining pressure* P_{dis}, so that the pressure at the free surface of a thin liquid film on a flat surface will be

$$P = P_0 + P_{dis}, \tag{2.40}$$

where P_0 is the pressure at the surface of a thick film under similar conditions. P_{dis} is negative for wetting fluids as a result of the attraction between the solid and liquid. In the system depicted in Fig. 2.8, a negative P_{dis} causes the liquid to flow from the deep container into the thin film (because of the apparent lower pressure in the film). The disjoining pressure not only affects the liquid film spreading but also alters the thermodynamic equilibrium conditions at the vapor–liquid interphase, as will be seen in the next section. The disjoining pressure increases with decreasing liquid film thickness.

A useful discussion of disjoining pressure can be found in Faghri and Zhang (2006), where it is shown that when the long-range molecular interaction potential can be represented as $\phi(r) \approx -1/r^n$, where r is the intermolecular distance, then

$$P_{dis}(\delta_F) \approx -\frac{2}{\pi(n-2)(n-3)}\delta_F^{3-n}. \tag{2.41}$$

In the Lennard-Jones potential model (see Eq. (1.87)), the second term represents the long-range molecular interactions. For a fluid that follows the Lennard-Jones potential model of Eq. (1.87), $n = 6$, and that leads to

$$P_{dis} = \frac{A_0}{\delta_F^3}. \tag{2.42}$$

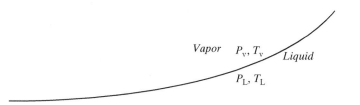

Figure 2.9. The vapor–liquid interphase at equilibrium.

This is a widely used representation of the disjoining pressure, where A_0 is a dispersion constant. The typical magnitude of A_0 can be demonstrated by the following two examples. For water, $A_0 = -2.87 \times 10^{-21}$ J (Park and Lee, 2003), and for ammonia $A_0 \approx -10^{21}$ J.

2.5 Liquid–Vapor Interphase at Equilibrium

We now consider the vapor–liquid interphase shown in Fig. 2.9, where mechanical and thermal equilibrium is assumed. First consider mechanical equilibrium. The Young–Laplace equation, Eq. (2.14), must be modified to consider the mechanisms that alter the surface tension, as well as the disjoining pressure, according to

$$P_v \cdot P_L = 2 \left(\sigma - \sum \Pi_j \right) K_{vL} - P_{dis}. \tag{2.43}$$

Equation (2.43) can now be called the augmented Young–Laplace equation.

Thermal equilibrium requires that $T_L = T_v = T_I$. For a flat, pure liquid–vapor interphase where disjoining pressure is absent, evidently $P_v = P_L = P$, and $T_L = T_v = T_{sat}(P)$ at the interphase. With $P_v \neq P_L$, however, both phases evidently cannot be at their normal saturated condition (i.e., saturation conditions corresponding to a flat interphase over a deep liquid layer). To find the relationship among P_v, P_L, and $P_{sat}(T_I)$, let us find the specific Gibbs free energy of the liquid and vapor.

Recall from thermodynamics that during any process involving vapor and liquid at equilibrium, the total Gibbs free energy in the system must remain unchanged. Also, recall that the chemical potential of a substance, ξ, is its partial specific (or molar) Gibbs free energy. Equilibrium thus requires that $\xi_v = \xi_L$. Using the definition $\xi = \mathbf{u} + P_v - T_s$, and noting that according to the Gibbs relation (the first Tds relation) $Tds = d\mathbf{u} + Pdv$, one can write

$$d\xi_v = -s_v dT + v_v dP_v \tag{2.44}$$

and

$$d\xi_L = -s_L dT + v_L dP_L. \tag{2.45}$$

The change in ξ_v when the vapor undergoes an isothermal process from $P_{sat}(T_I)$ to P_v will then be

$$\xi_v - \xi_g(T_I) = \int_{P_{sat}(T_I)}^{P_v} v_v dP = \left(\frac{R_u}{M} \right) T_I \ln \left(\frac{P_v}{P_{sat}(T_I)} \right), \tag{2.46}$$

where $\xi_g(T_I)$ is the chemical potential of saturated vapor at T_I. The vapor has been assumed to behave as an ideal gas. Likewise, the change in ξ_L when liquid undergoes an isothermal process from $P_{sat}(T_I)$ to P_L will be

$$\xi_L - \xi_f(T_I) = \int_{P_{sat}(T_I)}^{P_L} v_L dP = v_L\left[P_L - P_{sat}(T_I)\right], \qquad (2.47)$$

where $\xi_f(T_I)$ is the Gibbs free energy of saturated liquid. Now, $\xi_f(T_I) = \xi_g(T_I)$. Furthermore, equilibrium requires that $\xi_L = \xi_v$ at the interphase. Equations (2.46) and (2.47) then lead to

$$P_v = P_{sat}(T_I)\exp\left\{\frac{v_L\left[P_L - P_{sat}(T_I)\right]}{\frac{R_u}{M}T_I}\right\}. \qquad (2.48)$$

The substitution for P_L from Eq. (2.43) leads to

$$P_v = P_{sat}(T_I)\exp\left\{\frac{v_L\left[P_v - 2(\sigma - \sum \Pi_j)K_{vL} + P_{dis} - P_{sat}(T_I)\right]}{\frac{R_u}{M}T_I}\right\}. \qquad (2.49)$$

Often $P_v - P_{sat}(T_I) \ll 2(\sigma - \sum \Pi_j)K_{vL}$. When surface tension is a constant, $P_{dis} = 0$, and no interfacial force terms other than surface tension are present, this equation then reduces to the well-known Laplace–Kelvin relation

$$\ln\frac{P_v}{P_{sat}(T_I)} = -\frac{2\sigma v_L}{\frac{R_u}{M}T_I}K_{vL}. \qquad (2.50)$$

Equation (2.50) thus indicates that, when $K_{vL} > 0$ (e.g., inside a microbubble), the vapor pressure is in fact lower than the standard saturation pressure associated with the prevailing temperature. Thermal equilibrium between the bubble and the surrounding liquid thus requires the liquid to be slightly superheated. The opposite occurs on the surface of a microdroplet at equilibrium with its own vapor, where the vapor pressure must actually be higher than the saturation pressure.

This type of analysis can be extended to liquid–vapor mixtures where one or both phases contain an inert component. For a liquid at equilibrium with its own vapor mixed with a noncondensable gas, for example, Eq. (2.43) can be replaced with

$$P_v + P_n - P_L = 2\left(\sigma - \sum \Pi_j\right)K_{vL} - P_{dis}, \qquad (2.51)$$

where P_v represents the *vapor partial pressure* and P_n is the noncondensable partial pressure at the interphase. Equations (2.48) and (2.50) will apply, and Eq. (2.49) becomes

$$P_v = P_{sat}(T_I)\exp\left\{\frac{v_L\left[P_n + P_v - 2\left(\sigma - \sum \Pi_j\right)K_{vL} + P_{dis} - P_{sat}(T_I)\right]}{\frac{R_u}{M}T_I}\right\}. \qquad (2.52)$$

Also, for a solution composed of a solute (e.g., common salt) and a solvent (e.g., water) with a mole fraction of X_L, Eq. (2.50) can be applied provided that $P_{sat}(T_I)$ is replaced with P_s, with the latter defined as

$$P_s = P_{sat}(T_I)\gamma X_L, \qquad (2.53)$$

where γ is the activity coefficient. For ideal solutions, $\gamma = 1$.

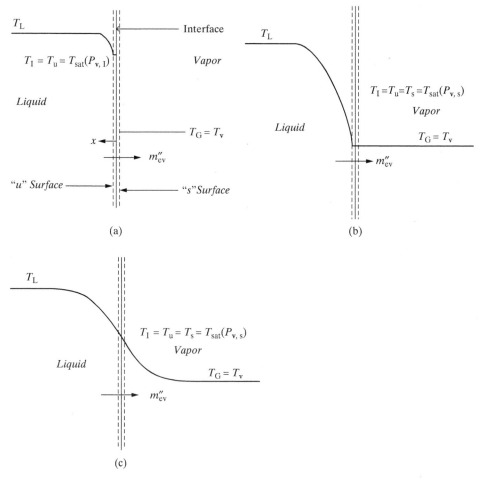

Figure 2.10. The temperature distribution near the liquid–vapor interphase: (a) early, during a very fast transient evaporation; (b) quasi-steady conditions with pure vapor; (c) quasi-steady conditions with a vapor–noncondensable mixture.

2.6 Attributes of Interfacial Mass Transfer

On the molecular scale, the interphase between a liquid and its vapor is always in violent agitation. Some liquid molecules that happen to be at the interphase leave the liquid phase (i.e., they evaporate), whereas some vapor molecules collide with the interphase during their random motion and join the liquid phase (i.e., they condense). The evaporation and condensation molecular rates are equal when the liquid and vapor phases are at thermal equilibrium. Net evaporation takes place when the molecules leaving the surface outnumber those that are absorbed by the liquid. When net evaporation or condensation takes place, the molecular exchange at the interphase is accompanied with a thermal resistance.

2.6.1 Evaporation and Condensation

For convenience of discussion, the interphase can be assumed to be separated from the gas phase by a surface (the s surface in Fig. 2.10(a)). When the interphase is flat and the disjoining pressure is negligible the temperature T_I and the vapor partial pressure at the interphase, $P_{v,I}$, are related according to

$$T_I = T_{\text{sat}}(P_{v,I}).\tag{2.54}$$

Equations (2.49) and (2.52) are the more general representations of the interphase, when P_v is replaced with $P_{v,I}$.

The conditions that lead to Eqs. (2.49), (2.52) or (2.54) are established over a time period that is comparable with molecular time scales and can thus be assumed to develop instantaneously for all cases of interest to us. Assuming that the vapor is at a temperature T_v in the immediate vicinity of the s surface, we can estimate the vapor molecular flux passing the s surface and colliding with the liquid surface from the molecular effusion flux as predicted by gaskinetic theory, when molecules are modeled as hard spheres (see Eq. (1.83)). If it is assumed that all vapor molecules that collide with the interphase join the liquid phase, then

$$j_{\text{cond}} = \frac{P_v}{\sqrt{2\pi \kappa_B m T_v}} = \frac{P_v}{\sqrt{2\pi (R_u/M_v)T_v}}.\tag{2.55}$$

The flux of molecules that leave the s surface and join the gas phase can be estimated from a similar expression where $P_{v,I}$ and T_I are used instead of P_v and T_v, respectively,

$$j_{\text{ev}} = \frac{P_{v,I}}{\sqrt{2\pi (R_u/M_v)T_I}}.\tag{2.56}$$

The net evaporation mass flux will then be

$$q_s'' = m_{\text{ev}}'' h_{\text{fg}} = h_{\text{fg}}\left[\frac{M_v}{2\pi R_u}\right]^{1/2}\left[\frac{P_{v,I}}{\sqrt{T_I}} - \frac{P_v}{\sqrt{T_v}}\right].\tag{2.57}$$

This expression is a theoretical maximum for the phase-change mass flux (the Knudsen rate). An interfacial heat transfer coefficient can also be defined according to

$$H_I = \frac{q_s''}{T_I - T_v}.\tag{2.58}$$

Equation (2.57) is known to deviate from experimental data. It has two important shortcomings, both of which can be remedied. The first is that it does not account for convective flows (i.e., finite molecular mean velocities) that result from the phase change on either side of the interphase. The second shortcoming is that Eq. (2.57) assumes that all vapor molecules that collide with the interphase condense, and none get reflected. Based on the predictions of gaskinetic theory when the gas moves with a finite mean velocity, Schrage (1953) derived

$$m_{\text{ev}}'' = \left[\frac{M_v}{2\pi R_u}\right]^{1/2}\left[\sigma_e \frac{P_{v,I}}{\sqrt{T_I}} - \Gamma\sigma_c \frac{P_v}{\sqrt{T_v}}\right],\tag{2.59}$$

where Γ is a correction factor that depends on the dimensionless mean velocity of vapor molecules that cross the s surface, namely $-m_{\text{ev}}''/\rho_v$, normalized with the mean molecular thermal speed $\sqrt{2R_u T_v/M_v}$, defined to be positive when net condensation takes place,

$$a = -\frac{m''_{ev}}{\rho_v}\left(\frac{2R_u T_v}{M_v}\right)^{-1/2} \approx -\frac{m''_{ev}}{P_v}\sqrt{\frac{R_u T_v}{2M_v}}, \tag{2.60}$$

and is given by

$$\Gamma = \exp(-a^2) + a\pi^{1/2}\left[1 + \mathrm{erf}(a)\right]. \tag{2.61}$$

The effect of mean molecular velocity only needs to be considered for vapor molecules that approach the interphase. No correction in needed for vapor molecules that leave the interphase, because there is no effect of bulk motion on them. Parameters σ_e and σ_c are the evaporation and condensation coefficients, and these are usually assumed to be equal, as would be required when there is thermostatic equilibrium. When $a < 10^{-3}$, as is often the case in evaporation and condensation, $\Gamma \approx 1 + a\pi^2$. Substitution into Eq. (2.59) and linearization then leads to

$$m''_{ev} = \left[\frac{M_v}{2\pi R_u}\right]^{1/2}\frac{2\sigma_e}{2-\sigma_e}\left[\frac{P_{v,I}}{\sqrt{T_I}} - \frac{P_v}{\sqrt{T_v}}\right]. \tag{2.62}$$

For $10^{-3} < a < 0.1$, the term $2\sigma_e/(2-\sigma_e)$ should be modified to $2\sigma_e/(2-1.046\sigma_e)$. This, equation, along with Eq. (2.58) and

$$q''_s = m''_{ev}h_{fg}$$

can now be used for the derivation of an expression for the interfacial heat transfer coefficient H_I.

The magnitude of the evaporation coefficient σ_e is a subject of some disagreement. For water, values in the $\sigma_e = 0.01$ to 1.0 range have been reported (Eames *et al.*, 1997). Careful experiments have shown that $\sigma_e \geq 0.5$ for water (Mills and Seban, 1967), however. Some investigators have obtained $\sigma_e = 1$ (Maa, 1967; Cammenga *et al.*, 1977) and have argued that smaller σ_e values measured by others were probably caused by experimental error.

EXAMPLE 2.2. A body of stagnant water is originally at a uniform temperature of 373 K. The surface of the body of water is instantaneously exposed to saturated water vapor at a pressure of 0.75 bar. Using Eq. (2.62) and assuming $\sigma_e = 1$, calculate the rate of evaporation with and without the interfacial thermal resistance included in the analysis.

SOLUTION. The heat transfer in the liquid can be modeled as diffusion in a semi-infinite medium if we neglect the motion of the interphase caused by evaporation, therefore

$$\rho_L C_{P,L}\frac{\partial T}{\partial t} = k_L \frac{\partial^2 T}{\partial x^2} \tag{a}$$

$$T(x,t) = T_0 = 373\,\mathrm{K} \quad \text{at} \quad t < 0 \tag{b}$$

$$T(\infty,t) = T_0 \quad \text{for} \quad t \geq 0. \tag{c}$$

An additional boundary condition, representing the conditions at the liquid–vapor interphase, is evidently needed.

If the interfacial thermal resistance is neglected, the situation will be similar to that depicted in Fig. 2.10(b), and we only need to use Eqs. (a)–(c), with $T_I = T_v = T_{sat}|_{0.75bar} = 364.9\,\text{K}$ as the boundary condition at the liquid–vapor interphase. The interphase temperature will be a constant and the solution will then be

$$\frac{T(x) - T_I}{T_0 - T_I} = \text{erf}\,\frac{x}{\sqrt{4\alpha_L t}}. \tag{d}$$

The heat flux from the liquid bulk to the interphase, the evaporation mass flux, and the interfacial heat transfer coefficient will then be, respectively,

$$q_s'' = -\left(-k_L \frac{\partial T}{\partial x}\right)_{x=0} = \frac{k_L}{\sqrt{\pi \alpha_L t}}(T_0 - T_I). \tag{e}$$

$$m_{ev}'' = q_s''/h_{fg} \tag{f}$$

$$H_I = \frac{\sqrt{\pi \alpha_L t}}{k_L}. \tag{g}$$

To include the effect of interfacial resistance (Fig. 2.10(a)), let us use the Clapeyron relation, Eq. (1.9), and replace v_{fg} with $v_{fg} \approx v_g \approx [P/(\frac{R_u}{M_v}T)]^{-1}$. One can then write

$$\ln \frac{P_{v,I}}{P_v} = \left(\frac{1}{T_v} - \frac{1}{T_I}\right)\frac{M_v h_{fg}}{R_u}. \tag{h}$$

Since $P_{v,I}/P_v \approx 1$ is expected, one can write

$$\ln \frac{P_{v,I}}{P_v} = \ln\left(1 + \frac{P_{v,I} - P_v}{P_v}\right) \approx \frac{P_{v,I} - P_v}{P_v}. \tag{i}$$

Therefore

$$P_{v,I} \approx P_v \cdot \left\{1 + \frac{h_{fg}(T_I - T_v)M_v}{R_u T_I T_v}\right\}. \tag{j}$$

One can now substitute for $P_{v,I}$ in Eq. (2.62) from this equation:

$$m_{ev}'' = 2\left[\frac{M_v}{2\pi R_u}\right]^{\frac{1}{2}} P_v \cdot \left\{\frac{1 + \frac{h_{fg}(T_I - T_v)M_v}{R_u T_I T_v}}{\sqrt{T_I}} - \frac{1}{\sqrt{T_v}}\right\}. \tag{k}$$

Combining with Eq. (f), this equation can be rewritten as:

$$k_L \frac{\partial T(x,t)}{\partial x}\bigg|_{x=0} = 2h_{fg}\left(\frac{M_v}{2\pi R_u}\right)^{1/2} P_v \left[\frac{1 + \frac{h_{fg}(T_I - T_v)M_v}{R_u T_I T_v}}{\sqrt{T_I}} - \frac{1}{\sqrt{T_v}}\right] \quad \text{for} \quad t \geq 0. \tag{l}$$

This equation represents the boundary condition at the liquid–vapor interphase. Equations (a)–(c) and (l) represent the rigorous solution of the problem, and require numerical solution. The numerical solution of course requires extremely fine nodalization in the vicinity of the liquid–vapor interphase, and extremely short time steps due to the extremely large temperature gradient and the extremely fast transient process. At any moment during the transient the interfacial heat transfer coefficient can be calculated from Eq. (2.58).

The calculation results are shown in Figs. 2.11 and 2.12, and are summarized in the following table. As noted, except for a very short period of time into the transient ($\approx 10^{-4}\,\text{s}$), the effect of interfacial thermal resistance is negligible.

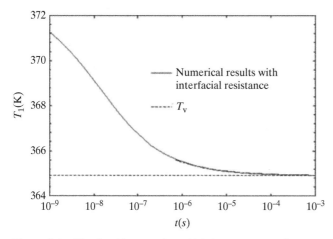

Figure 2.11. The liquid–vapor interfacial temperature for Example 2.2.

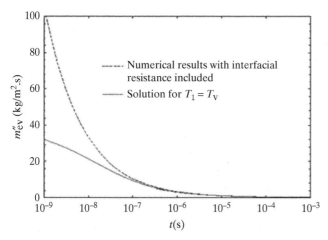

Figure 2.12. Evaporation mass flux for Example 2.2.

$t(s)$	$m''_{ev}(1)$ (kg/m^{2}s)	$m''_{ev}(2)$ (kg/m^{2}s)	$\sqrt{\pi \alpha_L t}/k_L$ (m^{2}K/W)	$1/H_1(2)$ (m^{2}K/W)
1×10^{-9}	104.972	32.243	3.42×10^{-8}	8.825×10^{-8}
1×10^{-8}	33.195	21.798	1.08×10^{-7}	8.748×10^{-8}
1×10^{-7}	10.497	9.860	3.42×10^{-7}	8.662×10^{-8}
1×10^{-6}	3.319	3.206	1.08×10^{-6}	8.615×10^{-8}
1×10^{-5}	1.050	1.014	3.42×10^{-6}	8.599×10^{-8}
1×10^{-4}	0.332	0.323	1.08×10^{-5}	8.595×10^{-8}
1×10^{-3}	0.105	0.103	3.42×10^{-5}	8.593×10^{-8}

(1) Interphase resistance neglected.
(2) Numerical solution with interphase resistance.

EXAMPLE 2.3. Now let us examine the effect of interfacial thermal resistance on evaporation of thin liquid films. An example is the case of nucleate boiling, where bubbles that grow on the heated surface are separated from the surface by a thin liquid film. Rapid evaporation takes place at the surface of the film, while the film is replenished by liquid flowing underneath the bubble. Estimate the evaporation rate at the surface of microlayers with film thicknesses in the 1–50 μm range during nucleate boiling of atmospheric water, and examine the effect of the interfacial thermal resistance on the calculation results. The wall is assumed to be at $T_w = 390$ K. For simplicity, treat the microlayer as a flat, quasi-steady liquid film.

SOLUTION. We deal with two thermal resistances in series; one represents heat conduction through the microlayer, and the other is associated with the liquid film–vapor interphase. With the interfacial thermal resistance and the effect of disjoining pressure neglected, one can write

$$q_w'' = m_{ev}'' h_{fg} = \frac{k_L}{\delta_F}(T_w - T_s).$$

When the interfacial thermal resistance and the effect of disjoining pressure are neglected, $T_I = T_s = T_{sat}(P_v) = 373.3$ K. With interfacial thermal resistance included, and under the assumption that T_w remains constant, the previous equation will be replaced with

$$q_w'' = m_{ev}'' h_{fg} = \frac{k_L}{\delta_F}(T_w - T_I). \tag{a}$$

This equation, along with Eq. (2.62), should now be solved with q_w'' (the surface heat flux) and T_I as the two unknowns. We need to account for the effect of the disjoining pressure, however. Substitution from Eq. (2.42) in Eq. (2.49) gives

$$P_{v,I} = P_{sat}(T_I) \exp \left\{ \frac{v_L \left[P_{v,I} + \frac{A_0}{\delta_F^3} - P_{sat}(T_I) \right]}{\frac{R_u}{M_v} T_I} \right\}, \tag{b}$$

where $A_0 = -2.87 \times 10^{-21}$ J. Equations (2.62), (a), and (b) should now be solved with q_w'', T_I, and $P_{v,I}$ as the unknowns. The iterative solution can be performed by using Antoine's equation for the saturation vapor pressure of water, according to which

$$\log_{10}\left[P_{sat}(T_I) \right] = a - \frac{b}{T_I + c}, \tag{c}$$

where

$$a = 7.96681$$

$$b = 1668.21$$

$$c = 228,$$

where T_I is in degrees Celsius and P_{sat} is in torr. Antoine's equation is a popular tool for curvefitting the vapor pressure of volatile substances, and has reasonable accuracy for water in the pressure range of 1 to about 200 kilopascals. The calculation results are summarized in the following table. The interfacial temperature T_I is shown with one decimal point precision.

$|\delta_F(\mu m)|$	$m_{ev}''(1)(kg/m^2 \cdot s)$	$m_{ev}''(2)(kg/m^2 \cdot s)$	$T_I(K)$
50	0.09946	0.1001	373.0
10	0.4973	0.4989	373.1
5	0.9946	0.9934	373.1
2	2.486	2.452	373.4
1	4.973	4.801	373.7

(1) Interphase resistance neglected.
(2) Interphase resistance included.

These two examples demonstrate that in common engineering calculations the interfacial thermal resistance can be comfortably neglected, and the interphase temperature profile will be similar to Fig. 2.10(b) or (c). When microsystems or extremely fast transients are dealt with, however, the interfacial thermal resistance may be important.

2.6.2 Sparingly Soluble Gases

The mass fraction profiles for a gaseous chemical species that is insoluble in the liquid phase (a "noncondensable") during rapid evaporation are qualitatively displayed in Fig. 2.13. For convenience, once again the interphase is treated as an infinitesimally thin membrane separated from the gas and liquid phases by two parallel planes "s" and "u," respectively. Noncondensable gases are not completely insoluble in liquids, however. For example, air is present in water at about 25 ppm by weight when water is at equilibrium with atmospheric air at room temperature. In many evaporation and condensation problems where noncondensables are present, the effect of the noncondensable that is dissolved in the liquid phase is small, and there is no need to keep track of the mass transfer process associated with the noncondensable in the liquid phase. There are situations where the gas released from the liquid plays an important role, however. An interesting example is the forced convection by a subcooled liquid in mini- and microchannels (Adams et al., 1999).

The release of a sparingly soluble species from a liquid that is undergoing net phase change is displayed in Fig. 2.14. Although an analysis based on the kinetic theory of gases may be needed for the very early stages of a mass transfer transient, such an analysis is rarely performed (Mills, 2001). Instead, equilibrium at the interphase with respect to the transferred species is often assumed. Unlike temperature, there is a significant discontinuity in the concentration (mass fraction) profiles at the liquid–gas interphase, even under equilibrium conditions. For convenience, we will follow the common practice of using X and Y to represent mole fractions in the liquid and gas phases, respectively. The equilibrium at the interphase with respect to a sparingly soluble inert species is governed by *Henry's law*, according to which

$$Y_{n,s} = He_n X_{n,u}, \tag{2.63}$$

where He_n is the *Henry number* for species n and the liquid and in general depends on pressure and temperature. The equilibrium at the interphase can also be presented in terms of *Henry's constant*, which is defined as $C_{He,n} = He_n P$, with P representing the

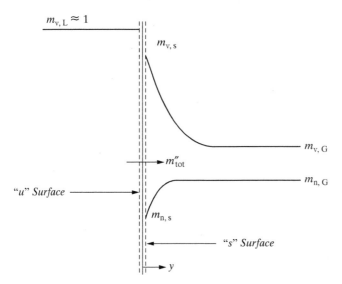

Figure 2.13. Mass fraction profiles near the liquid–vapor interphase during evaporation into a vapor–noncondensable mixture.

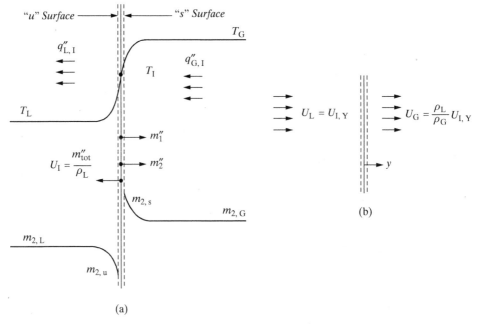

(a)

Figure 2.14. The gas–liquid interphase during evaporation and desorption of an inert species: (a) mass fraction profiles; (b) velocities when the coordinate is placed on the interphase.

total pressure. C_{He} is approximately a function of temperature only. If all the components of the gas phase are assumed to be ideal gases and the system is at equilibrium, then $X_{n,L} = X_{n,u}$ and $Y_{n,s} = Y_{n,G}$, and

$$C_{He,n}X_{n,u} = Y_{n,s}P = P_{n,s}, \tag{2.64}$$

where $P_{n,s}$ is the partial pressure of species n at the s surface. When the bulk gas and liquid phases are at equilibrium, then

$$C_{\mathrm{He},n}X_{n,\mathrm{L}} = Y_{n,\mathrm{G}}P = P_{n,\mathrm{G}}, \tag{2.65}$$

where now all parameters represent the gas and liquid bulk conditions. Evidently, C_{He} is related to the solubility of species n in the liquid. It is emphasized that these linear relationships only apply to sparingly soluble gases. When the gas phase is highly soluble in the liquid, Eq. (2.64) should be replaced with tabulated values of a nonlinear relation of the generic form $P_{n,s} = P_{n,s}(X_{n,\mathrm{u}}, T_{\mathrm{I}})$.

2.7 Semi-Empirical Treatment of Interfacial Transfer Processes

In most engineering problems the interfacial resistance for heat and mass transfer is negligibly small, and equilibrium at the interphase can be comfortably assumed. The interfacial transfer processes are then controlled by the thermal and mass transfer resistances between the liquid bulk and the interphase (i.e., the liquid-side resistances) and between the gas bulk and the interphase (i.e., the gas-side resistance).

Let us consider the situation where a sparingly soluble substance 2 is mixed with liquid represented by species 1. If the interphase is idealized as a flat surface, the configuration for a case when evaporation of species 1 and desorption of a dissolved species 2 occur simultaneously will be similar to Fig. 2.14(a). For simplicity, let us treat the mass flux of species 1 as known for now and focus on the transfer of species 2. The interfacial mass fluxes will then be

$$m_2'' = (1 - m_{1,\mathrm{u}})m_{\mathrm{tot}}'' - \rho_{\mathrm{L},\mathrm{u}}\boldsymbol{D}_{12,\mathrm{L}} \left.\frac{\partial m_2}{\partial y}\right|_{y=0} \tag{2.66}$$

and

$$m_{\mathrm{tot}}'' = m_1'' + m_2''. \tag{2.67}$$

Sensible and latent heat transfer can take place on both sides of the interphase. When the coordinate center is fixed to the interphase, as shown in Fig. 2.14(b), there will be fluid motion in the y direction on both sides of the interphase, where

$$U_{\mathrm{I},y} = \frac{m_{\mathrm{tot}}''}{\rho_{\mathrm{L}}}. \tag{2.68}$$

An energy balance for the interphase gives

$$m_1''h_{\mathrm{f}} + m_2''h_{2,\mathrm{LI}} + m_{\mathrm{tot}}''\frac{1}{2}U_{1,y}^2 - q_{\mathrm{LI}}'' = m_1''h_{\mathrm{g}} + m_2''h_{2,\mathrm{GI}} + m_{\mathrm{tot}}''\frac{1}{2}\left(\frac{\rho_{\mathrm{L}}U_{\mathrm{I}}}{\rho_{\mathrm{G}}}\right)^2 - q_{\mathrm{GI}}''. \tag{2.69}$$

If we neglect kinetic energy changes, this equation can be re-written as

$$q_{\mathrm{GI}}'' - q_{\mathrm{LI}}'' = m_1''h_{\mathrm{fg}} + m_2''h_{2,\mathrm{LG}}, \tag{2.70}$$

where $h_{2,\mathrm{LG}}$ is the specific heat of desorption for species 2.

The sensible heat transfer terms follow Fourier's law and can be represented by the convection heat transfer coefficients

$$q''_{GI} = k_G \left. \frac{\partial T_G}{\partial y} \right|_{y=0} = H_{GI}(T_G - T_I), \tag{2.71}$$

$$q''_{LI} = k_L \left. \frac{\partial T}{\partial y} \right|_{y=0} = H_{LI}(T_I - T_L). \tag{2.72}$$

The convection heat transfer coefficients must account for the distortion of the temperature profiles caused by the mass-transfer-induced fluid velocities, as described in Section 1.8. Mass transfer for species 2 can be represented as

$$m''_2 = (1 - m_{1,s})m''_{tot} - \rho_{G,s}\boldsymbol{D}_{12,G} \left. \frac{\partial m_2}{\partial y} \right|_s, \tag{2.73}$$

$$m''_2 = (1 - m_{1,u})m''_{tot} - \rho_{L,u}\boldsymbol{D}_{12,L} \left. \frac{\partial m_2}{\partial y} \right|_u. \tag{2.74}$$

These equations include advective and diffusive terms on their right-hand sides. Note that $\boldsymbol{D}_{12,G}$ and $\boldsymbol{D}_{12,L}$ are the binary mass diffusivity coefficients in the gas and liquid phases, respectively. Once again, for convenience the diffusion terms can be replaced by

$$-\rho_{G,s}\boldsymbol{D}_{12,G} \left. \frac{\partial m_2}{\partial y} \right|_s = \dot{K}_{GI}(m_{2,s} - m_{2,G}), \tag{2.75}$$

$$-\rho_{L,u}\boldsymbol{D}_{12,L} \left. \frac{\partial m_2}{\partial y} \right|_u = \dot{K}_{LI}(m_{2,L} - m_{2,u}), \tag{2.76}$$

where the mass transfer coefficients $\dot{K}_{GI}$ and $\dot{K}_{LI}$ must account for the distortion in the concentration profiles caused by the blowing effect of the mass transfer at the vicinity of the interphase.

The effect of mass transfer-induced distortions of temperature and concentration profiles can be estimated by the Couette flow film model. The liquid- and gas-side transfer coefficients are modified as (see Section 1.8)

$$\frac{\dot{H}_{GI}}{H_{GI}} = \frac{m''_{tot}C_{PG,t}/H_{GI}}{\exp\left(m''_{tot}C_{PG,t}/H_{GI}\right) - 1}, \tag{2.77}$$

$$\frac{\dot{H}_{LI}}{H_{LI}} = \frac{m''_{tot}C_{PL,t}/H_{LI}}{\exp\left(m''_{tot}C_{PL,t}/H_{LI}\right) - 1}, \tag{2.78}$$

$$\frac{\dot{K}_{GI}}{K_{GI}} = \frac{m''_{tot}/K_{GI}}{\exp\left(m''_{tot}/K_{GI}\right) - 1}, \tag{2.79}$$

$$\frac{\dot{K}_{LI}}{K_{LI}} = \frac{-m''_{tot}/K_{LI}}{\exp\left(-m''_{tot}/K_{LI}\right) - 1}, \tag{2.80}$$

where $C_{PG,t}$ and $C_{PL,t}$ are the specific heats of the transferred species in the gaseous and liquid phases, respectively, and H_{LI}, H_{GI}, K_{LI}, and K_{GI} are the convective transfer coefficients for the limit $m''_{tot} \to 0$. When the gas–liquid system is single-component (e.g., evaporation or condensation of a pure liquid surrounded by its own pure vapor), then $C_{PG,t} = C_{PG}$ and $C_{PL,t} = C_{PL}$.

Equations (2.77)–(2.80) are convenient to use when mass fluxes are known. The Couette flow film model results can also be presented in the following forms, which are convenient when the species concentrations are known:

$$\frac{\dot{H}_{GI}}{H_{GI}} = \ln(1 + B_{h,G})/B_{h,G}, \tag{2.81}$$

$$\frac{\dot{H}_{LI}}{H_{LI}} = \ln(1 + B_{h,L})/B_{h,L}, \tag{2.82}$$

$$\frac{\dot{K}_{GI}}{K_{GI}} = \ln(1 + B_{m,G})/B_{m,G}, \tag{2.83}$$

$$\frac{\dot{K}_{LI}}{K_{LI}} = \ln(1 + B_{m,L})/B_{m,L}, \tag{2.84}$$

where

$$B_{h,L} = \frac{-m''_{tot}C_{PL,t}}{\dot{H}_{LI}}, \tag{2.85}$$

$$B_{h,G} = \frac{m''_{tot}C_{PG,t}}{\dot{H}_{GI}}, \tag{2.86}$$

$$B_{m,G} = \frac{m_{2,G} - m_{2,s}}{m_{2,s} - m''_2/m''_{tot}}, \tag{2.87}$$

$$B_{m,L} = \frac{m_{2,L} - m_{2,u}}{m_{2,u} - m''_2/m''_{tot}}. \tag{2.88}$$

The transfer of species 1 can now be addressed. Since species 2 is only sparingly soluble, its mass flux at the interphase will be typically much smaller than the mass flux of species 1, when phase change of species 1 is in progress. The transfer of species 1 can therefore be modeled by disregarding species 2, in accordance with Section 1.8. The following example shows how.

EXAMPLE 2.4. A spherical 1.5-mm-diameter pure water droplet is in motion in dry air, with a relative velocity of 2 m/s. The air is at 25 °C. Calculate the evaporation mass flux at the surface of the droplet, assuming that at the moment of interest the droplet bulk temperature is 5 °C. For simplicity, assume quasi-steady state, and for the liquid-side heat transfer coefficient (i.e., heat transfer between the droplet surface and the droplet liquid bulk) use the correlation of Kronig and Brink (1950) for the internal thermal resistance of a spherical droplet that undergoes internal recirculation according to Hill's vortex flow:

$$Nu_{D,L} = \frac{H_{LI}d}{k_L} = 17.9. \tag{a}$$

SOLUTION. In view of the very low solubility of air in water, we can treat air as a completely passive component of the gas phase. The thermophysical and transport properties need to be calculated first. For simplicity, they will be calculated at 25 °C.

The results are as follows:

$$C_{PL} = 4{,}200\,\text{J/kg·K}, \; C_{Pv} = 1{,}887\,\text{J/kg·K}, \; D_{12} = 2.54 \times 10^{-5}\,\text{m}^2\text{/s},$$

$$k_G = 0.0255\,\text{W/m·K}, \; k_L = 0.577\,\text{W/m·K}, \; h_{fg} = 2.489 \times 10^6\,\text{J/kg},$$

$$\mu_G = 1.848 \times 10^{-5}\,\text{kg/m·s}, \; \rho_G = 1.185\,\text{kg/m}^3, \; \text{Pr}_G = 0.728.$$

We also have $M_n = 29\,\text{kg/kmol}$ and $M_v = 18\,\text{kg/kmol}$. We can now calculate the convective transfer coefficients. We will use the Ranz–Marshall correlation for the gas side. The following results are obtained:

$$\text{Re}_G = \rho_G U d/\mu_G = 192.3,$$

$$\text{Sc}_G = \frac{\mu_G}{\rho_G D_{12}} = 0.613,$$

$$\text{Nu}_G = H_{GI} d/k_G = 2 + 0.3\text{Re}_G^{0.6}\,\text{Pr}_G^{0.333} \Rightarrow H_{GI} = 141.7\,\text{W/m}^2\text{·K},$$

$$\text{Sh}_G = \frac{K_{GI} d}{\rho_G D_{12}} = 2 + 0.3\text{Re}_G^{0.6}\text{Sc}_G^{0.333} \Rightarrow K_{GI} = 0.1604\,\text{kg/m}^2\text{·s},$$

$$\frac{H_{LI} d}{k_L} = 17.9 \Rightarrow H_{LI} = 6{,}651\,\text{W/m}^2\text{·K}.$$

The following equations should now be solved iteratively, bearing in mind that $P = 1.013 \times 10^5\,\text{N/m}^2$ and $m_{v,\infty} = 0$:

$$X_{v,s} = P_{sat}(T_I)/P,$$

$$m_{v,s} = \frac{X_{v,s} M_v}{X_{v,s} M_v + (1 - X_{vs})M_n},$$

$$B_{hL} = -\frac{m'' C_{PL}}{\dot{H}_{LI}},$$

$$B_{hG} = \frac{m'' C_{Pv}}{\dot{H}_{LI}},$$

$$B_{mG} = \frac{m_{v,\infty} - m_{v,s}}{m_{v,s} - 1},$$

$$\dot{H}_{LI} = H_{LI}\ln(1 + B_{hL})/B_{hL}, \tag{b}$$

$$\dot{H}_{GI} = H_{GI}\ln(1 + B_{hG})/B_{hG}, \tag{c}$$

$$\dot{H}_{GI}(T_G - T_I) - \dot{H}_{LI}(T_I - T_L) = m'' h_{fg}, \tag{d}$$

$$m'' = K_{GI}\ln(1 + B_{mG}),$$

$$h_{fg} = h_{fg}\big|_{T_{sat} = T_I}. \tag{e}$$

The last equation can be dropped, by noting that the interface temperature will remain close to T_G, and therefore h_{fg} will approximately correspond to T_G. It is wise to first perform a scoping analysis by neglecting the effect of mass transfer on convection heat transfer coefficients to get a good estimate of the solution. In that case Eqs. (b) and (c) are avoided, and Eq. (d) is replaced with

$$H_{GI}(T_G - T_I) - H_{LI}(T_I - T_L) = m'' h_{fg}. \tag{f}$$

This scoping solution leads to $m'' = 8.595 \times 10^{-4}\,\text{kg/m}^2\cdot\text{s}$, $B_{hL} = -5.428 \times 10^{-4}$, and $B_{hG} = 0.01145$. Clearly, $B_{hL} \approx 0$, and there is no need to include Eq. (b) in the solution in other words, we can comfortably write $\dot{H}_{LI} = H_{LI}$ and solve this set of equations including Eq. (d). (With $B_{hL} \approx 0$, the inclusion of Eq. (b) may actually cause numerical stability problems.) The iterative solution of the aforementioned equations leads to

$$T_I = 278.1\,\text{K} \quad \text{and}$$
$$m'' = 8.594 \times 10^{-4}\,\text{kg/m}^2\cdot\text{s}.$$

The difference between the two evaporation mass fluxes is very small because this is a low-mass-transfer process to begin with.

EXAMPLE 2.5. In Example 2.4, assume that the droplet contains dissolved CO_2, at a bulk mass fraction of 20×10^{-5}. Calculate the rate of release of CO_2 from the droplet, assuming that the concentration of CO_2 in the air stream is negligibly small. Compare the mass transfer rate of CO_2 from the same droplet, if no evaporation took place.

SOLUTION. We have $M_{CO_2} = 44\,\text{kg/kmol}$. Also, $T_I \approx T_L = 5\,^\circ\text{C}$ and $C_{He} = 7.46 \times 10^7\,\text{Pa}$. Let us use subscripts 1, 2, and 3 to refer to H_2O, air, and CO_2, respectively. Then

$$\mathbf{D}_{31,L} = 1.77 \times 10^{-9}\,\text{m}^2/\text{s}.$$

For the diffusion of CO_2 in the gas phase, since the gas phase is predominantly composed of air, we will use the mass diffusivity of a CO_2–air pair at 15 °C. As a result,

$$\mathbf{D}_{32,G} = 1.49 \times 10^{-5}\,\text{m}^2/\text{s}.$$

The forthcoming calculations then follow:

$$\text{Sc}_G = \frac{\nu_G}{\mathbf{D}_{32,G}} = 1.04,$$

$$\text{Sh}_G = \frac{K_{GI}d}{\rho_G \mathbf{D}_{32,G}} = 0.2 + 0.3\text{Re}_G^{0.6}\text{Sc}_G^{0.333} \rightarrow \text{Sh}_G = 9.14;$$

$$K_{GI} = 0.108\,\text{kg/m}^2\cdot\text{s},$$

$$\text{Sh}_L = \frac{K_{LI}d}{\rho_L \mathbf{D}_{31,L}} = 17.9 \Rightarrow K_{LI} = 0.0212\,\text{kg/m}^2\cdot\text{s}.$$

The following equations must now be simultaneously solved, bearing in mind that $m_{3,G} = 0$ and $m_{3,L} = 20 \times 10^{-5}$:

$$m''_{\text{tot}} = m''_1 + m''_3, \tag{a}$$

$$m''_3 = m_{3,s}m''_{\text{tot}} + K_{GI}\frac{\ln(1 + B_{mG})}{B_{mG}}(m_{3,s} - m_{3,G}), \tag{b}$$

$$m''_3 = m_{3,u}m''_{\text{tot}} + K_{LI}\frac{\ln(1 + B_{mL})}{B_{mL}}(m_{3,L} - m_{3,u}), \tag{c}$$

$$X_{3,u} = \frac{PX_{3,s}}{C_{He}}, \tag{d}$$

$$m_{3,\mathrm{u}} \approx \frac{X_{3,\mathrm{s}}M_3}{X_{3,\mathrm{s}}M_3 + (1 - X_{3,\mathrm{s}})M_2}, \tag{e}$$

$$m_{3,\mathrm{u}} = \frac{X_{3,\mathrm{u}}M_3}{X_{3,\mathrm{u}}M_3 + (1 - X_{3,\mathrm{u}})M_1}, \tag{f}$$

$$B_{m\mathrm{G}} = \frac{m_{3,\mathrm{G}} - m_{3,\mathrm{s}}}{m_{3,\mathrm{s}} - \frac{m_3''}{m_{\mathrm{tot}}''}}, \tag{g}$$

$$B_{m\mathrm{L}} = \frac{m_{3,\mathrm{L}} - m_{3,\mathrm{u}}}{m_{3,\mathrm{u}} - \frac{m_3''}{m_{\mathrm{tot}}''}}. \tag{h}$$

Note that, from Example 2.4, $m_1'' = 8.594 \times 10^{-4}\,\mathrm{kg/m^2 \cdot s}$. The iterative solution of Eqs. (a)–(h) results in

$$m_{3,\mathrm{u}} = 8.73 \times 10^{-8},$$
$$m_{3,\mathrm{s}} = 3.99 \times 10^{-5},$$
$$m_3'' = 4.32 \times 10^{-6}\,\mathrm{kg/m^2 \cdot s}.$$

When evaporation is absent, the same equation set must be solved with $m_1'' = 0$. In that case,

$$m_{3,\mathrm{u}} = 8.37 \times 10^{-8},$$
$$m_{3,\mathrm{s}} = 3.82 \times 10^{-5},$$
$$m_3'' = 4.23 \times 10^{-6}\,\mathrm{kg/m^2 \cdot s}.$$

2.8 Multicomponent Mixtures

The brief discussion in this section is meant to prepare us for the discussion of boiling and condensation in binary mixtures. For detailed discussion textbooks on physical chemistry (e.g., Benson, 2009) can be used.

Chemical potential is the property that determines the transfer of a species from one phase to another. Species diffuse down their own chemical potential gradients. Consider a multicomponent mixture, composed of N moles and n components, whereby $N = \sum_{j=1}^{n} N_j$. For convenience, molar specific thermodynamics properties will be used in the forthcoming discussions. The formulation based on mass-based specific properties can easily be developed.

Let us recall that for a mixture of chemical species, the *partial molar property* $\tilde{p}_{j,\mathrm{mix}}$ is defined as:

$$\tilde{p}_{j,\mathrm{mix}} = \left(\frac{\partial \mathscr{P}}{\partial N_j}\right)_{T,P,N_i} \tag{2.89}$$

where $\mathscr{P}$ represents the total property of the system and $\tilde{p}$ represents the molar specific property.

The chemical potential of component j represents the partial specific Gibbs free energy of that species in the mixture, and is represented as:

$$\tilde{\mu}_j = \left(\frac{\partial G}{\partial N_j}\right)_{T,P,N_i,A_1} \tag{2.90}$$

where G is the mixture total Gibbs free energy, and A_I is the total interfacial area in the system. Equation (2.90) considers the possibility of an interphase surface in the system. It can be shown that, equivalently,

$$\tilde{\mu}_j = \left(\frac{\partial F}{\partial N_j}\right)_{T,V,N_i,A_\mathrm{I}} = \left(\frac{\partial \mathbf{U}}{\partial N_j}\right)_{V,S,N_i,A_\mathrm{I}} = \left(\frac{\partial H}{\partial N_j}\right)_{P,S,N_i,A_\mathrm{I}} \tag{2.91}$$

where F, H, $\mathbf{U}$, and S represent the total Helmholtz free energy, total enthalpy, total internal energy, and total entropy of the system, respectively.

For a multicomponent mixture undergoing phase change the TdS relations can be represented as:

$$TdS = d\mathbf{U} + PdV - \sum_{j=1}^{n}\tilde{\mu}_j dN_j - \sigma dA_\mathrm{I} \tag{2.92}$$

$$TdS = dH - VdP - \sum_{j=1}^{n}\tilde{\mu}_j dN_j - \sigma dA_\mathrm{I} \tag{2.93}$$

where σ is the interfacial tension. The last term in these equations represents the work associated with a change in the total interfacial area in the system. Using the definitions of functions H and F (see Eqs. (2.2) and (2.3)) it can also be shown that:

$$dG = -SdT + VdP + \sum_{j=1}^{n}\tilde{\mu}_j dN_j + \sigma dA_\mathrm{I} \tag{2.94}$$

$$dF = -SdT - PdV + \sum_{j=1}^{n}\tilde{\mu}_j dN_j + \sigma dA_\mathrm{I} \tag{2.95}$$

Consider now a constant-temperature process in an ideal gas mixture. There will evidently be no phase change and Eq. (2.94) then gives

$$d\tilde{g} = \tilde{v}dP = \frac{R_\mathrm{u}T}{P}dP, \quad P = \sum_{j=1}^{n}Y_j P, \tag{2.96}$$

where Y_j is the mole fraction of species j in the gas phase. Using Eq. (2.90), one then gets:

$$d\tilde{\mu}_j = R_\mathrm{u}Td\ln(Y_j P). \tag{2.97}$$

Non-ideal mixtures deviate from the above expression. To account for these deviations we define the property *fugacity* of species j in the mixture, $\hat{f}_{j,\mathrm{mix}}$, according to

$$d\tilde{\mu}_j = R_\mathrm{u}Td\ln\left(\hat{f}_{j,\mathrm{mix}}\right). \tag{2.98}$$

Given that at the limit of $P \to 0$ gases behave as ideal gases, evidently we must have

$$\lim_{P\to 0}\left(\frac{\hat{f}_{j,\mathrm{mix}}}{Y_j P}\right) = 1 \tag{2.99}$$

Furthermore, we must have

$$\lim_{Y_j \to 1} \hat{f}_{j,\text{mix}} = \hat{f}_j, \tag{2.100}$$

where $\hat{f}_j$ represents the fugacity of the pure substance. Integration of Eq. (2.98) for a $T = \text{const.}$ process, starting from a reference condition, then gives:

$$\tilde{\mu}_{j,\text{mix}} = \tilde{\mu}_{\text{ref}} + R_u T \ln \left(\frac{\hat{f}_{j,\text{mix}}}{\hat{f}_{j,\text{ref}}} \right). \tag{2.101}$$

One could choose the pure j at temperature T as the reference state, in which case $\tilde{\mu}_{\text{ref}}$ would be the chemical potential (or specific molar Gibbs free energy) of pure species j and $\hat{f}_{j,\text{ref}} = \hat{f}_j$, with $\hat{f}_j$ representing the fugacity of pure substance j.

A property closely related to fugacity is the *fugacity coefficient*, $\hat{\phi}$. For species j in the mixture, we write

$$\hat{\phi}_{j,\text{mix}} = \frac{\hat{f}_{j,\text{mix}}}{Y_j P}. \tag{2.102}$$

Evidently, we must have $\lim_{P \to 0} \hat{\phi}_{j,\text{mix}} = 1$.

For an ideal gas $\hat{f}_{j,\text{mix}} = Y_j P$ and $\hat{\phi}_{j,\text{mix}} = 1$ at all pressures and temperatures. For non-ideal mixtures it can be shown that

$$\ln \left(\frac{\hat{f}_{j,\text{mix}}}{Y_j P} \right) \Bigg|_{T,P,x_i} = \int_0^P (Z_j - 1) \frac{dP}{P} \Bigg|_{T=\text{const.}} \tag{2.103}$$

where Z_j is the compressibility factor of species j, namely,

$$Z_j = \frac{P \tilde{v}_{j,\text{mix}}}{R_u T}. \tag{2.104}$$

Note that the above integration must be performed at constant temperature. Using Eq. (2.103) one can obtain the following useful equation,

$$\frac{\hat{f}_{j,\text{mix}} (Y_j, P_2)}{\hat{f}_{j,\text{mix}} (Y_j, P_1)} = \int_{P_1}^{P_2} (Z_j - 1) \frac{dP}{P} \Bigg|_{T=\text{const.}} \tag{2.105}$$

Equations (2.103) and (2.105) are general and can be used for the calculation of fugacity in a mixture in any phase, provided that the compressibility of the substance j is known over the entire pressure rage of interest. It is widely used for gases at low and moderate pressures.

Similar expressions can be developed for pure substances, as well as mixtures (as opposed to single components in a mixture). For a pure substance j, for example, we can write

$$d\tilde{g}_j = R_u T d \ln \left(\hat{f}_j \right) \tag{2.106}$$

$$\lim_{P \to 0} \left(\frac{\hat{f}_j}{P} \right) = 1 \tag{2.107}$$

$$\ln\left(\frac{\hat{f}_j}{P}\right)\Bigg|_{T,P} = \int_0^P (Z_j - 1)\frac{dP}{P}\Bigg|_{T=\text{const.}} \tag{2.108}$$

$$\frac{\hat{f}_j(P_2)}{\hat{f}_j(P_1)} = \int_{P_1}^{P_2}(Z_j - 1)\frac{dP}{P}\Bigg|_{T=\text{const.}}. \tag{2.109}$$

Now, let us consider the liquid phase. For an ideal solution we will have

$$\hat{f}_{j,\text{mix}} = X_j P, \ \hat{f}_j = P, \hat{\phi}_j = 1 \tag{2.110}$$

where X_j is the mole fraction of species j in the liquid.

Equilibrium among phases in a system, multicomponent or otherwise, occurs when the chemical potential of each species is uniform throughout the system. It can be shown that this requirement leads to

$$\hat{f}_{j,\alpha} = \hat{f}_{j,\beta} = \hat{f}_{j,\gamma} = \ldots, \tag{2.111}$$

where subscripts $\alpha, \beta, \gamma, \ldots$ represent the phases in the system. This equation can be derived simply by applying Eq. (2.101) to the state of species j in all phases and equating the derived chemical potentials. Thus, at equilibrium, the fugacity of any species should be uniform in all the phases throughout the system.

The non-ideal behavior of a liquid mixture can also be presented by defining *activity coefficient* of a species j in the mixture, γ_j, a dimensionless parameter, whereby

$$\tilde{\mu}_j = \tilde{g}_j + R_u T \ln\left(\gamma_j X_j\right), \tag{2.112}$$

where $\tilde{g}_j$ is the specific molar Gibbs free energy of species j, when the species is in pure form. The activity coefficient is in general a function of pressure, temperature and composition, and must satisfy

$$\lim_{x_j \to 0} \gamma_j = 1. \tag{2.113}$$

For an ideal solution $\gamma_j = 1$ for all pressures, temperatures, and compositions. It can be shown that fugacity and activity coefficient are related according to

$$\hat{f}_{j,\text{mix}} = X_j \gamma_j \hat{f}_j. \tag{2.114}$$

When we mix different chemical species without chemical reaction, for any mixture property $\tilde{p}$ we can define the property change of mixing according to

$$\Delta\tilde{p} = \tilde{p}_{\text{mix}} - \sum_{j=1}^n X_j \tilde{p}_j, \tag{2.115}$$

where $\tilde{p}_{\text{mix}}$ is the mixture average specific molar property and $\tilde{p}_j$ is the specific molar property of pure substance j at the pressure and temperature of the mixture. Equation (2.115) thus represents the deviation of the average mixture property from the mole fraction-weighted average of that property. For an ideal mixture we have

$$\Delta\tilde{g} = R_u T \sum X_j \ln X_j \tag{2.116}$$

$$\Delta\tilde{s} = -R_u \sum X_j \ln X_j \tag{2.117}$$

$$\Delta\tilde{v} = \Delta\tilde{u} = \Delta\tilde{h} = 0. \tag{2.118}$$

For non-ideal solutions the above expressions are not accurate and deviations occur due to the interaction among the molecules of various species.

Let us now consider a multicomponent liquid–vapor mixture. In accordance with Eq. (2.111), at equilibrium we must have

$$\hat{f}_{j,\mathrm{G}} = \hat{f}_{j,\mathrm{L}}. \tag{2.119}$$

Applying Eq. (2.114) to the right side of this equation, one gets

$$Y_j \hat{\phi}_j P = X_j \gamma_j \hat{f}_{j,\mathrm{L}}. \tag{2.120}$$

Note that the properties on the left side all refer to the vapor phase, and properties on the right represent the liquid phase. Some simple manipulation of this relation (see Problem 2.20) leads to

$$\hat{f}_{j,\mathrm{G}} = X_i \gamma_j P_{j,\mathrm{sat}}(T) \exp\left[\int_0^{P_{j,\mathrm{sat}}(T)} (Z_j - 1)\frac{dP}{P} + \frac{1}{R_u T} \int_{P_{j,\mathrm{sat}}(T)}^{P} \tilde{v}_j dP \right], \tag{2.121}$$

where $P_{j,\mathrm{sat}}(T)$ is the partial pressure of species j at the system temperature and Z_j is the vapor phase compressibility factor. Evidently, $\tilde{v}_j$ on the right side represents the specific volume of pure component j in liquid state. This important equation relates the fugacity of species j in the mixture to the local saturation pressure of that species. It can be further simplified. Assuming that the vapor phase acts as an ideal gas ($Z_j = 1$), and treating the liquid phase as incompressible (a good assumption as long as we are away from the critical pressure), one gets

$$Y_j P = X_j \gamma_j P_{j,\mathrm{sat}}(T) \exp\left[\frac{\tilde{v}_{j,f}\left(P - P_{j,\mathrm{sat}}(T)\right)}{R_u T} \right]. \tag{2.122}$$

The argument of the exponential term on the right side is typically very small, and therefore the exponential term is close to unity. We thus get

$$Y_j P = X_j \gamma_j P_{j,\mathrm{sat}}(T). \tag{2.123}$$

If we further assume that the liquid phase is an ideal solution, we obtain *Raoult's law:*

$$Y_j P = X_j P_{j,\mathrm{sat}}(T). \tag{2.124}$$

The above derivations were performed for a system at equilibrium. These results can be applied to a system undergoing changes, provided that equilibrium is assumed at the liquid–gas interphase.

Now, referring to Fig. 1.6 for a binary mixture, we can write,

$$X_1 + X_2 = 1 \tag{2.125}$$

$$P = Y_1 P + Y_2 P = X_1 \gamma_1 P_{1,\mathrm{sat}}(T) + X_2 \gamma_2 P_{2,\mathrm{sat}}(T). \tag{2.126}$$

These equations then lead to

$$X_1 = \frac{P - \gamma_2 P_{2,\mathrm{sat}}(T)}{\gamma_1 P_{1,\mathrm{sat}}(T) - \gamma_2 P_{2,\mathrm{sat}}(T)} \tag{2.127}$$

$$Y_1 = \frac{\gamma_1 X_1 P_{1,\mathrm{sat}}(T)}{P}. \tag{2.128}$$

These two equations can be used to construct binary mixture phase diagrams similar to Figs. 1.6 and 1.7.

2.9 Interfacial Waves and the Linear Stability Analysis Method

Liquid and gas can exist in a multitude of patterns. Some of the simpler morphological configurations are important in natural and industrial processes. Examples include a horizontal layer of one phase overlaid on a layer of the other phase (stratified flow) and a cylindrical jet of one phase surrounded by the other. These simple configurations often can be sustained only in certain parameter ranges and would otherwise be disrupted by small random perturbations. Hydrodynamic stability theory seeks to define the conditions that are necessary for a gas–liquid interface to remain stable when exposed to small disturbances.

Hydrodynamic stability is the cornerstone of the *hydrodynamic theory of boiling* (Lienhard and Witte, 1985), according to which some of the most important pool boiling processes are hydrodynamically controlled. The theory has led to relatively successful mechanistic models, which will be reviewed in Chapter 11.

The discussion in this and other forthcoming sections dealing with interfacial instability will be limited to linear stability analysis, which examines the response of flow fields to interfacial disturbances that have infinitesimally small amplitudes. In reality, the interphase can be disrupted by large-amplitude disturbances that require nonlinear stability analysis. However, for a system to be stable in response to large-amplitude disturbances, it must also be stable in response to infinitesimally small disturbances. This is because disturbances in a stable system must decay and vanish, and before they completely disappear they will inevitably pass through the infinitesimally small disturbance amplitude range. As a result, a system that is found to be unstable based on linear stability considerations will not be stable in response to large-amplitude disturbances either.

The field of hydrodynamic instability and interfacial waves is vast. Useful treatises on the topic include those by Lamb (1932), Levich (1962), and Chandrasekhar (1961). In the forthcoming sections we will primarily be interested in instability phenomena that have direct applications in boiling and condensation.

Integration of Euler's Equation for an Inviscid Flow

Linear stability models often assume inviscid flows, and they utilize Euler's equation, which is now introduced. The Navier–Stokes equation for an inviscid flow is

$$\rho \frac{D\vec{U}}{Dt} = -\nabla P + \vec{F}_{\text{body}}. \tag{2.129}$$

The left side of this equation can be recast as

$$\rho \frac{D\vec{U}}{Dt} = \rho \left[\frac{\partial \vec{U}}{\partial t} + \nabla \left(\frac{1}{2} U^2 \right) - \vec{U} \times (\nabla \times \vec{U}) \right]. \tag{2.130}$$

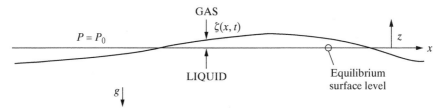

Figure 2.15. Disturbances on the surface of a quiescent liquid pool.

Therefore,

$$\rho \left[\frac{\partial \vec{U}}{\partial t} + \nabla \left(\frac{1}{2} U^2 \right) \right] = \rho \vec{U} \times (\nabla \times \vec{U}) - \nabla P + \rho \vec{g}. \qquad (2.131)$$

For irrotational flow, $\nabla \times \vec{U} = 0$. Furthermore, for irrotational flow a velocity potential ϕ can be defined so that

$$\vec{U} = \nabla \phi. \qquad (2.132)$$

Combining Eqs. (2.131) and (2.132), and integrating between two arbitrary points i and j, gives

$$\rho \left[\frac{\partial \phi}{\partial t} + \frac{1}{2} U^2 \right]_i^j = P_i - P_j - \rho g(z_j - z_i), \qquad (2.133)$$

where z_i and z_j are heights with respect to a reference plane in the gravitational field.

Note that Eq. (2.133) also applies along any streamline for an inviscid flow, even when the flow is rotational. To integrate Eq. (2.131) along a streamline, we should apply to both sides of the equation $\int_{\vec{r}_i}^{\vec{r}_j} \cdot d\vec{r}$, where $d\vec{r} = \vec{T} dl$, with $\vec{T}$ representing a unit tangent vector for the streamline and l representing distance along the streamline. Along a streamline, however, $\vec{U} \times (\nabla \times \vec{U}) \cdot \vec{T} = 0$, and the integration will lead to Eq. (2.133).

2.10 Two-Dimensional Surface Waves on the Surface of an Inviscid and Quiescent Liquid

The methodology for linear stability analysis is now demonstrated by the simple example displayed in Fig. 2.15. We would like to analyze the consequences of disturbances imposed on the surface of a deep and quiescent liquid pool.

The surface of the liquid is assumed to be disturbed by infinitesimally small two-dimensional disturbances. For an arbitrary point on the disturbed surface, and with the assumption that $\rho_G \ll \rho_L$, Eq. (2.133) leads to

$$\rho_L \frac{\partial \phi}{\partial t} \bigg|_{z=0}^{z=\zeta} = P_0 - P_1 - \rho_L g \zeta, \qquad (2.134)$$

where P_0 is the pressure at the interphase when the interphase is flat. Because of surface deflection the surface tension imposes a force that has to be balanced by

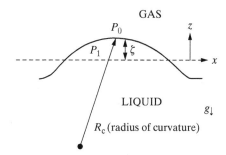

Figure 2.16. The disturbed interphase.

the pressure difference between the two sides of the interphase. This is depicted in Fig. 2.16, therefore,

$$P_1 - P_0 = \frac{\sigma}{R_c}, \tag{2.135}$$

where the radius of curvature is given by

$$R_c^{-1} = -\partial^2 \zeta / \partial x^2. \tag{2.136}$$

Substitution from Eqs. (2.135) and (2.136) in (2.134) gives

$$-\rho_L \left. \frac{\partial \phi}{\partial t} \right|_{z=\zeta} - \rho_L g \zeta + \sigma \frac{\partial^2 \zeta}{\partial x^2} = 0, \tag{2.137}$$

where $\rho_L \frac{\partial \phi}{\partial t}|_{z=0}^{z=\zeta} = \rho_L \frac{\partial \phi}{\partial t}|_{z=\zeta}$ has been used, because $\frac{\partial \phi}{\partial t}|_{z=0}$ here represents equilibrium and therefore corresponds to zero velocity.

We now would like to examine the response of the system to an arbitrary disturbance at the interphase. To be stable, the system must be stable for all arbitrary disturbances.

Assume that as a result of an arbitrary disturbance

$$\phi = A e^{kz} \text{Real} \left[\exp(-i [\omega t - kx]) \right], \tag{2.138}$$

where k is a real and positive number, as required by the boundary conditions. The reason for the selection of this general form is that any arbitrary ϕ can be represented as a Fourier integral (Chandrasekhar, 1961), whereby

$$\phi(x, z, t) = \int_{-\infty}^{+\infty} A_k(z, t) \exp(ikx) dk. \tag{2.139}$$

Given that the system equations will be linear, the response of the system to ϕ will thus depend on its response to disturbances of all wavelengths.

Equation (2.138) satisfies the following required conditions:

$$\nabla^2 \phi = 0 \quad \text{(mass continuity)}, \tag{2.140}$$

$$\lim_{z \to -\infty} \frac{\partial \phi}{\partial z} \to 0 \quad \text{(quiescent far field)}. \tag{2.141}$$

Kinematic consistency at the surface requires that the growth rate of the disturbance be the same as the liquid velocity at $z = \zeta$ in the y direction, therefore,

$$v = \left. \frac{\partial \phi}{\partial z} \right|_{z=\zeta} = \frac{\partial \zeta}{\partial t} + u \frac{\partial \zeta}{\partial x} = \frac{\partial \zeta}{\partial t}. \tag{2.142}$$

Since the disturbances are infinitesimally small, $\partial\phi/\partial z$ is found at $z = 0$, instead of $z = \zeta$. Substitution for ϕ from Eq. (2.138) in Eq. (2.142) leads to

$$\zeta = \zeta_0 \text{Real}\left\{i\exp(-i\left[\omega t - kx\right])\right\}, \qquad (2.143)$$

$$\zeta_0 = A\frac{k}{\omega}, \qquad (2.144)$$

where ζ_0 is the wave amplitude. The condition necessary for stability can now be seen in Eq. (2.143). The system will be stable as long as ω is real. Otherwise, the term $e^{-i\omega t}$ will grow indefinitely with time.

Now, substitution from Eqs. (2.138) and (2.143) into Eq. (2.137), and using the linear stability approximation of calculating everything at $z \approx 0$, gives a relation between frequency ω (in radians per second) and the wave number k:

$$\omega^2 = \sigma\frac{k^3}{\rho_L} + gk. \qquad (2.145)$$

Note that the wave number k is related to the wavelength λ according to

$$k = \frac{2\pi}{\lambda}. \qquad (2.146)$$

The system is evidently stable since ω is real for any value of k (or equivalently for any value of wavelength λ) and no unbounded growth with time can occur. The amplitude ζ_0 is arbitrary, as long as the condition $\zeta_0 \ll \lambda$ is satisfied. Also, the propagation velocity of waves is

$$c = \frac{\omega}{k}. \qquad (2.147)$$

Equation (2.145) can be re-written as

$$\omega^2 = \frac{8\pi^3\sigma}{\rho_L\lambda^3} + \frac{2\pi g}{\lambda}. \qquad (2.148)$$

It can be noted that for long waves, where $\lambda \gg 2\pi\sqrt{\sigma/\rho_L g}$, the second term on the right side is dominant, and $\omega \approx \sqrt{gk} = \sqrt{2\pi g/\lambda}$, and $c = \sqrt{g/k}$. These waves are called *gravity waves*, and their propagation velocity does not depend on surface tension. In contrast, waves with very short wavelengths ($\lambda \ll 2\pi\sqrt{\sigma/\rho_L g}$) are called *capillary waves* or *ripples*. For such waves, the second term on the right side of Eq. (2.148) can be neglected, leading to

$$\omega \approx \sqrt{\frac{\sigma k^3}{\rho_L}} \qquad (2.149)$$

and

$$c \approx \sqrt{\frac{\sigma k}{\rho_L}} = \sqrt{\frac{2\pi\sigma}{\rho_L\lambda}}. \qquad (2.150)$$

Thus, for ripples, the frequency and propagation speed depend primarily on surface tension and density, and not on the gravitational acceleration.

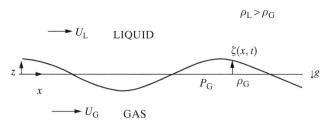

Figure 2.17. Interphase when liquid is overlaid on gas.

2.11 Rayleigh–Taylor and Kelvin–Helmholtz Instabilities

In this section two very important hydrodynamic instability types are discussed. The configurations are (1) a horizontal liquid layer superposed on a gas layer when both fluids are quiescent and (2) gas and liquid layers that have a relative velocity. The first is called the *Rayleigh–Taylor instability*, and the second is called the *Kelvin–Helmholtz instability*.

Consider two infinitely large inviscid, incompressible fluids separated by a horizontal interface. The fluids move at U_L and U_G when they are undisturbed (Fig. 2.17). The interface is disturbed by an arbitrary disturbance with an infinitesimally small amplitude.

Both fluids are inviscid and irrotational, and therefore velocity potentials can be defined for them. Mass continuity then requires that

$$\nabla^2 \phi_L = 0, \tag{2.151}$$

$$\nabla^2 \phi_G = 0. \tag{2.152}$$

Also, since velocities away from the interphase are known, the velocity potentials must satisfy

$$\lim_{z \to \infty} \nabla \phi_L = U_L \vec{e}_x, \tag{2.153}$$

$$\lim_{z \to -\infty} \nabla \phi_G = U_G \vec{e}_x, \tag{2.154}$$

where $\vec{e}_x$ is a unit vector in the x direction. These conditions are generally satisfied by

$$\phi_L = U_L x + \phi'_L, \tag{2.155}$$

$$\phi_G = U_G x + \phi'_G, \tag{2.156}$$

$$\phi'_L = b \exp\left[-kz + i(\omega t - kx)\right], \tag{2.157}$$

$$\phi'_G = b' \exp\left[kz + i(\omega t - kx)\right], \tag{2.158}$$

where k is real and positive. We can now apply the integral of Euler's equation to each phase, between the disturbed interphase and the neutral (flat) interphase, to get (for $i = L$ and G)

$$\rho_i \left[\left. \frac{\partial \phi'_i}{\partial t} \right|_0^\zeta + \frac{1}{2} \left\{ \left[U_i + \frac{\partial \phi'_i}{\partial x} \right]^2 + \left(\frac{\partial \phi'_i}{\partial z} \right)^2 \right\}_0^\zeta + g\zeta \right] = P_0 - P_i. \tag{2.159}$$

Expanding this equation and neglecting the second-order differential terms we will get

$$\rho i \left[\left.\frac{\partial \phi_i'}{\partial t}\right|_\zeta - \left.\frac{\partial \phi_i'}{\partial t}\right|_0 + \left[U_i \frac{\partial \phi_i'}{\partial x}\right]_0^\zeta + g\zeta \right] = P_0 - P_i. \qquad (2.160)$$

Subtracting Eq. (2.160) for gas from the same equation written for liquid gives

$$\rho_L \left(\frac{\partial \phi_L'}{\partial t} + U_L \frac{\partial \phi_L'}{\partial x} + g\zeta \right) - \rho_G \left(\frac{\partial \phi_G'}{\partial t} + U_G \frac{\partial \phi_G'}{\partial x} + g\zeta \right) = P_G - P_L. \qquad (2.161)$$

Kinematic conditions at the interphase require

$$\frac{\partial \zeta}{\partial t} + U_L \frac{\partial \zeta}{\partial x} \approx \left.\frac{\partial \phi_L'}{\partial z}\right|_{z=0}, \qquad (2.162)$$

$$\frac{\partial \zeta}{\partial t} + U_G \frac{\partial \zeta}{\partial x} \approx \left.\frac{\partial \phi_G'}{\partial z}\right|_{z=0}. \qquad (2.163)$$

Note that in writing the right-hand sides of these equations we have used the common linear stability analysis approximation. Both terms should really be calculated at $z = \zeta$; however, because the disturbance amplitude is infinitesimally small, the terms are instead calculated at $z = 0$.

Mechanical equilibrium at the interphase requires the following equation (which is similar to Eqs. (2.135) and (2.136)) to be satisfied:

$$P_G - P_L = \frac{\sigma}{R_c}. \qquad (2.164)$$

Now, assume that ζ is the real part of the arbitrary perturbation

$$\zeta = \zeta_0 \exp\left[i(\omega t - kx)\right], \qquad (2.165)$$

where $k > 0$. Substitution from Eqs. (2.157), (2.158), and (2.165) into Eqs. (2.162) and (2.163) gives

$$-bk = i\zeta_0(\omega - kU_L), \qquad (2.166)$$

$$b'k = i\zeta_0(\omega - kU_G). \qquad (2.167)$$

Equations (2.161), (2.136), (2.164), and (2.165) can now be combined to yield

$$\rho_L\left[i\omega b - iU_L bk - g\zeta_0\right] - \rho_G\left[i\omega b' - iU_G b'k + g\zeta_0\right] = \sigma\zeta_0 k^2. \qquad (2.168)$$

Substitution for b and b' from Eqs. (2.166) and (2.167) into Eq. (2.168) gives the *dispersion equation*:

$$c = \frac{\omega}{k} = \frac{\rho_L U_L + \rho_G U_G}{\rho_L + \rho_G} \pm \left[-\frac{g}{k}\frac{\rho_L - \rho_G}{(\rho_L + \rho_G)} - \frac{\rho_L \rho_G}{(\rho_L + \rho_G)^2}(U_L - U_G)^2 + \frac{\sigma k}{\rho_L + \rho_G} \right]^{1/2}. \qquad (2.169)$$

The derivations shown here dealt with infinitely thick layers of liquid and gas. When the layers have finite thicknesses equal to δ_L and δ_G, respectively, Eqs. (2.157) and (2.158) should be replaced with

$$\phi_L' = b \cosh\left[k(\delta_L - z)\right] \exp\left[i(\omega t - kx)\right], \qquad (2.170)$$

$$\phi_G' = b' \cosh\left[k(\delta_G + z)\right] \exp\left[i(\omega t - kx)\right]. \qquad (2.171)$$

It can then be shown that Eq. (2.169) applies, provided that ρ_L and ρ_G are replaced with $\rho'_L = \rho_L \coth(k\delta_L)$ and $\rho'_G = \coth(k\delta_G)$, respectively (see Problem 2.12).

The system is unstable when ω is complex. (Note that k is a real and positive number.) This can be seen from Eq. (2.165), because a complex ω would lead to an infinite growth of ζ.

A *neutral (critical)* condition results when the square-root term in Eq. (2.169) becomes equal to zero, and from there

$$\frac{g}{k_{cr}} \frac{\rho_L - \rho_G}{\rho_L + \rho_G} + \frac{\rho_L \rho_G}{(\rho_L + \rho_G)^2} (U_L - U_G)^2 - \frac{\sigma k_{cr}}{\rho_L + \rho_G} = 0. \tag{2.172}$$

Rayleigh–Taylor Instability

The Rayleigh–Taylor instability occurs when $U_L = U_G = 0$, in which case Eq. (2.169) leads to

$$\omega = \frac{k}{\sqrt{\rho_L + \rho_G}} \sqrt{\sigma k - g\Delta\rho/k}. \tag{2.173}$$

Whether the system is stable or not will depend on the sign of the square-root term. For neutral conditions, therefore

$$k_{cr} = \sqrt{g\Delta\rho/\sigma}. \tag{2.174}$$

The quantity $\lambda_L = 1/k_{cr} = \sqrt{\sigma/g\Delta\rho}$ is called the *Laplace length scale*. The neutral wavelength is therefore

$$\lambda_{cr} = \frac{2\pi}{k_{cr}} = 2\pi\lambda_L. \tag{2.175}$$

Waves with shorter wavelength are called ripples, and these do not cause the disruption of the interphase. Waves with wavelengths longer than λ_{cr} lead to the disruption of the interphase. The fastest growing wavelength (also sometimes referred to as the most dangerous wavelength!) can be found by applying $d\omega/dk = 0$ to Eq. (2.173), which results in

$$k_d = \sqrt{(g\Delta\rho)/(3\sigma)} = \frac{1}{\sqrt{3}\lambda_L}. \tag{2.176}$$

This is equivalent to

$$\lambda_d = 2\pi\sqrt{3}\lambda_L, \tag{2.177}$$

$$\omega_d = \left[\frac{4(\Delta\rho)^3 g^3}{27\sigma(\rho_L + \rho_G)^2}\right]^{0.25}. \tag{2.178}$$

Taylor instability analysis for a three-dimensional flow field (with a two-dimensional interphase on the (x, y) plane) can also be easily carried out (see Problem 2.10), whereby it can be shown that

$$\lambda_{cr2} = \sqrt{2}\lambda_{cr} = 2\pi\sqrt{2}\lambda_L, \tag{2.179}$$

$$\lambda_{d2} = \sqrt{2}\lambda_d = 2\pi\sqrt{6}\lambda_L. \tag{2.180}$$

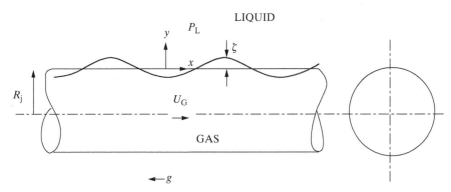

Figure 2.18. Schematic of a gaseous jet moving in a quiescent liquid.

Kelvin–Helmholtz Instability

The Kelvin–Helmholtz instability applies when the relative phase velocity $U_r = U_G - U_L$ is finite. Equation (2.169) indicates that complex ω, and therefore instability, occurs when

$$\frac{\rho_L \rho_G}{(\rho_L + \rho_G)^2}(U_L - U_G)^2 > \frac{\sigma k}{\rho_L + \rho_G} - \frac{g}{k}\frac{\rho_L - \rho_G}{\rho_L + \rho_G}. \tag{2.181}$$

Examination of the terms on the right side of this equation shows that surface tension attempts to stabilize the system, whereas gravity is destabilizing. When the gravitational term (the first term in Eq. (2.172) or the last term in Eq. (2.181)) is negligible, the critical wavelength will be

$$\lambda_{cr} = \frac{2\pi\sigma(\rho_L + \rho_G)}{\rho_L \rho_G U_r^2}. \tag{2.182}$$

An important application of Kelvin–Helmholtz instability theory is the critical (neutral) rise velocity of a gas or vapor jet in a quiescent liquid (see Fig. 2.18). Neglecting the effect of jet surface curvature, the jet critical rise velocity can be found by dropping the first term in Eq. (2.172), or equivalently the last term in Eq. (2.181) (because of the vertical configuration of the jet), and imposing $U_L = 0$. The result will be

$$U_{G,cr} = \left[\frac{\sigma k_H(\rho_G + \rho_L)}{\rho_G \rho_L}\right]^{1/2}, \tag{2.183}$$

where k_H is the critical wavelength for the jet, to be discussed shortly. For $\rho_G \ll \rho_L$, which is often true in boiling systems at low and moderate pressures, Eq. (2.183) gives

$$U_{G,cr} \approx (\sigma k_H/\rho_G)^{1/2} = (2\pi\sigma/\rho_G \lambda_H)^{1/2}, \tag{2.184}$$

where $\lambda_H = 2\pi/k_H$ is the wavelength corresponding to k_H (the jet neutral wavelength). If the rising jet attempts to move at a higher velocity than $U_{G,cr}$, it will become unstable and will break up.

Stability of a Gaseous Jet with Respect to Axisymmetric Disturbances

Consider a cylindrical jet with radius R_j moving parallel to $\vec{g}$, whose surface has been disturbed by a small axisymmetric disturbance. Assuming that the amplitude of the disturbances is infinitesimally small ($\zeta \ll R_j$), we can apply the analytical steps up to the derivation of Eq. (2.169), provided that the gravity term is dropped, and the mechanical equilibrium at the interphase (Eq. (2.164)) is replaced with

$$P_G - P_L = \sigma \left(\frac{1}{R_c} + \frac{1}{R_t} \right), \tag{2.185}$$

where R_c and R_t are the principal radii of curvature:

$$\frac{1}{R_c} = -\frac{\partial^2 \zeta}{\partial x^2}, \tag{2.186}$$

$$\frac{1}{R_t} \approx \frac{1}{R_j + \zeta} \approx \frac{1}{R_j} \left(1 - \frac{\zeta}{R_j} \right) = \frac{1}{R_j} - \frac{\zeta}{R_j^2}. \tag{2.187}$$

For the undisturbed condition we have

$$P_{G0} - P_{L0} = \frac{\sigma}{R_j}. \tag{2.188}$$

With this modification, the dispersion relation (Eq. (2.169)) becomes

$$c = \frac{\omega}{k} = \frac{\rho_L U_L + \rho_G U_G}{\rho_L + \rho_G}$$
$$\pm \left[-\frac{\rho_L \rho_G}{(\rho_L + \rho_G)^2} (U_L - U_G)^2 + \frac{\sigma k}{\rho_L + \rho_G} - \frac{\sigma}{k(\rho_L + \rho_G) R_j^2} \right]^{1/2}. \tag{2.189}$$

A stable jet can exist when the square-root term becomes equal to zero. Therefore,

$$(U_L - U_G)^2 \le \frac{\rho_L + \rho_G}{\rho_L \rho_G} \sigma \left[k - \frac{1}{k R_j^2} \right]. \tag{2.190}$$

Critical Wavelength of a Circular Jet (Rayleigh Unstable Wavelength)

Consider now an inviscid liquid jet issuing into an inviscid gas, with a constant velocity, parallel to $\vec{g}$. Since the gas phase is inviscid, the liquid velocity will remain uniform across the jet cross section. We can therefore attach the coordinate system to the jet, whereby we will essentially deal with disturbances at the surface of a liquid jet with stationary neutral conditions. Equation (2.133), when applied to the jet surface, gives

$$\rho_L \left. \frac{\partial \phi}{\partial t} \right|_0^\zeta = (P_{L0} - P_G) - (P_L - P_G), \tag{2.191}$$

where y is the displacement from neutral situation of the jet surface, $P_{L0} - P_G = \sigma / R_j$, and

$$(P_L - P_G) \approx \sigma \left(-\frac{\partial^2 \zeta}{\partial x^2} + \frac{1}{R_j} - \frac{\zeta}{R_j^2} \right).$$

Proceeding with an analysis similar to the previous cases, one can derive

$$\left(\frac{\omega}{k} \right)^2 = \frac{k\sigma}{(\rho_L + \rho_G)} - \frac{\sigma}{R_j^2 k(\rho_L + \rho_G)}. \tag{2.192}$$

Stability requires that the right side be positive (i.e., $k > 1/R_j$), and a neutrally stable jet will occur with $k_H = 1/R_j$, or

$$\lambda_H = 2\pi R_j. \tag{2.193}$$

This is the *critical Rayleigh unstable wavelength*. Disturbances with longer wavelengths would disrupt the jet. Lord Rayleigh derived this expression based on the argument that the jet is stable as long as the total capillary surface energy decreases as a result of surface disturbances (Lienhard and Witte, 1985).

A more rigorous treatment of the problem of liquid jet stability can be performed by considering asymmetric two-dimensional disturbances (Lamb, 1932; Chandrasekhar, 1961), whereby

$$\phi = \phi_1(r, \theta) \exp(i\omega t) \cos(kx), \tag{2.194}$$

where θ is the azimuthal angle. The continuity equation $\nabla^2 \phi = 0$ must be satisfied, and that leads to

$$\frac{\partial^2 \phi_1}{\partial r^2} + \frac{1}{r} \frac{\partial \phi_1}{\partial r} + \frac{1}{r^2} \frac{\partial^2 \phi_1}{\partial \theta^2} - k^2 \phi_1 = 0. \tag{2.195}$$

Following the separation of variables technique for the solution of linear partial differential equations, one can assume that ϕ_1 is the product of two functions: one a function of r only and the other one a function of θ only. It can then be shown that

$$\phi_1(r, \theta) = \phi_0 I_m(kr) \cos(m\theta), \tag{2.196}$$

where $m = 0, 1, 2, 3, \ldots$, and $I_m(x)$ is the modified Bessel function of the first kind and mth order. The resulting velocity potential ϕ will then be consistent with disturbances of the form

$$R = R_j - i\phi_0 k I'_m(kR_j) \cos(kx) \cos(m\theta) \frac{\exp(i\omega t)}{\omega}. \tag{2.197}$$

The details of the solution can be found in Lamb (1932) and Chandrasekhar (1961). The resulting dispersion relation is

$$\omega^2 = \frac{(kR_j) I'_m(kR_j)}{I_m(kR_j)} \left[(kR_j)^2 + m^2 - 1 \right] \frac{\sigma}{\rho_L R_j^3}, \tag{2.198}$$

where $I'_m(x) = \frac{d}{dx} I_m(x)$. For $m > 0$, we will have $\omega^2 > 0$ for any kR_j value. Therefore, the jet is stable with respect to asymmetric surface disturbances. For axisymmetric disturbances, $m = 0$, and one gets

$$\omega^2 = \frac{(kR_j) I'_0(kR_j)}{I_0(kR_j)} \left[(kR_j)^2 - 1 \right] \frac{\sigma}{\rho_L R_j^3}. \tag{2.199}$$

In this case, $\omega^2 > 0$ when $kR_j > 1$, and the critical wavelength will occur when $kR_j = 1$, and that leads to $\lambda_H = 2\pi R_j$. This result is the same as Eq. (2.193).

2.12 Rayleigh–Taylor Instability for a Viscous Liquid

Consider a horizontal layer of gas with the depth of h, underneath an infinitely thick liquid layer, similar to Fig. 2.17. Both phases are incompressible, and the gas phase is inviscid. The conservation equations for liquid are

$$\frac{\partial u_L}{\partial x} + \frac{\partial v_L}{\partial y} = 0, \tag{2.200}$$

$$\rho_L \frac{\partial u_L}{\partial x} = -\frac{\partial P_L}{\partial x} + \mu_L \left(\frac{\partial^2 u_L}{\partial x^2} + \frac{\partial^2 u_L}{\partial y^2} \right), \tag{2.201}$$

$$\rho_L \frac{\partial v_L}{\partial t} = -\frac{\partial P_L}{\partial y} + \mu_L \left(\frac{\partial^2 v_L}{\partial x^2} + \frac{\partial^2 v_L}{\partial y^2} \right) - \rho_L g. \tag{2.202}$$

The periodic motion of incompressible fluid surfaces obeying the Navier–Stokes equations can be represented by using a potential and a stream function (Lamb, 1932; Levich, 1962). Each velocity component of the liquid is assumed to consist of an inviscid term and a perturbation that represents the effect of viscosity:

$$u_L = u_L^0 + u_L', \tag{2.203}$$

$$v_L = v_L^0 + v_L', \tag{2.204}$$

$$P_L = P_L^0, \tag{2.205}$$

where parameters with superscript 0 represent ideal (inviscid) flow conditions. Thus, for liquid, we must have

$$\frac{\partial u_L^0}{\partial x} + \frac{\partial v_L^0}{\partial y} = 0, \tag{2.206}$$

$$\rho_L \frac{\partial v_L^0}{\partial t} = -\frac{\partial P_L^0}{\partial x}, \tag{2.207}$$

$$\rho_L \frac{\partial v_L^0}{\partial t} = -\frac{\partial P_L^0}{\partial y} - \rho_L g, \tag{2.208}$$

$$\frac{\partial u_L'}{\partial x} + \frac{\partial v_L'}{\partial y} = 0, \tag{2.209}$$

$$\rho_L \frac{\partial u_L'}{\partial t} = \mu_L \left(\frac{\partial^2 u_L'}{\partial x^2} + \frac{\partial^2 u_L'}{\partial y^2} \right), \tag{2.210}$$

$$\rho_L \frac{\partial v_L'}{\partial t} = \mu_L \left(\frac{\partial^2 v_L'}{\partial x^2} + \frac{\partial^2 v_L'}{\partial y^2} \right). \tag{2.211}$$

Because the gas phase is inviscid, only Eqs. (2.206)–(2.208), with subscript L replaced with G, will be needed for the gas. The ideal fluid velocities can be represented by the following velocity potentials:

$$\phi_L = A_L e^{-ky + \alpha t} \cos kx, \tag{2.212}$$

$$\phi_G = A_G \cosh\left[k(y + h)\right] e^{\alpha t} \cos kx, \tag{2.213}$$

where α is the wave growth parameter, and for each phase

$$\left(u_i^0, v_i^0\right) = \left(\frac{\partial \phi_i}{\partial x}, \frac{\partial \phi_i}{\partial y}\right),$$

$$P_i^0 = -\rho i \frac{\partial \phi_i}{\partial t} - \rho_i g y,$$

where the last expression results from Euler's equation. For the flow field to be stable, α should not have a positive real component. For the viscosity-induced rotational motion in the liquid phase, we can define a stream function ψ_L such that

$$\left(u_L', v_L'\right) = \left(-\frac{\partial}{\partial y}, \frac{\partial}{\partial x}\right) \psi_L. \tag{2.214}$$

Substitution from Eq. (2.214) into Eqs. (2.210) and (2.211) leads to

$$\frac{\partial}{\partial y}\left[\frac{\partial \psi_L}{\partial t} - \nu_L \nabla^2 \psi_L\right] = 0, \tag{2.215}$$

$$\frac{\partial}{\partial x}\left[\frac{\partial \psi_L}{\partial t} - \nu_L \nabla^2 \psi_L\right] = 0. \tag{2.216}$$

Clearly, we must have

$$\frac{\partial \psi_L}{\partial t} = \nu_L \left[\frac{\partial^2 \psi_L}{\partial x^2} + \frac{\partial^2 \psi_L}{\partial y^2}\right]. \tag{2.217}$$

Equation (2.217) and all boundary conditions can be satisfied by

$$\psi_L = B_L e^{-my+\alpha t} \sin kx, \tag{2.218}$$

where from Eq. (2.217)

$$m = \sqrt{k^2 + \frac{\alpha}{\nu_L}}. \tag{2.219}$$

The requirement $\partial \zeta / \partial t = v_L|_{y=0}$ leads to

$$\zeta = \frac{k}{\alpha}(B_L - A_L)e^{\alpha t} \cos kx. \tag{2.220}$$

A linear stability analysis can now be performed by applying the following conditions at the interphase ($y = 0$):

$$v_L^0 + v_L' = v_G, \tag{2.221}$$

$$-P_L + 2\mu_L \frac{\partial \left(v_L^0 + v'\right)_L}{\partial y} = -P_G - \sigma \frac{\partial^2 \zeta}{\partial x^2}, \tag{2.222}$$

$$\mu_L \left(\frac{\partial \left(u_L^0 + u'_L\right)}{\partial y} + \frac{\partial \left(v_L^0 + v'_L\right)}{\partial x}\right) = 0. \tag{2.223}$$

Equation (2.133) can now be applied to both phases, and one of the resulting equations can be subtracted from the other to get

$$P_L - P_G = -\rho_L \left.\frac{\partial \phi_L}{\partial t}\right|_{y\approx 0} + \rho_G \left.\frac{\partial \phi_G}{\partial t}\right|_{y\approx 0} - (\rho_L - \rho_G)g\zeta + \left(P_L^0\big|_{y=0} - P_G^0\big|_{y=0}\right). \tag{2.224}$$

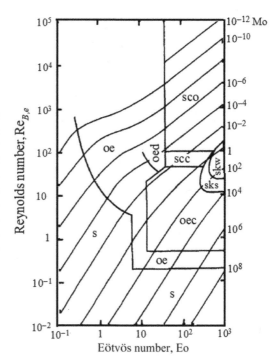

Figure 2.19. Shape regimes for bubbles in unhindered gravitational motion in liquids: s, spherical; oe, oblate ellipsoid; oed, oblate ellipsoidal (disk-like and wobbling); oec, oblate ellipsoidal cap; scc, spherical cap with closed, steady wake; sco, spherical cap with open, unsteady wake; sks, skirted with smooth, steady skirt; skw, skirted with wavy unsteady skirt. (From Bhaga and Weber, 1981.)

For the system of interest here, $P_L^0|_{y=0} = P_G^0|_{y=0}$ (Lamb, 1932; Levich, 1962) because they represent pressures at the interphase in neutral conditions and in the absence of any interfacial curvature. Equation (2.224) can now be used for eliminating $(P_L - P_G)$ from Eq. (2.222). Equations (2.221) through (2.223) then lead to

$$-A_L + B_L - A_G \sinh(kh) = 0, \tag{2.225}$$

$$2k^2 A_L - (m^2 + k^2) B_L = 0, \tag{2.226}$$

$$\left[2\mu_L k^2 + \frac{k^3 \sigma}{\alpha} + \rho_L \alpha - \frac{k}{\alpha}(\rho_L - \rho_G)g \right] A_L$$
$$+ \left[-2\mu_L km - \frac{k^3 \sigma}{\alpha} - \frac{k}{\alpha}(\rho_L - \rho_G)g \right] B_L - [\rho_G \alpha \cosh(kh)] A_G = 0. \tag{2.227}$$

A nontrivial solution for the unknowns A_L, B_L, and A_G is possible if the determinant of the coefficient matrix of Eqs. (2.225)–(2.227) is equal to zero. Equating the coefficient matrix to zero will lead to the dispersion relation for the system. The system will be unstable if α has a positive, real component.

2.13 Waves at the Surface of Small Bubbles and Droplets

Waves can develop at the surface of a bubble, leading to its deformation and oscillation and affecting its internal circulation flow, as well as the flow field of the surrounding liquid. Similar statements can be made about a droplet.

Experiment shows that a bubble moving in a stagnant liquid can acquire several different shapes, depending on the properties of the surrounding liquid, and most importantly on the volume of the bubble, V_B. Figure 2.19 depicts an empirical bubble shape regime map (Bhaga and Weber, 1981). The map is in terms of the bubble Eötvös number, Morton number, and Reynolds number, defined, respectively, as

$$\mathrm{Eo} = g(\rho_\mathrm{L} - \rho_\mathrm{G})d_{\mathrm{B,e}}^2/\sigma, \tag{2.228}$$

$$\mathrm{Mo} = g\mu_\mathrm{L}^4(\rho_\mathrm{L} - \rho_\mathrm{G})/\left(\rho_\mathrm{L}^2\sigma^3\right), \tag{2.229}$$

$$\mathrm{Re}_{\mathrm{B,e}} = \rho_\mathrm{L}U_\mathrm{B}d_{\mathrm{B,e}}/\mu_\mathrm{L}, \tag{2.230}$$

where $d_{\mathrm{B,e}} = (6V_\mathrm{B}/\pi)^{1/3}$. Evidently, except for very low bubble Reynolds numbers, deformation and shape oscillations are to be expected. Bubble and droplet deformation in fact can lead to breakup.

The discussions in this section will be primarily relevant to the interfacial waves and oscillations of bubbles and droplets that remain nearly spherical. This is an important regime for bubbles and is common in stirred mixing tanks and highly turbulent flow systems.

Generally speaking, however, in gas–liquid two-phase flows the interaction between the two phases is often more complicated than the case of bubbles rising in a stagnant liquid pool. Hydrodynamically induced bubble breakup phenomena often keep the size of the bubbles small, whereas interfacial drag and other forces limit the relative velocity between the two phases. Near-spherical droplets are probably more common than near-spherical bubbles and are best exemplified by spray droplets.

Bubbles and droplets can oscillate at their natural frequencies. Two different oscillation modes can be defined for bubbles: volume oscillations, where the bubble volume oscillates while its shape remains unchanged, and shape oscillations, where the geometric shape of the bubble undergoes periodic changes. The angular frequency of volume oscillations for a spherical bubble of an ideal gas surrounded by an incompressible liquid is (Shima, 1970)

$$\omega_0 = \frac{1}{R_0}\left\{\frac{3\gamma}{\rho_\mathrm{L}}\left[P_\mathrm{L} + \left(1 - \frac{1}{3\gamma}\right)\frac{2\sigma}{R_0}\right] - \left(\frac{2\mu_\mathrm{L}}{\rho_\mathrm{L}R_0}\right)^2\right\}^{1/2}, \tag{2.231}$$

where R_0 is the bubble static (equilibrium) radius, ρ_L and P_L are the density and the ambient pressure of the surrounding liquid, respectively, and $\gamma = C_\mathrm{P}/C_\mathrm{v}$ is the bubble specific heat ratio. If viscosity and surface tension effects are neglected, as is justified for example for air bubbles with $R_0 > 100$ µm (visible bubbles) in water (Plesko and Leutheusser, 1982), then

$$\omega_0 = \left[3\gamma P_\mathrm{L}/\left(\rho_\mathrm{L}R_0^2\right)\right]^{1/2}. \tag{2.232}$$

Volume oscillations have little effect on the interfacial transfer processes; however, bubble and droplet shape oscillations are more complex and significantly influence the transfer processes on both sides of the interphase.

By assuming two-dimensional (r, θ) flow (see Fig. 2.20), the natural oscillations of a near-spherical bubble or droplet can be analyzed by imposing disturbances of the following form on the bubble, when the bubble and the surrounding liquid have otherwise negligibly small relative velocity:

$$R = R_0 + \zeta, \quad \zeta \ll R_0, \tag{2.233}$$

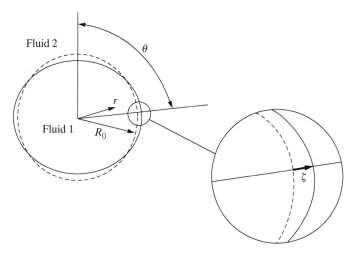

Figure 2.20. Schematic of an oscillating fluid particle.

where

$$\zeta = f(\theta)\sin(\omega t). \tag{2.234}$$

Assuming inviscid fluids inside and outside, then we must have, for the inside,

$$\nabla^2 \phi_1 = 0, \tag{2.235}$$

$$\vec{U}_1 = \nabla \phi_1, \tag{2.236}$$

and, for the outside,

$$\nabla^2 \phi_2 = 0, \tag{2.237}$$

$$\vec{U}_2 = \nabla \phi_2, \tag{2.238}$$

where ϕ_1 and ϕ_2 refer to velocity potentials inside and outside the sphere, respectively. Besides these equations, the solutions for inside and outside must of course be consistent at the interphase, where radial displacements and velocities must be equal for the two phases. The solution of the problem then leads to the superposition of an infinite number of solutions (oscillation modes), whereby

$$\zeta = \sum_{n=1}^{\infty} \zeta_{0,n} P_n(\cos\theta)\sin(\omega_n t), \tag{2.239}$$

$$\phi_1 = \phi_1^0 + \sum_{n=1}^{\infty} \frac{R_0\omega_n}{n}\zeta_{0,n}\left(\frac{r}{R_0}\right)^n P_n(\cos\theta)\cos(\omega_n t), \tag{2.240}$$

$$\phi_2 = \phi_2^0 + \sum_{n=1}^{\infty} \frac{R_0\omega_n}{n+1}\zeta_{0,n}\left(\frac{R_0}{r}\right)^{n+1} P_n(\cos\theta)\cos(\omega_n t), \tag{2.241}$$

where ϕ_1^0 and ϕ_2^0 represent the undisturbed velocity potentials and P_n is Legendre's polynomial of degree n. Given that velocities are infinitesimally small, integration of Euler's equation gives

$$-P_1 + P_{0,1} = -\rho_1 \left(\frac{\partial \phi_1}{\partial t}\right)_{r=R_0}, \tag{2.242}$$

$$-P_2 + P_{0,2} = -\rho_2 \left(\frac{\partial \phi_2}{\partial t}\right)_{r=R_0}, \tag{2.243}$$

where $P_{0,1}$ and $P_{0,2}$ are pressures for the undisturbed system, and the derivatives on the right side are to be calculated at $r = R_0 + \zeta \approx R_0$, in accordance with linear stability analysis. Evidently,

$$P_{0,1} - P_{0,2} = \frac{2\sigma}{R_0}. \tag{2.244}$$

To find the oscillation frequency for the nth oscillation mode, we will assume that only $\zeta_{0,n}$ is finite, and $\zeta_{0,i} = 0$ for $i \neq n$. Equations (2.240)–(2.244) can then be combined to yield

$$\left(P_1 - P_2 - \frac{2\sigma}{R_0}\right) = R_0 \omega_n^2 \left(\frac{\rho_2}{n+1} + \frac{\rho_1}{n}\right) \zeta_{0,n} P_n(\cos\theta) \sin(\omega_n t). \tag{2.245}$$

Also, based on the minimum surface Gibbs free energy requirement for equilibrium, it can be shown that (Levich, 1962)

$$\left(P_1 - P_2 - \frac{2\sigma}{R_0}\right) = \frac{-2\zeta_n \sigma}{R_0^2} - \frac{\sigma}{R_0^2}\left[\frac{1}{\sin\theta}\frac{\partial}{\partial\theta}\left(\sin\theta\frac{\partial\zeta_n}{\partial\theta}\right)\right], \tag{2.246}$$

where $\zeta_n = \zeta_{0,n} P_n(\cos\theta) \sin(\omega_n t)$. From the definition of Legendre polynomials, we have

$$\frac{1}{\sin\theta}\frac{\partial}{\partial\theta}\left(\sin\theta\frac{\partial\zeta_n}{\partial\theta}\right) + n(1+n)\zeta_n = 0. \tag{2.247}$$

We can now substitute from Eq. (2.247) in Eq. (2.246), and then equate the resulting expression for $(P_1 - P_2 - 2\sigma/R_0)$ with Eq. (2.245). The frequency of the nth mode of oscillations is then derived (see Problem 2.18):

$$\omega_n^2 = \frac{n(n+1)(n-1)(n+2)\sigma}{R_0^3[\rho_1(n+1) + \rho_2 n]}. \tag{2.248}$$

Equation (2.248) indicates that shape oscillations are possible for $n \geq 2$. (The volume oscillations can in fact be considered as the zeroth mode.) The predominant oscillation mode is for $n = 2$, which leads to

$$\omega_2^2 = \frac{24\sigma}{R_0^3[3\rho_1 + 2\rho_2]}. \tag{2.249}$$

EXAMPLE 2.6. Calculate the second-mode shape oscillation frequencies for an air bubble surrounded by water, and a water droplet surrounded by air, at a temperature of 25 °C temperature and a pressure of 1 bar. Assume $R_0 = 2$ mm.

SOLUTION. The properties are as follows: $\rho_G = 1.185$ kg/m^3, $\rho_L = 997$ kg/m^3, and $\sigma = 0.071$ N/m. For the bubble, Eq. (2.249) leads to $\omega = 326.5$ rad/s, and therefore $f = \omega/2\pi = 51.97$ Hz. For the droplet, $\omega = 266.7$ rad/s, leading to $f = 42.45$ Hz.

The results derived here, as mentioned before, are correct when the undisturbed flow field represents essentially zero relative tangential velocity between the two phases. When the gas phase is treated as essentially a void, however, the effect of relative velocity between a liquid sphere and the surrounding fluid disappears when the coordinate system is assumed to move with the sphere. With this approximation, these results can be applied to a droplet moving in a low-pressure gas by using $\rho_1 = \rho_L$ and $\rho_2 = 0$.

Experimental observations reported by Montes *et al.* (1999, 2002) have shown that the shape oscillations are complex for bubbles rising in stagnant liquid but can be approximated as a linear combination of the aforementioned shape oscillations of modes 2, 3, and 4. One possible combination that agreed well with their experiments was

$$\frac{\zeta}{R_0} = \frac{\zeta_0}{R_0} \sum_{n=2}^{4} C_n P_n(\cos\theta) \cos(\omega_n t), \tag{2.250}$$

where $C_2 = 0.2$, $C_3 = 0.5$, and $C_4 = 0.3$.

2.14 Growth of a Vapor Bubble in Superheated Liquid

Bubbles that nucleate in superheated liquids can undergo growth. In heterogeneous nucleate boiling the nucleated bubbles reside on wall crevices surrounded by a superheated liquid while growing, until they are released. The growth process of such bubbles is complicated by the presence of the wall and the nonuniformity of the temperature in the surrounding liquid. These will be discussed in Chapter 11. Homogeneously nucleated bubbles, in contrast, are free spherical microbubbles. Freely floating microbubbles are also common in bubble chambers and liquid droplet neutron detectors. The growth of free microbubbles that are surrounded by superheated liquid is discussed in this section.

Following nucleation, microbubbles that are surrounded by a superheated liquid undergo three phases of growth. The mathematical solution for the first phase is referred to as the Rayleigh solution. The bubble growth in this phase is hydrodynamically controlled, and at low pressures the bubble grows approximately at a constant rate ($\dot{R} \approx$ const.). The time duration of this phase is very short, however (typically a fraction of a microsecond or so). The second phase of bubble growth represents transition from hydrodynamically controlled growth to a thermally controlled growth. In the third phase, which typically accounts for most of the bubble growth, inertia and surface tension effects are insignificant and the bubble growth is thermally controlled.

Since the period of growth of a bubble surrounded by superheated liquid is short (about 10 ms in situations relevant to boiling), it may be reasonable to assume that the bubble is surrounded by a stagnant and infinitely large liquid field. Assuming inviscid liquid behavior, furthermore, we can apply potential flow theory to the liquid, thereby

$$\nabla^2 \phi = \frac{1}{r^2} \frac{d}{dr}\left(r^2 \frac{d\phi}{dr}\right) = 0, \tag{2.251}$$

where ϕ is the liquid velocity potential. The boundary conditions are

$$\frac{d\phi}{dr} = U_L = \dot{R} \quad \text{at} \quad r = R \tag{2.252}$$

and

$$\frac{d\phi}{dr} = 0 \quad \text{for} \quad r \to \infty. \tag{2.253}$$

The solution to Eq. (2.251) and its boundary conditions is

$$\phi = -\frac{R^2 \dot{R}}{r}. \tag{2.254}$$

Now, assuming irrotational flow and neglecting gravity, we can use Euler's equation

$$\frac{\partial \phi}{\partial t} + \frac{1}{2}U_L^2 + \frac{1}{\rho_L}P_L = \frac{1}{\rho_L}P_\infty, \tag{2.255}$$

where P_L is the liquid pressure at the surface of the bubble and P_∞ refers to the pressure in a far-field location in the liquid. We can now substitute for ϕ from Eq. (2.254) in Eq. (2.255), and use $U_L = \partial \phi / \partial r = R^2 \dot{R}/r^2$. The resulting equation when applied for $r = R$ (i.e., the surface of the bubble) leads to the *Rayleigh equation*

$$R\ddot{R} + \frac{3}{2}\dot{R}^2 = \frac{P_L - P_\infty}{\rho_L}. \tag{2.256}$$

For the special case of $P_L - P_\infty = \text{const.}$, the solution to Rayleigh's equation is

$$\dot{R}(t) = \left\{ \frac{2(P_L - P_\infty)}{3\rho_L} \left[1 - \left(\frac{R_0}{R} \right)^3 \right] \right\}^{1/2}, \tag{2.257}$$

where R_0 is the bubble radius at $t = 0$. For $R \gg R_0$, the second term in the bracket on the right side can be neglected, and the solution of what remains gives

$$R(t) \approx \left\{ \frac{2(P_L - P_\infty)}{3\rho_L} \right\}^{1/2} t. \tag{2.258}$$

Rayleigh's equation is purely hydrodynamic and addresses the liquid phase only. It represents the hydrodynamic and liquid-inertia-controlled bubble growth. Coupling with the gas or vapor phase (pure vapor in the present case) can be provided by noting that

$$P_L - P_\infty = (P_L - P_v) + (P_v - P_\infty), \tag{2.259}$$

$$P_L - P_v = -\frac{2\sigma}{R}, \tag{2.260}$$

$$P_v - P_\infty = \frac{h_{fg}(T_B - T_{sat})}{T_{sat}(v_v - v_L)}, \tag{2.261}$$

where T_B is the bubble temperature and T_{sat} corresponds to P_∞. Equation (2.261) is an approximation to Clapeyron's relation. By combining Eqs. (2.256) and (2.259)–(2.261), the equation of motion (the *extended Rayleigh equation*) is obtained:

$$R\ddot{R} + \frac{3}{2}\dot{R}^2 + \frac{2\sigma}{\rho_L R} - \frac{h_{fg}(T_B - T_{sat})}{\rho_L T_{sat}(v_v - v_L)} = 0. \tag{2.262}$$

Equation (2.262) contains two unknowns: T_B and R. It can be solved simultaneously with the energy conservation equation for the liquid (where we note that the radial liquid velocity is $U_L = R^2 \dot{R}/r^2$):

$$\frac{\partial T_L}{\partial r} + \frac{R^2}{r^2}\dot{R}\frac{\partial T_L}{\partial r} = \frac{\alpha_L}{r^2}\frac{\partial}{\partial r}\left(r^2 \frac{\partial T_L}{\partial r} \right). \tag{2.263}$$

The initial and boundary conditions for this equation are

$$T_L = \begin{cases} T_\infty & \text{at } t = 0, \\ T_B & \text{at } r = R, t > 0, \\ T_\infty & \text{for } r \to \infty. \end{cases}$$

The initial, hydrodynamically controlled growth period is isothermal and for conditions where $\rho_L \gg \rho_v$ the bubble expands according to (van Stralen and Cole, 1979)

$$R(t) \approx \left[\frac{2\rho_v h_{fg}(T_\infty - T_B)}{3\rho_L T_B} \right]^{1/2} t. \qquad (2.264)$$

This equation is similar to Eq. (2.258), when Clapeyron's relation is linearized and used.

For the thermally controlled growth phase, an approximate solution can be derived by writing

$$q'' = \rho_v h_{fg} \dot{R}. \qquad (2.265)$$

This equation is derived by performing a simple energy balance on the bubble, and noting that $T_B \approx T_{sat}$ since the bubble is relatively large during its thermally controlled growth. The heat flux at the bubble surface can be estimated based on the one-dimensional transient heat conduction into a semi-infinite medium:

$$q'' = \frac{k_L(T_\infty - T_{sat})}{\sqrt{\pi \alpha_L t}}. \qquad (2.266)$$

Substituting Eq. (2.266) into Eq. (2.265) and combining the solution of the resulting differential equation with initial condition $R = 0$ at $t = 0$ gives

$$R = C \frac{2k_L(T_\infty - T_{sat})}{\sqrt{\alpha_L}\rho_v h_{fg}} \sqrt{t}. \qquad (2.267)$$

This simple analysis gives $C = 1/\sqrt{\pi}$. Plesset and Zwick (1954) and Forster and Zuber (1954) have solved the extended Rayleigh equation for the aforementioned first (Rayleigh expansion) and third (asymptotic thermally controlled expansion) phases of bubble growth. Plesset and Zwick (1954) solved the asymptotic bubble growth equations based on a thin thermal boundary layer and obtained $C = \sqrt{3/\pi}$. The solution by Forster and Zuber (1954) gave $C = \sqrt{\pi/2}$.

Improvements to the approximate solutions of Plesset and Zwick (1954) and Forster and Zuber (1955) have been proposed by several authors (Birkhoff *et al.*, 1958; Scriven, 1959; Bankoff, 1963; Skinner and Bankoff, 1964; Riznic *et al.*, 1999). Scriven (1959) solved the problem of bubble growth in a superheated liquid, for a single-component situation (pure liquid and vapor), as well as a two-component case (vapor and an inert species that has a finite solubility in liquid), by removing the assumption of a thin thermal boundary layer. Scriven's exact solution shows that the temperature drop in the liquid occurs in the $R < r < 2R$ range for $R/(2\sqrt{\alpha_L t}) > 1$, giving credit to the thin thermal boundary layer assumption (Hsu and Graham, 1986).

EXAMPLE 2.7. In bubble chambers a charged particle passes through a superheated stagnant liquid and creates a thermal spike that leads to the formation of bubbles. Derive an expression for the minimum energy that must be deposited in a thermal spike to form a stable critical-size bubble.

SOLUTION. The deposited energy must generate a bubble that satisfies the Laplace–Kelvin equation. Therefore,

$$\ln \frac{P_v}{P} = -\frac{2\sigma v_L}{\frac{R_u}{M} T_L R_{cr}}. \tag{a}$$

The bubble chamber can be idealized as an infinitely large pool of liquid, and the chamber pressure can be assumed to remain constant during the nucleation process. The energy needed for the generation of the interfacial surface can be derived by noting that for a pure substance $\sigma = g^I$, where g^I is the excess Gibbs free energy associated with a unit interfacial surface area (see Eq. (2.11)). Starting from the definition of the Gibbs excess free energy, $g = h - Ts$, we can write

$$dg = dh - Tds - sdT. \tag{b}$$

However,

$$Tds = dh - vdP. \tag{c}$$

Combining these two equations, we get

$$dg = -sdT + vdP. \tag{d}$$

Using Eq. (d), we can write $s^I = -(\frac{\partial g}{\partial T})P$. The specific interphase enthalpy can then be written as

$$h^I = g^I + T_L s^I = \sigma - T_L \left(\frac{\partial \sigma}{\partial T} \right)_P. \tag{e}$$

The energy needed for the generation of the interphase is thus $4\pi R_{cr}^2(\sigma - T_L \frac{\partial \sigma}{\partial T})$.

Following the formation of the critical-size bubble, the surrounding liquid will acquire a kinetic energy E_k, where

$$E_k = \frac{1}{2}\rho_L \int_{R_{cr}}^{\infty} 4\pi r^2 u^2 dr = 2\pi \rho_L R_{cr}^3 \dot{R}_{cr}^2, \tag{f}$$

where $u(r) = R_{cr}^2 \dot{R}_{cr}/r^2$ has been used (see the discussion before Eq. (2.214)).

The total deposited energy must therefore be larger than E_{cr}, where

$$E_{cr} = \frac{4}{3}\pi R_{cr}^3 \rho_v h_{fg} + 4\pi R_{cr}^2 \left(\sigma - T_L \frac{\partial \sigma}{\partial T} \right) + \frac{4}{3}\pi R_{cr}^3 P_L + 2\pi \rho_L R_{cr}^3 \dot{R}_{cr}^2. \tag{g}$$

Note that the third term on the right side represents the reversible work from bubble expansion. Irreversible energy loss also occurs from the generation of sound waves and by viscous effects. These effects are often negligibly small, however. In fact, computations have shown that the first three terms on the right side of Eq. (g) typically account for more than 99% of the critical energy needed for nucleation (Harper and Rich, 1993).

PROBLEMS

2.1 Consider the two-dimensional shallow liquid in Fig. P2.1. The sides of the container at $x = 0$ and $x = l$ are at temperatures T_1 and T_2, so that $T_1 > T_2$. In steady state, the thermocapillary effect results in a circulatory flow as shown in the figure.

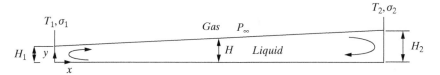

Figure P2.1. Shallow liquid film in Problem 2.1.

Assume constant liquid properties and negligible gas viscosity, and assume negligible inertial effects so that the momentum equation in the x direction reduces to

$$-\frac{dP}{dx} + \mu_{\mathrm{L}}\frac{d^2U}{dx^2} = 0.$$

Prove that the following expressions apply:

$$\sigma = \sigma_1 + \frac{1}{3}\rho_{\mathrm{L}}g(H^2 - H_1^2),$$

$$U(y) = \frac{y}{2\mu_{\mathrm{L}}}\left(\frac{3}{2}\frac{y}{H} - 1\right)\frac{d\sigma}{dx}.$$

2.2 Repeat the analysis of Example 2.1 for a complete spherical bubble.

2.3 The chopped microbubble displayed in Fig. P2.3 is axisymmetric and has a dry, circular base. The figure shows the projection of the bubble on a plane that is perpendicular to the bubble base and divides the bubble into two halves. The bubble is in steady state and is immersed in a quiescent liquid thermal boundary layer. The pressure inside the bubble is uniform. Because of the nonuniform temperature in the liquid thermal boundary layer, the bubble surface can have a nonuniform temperature distribution, which will result in the deformation of the bubble shape from a sphere. Prove that the shape of the bubble interphase follows

$$|Y'| = [1 + Y'^2]^{3/2}\left[\frac{2\sigma_{\mathrm{w}}}{\sigma} - \frac{1}{|X|\left(1 + 1/Y'^2\right)^{1/2}}\right],$$

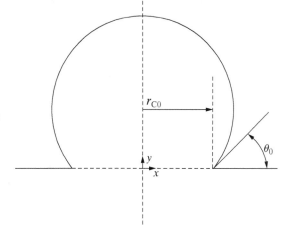

Figure P2.3. Bubble described in Problem 2.3.

where σ_w is the surface tension at the wall temperature, σ is the local surface tension, the derivatives are with respect to X, and

$$X = \frac{x}{r_{C0}/\sin\theta_0},$$
$$Y = \frac{y}{r_{C0}/\sin\theta_0}.$$

What are the boundary conditions for the above differential equation?

2.4 As of 2004, the concentration of CO_2 in Earth's atmosphere was about 377.4 parts per million (ppm) by volume (World Watch Institute, 2006).

(a) For water at equilibrium with atmospheric air at temperatures 290 and 300 K, calculate the liquid-side mass fraction of CO_2.

(b) The total volume of water in the oceans is approximately 1.35×10^9 km^3 (Emiliani, 1992). The global average land–ocean temperature was 14.48 °C in 2004. Assuming an average ocean temperature of 14.48 °C, estimate the amount of dissolved CO_2 in the ocean waters, if the oceans are at equilibrium with atmospheric air.

(c) During the year 2003, the concentration of CO_2 in Earth's atmosphere increased from 375.6 to 377.4 ppm by volume. During the same period, approximately 7.25×10^{12} kg of CO_2 was emitted into the atmosphere (World Watch Institute, 2006). Compare the total CO_2 emission during 2003 with what the oceans would be capable of dissolving had the oceans remained at equilibrium with the atmosphere. Assume a constant average ocean temperature of 14.48 °C.

2.5 A pure water droplet that is 0.55 mm in diameter is suspended in laboratory air that is at 25 °C. The relative humidity of the laboratory air is 60%, and the concentration of CO_2 in air is 500 ppm by volume. For simplicity, it is assumed that the droplet remains isothermal. It is also assumed that the concentration of CO_2 in the droplet remains uniform. It is also assumed that in the absence of mass transfer the heat transfer between the droplet and the surrounding air follows $Nu = Hd/k_G = 2$.

(a) Calculate the evaporation mass flux when the droplet temperature is 25 °C.

(b) Repeat part (a), this time assuming that the droplet temperature is 50 °C.

(c) For both cases (a) and (b) calculate the mass flux of CO_2 transferred between the droplet and the surrounding atmosphere.

2.6 According to Suo and Griffith (1964) a tube is considered to be a microchannel when $D \le 0.3\sqrt{\sigma/g\Delta\rho}$, where D is the tube diameter. Using this definition, find this threshold diameter for an atmospheric air–water mixture, a saturated water–steam mixture at 10 MPa, and a saturated refrigerant R-134a liquid–vapor mixture at 1.02 MPa.

2.7 For a microbubble surrounded and at thermal and mechanical equilibrium with liquid, prove that the liquid must be superheated according to:

$$P_{\text{sat}} - P_L \approx \frac{2\sigma}{R}\frac{\rho_L + \rho_v}{\rho_L}$$

where P_L is the pressure in the liquid phase and P_{sat} is the saturation pressure corresponding to the temperature of the temperature of the bubble and its surroundings.

Also, for a microdroplet surrounded and at thermal and mechanical equilibrium with its vapor, prove that the vapor must be supercooled according to

$$P_{\text{sat}} - P_{\text{v}} \approx -\frac{2\sigma}{R}\frac{\rho_{\text{v}}}{\rho_{\text{L}}}$$

where P_{v} is the pressure in the vapor phase and P_{sat} is the saturation pressure corresponding to the temperature of the temperature of the bubble and its surroundings.

2.8 Show that, in view of Kelvin's effect, vapor microbubbles can be stable in a quiescent liquid only when the liquid is slightly superheated. Estimate the liquid superheat needed for vapor bubbles with 2 and 5 μm radii to exist in pure, atmospheric water. Also, repeat the problem, this time for microdroplets with 2 and 5 μm radii in atmospheric pure water vapor.

2.9 Waves with an amplitude of 0.5 m and a wavelength of 100 m occur at the surface of a deep lake. Find the maximum fluid velocity in the vertical direction at the lake surface, and at 10 and 50 m below the surface.

2.10 Derive Eqs. (2.179) and (2.180) for Taylor instability in a three-dimensional flow field, where the interphase is on the (x, y) plane.

Hint: Equations (2.157) and (2.158) must be replaced with

$$\phi'_{\text{L}} = b \exp\left[-kz + i(\omega t - k_x x - k_y y)\right]$$

and

$$\phi'_{\text{G}} = b' \exp\left[kz + i(\omega t - k_x x - k_y y)\right],$$

respectively, and force balance at the interphase requires

$$P_{\text{G}} - P_{\text{L}} = \sigma\left(-\frac{\partial^2\zeta}{\partial x^2} - \frac{\partial^2\zeta}{\partial y^2}\right).$$

2.11 A network of steam jets rise from a horizontal plane submerged in saturated water under atmospheric conditions. The jets form a square network, with sides that are λ_{cr}, the critical wavelength according to Rayleigh–Taylor instability theory. The diameter of each jet is $\lambda_{\text{cr}}/2$. Calculate the vapor mass flow rate, per unit of plate surface area.

2.12 Derive a dispersion relation similar to Eq. (2.169), when the liquid and gas layers are δ_{L} and δ_{G} thick, respectively.

2.13 Using the results of Problem 2.11, find the neutral and fastest growing disturbance wavelength, for the limit $U_{\text{G}} \to 0$ and $U_{\text{L}} \to 0$.

2.14 Derive an expression similar to Rayleigh's equation for an infinitely long cylindrical bubble growing in an infinitely large stagnant liquid. Discuss the adequacy of the result.

2.15 The schematic of an ideal inverted annular two-phase flow regime in a pipe is depicted in Fig. P2.15. In this regime a cylindrical liquid core is separated from the wall by a thin gas film. For inverted annular flow in a horizontal pipe, when both phases are inviscid and incompressible, and assuming $\delta/R \ll 1$, show that the dispersion equation is

$$c = \frac{\omega}{k} = \frac{\rho_L U_L + \rho'_G U_G}{\rho_L + \rho'_G}$$

$$\pm \left[-\frac{\rho_L \rho'_G}{(\rho_L + \rho'_G)^2}(U_L - U_G)^2 + \frac{\sigma k}{\rho_L + \rho'_G} - \frac{\sigma}{k(\rho_L + \rho'_G)(R - \delta)^2} \right]^{1/2},$$

where $\rho'_G = \rho_G \coth(k\delta)$.

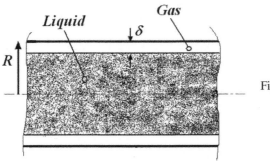

Figure P2.15.

2.16 Prove Eqs. (2.225)–(2.227).

2.17 Consider the pipe displayed in Fig. P2.17, which is assumed to be in microgravity conditions. Assuming that $h \ll R$, and assuming negligibly small axial fluid velocities, show that an analysis similar to the analysis in Section 2.12 leads to the following dispersion equation:

$$[\rho^* \coth(R^* K^* H^*) + 1]\Omega^2 - \frac{K^*}{R^{*2}(1 - H^*)^2} + K^{*3}$$

$$+ 4K^{*4}\left[1 - \left(1 + \frac{\Omega}{K^{*2}}\right)^{1/2}\right] + 4\Omega K^{*2} = 0,$$

where

$$\rho^* = \rho_G/\rho_L, \quad H^* = h/R,$$

$$R^* = R\frac{\sigma}{\rho_L \nu_L^2}, \quad K^* = k\left(\frac{\sigma}{\rho_L \nu_L^2}\right)^{-1}, \quad \Omega = \alpha\left(\frac{\rho_L^2 \nu_L^3}{\sigma^2}\right),$$

and α is the wave growth parameter.

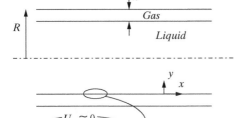

Figure P2.17. Inverted annular flow in Problem 2.17.

2.18 Calculate and compare the frequencies of volumetric and second-mode shape oscillations for air bubbles suspended in water at 25 °C and atmospheric pressure, with $R_0 = 10, 100,$ and 500 μm.

2.19 Using Eqs. (2.245), (2.246), and (2.247), prove Eqs. (2.248) and (2.249).

2.20 Prove Eq. (2.121).

3 Two-Phase Mixtures, Fluid Dispersions, and Liquid Films

3.1 Introductory Remarks about Two-Phase Mixtures

The hydrodynamics of gas–liquid mixtures are often very complicated and difficult to rigorously model. A detailed discussion of two-phase flow modeling difficulties and approximation methods will be provided in Chapter 5. For now, we can note that, although the fundamental conservation principles in gas–liquid two-phase flows are the same as those governing single-phase flows, the single-phase conservation equations cannot be easily applied to two-phase situations, primarily because of the discontinuities represented by the gas–liquid interphase and the fact that the interphase is deformable. Furthermore, a wide variety of morphological configurations (flow patterns) are possible in two-phase flow. Despite these inherent complexities, useful analytical, semi-analytical, and purely empirical methods have been developed for the analysis of two-phase flows. This has been done by adapting one of the following methods.

(1) Making idealizations and simplifying assumptions. For example, one might idealize a particular flow field as the mixture of equal-size gas bubbles uniformly distributed in a laminar liquid flow, with gas and liquid moving with the same velocity everywhere. Another example is the flow of liquid and gas in a channel, with the liquid forming a layer and flowing underneath an overlying gas layer (a flow pattern called stratified flow), when the liquid and gas are both laminar and their interphase is flat and smooth. It is possible to derive analytical solutions for these idealized flow situations. However, these types of models have limited ranges of applicability, and two-phase flows in practice are often far too complicated for such idealizations.

(2) Averaging parameters and conservation and transport equations. This approach is based on the realization that an essential first step in deriving workable analytical or empirical methods is establishing the definitions of workable two-phase flow properties. This is needed because gas–liquid two-phase flow is characterized by complicated spatial and temporal fluctuations. Furthermore, at any point and at any instant of time, only one of the phases can be present. At any point the flow parameters and properties such as velocity, density, and pressure all fluctuate. The local-instantaneous properties and velocities are not very useful or tractable (except in direct numerical simulations), whereas by averaging workable equations in terms of workable average properties are derived.

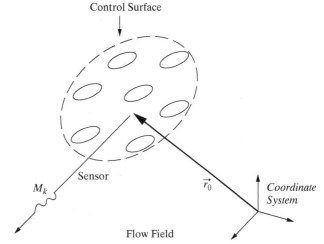

Figure 3.1. Schematic of a two-phase flow field.

Although other approaches also exist for defining workable properties and conservation equations, averaging is the most widely used among them. Workable flow properties (as well as simplified conservation equations) can be defined by performing some form of averaging (time, volume, flow area, etc.) on local and instantaneous properties and conservation equations. Averaging is equivalent to low-pass filtering to eliminate high-frequency fluctuations. In averaging, information about the details of fluctuations is lost in return for simplified and tractable properties and equations. Although fluctuation details are lost as a result of averaging, their statistical properties and their effects on the averaged balance equations can be accounted for. We can think of single-phase turbulent flow for a rough analogy. In single-phase turbulent flow we lose information about fluctuations by averaging, but we can include their effect on average momentum and energy equations by introducing eddy viscosity and heat transfer eddy diffusivity, or by using a turbulent transport model (such as the k–ε model). These additional models and correlations are often empirical.

The averaging techniques will be discussed only briefly here. Detailed discussions can be found in Ishii (1975), Nigmatulin (1979), Banerjee (1980), Bouré and Delhaye (1981), and Lahey and Drew (1988).

3.2 Time, Volume, and Composite Averaging

3.2.1 Phase Volume Fractions

Consider the sensor shown in Fig. 3.1, which generates a signal M_k, where

$$M_k = \begin{cases} 1 & \text{when the sensor tip is in phase } k, \\ 0 & \text{otherwise.} \end{cases}$$

The time-averaged local volume fraction of phase k at a location represented by the position vector $\vec{r}$ can then be defined as

$$\bar{\alpha}_k^t(t_0, \vec{r}) = \frac{1}{\Delta t} \int_{t_0 - \frac{\Delta t}{2}}^{t_0 + \frac{\Delta t}{2}} M_k \, dt. \tag{3.1}$$

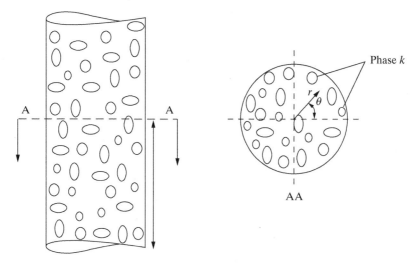

Figure 3.2. Schematic of a one-dimensional two-phase flow field.

In gas–liquid two-phase flow, and elsewhere in this book, unless otherwise stated, we use α for α_G, and we refer to it as the *void fraction*. Evidently, in a gas–liquid mixture

$$\alpha_L = 1 - \alpha. \tag{3.2}$$

We can also define an instantaneous volume fraction for phase k, at a location represented by $\vec{r}_0$, by considering a control volume surrounding the point of our interest (see Fig. 3.1) and writing

$$[\![\alpha_k(t_0, \vec{r}_0)]\!] = \frac{\Delta V_k}{\Delta V}, \tag{3.3}$$

where ΔV is the total volume of the control volume and ΔV_k is the volume occupied by phase k in the control volume when the flow field is frozen at t_0. Here also, the widely accepted convention is to use α for α_G and call it the void fraction.

Many two-phase flow problems involve flow in pipes and channels. A schematic is shown in Fig. 3.2. For these situations it is more convenient to replace volume averaging with flow-area averaging. Here again we can define an instantaneous flow-area-averaged phase volume fraction. The more useful definition is by composite time (or ensemble) and flow-area averaging. The double (composite) time and flow-area-averaged volume fraction of phase k is then defined as

$$\langle\bar{\alpha}_k^t(t_0, z_0)\rangle = \frac{1}{A\Delta t} \int\limits_{t_0-\frac{\Delta t}{2}}^{t_0+\frac{\Delta t}{2}} dt \int_A M_k(t, r, \theta)dA = \frac{1}{A\Delta t} \int\limits_{t_0-\frac{\Delta t}{2}}^{t_0+\frac{\Delta t}{2}} A_k(t)dt, \tag{3.4}$$

where $A_k(t)$ is the flow area occupied by phase k at time instant t and (t, r, θ) is meant to represent dependence on time and location on the channel cross section. For gas–liquid two-phase flow, following the aforementioned convention, $\langle \bar{\alpha}_G^t(t_0, z_0) \rangle$ is simply shown as $\langle \bar{\alpha}^t(t_0, z_0) \rangle$ and is called the void fraction.

3.2.2 Averaged Properties

For any property ξ, the local, time-averaged value is defined as

$$\bar{\xi}^t(t_0, \vec{r}_0) = \frac{1}{\Delta t} \int_{t_0 - \frac{\Delta t}{2}}^{t_0 + \frac{\Delta t}{2}} \xi(t, r_0) dt. \tag{3.5}$$

Time averaging is evidently meaningful when the following conditions are met.

(1) $\Delta t \gg$ the characteristic time scale of fluctuations desired to be filtered out (e.g., the time for passage of dispersed-phase parcels).
(2) $\Delta t \ll$ the characteristic time scale of the macroscopic system's transient processes (e.g., the time scale for significant changes in local pressure, temperature, etc.).

Time averaging is most appropriate for quasi-stationary processes. (For processes that involve fluctuations, such as turbulent flow, we use the term stationary instead of steady state. After all, these processes are never in steady state in the strict sense. In a stationary process, the statistical characteristics of properties do not vary with time.) However, for transient situations time averaging can be replaced with ensemble averaging. In ensemble averaging, we consider a large number of identical experiments in which measurements are repeated at specific locations and specific times after the initiation of each test. The ensemble-average property for ξ at time t_0 and location $\vec{r}$ is then defined as $\sum_{i=1}^{N} \xi(t_0, \vec{r}_0)/N$, where N is the total number of the identical experiments.

The instantaneous volume-averaged value of the same property ξ can be defined as

$$[\![\xi(t_0, \vec{r}_0)]\!] = \frac{1}{\Delta V} \int_{\Delta V} \xi(t_0, \vec{r}) dV, \tag{3.6}$$

where ΔV is an appropriately defined control volume. It is obvious that volume-averaged properties are useful when

(1) $\Delta V >$ the characteristic scale of spatial fluctuations that are to be filtered (e.g., the mixture volume associated with a single parcel of the dispersed phase),
(2) $\Delta V \ll$ physical system characteristic size, and
(3) $\Delta V <$ the characteristic size over which significant macroscopic flow field property variations can occur.

The averaging techniques represented by Eqs. (3.5) and (3.6) are appropriate for properties for which we do not need to distinguish the individual phases, for example the mixture velocity (rather than the velocity of each phase) or mixture temperature and pressure. In two-phase flow we often need phase-specific parameters. For example, we usually would like to keep track of liquid and gas phase velocities separately, because gas and liquid tend to move at different velocities. In these cases we define the time- and volume-averaged intrinsic phase properties as

$$\bar{\xi}^{t,k}(t_0, \vec{r}_0) = \frac{1}{\Delta t} \int_{t_0 - \frac{\Delta t}{2}}^{t_0 + \frac{\Delta t}{2}} \xi_k(t, \vec{r}_0) M_k dt, \tag{3.7}$$

$$[\![\xi_k(t_0, \vec{r}_0)]\!]_k = \frac{1}{\Delta V_k} \int_{\Delta V_k} \xi_k(t_0, \vec{r}) dV_k, \tag{3.8}$$

where $k = $ L or G. In Eq. (3.8), $\Delta V_k = [\![\alpha_k(t_0, \vec{r}_0)]\!]\Delta V$ is the volume occupied by phase k at time instant t.

Models and correlations are often based on double averaging, however. Double averaging is needed because single time, ensemble, space, or area averaging alone is not sufficient for removing all the discontinuities in two-phase flow. For modeling of two-phase flow we need a type of averaging that leads to properties and flow parameters that are continuous and have continuous derivatives. (Averaging should also separate signals from noise and lead to parameters that can be measured with realistic instrumentation.) Single averaging does not get rid of all the discontinuities in derivatives. For two- and three-dimensional flows, the composite time (or ensemble) and volume averaging is probably the most widely applied double-averaging concept, according to which

$$[\![\bar{\xi}^t(t_0, \vec{r}_0)]\!] = \frac{1}{(\Delta V)(\Delta t)} \int_{t_0 - \frac{\Delta t}{2}}^{t_0 + \frac{\Delta t}{2}} dt \int_{\Delta V} \xi(t, \vec{r}) dV, \tag{3.9}$$

$$[\![\bar{\xi}^t(t_0, \vec{r}_0)]\!]_k = \frac{1}{(\Delta V_k)(\Delta t)} \int_{t_0 - \frac{\Delta t}{2}}^{t_0 + \frac{\Delta t}{2}} dt \int_{\Delta V} M_k \xi_k(t, \vec{r}) dV. \tag{3.10}$$

Space and time averaging are, in principle, not commutative, and volume averaging should be performed first. For simplicity, however, we will not follow this requirement.

The composite time (or ensemble) and flow-area averaging can be considered the equivalent of composite time (or ensemble) and volume averaging when one deals with one-dimensional flow, and this topic will be discussed in some detail in the following.

3.3 Flow-Area Averaging

Volume averaging in a one-dimensional flow field leads to flow-area-averaged properties. The resulting equations are among the most widely used. The equivalent of Eq. (3.3) is

$$\langle \alpha_k(t_0, z_0) \rangle = \frac{A_k}{A}, \tag{3.11}$$

where z is the axial coordinate. The instantaneous flow-area-averaged ξ is defined as (see Fig. 3.2)

$$\langle \xi(t_0, z_0) \rangle = \frac{1}{A} \int_A \xi(t_0, z_0, r, \theta) dA, \tag{3.12}$$

where (r, θ) are meant to represent the coordinates on the cross section. The instantaneous flow-area-averaged intrinsic phase property $\langle \xi_k(t_0, z_0) \rangle_k$ is defined as

$$\langle \xi_k(t_0, z_0) \rangle_k = \frac{1}{A_k} \int_{A_k} \xi_k(t_0, z_0, r, \theta) dA_k. \tag{3.13}$$

The composite time- and flow-area-averaged property ξ is defined as

$$\langle \bar{\xi}^t(t_0, z_0) \rangle = \frac{1}{A \Delta t} \int_{t_0 - \frac{\Delta t}{2}}^{t_0 + \frac{\Delta t}{2}} dt \int_A \xi(t, z_0, r, \theta) dA. \tag{3.14}$$

In composite time and phase area-averaged form, we can write:

$$\langle \bar{\xi}_k^{t,k}(t_0, z_0) \rangle_k = \frac{1}{A \langle \bar{\alpha}_k^t \rangle} \int_A \overline{[\xi_k(t_0, r, \theta) \alpha_k(t_0, r, \theta)]}^t dA. \tag{3.15}$$

Thus, for example, the composite averaged mixture velocity will be:

$$\langle \bar{U}^t(t_0, z_0) \rangle = \frac{1}{A \Delta t} \int_{t_0 - \frac{\Delta t}{2}}^{t_0 + \frac{\Delta t}{2}} dt \int_A U(t, z_0, r, \theta) dA = \frac{1}{A} \int_A \overline{U(t_0, z_0, r, \theta)}^t dA. \tag{3.16a}$$

The composite liquid phase intrinsic velocity will be:

$$\langle \bar{U}_L^{t,L}(t_0, z_0) \rangle_L = \frac{1}{A \langle 1 - \bar{\alpha}^t \rangle} \int_A \overline{\{U_L(t_0, r, \theta)[1 - \alpha(t_0, r, \theta)]\}}^t dA. \tag{3.16b}$$

3.4 Some Important Definitions for Two-Phase Mixture Flows

Because we will primarily be dealing with composite-averaged properties and parameters in this book, unless otherwise stated, the forthcoming definitions are based on composite-averaged parameters.

3.4.1 General Definitions

The slip velocity is defined as

$$\vec{U}_r = \left[\kern-0.3em\left[\bar{\vec{U}}_G^{t,G} \right]\kern-0.3em\right]_G - \left[\kern-0.3em\left[\bar{\vec{U}}_L^{t,L} \right]\kern-0.3em\right]_L. \tag{3.17}$$

Intrinsic phase thermodynamic and transport properties will be simply shown as h_L, h_G, ρ_L, ρ_G, K_L, K_G, μ_L, μ_G, etc. The implicit assumption is that these properties remain uniform over the space where averaging is performed.

The mixture mass flux is defined as

$$\vec{G} = \rho_G \left[\kern-0.3em\left[\overline{\alpha \vec{U}}_G^t \right]\kern-0.3em\right] + \rho_L \left[\kern-0.3em\left[\overline{(1 - \alpha) \vec{U}_L}^t \right]\kern-0.3em\right]. \tag{3.18}$$

The phasic mass fluxes are defined as

$$\vec{G}_G = \rho_G \left[\kern-0.3em\left[\overline{\alpha \vec{U}_G}^t \right]\kern-0.3em\right], \tag{3.19}$$

$$\vec{G}_L = \rho_L \left[\kern-0.3em\left[\overline{(1 - \alpha) \vec{U}_L}^t \right]\kern-0.3em\right]. \tag{3.20}$$

The mixture (two-phase) density is defined as

$$\bar{\rho} = \rho_L(1 - [\![\bar{\alpha}^t]\!]) + \rho_G[\![\bar{\alpha}^t]\!]. \tag{3.21}$$

The in- situ mixture properties are defined based on a frozen flow field. They represent mixture properties when a unit volume of the frozen flow field is considered. In the per unit volume form, they are defined according to

$$\bar{h} = [h_L\rho_L(1 - [\![\bar{\alpha}^t]\!])] + h_G\rho_G[\![\bar{\alpha}^t]\!]/\bar{\rho}, \tag{3.22}$$

$$\bar{\mathbf{u}} = [\mathbf{u}_L\rho_L(1 - [\![\bar{\alpha}^t]\!])] + \mathbf{u}_G\rho_G[\![\bar{\alpha}^t]\!]/\bar{\rho}, \tag{3.23}$$

$$\bar{s} = [s_L\rho_L(1 - [\![\bar{\alpha}^t]\!]) + s_G\rho_G[\![\bar{\alpha}^t]\!]]/\bar{\rho}. \tag{3.24}$$

3.4.2 Definitions for Flow-Area-Averaged One-Dimensional Flow

Since double-averaged properties and parameters are of interest, in the forthcoming discussions all parameters are assumed to be time or ensemble averaged. Thus, by $\xi(t_0, z_0)$ we will mean $\bar{\xi}^t(t_0, z_0)$. Likewise, by $\xi_k(t_0, z_0)$ we will imply $\bar{\xi}_k^t(t_0, z_0)$, and by $\langle U_L \rangle_L$ we will mean $\langle \bar{U}_L^t \rangle_L$, etc.

For flow-area-averaged one-dimensional flow parameters the following expressions result from the continuity of phase volumes:

$$\langle U_L \rangle_L = \frac{Q_L}{A(1 - \langle\alpha\rangle)} \tag{3.25}$$

and

$$\langle U_G \rangle_G = \frac{Q_G}{A\langle\alpha\rangle}, \tag{3.26}$$

where Q_L and Q_G represent the liquid and gas volumetric flow rates (in cubic meters per second, for example).

The superficial velocity for each phase is defined as the mean velocity if just that phase flowed in the flow passage, therefore,

$$\langle j_L \rangle = Q_L/A, \tag{3.27}$$

$$\langle j_G \rangle = Q_G/A. \tag{3.28}$$

The total volumetric flux, or equivalently the mixture center-of-volume velocity, is

$$\langle j \rangle = \langle j_L \rangle + \langle j_G \rangle. \tag{3.29}$$

The superficial velocities are related to phase intrinsic velocities according to

$$\langle j_L \rangle = (1 - \langle\alpha\rangle)\langle U_L \rangle_L, \tag{3.30}$$

$$\langle j_G \rangle = \langle\alpha\rangle\langle U_G \rangle_G. \tag{3.31}$$

The volumetric quality is defined as

$$\beta = \frac{Q_G}{Q_L + Q_G} = \frac{\langle j_G \rangle}{\langle j \rangle}. \tag{3.32}$$

Clearly, $0 \leq \beta \leq 1$.

The mixture mass flux can obviously be written as

$$G = (\rho_L Q_L + \rho_G Q_G)/A,$$

and that leads to

$$G = \rho_G \langle U_G \rangle_G \langle \alpha \rangle + \rho_L \langle U_L \rangle_L (1 - \langle \alpha \rangle) = \rho_G \langle j_G \rangle + \rho_L \langle j_L \rangle. \quad (3.33)$$

The phasic mass fluxes follow:

$$G_G = \rho_G Q_G/A = \rho_G \langle U_G \rangle_G \langle \alpha \rangle = \rho_G \langle j_G \rangle, \quad (3.34)$$

$$G_L = \rho_L Q_L/A = \rho_L \langle U_L \rangle_L (1 - \langle \alpha \rangle) = \rho_L \langle j_L \rangle. \quad (3.35)$$

The flow quality is defined as

$$\langle x \rangle = m_G/(m_G + m_L), \quad (3.36)$$

where $\dot{m}$ represents the mixture mass flow rate (in kilograms per second, for example), and $\dot{m}_G$ and $\dot{m}_L$ represent the gas and liquid mass flow rates, respectively. The following identities can be easily proven:

$$\langle x \rangle = \rho_G \langle U_G \rangle_G \langle \alpha \rangle /[\rho_G \langle U_G \rangle_G \langle \alpha \rangle + \rho_L \langle U_L \rangle_L (1 - \langle \alpha \rangle)]$$
$$= \rho_G \langle j_G \rangle /(\rho_G \langle j_G \rangle + \rho_L \langle j_L \rangle)$$
$$= G_G/G. \quad (3.37)$$

Evidently, $0 \leq \langle x \rangle \leq 1$.

An important definition is the *slip ratio*:

$$S_r = \langle U_G \rangle_G / \langle U_L \rangle_L. \quad (3.38)$$

Using relation (3.37) for $\langle x \rangle$ and the definition of the slip ratio, one can easily derive the following relation, which is often referred to as the *fundamental void-quality relation*:

$$\frac{\langle x \rangle}{1 - \langle x \rangle} = \frac{\rho_G}{\rho_L} S_r \frac{\langle x \rangle}{1 - \langle x \rangle}. \quad (3.39)$$

The in situ mixture density defined in Eq. (3.21) applies with $\langle \alpha \rangle$ replacing $[\![\bar{\alpha}^t]\!]$. One can also define a homogeneous mixture density by writing

$$\rho_h = G/\langle j \rangle = \left[\frac{1 - \langle x \rangle}{\rho_L} + \frac{\langle x \rangle}{\rho_G} \right]^{-1} = [v_L + \langle x \rangle (v_G - v_L)]^{-1}. \quad (3.40)$$

The flow quality defined in Eq. (3.37) is purely hydrodynamic. For a pure vapor–liquid mixture (i.e., a single-component two-phase mixture), the equilibrium thermodynamic quality (often referred to simply as the thermodynamic quality) can be defined as

$$\langle x_{eq} \rangle = (h - h_f)/h_{fg}, \quad (3.41)$$

where h_f is the saturated liquid enthalpy and h is the specific mixed-cup enthalpy of the mixture, defined as

$$h = (\dot{m}_f h_f + \dot{m}_g h_g)/(\dot{m}_f + \dot{m}_g) = (1 - \langle x \rangle)h_f + \langle x \rangle h_g, \quad (3.42)$$

where we have used subscripts f and g instead of L and G, because saturated liquid and vapor are implied. Thermodynamic (equilibrium) quality is only applicable to single-component mixtures (e.g., water and steam). It can be negative (for a subcooled liquid mixed-cup state) or larger than 1 (for a superheated vapor mixed-cup state).

3.4.3 Homogeneous-Equilibrium Flow

The simplest method for modeling and analysis of the thermo-hydrodynamics of two-phase flow is to assume that the two phases are everywhere well mixed, are at thermodynamic equilibrium, and move with the same velocity. These are the characteristics of a homogeneous-equilibrium mixture (HEM) two-phase flow. For a HEM, $S_r = 1$. It can easily be shown that

$$\langle U_G \rangle_G = \langle U_L \rangle_L = \langle j \rangle, \tag{3.43}$$

$$\langle \alpha \rangle = \langle \beta \rangle, \tag{3.44}$$

$$\bar{\rho} = \rho_h. \tag{3.45}$$

For a single-component (pure vapor–liquid) mixture, HEM requires the mixture to be at saturation, and therefore

$$h = \bar{h}, \tag{3.46a}$$

$$\langle x \rangle = \langle x_{eq} \rangle. \tag{3.46b}$$

3.5 Convention for the Remainder of This Book

The averaging notations described in the previous sections are tedious and are usually not used. They are also often unnecessary, because averaged two-phase properties and parameters are assumed in most models and correlations anyway.

In the remainder of this book, unless otherwise stated, all parameters are assumed to be properly composite-averaged. However, Chapter 6, where the drift flux model is discussed, will be an exception because in that chapter the nonuniformities in the velocity and void fraction over the cross section of flow channels are the major subject of discussion. Thus, in one-dimensional flow, everywhere except Chapter 6, x will imply $\langle \bar{x}^t \rangle$, and U_G and U_L will stand for $\langle \bar{U}_G^{t,G} \rangle_G$ and $\langle \bar{U}_L^{t,L} \rangle_L$, respectively. Likewise, by $\xi(t_0, z_0)$ we will mean $\langle \bar{\xi}^t(t_0, z_0) \rangle$, and by $\xi_k(z_0, t_0)$ we will mean $\langle \bar{\xi}_k^t(z_0, t_0) \rangle_k$. In two- and three-dimensional flows, where volume averaging rather than flow-area averaging is needed, $\vec{U}_G$ and $\vec{U}_L$ will stand for $[\![\vec{\bar{U}}_G^{t,G}]\!]_G$ and $[\![\vec{\bar{U}}_L^{t,L}]\!]_L$, respectively.

EXAMPLE 3.1. Consider the flow of water in a 5.25-cm-inner-diameter pipe at a pressure of 10 bars. The mass flux is $G = 2100$ kg/m^2·s.

(a) Calculate the mixture mixed-cup enthalpies for $x_{eq} = -0.32, -0.1, 0.21$, and 1.13. Also, assuming that vapor and liquid remain saturated for $0 \leq x_{eq} \leq 1$, calculate the mixture temperatures.
(b) Assuming homogeneous-equilibrium flow conditions, calculate the liquid and vapor mass flow rates and the mixture velocities at $x_{eq} = 0.02$ and 0.7.

(c) For part (b), assuming that $S_r = (\rho_f/\rho_g)^{1/3}$, calculate the liquid and vapor mass flow rates, as well as the liquid and vapor phasic (intrinsic) velocities.

SOLUTION.

(a) At 10 bars of pressure we have $\rho_f = 886.6$ kg/m³, $\rho_g = 5.208$ kg/m³, $h_f = 765.3$ kJ/kg, $h_g = 2.778$ MJ/kg, and $h_{fg} = 2.013$ MJ/kg.

The mixture mixed-cup enthalpy values are found from $h = h_f + x_{eq}h_{fg}$ and are listed in the table at the end of this example. For cases where $x_{eq} < 0$, the fluid is in a subcooled liquid state. The temperature can be found from thermodynamic property tables, by noting that $T = T(h, P)$. The same can be said about the cases with $x_{eq} > 1$, where we deal with superheated vapor. When $1 \leq x_{eq} \leq 0$, the mixture is saturated, and the mixture temperature is $T_{sat}(P)$. These temperatures are also shown in the table.

(b) The phasic mass flow rate and the mixture velocity are found from

$$\dot{m}_f = \frac{\pi}{4}D^2 G(1 - x_{eq}),$$

$$\dot{m}_g = \frac{\pi}{4}D^2 G x_{eq},$$

$$\bar{U} = G/\bar{\rho} = G\left(\frac{x_{eq}}{\rho_g} + \frac{1 - x_{eq}}{\rho_f}\right).$$

The results are summarized in the table.

(c) Using the aforementioned values for ρ_g and ρ_f, we find $S_r = 5.54$. The liquid and vapor mass flow rates are the same as those calculated in part (b). Using the fundamental void-quality relation, Eq. (3.39), we can now calculate the void fraction, α. The phasic velocities are then obtained from

$$U_g = \frac{G x_{eq}}{\rho_g \alpha},$$

$$U_f = \frac{G(1 - x_{eq})}{\rho_f(1 - \alpha)}.$$

The results are summarized in the following table.

Part (a)	$x_{eq} = -0.32$	$x_{eq} = -0.1$	$x_{eq} = 0.21$	$x_{eq} = 1.13$
h (kJ/kg)	121.2	564.1	1188	3040
T (K)	301.9	407.2	453.6	568.3
			$(= T_{sat}$ at 10 bar)	
Part (b)		$x_{eq} = 0.02$		$x_{eq} = 0.7$
$\dot{m}_f$ (kg/s)		4.455		1.364
$\dot{m}_g$ (kg/s)		0.0909		3.182
$\bar{U}(= j)$ (m/s)		10.39		283
Part (c)				
$\dot{m}_f$ (kg/s)		4.455		1.364
$\dot{m}_g$ (kg/s)		0.0909		3.182
U_f (m/s)		3.78		20.93
U_g (m/s)		51.65		286.2

3.6 Particles of One Phase Dispersed in a Turbulent Flow Field of Another Phase

3.6.1 Turbulent Eddies and Their Interaction with Suspended Fluid Particles

Particles of one phase entrained in a highly turbulent flow of another phase (e.g., microbubbles in a turbulent liquid flow) are common in many two-phase flow systems.

Examples include agitated mixing vessels and notation devices. Turbulence determines the behavior of particles by causing particle dispersion, particle–particle collision, particle–wall impact, and coalescence and breakup when particles are fluid. Kolmogorov's theory of isotropic turbulence provides a useful framework for modeling these systems.

A turbulent flow field is *isotropic* when the statistical characteristics of the turbulent fluctuations remain invariant with respect to any arbitrary rotation or reflection of the coordinate system. A turbulent flow is called *homogeneous* when the statistical distributions of the turbulent fluctuations are the same everywhere in the flow field. In isotropic turbulence clearly $\overline{u_1'^2} = \overline{u_2'^2} = \overline{u_3'^2}$, where subscripts 1, 2, and 3 represent the three-dimensional orthogonal coordinates. Isotropic turbulence is evidently an idealized condition, although near-isotropy is observed in some systems, for example, in certain parts of a baffled agitated mixing vessel. However, in practice a *locally isotropic* flow field can be assumed in many instances, even in shear flows such as pipes, by excluding regions that are in the proximity of walls (Schlichting, 1968).

Highly turbulent flow fields are characterized by random and irregular variations of velocity at each point. These velocity fluctuations are superimposed on the base flow and are characterized by turbulent eddies. Eddies can be thought of as vortices that move randomly around and are responsible for velocity variation with respect to the mean flow. The size of an eddy represents the magnitude of its physical size. It can also be defined as the distance over which the velocity difference between the eddy and the mean flow changes appreciably (or the distance over which the eddy loses its identity).

The largest eddies are typically of the order of the turbulence-generating feature in the system. These eddies are too large to be affected by viscosity, and their kinetic energy cannot be dissipated. They produce smaller eddies, however, and transfer their energy to them. The smaller eddies in turn generate yet smaller eddies, and this cascading process proceeds, until energy is transferred to eddies small enough to be controlled by viscosity. Energy dissipation (or viscous dissipation, i.e., the irreversible transformation of the mechanical flow energy to heat) then takes place.

A turbulent flow whose statistical characteristics do not change with time is called *stationary*. (We do not use the term steady state because of the existence of time fluctuations.) A turbulent flow is in *equilibrium* when the rate of kinetic energy transferred to eddies of a certain size is equal to the rate of energy dissipation by those eddies plus the kinetic energy lost by those eddies to smaller eddies. Conditions close to equilibrium can (and often do) exist under nonstationary situations when the rate of kinetic energy transfer through eddies of a certain size is much larger than their rate of transient energy storage or depletion.

The distribution of energy among eddies of all sizes can be better understood by using the *energy spectrum* of the velocity fluctuations, and by noting that as eddies become smaller the frequency of velocity fluctuations that they represent becomes larger. Suppose we are interested in the streamwise turbulence fluctuations at a particular point. We can write

$$\int_0^\infty \mathbf{E}_1(k_1, t) dk_1 = \overline{u_1'^2}, \tag{3.47}$$

where $\mathbf{E}_1(k_1, t)$ is the one-dimensional energy spectrum function for velocity fluctuation u_1', in terms of the wave number k_1. The wave number is related to frequency according to $k_1 = 2\pi f / \bar{U}_1$, where f represents frequency. Instead of Eq. (3.47), one could write

$$\int_0^\infty \mathbf{E}_1^*(f, t) df = \overline{u_1'^2} \tag{3.48}$$

or

$$\mathbf{E}_1^*(f, t) = \mathbf{E}_1(k_1, t) \frac{dk_1}{df} = \frac{2\pi}{\bar{U}_1} \mathbf{E}_1(k_1, t), \tag{3.49}$$

where $\bar{U}_1$ is the mean streamwise velocity and $\mathbf{E}_1^*(f, t)$ is the one-dimensional energy spectrum function of velocity fluctuation u_1' in terms of frequency f. For an isotropic three-dimensional flow field, one can write (Hinze, 1975)

$$\int_0^\infty \mathbf{E}(k, t) dk = \frac{3}{2}\overline{u'^2}, \tag{3.50}$$

where $\mathbf{E}(k, t)$ is the three-dimensional energy spectrum function and k is the radius vector in the three-dimensional wave-number space. The qualitative distribution of the three-dimensional spectrum for isotropic turbulence is depicted in Fig. 3.3. The spectrum shows the existence of several important eddy size ranges. The largest eddies, which undergo little change as they move, occur at the lowest frequency range. The energy-containing eddies, so named because they account for most of the kinetic energy in the flow field, occur next. Eddies in the *universal equilibrium range* occur next, and their name arises because they have *universal* characteristics that do not depend on the specific flow configuration. These eddies do not remember how they were generated and are not aware of the overall characteristics of the flow field. As a result, they behave the same way, whether they are behind a turbulence-generating grid in a wind tunnel or in a flotation device. These eddies follow local isotropy, except very close to the solid surfaces.

The universal equilibrium range itself includes two important eddy size ranges: the *dissipation range* and the *inertial size range*. In the *dissipation range* the eddies are small enough to be viscous. Their behavior can only be affected by their size, fluid density, viscosity, and the turbulence dissipation rate (energy dissipation per unit mass) ε. (The dissipation rate actually represents the local intensity of turbulence.) A simple dimensional analysis using these properties leads to the *Kolmogorov microscale*:

$$l_D = (\nu^3/\varepsilon)^{1/4}. \tag{3.51}$$

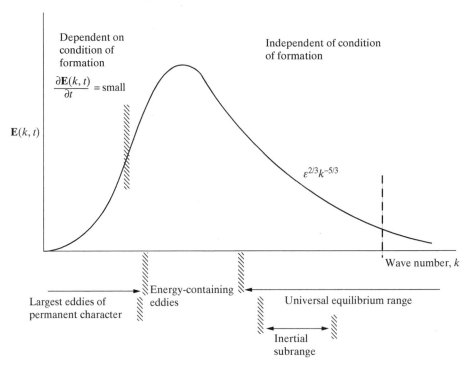

Figure 3.3. Schematic of the three-dimensional energy spectrum in isotropic turbulence.

Likewise, we can derive the following expressions for Kolmogorov's velocity and time scales:

$$u_{\mathrm{D}} = (\nu\varepsilon)^{1/4}, \tag{3.52a}$$

$$t_{\mathrm{c,D}} = (\nu/\varepsilon)^{1/2}. \tag{3.52b}$$

Eddies with dimensions less than about $10l_{\mathrm{D}}$ have laminar flow characteristics. Thus, when two points in the flow field are separated by a distance $\Delta r < 10l_{\mathrm{D}}$, they are likely to be within a laminar eddy. In that case, the variation of fluctuation velocities over a distance of Δr can be represented by (Schultze, 1984)

$$\sqrt{\overline{\Delta u'^2}} = 0.26\sqrt{\frac{\varepsilon}{\nu}}\,\Delta r. \tag{3.53}$$

The *inertial size range* refers to eddies with characteristic dimensions from about $20l_{\mathrm{D}}$ to about 0.05Λ, where Λ represents the turbulence macroscale. The macroscale of turbulence represents approximately the characteristic size of the largest vortices or eddies that occur in the flow field. The inertial eddies are too large to be affected by viscosity, and their behavior is determined by inertia. Their behavior can thus be influenced only by their size, the fluid density, and turbulent dissipation. The variation of fluctuation velocities across Δr, when Δr is within the inertial size range, can then be represented by (Schultze, 1984)

$$\sqrt{\overline{\Delta u'^2}} = (1.38)\varepsilon^{1/3}(\Delta r)^{1/3}. \tag{3.54}$$

Locally isotropic turbulent flow can occur in multiphase mixtures as well (Bose et al., 1997). An important characteristic of the inertial zone is that in that eddy scale range

$$\mathbf{E}(k) = 1.7\varepsilon^{2/3}k^{-5/3}, \tag{3.55}$$

where the coefficient 1.7 is due to Batchelor (1970). This relation (i.e., $\mathbf{E}(k) \sim k^{-5/3}$) provides a simple method for ascertaining the existence of an inertial eddy range in a complex turbulent flow field.

Bubbles, readily deformable particles, and their aggregates when they are suspended in highly turbulent liquids often have dimensions within the eddy scales of the inertial range. Their characteristics and behavior can thus be assumed to result from interaction with inertial eddies (Schultze, 1984; Coulaloglou and Tavlarides, 1977; Narsimhan et al., 1979; Tobin et al., 1990; Taitel and Dukler, 1976a, b).

The size of a dispersed fluid particle in a turbulent flow field is determined by the combined effects of breakup and coalescence processes. In dilute suspensions where breakup is the dominant factor, the maximum size of the dispersed particles can be represented by a critical Weber number, defined as

$$\mathrm{We_{cr}} = \frac{\rho_c \overline{\Delta u'^2} d_d}{\rho}, \tag{3.56}$$

where subscripts c and d represent the continuous and dispersed phases, respectively, and $\overline{\Delta u'^2}$ represents the magnitude of velocity fluctuations across the particle (i.e., over a distance of $\Delta r \approx d_d$). For particles that fall within the size range of viscous eddies, therefore, Eqs. (3.53) and (3.56) result in

$$d_{d,max} \approx \left(\frac{\nu\sigma}{\rho_c \varepsilon} \right)^{1/3} \mathrm{We_{cr}}^{1/3}. \tag{3.57}$$

For particles that fall within the inertial eddy size range in a locally isotropic turbulent field, Eqs. (3.54) and (3.56) indicate that the average equilibrium particle diameter should follow

$$d_{d,max} \approx \left(\frac{\sigma}{\rho_c} \right)^{3/5} \mathrm{We_{cr}}^{3/5} \varepsilon^{-2/5}. \tag{3.58}$$

The right-hand side of this equation also provides the order of magnitude of the particle Sauter mean diameter, $d_{d,32}$. In a pioneering study of the hydrodynamics of dispersions, Hinze (1955) noted that 95% of particles in an earlier investigation were smaller than

$$d_{d,max} = 0.725 \left(\frac{\sigma}{\rho_c} \right)^{3/5} \varepsilon^{-2/5}. \tag{3.59}$$

The validity of Eq. (3.58) has been experimentally demonstrated (Shinnar, 1961; Narsimhan et al., 1979; Bose et al., 1997; Tobin et al., 1990; Tsouris and Tavlarides, 1994).

EXAMPLE 3.2. In an experiment with a well-baffled stirred tank with some specific geometric characteristics (equal diameter and height, and impeller diameter 0.54 times the tank diameter), the mean turbulence dissipation could be estimated from the following expression:

$$\bar{\varepsilon} = 1.27 N^3 L_{\text{imp}}^2, \tag{a}$$

where N is the impeller rotational speed, L_{imp} is the impeller diameter, and $\bar{\varepsilon}$ is the specific energy dissipation rate in the tank. In an agitated lean liquid–liquid dispersion vessel, the average turbulence dissipation in the impeller zone is $\varepsilon_{\text{avg}} = 30\bar{\varepsilon}$. Estimate the maximum diameter of droplets of cyclohexane, for a dilute mixture of cyclohexane with distilled water at 25 °C, in a tank with $L_{\text{imp}} = 20$ cm and $N = 250$ rpm. The two phases are assumed to be mutually saturated, whereby $\rho_c = 997$ kg/m³, $\mu_c = 0.894 \times 10^{-3}$ kg/m·s, and $\rho_d = 761$ kg/m³, where subscripts c and d represent the continuous and dispersed phases, respectively. For the distilled water–cyclohexane mixture, when the two phases are mutually saturated, the interfacial tension is $\sigma = 0.0462$ N/m.

SOLUTION. The rotational speed will be $N = (250 \, \text{min}^{-1})(\text{min}/60 \, \text{s}) = 4.17 \, \text{s}^{-1}$. Using this value for the rotational speed in Eq. (a), we find $\bar{\varepsilon} = 3.675$ W/kg. As a result,

$$\varepsilon_{\text{avg}} = 30\bar{\varepsilon} = 110.2 \text{ W/kg}.$$

We can now use Eq. (3.59) to find $d_{\text{d,max}}$ by using ε_{avg} for ε, and that gives

$$d_{\text{d,max}} \approx 2.77 \times 10^{-4} \text{ m} = 0.277 \text{ mm}.$$

The Reynolds number can be calculated from $\text{Re} = \rho_c L_{\text{imp}}^2 N / \mu_c$. We thus get $\text{Re} \approx 1.86 \times 10^5$, implying fully turbulent flow in the impeller zone.

EXAMPLE 3.3. A dilute suspension of cyclohexane in distilled water at a temperature of 25 °C flows in a smooth pipe with 5.25-cm inner diameter. The mean velocity is 2.5 m/s. Estimate the size of the cyclohexane particles in the pipe.

SOLUTION. We will use the properties that were provided in Example 3.2. We can use Eq. (3.59), provided that we can estimate the turbulent dissipation in the pipe. We can estimate the latter from

$$\bar{\varepsilon} \approx \frac{1}{\rho_c} \bar{U} |(\nabla P)_{\text{fr}}|.$$

To find the frictional pressure gradient, let us write

$$\text{Re} = \rho_c \bar{U} D / \mu_c \approx 1.46 \times 10^5,$$
$$f' = 0.316 \, \text{Re}^{-0.25} \approx 0.0162,$$
$$|(\nabla P)_{\text{fr}}| = f' \frac{1}{D} \frac{1}{2} \rho_c \bar{U}^2 \approx 959 \text{ N/m}^3.$$

The dissipation rate will then be

$$\bar{\varepsilon} \approx 2.4 \text{ W/kg}.$$

Equation (3.59) then gives

$$d_{\text{max}} \approx 1.28 \times 10^{-3} \text{ m} = 1.28 \text{ mm}.$$

3.6.2 The Population Balance Equation

The size distribution of deformable particles suspended in a turbulent field of another fluid is determined by coalescence and breakup processes. The *population balance equation* (PBE), also referred to as the *general dynamic equation* in aerosol science (Friedlander, 2000), is a general mathematical representation of the processes that affect the particle number and size distributions in a mixture (Hulburt and Katz, 1964). The PBE has been used extensively for modeling and analysis of disperse-phase contactors, as well as aerosol populations.

In its general form, when particle growth (resulting, for example, from phase change) is not considered, the PBE can be written as

$$\frac{\partial n(V, t)}{\partial t} + \nabla \cdot [n(V, t)\vec{U}] = \nabla \cdot [\boldsymbol{D}(V)\nabla n(V, t)] + \dot{n}_{b} - \dot{n}_{d}, \qquad (3.60)$$

where $n(V, t)dV$ is the number density of particles in the V to $V + dV$ volume range, $\vec{U}$ represents the continuous phase velocity, $\boldsymbol{D}(V)$ represents the diffusion coefficient of particles, and $\dot{n}_{b}$ and $\dot{n}_{d}$ represent the birth and death rates for the particles with volume V in a unit mixture volume, respectively. The latter parameters are formulated as (Coulaloglou and Tavlarides, 1977; Tsouris and Tavlarides, 1994)

$$\dot{n}_{b} = \int_{0}^{V/2} n(V', t)n(V - V', t)h(V', V - V')\lambda(V', V - V')dV'$$

$$+ \int_{V}^{\infty} n(V', t)g(V')v(V')\beta(V, V')dV' \qquad (3.61)$$

and

$$\dot{n}_{d} = n(V, t) \left[g(V) + \int_{0}^{\infty} n'(V', t)h(V, V')\lambda(V, V')dV' \right], \qquad (3.62)$$

where $n(V - V', t)$ is the number density of particle with volume $V - V'$, $n(V', t)dV'$ is the number density of particles that have volumes in the V' to $V' + dV'$ range, $g(V)$ is the breakup frequency of particles with volume V, $h(V, V')$ is the collision frequency between particles with volumes V and V', and $\lambda(V, V')$ represents the coalescence (adhesion) efficiency. Also, $v(V')$ represents the number of particles formed from the breakup of particles with volume V', and $\beta(V, V')$ is the probability (or fraction) of breakup events of particles with volume V' that result in the generation of a particle with volume V. The first term on the right side of Eq. (3.61) represents the rate of generation of particles by coalescence of smaller particles, and the second term represents the generation of particles as a result of the breakup of larger particles.

Equation (3.60) is general and accounts for convection (the second term on the left side) and diffusion (the first term on the right side). Simplified forms of the PBE have been used for the derivation of correlations for coalescence of liquid dispersions (Das *et al.*, 1987; Konno *et al.*, 1988; Muralidhar *et al.*, 1988; Tobin *et al.*, 1990), droplet breakup in liquid dispersions (Narsimhan *et al.*, 1979; Konno *et al.*, 1983), and simultaneous coalescence and breakup of liquid droplet dispersions (Coulaloglou and Tavlarides, 1977; Sovova and Prochazka, 1981; Tsouris and Tavlarides, 1994). By imposing a sudden sharp reduction in turbulence intensity in a system that has been initially in steady state (e.g., by a sharp reduction in the impeller speed in an agitated mixing vessel), one can effectively shut down breakup while coalescence is under way. The temporal variation of the particle size distribution characteristics can then be used to deduce the coalescence parameters.

Analytical solution of the PBE is difficult, but has been derived for a few specific situations for aerosol fields where self-similar solutions were applicable (Friedlander, 2000). Several numerical methods have been proposed and demonstrated, however (see, e.g., Song et $al.$, 1997; Bennett and Rohani, 2001; Mahoney and Ramkrishna, 2002). The homogeneous flow field can be assumed in an agitated tank only if the recirculation time in the tank is significantly shorter than the characteristic time for coalescence or breakup. In a homogeneous tank the PBE can be solved by trial and error and the coalescence and/or breakup parameters can be adjusted until a particle size distribution is obtained that matches measurements. Under dynamic conditions, it is more convenient to use a discretized size distribution (Sovova and Prochazka, 1981). Spatial nonuniformities in an agitated vessel can also be accounted for by dividing the vessel into a number of nodes, and applying the PBE to each node separately (Tsouris and Tavlarides, 1994; Alopaeus et $al.$, 1999). Particle coalescence and breakup models are evidently needed for the solution of the PBE.

3.6.3 Coalescence

For dispersed fluid particles that are small enough to fall in the viscous eddy size range, an expression for the collision frequency is (Saffman and Turner, 1956)

$$h(V, V') = 0.31(V^{1/3} + V'^{1/3})^3(\varepsilon/\nu_c)^{1/2}, \qquad (3.63)$$

where subscript c refers to the continuous phase. In the forthcoming discussion, furthermore, subscript d will represent the dispersed phase.

Mechanistic coalescence models for agitated vessels often have the following assumptions in common.

(1) The coalescence frequency of two particles, $\theta(V, V')$, can be represented as the product of a collision frequency, $h(V, V')$, and a coalescence efficiency, $\lambda(V, V')$.
(2) The flow field is locally isotropic turbulent, and particles are within the inertial size distribution range of turbulent eddies.
(3) A particle has a characteristic turbulent velocity equal to the characteristic velocity of turbulent eddies of its size.

These assumptions lead to the following expression for collision frequency (Levich, 1962; Coulaloglou and Tavrialides, 1977; Tsouris and Tavlarides, 1994):

$$h(V, V') \approx \varepsilon^{1/3}(V^{1/3} + V'^{1/3})^2(V^{2/9} + V'^{2/9})^{1/2}. \qquad (3.64)$$

This expression considers two eddy velocities, one for each of the interacting particles. A slightly different form can be derived if, instead of two eddy velocities representing the diameters of the two particles, a single eddy representing the average diameter of the two particles is considered (Muralidhar et $al.$, 1988; Tobin et $al.$, 1990).

Aerosol population models often assume 100% coalescence efficiency (Friedlander, 2000). Following the collision between two particles or bubbles in a turbulent liquid flow field, however, coalescence requires the thinning and rupture of the liquid film that separates the two particles. This leads to an imperfect coalescence. Several models have been proposed for the coalescence efficiency, $\lambda(V, V')$. Coulaloglou and Tavlarides (1977), for example, proposed

$$\lambda(V, V') = \exp(-t_{c,coal}/t_{c,cont}), \tag{3.65}$$

where $t_{c,coal}$ is the average coalescence time and $t_{c,cont}$ is the average contact time. The coalescence time depends on the process of drainage of the liquid film separating the two colliding particles. Models that treat the film drainage as a stochastic process have been proposed by Das *et al.* (1987) and Muralidhar *et al.* (1988), among others. If the characteristic period of velocity fluctuation of eddies of the size $d + d'$ is used for $t_{c,cont}$, where $d = (6/\pi)^{1/3} V^{1/3}$ and $d' = (6/\pi)^{1/3} V'^{1/3}$, then (Coulaloglou and Tavlarides, 1977)

$$t_{c,cont} \approx \frac{(d + d')^{2/3}}{\varepsilon^{1/3}}. \tag{3.66}$$

An expression for λ is (Tsouris and Tavlarides, 1994; Alopaeus *et al.*, 1999)

$$\lambda(d, d') = \exp\left[-C\frac{\mu_c \rho_c \varepsilon}{\sigma^2 (1 + \alpha_d)^3}\left(\frac{dd'}{d + d'}\right)^4\right], \tag{3.67}$$

where C is a constant and α_d is the dispersed phase volume fraction. This and other forthcoming functions are in terms of the particle diameter, rather than particle volume. They can be used in equations similar to Eqs. (3.60) and (3.61), by changing the variable from V to d. Thus, if the integrand of an integral is $G(V)$, changing the variable to d is done by using

$$G(V) = G[V(d)]\frac{dd}{dV}; \quad V = \frac{\pi}{6}d^3. \tag{3.68}$$

When coalescence of bubbles is addressed, mechanisms other than turbulence-induced collision can also be significant. These include buoyancy and laminar shear, both of which cause faster moving particles to collide and coalesce with slower moving particles in their vicinity.

3.6.4 Breakup

Similar to coalescence, particle breakup can be considered to have two stages; particle–eddy collision and shattering of the particle. A particle–eddy collision frequency (not to be confused with particle–particle collision frequency in coalescence) and a breakage probability can thus be defined, with the product of the two representing the breakage frequency. Some investigators have modeled breakup by comparing the energy of eddies colliding with the particle with the particle surface energy (Coulaloglou and Tavlarides, 1977) or its increase as a result of the breakup of the particle (Narsimhan *et al.*, 1979). Another group of models have been derived based on assumed similarity in drop size distributions (Narsimhan *et al.*, 1980, 1984). The following expressions for breakage frequency g and the probability density $\beta(d, d')$

associated with the generation of particles with diameter d from the breakup of a particle with diameter d' have been used by Tsouris and Tavlarides (1994) and Alopaeus *et al.* (1999):

$$g(d') = C_1 \frac{\varepsilon^{1/3}}{(1 + \alpha_\mathrm{d}) d'^{2/3}} \exp \left[-C_2 \frac{\sigma (1 + \alpha_\mathrm{d})^2}{\rho_\mathrm{d} \varepsilon^{2/3} d'^{5/3}} \right], \tag{3.69}$$

where $C_1 = 0.00481$ and $C_2 = 0.08$ (Bapat and Tavlarides, 1985; Alopaeus *et al.*, 1999), and

$$\beta(d, d') = \frac{90 d_1^2}{d'^3} \left(\frac{d^3}{d'^3} \right)^2 \left(1 - \frac{d^3}{d'^3} \right)^2. \tag{3.70}$$

3.7 Conventional, Mini-, and Microchannels

3.7.1 Basic Phenomena and Size Classification for Single-Phase Flow

The fluid mechanics and heat transfer literature is primarily based on observations and experience dealing with systems of conventional sizes. We are faced with important questions when we deal with very small flow passages. Are the well-known phenomena, models, and correlations applicable for all size scales? If not, what is the size limit for their applicability, and what must be done for smaller systems?

For single-phase flow, the most obvious issue is the validity of the continuum assumption for the fluid. Strictly speaking, this assumption breaks down when the mean free path of the fluid molecules (or the mean intermolecular distance in the case of liquids) becomes comparable with the smallest important physical features of the system.

For liquids, there is little to be concerned about, because the intermolecular distance for liquids is of the order of 10^{-9} m or 10^{-3} μm, and the continuum assumption is valid for flow passages as small as about 1 μm. For gases we can define flow regimes using the *Knudsen number*

$$\mathrm{Kn} = \lambda_\mathrm{G}/l, \tag{3.71}$$

where λ_G is the gas molecular mean free path and l is the characteristic dimension of the flow path. By using statistical thermodynamics (Carey, 1999), the molecular mean free path can be calculated from

$$\lambda_\mathrm{G} = \frac{3}{2} \nu_\mathrm{G} \left(\frac{\pi M_\mathrm{G}}{2 R_\mathrm{u} T} \right)^{1/2}, \tag{3.72}$$

where ν_G and M_G represent the kinematic viscosity and the molecular mass of the gas, respectively. Alternatively, one can use simple gaskinetic theory to derive (Golden, 1964)

$$\lambda_\mathrm{G} = \frac{\sqrt{2}}{2\pi} \frac{\kappa_\mathrm{B} T}{P d^2}, \tag{3.73}$$

where κ_B is Boltzmann's constant and d is the range of the repulsive force around molecules. A typical value for d is $\approx 5 \times 10^{-10}$ m. Based on the magnitude of Kn, the following flow regimes are often defined for gas-carrying systems:

$$\begin{aligned}
&\text{continuum:} &&\text{Kn} \leq 10^{-3}, \\
&\text{velocity slip and temperature jump:} &&10^{-3} < \text{Kn} \leq 0.1, \\
&\text{transition regime:} &&0.1 < \text{Kn} \leq 10, \\
&\text{free molecular flow:} &&\text{Kn} > 10.
\end{aligned}$$

In the continuum regime, intermolecular collisions determine the behavior of the fluid. The continuum-based conservation equations, along with no-slip and thermal equilibrium boundary conditions at fluid–solid boundaries, can be used. In the velocity slip and temperature jump regime, however, the no-slip boundary condition as well as equality between wall and fluid temperatures at the solid–fluid interphase are inadequate. Intermolecular collisions still predominate in the velocity slip and temperature jump regime, however, and the predictions of continuum-based theory need to be corrected to account for near-wall phenomena. The behavior of the gas is determined by the wall–molecule collisions in the free molecular regime, and continuum-based methods are completely irrelevant.

EXAMPLE 3.4. For air flow in circular tubes at 300 K temperature, find the smallest tube diameter for the validity of the continuum regime for the following conditions: $P = 0.01$ bar, $P = 1$ bar, and $P = 1$ MPa.

SOLUTION. Let us use $\text{Kn} = \lambda_G/D = 10^{-3}$ as the criterion. At 300 K, $\mu_G = 1.857 \times 10^{-5}$ kg/m·s and is insensitive to pressure. The results of the calculations are summarized in the following table, where λ_G has been calculated by using Eq. (3.72).

P	ρ_g (kg/m^3)	ν_g (m^2/s)	λ_G (μm)	D (mm)
0.01 bar	0.0116	1.60×10^{-3}	10.2	10.2
1 bar	1.16	1.60×10^{-5}	0.102	0.102
1 MPa	11.61	1.60×10^{-6}	0.0102	0.0102

This example shows that, when the gas flow at moderate and high pressures is considered, channels with hydraulic diameters larger than about 100 pm conform to continuum treatment with no-slip conditions at solid surfaces. For liquid flow, as mentioned earlier, continuum treatment and no-slip conditions apply to much smaller channel sizes.

Single-phase flow and heat transfer in sub-millimeter channels have been studied rather extensively in the recent past. Note that for these channels there is no breakdown of continuum, and velocity slip and temperature jump are negligibly small. Some investigators have reported that well-established correlations for pressure drop and heat transfer and for laminar-to-turbulent flow transition deviate from the measured data obtained with such channels, suggesting the existence of unknown scale effects. It was also noted, however, that the apparent disagreement between conventional models and correlations on one hand and microchannel data on the other was relatively minor, indicating that conventional methods can be used at least for rough microchannel analysis. Basic theory does not explain the existence of an intrinsic scale effect, however. (After all, the Navier–Stokes equations apply to these flow channels as well.) The identification of the mechanisms responsible for the reported differences between conventional and microchannels and the development of predictive methods for microchannels have remained the foci of research.

There is now sufficient evidence that proves that in laminar flow the conventional theory agrees with microchannel data well and that the differences reported by some investigators in the past were likely due to experimental errors and misinterpretations (Sharp and Adrian, 2004; Kohl *et al.*, 2005; Herwig and Hausner, 2003; Tiselj *et al.*, 2004).

Some experimental investigations have also reported that the laminar-turbulent transition in microchannels occurred at a considerably lower Reynolds number than in conventional channels (Wu and Little, 1983; Stanley *et al.*, 1997). However, careful recent experiments by Kohl *et al.* (2005) using channels with $D_H = 25$–100 µm have shown that laminar flow theory predicts wall friction very well at least for $Re_D \leq 2000$, where Re_D is the channel Reynolds number, thus supporting the standard practice where laminar–turbulent transition is assumed to occur at $Re_D \approx 2300$. Sharp and Adrian (2004) have also reported that laminar to turbulent transition occurred in their experiments at $Re_D \approx 1800$–2000.

With respect to turbulent flow the situation is less clear. Measured heat transfer coefficients obtained by some investigators have been lower than what conventional correlations predict (Wang and Peng, 1994; Peng *et al.*, 1995; Peng and Peterson, 1995), whereas an opposite trend has been reported by others (Choi *et al.*, 1991; Yu *et al.*, 1995; Adams *et al.*, 1997, 1999). Nevertheless, the disagreement between conventional correlations and microchannel experimental data is relatively small, and the discrepancy is typically less than a factor of 2.

The following factors are likely to contribute to the reported differences between the behaviors of microchannels and conventional channels.

(1) Surface roughness and other configurational irregularities. The relative magnitudes of surface roughness in microchannels can be significantly larger than in large channels. Also, at least for some manufacturing methods (e.g., electron discharge machining), the cross-sectional geometry of a microchannel may slightly vary from one point to another (Mala and Li, 1999, Qu *et al.*, 2000).

(2) Suspended particles. Microscopic particles that are of little consequence in conventional systems can potentially affect the behavior of turbulent eddies in microchannels (Ghiaasiaan and Laker, 2001).

(3) Surface forces. Electrokinetic forces (i.e., forces arising from the electric double layer) can develop during the flow of a weak electrolyte (e.g., aqueous solutions with weak ionic concentrations), and these forces can modify the channel hydrodynamics and heat transfer (Yang *et al.*, 1998).

(4) Fouling and deposition of suspended particles. Fouling and deposition can change surface characteristics, smooth sharp corners, and cause local partial flow blockage.

(5) Compressibility. This is an issue for gas flows. Large local pressure and temperature gradients are common in microchannels. As a result, in gas flow, fully developed hydrodynamics does not occur.

(6) Conjugate heat transfer effects. Axial conduction in the fluid, as well as heat
conduction in the solid structure surrounding the channels, can be important
in microchannel systems. As a result, the local heat fluxes and transfer coefficients sometimes cannot be determined without a conjugate heat transfer analysis of the entire flow field and its surrounding solid structure system. Neglecting
the conjugate heat transfer effects can lead to misinterpretation of experimental
data (Herwig and Hausner, 2003; Tiselj et al., 2004).

(7) Dissolved gases. In heat transfer experiments with liquids, unless the liquid is
effectively degassed, dissolved noncondensables will be released from the liquid
as a result of depressurization and heating. The released gases, although typically
small in quantity (water at room temperature saturated with air contains about
10 ppm of dissolved air), can affect heat transfer by increasing the mean velocity,
disrupting the liquid velocity profile, and disrupting the thermal boundary layer
on the wall (Adams et al., 1999).

In summary, for single-phase laminar flow in mini- and microchannels of interest to
this book, the conventional models and correlations appear to be adequate. Transition from laminar to turbulent flow can also be assumed to occur under conditions
similar to those in conventional systems. In light of these discussions, furthermore,
conventional turbulent flow models and correlations may also be utilized for mini-
and microchannels, provided that the uncertainty with respect to the accuracy of such
correlations for application to mini- and microchannels is considered.

EXAMPLE 3.5. Consider steady and fully developed and turbulent flow of water in a
horizontal pipe, with $\mathrm{Re}_D = 4.0 \times 10^4$. The water temperature is 25 °C.

(a) Calculate the maximum wall roughness size for hydraulically smooth conditions.
Also, estimate the Kolmogorov microscale and the lower limit of the size range
of inertial eddies in the turbulent core of a tube with $D = 25$ mm.
(b) Repeat part (a) for a tube with $D = 0.8$ mm.

For both cases, for estimating the size of Kolmogorov's eddies, assume a hydraulically
smooth wall, and assume that conventional friction factor correlations apply.

SOLUTION.

(a) Using $\mathrm{Re}_D = \rho_c \bar{U} D / \mu_c$, we find $\bar{U} = 1.43$ m/s. Using the approach of
Example 3.3, we can then calculate the friction factor f', and use it for the calculation of the absolute value of the pressure gradient. The results will be

$$f' = 0.022$$

and

$$|(\nabla P)_{\mathrm{fr}}| \approx 916 \ \mathrm{N/m}^3.$$

The mean dissipation rate $\bar{\varepsilon}$ is then calculated following Example 3.3, with the result

$$\bar{\varepsilon} \approx 1.317 \text{ W/kg.}$$

The Kolmogorov microscale can now be calculated from Eq. (3.51), where $v_c = \mu_c/\rho_c = 8.96 \times 10^{-7} \text{ m}^2\text{/s}$ and $\bar{\varepsilon} = 1.317 \text{ W/kg}$ are used. The result will be

$$l_D \approx 2.7 \times 10^{-5} \text{ m} = 27 \text{ µm.}$$

The size range of viscous eddies will therefore be $l \leq 10 l_D \approx 270$ µm. The lower limit of the size range of inertial eddies will be $l \approx 20 l_D \approx 0.54$ mm. It is to be noted that these calculations are approximate, and the viscous dissipation rate is not uniform in a turbulent pipe.

(b) For the tube with $D = 0.8$ mm, the calculations lead to

$$|(\nabla P)_{\text{fr}}| \approx 2.8 \times 10^7 \text{ N/m}^3,$$

$$\bar{\varepsilon} \approx 1.26 \times 10^6 \text{ W/kg,}$$

$$l_D \approx 8.7 \times 10^{-7} \text{m} = 0.87 \text{ µm.}$$

The size range of viscous eddies will thus be $l \lesssim 8.7$ µm, whereas the lower limit of the inertial eddy size will be approximately 17 µm.

3.7.2 Size Classification for Two-Phase Flow

Gas–liquid two-phase flow is sensitive to the flow path physical size (scale). The sensitivity to scales is primarily due to the change in the relative magnitudes of the forces that are experienced by the two phases. Important dimensionless numbers that can characterize two-phase flow in a confined flow passage include:

$$\text{Eo} = \frac{\Delta \rho g l^2}{\sigma} \quad \text{(Eötvös number)} \tag{3.74}$$

$$\text{We}_i = \frac{j_i^2 l \rho_i}{\sigma} \quad \text{(Weber number)} \tag{3.75}$$

$$\text{Re}_i = j_i D_H / v_i \quad \text{(Reynolds number)} \tag{3.76}$$

$$\text{Bd} = l^2 \left/ \left(\frac{\sigma}{g \Delta \rho} \right) \right. \quad \text{(Bond number)} \tag{3.77}$$

$$\text{Ca} = \frac{j_L \mu_L}{\sigma} \quad \text{(Capillary number)}, \tag{3.78}$$

where l represents the characteristic dimension for the flow passage, and subscript i represents a phase i ($i = $ L or G). The hydraulic diameter is often chosen for the characteristic dimension for the flow passage, thus $l = D_H$. In the small channels of interest in high-performance miniature heat exchangers, Eo < 1 (or, equivalently, Bd < 1, where Bd is the Bond number); at least one of the Weber numbers is of the order of $1\sim10^2$; and $Re_L \geq 1$; whereby buoyancy is insignificant while inertial, viscous and capillary effects are all important. Dimensionless numbers representing the competition between buoyancy and viscous forces (Grashof and Rayleigh numbers) do not need to be included in the aforementioned list of the most relevant dimensionless numbers to small flow passages for this reason. With common (air–water and water–steam) fluids, flow passages with hydraulic diameters in the 100 μm – 1 mm range can meet these conditions. Applications of such flow passages in energy and process systems include miniature (meso) heat exchangers, cooling of high-powered electronic systems, three-phase catalytic reactors, cooling of plasma-facing components of fusion reactors, miniature refrigeration systems, fuel injection systems of some internal combustion equipment, and evaporator components of fuel cells, to name a few.

The Laplace length scale $\lambda_L = \sqrt{\sigma/g\Delta\rho}$ (also referred to as the capillary length) has been used as the basis for defining the threshold between mini- and macrochannel sizes. For channels smaller in characteristic size than the capillary length the Taylor instability-driven phenomena described in the previous chapter, which are crucial to many two-phase flow and change-of-phase processes in large channels, are likely to be irrelevant. Less obvious contributors to these differences are the different relative time and length scales in very small and common large channels. A comparison between the channel size and capillary length can evidently be represented by direct comparison between the channel size and λ_L, in terms of the dimensionless Bond number, or in terms of the dimensionless confinement number N_{con} introduced below. Suo and Griffith (1964) argued that the threshold for the size limit for small channels (minichannels in our terminology) should be the size below which stratified flow in horizontal and near-circular channels impossible, and based on their experimental results proposed

$$\sqrt{Bd} \leq 0.3. \tag{3.79}$$

Kew and Cornwell (1997) defined a *confinement number:*

$$N_{con} = (\lambda_L/D_H)^2, \tag{3.80}$$

the threshold for macro–micro size distinction. Accordingly, models and correlations developed for commonly applied macroscale systems will not do well when $N_{con} > 0.5$.

Triplett *et al.* (1999a) and Feng and Serizawa (2000) proposed the following threshold for aforementioned size limit:

$$D_H \leq \lambda_L \tag{3.81}$$

Li and Wang (2003) used the capillary length to propose a more detailed classification, which in the terms of the Bond number can be summarized as follows.

- $D_H < 0.224\lambda_L$ or Bd < 0.05 Surface tension dominates and renders gravity completely ineffective.

Table 3.1. *Laplace length scale values for saturated liquids of some fluids at 300 K.*

Fluid	λ_L (mm)
water	2.708
R-12	0.8181
R-22	0.8384
R-123	1.024
R-124	0.8524
R-134a	0.8278
R-236fa	0.8482
R-245fa	1.031
R-744	0.2944
R-1234ze	0.8873
R-404a	0.6349
FC-72	0.78
FC-87	0.76

- $0.224\lambda_L < D_H < 1.75\lambda_L$ or $0.05 < \mathrm{Bd} < 3.0$ Surface tension and gravity are both effective.
- $D_H < 0.224\lambda_L$ or $\mathrm{Bd} > 3.0$ Gravity is important while surface tension is ineffective.

Other size ranges can be defined for significantly smaller channels, however. With $D_H \leq O\ (10\ \mu\mathrm{m})$, for example, only large bubbles, comparabale in size to the channel diameter, form during low-flow boiling (Peles *et al.*, 2000; Qu and Mudawar, 2002).

The following classification will generally be used in this book when air–water, steam–water and other fluids with similar properties are considered. Such fluids constitute the majority of applications in miniature heat exchangers and process devices.

Microchannels: 10 μm $\lesssim D_H \lesssim$ 100 μm
Minichannels: 100 μm $\lesssim D_H \lesssim$ 1000 μm
Conventional channels: $D_H \gtrsim$ 3 mm

The classification obviously does not consider the effects of fluid properties. This classification, it must be emphasized, applies to air–water-like fluid mixtures. The demarcations among the categories are approximate, furthermore. The microchannel range is likely to represent two or more distinct scale ranges. The upper limit of the minichannel range for air–water-like fluids is probably slightly larger than 1 mm. The macrochannel and conventional channel size rages also are likely to have an overlap range.

Tables 3.1 and 3.2 depict the values of the capillary length for some widely used refrigerants and dielectric industrial fluids. As can be noted, the capillary length varies over three orders of magnitude, implying that the threshold sizes can be significantly different depending on the fluid. Channels with 200 μm $\lesssim D_H \lesssim$ 3 mm are used in miniature refrigeration systems and compact heat exchangers. As a result, for boiling and condensation in small channels, the common practice in the literature has been to discuss and correlate data for heated tubes in the range 200 μm $\lesssim D_H \lesssim$ 3 mm. In these cases, the term *small flow passage* will be used.

Table 3.2. *Laplace length scale for some saturated cryogens at 1 atm pressure.*

Fluid	λ_L (mm)
Helium	0.3214
Hydrogen	1.675
Methane	1.77
Krypton	0.5082
Neon	0.6206
Nitrogen	1.059
Oxygen	1.085
Argon	0.96
Fluorine	0.96

The flow passage size classifications discussed thus far did not consider the effect of mass flux and are suitable for adiabatic two-phase flow. In heated channels where bubble generation caused by nucleation, growth, and release from the heated wall is predominant the most important size threshold is the confinement effect on bubble growth. As a result often only one size threshold (sometimes referred to as micro–macro size transition) is considered. Experiments show that there is a distinct effect of mass flux on the micro–macro channel size threshold as well as the two-phase flow regimes when the flow passage can be considered as a microchannel. Based on an analysis of extensive flow boiling data representing standard and microgravity conditions, Li and Wu (2010b), proposed that the confinement effect on bubble growth and release is significant when either of the following equations applies:

$$\mathrm{Bd}^{0.5}\,\mathrm{Re_L} = \frac{1}{\mu_L}\left[\frac{g(\rho_L - \rho_G)}{\sigma}\right]^{0.5}G(1-x)D_H^2 < 200 \qquad (3.81a)$$

$$\mathrm{Bd} < 4. \qquad (3.81b)$$

Equation (3.81a) evidently reflects the effect of flow inertia, whereby higher flow inertia causes the departure of smaller bubbles from the channel surface. Equation (3.81b) is identical to $N_{con} > 0.5$ proposed earlier by Kew and Cornwell (1997), imposes the confinement effect at low liquid Reynolds numbers (Wu and Sundén, 2016).

Based on flow boiling data with the perfluorinated dialectric fluid FC-77, Harirchian and Garimella (2010) noted that bubble confinement depends on channel size as well as mass flux and developed the map shown in Figure 3.4. The boundary line that separates the confined and unconfined flow regimes can be represented as:

$$\mathrm{Bd}^{0.5}\,\mathrm{Re_{L0}} = \frac{1}{\mu_L}\left[\frac{g(\rho_L - \rho_G)}{\sigma}\right]^{0.5}GD_H^2 = 160. \qquad (3.82)$$

The left-hand side of the above equation has been termed the *convective confinement number* (Harirchian and Garimella, 2010).

The criterion evidently does not depend on heat flux. The criterion was shown to predict data from several sources representing water and a number of dielectric fluids and refrigerants.

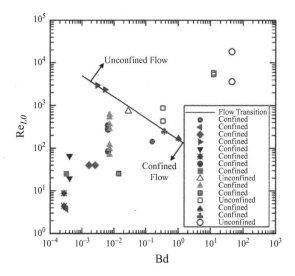

Figure 3.4. The flow regime map and confinement criterion of Harirchian and Garimella (2010).

3.8 Falling Liquid Films

Thin liquid films flowing on a solid surface are common in gas–liquid two-phase flows and change-of-phase heat transfer processes. Falling liquid films are in fact the preferred flow pattern in numerous industrial applications where heat or mass transfer between a liquid and a gas is sought, because of their very large gas–liquid interphase area. Falling film absorbers, which are the critical components of heat driven heat pumps, have been of great interest and have led to extensive research (Killian and Garimella, 2001). Passive cooling of heated surfaces, encountered for example in the passive containment cooling system (PCCS) of advanced pressurized water reactors (PWRs), is another recent application (Ambrosini *et al.*, 2002; Huang *et al.*, 2014). Despite their apparent simple configuration, liquid films can support a rich variety of flow regimes and support complex transport phenomena.

Dimensional analysis shows that the film hydrodynamics should depend on the surface angle of inclination, film Reynolds number Re_F and *Kapitza number*, Ka, where:

$$Re_F = 4\frac{\Gamma_F}{\mu_L} \tag{3.83}$$

$$Ka = \nu_L^4 \rho_L^3 g / \sigma^3. \tag{3.84}$$

The justification for the above definition of film Reynolds number will be presented in the next section. For a liquid falling film at least three major flow regimes can be identified: laminar, laminar-wavy, and turbulent (Fulford, 1964). The flow regime transitions are gradual, however. Ishigai *et al.* (1972) have suggested that for a falling liquid film on a vertical surface five flow regimes can be defined:

- Laminar flow

$$Re_F \leq 1.88\,Ka^{-0.1} \tag{3.85}$$

- First transition regime (from laminar to laminar-wavy)

$$1.88\,Ka^{-0.1} \leq Re_F < 8.8\,Ka^{-0.1} \tag{3.86}$$

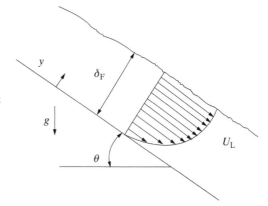

Figure 3.5. Laminar liquid film on a flat inclined surface.

- Stable wavy-laminar

$$8.8\,\mathrm{Ka}^{-0.1} \leq \mathrm{Re_F} < 300 \qquad (3.87)$$

- Second transition regime (from stable wavy-laminar to turbulent)

$$300 \leq \mathrm{Re_F} < 1600 \qquad (3.88)$$

- Fully turbulent

$$\mathrm{Re_F} \geq 1600 \qquad (3.89)$$

In the laminar regime the film is laminar and has a smooth and flat surface. With increasing $\mathrm{Re_F}$, however, waves appear at the film surface. These waves are of two types (Brauner, 1989): capillary waves and inertial (roll) waves. Capillary waves are small, low-amplitude, uniform, and symmetric. Capillary waves appear at the film surface with regime change from laminar to the first transition. The surface ripples have a negligible influence on the near-wall hydrodynamics (Karimi and Kawaji, 1998). With increasing $\mathrm{Re_F}$ capillary waves give way to inertial (roll) waves. These waves are high amplitude (typically several times the average thickness of the liquid film) and can move in solitary mode with adjacent waves separated from each other by long intervals of smooth film. The intervals increase in length with increasing $\mathrm{Re_F}$. These waves are steep at their front, and long on their back, their effect on the liquid hydrodynamics extends all the way to the wall, and they can cause significant recirculation in the liquid film (Karimi and Kawaji, 1999). The second transition regime occurs when adjacent inertial waves start interacting. The interaction of roll waves leads to three-dimensional wave patterns. The flow becomes more chaotic with increasing $\mathrm{Re_F}$, and eventually the film becomes fully turbulent. In the fully turbulent film regime the film surface is covered with small turbulent waves.

3.8.1 Laminar Falling Liquid Films

Consider a flat surface, inclined with respect to the horizontal plane by the angle θ, that supports a laminar, incompressible liquid film (Fig. 3.5). Assuming steady state gives the momentum equation for the liquid phase of

$$\frac{d^2 U_L}{dy^2} + \frac{g \sin \theta \, \Delta \rho}{\mu_L} = 0, \tag{3.90}$$

where $\Delta \rho = \rho_L - \rho_G$. The boundary conditions are

$$U_L = 0 \text{ at } y = 0, \tag{3.91}$$

$$\mu_L \frac{dU_L}{dy} = \tau_I \text{ at } y = \delta_F. \tag{3.92}$$

The interfacial shear stress τ_I is small when the liquid film flows in stagnant gas and can be neglected. The solution of this system will then give

$$U_F(y) = \frac{g \Delta \rho \delta_F^2 \sin \theta}{\mu_L} \left[\left(\frac{y}{\delta_F} \right) - \frac{1}{2} \left(\frac{y}{\delta_F} \right)^2 \right]. \tag{3.93}$$

The velocity profile can now be used in the following derivations:

$$\Gamma_F = \int_0^{\delta_F} \rho_L U_L(y) dy = \frac{g \sin \theta \delta_F^3 \Delta \rho}{3 \nu_L}, \tag{3.94}$$

$$\bar{U}_F = \frac{\Gamma_F}{\delta_F \rho_L} = \frac{g \sin \theta \delta_F^2 \Delta \rho}{3 \mu_L}. \tag{3.95}$$

The *film Reynolds number* is defined as

$$\mathrm{Re}_F = 4 \frac{\bar{U}_F \delta_F}{\nu_L} = 4 \frac{\Gamma_F}{\mu_L}. \tag{3.96}$$

The film properties can be represented in terms of Re_F as

$$\delta_F = \left[\frac{3}{4} \frac{\rho_L \nu_L^2}{g \sin \theta \, \Delta \rho} \right]^{1/3} \mathrm{Re}_F^{1/3}, \tag{3.97}$$

$$\tau_w = (\rho_L - \rho_G) g \delta_F. \tag{3.98}$$

These expressions can be easily modified for the case where τ_I is finite and known (Problem 3.5). In most applications $\rho_L \gg \rho_G$, and $\Delta \rho = \rho_L - \rho_G \approx \rho_L$ can be used. The steady-state heat and mass transfer rates through a laminar and smooth liquid film follow $q'' = k_L(T_w - T_I)/\delta_F = H_F(T_w - T_I)$ and $m_i'' = \rho_L D_{iL}(m_{i,w} - m_{i,u})/\delta_F = K_F(m_{i,w} - m_{i,u})$, where subscript u refers to the "u" surface, leading to

$$\mathrm{Nu}_F = \mathrm{Sh}_F = 1.1 \, \mathrm{Re}_F^{-1/3}, \tag{3.99}$$

where $\mathrm{Nu}_F = H_F l_F / k_F$ and $\mathrm{Sh}_F = K_F l_F / (\rho_L D_{iL})$. The length scale l_F is defined as

$$l_F = \left(\frac{\rho_L \nu_L^2}{g \sin \theta \Delta G \rho} \right)^{1/3}. \tag{3.100}$$

It must be noted that the mass transfer version of Eq. (3.99) (i.e., $\mathrm{Sh}_F = 1.1 \, \mathrm{Re}_F^{-1/3}$) is rarely applicable in practice. This equation applies to quasi-steady conditions. Mass transfer processes involving falling films are often controlled by the mass transfer resistance near the gas–liquid interphase, however, and are of entrance-effect type.

Smooth laminar films can be sustained only with small flow rates, however. Ripples and waves appear on the film surface at moderate flow rates. The waves enhance heat and mass transfer in both the film and the adjacent gas. Linear stability analysis has been applied for the development of a criterion to predict the onset of waviness (Kapitza, 1948; Benjamin, 1957; Hanratty and Hershman, 1961). According to the theory by Kapitza (1948), at the inception of waviness

$$\mathrm{Re}_{F,inc} = 2.44[\mathrm{Ka}\sin\theta]^{-1/11}, \tag{3.101}$$

where the *Kapitza number*, Ka, is defined according to Eq. (3.84).

Kapitza's theory predicts incipience of waves at a finite Re_F (about 6 for room-temperature water flowing on a vertical plane). Some linear stability models indicate that laminar and smooth falling films are unstable essentially at all flow rates (Benjamin, 1957; Hanratty and Hershman, 1961). The linear stability theory of Benjamin (1957), for example, suggests for the onset of ripples

$$\frac{\cos\theta}{\mathrm{Fr}^2} = 3.6 - \left(\frac{2\pi\delta_F}{\lambda}\right)^2 \mathrm{We}_F^2, \tag{3.102}$$

where λ is the wavelength. The film Froude and Weber numbers are defined, respectively, as

$$\mathrm{Fr}_F = \frac{\overline{U}_F}{(g\delta_F)^{1/2}} \tag{3.103}$$

and

$$\mathrm{We}_F = [\rho_L \overline{U}_F^2 \delta_F / \sigma]^{1/2}. \tag{3.104}$$

Linear stability methods, when compared with experimental data, do not appear to predict the data for all slopes, and Kapitza's theory does reasonably well for $\theta \geq 30°$ (Ganic and Mastanaiah, 1983). Experimental data show that wave inception may occur on vertical surfaces at $\mathrm{Re}_{F,inc} \approx 10$ (Fulford, 1964; Brauer, 1956; Binnie, 1957). For falling films on vertical surfaces, transition to waviness is usually assumed to take place at $\mathrm{Re}_{F,inc} \approx 30$ (Edwards *et al.*, 1979).

Two main types of waves occur on falling liquid films: small-amplitude waves that are almost sinusoidal and large-amplitude waves (also called roll waves). Small waves appear at film Reynolds numbers slightly higher than $\mathrm{Re}_{F,inc}$. Roll waves are asymmetrical, are neither periodic nor linear, have amplitudes that are typically 2 to 5 times the average film thickness, carry the bulk of the liquid, and enhance very significantly the mixing and transport speed of processes in the film. The roll waves sometimes interact with one another. The enhancement in transfer processes is primarily due to mixing caused by the waves, although the increase in the film surface area caused by the waves also makes a small contribution (typically a few per cent) to the enhancement.

An empirical correlation for wavy laminar falling films, which is applicable for the range $\mathrm{Re}_F \ll 30$ to 1000, is (Edwards *et al.*, 1979)

$$\mathrm{Nu}_F = 0.82\,\mathrm{Re}_F^{-0.22}. \tag{3.105}$$

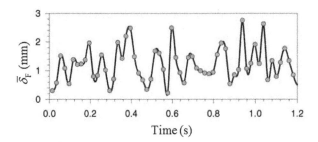

Figure 3.6. Fluctuations in the thickness of a liquid falling film on a vertical surface at $Re_F = 5275$ (Karimi and Kawaji, 1998).

Once again, the mass transfer equivalent of this equation, namely $Sh_F = 0.82\, Re_F^{-0.22}$, although in principle correct for a quasi-steady mass transfer process, is rarely applicable because of the often entrance-effect-dominated mass transfer processes in falling liquid films.

3.8.2 Turbulent Falling Liquid Films

The wavy laminar liquid film becomes turbulent at high film Reynolds numbers. Transition to turbulent film occurs over the $Re_F \ll 1000\text{–}1800$ range and is preceded by the development of a chaotic wave pattern.

The surface of a turbulent film is covered by waves and as a result the thickness of a turbulent falling liquid film is not uniform and fluctuates around an average value with amplitudes that are typically of the same order of magnitude as, or larger than, the average film thickness. The large amplitude waves can accommodate several eddies, and the resulting recirculation strongly augments the transport processes within the film (Jayanti and Hewitt, 1997). Figure 3.6 is a typical measured time trace of a turbulent falling liquid film thickness (Karimi and Kawaji, 1998). An average film thickness can be defined, however.

The average film thickness for falling films on vertical surfaces, when the interfacial gas–liquid shear stress is negligible, has been correlated by several authors. Among the widely referenced correlations are the following:

$$\bar{\delta}_F^+ = 0.0178\, Re_F \quad \text{(Brötz, 1954),} \tag{3.106}$$

$$\bar{\delta}_F^+ = 0.0947\, Re_F^{0.8} \quad \text{(Brauer, 1956),} \tag{3.107}$$

$$\bar{\delta}_F^+ = 0.051\, Re_F^{0.87} \quad \text{(Ganchev \textit{et al.}, 1972),} \tag{3.108}$$

$$\delta_F^+ = 0.109\, Re_F^{0.789} \quad \text{(Takahama and Kato, 1980),} \tag{3.109}$$

and

$$\bar{\delta}_F^+ = 1.657\, Ka^{-0.104}\, Re_F^{0.708} \quad \text{(Miller and Keyhani, 1989).} \tag{3.110}$$

Shear-driven liquid films also occur in the annular flow regime in internal flow situations, as will be discussed later in Chapter 4 and elsewhere. The behavior and thickness of such liquid films are influenced by the shear stress at the gas–liquid interphase, as expected. Several correlations have been developed for the prediction of the thickness of such liquid films. The correlations can be applied to falling films, however, when the interfacial shear stress is neglected. Some of these correlations are

$$\bar{\delta}_F^+ = 0.0512 \, \text{Re}_F^{0.875} \quad \text{(Kosky, 1971)} \tag{3.111}$$

$$\bar{\delta}_F^+ = 0.089 \, \text{Re}_F^{0.789} \quad \text{(Takahama and Kato, 1980)} \tag{3.112}$$

$$\bar{\delta}_F^+ = 0.34 \, \text{Re}_F^{0.6} \quad \text{(Asali } et \, al., 1985), \tag{3.113}$$

where $\bar{\delta}_F^+$ is in wall units

$$\bar{\delta}_F^+ = \frac{\bar{\delta}_F \sqrt{\tau_w/\rho_L}}{\nu_L}. \tag{3.114}$$

To apply these correlations, τ_w, the shear stress at wall surface, is needed. This can be calculated simply by performing a force balance on the film. For a free-falling film on a vertical surface, for example, we will have

$$\tau_w = \rho_L g \bar{\delta}_F.$$

The correlation obtained by Asali $et \, al.$ (1985) agreed with the experimental data of Ambrosini $et \, al.$ (2002), which represented water falling films on vertical as well as 45°-inclined flat surfaces.

The thickness of a turbulent falling film on an inclined surface can be found by replacing g with $g \sin \theta$ in correlations that address flow over vertical surfaces.

3.9 Heat Transfer Correlations for Falling Liquid Films

In light of the previous discussion, it should be clear that theoretical prediction of heat and mass transfer in wavy laminar and turbulent films is difficult. Some empirical correlations are listed in the following. The correlations are based on the average film thickness, $\bar{\delta}_F$, for the obvious reason that instantaneous film thickness varies because of the occurrence of waves.

The following correlations, proposed by Fujita and Ueda (1978), deal with wall-liquid film heat transfer for a falling film on a vertical surface:

$$\text{Nu}_F = \begin{cases} 1.76 \, \text{Re}_F^{-1/3} & \text{for } \text{Re}_F \leq 2460 \, \text{Pr}_L^{-0.646} & (3.115) \\ 0.0323 \, \text{Re}_F^{1/5} \text{Pr}_L^{0.344} & \text{for } 2460 \text{Pr}_L^{-0.646} < \text{Re}_F \leq 1600, & (3.116) \\ 0.00102 \, \text{Re}_F^{2/3} \, \text{Pr}_L^{0.344} & \text{for } 1600 < \text{Re}_F \leq 3200, & (3.117) \\ 0.00871 \, \text{Re}_F^{2/5} \, \text{Pr}_L^{0.344} & \text{for } \text{Re}_F > 3200, & (3.118) \end{cases}$$

where, in accordance with Eq. (3.100), $\text{Nu}_{Fw} = H_F (\nu_L^2/g)^{1/3}/k_L$.

Won and Mills (1982) performed gas absorption experiments and measured the mass transfer coefficients in falling liquid films. They developed the following correlation for the liquid-side mass transfer coefficient representing the resistance between the film–gas interphase and the film bulk:

$$\frac{K_{FI}}{\rho_L (\nu_L g)^{1/3}} = C \, \text{Re}_F^m \text{Sc}_L^{-n}. \tag{3.119}$$

Using the analogy between heat and mass transfer, their correlation can be used for calculating the heat transfer between the falling film bulk and its surface from

$$H_{FI}/\left[\rho_L C_{PL}(\nu_L g)^{1/3}\right] = C \, Re_F^m Pr_L^{-n}, \tag{3.120}$$

$$C = 6.97 \times 10^{-9} \, Ka^{-0.5}, \tag{3.121}$$

$$m = 3.49 \, Ka^{0.068}, \tag{3.122}$$

$$n = 0.137 \, Ka^{-0.055}. \tag{3.123}$$

Instead of Eq. (3.123), $n = 0.36 + 2.43\sigma$ can also be used, where a is in newtons per meter. The data base for the correlation is $1000 < Re_F < 10\,000, 80 < Sc < 2700$, and $5.06 \times 10^{-12} < Ka < 1.36 \times 10^{-8}$.

For falling film evaporation or condensation, when the total thermal resistance of the film is of interest, Edwards *et al.* (1979) recommend

$$Nu_F = \begin{cases} 1.10 \, Re_F^{-1/3} \text{ for } Re_F < 30 \text{ (laminar film)}, & (3.124) \\ 0.82 \, Re_F^{-0.22} \text{ for } 1,000 \geq Re_F \gtrsim 30 \text{ (wavy laminar film)}, & (3.125) \\ 3.8 \times 10^{-3} \, Re_F^{0.4} Pr_L^{0.65} \text{ for } Re_F > 1800 \text{ (turbulent film)}. & (3.126) \end{cases}$$

For the transition range $1000 < Re_F < 1800$, the larger of the wavy laminar and turbulent film correlations is recommended.

EXAMPLE 3.6. A heated flat vertical surface is cooled by a falling water film. At a particular location, the mean film mass flux is $\Gamma_F = 0.2767 \, kg/m^2 s$. The heated surface temperature is 107 °C, and the liquid film bulk temperature is 92 °C.

(a) Calculate the heat flux at the heated surface.
(b) Suppose the falling film is saturated liquid, occurs under atmospheric conditions, and is surrounded by pure saturated steam. Calculate the evaporation rate at the surface of the liquid film.

SOLUTION. Let us use water properties corresponding to a temperature of 98 °C:

$$\rho_L = 959 \, kg/m^3, \, k_L = 0.665 \, W/m \cdot K, \, \nu_L = 2.96 \times 10^{-7} m^2/s, \text{ and } Pr_L = 1.8.$$

The film Reynolds number can now be calculated by using Eq. (3.96), leading to $Re_F = 3900$. Since $Re_F > 1800$, the film is turbulent.

(a) The wall liquid heat transfer is evidently needed. From Eq. (3.118) we find

$$\overline{Nu}_F = 0.00871(3,900)^{2/5}(1.8)^{0.344} = 0.2912,$$

$$\overline{Nu}_F = \frac{\overline{H}_{Fw}(\nu_L^2/g)^{1/3}}{k_L} = 0.2912 \rightarrow \overline{H}_{Fw} = 9,233 \, W/m^2 \cdot K.$$

The wall–film heat flux can now be found:

$$q_w'' = \overline{H}_{Fw}(T_w - \overline{T}_L) = 9,233(107 - 92) = 1.385 \times 10^5 \, W/m^2.$$

(b) In this case, the properties should correspond to $T_{sat} = 100\ °C$, and that would introduce only a minor change in the properties calculated in part (a). The heat transfer coefficient is now found from Eq. (3.126), which gives

$$\overline{Nu}_F = 0.1514,$$

$$\overline{Nu}_F = \frac{\overline{H}_F (v_L^2/g)^{1/3}}{k_L} \Rightarrow \overline{H}_F = 4{,}823\ \text{W/m}^2\cdot\text{K}.$$

The wall heat flux and evaporation rate can now be found as

$$q_w'' = \overline{H}_F (T_w - T_{sat}) = 3.316 \times 10^4\ \text{W/m}^2,$$

$$m_{ev}'' = q_w''/h_{fg} = 1.47 \times 10^{-2}\ \text{kg/m}^2\cdot\text{s}.$$

3.10 Mechanistic Modeling of Liquid Films

It has been shown that it is possible to model a turbulent liquid film by assuming a constant film thickness equal to the average turbulent film thickness, and using an appropriate eddy diffusivity model. This approach can in fact be applied for turbulent liquid films in configurations other than falling films (e.g., in the annular flow regime).

For a steady-state, constant-property turbulent film on a flat surface the momentum equation will give

$$\frac{d}{dy}\left[(v_L + E)\frac{dU_L}{dy} \right] - \frac{1}{\rho_L}\frac{dP}{dz} + g\sin\theta = 0, \tag{3.127}$$

where E is the eddy diffusivity in the liquid film. Note that for a liquid film in a stagnant gas, $-dP/dz = -\rho_G g \sin\theta$, and therefore the last two terms combine into $\frac{\Delta\rho}{\rho_L} g\sin\theta$. By using the no-slip boundary condition at $y = 0$, and $dU_L/dy = 0$ at $y = \bar{\delta}_F$, integration of Eq. (3.127) twice will give

$$U_L^*(y^*) \int_0^{y^*} \frac{\left(1 - \frac{y^*}{\delta_F^*}\right)}{1 + \frac{E}{v_L}}\,dy^*, \tag{3.128}$$

where

$$y^* = y\frac{\sqrt{\bar{\delta}_F\left[-\frac{1}{\rho_L}\frac{dP}{dz} + g\sin\theta\right]}}{v_L} \tag{3.129}$$

and

$$U_L^* = U_L \left/ \sqrt{\bar{\delta}_F\left[-\frac{1}{\rho_L}\frac{dP}{dz} + g\sin\theta\right]}\right. . \tag{3.130}$$

Equation (3.94) in dimensionless form will become

$$\frac{\Gamma_F}{\mu_L} = \int_0^{\bar{\delta}_F^*} U_L^*\,dy^*. \tag{3.131}$$

This derivation assumed $\tau_I \approx 0$ at the film–gas interphase, which is a good approximation for falling films in stagnant gas. When the interfacial shear is important, Eq. (3.92) will be the boundary condition for the velocity profile at the liquid–gas interphase and provides coupling with the gas-side conservation equations. Knowing Γ_F, the iterative numerical solution of Eq. (3.128), along with an appropriate eddy diffusivity model will provide a complete representation of the film hydrodynamics, including the velocity profile and film average thickness. Some eddy diffusivity models will be discussed shortly.

Once the film hydrodynamics have been solved for, the heat transfer in the film can be dealt with by writing the steady-state energy conservation equation for the liquid:

$$\frac{\partial}{\partial y}\left[\left(\alpha_L + \frac{E}{Pr_{L,turb}}\right)\frac{\partial T_L}{\partial y}\right] = U_L(y)\frac{\partial T_L}{\partial z}, \qquad (3.132)$$

where α_L is the thermal diffusivity of liquid and $Pr_{L,turb}$ is the turbulent Prandtl number. For common substances $Pr_{L\,turb} \approx 1$. When one deals with an evaporating liquid film, or a condensate liquid film, the right-hand side of this equation can be neglected. When heating (or cooling) of a subcooled liquid film is considered, the concept of *thermally developed flow* for a constant wall heat flux boundary condition can be borrowed from convection heat transfer theory (see, e.g., Kays *et al.*, 2005), whereby $\partial T_L/\partial z = d\overline{T}_L/dz$. Either way, Eq. (3.132) becomes an ordinary differential equation and can be integrated with proper boundary conditions at the wall ($y = 0$) and the interphase ($y = \bar{\delta}_F$). The numerical solution of Eq. (3.132) is actually simple because the hydrodynamics of the film are already known.

The mass transfer of a species i in the liquid film can likewise be solved for by starting from

$$\frac{\partial}{\partial y}\left[\left(D_{iL} + \frac{E}{Sc_{iL,turb}}\right)\frac{\partial m_{i,L}}{\partial y}\right] = U_L(y)\frac{\partial m_{i,L}}{\partial z}, \qquad (3.133)$$

where $Sc_{iL,turb}$, which is typically of the order of 1, is the turbulent Schmidt number for species i that diffuses in the liquid of interest.

The turbulence in liquid films resembles the wall-bound turbulence elsewhere, except very close to the gas–liquid interphase. Thus, near the wall, the viscous, buffer, and fully turbulent layers occur, and the universal velocity profile applies. Turbulent eddies are damped by the gas–liquid interphase, however. The effect of this damping, while relatively unimportant with respect to hydrodynamics and momentum transfer, is significant for heat and mass transfer. For mass transfer, in particular, the phenomena near the liquid–gas interphase are crucial because of the typically very thin mass transfer boundary layers in liquids. Thus, the well-established diffusivity models (e.g., Reichardt, 1951a, b; Deissler, 1954; van Driest, 1956) are good for the bulk of the liquid film but need modification to account for the damping of eddies near the interphase. Eddy diffusivity models for falling liquid films have been proposed by many investigators (Chun and Seban, 1971; Mills and Chung, 1973; Sandal, 1974; Subramanian, 1975; Hubbard *et al.*, 1976; Seban and Faghri, 1976; Mudawar and El-Masri, 1986; Shmerler and Mudawar, 1988). The correlation of Mudawar and El-Masri is a modification of van Driest's eddy diffusivity model and reads

$$\frac{E}{\nu_L} = -\frac{1}{2} + \frac{1}{2}\left[1 + 4\kappa^2 y^{+2}\left(1 - \frac{y^+}{\delta_F^+}\right)F\right]^{1/2},\tag{3.134}$$

where $\kappa = 0.4$ is von Kármán's constant, $y^+ = y\sqrt{\tau_w/\rho_L/\nu_L}$, $\delta_F^+ = \delta_F\sqrt{\tau_w/\rho_L/\nu_L}$, and

$$F = \left\{1 - \exp\left[-\frac{y^+}{26}\left(1 - \frac{y^+}{\delta_F^+}\right)^{1/2}\left(1 - \frac{0.865\,\mathrm{Re}_{F,\mathrm{crit}}^{1/2}}{\delta_F^+}\right)\right]\right\}^2.\tag{3.135}$$

The critical film Reynolds number is to be calculated from $\mathrm{Re}_{F,\mathrm{crit}} = 0.04/\mathrm{Ka}^{0.37}$.

PROBLEMS

3.1 On a graph of $\langle j_G\rangle_G$ versus $\langle j_L\rangle_L$, assuming constant properties, show lines of constant $\langle j\rangle$, constant mass flux G, and constant quality $\langle x\rangle$. Can lines of constant void fraction $\langle\alpha\rangle$ be drawn?

3.2 A mixture of liquid and vapor R-134a is flowing in a tube with an inner diameter of 4 mm. The pressure is 13 bar. With local qualities of $x = 0.5\%$ and 5%, the measured void fractions are 40% and 95%, respectively.

(a) Calculate the slip ratios and mixture densities.

(b) For total mass fluxes of 150 and 800 kg/m²·s, calculate the phase velocities for liquid and vapor.

3.3 Consider the flow of saturated R-134a in a 4-mm-inner-diameter tube, at 7 bars pressure. The mass flux is $G = 230$ kg/m²·s. Assuming one-dimensional flow, calculate the void fraction for qualities of 0.01 and 0.08 by:

(a) assuming homogeneous equilibrium flow;

(b) using the following correlation for slip ratio (Xu and Fang, 2014):

$$\langle\alpha\rangle = \left[1 + \left(1 + 2Fr_{L0}^{-0.2}\beta^{3.5}\right)\left(\frac{1-x}{x}\right)\left(\frac{\rho_G}{\rho_L}\right)\right]^{-1},$$

where β is the volumetric quality, and the all-liquid Froude number is defined as

$$Fr_{L0} = \left(\frac{G}{\rho_L}\right)^2 \bigg/ (gD).$$

3.4 In Problem 3.2, suppose that for $x = 0.5\%$ and 5% qualities, the equilibrium qualities are estimated to be $x_{eq} = 0.35\%$ and 5.5%.

(a) Find the mixture enthalpies following the definition in Eq. (3.42) and the mixture internal energies when it is defined similarly to Eq. (3.42).

(b) Find the in-situ mixture enthalpies and internal energies, following the definition

$$\bar{h} = [\rho_L(1-\alpha)h_L + \rho_G\alpha h_G]/\bar{\rho},$$

where, consistent with Eq. (3.21), $\bar{\rho} = \rho_L(1 - \langle\alpha\rangle) + \rho_G\langle\alpha\rangle$.

(c) What are the likely states of liquid and vapor?

3.5 For turbulent flow in a pipe, assuming that the friction factor can be found from Blasius's correlation, show that the Kolmogorov microscale can be estimated from

$$\frac{l_D}{R} \approx \text{Re}^{-0.25} \, \text{Re}_\tau^{-0.5},$$

where $\text{Re}_\tau = U_\tau D / \nu$ and $U_\tau = \sqrt{\tau_w / \rho}$. For room-temperature water flowing in tubes with 0.8 and 2.5 mm diameters, calculate and plot the variation of l_D as a function of Re, for the $7000 < \text{Re} < 22\,000$ range.

3.6 For saturated liquid–vapor mixtures of the following fluids at 300 K temperature:

(a) Calculate the Laplace length scale. Also, find the diameter threshold for a capillary pipe at which buoyancy effect becomes insignificant.

(b) For capillary tube diameters equal to 0.5 mm and 1 mm, find the maximum mass flux at which confinement effects for bubble nucleation and growth are important.

R-123
R-124
R-134a
R-404a
FC-72
FC-87

3.7 Rederive Eqs. (3.97) and (3.98) for the case where the gas–liquid interfacial shear stress τ_I is known.

3.8 For room-temperature water flowing on a vertical, flat surface, calculate and compare $\Gamma_F, \bar{\delta}_F,$ and $\overline{U}_L$ at $\text{Re}_F = 2500$ and 5000, using the correlations of Brötz (1954) and Brauer (1956).

3.9 Shear-driven liquid films occur in the annular flow regime in internal flow situations. The behavior and thickness of such liquid films are influenced by the shear stress at the gas–liquid interphase, as expected.

Consider liquid falling film made of atmospheric and room-temperature water. For $\text{Re}_F = 2100$ and 3200 do the following.

(a) Find Γ_F, the liquid film mass flow rate per unit length, and determine the film flow regime.

(b) Find the mean liquid film thickness, using the above correlations.

(c) Compare the above liquid film thickness values with the predictions of some correlations that are commonly used for falling films.

Hint: For a falling liquid film in steady flow, the wall shear stress can be found by a simple momentum (or force) balance on the film.

3.10 Equations (3.127) through (3.130) represent liquid film flow on a flat surface. Modify these equations for a liquid film flowing inside a channel with circular cross section.

3.11 The inner surface of a vertical, 1-cm-diameter heated tube is being cooled by a subcooled falling film of water. The pressure is 300 kPa, the wall temperature is

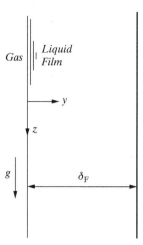

Figure P3.13. Schematic for Problem 3.13.

137 °C, and the mean film temperature is 85 °C. For $Re_F = 125$ and 1100 do the following.

(a) Calculate the liquid film thickness, assuming negligible gas–liquid interfacial shear.

(b) Calculate the heat transfer rate between the heated wall and the falling film.

(c) Find the liquid mass flow rates needed to cause a similar heat transfer rate, had the falling film been replaced with ordinary pipe flow with the same mean liquid temperature.

3.12 The eddy diffusivity model of van Driest (1956) is

$$\frac{E}{\nu} = -\frac{1}{2} + \left(1 + 4l_m^{+2}\right)^{1/2},$$

where $l_m^+ = l_m \sqrt{\tau_w/\rho}/\nu$ is the turbulent mixing length in wall units and is to be found from

$$l_m = \kappa y[1 - \exp(-y^+/A)], \kappa = 0.4, A = 26.$$

Specify and discuss the modifications that Mudawar and El-Masri (1986) have implemented on van Driest's model.

3.13 In gas absorption by laminar falling liquid films, because of the slow diffusion of the absorbed gas into the liquid film, the absorbed gas often penetrates only a small distance below the gas–liquid interphase. Consider a laminar and smooth falling film on a flat and vertical surface. An inert and sparingly soluble gas is absorbed by the liquid from the gas phase.

(a) Show that the mass species conservation equation for the transferred species in the liquid can be approximately represented by

$$\left(\frac{g\delta_F^2}{2\nu_L}\right)\frac{\partial m_1}{\partial z} = D_{12}\frac{\partial^2 m_1}{\partial y^2},$$

where y is now defined as the distance from the interphase (see Fig. P3.13).

(b) Solve the equation in part (a) for the following boundary conditions:

$$z = 0, m_1 = m_{1,in},$$
$$y = 0, m_1 = m_{1,u} = \text{const.},$$
$$y \to \infty, m_1 = m_{1,in}.$$

(c) Using the solution obtained in part (b), prove that the local liquid-side mass transfer coefficient between the liquid film surface and bulk is

$$K_{LI} = \rho_L \left(\frac{g}{2} \frac{\delta_F^2}{\nu_L} \frac{D_{12}}{\pi z} \right)^{1/2}.$$

4 Two-Phase Flow Regimes – I

4.1 Introductory Remarks

Gas–liquid two-phase mixtures can form a variety of morphological flow configurations. The two-phase flow regimes (flow patterns) represent the most frequently observed morphological configurations.

Flow regimes are extremely important. To get an appreciation for this, one can consider the flow regimes in single-phase flow, where laminar, transition, and turbulent are the main flow regimes. When the flow regime changes from laminar to turbulent, for example, it is as if the personality of the fluid completely changes as well, and the phenomena governing the transport processes in the fluid all change. The situation in two-phase flow is somewhat similar, only in this case there is a multitude of flow regimes. The flow regime is the most important attribute of any two-phase flow problem. The behavior of a gas–liquid mixture – including many of the constitutive relations that are needed for the solution of two-phase conservation equations – depends strongly on the flow regimes. Methods for predicting the ranges of occurrence of the major two-phase flow regimes are thus useful, and often required, for the modeling and analysis of two-phase flow systems.

Flow regimes are among the most intriguing and difficult aspects of two-phase flow and have been investigated over many decades. Current methods for predicting the flow regimes are far from perfect. The difficulty and challenge arise out of the extremely varied morphological configurations that a gas–liquid mixture can acquire, and these are affected by numerous parameters. Some of the physical factors that lead to morphological variations include the following:

(a) the density difference between the phases; as a result the two phases respond differently to forces such as gravity and centrifugal force;
(b) the deformability of the gas–liquid interphase that often results in incessant coalescence and breakup processes; and
(c) surface tension forces, which tends to maintain one phase dispersal.

Flow regimes and their ranges of occurrence are thus sensitive to fluid properties, system configuration/and orientation, size scale of the system, occurrence of phase change, etc. Nevertheless, for the most widely used configurations and/or relatively well-defined conditions (e.g., steady-state and adiabatic air–water and steam–water flow in uniform-cross-section long vertical pipes, or large vertical rod bundles with uniform inlet conditions) reasonably accurate predictive methods exist. The literature also contains data and correlations for a vast number of specific system

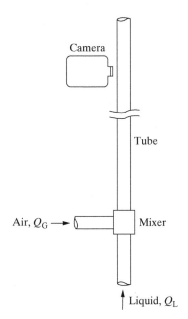

Figure 4.1. A simple flow-regime-observation experimental system for vertical pipes.

configurations, fluid types, etc. Although experiments are often needed when a new system configuration and/or fluid type is of interest, even in these cases the existing methods can be used for preliminary analysis and design calculations.

In this chapter the major flow regimes and the empirical predictive methods for adiabatic two-phase flow in straight channels and rod bundles will be discussed. The discussion of mechanistic models for regime transitions will be postponed to Chapter 7, so that the necessary background for understanding these mechanistic models is acquired in Chapters 5 and 6.

Also, in this chapter only conventional flow passages (i.e., flow passages with $D_H \geq 3$ mm) and rod bundles will be considered. There are important differences between commonly used channels and mini- or microchannels with respect to the gas–liquid two-phase flow hydrodynamics. Two-phase flow regimes and conditions leading to regime transitions in mini- and microchannels will be discussed in Chapter 10.

4.2 Two-Phase Flow Regimes in Adiabatic Pipe Flow

4.2.1 Vertical, Co-current, Upward Flow

Consider the simple experiment depicted in Fig. 4.1, where steady-state flow in a long tube with low or moderate liquid flow rate is considerede. The experiment proceeds with constant liquid volumetric flow rate Q_L, whereas the gas volumetric flow rate Q_G is started from a very low value and is gradually increased.

The major flow regimes that will be observed are depicted in Fig. 4.2. We will postpone for the moment discussion of the finely dispersed bubbly regime and focus on the others.

In *bubbly flow* (Fig. 4.2(a)) distorted-spherical and discrete bubbles move in a continuous liquid phase. The bubbles have little interaction at very low gas flows,

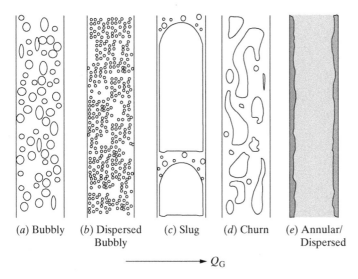

(a) Bubbly   (b) Dispersed   (c) Slug   (d) Churn   (e) Annular/
                  Bubbly   Dispersed

$$\longrightarrow Q_G$$

Figure 4.2. Major flow regimes in vertical upward pipe flow.

but they increase in number density as Q_G is increased. At higher Q_G rates, bubbles interact, leading to their coalescence and breakup.

Bubbly flow ends when discrete bubbles coalesce and produce very large bubbles. The *slug flow* regime (Fig. 4.2(c)) then develops; it is dominated by bullet-shaped bubbles (Taylor bubbles) that have approximately hemispherical caps and are separated from one another by *liquid slugs*. The liquid slug often contains small bubbles. A Taylor bubble approximately occupies the entire cross section and is separated from the wall by a thin liquid film. Taylor bubbles coalesce and grow in length until a relative equilibrium liquid slug length ($L_s/D \sim 16$) in common vertical channels (Taitel *et al.*, 1980) is reached.

At higher gas flow rates, the disruption of the large Taylor bubbles leads to *churn (froth) flow* (Fig. 4.2(d)), where chaotic motion of the irregular-shaped gas pockets takes place, with literally no discernible interfacial shape. Both phases may appear to be contiguous, and incessant churning and oscillatory backflow are observed. An oscillatory, time-varying regime where large waves moving forth in the flow direction are superimposed on an otherwise wavy annular-dispersed flow pattern involving a thick liquid film on the wall is also referred to as churn flow. Churn flow also occurs at the entrance of a vertical channel, before slug flow develops. This is a different interpretation of churn flow and represents the irregular region near the entrance of a long channel where eventually a slug flow pattern will develop.

Annular-dispersed (annular-mist) flow (Fig. 4.2(e)) replaces churn flow at higher gas flow rates. A thin liquid film, often wavy, sticks to the wall while a gas-occupied core, often with entrained droplets, is observed. In common pipe scales, the droplets are typically 10–100 μm in diameter (Jepsen *et al.*, 1989). The annular-dispersed flow regime is usually characterized by continuous impingement of droplets onto the liquid film and simultaneously an incessant process of entrainment of liquid droplets from the liquid film surface. Figure 4.3 depicts the cross section of a tube in the annular-dispersed regime (Srivastava, 1973).

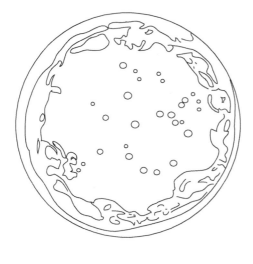

Figure 4.3. Cross-sectional view of annular-dispersed flow. (From Srivastava, 1973.)

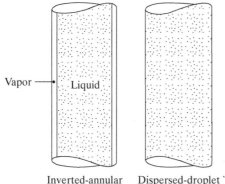

Vapor — Liquid

Inverted-annular   Dispersed-droplet

Figure 4.4. Inverted-annular and dispersed-droplet regimes.

The *inverted-annular regime* and *dispersed-droplet regime*, depicted schematically in Fig. 4.4, should also be mentioned here. These regimes are not observed in adiabatic gas–liquid flows. They do occur in boiling channels, however. In the inverted-annular flow regime a vapor film separates a predominantly liquid flow from the wall. The liquid flow may contain entrained bubbles. This flow regime takes place in channels subject to high wall heat fluxes and leads to an undesirable condition called the departure from nucleate boiling. In the dispersed-droplet regime an often superheated vapor containing entrained droplets flows in an otherwise dry channel. This regime can occur in boiling channels when massive evaporation has already caused the depletion of most of the liquid.

Flow regimes associated with very high liquid flow rates are now discussed. In these circumstances, in all flow regimes except annular (i.e., all flow regimes where the two phases are not separated), because of the very large liquid and mixture velocities the slip velocity between the two phases is often small in comparison with the average velocity of either phase, and the effect of gravity is relatively small. Furthermore, as long as the void fraction is small enough to allow the existence of a continuous liquid phase, the highly turbulent liquid flow does not allow the existence of large gas chunks and shatters the gas into small bubbles. Bubbly flow is thus replaced by a *finely dispersed bubbly flow* regime, where the bubbles are quite small and nearly spherical (Fig. 4.2(b)). No froth (churn) flow may take place;

furthermore, the transition from slug to annular-mist flow may only involve churn flow characterized by the oscillatory flow caused by the intermittent passing of large waves through a wavy annular-like base flow pattern.

It must be emphasized that the flow regimes shown in Fig. 4.2 are the *major and easily distinguishable* flow patterns. In an experiment similar to the one described here, transition from one major flow regime to another is never sudden, and each pair of major flow regimes are separated from one another by a relatively wide transition zone. Figure 4.5, borrowed from Govier and Aziz (1972), displays schematics of flow regimes and their range of phase superficial velocities for air–water flow in a 2.6-cm-diameter vertical tube.

4.2.2 Co-current Horizontal Flow

Let us now consider the simple experiment displayed in Fig. 4.6, where we establish a fixed liquid volumetric flow rate Q_L. We then start with a small gas volumetric flow rate Q_G, and increase Q_G while visually characterizing the flow regimes.

First, consider flow regimes at low liquid flow rates. For "low liquid flow rate" conditions assume Q_L is low enough that during drainage of liquid from the pipe when $Q_G = 0$, as shown in Fig. 4.7, the liquid occupies less than half of the pipe's cross-sectional height (i.e., $h_L < D/2$). The major flow regimes are shown in Fig. 4.8.

The *stratified-smooth flow* regime occurs at very low gas flow rates and is characterized by a smooth gas–liquid interphase. With increasing gas flow rate, the *stratified-wavy flow* regime is obtained, where hydrodynamic interactions at the gas–liquid interphase result in the formation of large-amplitude waves.

The *slug flow* regime occurs with further increasing gas flow rate. In comparison with the stratified-wavy regime, it appears as if the "waves" generated at the surface of the liquid grow large enough to bridge the entire channel cross section. The slug flow regime in horizontal channels is thus different from the slug flow defined for vertical channels. The gas phase is thus no longer contiguous. The liquid can contain entrained small droplets, and the gas phase may contain entrained liquid droplets.

The annular-dispersed (annular-mist) flow regime is established at higher gas flow rates. The flow regime resembles the annular-dispersed regime in vertical tubes, except that here gravity causes the liquid film to be thicker near the bottom.

The flow regimes at high liquid flow rates are now described. Referring to Fig. 4.7, we are now considering cases where, in the absence of a gas flow, liquid drainage out of the tube would result in $h_L > D/2$. The major flow regimes are depicted in Fig. 4.9.

The following flow regimes are observed as the gas flow rate is increased. In the *bubbly flow* regime, discrete bubbles tend to collect at the top of the pipe owing to the buoyancy effect. The *finely dispersed bubbly flow* regime is similar to the finely dispersed bubbly flow pattern in vertical flow channels. It occurs only at very high liquid flow rates. It is characterized by small spherical bubbles, approximately uniformly distributed in the channel. The *plug* or *elongated bubbles flow* regime is the equivalent of the slug flow regime in vertical channels. Finally, the *annular-dispersed (annular-mist) flow* regime is obtained at very high gas flow rates.

It is once again emphasized that the flow patterns in Fig. 4.9 only display the major flow regimes that are easily discernible visually and with simple photographic

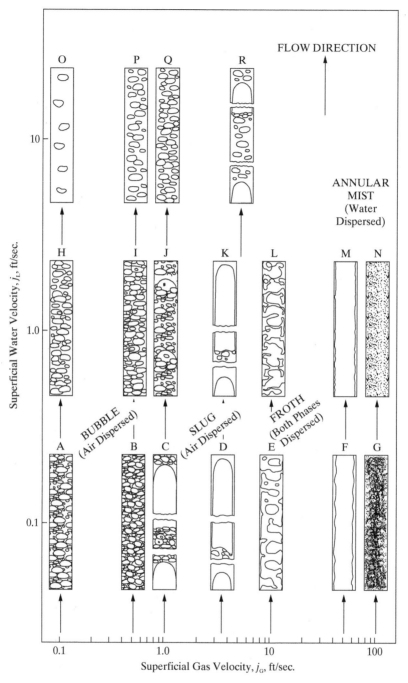

Figure 4.5. Flow regimes for air–water flow in a 2.6-cm-diameter vertical tube. (From Govier and Aziz, 1972.)

techniques and are commonly addressed in flow regime maps and transition models. Many subtle variations within some of the flow patterns can be recognized by using more sophisticated techniques (Spedding and Spence, 1993). Figure 4.10, borrowed from Govier and Aziz (1972), displays schematics of flow regimes and their range of phase superficial velocities for air–water flow in a 2.6-cm-diameter tube.

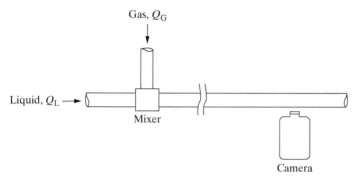

Figure 4.6. A simple flow-regime-observation experimental system for horizontal pipes.

Figure 4.7. Drainage of liquid out of a horizontal pipe.

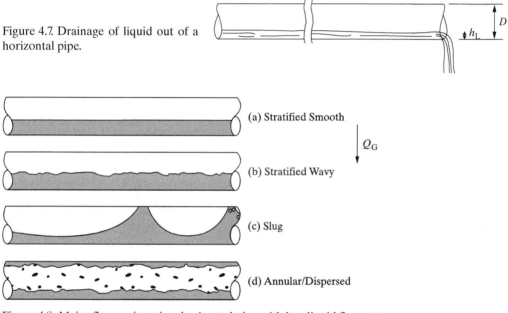

(a) Stratified Smooth

(b) Stratified Wavy

(c) Slug

(d) Annular/Dispersed

Figure 4.8. Major flow regimes in a horizontal pipe with low liquid flow rates.

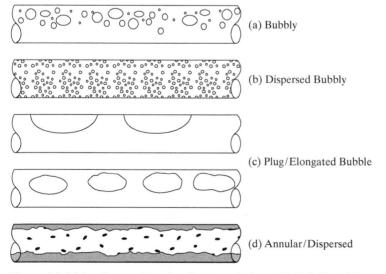

(a) Bubbly

(b) Dispersed Bubbly

(c) Plug/Elongated Bubble

(d) Annular/Dispersed

Figure 4.9. Major flow regimes in a horizontal pipe with high liquid flow rates.

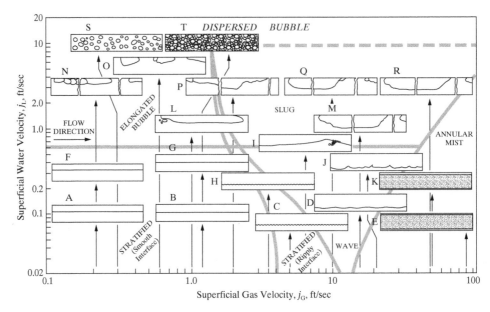

Figure 4.10. Flow regimes for air–water flow in a 2.6-cm-diameter horizontal tube. (From Govier and Aziz, 1972.)

With regard to the two-phase flow regimes, the following points should be borne in mind.

(1) Flow regimes and conditions leading to regime transitions are geometry dependent and are sensitive to liquid properties. The most important properties are surface tension, liquid viscosity, and liquid/gas density ratio. Important geometric attributes include orientation with respect to the gravitational vector, the size and shape of the flow channel, the aspect ratio (length to diameter) of the channel, and any feature that may cause flow disturbances.

(2) The basic flow regimes such as bubbly, stratified, churn, and annular-dispersed occur in virtually all system configurations, such as slots, tubes, and rod bundles. Details of the flow regimes of course vary according to channel geometry.

(3) The apparently well-defined flow regimes described here do not represent a complete picture of all possible flow configurations. In fact, by focusing on the flow regime intricate details, it is possible to define a multitude of subtle flow regimes (e.g., see Spedding and Spence, 1993). However, flow regime maps based on the basic regimes presented here have achieved wide acceptance over time. The regime change boundaries are generally difficult to define because of the occurrence of extensive "transitional" regimes.

(4) Bubbly, plug/slug, churn, and annular flow also occur in minichannels (i.e., channels with $100\,\mu\text{m} \le D_\text{H} \le 1$ mm).

(5) In adiabatic, horizontal flow, often for simplicity the regimes are divided into four zones:
   • stratified (smooth and wavy),
   • intermittent (plug, slug, and all subtle flow patterns between them),
   • annular-dispersed, and
   • bubbly.

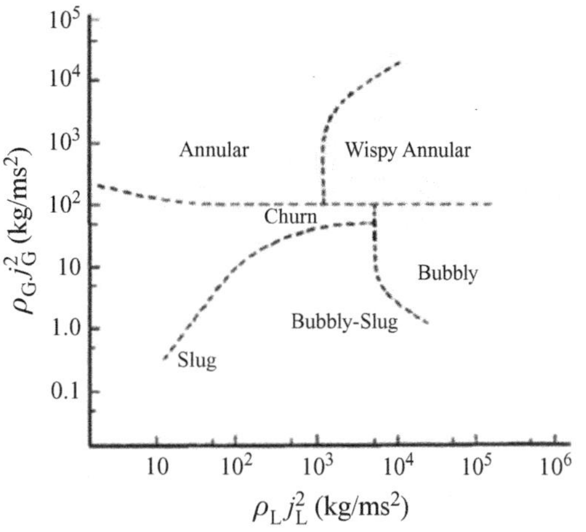

Figure 4.11. The flow regime map of Hewitt and Roberts (1969) for upward, cocurrent vertical flow.

(6) Flow regimes in boiling and condensing flows are significantly different than those in adiabatic channels. They will be discussed later.

4.3 Flow Regime Maps for Pipe Flow

Flow regime maps are the most widely used predictive tools for two-phase flow regimes. They are often empirical two-dimensional maps with coordinates representing easily quantifiable parameters. The coordinate parameters in the majority of widely used maps are either the phasic superficial velocities (Mandhane coordinates, after Mandhane et al., 1974) or include the phasic superficial velocities as well as some other properties. Most of the widely used regime maps are based on data for vertical or horizontal tubes with small and moderate diameters (typically $1 \le D \le 10$ cm) and for liquids with properties not too different from those of water. They also primarily represent "developed" conditions, with minimal channel end effects. Experimental data and regime maps for a wide variety of scales, geometric configurations, orientations, and properties can also be found in the open literature.

The flow regime map of Hewitt and Roberts (1969) is displayed in Fig. 4.11. This flow regime map is for co-current, vertical upward flow in pipes. The coordinates are defined as

$$\rho_G j_G^2 = \frac{(Gx)^2}{\rho_G}, \tag{4.1}$$

$$\rho_L j_L^2 = \frac{[G(1-x)]^2}{\rho_L}. \tag{4.2}$$

The flow regime map of Baker (1954), shown in Fig. 4.12, deals with co-current horizontal flow in pipes. The data base of this flow regime map is primarily air–water mixture. The property parameters are defined as

$$\lambda = \left[\frac{\rho_G}{\rho_a} \frac{\rho_L}{\rho_w} \right]^{1/2}, \tag{4.3}$$

$$\psi = \left(\frac{\sigma_w}{\sigma} \right) \left[\left(\frac{\mu_L}{\mu_w} \right) \left(\frac{\rho_w}{\rho_L} \right)^2 \right]^{1/3}. \tag{4.4}$$

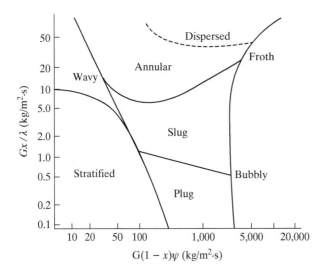

Figure 4.12. The flow regime map of Baker (1954) for co-current flow in horizontal pipes.

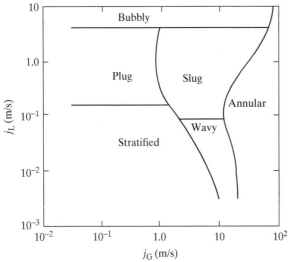

Figure 4.13. The flow regime map of Mandhane *et al.* (1974) for co-current flow in horizontal pipes.

and are meant to account for deviations from air and water properties. In these expressions the subscript a stands for air, W for water (both at normal conditions), G for the gas of interest, and L for the liquid of interest. In Eq. (4.4), σ_w represents the air–water surface tension and a is the surface tension of the gas–liquid pair of interest. Water and air properties are to be found at atmospheric pressure and temperature.

The flow regime map of Mandhane *et al.* (1974), displayed in Fig. 4.13, is probably the most widely accepted map for cocurrent flow in horizontal pipes. The range of its data base is as follows:

Pipe diameter	12.7–165.1 mm
Liquid density	705–1,009 kg/m^3
Gas density	0.80–50.5 kg/m^3
Liquid viscosity	3×10^{-4}–9×10^{-2} kg/m·s
Gas viscosity	10^{-5}–2.2×10^{-5} kg/m·s
Surface tension	0.024–0.103 N/m
Liquid superficial velocity	0.9×10^{-3}–7.31 m/s
Gas superficial velocity	0.04–171 m/s

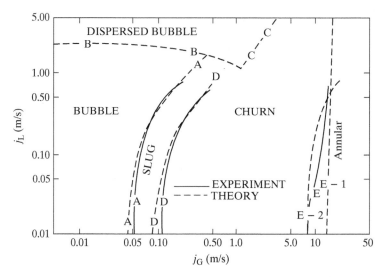

Figure 4.14. The rod bundle flow regime data of Venkateswararao *et al.* (1982).

4.4 Two-Phase Flow Regimes in Rod Bundles

The thermal-hydraulics of rod bundles is important because the cores of virtually all existing power-generating nuclear reactors consist of rod bundles. Two-phase flow occurs in the core of boiling water reactors (BWRs) during normal operations and in pressurized water reactors (PWRs) during many accident scenarios.

Adiabatic experimental studies (i.e., experiments without phase change) using a 20-rod bundle (16 complete and 4 half-rods) with near-prototypical bundle height, rod diameter, and pitch have indicated that the flow patterns include bubbly, slug, churn, annular, and possibly dispersed-bubbly (Venkateswararao *et al.*, 1982). In bubbly flow, the bubbles are typically small enough to move within a subchannel defined by four rods in bundles with rectangular pitch and three rods in bundles with triangular pitch. The slug flow regime can have at least three configurations (Venkateswararao *et al.*, 1982): Taylor bubbles moving within subchannels (cell-type slug flow); large-cap bubbles occupying more than a subchannel; and Taylor-like bubbles occupying the test section's entire flow area in a 20-rod bundle (shroud-type Taylor bubbles). The churn flow regime is characterized by irregular and alternating motion of liquid and can result from the instability of "cell-type" slug flow. Figure 4.14 displays the experimental flow regime map of Venkateswararao *et al.* (1982). These authors showed that their data could be predicted by the flow regime transition models of Taitel *et al.* (1980) (designated as theory in the figure), to be described in Chapter 7, with modifications to account for the rod bundle geometric configuration.

Paranjape *et al.* (2011) performed air–water experiments in an 8 × 8 square-lattice rod bundle test section that simulated a prototypical BWR rod bundle (12.7-mm rod diameter, 16.7-mm pitch), using an artificial neural network-based technique for flow regime identification. The flow area averaged void fractions were measured by an impedance void meter. The cumulative probability distribution functions (CPDFs) of the signals from the impedance meters were used for flow regime transition identification. The flow regime identification techniques allowed the identification of four distinct regimes.

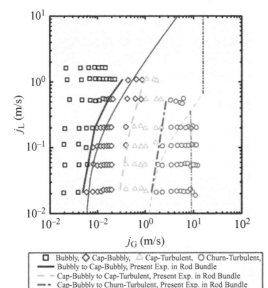

Fig. 4.15. The experimental flow regime map of Paranjape *et al.* (2011).

- Bubbly Flow. The flow regime consisted of dispersed spherical and distorted-spherical bubbles throughout the test section, ranging in size from 2 mm to 10 mm approximately. Bubbles were present in the subchannels as well as in the gap between the rods. Most of the bubbles moved in the flow direction in a subchannel, and often migrated among neighboring channels.
- Cap-Bubbly Flow. Cap-shaped bubbles which spanned one or two subchannels resulted as bubbles grew. The cap-bubble motions were found to be relatively steady.
- Cap-Turbulent Flow. This flow pattern occurred as cap bubbles grew sufficiently large to occupy more than two subchannels, with motions that appeared to be turbulent.
- Churn-Turbulent Flow. Large bubbles spanning five to six subchannels, with highly agitated motion, were observed. These bubbles often moved in a zigzag manner across the entire test section.

Figure 4.15 displays the flow regime map of Paranjape *et al.* (2011), where the flow regime transition lines are compared with the predictions of flow regime transition models that will be discussed in the forthcoming Chapter 7. The flow regimes developed over a short distance from the rod bundle inlet, indicating a relatively short entrance effect. The effect of spacer grids (which were included in the test section) on the flow regimes was also significant only for a short distance and disappeared within about 16 hydraulic diameters downstream of a spacer grid. The transition from bubbly to cap-bubbly regime in the rod bundle experiments agrees with the bubbly-slug flow regime transition model of Mishima and Ishii (1984) for circular pipes (see Eq. (7.13) in Chapter 7). This appears to suggest that the two flow regime transitions occur at a void fraction of about 30%. Furthermore, for low liquid superficial velocities ($\langle j_L \rangle \lesssim 0.1$ m/s) the cap-turbulent to churn-turbulent flow regimes in the rod

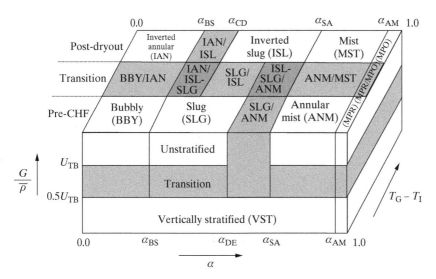

Figure 4.16. Schematic of RELAP5–3D vertical flow regime map (RELAP5–3D Code Development Team, 2012). Shaded areas indicate transitions.

bundle agreed with the predictions of the slug to churn-turbulent flow regime transition model of Mishima and Ishii (1984) for circular pipes (see Eqs. (7.14) and (7.15) in Chapter 7).

Flow regime maps and models that are used in reactor thermal-hydraulic computer codes usually assume that the basic flow regimes include bubbly, slug/churn, and annular, and they often include relatively large regime transition regions as well. For thermal-hydraulic codes, the following points should be noted. First, hydrodynamic parameters that are not easily measurable can be readily used in the development of regime models because these parameters are calculated and therefore "known" by the code. Second, what is really important for reactor codes is the correct prediction of regime-dependent parameters such as interfacial friction, heat transfer rates, etc.

The two-phase flow regime models of a well-known thermal-hydraulic code are now briefly discussed as examples. These models utilize the void fraction and volumetric fluxes, based on the argument that in transient and multi-dimensional situations they are the appropriate parameters that determine the two-phase flow morphology (Mishima and Ishii, 1984).

The *RELAP5–3D code* (RELAP5–3D Code Development Team, 2012) uses separate flow regime maps for vertical and horizontal flow configurations. The vertical flow regime map is used when $60° \leq |\theta| \leq 90°$, the horizontal flow regime map is applied when $0° \leq |\theta| \leq 30°$, and interpolation is applied when $30° < |\theta| < 60°$, where θ is the angle of inclination with respect to the horizontal plane. (Note that regime maps are primarily used for the calculation of parameters such as interfacial area concentration, interfacial heat transfer coefficients, etc. Interpolation is used for the calculation of these parameters.) The flow regime map for vertical flow is shown in Fig. 4.16. Distinction is made between precritical heat flux (pre-CHF) and post-CHF (post-dryout) regimes. Flow boiling and critical heat flux and postcritical heat flux (post-CHF) will be discussed in Chapters 13 and 14, respectively. The post-CHF regimes occur when, because of boiling, sustained or macroscopic physical contact between the surface and the liquid is interrupted. Post-CHF regimes are assumed when $T_G - T_I > 1$ K. The parameters in Fig. 4.16 are defined as follows:

$$U_{TB} = 0.35\sqrt{gD\Delta\rho/\rho_L} \text{ (Taylor bubble rise velocity in vertical tubes)} \quad (4.5)$$

$$\alpha_{BS} = \begin{cases} \alpha_{BS}^* & \text{for } G \leq 2000\,\text{kg/m}^2\cdot\text{s}, & (4.6) \\ \alpha_{BS}^* + (0.5 - \alpha_{BS}^*)\dfrac{G - 2000}{1000} & \text{for } 2000 < G < 3000\,\text{kg/m}^2\cdot\text{s}, & (4.7) \\ 0.5 & \text{for } G \geq 3000\,\text{kg/m}^2\cdot\text{s}, & (4.8) \end{cases}$$

$$\alpha_{BS}^* = 0.25\min[1, (0.045D^*)^8], \quad (4.9)$$

$$D^* = D/\sqrt{\sigma/g\Delta\rho}, \quad (4.10)$$

$$\alpha_{CD} = \alpha_{BS} + 0.2, \quad (4.11)$$

$$\alpha_{SA} = \max\left[\alpha_{AM}^{\min}, \min\left(\alpha_{crit}^f, \alpha_{crit}^e, \alpha_{BS}^{\max}\right)\right], \quad (4.12)$$

$$\alpha_{crit}^f = \begin{cases} \min\{[\sqrt{gD\Delta\rho/\rho_G}/U_G], 1\} & \text{for upward flow,} & (4.13) \\ 0.75 & \text{for downward or countercurrent flow,} & (4.14) \end{cases}$$

$$\alpha_{crit}^e = \min\left\{\frac{3.2}{U_G}[\sigma g\Delta\rho/\rho_G^2]^{1/4}, 1\right\}, \quad (4.15)$$

$$\alpha_{AM}^{\min} = \begin{cases} 0.5 & \text{for pipes,} & (4.16) \\ 0.8 & \text{for bundles,} & (4.17) \end{cases}$$

$$\alpha_{BS}^{\max} = 0.9, \quad (4.18)$$

$$\alpha_{DE} = \max(\alpha_{BS}, \alpha_{SA} - 0.05), \quad (4.19)$$

$$\alpha_{AM} = 0.9999. \quad (4.20)$$

For a vertically stratified flow regime to occur at a point in the computational domain (i.e., in a control volume), the void fraction above that point (i.e. in that control volume) should be greater than 0.7, and there must be at least a void fraction difference of 0.2 across the control volume.

The RELAP5–3D horizontal flow regime map is displayed in Fig. 4.17. The parameters in the flow regime map are defined as follows:

$$U_{crit} = \frac{1}{2}\left[\frac{\Delta\rho g\alpha A}{\rho_g D\sin\theta'}\right]^{1/2}(1 - \cos\theta'), \quad (4.21)$$

$$\alpha_{BS} = \begin{cases} 0.25 & \text{for } G \leq G_1^*, & (4.22) \\ 0.25 + 0.00025(G - G_1^*) & \text{for } G_1^* < G < G_2^*, & (4.23) \\ 0.5 & \text{for } G \geq G_2^*, & (4.24) \end{cases}$$

where $G_1^* = 2000\,\text{kg/m}^2\cdot\text{s}$, $G_2^* = 3000\,\text{kg/m}^2\cdot\text{s}$, $\alpha_{DE} = 0.75$, $\alpha_{SA} = 0.8$, and θ' is the angle defined in Fig. 4.18 when stratified flow is assumed.

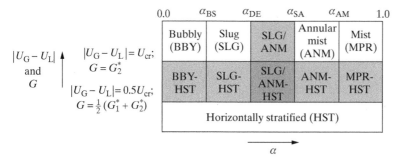

Figure 4.17. Schematic of RELAP5–3D horizontal flow regime map (RELAP5–3D Code Development Team, 2012). Shaded areas indicate transitions.

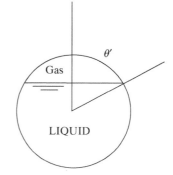

Figure 4.18. Definition of the angle θ'.

These flow regime transition models are sometimes modified and improved for various conditions (see, e.g., Hari and Hassan, 2002).

4.5 Two-Phase Flow in Curved Passages

Flow through curved flow passages is a common occurrence. Short flow passages include bends and various fittings, which are commonly treated as flow discontinuities (flow disturbances). Empirical methods for the prediction of pressure losses in bends will be reviewed in Section 8.6. Longer flow passages include spiral and helicoidally coiled tubes, which are important features of many process systems. We will focus on helicoid ducts, and to a lesser extent on spiral ducts, in this section and the forthcoming Sections 8.7. These types of curved flow passages are used in heat exchangers, boilers/steam generators, cryogenic medical devices, and thermosyphons, among other applications. The application of helicoidally coiled tubes, in particular, is rapidly increasing in power and refrigeration systems.

The geometric characteristics and their related terminology for helicoid and spiral flow passages are displayed in Fig. 4.19. For a circular cross-section tube R_i and R_o are the inner and outer radii of the tube, R_{cl} is the coil radius (pitch radius), and l_{pch} is the coil pitch. Helicoidally coiled flow passages are often tightly wound for compactness. The terminology about the orientation of helicoid coils is rather confusing. A helicoid coil is sometimes referred to as horizontal if the coil axis is vertical (Shah and Joshi, 1987), because for such a helicoid coil the flow passage itself is typically only slightly inclined with respect to the horizontal plane. Accordingly, a helicoid coil

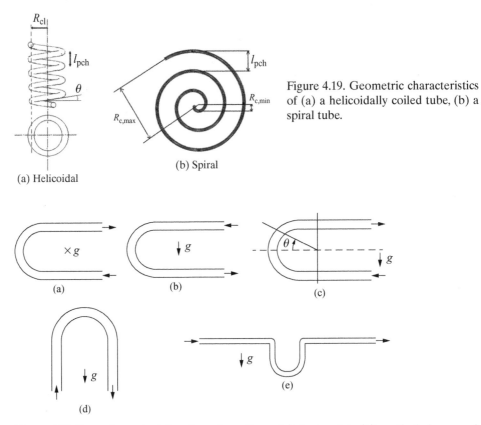

Figure 4.19. Geometric characteristics of (a) a helicoidally coiled tube, (b) a spiral tube.

Figure 4.20. Some important bend configurations; (a) horizontal, (b) vertical downward, (c) vertical upward, (d) inverted U-tube, (e) U-tube.

is called vertical if its axis is horizontal, because in this case the flow passage itself is typically only slightly inclined with respect to a vertical plane (Shah and Joshi, 1987). However, most recent literature uses the orientation of the helix's axis as the basis for defining the orientation. In this book, to avoid confusion, orientation will always be based on the axis of the flow passage's curvature. Thus in a *vertical helicoidal coil* the coil's axis is vertical, and in a *horizontal helicoidal coil* the coil's axis is horizontal.

Two-Phase Flow through Bends

Bends, and in particular 180° bends (U-bends), are among the most common fittings in piping systems. Two-phase flow in bends has been studied rather extensively, mostly with the purpose of understanding their effect on two-phase flow regimes and, more importantly, their pressure loss. Pressure drop in bends (and other flow passage features) will be discussed later in Chapter 8. In this section the two-phase flow regimes and their response to bends will be discussed.

With respect to the flow regime an important factor is the orientation of the U-bend. Figure 4.20 depicts some important orientations. Most U-bends that are used in industry are either *horizontal* or *vertical*, however, and as a result past investigations have focused on these two configurations. For a horizontal bend

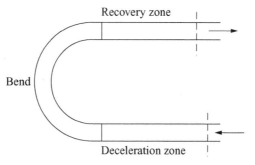

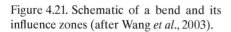

Figure 4.21. Schematic of a bend and its influence zones (after Wang *et al.*, 2003).

(Fig. 4.20(a)) the entire bend and the two straight flow passages connected to it are in a horizontal plane, and for a vertical bend (Figs. 4.20(b) and (c)) the bend itself is in a vertical plane while the two straight tubes connected to its two ends are horizontal. For a vertical bend, furthermore, we may deal with a co-current (Fig. 4.20(b)) or countercurrent (Fig. 4.20(c)) flow situation. Inverted U-tubes, as in Fig. 4.20(d), are encountered in U-tube steam generators in most pressurized water reactors. The bend configuration in Fig. 4.20(e), where a vertical U-shaped bend is connected to two horizontal straight flow passages, is often an undesirable flow loop feature because it can lead to the *loop seal effect*. The loop seal effect is encountered under low-flow two-phase flow conditions when liquid fills the bottom portion of the bend and effectively blocks the flow passage. Clearing of the blockage (loop seal clearing) can then be achieved only when the pressure difference between the two sides of the bend increases sufficiently to cause the ejection of the trapped liquid. The loop seal effect can cause flow intermittency, or even complete flow blockage. The formation and clearing of loop seals are parts of the phenomenology of loss of coolant accidents in some pressurized water reactors (Iguchi *et al.*, 1988; Kukita *et al.*, 1990; Lee and Kim, 1992). Analysis has shown that during the Three Mile Island incident a loop seal in the surge line prevented the flow of about four tons of liquid water from the pressurizer into the reactor core while the core was undergoing dryout and meltdown (Munis *et al.*, 1987).

For a bend connected to straight flow passages three zones can be defined, as shown in Fig. 4.21. Centrifugal forces resulting from the flow passage curvature, and buoyancy force affect the flow pattern in bends. Furthermore, the bend may affect the flow pattern in a short zone upstream of the bend (the *deceleration zone* (Wang *et al.*, 2003)). The effect of the bend disappears in the *recovery zone* downstream from the bend. The extent of the deceleration and recovery zones depend on a number of parameters including the flow rate, quality, channel size, bend radius of curvature, and the bend orientation. The acceleration zone is likely to be very short, even though some investigators did not observe it at all (Da Silva Lima and Thome, 2012). The recovery zone length is also relatively short, and is typically about 10 to 20 times the diameter of the flow channel (Padilla *et al.*, 2013).

The liquid and gas flow patterns are affected by centrifugal and gravitational forces, as noted earlier. The centrifugal force tends to separate the phases if it acts alone. The gravitational force, whose orientation varies along the flow passage, complicates the flow patterns, however. The phase separation that is caused by the centrifugal force depends on the phase velocities. At low gas velocities the liquid phase

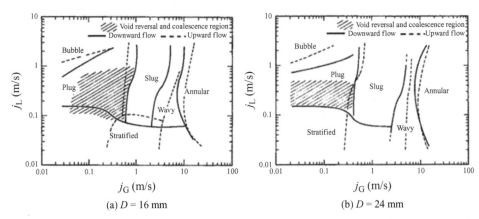

(a) $D = 16$ mm   (b) $D = 24$ mm

Figure 4.22. Flow regime map for air–water flow in a vertical C-shaped tube (Usui *et al.*, 1981).

is pushed towards the outer surface of the pipe due to its higher density. When the gas velocity is significantly higher than the liquid velocity, however, the trend is reversed and a phenomenon termed *flow reversal* occurs (Banerjee *et al.*, 1967).

Usui *et al.* (1980, 1981) carried out a detailed investigation of flow regimes, void fraction and pressure drop in a vertical bend connected to horizontal pipes at its end (similar to Fig. 4.20(c)), using room temperature air–water mixtures at near-atmospheric pressure. The range of parameters in their study was: $D = 16$ and 24 mm; $R_c = 90$–180 mm; $j_L = 0.09$–1.7 m/s; $j_G = 0.04$–30 m/s.

In upward flow (Usui *et al.*, 1980) the flow regimes resembled the flow regimes in straight horizontal tubes, and included bubbly, stratified, plug, slug, stratified-wavy, and annular. Figure 4.22 depicts their flow regime map for upward and downward flow in their vertical C-shaped test section. The combined effects of centrifugal and gravitational forces cause complications, however. In the slug flow regime liquid flow reversal occurred, and the resulting local countercurrent flow was sometimes accompanied by flooding. In stratified, plug, and slug flow regimes the aforementioned flow inversion occurred in some cases whereby in the downstream half of the bend the liquid would tend to collect on the inner surface of the bend. In the annular flow regime secondary flows similar to those observed in single-phase flow through curved pipes take place in the gas phase, and the resulting interfacial stress caused a tangential flow in the liquid film.

The centrifugal and gravitational forces in combination determine the distribution of the two phases in a bend. As long as the centrifugal force does not have a component that would oppose gravity the liquid phase moves to the outer surface of the bend due to its larger inertia. However, when the centrifugal force has a component that is in the opposite direction to gravity, the phase for which the net per-unit-mass force in the outward direction is larger tends to move to the outer surface of the tube where the radius of curvature is largest. If the net per-unit-mass outward force is larger for the gas phase the aforementioned flow inversion is observed. Usui *et al.* (1980) derived a method for the prediction of conditions that lead to flow inversion by assuming: (1) one-dimensional flow; (2) negligible secondary flows; and (3) $R_c/D \gg 1$. Since inversion occurs only when inertia has a component that opposes gravity, in upward flow in a vertical bend flow inversion is possible only in the second (downstream) half of the bend. The per-unit-mass forces acting on the liquid and gas phases are then equal when

$$\rho_L \frac{U_L^2}{R_c} + \rho_L g \cos \theta' = \rho_G \frac{U_G^2}{R_c} + \rho_G g \cos \theta', \tag{4.25}$$

where θ' is the local angle of inclination of the flow passage with respect to the vertical direction. The above balance between the forces can be recast as

$$\text{Fr}(\theta') = 1 \tag{4.26}$$

where the local Froude number is defined as

$$\text{Fr}(\theta') = \frac{U_L^2}{\frac{\rho_L - \rho_G}{\rho_L} R_c g |\cos \theta'|} \left[1 - \frac{\rho_G U_G^2}{\rho_L U_L^2} \right]. \tag{4.27}$$

Usui *et al.* (1980) noted that the liquid phase moves to the outer surface of bend when $\text{Fr}(\theta') > 1$, and the liquid phase moves to the inner surface of the bend (i.e., flow inversion occurs) when $\text{Fr}(\theta') < 1$.

To apply the above criterion of Usui *et al.* (1980) one needs to calculate the phase velocities U_L and U_G, and void fraction will be needed for that purpose. Usui *et al.* (1980) noted that the following correlation of Smith (1969) could predict their measured void fractions well,

$$\alpha = \left\{ 1 + \left[0.4 + 0.6 \sqrt{\frac{(\rho_L/\rho_G) + 0.4(1-x)/x}{1 + 0.4(1-x)/x}} \right] (\rho_G/\rho_L)(1-x)/x \right\}^{-1}. \tag{4.28}$$

Usui *et al.* (1981) studied downward two-phase flow in their test apparatus. The flow regime maps for downward flow are also shown in Fig. 4.22. As noted, except for stratified flow, the flow regime transition lines, are not significantly different for upward or downward flow. For both flow directions the flow regimes are similar to the flow regimes that occur in straight horizontal pipes.

The similarity of flow regimes in a bend to the flow regimes in a horizontal pipe, when the bend is connected to horizontal flow passages at its ends, has been confirmed in other investigations (Wang *et al.*, 2003; Padilla *et al.*, 2012; Da Silva Lima and Thome, 2012).

More recent investigations have been focused on flow of refrigerants in bends, with interest primarily in pressure drop (Padilla *et al.*, 2009, 2012, 2013; Tammaro *et al.*, 2013; Da Silva Lima and Thome, 2012). Wang *et al.* (2003) used air–water mixtures in horizontal bends (similar to Fig. 4.20(a)) with tube diameters in the 3 mm to 6.9 mm range, and reported the existence of a deceleration zone upstream of the bend. Da Silva Lima and Thome (2012) performed a detailed experimental study with R-134a refrigerant using horizontal and vertical bends connected to horizontal tubes, with tube diameters in the 8 mm to 11 mm range. With the vertical bend they examined both downward and upward flows. The centrifugal force dominated in their experiments and they did not observe a flow deceleration region. They emphasized that a flow regime map should include θ, the local angle of inclination of the flow path. Padilla *et al.* (2013) applied the aforementioned criterion of Usui *et al.* (Eq. (4.27)) for assessing some parametric effect on the vapor centrifugal force in their experiments with vertical-downward bends.

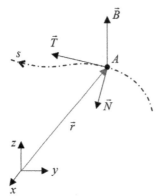

Figure 4.23. Definitions for fluid particle motion on a space curve.

Two-Phase Flow through Helicoidally Coiled Passages

Most helicoidally coiled flow passages in which phase change is expected have their axes vertically oriented. In boiling systems the flow is usually upwards, furthermore, so that buoyancy helps the flow and countercurrent flow limitation does not occur. Coils with horizontally oriented axes are also usually avoided because in every turn part of the flow passage is close to vertical, and for approximately half of each turn the flow will move against gravity and for the other half of the turn it moves in the direction of gravity. Flow in the direction opposite to gravity should be avoided wherever possible for condensing systems because the liquid would tend to move downwards by gravity and there will be the possibility of countercurrent flow limitation. Likewise, flow in the direction of gravity is not advisable for evaporating flows because of the same reason, the vapor would tend to move upwards by buoyancy and countercurrent flow limitation may occur.

Let us focus on helically coiled flow passages with vertically oriented axes, which represent the vast majority of curved flow passages in use in two-phase flow applications. The flow passage in this case has an angle of inclination with respect to the horizontal plane that can be represented as

$$\theta = \tan^{-1}[l_{pch}/(2\pi R_{cl})] = \sin^{-1}[l_{pch}/(2\pi R_c)]. \tag{4.29}$$

where R_c is the radius of curvature of the flow passage, to be discussed shortly. Helicoidally coiled flow passages are often tightly wound for compactness. As a result, θ is often small, typically not larger than about 10°.

The fluid motion in curved flow passages takes place on a space curve. Consider Fig. 4.23, where a space curve is depicted on which the fluid particle located at point A is moving. In this figure s represents the arc length, and $\vec{r}$ is the position vector defined based on an arbitrarily defined coordinate system. Then, recall from vector analysis that

$$\frac{d\vec{r}}{ds} = \vec{T} \tag{4.30}$$

$$\frac{d\vec{T}}{ds} = K\vec{N} \tag{4.31}$$

$$\frac{d\vec{N}}{ds} = T\vec{B} - K\vec{T}, \tag{4.32}$$

where $\vec{T}$, $\vec{N}$, and $\vec{B}$ are, respectively, the unit tangent vector, the principal normal unit vector, and the binormal unit vector. Parameters K and T are the *curvature* and *torsion*, respectively. These unit vectors are related by

$$\vec{B} = \vec{T} \times \vec{N}. \tag{4.33}$$

Furthermore,

$$\frac{d\vec{B}}{ds} = -T\vec{N}. \tag{4.34}$$

The plane that passes through point A and contains $\vec{T}$ and $\vec{N}$ is called the *osculating plane*. The plane that passes through point A and is perpendicular to $\vec{T}$ (and therefore contains $\vec{B}$ and $\vec{N}$) is called the *normal plane*. The *rectifying plane* is a plane that passes through point A and is perpendicular to $\vec{N}$. The radius of curvature, R_c, and the radius of torsion, R_t, are defined as

$$R_c = \frac{1}{K} \tag{4.35}$$

$$R_t = \frac{1}{T} \tag{4.36}$$

For a helicoid tube with circular cross section, the curvature and torsion of the centerline of the tube are, respectively,

$$K = \frac{R_{cl}}{R_{cl}^2 + \left(\frac{l_{pch}}{2\pi}\right)^2} \tag{4.37}$$

$$T = \frac{\left(\frac{l_{pch}}{2\pi}\right)}{R_{cl}^2 + \left(\frac{l_{pch}}{2\pi}\right)^2}. \tag{4.38}$$

The radius of curvature for a helicoid tube can thus be found from (Truesdell and Adler, 1970)

$$R_c = R_{cl}\left[1 + \left(\frac{l_{pch}}{2\pi R_{cl}}\right)^2\right]. \tag{4.39}$$

For a helicoid tube the radius of curvature is a constant. For a spiral tube evidently the radius of curvature is not a constant. A spiral tube with circular cross section is called a *simple* or *Archimedean spiral coil* if its pitch l_{pch} is a constant. For these flow passages the minimum and maximum radii of curvature, $R_{c,min}$ and $R_{c,max}$, represent the radii of curvature at the beginning and end of the flow passage.

Centrifugal forces that result from the flow path curvature influence the flow and heat transfer behavior of curved flow passages. In single-phase flow these forces lead to secondary flows that cause fluid mixing, move the location of maximum axial velocity towards the outer surface, and thus lead to complicated flow patterns. As a result, in comparison with flows in straight flow passages with similar cross section and boundary conditions, in curved flow passages the transition from laminar to turbulent flow regimes occurs at a higher Reynolds number when the cross section does not include sharp angles, the entrance lengths (both hydrodynamic and thermal) are

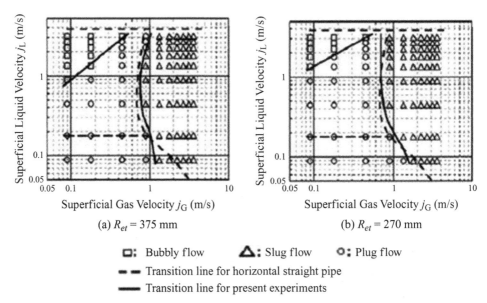

(a) $R_{et} = 375$ mm   (b) $R_{et} = 270$ mm

□: Bubbly flow   ▲: Slug flow   ○: Plug flow
– – Transition line for horizontal straight pipe
—— Transition line for present experiments

Figure 4.24. Flow regime maps for air–water flow in a 20-mm-inner-diameter vertically oriented helically coiled tube (Murai *et al.*, 2006).

shorter, and the circumferential-average friction factor and heat transfer coefficient both higher. Friction factor and heat transfer coefficient have non-uniform distributions around the circumference, and their nonuniformity becomes stronger as the Dean number (to be defined shortly) is increased.

In two-phase flow the centrifugal forces also complicate the flow and transport processes. Centrifugal force causes the phase with higher density to move outwards with respect to the curvature axis. In short flow passages such as bends this tendency of the heavier (liquid) phase to move outwards helps the phases to separate. In helically coiled passages, however, secondary flows occur at a level that further complicates the flow field.

For two-phase flow in helicoidally coiled tubes with vertically oriented axes, when the flow passage angle of inclination with respect to the horizontal plane is small, and as long as $l_{pch}/R_{cl} \ll 1$, the two-phase flow regimes that occur are similar to those observed in horizontal flow passages (Banerjee *et al.*, 1969; Chen and Zhang, 1984; Kaji *et al.*, 1984). In air–water experiments with θ up to 12°, Chen and Zhang (1984) observed plug, slug, annular-dispersed, stratified-wavy, and dispersed bubbly flow regimes. More recent experimental studies also confirm the aforementioned similarity between flow regimes in a vertically oriented helicoid coil and a horizontal tube as long as the angle of inclination is not larger than about 15° or so. The flow passage curvature causes some deviations from flow regime transition boundaries in comparison with horizontal flow passages, however.

Figure 4.24 displays the flow regimes observed by Murai *et al.* (2006) in their air–water experiments with a 20-mm-inner-diameter vertically oriented helically coiled tube. Their test section included a straight horizontal segment, thus making direct comparison between flow patterns in a straight pipe and a helically coiled pipe possible. The range of parameters in the tests allowed for stratified, bubbly, plug, and slug flow regimes (no annular flow). They noted that stratified flow did not occur in their helicoidally coiled tube, while it did take place in the horizontal tube. Furthermore, compared with the horizontal flow passage, the flow passage curvature enhanced bubble coalescence and therefore reduced the range over which bubbly flow was observed. In the air–water experiments of Kaji *et al.* (1984) the angle of inclination of the helicoidally coiled tube was 8°. The slug–annular flow regime transition was predicted well by the flow regime map of Mandhane *et al.* (1974), while the annular–wavy stratified flow regime transition boundary in their experiments was better predicted by the flow regime map of Baker (1954).

Much attention has been paid to the "film inversion" phenomenon in the annular flow regime, where in certain conditions, contrary to intuition, the liquid film is thicker towards the inner side of the helicoid tube. The annular flow regime is very important in boiling channels, because of its prevalence in the flow regime map, and more importantly, because of the occurrence of dryout-type critical heat flux (see Chapter 13). The centrifugal force on a unit mass of either phase can be represented as $\rho u^2 / R_c$, where u represents the local phasic velocity in the main flow direction (the direction tangent to the centerline of the flow passage). Clearly, for the same velocity, the heavier phase (the liquid phase in the case of a gas–liquid flow) experiences a larger force per unit volume, and therefore will tend to move outward. Under normal operating conditions of steam generators this leads to a thicker liquid film layer on the outer surface of the helical tubes (Hewitt and Jayanti, 1992). In some experiments, however, the opposite trend with respect to the thickness of the liquid film has been observed (Whalley, 1980; Banerjee *et al.*, 1967). Banerjee *et al.* (1967) have argued that film inversion occurs when the centrifugal force for gas exceeds the centrifugal force for liquid. This can happen because the two phases move with different velocities, and the velocity differential between the two phases is particularly significant in the annular flow regime.

Two-phase flow in helicoidally coiled tubes with horizontally oriented axes has not been extensively studied, probably due to the fewer applications of such coiled tubes. An experimental study with air and water was reported by Awwad *et al.* (1995a), who used 25.4-mm-inner-diameter tubes with 350 mm and 660 mm coil diameters. They identified six flow patterns, which are displayed in Fig. 4.25, and divided them into three groups: unsteady flow patterns (Figs. 4.25(a) and (b)); transition flow patterns (Figs. 4.25(c) and (d)), and steady flow patterns (Figs. 4.25(e) and (f)). In "steady" flow patterns, which occur when the mixture mass flux is high, the flow is stable and the two phases are each contiguous in the flow field. This is, of course, possible if the flow pattern everywhere is either annular or stratified. In "unsteady" flow patterns, on the other hand, in different parts of the tube one of the phases is in the form of bullet-shaped bubbles or slugs and is therefore discontinuous. Both phases are thus discontinuous in the tube and intermittent flow oscillations occur. The "transition" flow patterns show the characteristics of both of the aforementioned flow pattern types at different times and locations. The unsteady flow conditions are evidently undesirable, particularly in boiling and condensation applications.

The hydrodynamics of single- and two-phase flows in helically coiled flow channels will be further discussed in Section 8.7.

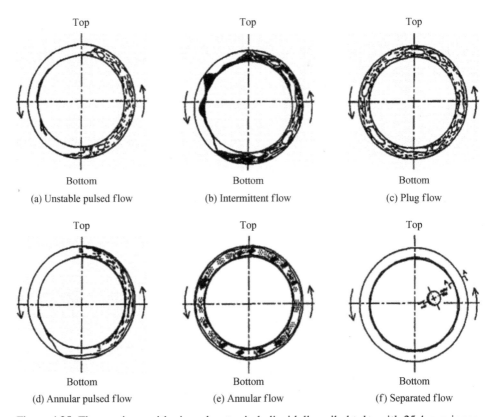

Figure 4.25. Flow regimes with air and water in helicoidally coiled tube with 25.4-mm-inner-diameter tubes with 350 mm and 660 mm coil diameters (Awwad *et al.*, 1995a).

4.6 Comments on Empirical Flow Regime Maps

Empirical flow regime maps have been in use for decades. They suffer from several shortcomings, however. Some of their major shortcomings are as follows.

(1) These flow regime maps generally address "developed" flow conditions and are not very accurate for short flow passages.

(2) Empirical flow regime maps often attempt to specify parameter ranges for various flow regimes using a common set of coordinates. Since the mechanisms that cause various regime transitions are different, a common set of coordinates may not be appropriate for the entire flow regime map.

(3) Most flow regime maps are based on data obtained with water, or liquids whose properties are not significantly different from those of water, in channels with diameters in the 1–10 cm range. The maps may not be useful for significantly different channel sizes or fluid properties.

(4) Closure relations are necessary for the solution of conservation equations (e.g., interfacial area concentration, interfacial forces, and transfer process rates, etc.) and these closure relations depend on flow regimes. A flow regime change thus implies switching from one set of correlations and models to another. This can introduce discontinuities and can cause numerical difficulties. This difficulty is mitigated to some extent by defining flow regime transition zones.

Two-phase flow regimes will be further discussed in Chapter 7, after the two-phase model conservation equations are discussed in the next chapter.

PROBLEMS

4.1 Saturated liquid R-134a is flowing in a vertical heated tube that has a diameter of 1 cm. The pressure is 16.8 bar, which remains approximately constant along the tube. A heat flux of 100 kW/m^2 is imposed on the tube.

(a) Assuming that friction and changes in kinetic and potential energy are negligible, prove that the first law of thermodynamics leads to

$$G\frac{dx_{eq}}{dz} = \frac{4q''_w}{Dh_{fg}},$$

where z is the axial coordinate.

(b) Assuming the flow regime maps based on adiabatic flow apply, using the flow regime map of Hewitt and Roberts (1969) determine the sequence of two-phase flow regimes and the axial coordinate where each regime is established for the following mass fluxes: $G = 200, 500$, and 1500 kg/m^2·s.

4.2 Repeat Problem 4.1, this time assuming that the tube is horizontal, and use the flow regime maps of Baker (1954) as well as Mandhane et al. (1974). Compare and discuss the predictions of the two flow regime maps.

4.3 A horizontal pipeline that is 15 cm in diameter is at 20 °C and carries a mixture of kerosene ($\rho_L = 804$ kg/m^3; $\mu_L = 1.92 \times 10^{-3}$ kg/m·s) and methane gas ($M = 16$ kg/kmol; $\mu_G = 1.34 \times 10^{-5}$ kg/m·s). Because of pressure drop considerations, it is important that the flow regime remains stratified or wavy, but not intermittent. The pressure along the pipeline varies in the 1–10 bar range. Using the flow regime map of Mandhane et al. (1974), determine the allowable range of methane mass flux for the following kerosene mass fluxes: $G_L = 10, 35$, and 75 kg/m^2·s. Discuss the validity of the flow regime map of Mandhane et al. for the described system.

4.4 The fuel rods in a PWR are 1.1 cm in diameter and 3.66 m long. The rods are arranged in a square lattice, as shown in Fig. P4.4, with a pitch-to-diameter ratio of 1.33. For a period of time during a particular core uncovery incident, the core remains at 40 bar pressure, while saturated liquid water enters the bottom of the core. The heat flux along one of the channels is assumed to be uniform and equal to 6.0×10^3 W/m^2. The flow is assumed to be one dimensional and the equilibrium quality at the exit of the channel is 0.12.

(a) Assuming that quality varies along the channel according to $AG\, dx_{eq}/dz = p_{heat}q''_w/h_{fg}$ (where A is the flow area and p_{heat} is the heated perimeter), calculate the coolant mass flux.

(b) Using the flow regime map of Hewitt and Roberts (1969), determine the sequence of flow regimes and the approximate axial location of regime transitions.

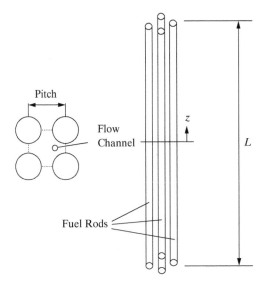

Figure P4.4. Figure for Problem 4.4.

4.5 In an experiment with a test section that includes a vertical pipe with $D = 5.25$ cm inner diameter, co-current upward two-phase flow regimes are to be studied. Liquid superficial velocities are set at $j_f = 0.2, 1.0,$ and 2.5 m/s, j_g is varied from 0.1 to 10 m/s, and the flow regimes and their transition conditions are recorded. Using the flow regime map of Hewitt and Roberts (1969), and the flow regime map of the RELAP5–3D code, find the flow regimes and the conditions when they are established for saturated steam–water mixtures at 1 and 5 bar pressures. Compare the predictions of the two methods, and comment on the results. For void fraction calculation, when needed, use the following correlation for the slip ratio:

$$S_r = \sqrt{1 - x\left(1 - \frac{\rho_f}{\rho_g}\right)}.$$

4.6 Consider the flow of liquid methane in a horizontal pipe that has a 10 cm inner diameter. The pressure is 5 bars. The total mass flux is 1.6×10^3 kg/m²·s. Because of heat transfer, evaporation takes place along the tube.

(a) Using the flow regime map of Mandhane *et al.* (1974) find the major flow regimes that will occur along the pipe, and specify the vapor quality where flow regime transitions take place.

(b) Repeat part (a), this time using the flow regime map of Baker (1954)

(c) Where possible, compare the results of parts (a) and (b) with the predictions of the flow regime map of RELAP5–3D.

For simplicity assume that the pressure remains constant, and assume that the liquid–vapor mixture is always saturated. Also, where needed, you can use the void-quality relation of Cioncolini and Thome (2012) for calculating the void fraction:

$$\langle \alpha \rangle = \frac{c\langle x \rangle^n}{1 + (c - 1)\langle x \rangle^n}$$

$$c = -2.129 + 3.129\left(\frac{\rho_G}{\rho_L}\right)^{-0.2186}$$

$$n = 0.3487 + 0.6513\left(\frac{\rho_G}{\rho_L}\right)^{0.5150}$$

$$0 < \langle x \rangle < 1; \ 10^{-3} < \rho_G/\rho_L < 1.0.$$

4.7 Consider the concurrent flow of a saturated liquid–vapor mixture of lique-fied petroleum gas (LPG) in an insulated 4-cm-inner-diameter pipe. For simplicity assume that LPG is pure propane. The local pressure is 4 bars, the mixture mass flux is 900 kg/m²·s, and the local quality is 3%.

(a) What is the most likely flow regime according to the flow regime map of Mand-hane *et al.* (1974)?

(b) What is the most likely flow regime according to the flow regime map of Baker (1954)? Comment on the applicability of the predicted flow regimes.

(c) Suppose the same pipe was vertical, so that a concurrent upward two-phase flow occurred. What would the flow regime be?

4.8 Refrigerant R-123 at a pressure of 4 bars flows in an adiabatic and circular cop-per tube that has 10.0 mm inner diameter. The tube is helically coiled and has a ver-tical axis with $R_{cl} = 35$ cm and $l_{pch} = 3$ cm. The flow is upwards. For a mass flux of 400 kg/m²·s and for qualities of 0.01, 0.25, and 0.5 determine the flow regime using an appropriate flow regime map that represents two-phase flow regimes in a horizontal tube. Also determine whether film inversion takes place. Discuss the validity of the flow regimes. Where needed, you can use the void-quality relation of Smith (1969) for calculating the void fraction (see Eq. (4.28)).

4.9 Refrigerant R-134a at a temperature of 233 K flows in a circular copper tube that has 4.0 mm inner diameter. The tube is adiabatic, has a vertical axis and is heli-cally coiled with $R_{cl} = 6$ cm and $l_{pch} = 1.5$ cm. The flow is upwards. For a mass flux of 300 kg/m²·s and for qualities of 0.01, 0.4, and 0.8 determine the flow regime by using an appropriate method applicable to a straight horizontal pipe. Also determine whether film inversion takes place. Discuss the validity of the flow regimes. Where needed, you can use the void-quality relation of Smith (1969) (see the previous prob-lem) for calculating the void fraction.

5 Two-Phase Flow Modeling

5.1 General Remarks

The design and analysis of systems often require the solution of mass, momentum, and energy-conservation equations. This is routinely done for single-phase flow systems, where the familiar Navier–Stokes equations are simplified as far as possible and then solved. The situation for two-phase flow systems is more complicated, however. The solution of the rigorous differential conservation equations is impractical, and a set of tractable conservation equations is needed instead. To derive tractable and at the same time reasonably accurate conservation equations, one needs deep physical insight (to make sensible simplifying assumptions) and mathematical skill. Fortunately, the subject has been investigated for decades, and at this time we have well-tested sets of tractable two-phase conservation equations that have been shown to do well in comparison with experimental data.

Generally speaking, conservation equations can be formulated and solved for multiphase flows in two different ways. In one approach, every phase is treated as a continuum, and all the conservation equations are presented in the Eulerian frame (i.e., a frame that is stationary with respect to the laboratory). This approach is quite general and can be applied to all flow configurations. In another approach, which is applicable when one of the phases is dispersed while the other phase is contiguous (e.g., in dispersed-droplet flow), the contiguous phase (the gas phase in the dispersed-droplet flow example) is treated as a continuum and its conservation equations are formulated and solved in the Eulerian frame. The dispersed phase, however, is treated by tracking the trajectories of a sample population of the dispersed-phase particles in the Lagrangian frame (i.e., a frame that moves with the particle). Iterative solutions of the two sets of equations are often needed to account for the interactions between the two phases. This powerful and computation-intensive method, often referred to as Eulerian–Lagrangian, is nowadays routinely used for the analysis of sprays and particle-laden flows. In fact, many commercial computational fluid dynamic (CFD) codes are capable of performing this type of Eulerian–Lagrangian analysis. With the exception of condensation on spray droplets, the Eulerian–Lagrangian method is not appropriate for boiling and condensation systems, and it is rarely used for the analysis of such systems, because most of the flow patterns in these systems do not involve dispersed particles. Even in some flow regimes where one of the phases is particulate (e.g., bubbly flow), the interparticle interactions are too complicated for a Lagrangian–Eulerian simulation. In light of these attributes, everywhere in this book we will limit our discussion of conservation equations to the Eulerian frame, and we will treat each phase as a continuum.

Table 5.1. *Summary of parameters for Eq. (5.1) in phase k.*

Conservation/transport law	ψ_k	φ_k	J_k^{**}
Mass	1	0	0
Momentum	$\vec{U}_k$	$\vec{g}$	$P_k \bar{\bar{I}} - \bar{\bar{\tau}}_k$
Energy	$\mathbf{u}_k + \frac{U_k^2}{2}$	$\vec{g} \cdot \vec{U}_k + \dot{q}_{v,k}/\rho_k$	$\vec{q}_k'' + (P_k \bar{\bar{I}} - \bar{\bar{\tau}}_k) \cdot \vec{U}_k$
Thermal energy in terms of enthalpy	h_k	$\frac{1}{\rho_k}\left(\dot{q}_{v,k} + \frac{DP_k}{Dt} + \bar{\bar{\tau}}_k : \nabla \vec{U}_k\right)$	$\vec{q}_k''$
Thermal energy in terms of internal energy	$\mathbf{u}_k$	$\frac{1}{\rho_k}(\dot{q}_{v,k} - P_k \nabla \cdot \vec{U}_k + \bar{\bar{\tau}}_k : \nabla \vec{U}_k)$	$\vec{q}_k''$
Species l, mass-flux-based	$m_{l,k}$	$\dot{r}_{l,k}/\rho_k$	$\vec{j}_{l,k}$
Species l, molar-flux-based*	$X_{l,k}$	$\dot{R}_{l,k}/\mathbf{C}_k$	$\vec{J}_{l,k}$

* In Eq. (5.1), ρ_k must be replaced with $\mathbf{C}_k$ everywhere.

In this chapter we will first discuss the differential balance laws and their relationship to two-phase flows. One-dimensional conservation equations are then presented. Rather than simplifying multi-dimensional conservation equations and presenting them in one-dimensional form, a set of one-dimensional conservation equations are derived in a heuristic manner by performing mass, momentum, and energy balances on a slice control volume in a channel. This is a useful exercise, since it clearly shows how the phase interactions and transport phenomena are accounted for in the conservation equations. The two-fluid conservation equations, in their general and multi-dimensional forms, are then presented and discussed.

5.2 Local Instantaneous Equations and Interphase Balance Relations

A gas–liquid two-phase flow field is always made of regions that contain one of the phases only and are separated from one another by an interphase. Given the general nature of the single-phase conservation equations there is no reason why they should not be applicable to the single-phase regions in a multiphase flow field. In fact, two-phase conservation equations, no matter how they are derived, should be consistent with the requirement that single-phase conservation equations must not be violated anywhere in the flow field.

Let us first revisit the single-phase conservation equations and their important attributes. The equations for phase k, in their local instantaneous form, can all be presented in the following shorthand expression:

$$\frac{\partial \rho_k \psi_k}{\partial t} + \nabla \cdot (\vec{U}_k \rho_k \psi_k) = -\nabla \cdot J_k^{**} + \rho_k \varphi_k, \qquad (5.1)$$

where k is a phase index (e.g., $k = 1$ for liquid and $k = 2$ for gas). This equation is of course identical to Eq. (1.14), except for the subscript k that has now appeared for all phase-specific parameters. The parameters summarized in Table 1.1 and the constitutive and closure relations thereof (Eqs. (1.15)–(1.26)) all apply, provided that they are in the phase-specific forms (see Table 5.1). Thus $\dot{q}_{v,k}$ is the volumetric energy generation rate in phase k; q_k'' is the heat flux in phase k; $\mathbf{u}_k$ is the specific internal energy of phase k; h_k is the specific enthalpy of phase k; $m_{l,k}$ is the mass fraction of species l in phase k; and $X_{l,k}$ is the mole fraction of species l in phase k.

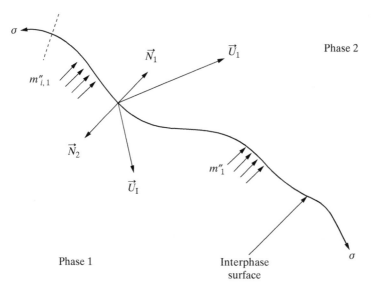

Figure 5.1. Schematic of the gas–liquid interphase.

The state variables (i.e., the unknowns that, if calculated, fully define the state of the fluid) in these equations typically are P_k, $\vec{u}_k$, $\mathbf{u}_k$, and $m_{l,k}$ or $X_{l,k}$. If we assume a Newtonian fluid and that molecular diffusion of species l in phase k is governed by Fick's law, the rate equations are

$$\bar{\bar{\tau}}_k = (\tau_{ij}e_ie_j)_k = \mu_k\left(\frac{\partial u_i}{\partial x_j} + \frac{\partial u_j}{\partial x_i}\right)_k - \frac{2}{3}\mu_k\delta_{ij}\nabla \cdot \vec{U}_k, \tag{5.2}$$

$$\vec{q}''_k = -k_k\nabla T_k + \sum_l \vec{j}_{l,k}h_{l,k}, \tag{5.3}$$

$$\vec{j}_{l,k} = -\rho_k\mathbf{D}_{lk}\nabla m_{l,k}, \tag{5.4}$$

$$\vec{J}_{l,k} = -C_k\mathbf{D}_{lk}\nabla X_{l,k}, \tag{5.5}$$

where $\mathbf{D}_{lk}$ represents the *mass diffusivity* of species l with respect to the liquid phase. Note that, as mentioned in Chapter 1, the mass diffusion is assumed to comply with Fick's law, which is true for a binary gaseous system and when the species k is only sparingly soluble in the liquid.

The constitutive relations provide for the thermophysical properties and include

$$\rho_k = \rho_k(P_k, \mathbf{u}_k, m_1, m_2, \ldots, m_{n-1}), \tag{5.6}$$

$$T_k = f(P_k, \mathbf{u}_k, m_1, m_2, \ldots, m_{n-1}), \tag{5.7}$$

where n is the total number of species in phase k. In these two equations, the mass fractions $m_1, m_2, \ldots, m_{n-1}$ can alternatively be replaced with mole fractions $X_1, X_2, \ldots, X_{n-1}$. For pure substances $n = 1$, and there is of course no need to include mass fractions.

An additional closure relation for a two-phase flow field is the topological constraint, which states that at any point in the flow field, and at any instant, only one of the phases can be present.

Equation (5.1) evidently does not apply to the gas–liquid interphase itself. A schematic of the interphase is shown in Fig. 5.1, where $\vec{N}_1$ represents the unit normal vector and $\dot{m}_1''$ is the mass flux of phase 1 moving toward the interphase. The interphase can be treated as an infinitely thin membrane that, by virtue of its essentially zero volume, is always at steady state with respect to all transfer processes. It is also at thermal equilibrium, equilibrium with respect to the concentration of species i in the two phases, and at mechanical equilibrium. The thermal equilibrium assumption is valid except in extremely fast transients. Mechanical equilibrium requires that the forces that act on an element of the interphase balance each other out. These arguments lead to the following, simplified *interphase jump conditions*:

$$m_1'' = \rho_1(\vec{U}_1 - \vec{U}_I) \cdot \vec{N}_1, \tag{5.8}$$

$$T_1 = T_2 = T_1, \tag{5.9}$$

$$m_1'' + m_2'' = 0, \tag{5.10}$$

$$m_1''(\vec{U}_1 - \vec{U}_2) + (P_1\bar{\bar{I}} - \bar{\bar{\tau}}_1 - P_2\bar{\bar{I}} + \bar{\bar{\tau}}_2) \cdot \vec{N}_1 - 2\sigma K_{12}\vec{N}_1 = 0, \tag{5.11}$$

$$K_{12} = \frac{1}{2}\left(\frac{1}{R_{C1}} + \frac{1}{R_{C2}}\right), \tag{5.12}$$

$$m_1''\left(h_1 + \frac{1}{2}(U_1^2 - U_1^2)\right) + m_2''\left(h_2 + \frac{1}{2}(U_2^2 - U_1^2)\right) + \vec{q}_1'' \cdot \vec{N}_1 + \vec{q}_2'' \cdot \vec{N}_2$$
$$- (\vec{N}_1 \cdot \bar{\bar{\tau}}) \cdot (\vec{U}_1 - \vec{U}_I) - (N_2 \cdot \bar{\bar{\tau}}_2) \cdot (\vec{U}_2 - \vec{U}_I) = 0, \tag{5.13}$$

$$m_1'' m_{i,1} + \vec{j}_{i,1} \cdot \vec{N}_1 + m_2'' m_{i,2} + \vec{j}_{i,2} \cdot \vec{N}_2 = 0, \tag{5.14}$$

and

$$N_1'' + X_{i,1} \cdot \vec{J}_{i,1} \cdot \vec{N}_1 + N_2'' X_{i,2} + \vec{j}_{i,2} \cdot \vec{N}_2 = 0. \tag{5.15}$$

Equation (5.8) is a kinematic consistency requirement, Eq. (5.9) represents thermal equilibrium, Eq. (5.10) satisfies mass continuity, Eq. (5.11) represents the balance of linear momentum, Eq. (5.12) defines the mean curvature, and Eq. (5.13) represents the conservation of energy. In Eq. (5.11) the surface tension σ has been assumed to be constant, R_{C1} and R_{C2} are the interphase principal radii of curvature, and K_{12} is therefore the average interphase curvature. Equations (5.14) and (5.15) are equivalent and represent the interphase balance conditions for an inert species i, in terms of mass and molar fluxes, respectively, assuming that no accumulation of that species at the interphase takes place. Only one of them is therefore used.

This set of equations along with their appropriate closure relations can, in principle, be solved by direct numerical simulation, or by using any of several discretization methods (finite-difference, finite-volume, finite-element, etc.) provided that the flow field boundary conditions are known and, more importantly, that the exact location of the gas–liquid interfacial surface is also known at any time. Direct numerical simulation would require the use of time and spatial steps small enough to capture the smallest important fluctuations, over a domain large enough to capture the largest important flow features. However, a major difficulty is that the whereabouts of the interphase is not known *a priori*, and in fact it has to be found as part of the solution. This makes the numerical solution of these equations difficult. Solution of multiphase conservation equations with deformable interphase surfaces is an active research area, and a handful of techniques for resolving and modeling the interphase boundary motion are now available (Nichols *et al.*, 1980; Sussman and Fatemi, 2003; Tryggvason *et al.*, 2001; Osher and Fedkiw, 2003). These techniques are computationally intensive and evolving. They are currently used primarily for research purposes and are not yet convenient for typical design and analysis applications.

Rigorous modeling of gas–liquid two-phase flow based on the solution of local and instantaneous conservation principles is thus generally not feasible. Simplified models that are based on idealization and time, volume, and ensemble averaging are usually used instead. Simplified multiphase flow conservation equations can be obtained in several ways, including the following:

(a) assuming that each point in the mixture is simultaneously occupied by both phases and deriving a mixture model;
(b) developing control-volume-based balance equations;
(c) performing some form of averaging (time, volume, flow area, ensemble, or composite) on local and instantaneous conservation equations; or
(d) postulating a set of conservation equations based on physical and mathematical insight.

Among these, the most widely used is the averaging method, which can lead to flow parameters that are measurable with available instrumentation, are continuous, and in the case of double averaging have continuous first derivatives. Good discussions about various types of averaging can be found in Ishii (1975), Bouré and Delhaye (1981), Banerjee and Chan (1980), and Lahey and Drew (1988), among others. Averaging is in fact equivalent to low-pass filtering to eliminate high-frequency fluctuations. By averaging, we lose information about details of fluctuations, and in return we get simplified and tractable model equations. This is not a hopeless loss of information, however. Although fluctuation details are lost, their statistical properties and macroscopic effects on balance equations can be accounted for by using appropriate closure relations. The situation is somewhat similar to turbulent flow, for which, by using time-averaged equations, we lose information about velocity fluctuations but include the effect of these fluctuations in the macroscopic conservation equations by introducing Reynolds stresses and fluxes or using momentum and thermal eddy diffusivities.

5.3 Two-Phase Flow Models

The objective of modeling a physical process is to devise a mathematical model that is tractable and represents the behavior of the flow field of interest with a satisfactory approximation. The mathematical model in our case will include the conservation equations, the transport (rate) equations, expressions for the rates of interphase transfer processes, thermophysical and transport properties (constitutive relations), and topological constraint. The crucial step is evidently the development of tractable conservation equations.

As mentioned earlier, tractable conservation equations can be derived based on averaging, by first dividing the flow field into a number of domains, while accounting for the flow structure and making assumptions about the nature of phase interactions. For example, one can define a single flow domain and assume that at any location the gas and liquid phases are at equilibrium in all respects. Alternatively, one can divide each phase into several domains to account for the various possible nonuniformities. Intuition suggests that model accuracy can be increased by increasing the number of domains. That is not always true, however, because of the unavailability of good models for phase interactions and closure relations.

Two-phase flow models can be divided into three main categories.

(1) *Homogeneous mixture model*: In this model, the two phases are assumed to be well mixed and have the same velocity at any location. Thus, only one momentum equation is needed. Furthermore, if in a single-component flow thermodynamic equilibrium is also assumed between the two phases everywhere, the homogeneous-equilibrium mixture model results. The two phases do not need to be at thermodynamic equilibrium, however. Examples include flashing of liquids and condensation of vapor bubbles surrounded by subcooled liquid. The HEM model is the simplest two-phase flow model, and it essentially treats the two-phase mixture as a single fluid. The solution of conservation equations is more complicated than single-phase flow, however. After all, the fluid mixture is compressible, with thermophysical properties that can vary significantly with time and position.

(2) *Multifluid models*: In this case, the flow field is divided into at least two (liquid and gas) domains, and each domain is represented by one momentum equation. A good example is the two-fluid model (2FM), which is currently the most widely used two-phase flow model. In the 2FM, gas and liquid phases are each represented by one complete set of differential conservation equations (for mass, momentum, and energy). The assumptions of thermodynamic equilibrium between the two phases or saturation state for one of the phases are sometimes made. Either of these assumptions will lead to the redundancy and elimination of one of the energy equations.

(3) *Diffusion models*: In these models the liquid and gas phases constitute the two domains. Only a single momentum equation is used, however. This is made possible by obtaining the relative (slip) velocity between the two phases, or the relative velocity of one phase with respect to the mixture, from a model or correlation. The slip velocity relation is usually algebraic (rather than a differential equation). The drift flux model (DFM) is the most widely used diffusion model. The DFM (the Zuber–Findlay model) is more often used for void fraction calculations, however.

5.4 Flow-Area Averaging

Conservation equations will be heuristically derived in the following sections for a one-dimensional flow in a channel whose flow area changes along the channel axis only slowly. The derivations will be based on a simple control volume analysis

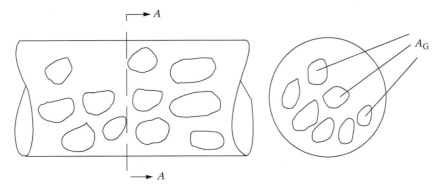

Figure 5.2. Schematic of a channel.

with a slice of the channel as the control volume, following a methodology similar to Yadigaroglu and Lahey (1976) and Lahey and Moody (1993). For simplicity of derivations, the two phases are displayed as if they are completely separated (as, for example, in stratified or annular flow). The resulting differential equations are much more general, however, as long as the one-dimensional flow assumption makes sense and we are satisfied with having only two domains (liquid and gas).

Let us first review a few additional definitions and rules dealing with flow-area cross-sectional averaging. Figure 5.2 shows a schematic of a gas–liquid two-phase flow in a channel. At any instant, a cross section is partially covered by gas. Assume that A_G is the time- or ensemble-average area covered by gas. The average void fraction is defined as

$$\langle \alpha \rangle = A_G/A. \tag{5.16}$$

Assume also that all other parameters are time- or ensemble-averaged. The average properties to be defined will thus be double (composite) averaged. An in-situ flow-area-average value for any property ξ will be

$$\langle \xi \rangle = \frac{1}{A} \int_A \xi \, dA. \tag{5.17}$$

Some properties are phase specific (e.g., the density of the gas phase) and should only be averaged over their corresponding phases. Therefore (see Eqs. (3.15) and (3.16))

$$\langle \xi_G \rangle_G = \frac{1}{A_G} \int_{A_G} \xi_G \, dA_G = \frac{1}{A\langle \alpha \rangle} \int_A \xi_G \alpha \, dA = \frac{\langle \xi_G \alpha \rangle}{\langle \alpha \rangle}, \tag{5.18}$$

$$\langle \xi_L \rangle_L = \frac{1}{A_L} \int_{A_L} \xi_L \, dA_L = \frac{1}{A(1-\langle \alpha \rangle)} \int_{A_L} \xi_L (1-\alpha) \, dA = \frac{\langle \xi_L(1-\alpha) \rangle}{\langle 1-\alpha \rangle}. \tag{5.19}$$

In these and other expressions elsewhere in this section, the right side of the first equal sign is the definition of the averaged parameter, and the right sides of the other remainder equal signs represent identities that can be easily proved. Phasic superficial velocities are defined as

$$\langle j_G \rangle = \frac{Q_G}{A} = \langle \alpha U_G \rangle = \langle \alpha \rangle \langle U_G \rangle_G, \tag{5.20}$$

$$\langle j_L \rangle = \frac{Q_L}{A} = \langle (1 - \alpha) U_L \rangle = (1 - \langle \alpha \rangle) \langle U_L \rangle_L, \tag{5.21}$$

where Q_G and Q_L are the total volumetric flow rates of gas and liquid, respectively. The total volumetric flux (mixture center-of-volume velocity) is

$$\langle j \rangle = (Q_G + Q_L)/A = \langle j_G \rangle + \langle j_L \rangle. \tag{5.22}$$

The mixture mass flux is

$$\begin{aligned} \langle G \rangle &= (\rho_G Q_G + \rho_L Q_L)/A = \rho_G \langle j_G \rangle + \rho_L \langle j_L \rangle \\ &= \rho_G \langle \alpha \rangle \langle U_G \rangle_G + \rho_L \langle 1 - \alpha \rangle \langle U_L \rangle_L. \end{aligned} \tag{5.23}$$

Phase mass fluxes follow as

$$\langle G_G \rangle = \rho_G Q_G/A = \rho_G \langle \alpha \rangle \langle U_G \rangle_G = \rho_G \langle j_G \rangle, \tag{5.24}$$

$$\langle G_L \rangle = \rho_L Q_L/A = \rho_L \langle 1 - \alpha \rangle \langle U_L \rangle_L = \rho_L \langle j_L \rangle. \tag{5.25}$$

Now that we have defined the flow-area-averaged parameters, for convenience let us drop all averaging notation, and from now on assume that all parameters are composite flow-area time or flow-area ensemble averaged. Thus, for example, wherever U_G is used, it implies $\langle U_G \rangle_G$, α everywhere implies $\langle \alpha \rangle$, j and j_G imply $\langle j \rangle$ and $\langle j_G \rangle$, respectively, and G implies $\langle G \rangle$. We can now proceed with the derivations.

5.5 One-Dimensional Homogeneous-Equilibrium Model: Single-Component Fluid

By single-component fluid, we mean a pure liquid mixed with its own pure vapor. The HEM is the simplest of all two-phase flow models. The two phases are everywhere assumed to be well mixed, have the same velocity, and to be at thermal equilibrium. For a single-component two-phase flow (e.g., water and steam), liquid–vapor thermodynamic equilibrium obviously implies a saturated mixture. The following relations then apply:

$$\bar{\rho} = \rho_h = G/j = \alpha \rho_g + (1 - \alpha) \rho_f = [v_f + x(v_g - v_f)]^{-1}, \tag{5.26}$$

$$\frac{x}{1-x} = \frac{\rho_g \alpha}{\rho_g (1 - \alpha)}, \tag{5.27}$$

$$h = \bar{h} = [\rho_g \alpha h_g + \rho_f (1 - \alpha) h_f]/\bar{\rho} = h_f + x(h_g - h_f), \tag{5.28}$$

$$x = x_{eq} = (h - h_f)/h_{fg}, \tag{5.29}$$

where x_{eq} is the equilibrium quality.

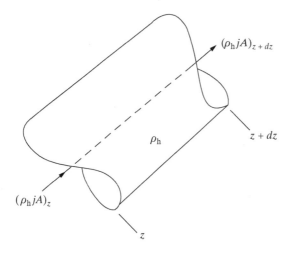

Figure 5.3. Control volume for the derivation of the mass conservation equation.

Mass Conservation

The differential mass conservation equation can be derived by performing a mass balance on a slice of the channel, as shown in Fig. 5.3:

$$\frac{\partial}{\partial t}(A\delta z \rho_{\mathrm{h}}) = (\rho_{\mathrm{h}} A j)_z - \left[(\rho_{\mathrm{h}} A j)_z + \frac{\partial}{\partial z}(\rho_{\mathrm{h}} A j)\delta z + \cdots \right], \tag{5.30}$$

where the second (bracketed) term on the right side represents the mass flow rate out of the control volume. In the limit of $\delta z \to 0$, and with a fixed geometry assumed,

$$\frac{\partial \rho_{\mathrm{h}}}{\partial t} + \frac{1}{A}\frac{\partial}{\partial z}(A\rho_{\mathrm{h}} j) = 0. \tag{5.31}$$

This can be recast as

$$\frac{Dj}{Dt}(A\rho_{\mathrm{h}}) + \rho_{\mathrm{h}} A \frac{\partial j}{\partial z} = 0, \tag{5.32}$$

where the material derivative is defined as

$$\frac{Dj}{Dt} = \frac{\partial}{\partial t} + j\frac{\partial}{\partial z}. \tag{5.33}$$

For a channel with uniform flow area, Eq. (5.32) gives

$$\frac{Dj}{Dt}\rho_{\mathrm{h}} + \frac{\partial j}{\partial z} = 0. \tag{5.34}$$

Momentum Conservation

Consider the forces that act on the fluid mixture in a slice of the flow channel, as shown in Fig. 5.4. The force term F_{w} represents the frictional force and can be cast as $F_{\mathrm{w}} = p_{\mathrm{f}}\tau_{\mathrm{w}}\delta z$, where p_{f} is the channel wetted perimeter and τ_{w} is the wall shear stress. The net force exerted by the channel wall on the fluid in the positive z direction is

$$F_{\mathrm{A}} = P\frac{\partial A}{\partial z}\delta z = (PA)_{z+\delta z} - (PA)_z - A\frac{\partial P}{\partial z}\delta z. \tag{5.35}$$

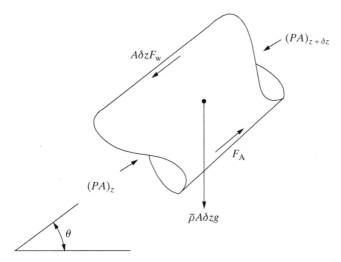

Figure 5.4. Forces acting on a control volume in homogeneous flow.

Applying Newton's second law of motion to the control volume shown in Fig. 5.4, and taking the limit of $\delta z \to 0$, we get

$$\frac{\partial G}{\partial t} + \frac{1}{A}\frac{\partial}{\partial z}(\rho_h j^2 A) = -\frac{\partial P}{\partial z} - g\rho_h \sin\theta - F_w, \qquad (5.36)$$

where F_w is the dissipative wall friction force, per unit mixture volume. In terms of wall shear stress, $F_w = p_f \tau_w / A$, where τ_w is the shear stress imposed on the flow by the wall. In addition to friction, so-called minor losses also occur owing to flow-area variations, flow-path curvature, or any type of flow disturbance. Minor losses will be discussed in Chapter 9. For now, let use assume that friction is the only significant dissipative wall–fluid interaction force.

Since $j = G/\rho_h$, Eq. (5.36) can be re-written as

$$\frac{\partial}{\partial t}(\rho_h j) + \frac{1}{A}\frac{\partial}{\partial z}(\rho_h j^2 A) = -\frac{\partial P}{\partial z} - g\rho_h \sin\theta - F_w. \qquad (5.37)$$

When the channel flow area is uniform,

$$\frac{\partial}{\partial t}(\rho_h j) + \frac{\partial}{\partial z}(\rho_h j^2) = -\frac{\partial P}{\partial z} - g\rho_h \sin\theta - F_w. \qquad (5.38)$$

Equations (5.36), (5.37), and (5.38) are in conservative form. They can be recast in nonconservative form by using the mass conservation equation. For example, the left-hand side of Eq. (3.37) can be written as

$$\rho_h \frac{\partial j}{\partial t} + \rho_h j \frac{\partial j}{\partial z} + j\left[\frac{\partial \rho_h}{\partial t} + \frac{1}{A}\frac{\partial}{\partial z}(\rho_h jA)\right]. \qquad (5.39)$$

The last (bracketed) term in this expression is identically equal to zero because of the conservation of mass. Therefore, Eq. (5.37) can be written as $v_g = 1/\rho_g$

$$\rho_h \frac{D_j}{Dt} j = -\frac{\partial P}{\partial z} - \rho_h g \sin\theta - F_w. \qquad (5.40)$$

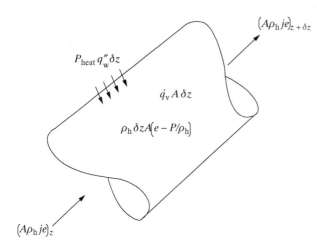

Figure 5.5. Energy terms for a control volume.

The wall force term, F_w, is in fact the frictional pressure gradient and can be shown as

$$F_w = \left(-\frac{dP}{dz}\right)_{fr} = \tau_w p_f/A. \qquad (5.41)$$

A popular form of the steady-state HEM momentum equation fo a single-component mixture in a uniform-flow-area pipe is obtained by expanding Eq. (5.36), using $v_f = 1/\rho_f = $ const., and writing

$$\frac{\partial v_g}{\partial z} = \left(\frac{dv_g}{dP}\right)\left(\frac{dP}{dz}\right), \qquad (5.42)$$

$$\tau_w = f_{TP}\frac{G^2}{2\rho_h}. \qquad (5.43)$$

The result will be (see Problem 5.2)

$$-\frac{dP}{dz} = \frac{\frac{2f_{TP}G^2 v_f}{D}\left[1 + x\frac{v_{fg}}{v_f}\right] + G^2 v_f\left(\frac{v_{fg}}{v_f}\right)\frac{dx}{dz} + \frac{g\sin\theta}{v_f\left[1 + x\frac{v_{fg}}{v_f}\right]}}{1 + G^2 x\left(\frac{dv_g}{dP}\right)}. \qquad (5.44)$$

Energy Conservation

The control volume energy storage and transport terms are shown in Fig. 5.5, where the total specific convected energy is defined as

$$e = h + \frac{1}{2}j^2 + gz\sin\theta. \qquad (5.45)$$

The parameter p_{heat} is the heated perimeter, q_w'' is the wall heat flux, and $\dot{q}_v$ is the volumetric energy generation rate. We can now apply the first law of thermodynamics to the control volume and take the limit of $\delta z \to 0$ to get

$$A\frac{\partial}{\partial t}(\rho_h e - P) + \frac{\partial}{\partial z}(A\rho_h je) = p_{heat}q_w'' - A\dot{q}_v. \qquad (5.46)$$

The nonconservative form of this equation can be derived by expanding the left side and using mass conservation:

$$\rho_h \frac{Dj}{Dt} h + \rho_h \frac{Dj}{Dt}\left(\frac{1}{2}j^2\right) + gj\rho_h \sin\theta = q''_w p_{\text{heat}}/A + \dot{q}_v + \frac{\partial P}{\partial t}. \tag{5.47}$$

Thermal and Mechanical Energy Equations

Equations (5.46) and (5.47) contain thermal and mechanical (kinetic and potential) energy terms. Although there is nothing wrong with solving these equations, sometimes it is more convenient to remove the redundant mechanical energy terms from energy equations before solving them. To do this, we first multiply Eq. (5.40) by j (equivalent to getting the dot product of the momentum equation with the velocity vector). The result will be

$$\rho_h \frac{Dj}{Dt}\left(\frac{1}{2}j^2\right) = -\frac{\partial P}{\partial z} - \rho_h gj \sin\theta - \frac{\tau_w p_f}{A}j. \tag{5.48}$$

This is a transport equation for mechanica lenergy. The last term on the right side is the familiar frictional (viscous) dissipation and represents the irreversible transformation of mechanical energy into heat. We can now subtract Eq. (5.48) from Eq. (5.47) to derive the thermal energy equation:

$$\rho_h \frac{Dj}{Dt} h = \frac{q''_w p_{\text{heat}}}{A} + \dot{q}_v + \frac{DjP}{Dt} + \frac{\tau_w p_f}{A}j. \tag{5.49}$$

As expected, the viscous dissipation term appeared on the right side of Eq. (5.49), only with a positive sign. This term is of course always positive.

Summary and Comments

The three differential conservation equations are the following:

mass conservation: Eq. (5.31) or (5.34),
momentum conservation: Eq. (5.37) or (5.40), and
energy conservation: Eq. (5.46), (5.47), or (5.49).

The unknowns (state variables) are (P, j, h). Note that since h and x are related (see Eq. (5.28)), one can treat x as a state variable, instead of h.

The HEM model essentially reduces a two-phase flow problem to an idealized single-phase flow, where compressibility and variability of fluid properties are significant. The model is simple, consistent, and unambiguous. Besides the assumption of homogeneity and equilibrium, few other assumptions are needed.

The model is reasonably accurate for many applications involving flow regimes such as bubbly and dispersed-droplet flows. However, even for these regimes, the model is good only when gas–liquid velocity slip is insignificant (e.g., in high mixture-velocity flows). It is inaccurate for most flow regimes and for any flow configuration that could lead to phase separation.

5.6 One-Dimensional Homogeneous-Equilibrium Model: Two-Component Mixture

The HEM conservation equations are now presented for a liquid mixed with a vapor–noncondensable gas mixture. However, it is assumed that the solubility of the noncondensable gas in the liquid phase is small, so that the thermodynamic and transport properties of the liquid phase are essentially the same as the properties of a pure liquid. The equilibrium assumption requires that everywhere (a) the gas and liquid phases be at the same temperature, (b) the vapor–noncondensable mixture be saturated, and (c) the liquid and gas phases be at equilibrium with respect to the concentration of the noncondensable. The thermodynamic property relations discussed in Section 1.3 apply. The liquid is generally subcooled with respect to the local pressure; therefore $\rho_L = \rho_L(P, T)$ and $h_L = h_L(P, T)$. Assuming that the noncondensable can be treated as a single species with ideal gas behavior, one can write

$$\rho_G = \rho_n + \rho_g(T), \tag{5.50}$$

$$\rho_n = \rho_n(\rho_n, T) = \frac{P_n}{\frac{R_u}{M_n}T}, \tag{5.51}$$

$$P_n = P - P_v, \tag{5.52}$$

$$P_v = P_{sat}(T), \tag{5.53}$$

$$h_G = (\rho_n h_n + \rho_v h_v)/\rho_G = m_v h_v + (1 - m_v)h_n, \tag{5.54}$$

$$h_n = h_n(T), \tag{5.55}$$

where $h_v = h_g(T)$ and m_v is the mass fraction of vapor in the vapor–noncondensable mixture. The gas–liquid mixture relations are now

$$\bar{\rho} = \rho_h = G/j = \alpha\rho_G + (1 - \alpha)\rho_L = [v_G + x(v_G - v_L)]^{-1}, \tag{5.56}$$

$$\frac{x}{1 - x} = \frac{\rho_G \alpha}{\rho_L(1 - \alpha)}, \tag{5.57}$$

$$h = [\rho_G \alpha h_G + \rho_L(1 - \alpha)h_L]/\bar{\rho} = h_L + x(h_G - h_L). \tag{5.58}$$

The conservation equations for the mixture mass, momentum, and energy derived in the previous section all apply, provided that the liquid and gas properties given here are used in those equations

An additional equation representing the conservation of the noncondensable species in the mixture should also be included. Neglecting the diffusion of the noncondensable, in comparison with its advection (an assumption that is valid in the vast majority of problems), one can show that

$$\frac{\partial}{\partial t}[\alpha\rho_G m_{n,G} + (1 - \alpha)\rho_L m_{n,L}] + \frac{1}{A}\frac{\partial}{\partial z}\{A j[\alpha\rho_G m_{n,G} + (1 - \alpha)\rho_L m_{n,L}]\} = 0. \tag{5.59}$$

The condition of equilibrium between the liquid and gas phases with respect to the concentration of the noncondensable leads to (see Eq. (2.64))

$$X_{n,L} = \frac{X_{n,G}P}{C_{He,n}}, \tag{5.60}$$

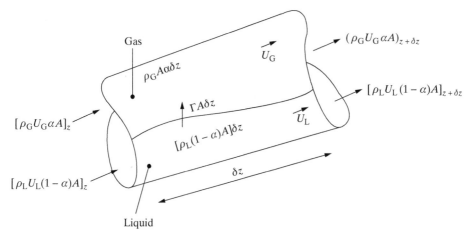

Figure 5.6. Control volume for mass conservation in separated flow.

where $X_{n,L}$ and $X_{n,G}$ are the mole fractions of the species n in the liquid and gas, respectively. They are related to mass fractions according to Eq. (1.45).

There are now four conservation equations: mixture mass (modified Eq. (5.31) or (5.34)), mixture momentum (modified Eq. (5.37) or (5.40)), mixture energy (modified Eq. (5.46), (5.47), or (5.49)), and noncondensable species (Eq. (5.59)). The unknowns are P, T, j, and either $m_{n,G}$ or $m_{n,L}$. In problems dealing with boiling and condensation in the presence of a noncondensable, the solubility of the noncondensable in the liquid phase is often negligibly small and can be neglected (i.e., $C_{He,n} = \infty$). In that case, $m_{n,L} = 0$, terms containing $m_{n,L}$ in Eq. (5.59) are all dropped, and Eq. (5.60) becomes irrelevant. The unknowns will be P, T, j, and $m_{n,G}$.

5.7 One-Dimensional Separated-Flow Model: Single-Component Fluid

In separated flow modeling we derive a mass and momentum equation for each phase. The number of energy equations can be one or two. When a single-component liquid–vapor mixture is considered, often one of the phases (liquid in bulk boiling and vapor in condensation, for example) can be assumed to be saturated with respect to the local pressure. In these cases only one energy equation is needed. When conditions involving a subcooled liquid and superheated vapor are to be considered, then two energy conservation equations are needed, one for each phase.

In the following derivations, for simplicity, a stratified or annular flow pattern is used in figures for demonstration of various terms. The results of the derived equations are more general, however. Because we deal with pure vapor, the following apply everywhere in the equations: $\rho_G = \rho_v, h_G = h_v$, etc.

Mass Conservation Equation

Let us define V as the rate of phase change, per unit mixture volume (in kilograms per meter cubed per second, for example); with positive values for evaporation. The mass flow terms are depicted in Fig. 5.6 for a slice of the flow channel. Phase conservation equations can be derived by applying the mass continuity principle to

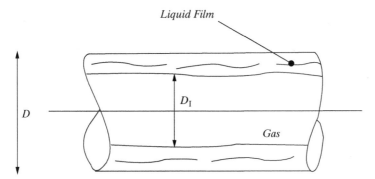

Figure 5.7. Ideal annular flow.

gas- and liquid-occupied portions of the control volume and taking the limit of $\delta z \to 0$. The resulting equations will be:

$$\frac{\partial}{\partial t}[\rho_L(1-\alpha)] + \frac{1}{A}\frac{\partial}{\partial z}[A\rho_L U_L(1-\alpha)] = -\Gamma \qquad (5.61)$$

$$\frac{\partial}{\partial t}(\rho_G\alpha) + \frac{1}{A}\frac{\partial}{\partial z}(A\rho_G U_G\alpha) = \Gamma, \qquad (5.62)$$

for liquid and gas (vapor) mass conservation, respectively. The mixture mass conservation equation can be obtained by adding Eqs. (5.61) and (5.62):

$$\frac{\partial}{\partial t}[\rho_L(1-\alpha) + \rho_G\alpha] + \frac{1}{A}\frac{\partial}{\partial z}\{A[\rho_L(1-\alpha)U_L + \rho_G\alpha U_G]\} = 0. \qquad (5.63)$$

Equation (5.63) can be recast as

$$\frac{\partial}{\partial t}\bar{\rho} + \frac{1}{A}\frac{\partial}{\partial z}(AG) = 0. \qquad (5.64)$$

Note that out of Eqs. (5.61)–(5.63), only two are independent.

Now let us briefly discuss the volumetric phase change rate Γ. In general we can write

$$\Gamma = m_I'' a_I'' = \frac{m_I'' p_I}{A}, \qquad (5.65)$$

where m_I'' represents the interfacial mass flux (in kilograms per meter squared per second) defined here to be positive for evaporation, p_I is the gas–liquid interfacial perimeter (total interfacial surface area per unit channel length), and $a_I'' = p_I/A$ is the interfacial area concentration (interfacial surface area per unit mixture volume). Evidently p_I depends on the two-phase flow regime.

Two examples for the interfacial perimeter are now given.

(1) Ideal annular flow in a pipe: Referring to Fig. 5.7, the following relations apply:

$$D_I/D = \sqrt{\alpha},$$
$$p_I = \pi D\sqrt{\alpha},$$
$$\Gamma = 4m_I''\sqrt{\alpha}/D.$$

(2) Ideal bubbly flow: Assume spherical, uniform–sized bubbles. Then

$$p_{\mathrm{I}} = A\pi d_{\mathrm{B}}^2 N_{\mathrm{B}},$$
$$N_{\mathrm{B}} = \alpha/\left(\tfrac{\pi}{6}d_{\mathrm{B}}^3\right)$$

where d_{B} is the bubble diameter and N_{B} is the number of bubbles in a unit mixture volume.

Momentum Conservation Equation

The momentum transfer terms for the two phases are shown in Fig. 5.8(a) for a slice of the flow channel, where U_{I} represents the axial velocity at the interphase. Forces acting on the liquid and gas phases are depicted in Figs. 5.8(b) and (c), respectively.

The wall friction force acting on the liquid is $A F_{\mathrm{wL}} \delta z$ (where F_{wL} is the force acting on the liquid phase, per unit mixture volume). Often in two-phase flow the wall friction is found from pressure drop correlations that only address the two-phase mixture as a whole. In that case, the force on a unit mixture volume has to be distributed between the two phases. For example, we can write

$$F_{\mathrm{wL}} = \frac{\tau_{\mathrm{w}} p_{\mathrm{f}}}{A}[1 - f(\alpha)], \tag{5.66}$$

where τ_{w} is the wall shear stress on the two-phase mixture, p_{f} is the flow passage wetted perimeter, and $f(\alpha)$ is the fraction of wall shear force directly imposed on the gas.

The parameter F_{I} represents the interfacial force per unit mixture volume. Interfacial drag and friction both contribute to this force. When friction is dominant, one can write

$$F_{\mathrm{I}} = \frac{\tau_{\mathrm{I}} p_{\mathrm{I}}}{A}, \tag{5.67}$$

where τ_{I} is the interfacial shear stress (positive when $U_{\mathrm{G}} > U_{\mathrm{L}}$) and p_{I} is the interfacial perimeter. The parameter F_{VM} is *the virtual mass force* and will be discussed later.

The forces acting on the gas phase, shown in Fig. 5.8(c), are similar to the ones that act on the liquid phase. All the forces associated with interaction between the two phases have exactly the same magnitudes, only in the opposite direction. Also, the wall friction force consistent with Eq. (5.66) will be

$$F_{\mathrm{wG}} = \frac{\tau_{\mathrm{w}} p_{\mathrm{f}}}{A} f(\alpha). \tag{5.68}$$

To derive the momentum equation for each phase, one applies Newton's law of motion to that phase's control volume and takes the limit of $\delta z \to 0$. The phasic momentum equations will be

$$\frac{\partial}{\partial t}[\rho_{\mathrm{L}}(1-\alpha)U_{\mathrm{L}}] + \frac{1}{A}\frac{\partial}{\partial z}[A\rho_{\mathrm{L}}(1-\alpha)U_{\mathrm{L}}^2 + \Gamma U_{\mathrm{I}}]$$
$$= -(1-\alpha)\frac{\partial P}{\partial z} - F_{\mathrm{wL}} - \rho_{\mathrm{L}}g(1-\alpha)\sin\theta + F_{\mathrm{I}} - F_{\mathrm{VM}}, \tag{5.69}$$

$$\frac{\partial}{\partial t}(\rho_{\mathrm{G}}\alpha U_{\mathrm{G}}) + \frac{1}{A}\frac{\partial}{\partial z}(A\rho_{\mathrm{G}}\alpha U_{\mathrm{G}}^2) - \Gamma U_{\mathrm{I}} = -\alpha\frac{\partial P}{\partial z} - \rho_{\mathrm{G}}g\alpha\sin\theta - F_{\mathrm{wG}} - F_{\mathrm{I}} + F_{\mathrm{VM}}. \tag{5.70}$$

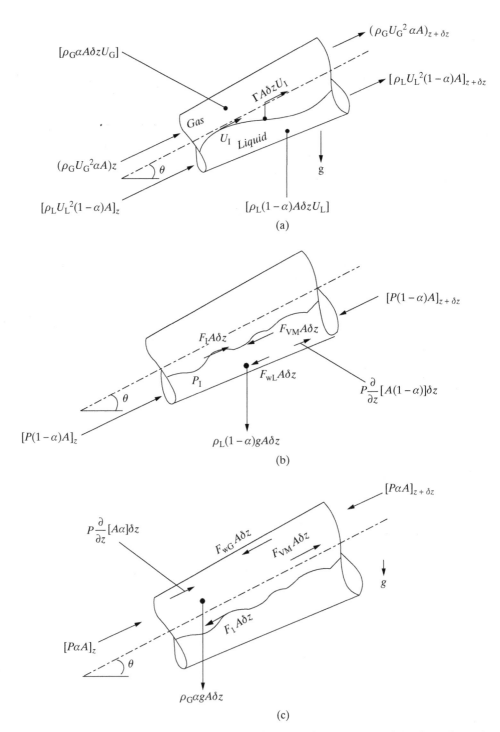

Figure 5.8. Separate flow: (a) momentum transfer terms for a segment of the flow channel; (b) forces acting on the liquid phase; (c) forces acting on the gas phase.

The mixture momentum equation can be obtained by adding Eqs. (5.69) and (5.70). As expected, all the interfacial force terms cancel out, leaving

$$\frac{\partial}{\partial t}[\rho_L(1-\alpha)U_L + \rho_G\alpha U_G] + \frac{1}{A}\frac{\partial}{\partial z}[A\rho_L(1-\alpha)U_L^2 + A\rho_G\alpha U_G^2]$$

$$= -\frac{\partial P}{\partial z} - [\rho_G\alpha + \rho_L(1-\alpha)]g\sin\theta - F_w, \tag{5.71}$$

where $F_w = \tau_w p_f/A$. Equation (5.71) can be recast in the following form:

$$\frac{\partial G}{\partial t} + \frac{1}{A}\frac{\partial}{\partial z}\left(A\frac{G^2}{\rho'}\right) = -\frac{\partial P}{\partial z} - \bar{\rho}g\sin\theta - F_w, \tag{5.72}$$

where $\bar{\rho} = \rho_L(1-\alpha) + \rho_G\alpha$, and the "momentum density" is defined as

$$\rho' = \left[\frac{(1-x)^2}{\rho_L(1-\alpha)} + \frac{x^2}{\rho_G\alpha}\right]^{-1}. \tag{5.73}$$

The interfacial velocity U_I is flow-regime-dependent. A simple and widely used choice is (Wallis, 1969)

$$U_I = \frac{1}{2}(U_L + U_G). \tag{5.74}$$

The interfacial force f_I is also flow-regime-dependent. For separated regimes such as stratified or annular flow, we can write

$$\tau_I = f_I\frac{1}{2}\rho_G|U_G - U_L|(U_G - U_L), \tag{5.75}$$

where the parameter f_I is the skin friction factor only when the interphase is smooth (e.g., in stratified-smooth flow regime). Since ripples and waves are rampant, however, f_I must account for their effect. The situation of the gas phase is somewhat similar to flow in a rough channel. A widely used correlation for the annular flow regime, for example, is (Wallis, 1969)

$$f_I = 0.005[1 + 75(1-\alpha)]. \tag{5.76}$$

For regimes such as bubbly and dispersed-droplet, the drag force may be predominant. For bubbly flow, for example, assuming spherical and uniform-sized bubbles, one can write

$$F_I = F_D N, \tag{5.77}$$

$$N = \alpha/(\pi d_B^3/6), \tag{5.78}$$

$$F_D = C_D\rho_L\frac{\pi d_B^2}{4}\frac{1}{2}|U_G - U_L|(U_G - U_L), \tag{5.79}$$

where C_D is the drag coefficient, which can be estimated from standard drag coefficient laws when the number density of particles of the dispersed phase is small. Otherwise, the hydrodynamic effect of the dispersed phase particles on each other will be important, and appropriate correlations should be used.

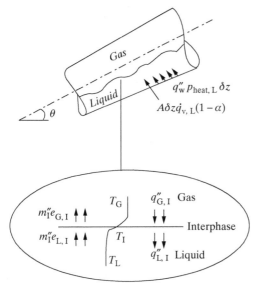

Figure 5.9. Liquid and vapor phase control volumes and the interphase.

Virtual Mass (Added Mass) Force Term

The virtual mass force occurs only when one of the phases accelerates with respect to the other phase. It results from the fact that the motion of the discontinuous phase results in the acceleration of the continuous phase as well. A more detailed discussion of this force will be given in Section 5.9. A simple and widely used expression for one-dimensional separated flow is

$$F_{VM} = -C_{VM}\left[\frac{\partial U_G}{\partial t} + U_G\frac{\partial U_G}{\partial z} - \frac{\partial U_L}{\partial t} - U_L\frac{\partial U_L}{\partial z}\right]. \tag{5.80}$$

Watanabe *et al.* (1990) have suggested

$$C_{VM} = C'\alpha(1-\alpha)\bar{\rho} \tag{5.81}$$

with

$$C' \approx 1. \tag{5.82}$$

In terms of magnitude, F_{VM} is significant only if the gas phase is dispersed, and even then only in rather extreme flow acceleration conditions (e.g., choked flow). Despite the insignificant quantitative effect, however, the virtual mass term is important because it modifies the mathematical properties of the momentum conservation equation and improves the numerical stability of the conservation equation set.

Energy Conservation Equations

A control volume composed of a short segment of the flow channel is depicted in Fig. 5.9, where T_L and T_G are the bulk liquid and gas phase temperatures, T_I is the temperature at the interphase, and m_I'' is the interfacial mass flux (defined here to be positive for evaporation). Other parameters are defined as follows:

$p_{\text{heat,L}}$ = part of the flow passage heated perimeter directly in contact with liquid,

$\dot{q}_{\text{v,L}}, \dot{q}_{\text{v,G}}$ = volumetric energy generation rates in the liquid and vapor phases, respectively,

$$e = h + \frac{1}{2}U^2 + gz\sin\theta, \tag{5.83a}$$

$$e_{\text{LI}} = h_{\text{LI}} + \frac{1}{2}U_I^2 + gz\sin\theta, \tag{5.83b}$$

$$e_{\text{GI}} = h_{\text{GI}} + \frac{1}{2}U_I^2 + gz\sin\theta, \tag{5.83c}$$

$h_{\text{L,I}}, h_{\text{G,I}}$ = liquid and vapor specific enthalpies at the interphase, respectively, and $q''_{\text{L,I}}, q''_{\text{G,I}}$ = heat fluxes between liquid and gas phases and the interphase (in watts per meter squared, for example).

The inclusion of $gz\sin\theta$ in the definitions of e_{LI} and e_{GI} is for generality. We often deal with $e_{\text{GI}} - e_{\text{LI}}$, whereby the $gz\sin\theta$ term cancels out.

To derive the phasic energy conservation equations, one should apply the first law of thermodynamics to the liquid and gas control volumes and take the limit of $\delta z \to 0$. The liquid phase energy conservation equation will be

$$\frac{\partial}{\partial t}\left[\rho_L(1-\alpha)\left(e_L - \frac{P}{\rho_L}\right)\right] + \frac{1}{A}\frac{\partial}{\partial z}[\rho_L A(1-\alpha)e_L U_L] + \Gamma e_{\text{LI}} - P\frac{\partial\alpha}{\partial t}$$
$$- \frac{p_{\text{heat,L}}}{A}q''_w - \frac{p_I}{A}q''_{\text{LI}} - \dot{q}_{\text{v,L}}(1-\alpha) - [F_I - F_{\text{VM}}]U_I = 0, \tag{5.84}$$

And, for the gas phase energy, we get

$$\frac{\partial}{\partial t}\left[\rho_G\alpha\left(e_G - \frac{P}{\rho_G}\right)\right] + \frac{1}{A}\frac{\partial}{\partial z}[\rho_G A\alpha e_G U_G] - \Gamma e_{\text{GI}} + P\frac{\partial\alpha}{\partial t}$$
$$- \frac{p_{\text{heat,G}}}{A}q''_w + \frac{p_I}{A}q''_{\text{GI}} - \dot{q}_{\text{v,G}}\alpha + [F_I - F_{\text{VM}}]U_I = 0. \tag{5.85}$$

The last term in both equations represents energy dissipation caused by the interfacial forces.

The mixture energy can be derived by simply adding Eqs. (5.84) and (5.85). All the interfacial transfer terms should of course disappear when we add the two equations. The resulting mixture energy equation can be recast in the following form, where thermal and mechanical energy terms have been separated:

$$\frac{\partial}{\partial t}(\bar{\rho}\bar{h}) + \frac{\partial}{\partial t}\left(\frac{G^2}{2\rho'}\right) + \frac{1}{A}\frac{\partial}{\partial z}(AGh) + \frac{1}{A}\frac{\partial}{\partial z}\left(A\frac{G^3}{2\rho'''^2}\right) + g\sin\theta G - \frac{\partial P}{\partial t}$$
$$= p_{\text{heat}}q''_w/A + [\dot{q}_{\text{v,L}}(1-\alpha) + \dot{q}_{\text{v,G}}\alpha]. \tag{5.86}$$

Two different definitions for mixture enthalpy have been used in this equation. The in-situ mixture specific enthalpy is defined as

$$\bar{h} = [\rho_L(1-\alpha)h_L + \rho_G\alpha h_G]/\bar{\rho}. \tag{5.87}$$

The mixed-cup enthalpy h is defined in Eq. (3.42). The mixture density is also defined as an in-situ mixture property (see Eq. (3.21)), according to

$$\bar{\rho} = \alpha\rho_G + (1-\alpha)\rho_L. \tag{5.88}$$

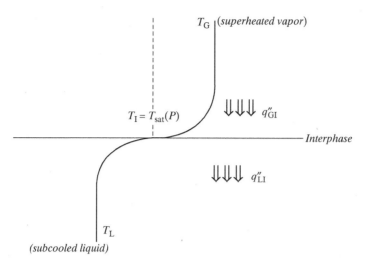

Figure 5.10. Temperature profiles at the vicinity of the interphase when vapor is superheated and liquid is subcooled.

Also,

$$\rho'''^2 = \left[\frac{(1-x)^3}{\rho_L^2 (1-\alpha)^2} + \frac{x^3}{\rho_G^2 \alpha^2} \right]^{-1}. \tag{5.89}$$

Out of the latter three energy equations, of course, only two are independent. Furthermore, if the state of one of the phases is known (e.g., when the vapor is saturated, $h_G = h_g(P)$), then one of the energy equations (Eq. (5.84) when the liquid is saturated and Eq. (5.85) when the vapor is saturated) becomes redundant and only one energy equation (usually Eq. (5.86)) need be used.

More about Interphase Mass and Energy Transfer

In a single-component liquid–vapor mixture (e.g., a water–steam mixture, without any noncondensables), we always have

$$T_I = T_{sat}(P), \tag{5.90}$$

where P is the local pressure. The mass and heat fluxes at the interphase are not independent. The temperature profiles depicted in Fig. 5.10 represent the general situation where the vapor phase is superheated while the liquid phase is subcooled. The energy balance at the interphase gives

$$m_I'' \left(h_L + \frac{1}{2} U_{LI}^2 \right)_I - q_{LI}'' = m_I'' \left(h_G + \frac{1}{2} U_{GI}^2 \right)_I - q_{GI}'', \tag{5.91}$$

where, in accordance to Eqs. (2.71) and (2.72), we have $q_{GI}'' = \dot{H}_{GI}(T_G - T_I)$ and $q_{LI}'' = \dot{H}_{LI}(T_I - T_L)$. (Note that in this section we follow the convention that $m_I'' > 0$ for evaporation.) The parameters $\dot{H}_{LI}$ and $\dot{H}_{GI}$ are the convective heat transfer coefficients between the interphase and the liquid and gas bulks, respectively. These heat transfer coefficients should account for the effect of mass transfer (see Eqs. (2.77) and (2.78)).

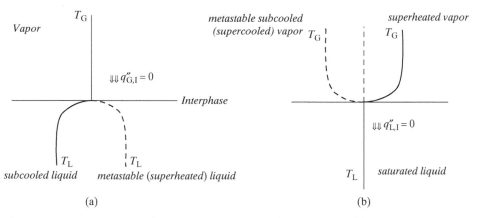

Figure 5.11. Temperature profiles at the vicinity of the interphase for (a) saturated vapor and (b) saturated liquid.

The kinetic energy change during the phase change process is typically negligible, and Eq. (5.91) therefore reduces to

$$m_I'' = \frac{\dot{H}_{GI}(T_G - T_I) - \dot{H}_{LI}(T_I - T_L)}{h_{fg}}. \tag{5.92}$$

Knowing m_I'' and a_I'' (the latter parameter to be found by using information about the flow regime), Γ is calculated from Eq. (5.65). Equation (5.92) in fact determines whether evaporation ($m_I'' > 0$) or condensation ($m_I'' < 0$) takes place.

Figure 5.11(a) is a schematic for the interphase when the vapor is saturated, a situation that is common during condensation. In this case,

$$T_G = T_I = T_{sat}(P).$$

When the liquid phase is saturated, as is often the case during evaporation, the temperature profiles are similar to Fig. 5.11(b). Then

$$T_L = T_I = T_{sat}(P).$$

The vapor phase is saturated vapor in most cases. Superheated vapor can be encountered in some situations, however. An example is the steam conditioning equipment in steam power plants where subcooled liquid is sprayed into superheated vapor. The spray droplets reach saturation rapidly and continue evaporating while surrounded in superheated vapor. Another example is the dispersed-droplet flow regime in heated flow channels under the postcritical heat flux regime (to be discussed in Chapter 13), where thermodynamic nonequilibrium conditions are encountered (Groeneveld and Delorme, 1976; Chen et al., 1979; Nijhawan et al., 1980). The condition where a metastable supercooled vapor phase is in contact with liquid is rarely encountered.

Summary and Comments

The phasic properties are in general thermodynamic functions of state variables, and they are usually represented as functions of pressure and temperature, that is, $\rho_L = \rho_L(P, T_L), h_L = h_L(P, T_L), \rho_G = \rho_v(P, T_G)$, and $h_G = h_v(P, T_G)$, etc.

The conservation equations are as follows:

mass: two out of Eqs. (5.61)–(5.63),
momentum: two out of Eqs. (5.69)–(5.71), and
energy: two out of Eqs. (5.84)–(5.86),

with unknowns $U_L, U_G, h_L, h_G, P,$ and α.
The following should be pointed out.

(1) The phasic enthalpies h_G and h_L can be equivalently replaced with T_G and T_L, respectively, as state variables.

(2) Knowing $\alpha, U_L,$ and U_G, x can be found from the fundamental void-quality relation, Eq. (3.39). It is also possible to use x as a state variable, instead of α.

(3) If one of the phases is saturated (e.g., saturated vapor, for which $\rho_G = \rho_g(P)$ and $T_G = T_{sat}(P)$), only one energy equation can be used. If both phases remain saturated, furthermore, then $h_G = h_g$ and $h_L = h_f$, and the unknowns will be P, $\alpha, U_L, U_G,$ and Γ. No interfacial heat transfer model can be used, because the assumption of equilibrium between the two phases implies an infinitely fast heat transfer at the interphase.

5.8 One-Dimensional Separated-Flow Model: Two-Component Fluid

Two-component separated-flow conservation equations are now presented and discussed. We will limit the discussion to conditions of interest in boiling/evaporation and condensation. A liquid mixed with a vapor–noncondensable gas mixture is considered, and it is assumed that the solubility of the inert, noncondensable gas in the liquid phase is small enough that it can be totally neglected. The liquid phase is thus impermeable to the inert gas, and the thermodynamic and transport properties of the liquid phase are those of a pure liquid. The liquid and gas phases can in general be subcooled, saturated, or superheated with respect to the local pressure. For the liquid, $\rho_L = \rho_L(P, T_L)$ and $h_L = h_L(P, T_L)$. Assuming that the noncondensable can be treated as a single species with ideal gas behavior, we have

$$P = P_n + P_v, \tag{5.93}$$

$$\rho_G = \rho_n + \rho_v, \tag{5.94}$$

$$\rho_v = \rho_v(P_v, T_G), \tag{5.95}$$

$$\rho_n = \rho_n(\rho_n, T_G) = \frac{P_n}{\frac{R_u}{M_n} T_G}, \tag{5.96}$$

$$h_G = (\rho_n h_n + \rho_v h_v)/\rho_G = m_v h_v + (1 - m_v) h_n, \tag{5.97}$$

$$h_n = h_n(T_G). \tag{5.98}$$

These equations are of course similar to Eqs. (5.50)–(5.55), with the difference that the vapor is no longer saturated, and the gas and liquid temperatures are not necessarily the same. The conservation equations of the previous section all apply. The parameter Γ, however, must include the phase change (evaporation or condensation) as well as the interfacial transfer of the inert gas component if such transfer indeed takes place. In dealing with phase change in the presence of a noncondensable, however, the effect of the absorption or desorption of the inert gas is often negligible, and only the phase change of the condensable component needs to be considered. An additional equation representing the conservation of mass for the noncondensable component is also needed. Neglecting the diffusion of the noncondensable, in comparison with its advection (an assumption that is valid in the vast majority of problems), the conservation of mass for the noncondensable species can be written as

$$\frac{\partial}{\partial t}(\rho_G \alpha m_{n,G}) + \frac{1}{A}\frac{\partial}{\partial z}(A\rho_G U_G \alpha m_{n,G}) = 0. \tag{5.99}$$

In comparison with the single-component separated-flow equations, we have added one new unknown ($m_{n,G}$) and a new conservation equation.

The interfacial mass transfer, representing evaporation or condensation, should now be discussed. Equation (5.91) applies. Neglecting the contribution of absorption or desorption of the noncondensable by the liquid to the energy transfer, furthermore, we can replace $h_G - h_L$ in that equation with $h_{fg}(T_I)$. The equilibrium conditions at the interphase will be

$$T_I = T_{sat}(P_{v,s}), \tag{5.100}$$

where $P_{v,s}$ is the vapor pressure at the interphase (the "s" surface; see Fig. 2.10). When the process is sufficiently slow such that the film model described in Section 2.7 can be used, then $q''_{LI} = \dot{H}_{LI}(T_I - T_L)$ and $q''_{GI} = \dot{H}_{GI}(T_G - T_I)$. The interphase temperature T_I, or, equivalently, $P_{v,s}$, is an additional unknown. Therefore, additional equations are needed to close the set of equations. The film model provides the following expressions that result in the closure of the equation set:

$$P_{v,s} = PX_{v,s}, \tag{5.101}$$

$$m''_I = K_{GI} \ln \frac{m_{n,G}}{1 - m_{v,s}}, \tag{5.102}$$

$$m_{v,s} = \frac{X_{v,s}M_v}{X_{v,s}M_v + (1 - X_{v,s})M_n}. \tag{5.103}$$

5.9 Multi-dimensional Two-Fluid Model

In two-fluid modeling, the local instantaneous phasic mass, momentum, and energy conservation equations are averaged, with each phase being treated as a "fluid." The averaging process follows what was described in Chapter 4. The resulting conservation equations of the two "fluids" are coupled via their interfacial interactions.

The composite volume and time/ensemble-averaged two-fluid model equations for mass, momentum, and energy (enthalpy) are (Ishii and Mishima, 1984)

$$\frac{\partial}{\partial t}(\alpha_k \rho_k) + \nabla \cdot (\alpha_k \rho_k \vec{U}_k) = \Gamma_k, \tag{5.104}$$

$$\frac{\partial}{\partial t}(\alpha_k \rho_k \vec{U}_k) + \nabla \cdot (\alpha_k \rho_k \vec{U}_k \vec{U}_k) = -\alpha_k \nabla P_k + \nabla \cdot \alpha_k(\bar{\bar{\tau}}_k + \bar{\bar{\tau}}_{k,turb})$$
$$+ \vec{U}_{kI}\Gamma_k + \vec{M}_{Ik} - (\nabla \alpha_k) \cdot \bar{\bar{\tau}}_I + \alpha_k \rho_k \vec{g}, \tag{5.105}$$

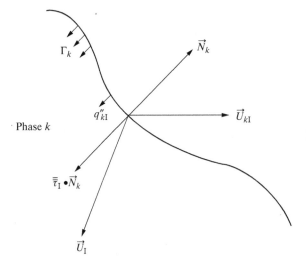

Figure 5.12. Schematic of the interphase.

and

$$\frac{\partial}{\partial t}(\alpha_k \rho_k h_k) + \nabla \cdot (\alpha_k \rho_k h_k \vec{U}_k) = -\nabla \cdot [\alpha_k(\vec{q}_k'' + \vec{q}_{k,\text{turb}}'')] + \alpha_k \left(\frac{\partial}{\partial t} + \vec{U}_k \cdot \nabla\right) P_k$$
$$+ h_{kI}\Gamma_k + a_I'' q_{kI}'' + (\vec{U}_I - \vec{U}_k) \cdot \vec{M}_{Ik} + (\mu\Phi)_k,$$

(5.106)

respectively, where

$$(\mu\Phi)_k = \alpha_k \bar{\bar{\tau}}_k : (\nabla \vec{U}_k) - \vec{U}_k \cdot \nabla \cdot (\alpha_k \bar{\bar{\tau}}_{k,\text{turb}}).$$ (5.106a)

Figure 5.12 defines the interfacial transport terms. The parameters in these equations are refined as

Γ_k = volumetric generation rate of phase k (in kilograms per meter cubed per second, for example) per unit mixture volume,

$\bar{\bar{\tau}}_k$ = viscous stress tensor in phase k,

$\bar{\bar{\tau}}_{k,\text{turb}}$ = turbulent (Reynolds) stress tensor,

P_k = pressure within phase k,

h_k = enthalpy of phase k,

$\vec{q}_k''$ = molecular heat conduction (diffusion) flux,

α_k = in situ volume fraction of phase,

$\vec{q}_{k,\text{turb}}''$ = turbulent heat diffusion flux,

$\vec{M}_{Ik}$ = generalized interfacial drag force (in newtons per meter cubed), per unit mixture volume, exerted on phase k,

$(\mu\Phi)_k$ = viscous and turbulent dissipation per unit mixture volume (in watts per meter cubed),

q_{kI}'' = heat flux from the interphase into phase k, and

h_{kI} = enthalpy of phase k at the interphase.

These equations must be written for both $k = $ L and $k = $ G. Adding the two phasic conservation equations for mass, momentum, or energy would give the corresponding mixture equation. In the mixture equations the interfacial transfer terms all vanish, that is,

$$\sum_{k=1}^{2} \Gamma_k = 0, \tag{5.107}$$

$$\sum_{k=1}^{2} \vec{M}_{Ik} = 0, \tag{5.108}$$

$$\sum_{k=1}^{2} (\Gamma_k h_{kI} + a_I'' q_{kI}'') = 0. \tag{5.109}$$

Generalized Drag Force

Consider the flow regimes where one phase is dispersed, while the other phase is continuous (e.g., bubbly, plug, or slug flows). For the dispersed phase, the interfacial force has two components – the drag force and the virtual mass force, and so

$$\vec{M}_{Id} = \vec{M}_{ID,d} + \vec{M}_{IV,d}. \tag{5.110}$$

The standard drag force term, $\vec{M}_{ID,d}$, can be represented as

$$\vec{M}_{ID,d} = -\frac{\alpha_d}{B_d} C_D \frac{1}{2} \rho_c |\vec{U}_d - \vec{U}_c|(\vec{U}_d - \vec{U}_c)A_d, \tag{5.111}$$

where subscripts c and d represent the continuous and dispersed phases, respectively; B_d is the volume of an average dispersed phase particle; α_d is the volume fraction of the dispersed phase; A_d is the frontal area of a dispersed phase particle; and C_D is the drag coefficient found from correlations with the generic form

$$C_D = f(\text{Re}_d), \tag{5.112}$$

where

$$\text{Re}_d = \rho_c |\vec{U}_c - \vec{U}_d| d_d / \mu_c \tag{5.113}$$

with d_d the diameter of dispersed-phase particles.

The virtual mass term, $\vec{M}_{IV,d}$ which is the same as F_{VM}, now in three-dimensional form, arises when a dispersed phase accelerates with respect to its surrounding continuous phase. The dispersed phase appears to impose its acceleration on some of the surrounding continuous fluid. An example, where theory predicts the virtual mass effect, is an accelerating rigid spherical particle in a quiescent fluid in the creep flow regime, starting from rest at $t = 0$, as shown in Fig. 5.13. The force needed to sustain the motion is (Michaelides, 1997)

$$\vec{F} = \rho_P V_P \frac{d\vec{U}}{dt} + \rho \frac{1}{2} V_P \frac{d\vec{U}}{dt} + 3\pi \mu d\vec{U} + \frac{3}{2} d^2 \sqrt{\pi\mu\rho} \int_0^t \frac{\dot{\vec{U}}(t')}{\sqrt{t-t'}} dt' \tag{5.114}$$

(the Boussinesq–Basset expression), where the first term on the right side represents the force needed to accelerate the particle itself, and the remaining three terms represent the transient hydrodynamic force. The second term on the right side is due to the virtual (added) mass, and the third term on the right side is the drag force. The last term on the right side represents the Basset force.

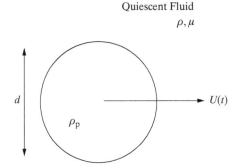

Quiescent Fluid
ρ, μ

Figure 5.13. Spherical particle accelerating in a quiescent fluid.

The exact form of the virtual mass force term is only known from theory for some simple and idealized conditions (Zuber, 1964; Van Wijngaarden, 1976; Wallis, 1990). The general form of virtual mass force in two-phase flow has been a subject of considerable discussion. Drew *et al.* (1979) derived a general form for the force term based on the argument that the force must be objective (frame-independent). Their proposed general form includes regime-dependent parameters that can only be obtained from theory for some idealized configurations, however. One suggested and widely used expression for the virtual mass term is from Ishii and Mishima (1984):

$$\vec{M}_{IV,d} = -\frac{1}{2}\alpha_d \frac{1+2\alpha_d}{1-\alpha_d}\rho_c\left[\frac{D_d}{Dt}(\vec{U}_d - \vec{U}_c) - (\vec{U}_d - \vec{U}_c)\cdot\nabla\vec{U}_c\right], \qquad (5.115)$$

where the *material derivative* is defined as

$$\frac{D_d}{Dt} = \frac{\partial}{\partial t} + (\vec{U}_d \cdot \nabla). \qquad (5.116)$$

Drew and Lahey (1987) have shown that for a single sphere accelerating in an incompressible inviscid fluid the total force exchanged between the sphere and the surrounding fluid is objective when the virtual mass force term is represented by the expression given here.

The effect of the virtual mass force term on the model predictions for most two-phase flow processes is usually small and unimportant. However, the mathematical form of the term significantly improves the stability of the numerical solution algorithm.

5.10 Numerical Solution of Steady, One-Dimensional Conservation Equations

Advanced thermal-hydraulics codes for nuclear reactor safety analysis often need to numerically solve the two-phase flow model conservation equations to simulate the flow and heat transfer processes in large and complex flow loops. As a result of this need, efficient and robust methods for the numerical solution of transient, one- and multi-dimensional two-phase conservation equations, for two- and three-fluid models, have been developed, evolved, and extensively applied in the past three decades (Mahafy, 1982; Taylor *et al.*, 1984; Spalding, 1980, 1983; Ren *et al.*, 1994a, b; Yao and Ghiaasiaan, 1996a, b; RELAP5–3D Code Development Team, 2005). The numerical solution methods for transient and multi-dimensional model conservation equations are complicated and typically need sophisticated algorithms incorporated in large computer programs. Detailed discussion of these numerical methods reside outside the scope of this book. Useful reviews and discussions can be found in Wulff (1990) and Yao and Ghiaasiaan (1996a, b).

A large variety of applications (e.g., boiler tubes, in-tube condensers, refrigeration loops, and pipelines), however, can be adequately treated using steady-state, one-dimensional model equations. Unlike the case of transient and/or multi-dimensional flow conditions, the numerical solution of steady, one-dimensional model equations is relatively simple and straightforward. Steady-state and one-dimensional model equations can in general be cast as a set of coupled ordinary differential equations (ODEs). The system of ODEs can then be numerically solved by using a numerical integration algorithm. Numerical integration tools are in fact readily available from numerous commercial and other sources. The method for the numerical solution of steady, one-dimensional two-phase model equations will be discussed in this section.

5.10.1 Casting the One-Dimensional ODE Model Equations in a Standard Form

The set of model equations generally includes a number of differential conservation equations and a number of closure and constitutive relations that are often algebraic. To have a unique solution, in addition to proper initial conditions, the system of equations must form a closed equation set (i.e., the number of unknowns must be equal to the total number of differential conservation equations and algebraic closure and constitutive relations). Among the unknowns, the *state variables* are equal in number to the total number of model conservation equations. The state variables are variables that together fully define the state of the physical system.

By expanding the differential terms in the model equations, making proper use of the chain rule, and applying thermodynamic property relations, it is possible to transform a set of steady, one-dimensional model conservation equations into the following generic form:

$$\mathbf{A}\frac{d\mathbf{Y}}{dz} = \mathbf{C}, \tag{5.117}$$

where z is the coordinate along the flow direction and $\mathbf{Y}$ is a column vector, the elements of which are the state variables. The square matrix $\mathbf{A}$ is the coefficient matrix for the set of ODEs, and $\mathbf{C}$ is a column vector containing known quantities.

The state variables and their number depend on the two-phase model. The pressure, mixture or phasic velocity (depending on the two-phase modeling approach), mixture or phasic enthalpy or internal energy, and species mass or mole fractions (when mass transfer is of interest) are typically among the state variables. In expanding the terms in the model conservation equations one needs to remember that thermophysical properties are not always constants. They vary along the flow path as a result of variations in pressure and temperature. However, because thermophysical properties are often not among the state variables in numerical solutions (being usually provided by closure relations), their spatial variations must be presented in terms of the spatial derivatives of the state variables. This can be done by applying the chain rule and thermodynamic relations to the closure relations and sometimes needs lengthy and careful algebra. The following examples will help clarify these points.

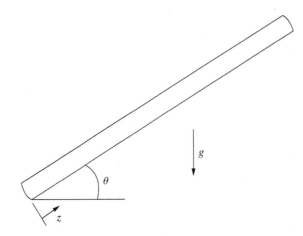

Figure 5.14. The inclined, heated tube for Examples 5.1 through 5.3.

EXAMPLE 5.1. A pure, subcooled liquid flows through the heated inclined tube displayed in Fig. 5.14 . Steady state can be assumed. The tube receives a uniform wall heat flux q''_w. Near the inlet, where subcooling is large, no boiling takes place and the flow is essentially a single-phase liquid. Derive a set of ODEs based on the time- and flow-area-averaged conservation equations, in the standard form of Eq. (5.117). Include the thermal expansion of the liquid in the derivations.

SOLUTION. For single-phase liquid flow, the momentum and energy conservation equations will be

$$\frac{d}{dz}\left(\frac{G^2}{\rho_L}\right) = -\frac{dP}{dz} - \rho_L g \sin\theta - 4\tau_w/D, \tag{5.118}$$

$$\frac{d}{dz}(Gh_L) = 4q''_w/D - gG\sin\theta. \tag{5.119}$$

Let us use P and h_L as the state variables. Steady state implies that $G = \text{const}$. We can manipulate the left side of the Eq. (5.118), using simple thermodynamic relations and the chain rule, as

$$\frac{d}{dz}\left(\frac{G^2}{\rho_L}\right) = -\frac{G^2}{\rho_L^2}\left[\left(\frac{\partial\rho_L}{\partial P}\right)_{h_L}\frac{dP}{dz} + \left(\frac{\partial\rho_L}{\partial h_L}\right)_P\frac{dh_L}{dz}\right]. \tag{5.120}$$

For common liquids at pressures considerably below their critical pressures, $(\partial\rho_L/\partial P)_{h_L} \approx 0$. Furthermore, since $h_L = h_L(P, T)$, we have

$$\left(\frac{\partial\rho_L}{\partial h_L}\right)_P = \left(\frac{\partial\rho_L}{\partial T}\right)_P\frac{\partial T}{\partial h_L} = (-\rho_L\beta)\frac{1}{C_{PL}}, \tag{5.121}$$

where

$$\beta = -\frac{1}{\rho_L}\left(\frac{\partial\rho_L}{\partial T}\right)_P$$

is the volumetric thermal expansion coefficient of the liquid. Substitution in Eq. (5.118) results in

$$\frac{G^2\beta}{\rho_L C_{PL}}\frac{dh_L}{dz} + \frac{dP}{dz} = -\rho_L g\sin\theta - 4\tau_w/D. \tag{5.122}$$

Equations (5.119) and (5.122) are now in the right form, if we note that the left side of Eq. (5.119) can be written as $G dh_L/dz$. In accordance with Eq. (5.117), we thus have

$$y_1 = h_L,$$
$$y_2 = P,$$
$$A_{1,1} = G,$$
$$A_{1,2} = 0,$$
$$A_{2,1} = \frac{G^2 \beta}{\rho_L C_{PL}},$$
$$A_{2,2} = 1,$$
$$C_1 = 4q_w''/D - gG\sin\theta,$$

and

$$C_2 = -\rho_L g \sin\theta - 4\tau_w/D.$$

EXAMPLE 5.2. For the situation of Example 5.2, subcooled boiling takes place when the liquid mean temperature approaches saturation. Nonequilibrium two-phase flow takes place during subcooled boiling, where subcooled liquid and saturated vapor coexist. Assuming that (a) the two-phase flow is one-dimensional and the two phases have equal velocities at any point and (b) the equilibrium and flow qualities are related according to $x = f(x_{eq})$, where $f(x_{eq})$ is a known continuous function of x_{eq}, derive a set of ODEs based on the time- and flow-area-averaged conservation equations, in the form of Eq. (5.117).

SOLUTION. Equations (5.31), (5.40), and (5.49), in steady state, reduce to

$$\frac{d}{dz}(\rho_h j) = 0, \tag{5.123}$$

$$\rho_h j \frac{dj}{dz} = -\frac{dP}{dz} - \rho_h g \sin\theta - \frac{4\tau_w}{D}, \tag{5.124}$$

$$\rho_h j \frac{dh}{dz} = 4q_w''/D + j\frac{dP}{dz} + 4j\tau_w/D, \tag{5.125}$$

where ρ_L and ρ_g represent the local densities of liquid and saturated vapor, respectively, and ρ_h is defined as

$$\rho_h = \left(\frac{1-x}{\rho_L} + \frac{x}{\rho_g}\right)^{-1}, \tag{5.126}$$

$$x = f(x_{eq}), \tag{5.127}$$

$$x_{eq} = (h - h_f)/h_{fg}. \tag{5.128}$$

All properties in these relations are local. It is reasonable to assume, for simplicity, an incompressible liquid phase; therefore $\rho_L = $ const. Let us use j, x_{eq}, and P as state variables. Equation (5.124) can now be manipulated as

$$\frac{d}{dz}(\rho_h j) = \rho_h \frac{dj}{dz} + j\frac{d\rho_h}{dz}. \tag{5.129}$$

The term $d\rho_h/dz$ gives

$$\frac{d\rho_h}{dz} = \frac{d\rho_h}{dp}\frac{dP}{dz} + \frac{\partial \rho_h}{\partial x}\frac{\partial x}{\partial x_{eq}}\frac{dx_{eq}}{dz}. \tag{5.130}$$

For convenience, let us from now on use the partial derivative notation without specifying the variables that are kept constant; for example, let us use $\partial\rho_h/\partial\rho_L$ for $(\partial\rho_h/\partial\rho_L)_{\rho g}$. (These subscripts are actually redundant and are sometimes used in thermodynamics for clarity of discussions.)

From Eq. (5.126), we get

$$\frac{\partial\rho_h}{\partial P} = \frac{\partial\rho_h}{\partial\rho_L}\frac{\partial\rho_L}{\partial P} + \frac{\partial\rho_h}{\partial\rho_g}\frac{d\rho_g}{dP} \approx \frac{\partial\rho_h}{\partial\rho_g}\frac{d\rho_g}{dP} = f(x_{eq})(\rho_h/\rho_g)^2\frac{d\rho_g}{dP}. \tag{5.131}$$

Also, using (5.126) gives

$$\frac{\partial\rho_h}{\partial x} = \rho_h^2\left(\frac{1}{\rho_L} - \frac{1}{\rho_g}\right). \tag{5.132}$$

Substitution in Eq. (5.129) then gives

$$\rho_h\frac{dj}{dz} + j\frac{f(x_{eq})}{\rho_g^2}\rho_h^2\frac{d\rho_g}{dP}\left(\frac{dP}{dz}\right) + j\left(\frac{1}{\rho_L} - \frac{1}{\rho_g}\right)\rho_h^2\frac{df(x_{eq})}{dx_{eq}}\left(\frac{dx_{eq}}{dz}\right) = 0. \tag{5.133}$$

This is the final form of the mass conservation equation, in accordance with Eq. (5.117). The momentum equation (Eq. (5.125)) only needs rearrangement of the terms:

$$\rho_h j\frac{dj}{dz} + \frac{dP}{dz} = -\rho_h g\sin\theta - 4\tau_w/D. \tag{5.134}$$

For the energy equation, we note that $h = h_f + x_{eq}h_{fg}$; therefore,

$$\frac{dh}{dz} = \frac{\partial h}{\partial P}\frac{dP}{dz} + \frac{\partial h}{\partial x_{eq}}\frac{dx_{eq}}{dz} = \left[(1 - x_{eq})\frac{dh_f}{dP} + x_{eq}\frac{dh_g}{dP}\right]\frac{dP}{dz} + h_{fg}\frac{dx_{eq}}{dz}. \tag{5.135}$$

Substitution from Eq. (5.135) in Eq. (5.125) and rearranging gives

$$\rho_h h_{fg} j\frac{dx_{eq}}{dz} + \left\{\rho_h j\left[(1 - x_{eq})\frac{dh_f}{dP} + x_{eq}\frac{dh_g}{dP}\right] - j\right\}\frac{dP}{dz} = 4q_w''/D + 4j\tau_w/D. \tag{5.136}$$

We thus get

$$y_1 = j, \quad y_2 = x_{eq}, \quad y_3 = P,$$

$$A_{1,1} = \rho_h,$$

$$A_{1,2} = j\left(\frac{1}{\rho_L} - \frac{1}{\rho_g}\right)\rho_h^2\frac{df(x_{eq})}{dx_{eq}},$$

$$A_{1,3} = j\left(\frac{d\rho_g}{dP}\right)\rho_h^2 f(x_{eq})/\rho_g^2,$$

$$C_1 = 0,$$

$$A_{2,1} = \rho_h j, \ A_{2,2} = 0, \ A_{2,3} = 1,$$

$$C_2 = -\rho_h g \sin\theta - 4\tau_w/D,$$

$$A_{3,1} = 0, \ A_{3,2} = \rho_h j h_{fg},$$

$$A_{3,3} = \rho_h j \left[(1 - x_{eq}) \frac{dh_f}{dP} + x_{eq} \frac{dh_g}{dP} \right] - j,$$

$$C_3 = 4q_w''/D + 4j\tau_w/D.$$

It should be noted that thermodynamic property derivatives such as $d\rho_g/dP, dh_f/dP$, and dh_g/dP are all in general calculable from the thermodynamic property tables.

Closure relations are needed for τ_w and $f(x_{eq})$. Methods for calculating τ_w are discussed in Chapter 8. The function $f(x_{eq})$ is related to the hydrodynamics and boiling in subcooled boiling, which will be discussed in Chapter 12.

EXAMPLE 5.3. Based on the 2FM equations, derive the set of ODEs in the form of Eq. (5.117) for steady, saturated flow boiling in the tube shown in Fig. 5.14.

SOLUTION. For flow of a saturated liquid–vapor mixture of a pure substance, according to the discussion at the end of Section 5.7, the unknowns can be chosen as P, α (or x), U_f, U_g, and Γ. The five conservation equations that are needed should include two mass equations, two momentum equations, and one energy equation. Let us choose the liquid and mixture mass (Eqs. (5.61) and (5.63)), liquid momentum (Eq. (5.69)), mixture energy (Eq. (5.86), or the equation found by adding Eqs. (5.84) and (5.85)), and the mixture momentum (Eq. (5.71)). Simplified for steady-state flow in a tube, these equations reduce to

$$\frac{d}{dz}[\rho_f U_f(1 - \alpha)] = -\Gamma, \tag{5.137}$$

$$\frac{d}{dz}[\rho_f U_f(1 - \alpha) + \rho_g U_g \alpha] = 0, \tag{5.138}$$

$$\frac{d}{dz}[\rho_f(1 - \alpha)U_f^2] = -(1 - \alpha)\frac{dP}{dz} - \rho_f g(1 - \alpha)\sin\theta + F_I - F_{wL}$$
$$- \Gamma U_I + C_{VM}\left(U_g \frac{dU_g}{dz} - U_f \frac{dU_f}{dz} \right), \tag{5.139}$$

$$\frac{d}{dz}\left[\rho_f U_f(1 - \alpha)\left(h_f + \frac{U_f^2}{2} + gz\sin\theta \right) + \rho_g U_g \alpha \left(h_g + \frac{U_g^2}{g} + gz\sin\theta \right) \right] = 4q_w''/D, \tag{5.140}$$

$$\frac{d}{dz}[\rho_f(1 - \alpha)U_f^2 + \rho_g \alpha U_g^2] = -\frac{dP}{dz} - [\rho_f(1 - \alpha) + \rho_g \alpha]g\sin\theta - F_w. \tag{5.141}$$

We can now eliminate Γ from Eq. (5.139) using Eq. (5.137) to get

$$\frac{d}{dz}[\rho_f(1-\alpha)U_f^2] = -(1-\alpha)\frac{dP}{dz} - \rho_f g(1-\alpha)\sin\theta + F_I - F_{wL}$$

$$+ U_I\frac{d}{dz}[\rho_f U_f(1-\alpha)] + C_{VM}\left(U_g\frac{dU_g}{dz} - U_f\frac{dU_f}{dz}\right). \qquad (5.142)$$

Equations (5.138), (5.141), (5.140), and (5.142) now form a set of four ODEs, with state variables P, U_g, U_f, and α. (Note that this order of equations is consistent with the order of the elements of the forthcoming coefficient matrix.) Equation (5.137) does not need to be included among the ODEs for integration. It can be used instead for calculating Γ. By expanding the derivative terms in these equations and using the chain rule and thermodynamic properties, these ODEs can be cast in the form of Eq. (5.117) with

$$y_1 = P, \quad y_2 = U_g, \quad y_3 = U_f, \quad y_4 = \alpha,$$

$$A_{1,1} = U_f(1-\alpha)\frac{d\rho_f}{dP} + U_f\alpha\frac{d\rho_g}{dP},$$

$$A_{1,2} = \alpha\rho_g,$$

$$A_{1,3} = \rho_f(1-\alpha),$$

$$A_{1,4} = \rho_g U_g - \rho_f U_f,$$

$$C_1 = 0,$$

$$A_{2,1} = (1-\alpha)U_f^2\frac{d\rho_f}{dP} + \alpha U_g^2\frac{d\rho_g}{dP} + 1,$$

$$A_{2,2} = 2\rho_g\alpha U_g,$$

$$A_{2,3} = 2\rho_f(1-\alpha)U_f,$$

$$A_{2,4} = \rho_g U_g^2 - \rho_f U_f^2,$$

$$C_2 = -[\rho_f(1-\alpha) + \rho_g\alpha]g\sin\theta - F_w,$$

$$A_{3,1} = \rho_f U_f(1-\alpha)\frac{dh_f}{dP} + U_f(1-\alpha)e_f\frac{d\rho_f}{dP} + \rho_g U_g\alpha\frac{dh_g}{dP} + U_g\alpha e_g\frac{d\rho_g}{dP},$$

$$A_{3,2} = \rho_g\alpha(e_g + U_g^2)$$

$$A_{3,3} = \rho_f(1-\alpha)(e_f + U_f^2), \quad C_3 = 4q_w''/D - [\rho_g U_g\alpha + \rho_f U_f(1-\alpha)]g\sin\theta,$$

$$A_{3,4} = \rho_g U_g e_g - \rho_f U_f e_f,$$

$$A_{4,1} = (1-\alpha)\left[1 + U_f(U_f - U_I)\frac{d\rho_f}{dP}\right],$$

$$A_{4,2} = -C_{VM}U_g,$$

$$A_{4,3} = [C_{VM}U_f + \rho_f(1-\alpha)(2U_f - U_I)],$$

$$A_{4,4} = -\rho_f U_f(U_f - U_I),$$

$$C_4 = -\rho_f g(1-\alpha)\sin\theta + F_I - F_{wL}.$$

5.10.2 Numerical Solution of the ODEs

Once the model conservation equations are cast in the form of a system of coupled ODEs, their numerical integration becomes straightforward. The derivatives of the state variables are found from

$$\frac{d\mathbf{Y}}{dz} = \mathbf{A}^{-1}\mathbf{C}, \tag{5.143}$$

where $\mathbf{A}^{-1}$ is the inverse of the matrix $\mathbf{A}$. The set of ODEs represented by Eq. (5.117) or (5.143) can in principle be solved by various integration algorithms, for example, the fourth-order Runge–Kutta method, or even the Euler method. However, for boiling and condensing channels the set of ODEs is usually *stiff*, and its numerical solution with commonly used integration methods over a large range of the independent variable (a large range of z for our case) may require an excessive amount of computation time. Detailed discussion of ODEs and their properties can be found in *Numerical Recipes* (Press *et al.*, 1992). Stiffness is encountered in a problem when the scales of the independent variable (z in our case) over which two or more dependent variables vary are significantly different in magnitude. Put differently, a problem is stiff when the physical system it represents has a multitude of degrees of freedom that have significantly different rates of responses. In Example 5.3, for instance, in SI units, over a finite length of the boiling channel, P may change by $\approx 10^5$ Pa or more, whereas α may only vary by ~ 0.1 or less. The fast response of P requires small integration steps, even when solutions for long segments of the channel are of interest. Efficient numerical solution of stiff ODE systems requires an implicit numerical solution technique, small and adjustable integration steps, and high integration orders. Fortunately, efficient and robust algorithms are available for this purpose. LSODE or LSODI (Hindmarsh, 1980; Sohn *et al.*, 1985) integration packages are good examples. These packages use implicit, variable-step, and variable-integration-order algorithms and are easily accessible (see www.netlib.org/). Other easily accessible stiff ODE solvers include MATLAB© ode23s and the *stiff* and *stifbs* algorithms in *Numerical Recipes*.

PROBLEMS

5.1 Gas flows through the column shown in the figure at steady state. There is no liquid through-flow, and the gas bubbles are assumed to move at their terminal velocities. Starting from an appropriate mixture momentum equation, prove that

$$\Delta P = [\rho_L(1 - \alpha) + \rho_G\alpha]gH.$$

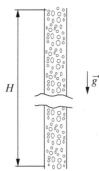

Figure P5.1. Schematic for Problem 5.1.

5.2 Starting from Eq. (5.40), derive Eq. (5.44). What does the condition

$$\left(\frac{dv_g}{dP}\right) \rightarrow -\left(\frac{1}{G^2 x}\right)$$

imply?

5.3 For steady-state annular flow in a vertical tube, prove that

$$\frac{dP}{dz} = -\rho_L g - \frac{4\tau_w}{D(1-\alpha)} \pm \frac{4\sqrt{\alpha}\tau_I}{D(1-\alpha)},$$

where D represents the tube diameter.

5.4 According to the DFM, the slip ratio in a one-dimensional flow can be represented as

$$S_r = C_0 + \frac{x(C_0 - 1)\rho_L}{(1-x)\rho_G} + \frac{\rho_L V_{gj}}{G(1-x)},$$

where C_0 and V_{gj} are empirical parameters.

(a) What implication does this ratio have on the closure issue for the separated-flow conservation equations?

(b) Using this relation, manipulate the one-dimensional separated-flow mixture momentum equation, Eq. (5.71), and cast that equation in terms of the mixture velocity defined as $U_m = G/\bar{\rho}$.

5.5 A steady two-phase mixture consisting of superheated liquid water and saturated steam bubbles flows in an adiabatic channel. The bubbles are uniformly sized, with radius R_B. The evaporation mass flux at the liquid–vapor interphase is m''. Prove that the liquid temperature varies along the channel according to

$$\frac{dT_L}{dz} = \frac{\left\{\frac{j}{D}\frac{dP}{dz} - [h_{fg} + C_{PL}(T_{sat} - T_L)]p_I m''/A\right\}}{C_{PL}(1-x)}.$$

Also, derive expressions for p_I in terms of x and R_B.

5.6 In Problem 5.3, assume that $\tau_I = f_I \frac{1}{2}\rho_G \overline{U}_G^2$ and $\tau_w = f_w \frac{1}{2}\rho_L \overline{U}_L^2$, where $\overline{U}_G$ is the mean gas velocity in the gaseous core and $\overline{U}_L$ is the liquid mean velocity in the tube assuming that the liquid film extends to the tube center.

Assuming that $\overline{U}_L = j_L/(1-\alpha)^{8/7}$, prove that

$$(1-\alpha)^{23/7} - 2f_w \text{Fr}_L^2\left[1 \pm \frac{f_I}{f_w}\frac{(1-\alpha^{16/7})}{\alpha^{5/2}}\frac{\rho_G}{\rho_L}\left(\frac{j_G}{j_L}\right)^2\right] = 0,$$

where

$$\text{Fr}_L = \frac{j_L}{\sqrt{gD(\rho_L - \rho_G)/\rho_L}}.$$

5.7 Repeat Problem 5.6, assuming that the annular film is turbulent and the turbulent velocity profile can be represented as

$$U_L(r) = \frac{60}{49}\left[1 - \frac{2r}{D}\right]^{1/7}.$$

5.8 Consider separated (two-fluid) momentum equations for adiabatic one-dimensional two-phase flow. The interfacial force sometimes is represented in the following form

$$F_I = \alpha(1-\alpha)\rho_L\rho_G(U_G - U_L)F_{FI}.$$

For the ideal annular flow regime, and for ideal bubbly flow regime (where bubbles are uniform-sized) formulate the parameter F_{FI} in terms of parameters such as the friction factor and the drag coefficient. Assume that the void fraction is known.

5.9 Consider a steady flow of pure superheated steam in a uniformaly heated tube. Derive the appropriate ordinary differential equations in the form of Eq. (5.117).

5.10 A steady gas–liquid mixture flows in an adiabatic, horizontal tube with diameter $D = 5$ cm. No phase change takes place in the channel. At the inlet, $P = 1$ MPa, $U_L = 5$ m/s, $S_r = 1.1$, and $\alpha = 0.5$. For simplicity, assume the following.

(a) The interfacial force varies according to

$$F_I = A_0\alpha(U_G - U_L)^2,$$

with

$$A_0 = 7.5 \times 10^8 \text{ kg/m}^4.$$

(b) Both phases are incompressible with $\rho_L = 887$ kg/m^3 and $\rho_G = 5.14$ kg/m^3.
(c) Wall friction follows $\tau_w = f_w\frac{1}{2}\rho_L j^2$, with $f_w = 0.05$.

Write the one-dimensional conservation equations that are sufficient for the hydrodynamic solution of the flow evolution in the pipe. Cast the equations in the form of Eq. (5.117).

5.11 In Problem 5.10, using a numerical integration of your choice, solve the derived ordinary differential equations, and calculate the tube length necessary for the velocity slip to reduce to $S_r = 1.01$.

5.12 Write Eq. (5.59) for steady flow in a channel with uniform cross section. Then show that the equation can be manipulated to

$$\frac{\partial}{\partial z}\left[j\left\{\alpha\rho_G\frac{m_{n,L}}{\frac{P}{C_{He}} + \left(1 - \frac{M_v}{M_n}\right)m_{n,L}} + (1-\alpha)\rho_L m_{n,L}\right\}\right] = 0.$$

5.13 Consider the upward flow of saturated water–steam mixture in a 5 cm – inner diameter pipe. The local pressure is 6 bars, the mixture mass flux is 3100 kg/m^2·s, and local quality is 0.5%.

(a) What is the most likely flow regime?
(b) Assume that the flow regime is bubbly, the bubbles are of uniform size, and the bubble–liquid relative velocity is 0.15 m/s. Find the local void fraction.
(c) Assume that bubbles are 1.5 mm in diameter. Work out the terms representing fluid–wall and interfacial force terms in the separated-flow phase momentum equations (Eqs. (5.69) and (5.70)). Use relevant empirical correlations wherever necessary.

(d) Assume that a heat flux equal to 5000 W/m^2 is imposed on the wall. Work out the terms representing the interfacial heat and mass transfer terms in the separated flow energy equations. For simplicity, neglect the effect of pressure drop on saturation temperature and properties, and assume that evaporation leads only to an increase in the number density of bubbles.

Hint: For part (d) you need to find the evaporation rate in the tube by solving the simplified version of the mixture energy equation.

5.14 Highly subcooled water flows into the small-diameter horizontal tube shown in the Fig. P5.14. The water is saturated with air at the inlet, and the process is in a steady state. The tube receives a uniform heat flux, but water remains subcooled throughout the depicted segment of the tube. As the water flows along the tube, the reduction of pressure and rising liquid temperature cause the release of dissolved air from the water, and a two-phase flow develops. For simplicity it is assumed that (1) the bubbles resulting from the release of air from the water remain saturated with water vapor, and at thermal equilibrium with the surrounding water; (2) the noncondensable–vapor mixture is at equilibrium with the surrounding water with respect to the concentration; (3) the developed two-phase flow is homogeneous; and (4) air acts as an ideal gas. It is also assumed that, because of the small solubility of air in water, the properties of water are not affected by the dissolved air.

(a) Simplify Eqs. (5.49) and (5.59) for application to the described problem.

(b) Prove that the equations derived in part (a), when integrated between the inlet and an arbitrary point 2, lead to

$$\left\{ \frac{G}{\rho_h}[\rho_L(1-\alpha)]h_L + \rho_v\alpha h_v + \frac{G^2}{2\rho_h} \right\}_2 - G\left[h_L + \frac{G^2}{2\rho_L^2} \right]_{in} = \frac{4q_w''}{D}(z_2 - z_{in}),$$

$$\frac{1}{\rho_h}\left[\rho_L(1-\alpha)\frac{M_n}{M_v}\frac{P - P_{sat}}{C_{He}} + \alpha\frac{P - P_{sat}}{\frac{R_u}{M_n}T_L} \right]_2 = m_{n,in},$$

where $m_{n,in}$ is the mass fraction of dissolved air in water at the inlet.

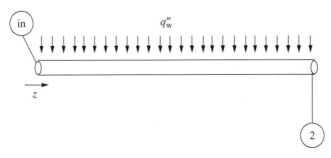

Figure P5.14. Schematic for Problem 5.14.

6 The Drift Flux Model and Void-Quality Relations

6.1 The Concept of Drift Flux

The drift flux model (DFM) is the most widely used diffusion model for gas–liquid two-phase flow. It provides a semi-empirical methodology for modeling the gas–liquid velocity slip in one-dimensional flow, while accounting for the effects of lateral (cross-sectional) nonuniformities. In its most widely used form, the DFM needs two adjustable parameters. These parameters can be found analytically only for some idealized cases and are more often obtained empirically. These empirically adjustable parameters in the model turn out to have approximately constant values or follow simple correlations for large classes of problems, however.

Recall that the diffusion models for two-phase flow need only one set of momentum conservation equations, often representing the mixture. Knowing the velocity for one of the phases (or the mixture), one can use the model's slip velocity relation (or its equivalent) to find the other phasic velocity. When used in the cross-section-average phasic momentum equations, the DFM thus leads to the elimination of one momentum equation. The mixture momentum equation can be recast in terms of mixture center-of-mass velocity. The elimination of one momentum equation leads to a significant saving in computational cost. Also, using the DFM, some major difficulties associated with the 2FM (e.g., the interfacial transport constitutive relations, the difficulty with flow-regime-dependent parameters, and numerical difficulties) can be avoided. These advantages of course come about at the expense of precision and computed process details.

Consider a one-dimensional flow, as shown in Fig. 6.1. Assume all parameters are time-averaged. In terms of local properties, we can write

$$U_G = j + (U_G - j). \tag{6.1}$$

Note that $j = j_L + j_G$, the total volumetric flux, is also the velocity of mixture center of volume. The term $(U_G - j)$ on the right side of Eq. (6.1) is thus the gas velocity with respect to the mixture center of volume.

Let us multiply both sides of Eq. (6.1) by α, the local void fraction, to get

$$j_G = \alpha j + \alpha(U_G - j). \tag{6.2}$$

Now, we can apply flow-area averaging to all the terms in the equation, bearing in mind the definition $\langle \xi \rangle = \frac{1}{A} \int_A \xi \, dA$, to obtain

$$\langle j_G \rangle = \langle \alpha j \rangle + \langle \alpha(U_G - j) \rangle. \tag{6.3}$$

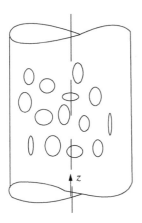

Figure 6.1. Schematic of a one-dimensional flow field.

The second term on the right side of Eq. (6.3) is the gas *drift flux*. Both terms on the right side involve averages of products of two quantities and are difficult to handle. We therefore define two parameters, which need to be obtained empirically:

$$C_0 = \frac{\langle \alpha j \rangle}{\langle \alpha \rangle \langle j \rangle},$$ (6.4)

$$V_{gj} = \frac{\langle \alpha (U_G - j) \rangle}{\langle \alpha \rangle}.$$ (6.5)

The parameter C_0 is called the *two-phase distribution coefficient*, or *concentration parameter*. It is a measure of global or overall interphase slip resulting from flow-area-averaging. The parameter V_{gj} is the *gas drift velocity*, and it represents the local slip. Equation (6.3) can now be recast as

$$\langle J_G \rangle = C_0 \langle \alpha \rangle \langle j \rangle + \langle \alpha \rangle V_{gj}$$ (6.6)

with

$$\langle \alpha \rangle = \frac{\langle j_G \rangle}{C_0 \langle j \rangle + V_{gj}}.$$ (6.7)

When C_0 and V_{gj} are empirically known, Eq. (6.7) can be used for calculating $\langle \alpha \rangle$ by noting that

$$\langle j_G \rangle = G \langle x \rangle / \rho_G$$ (6.8)

and

$$\langle j_L \rangle = G (1 - \langle x \rangle) / \rho_L.$$ (6.9)

Substitution of Eqs. (6.8) and (6.9) into Eq. (6.7) then gives

$$\langle \alpha \rangle = \frac{\langle x \rangle}{C_0 \left[\langle x \rangle + \frac{\rho_G}{\rho_L} (1 - \langle x \rangle) \right] + \frac{\rho_G V_{gj}}{G}}.$$ (6.10)

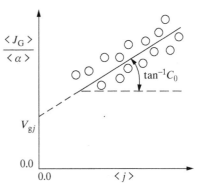

Figure 6.2. Estimation of DFM parameters from experimental data.

It can also be easily shown that the *slip ratio* S_r and the *slip velocity* U_r are related to C_0 and V_{gj} according to

$$S_r = \frac{\langle U_G \rangle_G}{\langle U_L \rangle_L} = C_0 + \frac{\langle x \rangle (C_0 - 1) \rho_L}{\rho_G (1 - \langle x \rangle)} + \frac{\rho_L V_{gj}}{G(1 - \langle x \rangle)}, \tag{6.11}$$

$$U_r = \langle U_G \rangle_G - \langle U_L \rangle_L = \frac{V_{gj} + (C_0 - 1) \langle j \rangle}{(1 - \langle \alpha \rangle)}. \tag{6.12}$$

Some other useful relations are

$$\langle j \rangle = \frac{G}{\langle \bar{\rho} \rangle} + \frac{\langle \alpha \rangle (\rho_L - \rho_G)}{\langle \bar{\rho} \rangle} V'_{gj}, \tag{6.13}$$

$$\langle U_L \rangle_L = \frac{G}{\langle \bar{\rho} \rangle} - \frac{\langle \alpha \rangle}{1 - \langle \alpha \rangle} \frac{\rho_G}{\langle \bar{\rho} \rangle} V'_{gj}, \tag{6.14}$$

$$\langle U_G \rangle_G = \frac{G}{\langle \bar{\rho} \rangle} + \frac{\rho_L}{\langle \bar{\rho} \rangle} V'_{gj}, \tag{6.15}$$

where $\langle \bar{\rho} \rangle = \rho_L (1 - \bar{\alpha}) + \rho_G \langle \bar{\alpha} \rangle$ is the mixture density, $G/\langle \bar{\rho} \rangle$ is the mixture center-of-mass velocity, and the *mean transport drift velocity* is defined as

$$V'_{gj} = V_{gj} + (C_0 - 1) \langle j \rangle. \tag{6.16}$$

An easy way to check the suitability of the DFM for a system, and thereby experimentally calculate C_0 and V_{gj}, is as follows. Equation (6.6) can be cast as

$$\frac{\langle j_G \rangle}{\langle \alpha \rangle} = C_0 \langle j \rangle + V_{gj}. \tag{6.17}$$

Using experimental data, one can then plot $\langle j_G \rangle / \langle \alpha \rangle$ versus $\langle j \rangle$. A curve fit to the data points can then be performed, as shown in Fig. 6.2. The ordinate intercept of the curve will provide V_{gj}, and C_0 will be the slope of the curve. A linear curve would imply a constant C_0 and V_{gj}.

The DFM parameters have been extensively studied. Experimental data representing a wide variety of flow situations approximately follow a linear profile and confirm the usefulness of the model. Given that C_0 and V_{gj} are both functions of the void fraction distribution, furthermore (see Eqs. (6.4) and (6.5)), one could expect a separate (C_0, V_{gj}) pair for each flow regime.

Notwithstanding its simplicity and convenience, the DFM has important limitations and is inadequate for many applications. The most important limitations of the DFM are as follows.

(1) The DFM is best applicable to one-dimensional flows. A one-dimensional flow can be inside a channel, or it can be in a vertical column or even inside the rod bundles in a nuclear reactor core.
(2) The DFM is not recommended for flow patterns where large slip velocities occur. It is thus best applicable to bubbly, slug, and churn flow regimes.

6.2 Two-Phase Flow Model Equations Based on the DFM

The separated-flow model equations described in Section 5.7 can be simplified by substituting for $\langle U_L \rangle_L$ and $\langle U_G \rangle_G$ using Eqs. (6.14) and (6.15) (see Problem 6.2). Given that $\langle U_L \rangle_L$ and $\langle U_G \rangle_G$ are not independent in the DFM, only one momentum equation (usually the mixture momentum equation) will be needed, and therefore a phasic momentum equation can be eliminated. The resulting mixture momentum equation can be presented as (Lahey and Moody, 1993)

$$\frac{\partial G}{\partial t} + \frac{1}{A}\frac{\partial}{\partial z}\left(A\frac{G^2}{\langle \bar{\rho} \rangle}\right) = -\frac{\partial}{\partial z} - \frac{1}{A}\frac{\partial}{\partial z}\left[A\frac{(\rho_L - \langle \bar{\rho} \rangle)}{(\langle \bar{\rho} \rangle - \rho_G)}\frac{\rho_L \rho_G}{\langle \bar{\rho} \rangle}V_{jg}^{\prime 2}\right] - \frac{\tau_w p_f}{A} - \langle \bar{\rho} \rangle g \sin\theta. \tag{6.18}$$

(Note that flow-area-averaging notations similar to those described in Chapter 3 have been used in this equation for consistency with the remainder of this chapter. In Chapter 5, where one-dimensional flow-area-averaged conservation equations were discussed, these notations were left out for convenience.) Also, although thermal nonequilibrium between the two phases can be accounted for in the DFM, often the two phases are assumed to be in thermal equilibrium. The mixture energy equation (e.g., Eq. (5.86)) will then be applicable provided that Eqs. (6.10), (6.14), and (6.15) are used in order to eliminate $\langle \alpha \rangle$, $\langle U_L \rangle_L$, and $\langle U_G \rangle_G$ from Eq. (5.86).

Alternatively, the separated flow model (or 2FM) equations can be solved by using only one momentum equation (preferably the mixture momentum equation, e.g., Eq. (5.71)) and treating $\langle \alpha \rangle$ not as an unknown state variable but as a parameter provided by the "closure relation" of Eq. (6.10).

As mentioned earlier, the DFM-based model equations have the following advantages over the 2FM.

(a) They are simpler and have better numerical robustness.
(b) They involve considerably fewer computations.
(c) They avoid the numerous, often inaccurate interfacial transfer models and correlations.

The last item of course may only mask the real problem.

The main disadvantages of the DFM-based method are the following.

(a) The DFM is inadequate for flow fields that are not one-dimensional or for those that involve significant interfacial slip.
(b) There is a loss of information about the flow field details that the 2FM can provide.

Empirical DFM parameters are now presented for pipe flow in the next section and for rod bundles in Section 6.4.

6.3 DFM Parameters for Pipe Flow

Many investigators have proposed DFM parameters for flow in channels, for adiabatic, boiling, and condensing flow conditions. Table 6.1 summarizes some of the more popular models. Most of the older DFM models were concerned with air–water and steam–water mixtures. Recently, however, refrigerant liquid–vapor mixtures have been the subject of interest. In the forthcoming paragraphs some of the most widely applied DFM parameters will be discussed.

For *co-current slug flow in vertical pipes*, Nicklin *et al.* (1962) noted that the average gas velocity could be correlated according to

$$U_G = C(j_G + j_L) + U_{B,\infty}, \tag{6.19}$$

where $U_{B,\infty}$ is the rise velocity of a Taylor bubble in a quiescent liquid and $C \approx 1.2$ is approximately equal to the ratio between maximum and mean velocity of the liquid phase in fully developed turbulent pipe flow. This is evidently equivalent to $C_0 = 1.2$ and $V_{gj} = U_{B,\infty}$. When $\rho_G/\rho_L \ll 1$, the latter parameter can be predicted from (Dumitrescu, 1943; Davis and Taylor, 1950)

$$V_{gj} = U_{B,\infty} = C_1\sqrt{gD}, \tag{6.20}$$

$$C_1 = 0.35. \tag{6.21}$$

These expressions were found to do very well for *countercurrent slug flow in vertical pipes* as well (Ghiaasiaan *et al.*, 1997a, b; Welsh *et al.*, 1999). Sadatomi *et al.* (1982) have examined the applicability of these relations to noncircular channels (rectangular, triangular, and annular, with centimeter-range hydraulic diameters). They noted that Eqs. (6.20) and (6.21) were valid, with $C_0 \approx 1.20$–1.24 for their rectangular channels, $C_0 \approx 1.30$ for their annular test section, and $C_0 \approx 1.34$ for their triangular test section. The measured values of C_0 corresponded approximately to the ratio between maximum and mean velocity of the liquid phase in fully developed turbulent flow in each channel.

For vertical, upward two-phase flow in pipes with $D = 25$–50 mm diameters, Ishii (1977) proposed the following correlations, which have been widely used for boiling channels. The distribution coefficient is to be found from

$$C_0 = \left[1.2 - 0.2\sqrt{\rho_G/\rho_L}\right][1 - \exp(-18\langle\alpha\rangle)]. \tag{6.22}$$

The gas drift velocity depends on the flow regime. For bubbly flow,

$$V_{gj} = \sqrt{2}\left(\frac{\sigma g \Delta\rho}{\rho_L^2}\right)^{1/4}(1 - \langle\alpha\rangle)^{1.75}. \tag{6.23}$$

For slug flow,

$$V_{gj} = 0.35\sqrt{\frac{gD\Delta\rho}{\rho_L}}. \tag{6.24}$$

Table 6.1 *Summary of some drift flux models.*

Author	Distribution coefficient	Drift velocity	Comments
Wallis (1969)	$C_0 = 1.0$	$V_{gj} = 1.53\left[\frac{\sigma g(\rho_L - \rho_G)}{\rho_L^2}\right]^{1/4}$	Isolated bubbles without coalescence
Zuber and Findlay (1965)	$C_0 = 1.2$	$V_{gj} = 1.53\left[\frac{\sigma g(\rho_L - \rho_G)}{\rho_L^2}\right]^{1/4}$	Churn-turbulent flow regime in a vertical tube
Dix (1971)	Eqs. (6.27) and (6.28)	$V_{gj} = 1.18\,(1 - \langle x \rangle)\left[\frac{\sigma g \Delta\rho}{\rho_f^2}\right]^{1/4}$	Low-flow boiling in vertical rod bundles
Bonnecaze et al. (1971)	$C_0 = 1.2$	$V_{gj} = 0.35\left[\frac{gD(\rho_L - \rho_G)}{\rho_L}\right]^{1/2}$	Slug flow regime in a vertical tube
Rouhani and Axelsson (1970)	$C_0 = 1 + 0.12\,(1 - \langle x \rangle)$ (version I) $C_0 = 1 + 0.2\,(1 - \langle x \rangle)\,(gD)^{1/4}\left(\frac{\rho_L}{G}\right)^{1/2}$ (version II)	$V_{gj} = 1.18\left[\frac{\sigma g(\rho_L - \rho_G)}{\rho_L^2}\right]^{0.25}$ (vertical)	Subcooled and saturated boiling in tubes; valid for $\langle \alpha \rangle > 0.1$
Ishii (1977)	Eq. (6.22)	Eqs. (6.23)–(6.26)	Boiling in vertical tubes
Sun et al. (1980)	$C_0 = [0.82 + 0.18\,(P/P_{cr})]^{-1}$	$V_{gj} = 1.41\left[\frac{\sigma g(\rho_L - \rho_G)}{\rho_L^2}\right]^{0.25}$	Low-flow boiling of water in rod bundles
Shipley (1982)	$C_0 = 1.2$ (V_{gj} in m/s)	$V_{gj} = 0.24 + 0.35\beta^2\sqrt{gD}\,\langle\alpha\rangle$	Two-phase flow in large diameter tubes
Pearson et al. (1984)	$C_0 = 1 + 0.796\exp\left(-0.061\sqrt{\rho_L/\rho_G}\right)$	$V_{gj} = 0.034\left[\sqrt{\rho_L/\rho_G} - 1\right]$	Level swell

Reference	C_0	V_{gj}	Applicability
Kataoka and Ishii (1987); Hibiki and Ishii (2003a)	$C_0 = 1 - 0.2\sqrt{\rho_G/\rho_L}$	Low viscosity: $N_{\mu_L} \le 2.25 \times 10^{-3}$ $V_{gj}^* = 0.0019 D_H^{*0.809}(\rho_G/\rho_L)^{-0.157} N_{\mu_L}^{-0.562}$ for $D_H^* \le 30$ $V_{gj}^* = 0.030(\rho_G/\rho_L)^{-0.157} N_{\mu_L}^{-0.562}$ for $D_H^* > 30$ High viscosity: $N_{\mu_L} > 2.25 \times 10^{-3}$ $V_{gj}^* = 0.92(\rho_G/\rho_L)^{-0.157}$ for $D_H^* > 30$ $D_H^* = D_H/\left\{\sigma/[g(\rho_L - \rho_G)]\right\}^{1/2}$; $V_{gj}^* = V_{gj}/\left\{\sigma g(\rho_L - \rho_G)/\rho_L^2\right\}^{1/4}$ $N_{\mu_L} = \mu_L/\left\{\rho_L\sigma\sqrt{\sigma/[g(\rho_L - \rho_G)]}\right\}^{1/2}$	Large-diameter pipes and bubbling or boiling pools
Steiner (1993)	$C_0 = 1 + 0.12(1 - \langle x \rangle)$	$V_{gj} = 1.18(1 - \langle x \rangle)\left[\frac{\sigma g(\rho_L - \rho_G)}{\rho_L^2}\right]^{0.25}$	Horizontal tube
Gomez (2000)	$C_0 = 1.15$	$V_{gj} = 1.53\left[\frac{\sigma g(\rho_L - \rho_G)}{\rho_L^2}\right]^{0.25}\sqrt{1 - \langle \alpha \rangle}\,\sin\theta$	Vertical and inclined tubes
Kataoka and Ishii (1987); Hibiki and Ishii (2003b)	$C_0 = \left[1 - 0.2\sqrt{\rho_G/\rho_L}\right]$ $\times [1 - \exp(-18\langle\alpha\rangle)]$ (bubbly) $C_0 = 1 - 0.2\sqrt{\rho_G/\rho_L}$ (slug and churn) $C_0 = 1 + \dfrac{1 - \langle\alpha\rangle}{\langle\alpha\rangle + \left[\frac{1+75(1-\langle\alpha\rangle)}{\sqrt{\langle\alpha\rangle}}\frac{\rho_G}{\rho_L}\right]^{\frac12}}$ $\times\left[1 + \dfrac{\sqrt{\frac{gD(\rho_L-\rho_G)(1-\langle\alpha\rangle)}{0.015\rho_L}}}{j}\right]$ (annular)	$V_{gj} = 1.41[\sigma g(\rho_L - \rho_G)/\rho_L^2]^{1/4}$ $\times (1 - \langle\alpha\rangle)^{1.75}$ (bubbly) $V_{gj} = 0.35[gD(\rho_L - \rho_G)/\rho_L]^{1/2}$ (slug) $V_{gj} = 1.41[\sigma g(\rho_L - \rho_G)/\rho_L^2]^{1/4}$ (churn) $V_{gj} = \dfrac{1 - \langle\alpha\rangle}{(\langle\alpha\rangle) + \left[\frac{1+75(1-\langle\alpha\rangle)}{\sqrt{\langle\alpha\rangle}}\frac{\rho_G}{\rho_L}\right]^{\frac12}}$ $\times\left[j + \sqrt{\frac{gD(\rho_L-\rho_G)(1-\langle\alpha\rangle)}{0.015\rho_L}}\right]$ (annular)	Large-diameter vertical pipes and pools
Woldesemayat and Ghajar (2007)	Eqs. (6.27) and (6.28)	Eq. (6.29)	See discussion preceding Eq. (6.27)

(cont.)

Table 6.1 (cont.)

Author	Distribution coefficient	Drift velocity	Comments
Choi et al. (2012)	$C_0 = \dfrac{2}{1+(\mathrm{Re}/1,000)^2} + \dfrac{1.2-0.2\sqrt{\rho_G/\rho_L}[1-\exp(-18\langle\alpha\rangle)]}{1+(1,000/\mathrm{Re})^2}$ $\mathrm{Re} = \rho_L j D/\mu_L$	$V_{gj} = 0.0246\cos\theta$ $\quad + 1.606\big[\sigma g\,(\rho_L - \rho_G)/\rho_L^2\big]^{1/4}\sin\theta$	0.05–0.15 m pipe diameters, $-10° < \theta < 10°$
Bhagwat and Ghajar (2014)	Eqs. (6.30)–(6.32)	Eqs. (6.36)–(6.39)	Extensive data base, $-90° \le \theta \le 90°$, various fluids
Takeuchi et al. (1992)	Eqs. (6.42)	Eqs. (6.43)	Based on data representing PWR rod bundles
Bestion (1990)	$C_0 = 1.0$	$V_{gj} = 0.188[g\,(\rho_L - \rho_G)\,D_H/\rho_G]^{1/2}$	Bubbly, slug and churn-turbulent regimes in PWR rod bundles and secondary sides of steam generators during boil off
Svetlov et al. (1999)	$C_0 = \max\left[\dfrac{0.675(\rho_L/\rho_G)^{0.1}}{1-0.6\exp(-18\langle\alpha\rangle)},\ 1.0\right]$	$V_{gj}/\{\sigma g\,(\rho_L - \rho_G)/\rho_L^2\}^{1/4}$ $= 2.19(\rho_L - \rho_G)^{0.25}\,Bd^{-0.25}$ $\times\,[1 - 0.6\exp(-18\langle\alpha\rangle)]$	Rod bundles, $\langle\alpha\rangle \le 0.75$
Julia et al. (2009)	$C_0 = \begin{cases} (1.03 - 0.03\sqrt{\rho_G/\rho_L}) \\ \quad \times[1 - \exp(-26.3\langle\alpha\rangle^{0.78})] \\ \quad \times D/P_0 = 0.3 \\ (1.04 - 0.04\sqrt{\rho_G/\rho_L}) \\ \quad \times[1 - \exp(-21.2\langle\alpha\rangle^{0.762})] \\ \quad \times D/P_0 = 0.5 \\ (1.05 - 0.05\sqrt{\rho_G/\rho_L}) \\ \quad \times[1 - \exp(-34.1\langle\alpha\rangle^{0.925})] \\ \quad \times D/P_0 = 0.7 \end{cases}$ $P_0 = $ pitch $D = $ rod diameter	$V_{gj}/\{\sqrt{\sigma g\,(\rho_L - \rho_G)/\rho_L^2}\}^{1/4}$ $= \dfrac{\sqrt{2}}{\sqrt{2}}(1 - \langle\alpha\rangle)^{1.75}\,B_{sf}$ $B_{sf} = \begin{cases} \dfrac{1 - d_B/(0.9L_{max})}{1 - d_B/L_{max} < 0.6} \\ \text{for } d_B/L_{max} < 0.6 \\ 0.12(d_B/L_{max})^{-2} \\ \text{for } d_B/L_{max} \ge 0.6 \end{cases}$ $L_{max} = \sqrt{2}.P_0 - D$	Bubbly flow in rod bundles, $\langle\alpha\rangle \le 0.20$

| Chen et al. (2012) | $C_0 = 4.79 j_G^* + 1$ for $j_G^* \leq 0.5 \cdot \sqrt{\rho_G/\rho_L}$
 $C_0 = C_\infty - (C_\infty - 1)\sqrt{\rho_G/\rho_L}$
 for $j_G^* > 0.5$
 $C_\infty = 3.45 \exp(-0.52 j_G^{*0.51}) + 1$
 $j_G^* = \langle j_G \rangle / \{\sigma g (\rho_L - \rho_G)/\rho_L^2\}^{1/4}$ | $V_{gj}^* = V_{gj,B}^* \cdot \exp(-1.39 j_G^*) + V_{gj,C}^* [1 - \exp(-1.39 j_G^*)]$
 $V_{gj,B}^* = \sqrt{2}(1 - \langle\alpha\rangle)^{1.75}$
 For $N_{\mu L} \leq 2.25 \times 10^{-3}$:
 $V_{gj,C}^* = 0019 D_C^{*0.809}(\rho_G/\rho_L)^{-0.157} N_{\mu L}^{-0.562}$ for $D_C^* \leq 30$
 $V_{gj,C}^* = 0.030(\rho_G/\rho_L)^{-0.157} N_{\mu L}^{-0.562}$ for $D_C^* > 30$
 For $N_{\mu L} > 2.25 \times 10^{-3}$:
 $V_{gj,C}^* = 0.92(\rho_G/\rho_L)^{-0.157}$ for $D_C^* \geq 30$

 $V_{gj}^* = V_{gj}/\{\sigma g (\rho_L - \rho_G)/\rho_L^2\}^{1/4}$
 $D_C^* = D_C/\sqrt{\sigma/[g(\rho_L - \rho_G)]}$
 D_C = Casing diameter or width of the rectangular casing | Adiabatic (air-water) and steam-water boiling in rod bundles; experimental rod bundles with 8×8, 4×4, 2×2, 6×22 rods. |

For churn flow,

$$V_{gj} = \sqrt{2}\left(\frac{\sigma g \Delta\rho}{\rho_L^2}\right)^{1/4}. \tag{6.25}$$

For annular flow,

$$V_{gj} = -(C_0 - 1)\langle j \rangle + \frac{1 - \langle\alpha\rangle}{\langle\alpha\rangle + \left[\frac{1+75(1-\langle\alpha\rangle)}{\sqrt{\langle\alpha\rangle}}\frac{\rho_G}{\rho_L}\right]^{1/2}} \cdot \left[\langle j \rangle + \sqrt{\frac{g\Delta\rho D(1-\langle\alpha\rangle)}{0.015\rho_L}}\right]. \tag{6.26}$$

Hibiki and Ishii (2003a, b; 2005) have modified these expressions for DFM parameters in vertical channels, introducing corrections that account for the effects of wall friction, interfacial geometry, and body force, on the velocity slip between the two phases.

Hibiki and Ishii (2003a, b) have also proposed DFM parameters for large-diameter flow passages. The flow phenomena in large-diameter flow passages differ from those in smaller channels in several respects. The height-to-diameter ratio in large-diameter flow passages is seldom large enough to justify the developed-flow assumption, and consequently strong end (entrance) effects are often present. The two-phase flow regimes in large-diameter flow passages are also different than in small channels. For example, slug flow may not be sustainable when $D/\sqrt{\sigma/g\Delta\rho} \geq 40$, in which case the Taylor bubbles that are common in small-diameter tubes are replaced with large bubble caps. Cocurrent annular-dispersed flow is also unlikely to happen in large-diameter flow passages since it requires exceedingly high gas flow rates. Other differences include the occurrence of multi-dimensional flow effects in large channels and recirculation patterns with downward liquid flow near the walls. For vertical, co-current upward flow in large-diameter pipes, Hibiki and Ishii (2003a, b) have proposed correlations for DFM parameters, based on air–water, N$_2$–water, and steam–water data covering the following parameter range: $10.2 \leq D \leq 48$ cm, $4.2 \leq z/D \leq 108$, and $0.1 \leq P \leq 15$ bar.

Woldesemayat and Ghajar (2007) performed a detailed review of the existing void fraction data covering two-phase flow in vertical-upward, horizontal, and inclined tubes and examined the accuracy of 68 correlations. The experimental data covered the following range: $12.7 \leq D \leq 102.26$ mm and $0.0° \leq \theta \leq 90°$, where θ represents the angle of inclination with respect to the horizontal plane. The fluids included air–water, water–natural gas, and air–kerosene. Overall, the DFM correlation of Dix (1971), to be discussed later, performed relatively well. Based on their entire data base, they introduced the following modification into the latter correlation. Accordingly, C_0 is to be found from (Dix, 1971)

$$C_0 = \frac{\langle j_G \rangle}{\langle j \rangle}\left[1 + \left(\frac{\langle j \rangle}{\langle j_G \rangle} - 1\right)^b\right], \tag{6.27}$$

with

$$b = (\rho_G/\rho_L)^{0.1}. \tag{6.28}$$

The drift velocity is found from

$$V_{gj} = 2.9 \left[\frac{gD\sigma(1+\cos\theta)(\rho_L - \rho_G)}{\rho_L^2} \right]^{0.25} (1.22 + 1.22\sin\theta)^{1/a}, \qquad (6.29)$$

where $a = (P/P_{atm})$ is the system nondimensionalized pressure and P_{atm} is the standard atmospheric pressure. The coefficient 2.9 in this equation is in $m^{-0.25}$ units in the SI unit system.

Chexal *et al.* (1991, 1997) have attempted to develop a series of correlations that together have a very wide range of applicability. Their main incentive was to eliminate the trouble and uncertainties associated with the two-phase flow regime map, and there is no mention of flow regimes in their correlations. The correlations address horizontal and vertical, and co-current as well as countercurrent flows. Two-phase flow in tubes or rod bundles and various property effects are all considered. Although the correlations are long and somewhat tedious, they appear to be remarkably accurate.

Bhagwat and Ghajar (2014) recently performed a comprehensive review of the available data for two-phase flow in flow channels with circular and rectangular cross-sections with hydraulic diameters in the 0.5–305 mm range, and developed the DFM correlations below. The angles of inclination with respect to the horizontal plane covered the entire range ($-90° \leq \theta \leq 90°$). Their data base covered the following.

- Gas–liquid mixtures: air–water, argon–water, natural gas–water, air–kerosene, air–glycerin, argon–acetone, argon–ethanol, argon–alcohol, steam–water, air–oil, and liquid–vapor mixtures of various refrigerants (R-11, R-12, R-22, R-134a, R-114, R-410A, R-290, R-1234yf).
- Liquid viscosity: 1.0×10^{-4} kg/m·s $\leq \mu_L \leq 0.6$ kg/m·s.
- System pressure: 0.1 MPa $\leq P \leq 18.1$ MPa.
- Two-phase Reynolds number (defined here as $Re_{TP} = \rho_L (j_L + j_G) D_H / \mu_L$): $10 \leq Re_{TP} \leq 5 \times 10^6$.

For the two-phase distribution coefficient Bhagwat and Ghajar (2014) derived,

$$C_0 = \frac{2 - (\rho_G/\rho_L)^2}{1 + (Re_{TP}/1000)^2} + \frac{\left\{ \left[\sqrt{[1 + (\rho_G/\rho_L)^2 \cos\theta]/(1+\cos\theta)} \right]^{(1-\langle\alpha\rangle)} \right\}^{0.4} + C_{0,1}}{1 + (1000/Re_{TP})^2}$$

$$(6.30)$$

where

$$C_{0,1} = \begin{cases} 0 & \text{for } -50° \leq \theta \leq 0 \text{ and } Fr \leq 0.1 \\ C_1 \left(1 - \sqrt{\rho_G/\rho_L}\right) \left[(2.6 - \beta)^{0.15} - \sqrt{f_{TP}} \right] (1 - \langle x \rangle)^{1.5} & \text{otherwise} \end{cases} \qquad (6.31)$$

$$C_1 = \begin{cases} 0.2 & \text{for circular and annular cross section} \\ 0.4 & \text{for rectangular cross section.} \end{cases} \qquad (6.32)$$

The Froude number and the two-phase Reynolds number are defined as

$$\mathrm{Fr} = \sqrt{\frac{\rho_G}{\rho_L - \rho_G}} \frac{\langle j_G \rangle}{\sqrt{gD\cos\theta}} \tag{6.33}$$

$$\mathrm{Re_{TP}} = \frac{\rho_L D_H \langle j \rangle}{\mu_L}. \tag{6.34}$$

The two-phase Fanning friction factor f_{TP} is to be found using the following correlation of Colebrook, where ε_D is the flow channel surface roughness

$$\frac{1}{\sqrt{f_{TP}}} = -4.0\log_{10}\left(\frac{\varepsilon_D/D_H}{3.7} + \frac{1.256}{\mathrm{Re_{TP}}\sqrt{f_{TP}}}\right). \tag{6.35}$$

For the drift velocity, Bhagwat and Ghajar (2014) derived,

$$V_{gj} = C_2 C_3 C_4 \left(0.35\sin\theta + 0.45\cos\theta\right)\sqrt{\frac{gD_H\Delta\rho}{\rho_L}}(1 - \langle\alpha\rangle)^{0.5} \tag{6.36}$$

where

$$C_2 = \begin{cases} \left[\dfrac{0.434}{\log_{10}(\mu_L/\mu_{\mathrm{ref}})}\right]^{0.15} & \text{for } (\mu_L/\mu_{\mathrm{ref}}) > 10 \\ 1 & \text{for } (\mu_L/\mu_{\mathrm{ref}}) \le 10 \end{cases} \tag{6.37}$$

$$C_3 = \begin{cases} \left[\dfrac{D_H^*}{0.025}\right]^{0.9} & \text{for } D_H^* < 0.025 \\ 1 & \text{for } D_H^* \ge 0.025 \end{cases} \tag{6.38}$$

$$C_4 = \begin{cases} -1 & \text{for } -50° \le \theta \le 0° \text{ and } \mathrm{Fr} \le 0.1 \\ +1 & \text{otherwise} \end{cases} \tag{6.39}$$

$$D_H^* = \sqrt{\frac{\sigma}{g(\rho_L - \rho_G)}}\Big/ D_H \tag{6.40}$$

$$\mu_{\mathrm{ref}} = 0.001\,\mathrm{kg/m\cdot s} \tag{6.41}$$

A correlation for void fraction in countercurrent flow in vertical channels has also been proposed by Yamaguchi and Yamazaki (1982).

6.4 DFM Parameters for Rod Bundles

Most of the current power-generating water-cooled nuclear reactors utilize vertical rod bundles in their cores. (Some CANDU reactors use horizontal rod bundles.) Liquid–vapor two-phase flow occurs during the normal operations in the core of BWRs and during accidents in other reactor types. Although the current state-of-the-art reactor thermal-hydraulics codes mostly use two-fluid modeling, the DFM is also attractive for slow processes where long real-time simulations are needed. The core uncovery/boiloff transient is among the processes most convenient for the application of the DFM. This transient follows a small-break loss of coolant accident (SB-LOCA) in PWRs. The primary coolant pumps stop, and the slow depletion of primary coolant leads to the formation of a swollen two-phase pool in the reactor core. Although the nuclear chain reaction is terminated early in the transient, heat generation in fuel rods continues from radioactive decay. The swell liquid pool undergoes boiloff, and the swell level in the reactor gradually recedes, leading to the uncovery of the fuel rods. Extensive experimental water–steam data are available for boiloff/uncovery processes in heated rod bundles, and several DFM correlations are available for rod bundles. A few, widely used correlations are reviewed in the following. A useful and concise review of the most accurate available correlations published before around the year 2000 for uncovery/boiloff conditions can be found in Coddington and Macian (2002). Reviews of more recent DFM parameters for rod bundles can be found in Ishii and Hibiki (2011) and Chen *et al.* (2012). Some of the recent DFM models for rod bundles are included in Table 6.1, and a few are reviewed below.

Two-phase flow regimes that occur in rod bundles were discussed in Section 4.4. As noted there, rod bundles are of two types: open lattice and closed lattice. In open lattice rod bundles each rod bundle is held together by spacer grids but otherwise does not have a physical boundary. Such rod bundles, when arranged next to each other in a nuclear reactor core, create a large array of parallel rods (typically tens of thousands in large pressurized water reactors) that are open to cross-flow everywhere. Each rod bundle is typically made of 15×15 or 17×17 rods. Closed lattice bundles are used in boiling water reactors. These rod bundles, which are typically 7×7 or 8×8, are separated from their surroundings by a shroud. As a result each rod bundle represents a separate flow passage, and the rod bundles in the core of a boiling water reactor represent a large number of parallel flow passages that are connected to common inlet (bottom) and outlet (top) plenums. In a rod bundle there are thus two distinct length scales: the length scale associated with the flow in an interstitial (subchannel) passage, and a length scale representing the rod bundle envelope (casing) (Chen *et al.*, 2012). The flow regimes in adiabatic two-phase flow in a rod bundle typically include dispersed bubbly, bubbly, cap bubbly, cap-turbulent flow, and churn-turbulent flow (Paranjape *et al.*, 2011). Dispersed bubbly and bubbly flow regimes are primarily controlled by the length scale associated with the subchannel. The flow in this case is essentially one-dimensional, bubbles are small enough to remain in the subchannels and they primarily move inside subchannels, even though cross-flow does occur among neighboring subchannels and occasionally bubbles even move from one subchannel to another. Cap bubbles, on the other hand, can occupy two or more subchannels, their maximum size is limited by the length scale associated with the casing of the rod bundle. The two-phase flow phenomena are thus more complicated than pipe flow.

Models originally developed based on two-phase flow and boiling in tubes are sometimes used for modeling flow and boiling in rod bundles as well. The aforementioned DFM expressions by Ishii (1977), for example, have been applied rather extensively to rod bundles.

Models originally developed and tested based on internal channel flow data are often used for rod bundles. The following correlation of Takaeuchi *et al.* (1992), based on tube data, appears to underpredict the void fraction for $P > 10$ MPa and for data where $\langle \alpha \rangle < 0.35$:

$$C_0 = 1.11775 + 0.45881 \langle \alpha \rangle - 0.57656 \langle \alpha \rangle^2, \tag{6.42}$$

$$V_{gj} = \sqrt{(\mathrm{Ku})^2/D^*} \frac{C_0 (1 - C_0 \langle \alpha \rangle)}{m^2 + C_0 \langle \alpha \rangle (\sqrt{\rho_G/\rho_L} - m^2)} \sqrt{g D_H \Delta \rho / \rho_L}, \tag{6.43}$$

where

$$m = 1.367, \tag{6.44}$$

$$D^* = D_{\mathrm{H}}\sqrt{\frac{g\Delta\rho}{\sigma}}. \tag{6.45}$$

The Kutateladze number Ku is found from

$$\mathrm{Ku} = \left[D^* \cdot \min\left(\frac{1}{2.4}, \frac{10.24}{D^*} \right) \right]. \tag{6.46}$$

The correlation of Dix (1971), based on rod bundle water–steam data, appears to underpredict void fractions for $P > 10$ MPa, for low pressures ($P < 1$ MPa), and for low mass fluxes ($G < 10^2$ kg/m^2·s). In this correlation C_0 is found from Eqs. (6.27) and (6.28), and the drift velocity is found from

$$V_{gj} = 2.9\left(\frac{\sigma g \Delta\rho}{\rho_{\mathrm{L}}^2} \right)^{1/4}. \tag{6.47}$$

The DFM of Julia *et al.* (2009) (see Table 6.1) explicitly accounts for the effect of sub-channel geometric characteristics. The model is applicable to bubbly flow, however, and it involves an estimate of bubble diameter. Julia *et al.* (2009) assumed an average bubble diameter of 1.3 mm in validating their model against experimental data. Paranjape *et al.* (2011), on the other hand, assumed a bubble diameter of 3 mm in applying the model to data representing typical BWR boiloff conditions. The DFM of Chen *et al.* (2012) is based on data obtained with various experimental rod bundles sizes and has a wide range of validity parameters for steam–water.

6.5 DFM in Minichannels

Experiments show that because of the predominance of surface tension and viscous effects, the slip velocity in minichannels is small in all flow regimes except for annular flow. Therefore, $V_{gj} \approx 0$ should be expected. (Recall that, for water-like liquids, stratified flow does not happen for $D \lesssim 1$ mm mini- and microchannels.)

Mishima and Hibiki (1996) have proposed that for bubbly and slug flow of air–water mixtures in a vertical minichannel ($D \lesssim 1$ mm) $V_{gj} = 0$ and

$$C_0 = 1.2 + 0.510e^{-0.692D}. \tag{6.48}$$

Kawahara *et al.* (2002, 2005) performed adiabatic two-phase flow experiments in a 250-μm-inner-diameter channel with water, methanol, and water–methanol mixtures as liquid and nitrogen as gas. Their void fraction data indicated $C_0 = 1.10$–1.22; this is significantly smaller than $C_0 = 1.6$ which is the value predicted by Eq. (6.48).

More recently, Zhang *et al.* (2010) have refined the aforementioned correlation to

$$C_0 = 1.2 + 0.380 \exp\left(-1.39/D_{\mathrm{H}}^* \right), \tag{6.49}$$

where $D_H^* = D_H / \sqrt{\frac{\sigma}{g(\rho_L - \rho_G)}}$. The correlation is based on atmospheric and room temperature air–water experimental data, and the range of the data base for its derivation is

$$0.20 < D_H < 4.90 \text{ mm}, \text{Re}_L \leq 2000, \text{Re}_G \leq 1000.$$

6.6 Void-Quality Correlations

As mentioned before, a correlation of the form $f(\langle \alpha \rangle, \langle x \rangle) = 0$ can be a very useful tool. In one-dimensional flow, the correlation makes it possible to solve the mixture momentum equation without the need for another momentum equation. Equation (6.17), along with the correlations for the DFM parameters already discussed can in fact be considered as void-quality correlations.

The fundamental void-quality relation for one-dimensional flow, Eq. (3.39), will serve the same purpose if a correlation for the slip ratio, S_r, is available. Many void-quality correlations are in fact expressed in terms of the slip ratio. Most of the void-quality correlations are for near-equilibrium flow conditions, however. Over the past several decades a large number of void-quality or slip ratio correlations have been proposed. Reviews can be found in Woldesemayat and Ghajar (2007), Pasch and Anghaie (2008), Winkler *et al.* (2012), and Xu and Fang (2014). Table 6.2 is a summary of some of the widely referenced correlations. Some important and widely used correlations are now discussed.

Equation (6.17) can be re-written as

$$\langle \alpha \rangle = \frac{\langle j_G \rangle}{C_0 \langle j \rangle + V_{gj}}. \tag{6.50}$$

When the drift velocity is low (i.e., $V_{gj} \ll \langle j \rangle$), one can write

$$\langle \alpha \rangle \approx \frac{\langle j_G \rangle}{C_0 \langle j \rangle} = K\beta, \tag{6.51}$$

where $\beta = \langle j_G \rangle / \langle j \rangle$ is the volumetric quality and K is *Armand's flow parameter* (Armand, 1959). Evidently $K = 1/C_0$ for low-drift-flux conditions.

Based on the principle of minimum entropy generation in equilibrium and ideal annular flow (no droplet entrainment), Zivi (1964) has derived

$$S_r = (\rho_f / \rho_g)^{1/3}. \tag{6.52}$$

A simple correlation for steam–water flow due to Chisholm (1973), which has shown good accuracy for steam–water data (Whalley, 1987), is

$$S_r = \sqrt{1 - \langle x \rangle \left(1 - \frac{\rho_L}{\rho_G}\right)}. \tag{6.53}$$

One of the most accurate correlations available is the CISE correlation (Premoli *et al.*, 1970):

$$S_r = 1 + E_1 \left(\frac{y}{1 + yE_2} - yE_2\right)^{1/2}, \tag{6.54}$$

Table 6.2. *Summary of some slip ratio and void-quality correlations.*

Author/source	Correlation	Comments
Homogeneous flow	$\langle\alpha\rangle = \alpha_h = \beta = \dfrac{1}{1+\left(\frac{1-\langle x\rangle}{\langle x\rangle}\right)\left(\frac{\rho_G}{\rho_L}\right)}; S_r = 1$	
Armand and Treschev (1946), Armand (1959), Bankoff (1960)	$\langle\alpha\rangle = K\beta, K =$ Armand flow parameter	Armand-type correlations; applicable to high flow, and some microchannel flow conditions. $K = 0.71+0.0145$ (P in MPa) (Bankoff, 1960), $K = 0.833$ (Armand and Treschev, 1946; Chisholm, 1983); $K = 0.8$ (Ali *et al.*, 1993).
Chisholm (1983)	$\langle\alpha\rangle = \dfrac{\beta}{\beta+(1-\beta)^{1/2}}$	
Kawahara *et al.* (2002, 2009)	$\langle\alpha\rangle = \dfrac{C_1\beta^{1/2}}{1-C_2\beta^{1/2}}$	Based on water–N_2, ethanol/water–N_2 data in minichannels. See Chapter 10 for the values of constants.
Zivi (1964)	Eq. (6.52)	Derived based on minimum entropy generation in annular flow regime; widely used in various models including critical (choked) two-phase flow
Thom (1964)	$S_r = \left[\left(\dfrac{\rho_L}{\rho_G}\right)^{0.112}\left(\dfrac{\mu_L}{\mu_G}\right)^{0.178}\right]$	Based on steam–water experimental data in a horizontal pipe
Smith (1969)	$S_r = e + (1-e)\left[\dfrac{\frac{\rho_L}{\rho_G}+\frac{1-\langle x\rangle}{\langle x\rangle}}{1+e\frac{1-\langle x\rangle}{\langle x\rangle}}\right]^{1/2}; e \approx 0.4$ $e =$ fraction of liquid in the form of entrained droplets	Theoretical derivation for annular-dispersed flow regime; $e =$ fraction of liquid in the form of entrained droplets $e \approx 0.4$ agreed with steam–water and air–water horizontal and vertical tube data
Chisholm (1973)	Eq. (6.53)	Steam–water flow
Premoli *et al.* (1970)	Eqs. (6.54)–(6.57)	Also known as CISE correlation, a widely used and very accurate empirical correlation
Osmachkin and Borisov (1970)	$S_r = 1 + \dfrac{0.6+1.5\beta^2}{Fr_{L0}^{0.25}}\left(1-\dfrac{P}{P_{cr}}\right);$ $Fr_{L0} = \dfrac{G^2}{gD_H\rho_L^2}$	Flow boiling in vertical rod bundles
Butterworth (1975)	$\langle\alpha\rangle = \left(1 + 0.28X_{tt}^{0.71}\right)^{-1}$ $X_{tt} = \left(\dfrac{\mu_L}{\mu_G}\right)^{0.25}\left(\dfrac{1-\langle x\rangle}{\langle x\rangle}\right)^{1.75}\dfrac{\rho_G}{\rho_L}$	Based on air–water and air–oil flow data in pipes of Lockhart and Martinelli (1949)

Table 6.2. *(cont.)*

Author/source	Correlation	Comments
Wallis (1969)	$\langle\alpha\rangle = \left(1 + X^{0.8}\right)^{-0.378}$; $X^2 = (-\partial P/\partial z)_{\mathrm{fr},L}/(-\partial P/\partial z)_{\mathrm{fr},G}$	Based on air–water and air–oil flow data in pipes of Lockhart and Martinelli (1949)
Baroczy (1965); Butterworth (1975)	$\langle\alpha\rangle = \left[1 + X_{\mathrm{tt}}^{0.82}\left(\frac{\rho_G}{\rho_L}\right)^{0.24}\left(\frac{\mu_L}{\mu_G}\right)^{0.05}\right]^{-1}$	Based on experimental data with water, R-22, sodium, potassium, rubidium, and mercury
Tandon *et al.* (1985)	For $50 < \mathrm{Re}_L < 1125$ $\langle\alpha\rangle = 1 - 1.928\mathrm{Re}_L^{-0.315}[F(X_{\mathrm{tt}})]^{-1} + 0.9293\mathrm{Re}_L^{-0.63}[F(X_{\mathrm{tt}})]^{-2}$ $\mathrm{Re}_L > 1125$ $\langle\alpha\rangle = 1 - 0.38\mathrm{Re}_L^{-0.088}[F(X_{\mathrm{tt}})]^{-1} + 0.0361\mathrm{Re}_L^{-0.176}[F(X_{\mathrm{tt}})]^{-2}$ $F(X_{\mathrm{tt}}) = 0.15\left[X_{\mathrm{tt}}^{-1} + 2.85X_{\mathrm{tt}}^{-0.476}\right]$	Based on analysis using turbulent boundary layer theory for annular flow regime
Yashar *et al.* (2001)	$\langle\alpha\rangle = \left[1 + \left(\sqrt{\frac{\langle x\rangle}{1-\langle x\rangle}}\mathrm{Fr}_L\right)^{-1} + X_{\mathrm{tt}}\right]^{-0.321}$	Based on data using refrigerants (R-134a and R-410A) in microfin tubes
Cioncolini and Thome (2012)	$\langle\alpha\rangle = \frac{c\langle x\rangle^n}{1+(c-1)\langle x\rangle^n}$ $c = -2.129 + 3.129\left(\frac{\rho_G}{\rho_L}\right)^{-0.2186}$ $n = 0.3487 + 0.6513\left(\frac{\rho_G}{\rho_L}\right)^{0.5150}$ $0 < \langle x\rangle < 1; 10^{-3} < \rho_G/\rho_L < 1.0$	Based on data with water–steam, R410a, water–air, water–argon, water–nitrogen, water plus alcohol–air, alcohol–air and kerosene–air; 1.05 mm to 45.5 mm tube hydraulic diameters (circular and noncircular)
Xu and Fang (2014)	$\langle\alpha\rangle =$ $\left[1 + \left(1 + 2\mathrm{Fr}_{L0}^{-0.2}\beta^{3.5}\right)\left(\frac{1-\langle x\rangle}{\langle x\rangle}\right)\left(\frac{\rho_G}{\rho_L}\right)\right]^{-1}$	Based on data for 5 refrigerants; 0.5 mm to 10.0 mm hydraulic diameters; circular, rectangular and flat channels; $40 < G < 1000$ kg/m$_2$·s; $0 < x < 1$; $0.02 < \mathrm{Fr}_{L0} < 145$; $0.004 < \rho_G/\rho_L < 0.153$

where

$$y = \frac{\beta}{1-\beta}, \tag{6.55}$$

$$E_1 = 1.578\mathrm{Re}_{L0}^{-0.19}\left(\frac{\rho_L}{\rho_G}\right)^{0.22}, \tag{6.56}$$

$$E_2 = 0.0273\,\mathrm{We}\,\mathrm{Re}_{L0}^{-0.51}\left(\frac{\rho_L}{\rho_G}\right) - 0.08, \tag{6.57}$$

Table 6.3. *Constants in various slip ratio correlations (Butterworth, 1975).*

Correlation	A	p	q	r
Homogeneous flow model	1	1	1	0
Zivi (1964)	1	1	0.67	0
Turner and Wallis (1965)	1	0.72	0.40	0.08
Lockhart and Martinelli 1949)	0.28	0.64	0.36	0.07
Thom (1964)	1	1	0.89	0.18
Baroczy (1963)	1	0.74	0.65	0.13

with

$$\text{We} = \frac{G^2 D_{\text{H}}}{\sigma \rho_{\text{L}}} \tag{6.58}$$

and

$$\text{Re}_{\text{L0}} = G D_{\text{H}} / \mu_{\text{L}}. \tag{6.59}$$

In these equations β is the volumetric quality. Several void-quality correlations, including Zivi's, can be represented in the following generic form (Butterworth, 1975):

$$\langle \alpha \rangle = \frac{1}{1 + A\left(\frac{1-\langle x \rangle}{\langle x \rangle}\right)^p \left(\frac{\rho_{\text{G}}}{\rho_{\text{L}}}\right)^q \left(\frac{\mu_{\text{L}}}{\mu_{\text{G}}}\right)^r}. \tag{6.60}$$

The constants in this relation are summarized in Table 6.3. The correlation of Lockhart and Martinelli (1949) has been found to predict the experimental data for two-phase flow in vertically oriented helicoidally coiled tubes (Banerjee *et al.*, 1969; Kasturi and Stepanek, 1972a, 1972b).

$$V_{gj} = 1.18 \left(1 - \langle x \rangle\right) \left[\frac{\sigma g \Delta \rho}{\rho_{\text{f}}^2}\right]^{1/4}.$$

EXAMPLE 6.1. A large fuel rod bundle that simulates the core of a PWR is made of 1.1-cm-diameter rods that are 3.66 m long. The rods are arranged in a square lattice, as shown in Fig. P4.4 (Problem 4.4), with a pitch-to-diameter ratio of 1.33. The tubes are uniformly heated. During an experiment, the rod bundle remains at 4 bar pressure, while saturated liquid enters the bottom of the bundle with a mass flux of 52 kg/m^2 s. The heat flux at the surface of the simulated fuel rods is 5×10^4 W/m^2. Calculate the equilibrium quality and the void fraction at the center of the rod bundle.

SOLUTION. The properties that are needed are as follows: $\rho_{\text{f}} = 798.5$ kg/m^3, $\rho_g = 20.1$ kg/m^3, $h_{\text{f}} = 1.087 \times 10^6$ J/kg, $h_{\text{fg}} = 1.713 \times 10^6$ J/kg, and $T_{\text{sat}} = 523.5$ K. Also, using Eq. (2.17) with $T_{\text{cr}} = 647.2$ K, we get $\sigma = 0.0264$ N/m. The flow area and the heated perimeter of a channel, as defined in Fig. P4.4 (Problem 4.4), are found by writing

$$A_{\text{c}} = (1.33D)^2 - \frac{\pi}{4}D^2 = 1.19 \times 10^{-4}\,\text{m}^2$$

and

$$p_{heat} = \pi D = 0.0346 \text{ m}.$$

In view of the high pressure, it is assumed that the properties remain constant along the flow channel. This assumption is reasonable since the pressure variations that can be expected will have a small effect on fluid properties. The quality at the center of the rod bundle can be estimated by writing

$$\langle x_{eq} \rangle = \langle x \rangle = \frac{p_{heat} q''_w L_{heat}/2}{A_c G h_{fg}} = 0.298,$$

where we have assumed thermodynamic equilibrium between the vapor and liquid phases. We can now calculate the superficial velocities at the center of the bundle:

$$\langle j_g \rangle = G \langle x \rangle / \rho_g = 0.772 \text{ m/s},$$
$$\langle j_f \rangle = G(1 - \langle x \rangle)/\rho_f = 0.046 \text{ m/s},$$
$$\langle j \rangle = \langle j_g \rangle + \langle j_f \rangle = 0.818 \text{ m/s}.$$

We can estimate the void fraction at the bundle center based on the DFM model, using the correlation of Dix (1971). Accordingly,

$$b = (\rho_g/\rho_f)^{0.1} = 0692.$$

Using Eq. (6.27), we will then get $C_0 = 1.078$, and from Eq. (6.47) we get $V_{gj} = 0.387$ m/s. Equation (6.7) now gives

$$\langle \alpha \rangle = \frac{\langle j_g \rangle}{C_0 \langle j \rangle + V_{gj}} \approx 0.61.$$

EXAMPLE 6.2. For a steady air–water two-phase flow in an upward, 7.37-cm-diameter tube, estimate the void fraction and phase velocities, using the DFM and the correlation of Woldesemayat and Ghajar (2007). The mixture mass flux is $G = 520 \text{ kg/m}^2 \cdot \text{s}$, and air constitutes 2% of the total mass flow rate. Assume that the water–air mixture is under atmospheric pressure and at room temperature (25 °C).

SOLUTION. The properties that are needed are $\rho_L = 997.1 \text{ kg/m}^3$, $\rho_G = 1.18 \text{ kg/m}^3$, and $\sigma = 0.071$ N/m. Knowing $\langle x \rangle = 0.02$, we find the superficial velocities by writing

$$\langle j_G \rangle = G \langle x \rangle / \rho_G = 8.78 \text{ m/s},$$
$$\langle j_L \rangle = G(1 - \langle x \rangle)/\rho_L = 0.51 \text{ m/s},$$
$$\langle j \rangle = \langle j_G \rangle + \langle j_L \rangle = 9.29 \text{ m/s}.$$

The calculations then proceed as follows:

$$b = (\rho_G/\rho_L)^{0.1} = 0.51,$$
$$a = (P/P_{atm}) = 1.$$

Also, $\theta = \pi/2$; therefore Eq. (6.29) gives

$$V_{gj} = 2.9 \left[\frac{gD\sigma(\rho_L - \rho_G)}{\rho_L^2} \right]^{0.25} (1.22 + 1.22)^1 = 0.60 \text{ m/s}.$$

Equation (6.27) leads to $C_0 = 1.17$. Finally, Eq. (6.7) gives

$$\langle \alpha \rangle = \frac{\langle j_G \rangle}{C_0 \langle j \rangle + V_{gj}} = 0.77.$$

PROBLEMS

6.1 Prove the identities in Eqs. (6.14) and (6.15).

6.2 Starting from the one-dimensional mixture momentum equation for separated flow in Chapter 5, prove Eq. (6.18).

6.3 In an experiment dealing with the flow of a gas–liquid mixture in a 5.08-mm-diameter vertical column, where the liquid was an aqueous suspension of cellulose fibers with 0.5% fiber and the gas was air, the data shown in Table P6.1 were recorded.

Table P6.1. *Data for Problem 6.3.*

$\langle j_L \rangle$ (cm/s)	$\langle j_G \rangle$ (cm/s)	$\langle \alpha \rangle$	Flow regime
23.4	8.86	0.201	Slug
22.8	26.2	0.031	Slug
23.2	23.8	0.36	Slug
33	11.3	0.216	Slug
33	18.1	0.282	Slug
41	14.4	0.23	Slug
41.7	19.1	0.285	Slug
23.3	1.83	0.07	Churn
22.2	3.73	0.123	Churn
22.6	6.0	0.163	Churn
32.3	3.6	0.113	Churn
32.5	5.9	0.125	Churn
32.8	8.6	0.168	Churn
42.1	5.17	0.111	Churn
41.9	8.3	0.150	Churn
42	11.9	0.210	Churn

(a) Treating the two flow regimes separately, examine the applicability of the DFM, and develop DFM parameters for the data.

(b) Repeat part (a), this time using the entire data set.

(c) Calculate the slip ratios for all the data points, and examine the feasibility of correlating them in terms of $\langle \alpha \rangle$ and $\langle j \rangle$.

6.4 A long vertical tube that is 0.06 m in diameter initially contains water at room temperature up to a height of $H = 1$ m. Air is injected into the bottom of the tube, leading to a steady swollen two-phase region. No water carry-over takes place. The air volumetric flow rate is 7.5×10^{-3} m³/s. Assume that the void fraction in the swollen two-phase region is uniform.

(a) Calculate the quality $\langle x \rangle$ and the gas and liquid superficial velocities $\langle j_L \rangle$ and $\langle j_G \rangle$.

(b) Using the DFM, calculate the swollen two-phase level height H_2 in the channel, assuming that the flow regime in the column is churn-turbulent.

(c) If the gas is assumed to be composed of uniform-size bubbles with diameter $d = \sqrt{\sigma / g \Delta \rho}$, calculate the total interfacial surface area and the interfacial surface area concentration in the pipe.

(d) What is the highest gas volumetric flow rate for which the steady swollen two-phase configuration can be maintained?

6.5 The Cunningham correlation, which is suitable for prediction of the void fraction profile in rod bundles during water boiloff processes, is (Wong and Hochreiter, 1981)

$$\langle \alpha \rangle = 0.925 \left(\frac{\rho_g}{\rho_f} \right)^{0.239} \left(\frac{\langle j_g \rangle}{j_{B,cr}} \right)^a \alpha_h^{0.6}, \quad \langle \alpha \rangle \leq 1,$$

where α_h is the homogeneous void fraction and

$$a = \begin{cases} 0.67, & \langle j_g \rangle / j_{B,cr} < 1, \\ 0.47, & \langle j_g \rangle / j_{B,cr} \geq 1, \end{cases}$$

$$j_{B,cr} = \frac{2}{3} \sqrt{g R_{B,cr}},$$

$$R_{B,cr} = 5.27 \sqrt{\frac{\sigma}{g \rho_f}}.$$

A vertical rod bundle in an experiment is composed of simulated nuclear fuel rods that are 0.9 cm in diameter and have a pitch-to-diameter ratio of 1.3 (see Fig. P6.5). Saturated water at 10 bars is subject to the flow of saturated water vapor in the experiment, and the collapsed liquid level height (i.e., the level the water reaches if steam flow is completely stopped) is 1.25 m. For steam superficial velocities in the range of 0.15–1.45 m/s, calculate and plot the swollen two-phase level in the bundle using the correlation of Cunningham given here, as well as the correlations of Takaeuchi *et al.* (1992).

Hint: Define a subchannel composed of a unit cell containing four quarter channels, and assume one-dimensional flow in the subchannel.

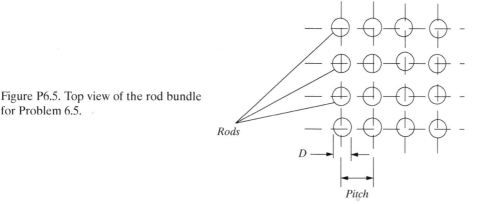

Figure P6.5. Top view of the rod bundle for Problem 6.5.

6.6 In an experiment using the rod bundle of Problem 4.4, while all other parameters are maintained the same as those in Problem 4.4, vapor is generated by imposing a uniform electric power of 5 kW/m on each rod, while sufficient saturated water is injected into the bundle to make up for the evaporated water. Calculate the two-phase swollen level height using the correlation of Cunningham (see Problem 6.5).

6.7 Consider steady-state, developed two-phase flow in a channel.

(a) Prove that the interfacial force follows

$$F_I = (1 - \langle \alpha \rangle) F_{wG} - \langle \alpha \rangle F_{wL} + \langle \alpha \rangle (1 - \langle \alpha \rangle)(\rho_L - \rho_G) g \sin \theta.$$

(b) Assume that F_I is related to the slip velocity U_r according to

$$F_I = C_I |\langle U_G - U_L \rangle| \langle U_G - U_L \rangle.$$

Prove that

$$C_I = \frac{(1 - \langle \alpha \rangle)^2}{V_{gj}{}^2} [\langle 1 - \alpha \rangle F_{wG} - \langle \alpha \rangle F_{wL}] + \frac{\langle 1 - \alpha \rangle^3 \langle \alpha \rangle}{V_{gj}^2} (\rho_L - \rho_G) g \sin \theta \cdot \rho.$$

(c) What assumptions are needed to justify the application of the expression in part (b) in a one-dimensional flow condition where transient effects and phase change occur?

7 Two-Phase Flow Regimes – II

7.1 Introductory Remarks

In Chapter 4 the basic gas–liquid two-phase flow regimes along with flow regime maps were reviewed. The discussion of flow regimes was limited to empirical methods applicable to commonly applied pipes and rod bundles. In this chapter mechanistic two-phase flow regime models will be discussed.

Empirical flow regime models suffer from the lack of sound theoretical or phenomenological bases. Mechanistic methods, in contrast, rely on physically based models for each major regime transition process. These models are often simple and rather idealized. However, since they take into account the crucial phenomenological characteristics of each transition process, they can be applied to new parameter ranges with better confidence than purely empirical methods. Some important investigations where regime transition models for the entire flow regime map were considered include the works of Taitel and Dukler (1976a, b), Taitel *et al.* (1980), Weisman and co-workers (1979, 1981), Mishima and Ishii (1984), and Barnea and co-workers (1986, 1987). The derivation of simple mechanistic regime transition models often involves insightful approximations and phenomenological interpretations. The review of the major elements of the successful models can thus be a useful learning experience.

In this chapter only conventional flow passages (i.e., flow passages with $D_H \gtrsim 3$ mm) will be considered. There are important differences between commonly applied channels and mini- or microchannels with respect to the gas–liquid two-phase flow hydrodynamics. Two-phase flow regimes and conditions leading to regime transitions in mini- and microchannels will be discussed in Chapter 10.

It is emphasized that for convenience in the remainder of this chapter and other chapters, with the exception of Chapters 3 and 6, all flow properties are assumed to be cross-section- and time/ensemble-averaged unless otherwise stated. Thus, everywhere U_L and U_G stand for $\langle U_L \rangle_L$ and $\langle U_G \rangle_G$, respectively, α and x represent $\langle \alpha \rangle$ and $\langle x \rangle$, respectively, and j means $\langle j \rangle$.

7.2 Upward, Co-current Flow in Vertical Tubes

7.2.1 Flow Regime Transition Models of Taitel *et al.*

For the Taitel *et al.* (1980) models the main flow patterns and the shape of their boundaries are as shown in Fig. 7.1, for air–water flow in a 5-cm-diameter tube. It

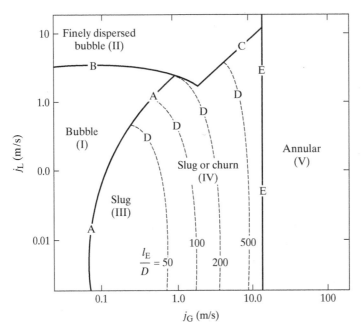

Figure 7.1. Flow regime transition lines in a tube 5 cm in diameter as predicted by the models of Taitel *et al.* (1980).

is important to remember that the figure is good only for that specific tube size and fluid pair. The positions of the transition lines change once a parameter (channel diameter or fluid properties) is changed. Thus, the correct way of using these and other mechanistic models is to directly apply the mathematical expression for each transition, rather than relying on graphical representations.

Line A in Fig. 7.1 represents the transition from the bubbly to the slug regime and is assumed to happen when bubbles become so numerous that they can no longer avoid coalescing and forming larger bubbles, eventually forming Taylor bubbles. Experience shows that the transition from bubbly to slug flow happens at $\alpha \approx 0.25$. Also, because the rise velocity of a typical bubble with respect to liquid in fact represents the gas–liquid velocity slip, then

$$U_{\text{G}} - U_{\text{L}} = U_{\text{B}}. \tag{7.1}$$

For typical bubbles encountered in the bubbly flow regime, the rise velocity of bubbles can be found from (Harmathy, 1960)

$$U_{\text{B}} = 1.53 \left[\frac{g \Delta \rho \sigma}{\rho_{\text{L}}^2} \right]^{1/4}, \Delta \rho = \rho_{\text{L}} - \rho_{\text{G}}. \tag{7.2}$$

Substituting for U_{B} in Eq. (7.1), and replacing the phasic velocities with easily quantifiable superficial velocities from $U_{\text{G}} = j_{\text{G}}/\alpha$ and $U_{\text{L}} = j_{\text{L}}/(1 - \alpha)$, we obtain the following equation for line A in Fig. 7.1:

$$j_{\text{L}} = 3 j_{\text{G}} - 1.15 \left[\frac{\sigma g \Delta \rho}{\rho_{\text{L}}^2} \right]^{1/4}. \tag{7.3}$$

Bubbly flow holds as long as j_L > the right side of Eq. (7.3), and slug flow develops once $j_L \leq$ the right side of Eq. (7.3).

We will now discuss the conditions that are necessary for the existence of bubbly flow. Experiments have shown that developed bubbly flow cannot be sustained in small tubes and is eventually replaced by slug flow. Taitel *et al.* (1980) argued that bubbly flow becomes impossible when the rise velocity of a Taylor bubble is lower than the rise velocity of regular bubbles, in which case the sporadic occurrence of a Taylor bubble into a vertical tube would cause the bubbles that pursue it to coalesce. Therefore, when $\rho_L \gg \rho_G$, bubbly flow would be impossible when $0.35\sqrt{gD} \leq 1.53[g\Delta\rho\sigma/\rho_L^2]^{1/4}$, where the left side represents the rise velocity of Taylor bubbles (Nicklin *et al.*, 1962; Davidson and Harrison, 1971). This expression can be recast as

$$\left[\frac{\rho_L^2 gD^2}{\sigma\Delta\rho}\right]^{1/4} \leq 4.36. \tag{7.4}$$

Line B in Fig. 7.1 represents the model for transition to finely dispersed bubbly flow. In the finely dispersed bubbly regime we deal with small, nearly spherical bubbles that remain discrete because of strong turbulence. Turbulent velocity fluctuations impose a hydrodynamic force on a bubble that can break up the bubble, should the bubble be larger than a certain critical size. The critical size depends on the level of turbulence energy dissipation. Taitel *et al.* (1980) assumed the following in the finely dispersed bubbly regime.

(a) The flow must be fully turbulent.
(b) The size of the finely dispersed bubbles is within the inertial turbulent eddy size range and is controlled by turbulence-induced aerodynamic breakup. (The inertial turbulent eddies are locally isotropic, and their properties depend on local turbulent energy dissipation, but not on liquid viscosity. See Section 3.6.)
(c) Dispersed bubbles must remain spherical since distorted bubbles have a higher chance of coalescence.

The equation defining line B then becomes

$$j_L + j_G = 4\left\{\frac{D^{0.429}\left(\frac{\sigma}{\rho_L}\right)^{0.089}}{\nu_L^{0.072}}\left(\frac{g\Delta\rho}{\rho_L}\right)^{0.446}\right\}. \tag{7.5}$$

Therefore, when $j_L + j_G$ > the right side of Eq. (7.5), dispersed bubbly flow occurs. Dispersed bubbly flow can be sustained up to a void fraction at which spherical bubbles become close-packed. A higher void fraction would "press" the bubbles against one another and cause coalescence. The maximum void fraction is in fact equal to the maximum packing for equal-size spherical bubbles in a large vessel. For the simple cubic configuration of spheres, this would give a maximum void fraction equal to $\alpha_{max} = \frac{\pi}{6}d_B^3/d_B^3 \approx 0.52$. (The highest theoretical void fraction would actually be 0.74, which corresponds to the face-centered cubic configuration.) Also, in this high-velocity regime the gas–liquid velocity slip is typically quite small in comparison with phasic superficial velocities; therefore $\alpha \approx \beta = j_G/(j_L + j_G)$. The upper limit for the existence of the dispersed bubbly regime then becomes

$$\frac{j_G}{j_L + j_G} = 0.52. \tag{7.6}$$

Thus, $j_G/(j_L + j_G) > 0.52$ would lead to slug flow.

Since transition to finely dispersed bubbly flow occurs typically at high liquid superficial velocities, Eq. (7.5) is not limited to upward vertical flow. This is true despite the presence of g in Eq. (7.5), which appears because the following correlation, representing the maximum diameter at which bubble shape distortion from a perfect sphere begins, has been used in its derivation (Brodkey, 1967):

$$d_{cr} = \left[\frac{0.4\sigma}{(\rho_L - \rho_G)g}\right]^{1/2}. \tag{7.7}$$

Lines D in Fig. 7.1 represent the churn-to-slug flow regime transition. The churn flow defined by Taitel *et al.* is in fact the entrance regime for the development of slug flow. This type of churn flow would eventually result in slug flow at a distance l_E from entrance. The model assumes that short Taylor bubbles and slugs are generated at the inlet. Consecutive short Taylor bubbles approach one another, however, and coalesce two by two, until the length of the liquid slugs separating them reaches $16D$, the latter representing the typical length of the liquid slugs in the stable slug regime. The model leads to the following expression for the distance from the entrance that is needed for the development of slug flow:

$$\frac{l_E}{D} = 40.6 \left(\frac{j}{\sqrt{gD}} + 0.22\right). \tag{7.8}$$

Thus, when Eq. (7.3) or (7.6) indicates that conditions necessary for slug flow are present, Eq. (7.8) must be tested. The flow regime will be slug only at distances from the inlet larger than l_E. Otherwise, the flow regime will be churn.

Line E in Fig. 7.1 represents the transition to the annular-dispersed flow regime. This transition is assumed to happen when the gas velocity is sufficient to shatter the liquid core in the pipe into dispersed droplets, so that the drag force imposed on the droplets overcomes their weight. It is assumed that (a) the droplet diameter d is governed by a critical Weber number as $\text{We}_{cr} = \rho_G j_G^2 d/\sigma = 30$ and (b) at the onset of annular-dispersed flow, the drag force on the droplet just balances the droplet's weight, therefore,

$$C_D \frac{\pi d^2}{4} \frac{1}{2} \rho_G j_G^2 = \frac{\pi}{6} d^3 g \Delta\rho. \tag{7.9}$$

Using $C_D = 0.44$ and eliminating the droplet diameter d between the expression $\rho_G j_G^2 d/\sigma = 30$ and Eq. (7.9) leads to

$$\frac{j_G \rho_G^{1/2}}{(\sigma g \Delta\rho)^{1/4}} = 3.1. \tag{7.10}$$

Annular-dispersed flow thus occurs when $j_G \rho_G^{1/2}/(\sigma g \Delta\rho)^{1/4} > 3.1$. This criterion coincides with the condition for zero liquid penetration rate (complete flooding) according to the Tien–Kutateladze (Tien, 1977) countercurrent flow limitation (flooding) correlation, to be discussed in Chapter 9.

7.2.2 Flow Regime Transition Models of Mishima and Ishii

The regime transition models of Mishima and Ishii (1984) are based on the argument that the void fraction is the most important geometric parameter affecting flow regime transition. Four major flow regimes are considered: bubbly, slug, churn-turbulent, and annular. Except for the transition from churn-turbulent to annular, other flow regime transition models are all based on critical void fraction thresholds. The channel-average void fraction is predicted by using the DFM with parameters proposed by Ishii (1977), and it is compared with the aforementioned critical void fraction thresholds to determine the flow regime transitions. According to Ishii (1977) (see Chapter 6),

$$
C_0 = \begin{cases} 1.2 - 0.2\sqrt{\frac{\rho_G}{\rho_L}} & \text{for round tubes,} \\ 1.35 - 0.35\sqrt{\frac{\rho_G}{\rho_L}} & \text{for rectangular ducts.} \end{cases}
$$

(7.11)

(7.12)

Transition from bubbly to slug flow is assumed to occur when $\alpha = 0.3$, and the void fraction is predicted by using the DFM, with V_{gj} found from Eq. (6.23), leading to

$$
j_L = \left(\frac{3.33}{C_0} - 1\right) j_G - \frac{0.76}{C_0}\left(\frac{\sigma g \Delta \rho}{\rho_L^2}\right)^{1/4}.
$$

(7.13)

Thus, the transition from bubbly to slug flow occurs when j_L is smaller than the right side of Eq. (7.13); otherwise bubbly flow would occur. Transition from the slug to the churn-turbulent flow regime is assumed to take place when the pipe-average void fraction surpasses the mean void fraction over an entire Taylor bubble (i.e., $\alpha \geq \alpha_B$), where

$$
\alpha = \frac{j_G}{C_0 j + 0.35\sqrt{\frac{\Delta \rho g D}{\rho_L}}},
$$

(7.14)

$$
\alpha_B = 1 - 0.813\left\{\frac{(C_0 - 1)j + 0.35\sqrt{\frac{\Delta \rho g D}{\rho_L}}}{j + 0.75\sqrt{\frac{\Delta \rho g D}{\rho_L}}\left[\frac{\Delta \rho g D^3 \rho_L}{\mu_L^2}\right]^{1/18}}\right\}^{3/4}.
$$

(7.15)

The mechanism causing transition from churn-turbulent flow to the annular flow regime depends on the channel diameter. For small-diameter tubes, flow regime transition occurs when flow reversal takes place in the liquid film surrounding the Taylor bubbles. Analysis based on this assumption leads to

$$
j_G = \sqrt{\frac{\Delta \rho g D}{\rho_G}}\alpha^{1.25}\left\{\frac{1 - \alpha}{0.015[1 + 75(1 - \alpha)]}\right\}^{1/2}.
$$

(7.16)

This criterion is to be used when

$$
D < \frac{\sqrt{\frac{\sigma}{g \Delta \rho}} N_{\mu L}^{-0.4}}{[(1 - 0.11 C_0)/C_0]^2}.
$$

(7.17)

For larger tube diameters, however, regime transition should not depend on the tube diameter. It is assumed that the destruction of the liquid slug and the entrainment of generated droplets cause the flow regime transition in this case. Using a criterion for the onset of liquid droplet entrainment in annular flow (Ishii, 1977), Mishima and Ishii (1984) derived

$$j_G \geq \left(\frac{\sigma g \Delta \rho}{\rho_G^2} \right)^{1/4} N_{\mu L}^{-0.2}, \tag{7.18}$$

where the viscosity number is defined as

$$N_{\mu L} = \frac{\mu_L}{\left[\rho_L \sigma \sqrt{\frac{\sigma}{g \Delta \rho}} \right]^{1/2}}. \tag{7.19}$$

Jayanti and Hewitt (1992) have indicated that the aforementioned mechanism for the regime transition from slug to churn-turbulent flow is unreasonable since it may require that the average void fraction in the liquid slug (representing small bubbles that are typically present in the liquid slug) be larger than the average void fraction over an entire Taylor bubble. Models based on flooding of the liquid film surrounding the Taylor bubbles have been proposed by a number of investigators, including McQuillan and Whalley (1985a, b). Jayanti and Hewitt (1992) have improved the model of McQuillan and Whalley.

EXAMPLE 7.1. Based on the flow regime models of Taitel *et al.* (1980), for an air–water mixture upward flow in a 2-m-long tube with 5-cm inner diameter, under atmospheric pressure and at room temperature, determine the flow regime for the following conditions:

(a) $j_L = 0.9$ m/s, $j_G = 8$ m/s;
(b) $j_L = 1.1$ m/s, $j_G = 0.4$ m/s.

SOLUTION. With respect to the relevant properties, we have $\rho_L = 997$ kg/m^3, $\rho_G = 1.185$ kg/m^3, $\mu_L = 8.94 \times 10^{-4}$ kg/m·s, $\mu_G = 10^{-5}$ kg/m·s, and $\sigma = 0.071$ N/m. Let us start with case (a), namely, $j_L = 0.9$ m/s and $j_G = 8$ m/s. Given the relatively high gas superficial velocity, we should use Fig. 7.1 as a guide, and start from the right side of the map. First check Eq. (7.10). For the given conditions, we find

$$\frac{j_G \rho_G^{1/2}}{(\sigma g \Delta \rho)^{1/4}} = 1.697 < 3.1.$$

The annular-dispersed regime is therefore not applicable. We next will check the finely dispersed bubbly and slug/churn regimes. Let us check Eq. (7.6). Accordingly,

$$\frac{j_G}{j_G + j_L} = 0.899 > 0.52.$$

Thus, finely dispersed bubbly flow is also not possible. A check of Eq. (7.3) would show that

$$3 j_G - 1.15 \left[\frac{\sigma g \Delta \rho}{\rho_L^2} \right]^{0.25} = 23.8.$$

Clearly, then, the flow regime is not bubbly. It must therefore be slug or churn. To determine which one, let us use Eq. (7.8), according to which, for our case, $l_E = 3.06$ m.

Since the total length of our tube is smaller than l_E, our entire tube will remain in churn flow.

We will now consider case (b), namely, $j_L = 1.1$ m/s and $j_G = 0.4$ m/s. Starting with Eq. (7.10), we find

$$\frac{j_G \rho_G^{1/2}}{(\sigma g \Delta \rho)^{1/4}} = 0.085 < 3.1.$$

Therefore annular-dispersed flow does not apply. (This was actually obvious, given that $j_G = 8$ m/s in case (a), which was larger than j_G in case (b), did not lead to annular-dispersed flow.)

We now examine the finely dispersed flow. Accordingly, $j_G/(j_G + j_L) = 0.267$. The right side of Eq. (7.5) is found to be 252.3, which is clearly larger than $j_G + j_L$. The flow regime cannot be finely dispersed bubbly. We should now check Eq. (7.3). For the given conditions we find

$$3 j_G - 1.15 \left[\frac{\sigma g \Delta \rho}{\rho_L^2} \right]^{0.25} = 1.01 < j_L.$$

The flow regime is therefore bubbly.

EXAMPLE 7.2. Repeat the problem in Example 7.1, this time using the flow regime transition models of Mishima and Ishii (1984). Compare the results with those obtained in Example 7.1.

SOLUTION. The properties calculated in Example 7.1 apply. From Eq. (7.12), we get $C_0 = 1.193$; from Eq. (6.25) for churn flow we get $V_{gj} = \sqrt{2}[\sigma g \Delta \rho / \rho_L^2]^{0.25} = 0.23$ m/s; and from Eq. (7.19), we get $N_{\mu L} = 2.04 \times 10^{-3}$. Also, with respect to the criterion of Eq. (7.17), we get

$$\frac{\sqrt{\frac{\sigma}{g \Delta \rho}} N_{\mu L}^{-0.4}}{[(1 - 0.11 C_0)/C_0]^2} = 0.06 \, \text{m} > D.$$

Let us now focus on the conditions of case (a), where $j_G = 8$ m/s and $j_L = 0.9$ m/s. First, we will check the possibility of annular-dispersed flow, given the relatively high value of j_G. Since the criterion of Eq. (7.17) is satisfied, we must calculate α from $\alpha = j_G/(C_0 j + V_{gj})$ and then check Eq. (7.16). The expression for the void fraction gives $\alpha = 0.737$. With this value of α, the right side of Eq. (7.16) is found to be 12.76 m/s, which is evidently larger than j_G. The flow regime, therefore, is not annular-dispersed. In other words, conditions for the transition from churn to annular have not been met.

Next, we will calculate α from Eq. (7.14) and α_B from Eq. (7.15). We will get $\alpha = 0.736$ and $\alpha_B = 0.77$. Since $\alpha < \alpha_B$, the flow regime cannot be churn. We are left with bubbly or slug. We should use Eq. (7.13) to decide which one of these two regimes applies. The right side of Eq. (7.13) is calculated to be 14.22 m/s, which is evidently larger than j_L. The flow regime is therefore slug.

We should now consider case (b), namely, $j_L = 1.1$ m/s and $j_G = 0.4$ m/s. Repetition of the previous calculations will result in the elimination of annular-dispersed and churn flow regimes. The right side of Eq. (7.13) is found to be 0.613 m/s, which is actually smaller than j_L. The flow pattern is therefore bubbly.

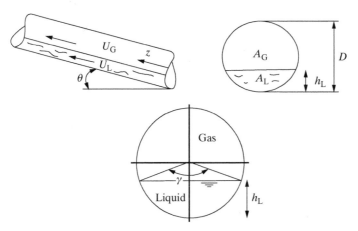

Figure 7.2. Equilibrium stratified flow in a slightly inclined pipe.

Comparison between the results of this and the previous example indicates that the predictions of the two-flow regime transition models were similar.

7.3 Co-current Flow in a Near-Horizontal Tube

In their pioneering work, Taitel and Dukler (1976a, b) divided the entire flow regime map for a horizontal or near-horizontal pipe into the following zones:

- stratified (smooth and wavy),
- intermittent (slug, plug/elongated bubbles),
- dispersed bubbly, and
- annular-dispersed.

They then proposed mechanistic models for all the relevant regime transitions. Among the transition regimes proposed by Taitel and Dukler, their models for stratified-to-wavy and stratified-to-intermittent have been the most successful. The stratified-to-intermittent regime transition is particularly important for pipelines and has been extensively investigated because intermittent flow regimes have a higher frictional pressure drop than stratified flow (Kordyban and Ranov, 1970; Mishima and Ishii, 1980). Intermittency also leads to countercurrent flow limitation (CCFL), or flooding, in channels with countercurrent gas–liquid flow. Semi-analytical models for various regime transitions have also been proposed by Weisman *et al.* (1979) (see Problem 7.9) as well.

A key element in the models dealing with regime transitions for horizontal flow is the flow conditions under an equilibrium stratified flow pattern, shown schematically in Fig. 7.2. To this end, first the steady-state and fully developed "separated-flow" phasic momentum equations are written as

$$-\frac{dP}{dz} - \frac{\tau_{wL}p_L - \tau_I p_I}{A(1-\alpha)} - \rho_L g \sin\theta = 0, \qquad (7.20)$$

$$-\frac{dP}{dz} - \frac{\tau_{wG}p_G + \tau_I p_I}{A\alpha} - \rho_G g \sin\theta = 0. \qquad (7.21)$$

Now, eliminating dP/dz between the two equations we obtain

$$\frac{\tau_{\mathrm{wG}}p_{\mathrm{G}}}{A\alpha} - \frac{\tau_{\mathrm{wL}}p_{\mathrm{L}}}{A(1-\alpha)} + \frac{\tau_{\mathrm{I}}p_{\mathrm{I}}}{A}\left(\frac{1}{1-\alpha} + \frac{1}{\alpha}\right) - (\rho_{\mathrm{L}} - \rho_{\mathrm{G}})g\sin\theta = 0, \quad (7.22)$$

where p_{L}, p_{G}, and p_{I} represent the wall–liquid, wall–gas, and gas–liquid interfacial perimeters. The fluid–surface and interfacial shear stresses can be estimated using friction factors as

$$\tau_{\mathrm{wL}} = f_{\mathrm{L}}\frac{1}{2}\rho_{\mathrm{L}}U_{\mathrm{L}}^2, \quad (7.23)$$

$$\tau_{\mathrm{wG}} = f_{\mathrm{G}}\frac{1}{2}\rho_{\mathrm{G}}U_{\mathrm{G}}^2, \quad (7.24)$$

and

$$\tau_{\mathrm{I}} = f_{\mathrm{I}}\frac{1}{2}\rho_{\mathrm{G}}|U_{\mathrm{G}} - U_{\mathrm{L}}|(U_{\mathrm{G}} - U_{\mathrm{L}}). \quad (7.25)$$

For simplicity, it is assumed that $f_{\mathrm{I}} = f_{\mathrm{G}}$. The gas friction factor is found from $f_{\mathrm{G}} = C_{\mathrm{G}}\mathrm{Re}_{\mathrm{G}}^{-m}$, where $\mathrm{Re}_{\mathrm{G}} = U_{\mathrm{G}}D_{\mathrm{G}}/\nu_{\mathrm{G}}$ and D_{G} represents the hydraulic diameter of the gas-occupied part of the pipe cross section (see Fig. 7.2). For turbulent gas flow $C_{\mathrm{G}} = 0.046$, $m = 0.2$, and for laminar flow $C_{\mathrm{G}} = 16$, $m = 1$. The liquid friction factor f_{L} is obtained by using the same expressions with subscript G replaced with L everywhere.

Knowing j_{G} and j_{L}, we can solve Eq. (7.22) numerically using geometric characteristics of the channel cross section to calculate α and h_{L} (the liquid level height). (Remember that $j_{\mathrm{L}} = U_{\mathrm{L}}(1 - \alpha)$ and $j_{\mathrm{G}} = U_{\mathrm{G}}\alpha$.) For circular pipes, for example, the following geometric relations apply (see Fig. 7.2):

$$\gamma = 2\cos^{-1}\left(1 - 2\frac{h_{\mathrm{L}}}{D}\right), \quad (7.26)$$

$$\alpha = 1 - \frac{1}{2\pi}(\gamma - \sin\gamma). \quad (7.27)$$

The transition from stratified-smooth to stratified-wavy flow, according to Taitel and Dukler (1976a, b), is associated with wave generation at the liquid–gas interphase, and occurs when

$$U_{\mathrm{G}} \geq \left[\frac{4\nu_{\mathrm{L}}\Delta\rho g\cos\theta}{S\rho_{\mathrm{L}}U_{\mathrm{L}}}\right]^{1/2}, \, S = 0.01, \quad (7.28)$$

where S is the sheltering coefficient.

Regime transition out of stratified flow can lead to bubbly, intermittent, or annular flow. Transition out of stratified flow was modeled by Taitel and Dukler (1976a, b) using an extended Kelvin–Helmholtz instability, and is assumed to occur when infinitesimally small waves at the interphase grow as a result of the aerodynamic force caused by the reduction in the gas-occupied flow area:

$$\mathrm{Fr}^2\left[\frac{1}{c_2^2}\frac{d\tilde{A}_{\mathrm{L}}/d\tilde{h}_{\mathrm{L}}}{\alpha\tilde{A}_{\mathrm{G}}}\right] \geq 1, \quad (7.29)$$

with

$$\mathrm{Fr} = \sqrt{\frac{\rho_{\mathrm{G}}}{\rho_{\mathrm{L}} - \rho_{\mathrm{G}}}}\frac{j_{\mathrm{G}}}{\sqrt{gD\cos\theta}}, \quad (7.30)$$

where $\tilde{h}_L = h_L/D$, $\tilde{A}_G = A_G/D^2$, and $\tilde{A}_L = A_L/D^2$. Annular-dispersed flow is assumed when Eq. (7.29) holds and $h_L/D < 0.5$, and intermittent flow is assumed when $h_L/D > 0.5$. For near-horizontal circular tubes, the experimental data indicated that

$$c_2 = 1 - \frac{h_L}{D}.$$

In dimensional form, the criterion of Eq. (7.29) for circular channels gives

$$U_G > \left(1 - \frac{h_L}{D}\right)\left[\frac{\Delta\rho g \cos\theta A_G}{\rho_G dA_L/dh_L}\right]^{1/2}. \tag{7.31}$$

This model has been found to be quite general, provided that c_2 is treated as an empirically adjustable parameter for flow configurations that are different from near-horizontal pipes. The application of the criterion presented here is rather tedious, however. Cheng et al. (1988) have curve fitted the predictions of Eqs. (7.29) and (7.30) for a horizontal channel (i.e., $\theta = 0$), apparently to a reasonable accuracy (Wong et al., 1990), according to

$$\mathrm{Fr} = \left(\frac{1}{0.65 + 1.11 X_{tt}^{0.6}}\right)^2, \tag{7.32}$$

where $X_{tt} = [(1-x)/x]^{0.9}(\mu_L/\mu_G)^{0.1}(\rho_G/\rho_L)^{0.5}$ is the turbulent–turbulent Martinelli parameter.

A simpler expression for the limit of stratification in horizontal channels is (Mishima and Ishii, 1980)

$$U_G - U_L = 0.487\sqrt{\frac{g(\rho_L - \rho_G)}{\rho_G}}(D_H - h_L). \tag{7.33}$$

This expression is the outcome of a theoretical analysis dealing with the growth of waves with finite amplitude. A larger $U_G - U_L$ value than what Eq. (7.33) sets thus leads to the development of intermittent flow.

The disruption of stratified flow, as mentioned, can lead to bubbly, intermittent, or annular-dispersed flow. Let us now discuss the conditions that dictate the occurrence of each of these regimes.

Taitel and Dukler argued that whether the disruption of stratified flow regime leads to intermittent or annular flow depends uniquely on the liquid level height in the equilibrium stratified flow. They thus suggest that annular flow will occur if $h_L/D > 0.5$, and intermittent flow occurs when $h_L/D < 0.5$. For the transition from intermittent to bubbly flow, one should notice that small bubbles tend to collect near the top of the channel because of buoyancy and tend to coalesce. The coalescence, if unchecked, would lead to the intermittent flow pattern. Taitel and Dukler (1976a, b) assumed that the transition to dispersed bubby flow occurs when forces caused by turbulence overwhelm buoyancy and therefore prevent coalescence. The argument leads to

$$U_L \geq \left[\frac{4A_G}{p_I}\frac{g\cos\theta}{f_L}\left(1 - \frac{\rho_G}{\rho_L}\right)\right]^{1/2}. \tag{7.34}$$

EXAMPLE 7.3. Water and air under atmospheric pressure and room temperature conditions flow co-currently in a long horizontal pipe that is 5 cm in diameter, under equilibrium conditions. The superficial velocities are $j_L = 0.1$ m/s and $j_G = 1.0$ m/s. Determine the two-phase flow regime in the pipe.

SOLUTION. The properties are similar to those calculated in Example 7.1. Since equilibrium conditions apply, we need to find the equilibrium stratified flow parameters first. The following equations are therefore solved simultaneously by trial and error: (7.22), (7.23), (7.24), (7.25) with f_I replaced with f_G, (7.26), and (7.27). Other equations are $j_L = U_L(1 - \alpha), j_G = U_G \alpha, f_G = C_G \text{Re}_G^{-m}, f_L = C_L \text{Re}_L^{-m}$, and

$$\text{Re}_G = \rho_G D_G U_G / \mu_G,$$
$$\text{Re}_L = \rho_L D_L U_L / \mu_L,$$
$$D_G = \frac{2\pi - (\gamma - \sin \gamma)}{2\pi - \gamma + 2 \sin(\gamma/2)} D,$$
$$D_L = \frac{\gamma - \sin \gamma}{\gamma + 2 \sin(\gamma/2)} D,$$
$$p_L = \gamma D/2,$$
$$p_G = (2\pi - \gamma) D/2,$$

and

$$p_I = D \sin(\gamma/2).$$

The iterative solution of these equations leads to

$$h_L = 0.036 \, \text{m},$$
$$\alpha = 0.227,$$
$$U_G = 4.14 \, \text{m/s},$$
$$U_L = 0.129 \, \text{m/s}.$$

We can now examine the criterion of Mishima and Ishii, Eq. (7.33). The right-hand side of the latter equation is found to be 5.21 m/s, which is clearly larger than $U_G - U_L$. A regime transition out of stratified flow does not occur, and therefore the flow pattern is stratified.

The right-hand side of Eq. (7.28) is calculated to be 0.165 m/s. Since $U_G > 0.165$ m/s, therefore the flow pattern is stratified wavy.

An alternative to Eq. (7.33) is Eq. (7.31). Instead of Eq. (7.31), however, we will use the criterion of Eq. (7.32), which is essentially a curve fit to the results of Eq. (7.31) for the critical conditions for horizontal flow. Thus,

$$x = \frac{\rho_G j_G}{\rho_G j_G + \rho_L j_L} = 0.0117,$$
$$X_{tt} = \left(\frac{1-x}{x}\right)^{0.9} \left(\frac{\mu_L}{\mu_G}\right)^{0.1} \left(\frac{\rho_G}{\rho_L}\right)^{0.5} = 2.745.$$

From Eq. (7.30), we get Fr = 0.0493. The right-hand side of Eq. (7.32) is calculated to be 0.1388. We thus have the following condition, which implies that the flow regime is stratified:

$$\text{Fr} < \left(\frac{1}{0.65 + 1.11 X_{\text{tt}}^{0.6}} \right)^2.$$

7.4 Two-Phase Flow in an Inclined Tube

Barnea, Taitel, and co-workers (Barnea *et al.*, 1985; Barnea, 1986, 1987; Taitel, 1990) studied extensively the two-phase flow regimes in inclined pipes and proposed a *unified model*, meant to predict the two-phase flow regimes for all pipe angles of inclination. Most of the transition models are modifications of the aforementioned models for vertical (Taitel *et al.*, 1980) or near-horizontal (Taitel and Dukler, 1976a, b) tubes. A brief review of these models is now presented.

The regime transition out of the stratified regime in inclined channels follows Eq. (7.29). The developed bubbly flow regime is not possible in small vertical tubes, as discussed earlier in Section 7.2 (see Eq. (7.4)). A similar observation has been made in inclined tubes. The phenomenon causing the disruption of bubbly flow is similar to what was described for vertical channels; namely, a stable bubbly flow becomes impossible if the rise velocity of Taylor bubbles is lower than the velocity of regular bubbles. This argument leads to

$$0.35\sqrt{gD}\sin\theta + 0.54\sqrt{gD}\cos\theta > 1.53\left[\frac{\sigma g \Delta \rho}{\rho_{\text{L}}^2}\right]^{1/4}\sin\theta, \tag{7.35}$$

where the left side is the axial velocity of elongated (Taylor) bubbles in an inclined pipe, and the right side is simply the axial component of the bubble rise velocity of Harmathy (1960) (see Eq. (7.2)).

The phenomenology of regime transition from bubbly to slug flow in inclined tubes is similar to that in vertical tubes. Equation (7.3) thus applies provided that U_{B} is replaced by $U_{\text{B}} \sin \theta$. The phenomenology of transition to the finely dispersed bubbly flow regime is also similar to that in vertical channels, with the additional requirement that the turbulence fluctuations must overwhelm buoyancy as well, so that crowding of bubbles near the top (creaming) is avoided. The necessary conditions are met when $d_{\text{B}} < d_{\text{cb}}$ and $d_{\text{B}} < d_{\text{cr}}$, where the bubble diameter d_{B} and the critical bubble diameters d_{cr} and d_{cb} are to be found, respectively, from (Barnea *et al.*, 1982; Taitel, 1990)

$$d_{\text{B}} = \left(0.725 + 4.15\alpha^{1/2}\right)\left(\frac{\sigma}{\rho_{\text{L}}}\right)^{3/5}\varepsilon^{-2/5}, \tag{7.36}$$

$$d_{\text{cr}} = 2\left[\frac{0.4\sigma}{\Delta\rho g}\right]^{1/2}, \tag{7.37}$$

$$d_{\text{cb}} = \frac{3}{8}\frac{\rho_{\text{L}}}{\Delta\rho}\frac{f_{\text{M}}j^2}{g\cos\theta}. \tag{7.38}$$

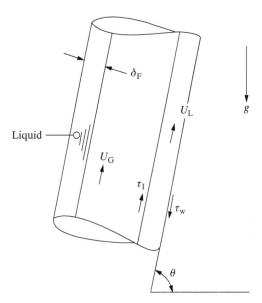

Figure 7.3. Annular flow regime in an inclined pipe.

In Eq. (7.36), ε represents the turbulent dissipation rate and can be estimated from

$$\varepsilon = -\left(\frac{dP}{dz}\right)_{\text{fr}} j = \frac{2f_M}{D} j^3, \tag{7.39}$$

where the turbulent Fanning friction coefficient is

$$f_M = 0.046(jD/\nu_L)^{-0.2}. \tag{7.40}$$

Two different mechanisms can disrupt the annular flow regime: (a) the formation of lumps of liquid (likely to happen when liquid film is very thick) and (b) film instability. First, assume steady-state and equilibrium (fully developed) flow, write the two-phase momentum equations, and eliminate the pressure gradient between them to get (see Fig. 7.3)

$$\frac{\tau_w p_f}{A} + \frac{\tau_I p_I}{A}\left(\frac{1}{1-\alpha} + \frac{1}{\alpha}\right) - \Delta\rho g \sin\theta = 0, \tag{7.41}$$

where for a circular channel

$$A = \pi D/4, \quad p_f = \pi D, \quad p_I = \pi D\sqrt{\alpha}, \quad \alpha = 1 - \frac{2\delta_F}{D}, \tag{7.42}$$

$$\tau_I = f_I \rho_G \frac{1}{2} \frac{j_L^2}{(1-\alpha)^2}, \tag{7.43}$$

$$f_I = f_G\left(1 + \frac{300\delta_F}{D}\right), \tag{7.44}$$

$$\tau_w = f_w \frac{1}{2} \rho_L \frac{j_L^2}{(1-\alpha)^2}. \tag{7.45}$$

Parameters f_G and f_w can be calculated from common channel single-flow correlations. For a known (j_G, j_L) pair, Eq. (7.41) can be solved to obtain δ_F or α. Mechanism

(a) is assumed to disrupt the annular flow regime when the void fraction calculated from Eq. (7.41) satisfies

$$1 - \alpha > \frac{1}{2}(1 - \alpha)_{\max}, \ (1 - \alpha)_{\max} = 0.48. \tag{7.46}$$

To model mechanism (b), algebraically solve Eq. (7.41) for τ_I, and apply the following to obtain $\delta_{F, \text{crit}}$:

$$\frac{\partial \tau_I}{\partial \delta_F} = 0. \tag{7.47}$$

The annular flow regime is disrupted when $\delta_F \geq \delta_{F, \text{crit}}$, where δ_F represents the prediction of Eq. (7.4).

7.5 Dynamic Flow Regime Models and Interfacial Surface Area Transport Equation

In multi-fluid modeling, separate sets of conservation equations are used, with each set representing one "fluid." In the 2FM, for example, each of the liquid and gas phases is represented by a set of conservation equations. The "fluids" interact with one another through their common interfacial areas. The rate of interfacial transport processes thus depends strongly on both the magnitude and configuration of the interphase. In the past, the most common approach to modeling the interphase has been to use flow regime maps or regime transition models, along with flow-regime-dependent constitutive relations. For example, we can use separate correlations for interfacial surface area concentration, interfacial drag, and heat transfer for bubbly, slug, churn, and annular flow regimes.

The application of the essentially static flow regime transition models with multi-fluid conservation equations is, however, in principle problematic. This is because the static flow regime transition models do not capture the dynamic variations of the interphase and can lead to instantaneous flow regime changes during simulations. Not only are these changes unphysical, but they can introduce mathematical discontinuities and cause spurious numerical oscillations.

It has been argued that the interfacial area concentration in gas–liquid two-phase flow is in fact a transported property. The theoretically correct way of treating the interfacial area is thus by an appropriate transport equation. Ishii (1975) proposed a transport equation for the local volumetric surface area concentration a_I'' (interfacial surface area per unit mixture volume). Studies addressing statistical, averaging, and other issues have since been published (Revankar and Ishii, 1992; Kocamustafaogullari and Ishii, 1995; Millies *et al.*, 1996; Morel *et al.*, 1999; Wu *et al.*, 1998; Kim *et al.*, 2002; Hibiki and Ishii, 2001; Ishii *et al.*, 2002; Sun *et al.*, 2004a, b; Ishii and Hibiki, 2011). The methodology of using an interfacial transport equation has been applied in some thermal-hydraulics codes. The VIPRE-02 code (Kelly, 1994), for example, is a thermal-hydraulics code that uses a dynamic flow regime model developed by Stuhmiller (1986, 1987). CULDESAC, a three-fluid model for vapor explosion analysis, is another example (Fletcher, 1991). The method has been used for modeling of critical two-phase flow (Geng and Ghiaasiaan, 2000).

A brief discussion of the methodology developed by Ishii and co-workers (Wu et al., 1998; Ishii et al., 2002; Sun et al., 2004a, b; Ishii and Hibiki, 2011) is presented in the following. It must be emphasized, however, that the method is still in development.

7.5.1 The Interfacial Area Transport Equations

Consider a flow regime where one of the phases is dispersed (e.g., bubbly flow), and define

$f(V_P, \vec{x}, \vec{U}_P, t) =$ distribution function of particles of the dispersed phase
$$[\text{in particles/m}^6(\text{m/s})^3],$$

where V_P is the particle volume, $\vec{x}$ is the position vector, $\vec{U}_P$ is the particle velocity, and t is time. The total number of particles per unit mixture volume at time t and location $\vec{x}$, $n_P(\vec{x}, t)$, can then be represented as

$$n_P(\vec{x}, t) = \int_{V_{P,\min}}^{V_{P,\max}} \int_{U_{P,x,\min}}^{U_{P,x,\max}} \int_{U_{P,y,\min}}^{U_{P,y,\max}} \int_{U_{P,z,\min}}^{U_{P,z,\max}} f(V_P, \vec{x}, \vec{U}_P, t) dV_P dU_{P,x} dU_{P,y} dU_{P,z}. \quad (7.48)$$

Assuming for simplicity that $f = f(V_P, \vec{x}, t)$ (meaning that f now has the dimensions of particles/m^6), we can write the transport equation for the distribution function as

$$\frac{\partial f}{\partial t} + \nabla \cdot (f \vec{U}_P) + \frac{\partial}{\partial V_P}\left(f \frac{dV_P}{dt}\right) = \sum_j S_j + S_{\text{ph}}, \quad (7.49)$$

where dV_P/dt represents the Lagrangian rate of change of particle volume; S_j are the source and sink terms for particles from collapse, breakup, and coalescence; and S_{ph} represents the source term from phase change (e.g., bubble nucleation).

The transport equation for number of particles can be obtained by applying $\int_{V_{P,\min}}^{V_{P,\max}} dV_P$ to all terms in Eq. (7.49); this results in

$$\frac{\partial n_P}{\partial t} + \nabla \cdot (n_P \vec{U}_{P,m}) = \sum_j R_j + R_{\text{ph}}, \quad (7.50)$$

where R_j is the generation rate of particles, per unit mixture volume (in particles per meter cubed per second) from mechanism j; R_{ph} is the generation rate of particles, per unit mixture volume, from nucleation; and the particle mean velocity is defined as

$$\vec{U}_{P,m} = \frac{1}{n_P} \int_{V_{P,\min}}^{V_{P,\max}} f(V_P, \vec{x}, t) \vec{U}_P(V_P, \vec{x}, t) dV_P. \quad (7.51)$$

A transport equation for the void fraction can also be derived, by bearing in mind that

$$\alpha(\vec{x}, t) = \int_{V_{P,\min}}^{V_{P,\max}} f(V_P, \vec{x}, t) V_P dV_P. \quad (7.52)$$

Then, based on conservation of volume principles, we can directly derive

$$\frac{\partial}{\partial t}(\alpha \rho_G) + \nabla \cdot (\alpha \rho_G \vec{U}_G) - \Gamma_G = 0, \tag{7.53}$$

where Γ_G is the total rate of generation of the gas phase (from phase change), per unit mixture volume (in kilograms per meter cubed per second). This equation can also be proved from Eq. (7.52).

The interfacial area transport equation can now be derived by multiplying Eq. (7.49) by the particle surface area, and integrating the product over the entire distribution function f. When the dispersed particles are gaseous spheres (e.g., in bubbly flow), this leads to

$$\frac{\partial a_I''}{\partial t} + \nabla \cdot (a_I'' \vec{U}_I) = \frac{2}{3} \left(\frac{a_I''}{\alpha} \right) \left[\frac{\partial \alpha}{\partial t} + \nabla \cdot (\alpha \vec{U}_G) - \dot{Q}_{ph} \right] + \int_{V_{P,min}}^{V_{P,max}} \left(\sum_j S_j + S_{ph} \right) A_P dV_P, \tag{7.54}$$

where $\dot{Q}_{ph}$ represents the total volumetric gas generation rate from nucleation, etc., per unit mixture volume; A_P is the average surface area of the fluid particles (bubbles) that have volume V_P; and $\vec{U}_I$ represents the velocity of the interphase and is defined as

$$\vec{U}_I(\vec{x}, t) = \frac{\int_{V_{P,min}}^{V_{P,max}} f(V_P, \vec{x}, t) A_P(V_P) \vec{U}_P(V_P, \vec{x}, t) dV_P}{\int_{V_{P,min}}^{V_{P,max}} (V_P, \vec{x}, t) A_P(V_P) dV_P}. \tag{7.55}$$

7.5.2 Simplification of the Interfacial Area Transport Equation

The interfacial area transport equation derived in Section 7.5.1 is difficult to use despite the aforementioned assumptions and restrictions, because of the complexity of the source and sink terms. Theoretical and semi-empirical formulation of these terms poses the major challenge for the application of the interfacial area transport method.

The terms in the transport equation can be modeled for some simple flow configurations when additional assumptions are made. For the bubbly regime, for example, one can represent source and sink terms, respectively, as

$$\int_{V_{P,min}}^{V_{P,max}} \sum_j S_j dV_P = \sum_j R_j, \tag{7.56}$$

and

$$\int_{V_{P,min}}^{V_{P,max}} \sum_j S_j A_P dV_P = \sum_j R_j \Delta A_P, \tag{7.57}$$

where ΔA_P represents the extra surface area, per particle, resulting from mechanism j. Now, assuming that (a) the coalescence of two equal-volume bubbles leads to a single bubble, and breakup of a bubble leads to two-equal volume bubbles, and (b) the bubbles resulting from nucleation have a diameter of d_{Bc} at birth, one can show that

$$n_P = \psi \frac{(a_I'')^3}{\alpha^2}, \tag{7.58}$$

$$\psi = \frac{1}{36\pi}(d_{Sm}/d_C)^3, \tag{7.59}$$

$$\Delta A_P = \begin{cases} -0.413 A_P & \text{for coalescence,} \\ 0.260 A_P & \text{for breakup,} \end{cases}$$

$$d_{Sm} = \frac{6\alpha}{a_I''} \quad \text{(Sauter mean diameter)},$$

$$d_C = \left(\frac{6V_P}{\pi}\right) \text{(volume-equivalent diameter).} \tag{7.60}$$

The interfacial surface area transport equation then becomes

$$\frac{\partial a_I''}{\partial t} + \nabla.(a_I'' \vec{U}_I) = \frac{2}{3}\frac{a_I''}{\alpha}\left[\frac{\partial \alpha}{\partial t} + \nabla.(\alpha \vec{U}_G) - \dot{Q}_{Ph}\right]$$

$$+ \frac{1}{3\psi}\left(\frac{\alpha}{a_I''}\right)^2 \sum_j R_j + \pi d_{Bc}^2 R_{Ph}, \tag{7.61}$$

where R_{Ph} is the rate of appearance of nucleation-generated bubbles, per unit mixture volume.

The one-group interfacial area transport equation represents the simplest method for the derivation of workable closure relations for Eq. (7.61). For the dispersed bubbly flow regime, assume (a) (approximately) spherical and uniform bubble size, (b) uniform nucleation bubble size, and (c) nucleation-generated bubbles that are much smaller than regular bubbles. As a result of these assumptions, $d_{Sm} = d_C$ and $\psi = \frac{1}{36\pi}$.

When flow-area averaging is performed, furthermore,

$$\langle \vec{U}_I \rangle \equiv \frac{\langle \vec{U}_I a_I'' \rangle}{\langle a_I'' \rangle} \approx \langle \vec{U}_G \rangle_G. \tag{7.62}$$

The terms R_j are evidently needed for the solution of Eq. (7.61). For dispersed bubbly flow, the following expressions have been derived by using simple mechanistic models, with constants that were quantified in steady-state adiabatic air–water experiments in vertical test sections (Wu *et al.*, 1998; Ishii *et al.*, 2002).

For disintegration resulting from impaction by turbulent eddies,

$$R_{TI} = C_{TI}\left(\frac{n_P u_t}{d_P}\right)\exp\left(-\frac{We_{cr}}{We}\right)\sqrt{1 - \frac{We_{cr}}{We}}, \tag{7.63}$$

with

$$C_{TI} = 0.085, \tag{7.64}$$

$$We_{cr} = 6.0 \quad \text{(critical Weber number)} \tag{7.65}$$

and

$$We = \rho_L d_P u_t^2/\sigma \quad \text{(bubble Weber number),} \tag{7.66}$$

where u_t is the root mean square of turbulent velocity fluctuations separated by the distance d_P. For inertial eddies in a locally isotropic turbulent field (see Section 3.6,

Eq. (3.54)),

$$u_t = \sqrt{\overline{\Delta u'^2}} \approx 1.38\varepsilon^{1/3}d_P^{1/3}, \tag{7.67}$$

where ε represents the turbulent energy dissipation, per unit mass.

For collision-induced coalescence resulting from random turbulent motion,

$$R_{RC} = -C_{RE}\left[\frac{n_P^2 u_t d_P^2}{\alpha_{max}^{1/3}(\alpha_{max}^{1/3} - \alpha^{1/3})}\right]\cdot\left[1 - \exp\left(-C\frac{\alpha_{max}^{1/3}\alpha^{1/3}}{\alpha_{max}^{1/3} - \alpha^{1/3}}\right)\right], \tag{7.68}$$

where

$$C_{RE} = 0.004, \tag{7.69}$$

$$C = 3.0, \tag{7.70}$$

and

$$\alpha_{max} = 0.75. \tag{7.71}$$

For coalescence resulting from the acceleration of a bubble caused by the wake of a preceding bubble,

$$R_{WE} = C_{WE}C_D^{1/3}n_P^2 d_P^2|U_G - U_L|, \tag{7.72}$$

where

$$C_{WE} = 0.002, \tag{7.73}$$

and C_D is the bubble drag coefficient.

Also,

$$\dot{Q}_{Ph} = \frac{\pi}{6}d_{Bc}^3 R_{Pc}. \tag{7.74}$$

The parameters d_{Bc} and R_{Pc} should be modeled separately for the processes of interest.

Equation (7.61) should be solved along with Eq. (7.53), which simply represents mass conservation for the gas phase, for the transport of void fraction. These expressions are valid for co-current, upward flow in a vertical channel. The transport equations are applicable for co-current, downward flow with some modifications to the closure relations (Ishii *et al.*, 2004).

7.5.3 Two-Group Interfacial Area Transport Equations

An obvious shortcoming of the one-group interfacial area transport equation (ITAE) is that it lumps the entire bubble population into a single group. In reality, depending on the flow regime the bubbles cover a vast range of sizes, and their shapes, hydrodynamic behavior and transport characteristics are not uniform. In most flow regimes where liquid is the continuous phase and gas is discontinuous, however, bubbles are predominantly of five groups: spherical, distorted, cap, Taylor, and irregular-shaped characteristic of the churn-turbulent regime. In the *two-group interfacial area transport* model the bubbles are divided into two groups: Group 1 includes bubbles that remain spherical or distorted-spherical, and Group 2 includes other larger bubbles (cap bubbles, Taylor bubbles, and irregular-shaped bubbles that occur in the churn-turbulent regime) (Ishii *et al.*, 2002; Sun *et al.*, 2004a, b; Ishii and Hibiki, 2011). The interfacial geometric variables that characterize the two-phase flow structure will be:

Group 1 volume fraction and interfacial area concentration: $a_{I,1}''$ and α_1, respectively.

Group 2 volume fraction and interfacial area concentration: $a_{I,2}''$ and α_2, respectively.

The constitutive requirement for the volume fractions is

$$\alpha_1 + \alpha_2 = \alpha. \tag{7.75}$$

Separate transport equations are developed for the interfacial area concentration associated with each group, as well as their volume fractions, with terms that account for the transfer of interfacial area from one group into another due to bubble coalescence or breakup, and caused by a multitude of mechanisms (Sun et al., 2004a).

In this section the two-group transport equations are presented and briefly discussed. It will be noted that the model needs a large number of closure relations, however, that in general depend on scale, geometric configuration, flow orientation, turbulence level, relative significance of buoyancy, etc. Currently, closure relations are in a reasonably developed state for vertical, upward flow. A detailed account about the derivation of these equations can be found in Ishii and Hibiki (2011).

The bubbles are divided into two groups, as mentioned earlier. The bubble volume that represents the boundary between distorted bubbles and bubble caps, according to Ishii and Zuber (1979) is

$$V_{B,c} = \frac{\pi}{6} d_{B,c}^3 \tag{7.76}$$

where

$$d_{B,c} = 4\sqrt{\frac{\sigma}{g(\rho_L - \rho_G)}}. \tag{7.77}$$

Bubbles with volumes smaller than $V_{B,c}$ are included in Group 1 and larger bubbles constitute Group 2. The interfacial area transport equations for the two groups are (Ishii and Hibiki, 2011)

$$\frac{\partial a_{I,1}''}{\partial t} + \nabla \cdot (a_{I,1}'' \vec{U}_{I,1}) = \left[\frac{2}{3} - \chi \left(\frac{d_{sc}}{d_{Sm,1}} \right)^2 \right] \frac{a_{I,1}''}{\alpha_1} \left[\frac{\partial \alpha_1}{\partial t} + \nabla \cdot (\alpha_1 \vec{U}_{G,1}) - \dot{Q}_{ph,1} \right]$$
$$+ \sum_j \phi_{j,1} + \phi_{ph,1} \tag{7.78}$$

$$\frac{\partial a_{I,2}''}{\partial t} + \nabla \cdot (a_{I,2}'' \vec{U}_{I,2}) = \frac{2}{3} \frac{a_{I,2}''}{\alpha_2} \left[\frac{\partial \alpha_2}{\partial t} + \nabla \cdot (\alpha_2 \vec{U}_{G,2}) - \dot{Q}_{ph,2} \right]$$
$$+ \chi \left(\frac{d_{sc}}{d_{Sm,1}} \right)^2 \frac{a_{I,1}''}{\alpha_1} \left[\frac{\partial \alpha_1}{\partial t} + \nabla \cdot (\alpha_1 \vec{U}_{G,1}) - \dot{Q}_{ph,1} \right]$$
$$+ \sum_j \phi_{j,2} + \phi_{ph,2}, \tag{7.79}$$

where $d_{Sm,1}$ represents the Sauter mean diameter of Group 1 bubbles. In writing these equations we have assumed that phase change (nucleation or condensation) only influence bubbles in Group 1. The coefficient χ is related to intergroup transfer and is neglected in the simplified model. The source and sink terms representing bubble–bubble interaction mechanism j and the phase change are:

$$\phi_{j,1} = \int_{V_{B,\min}}^{V_{sc}} S_j A_\mathrm{P} dV \tag{7.80}$$

$$\phi_{j,2} = \int_{V_{sc}}^{V_{B,\max}} S_j A_\mathrm{P} dV \tag{7.81}$$

$$\phi_{ph,1} = \int_{V_{B,\min}}^{V_{sc}} S_{ph} A_\mathrm{P} dV \tag{7.82}$$

$$\phi_{ph,2} = \int_{V_{sc}}^{V_{B,\max}} S_{ph} A_\mathrm{P} dV. \tag{7.83}$$

The transport equations for group-specific volume fractions are

$$\frac{\partial}{\partial t}(\alpha_1 \rho_\mathrm{G}) + \nabla \cdot (\alpha_1 \rho_\mathrm{G} \vec{U}_{\mathrm{G},1}) = \Gamma_1 - \Gamma_{1,2} \tag{7.84}$$

$$\frac{\partial}{\partial t}(\alpha_2 \rho_\mathrm{G}) + \nabla \cdot (\alpha_2 \rho_\mathrm{G} \vec{U}_{\mathrm{G},2}) = \Gamma_2 + \Gamma_{1,2} \tag{7.85}$$

where

$\Gamma_1, \Gamma_2 =$ volumetric vapor mass generation rate associated with Group 1 and 2 bubbles, respectively,

$\Gamma_{1,2} =$ volumetric mass transfer rate from Group 1 bubbles to Group 2 bubbles.

We need expressions that relate the group-specific parameters to gas phase parameters that can be used in two-fluid model equations. Evidently, we must have

$$\Gamma_1 + \Gamma_2 = \Gamma. \tag{7.86}$$

The vapor phase velocity, furthermore, follows

$$\vec{U}_\mathrm{G} = (\vec{U}_{\mathrm{G},1} \alpha_1 + \vec{U}_{\mathrm{G},2} \alpha_2)/\alpha. \tag{7.87}$$

It is possible to derive separate momentum conservation equations for the two bubble groups. It is simpler to treat the mixture as a three-phase flow composed of liquid, Group 1 gas and Group 2 gas, and assume that Group 1 and Group 2 bubbles do not directly interact with respect to momentum exchange, which leads to (Ishii and Hibiki, 2011)

$$\frac{\partial}{\partial t}(\alpha \rho_\mathrm{G} \vec{U}_\mathrm{G}) + \nabla \cdot (\alpha \rho_\mathrm{G} \vec{U}_\mathrm{G} \vec{U}_\mathrm{G}) = -\alpha \nabla P_\mathrm{G} + \alpha \nabla \cdot (\bar{\bar{\tau}}_\mathrm{G} + \bar{\bar{\tau}}_{\mathrm{G,\,turb}})$$
$$- \nabla \cdot \left[\rho_\mathrm{G} \frac{\alpha_1 \alpha_1}{\alpha} |\vec{U}_{\mathrm{G},1} - \vec{U}_{\mathrm{G},2}|^2 \right]$$
$$+ [\Gamma_1 \vec{U}_{\mathrm{IG},1} + \Gamma_2 \vec{U}_{\mathrm{IG},2} + \Gamma_{12}(\vec{U}_{\mathrm{IG},2} - \vec{U}_{\mathrm{IG},1})]$$
$$+ (P_{\mathrm{G,\,I}} - P_\mathrm{G})\nabla \alpha + \vec{\mathrm{M}}_{\mathrm{IG},1} + \vec{\mathrm{M}}_{\mathrm{IG},2} + \alpha \rho_\mathrm{G} \vec{g} \tag{7.88}$$

$$\frac{\partial}{\partial t}[(1-\alpha)\rho_\mathrm{L} \vec{U}_\mathrm{L}] + \nabla \cdot [(1-\alpha)\rho_\mathrm{L} \vec{U}_\mathrm{L} \vec{U}_\mathrm{L}] = -(1-\alpha)\nabla P_\mathrm{L} + (1-\alpha)\nabla \cdot (\bar{\bar{\tau}}_\mathrm{L} + \bar{\bar{\tau}}_{\mathrm{L,turb}})$$
$$+ (P_{\mathrm{L,I}} - P_\mathrm{L})\nabla(1-\alpha) + \Gamma \vec{U}_\mathrm{IL} + \vec{\mathrm{M}}_\mathrm{IL} + (1-\alpha)\rho_\mathrm{L} \vec{g}. \tag{7.89}$$

As noted, the gas phase pressure has been assumed to be identical for the two bubble groups. The generalized drag force terms must satisfy the requirement of tangential force balance for the interphase, which leads to

$$\vec{M}_{IL} + \vec{M}_{IG,1} + \vec{M}_{IG,2} = 0. \tag{7.90}$$

The thermal energy equations are derived by assuming that the two bubble groups have the same temperature, and assuming that the average stresses in the liquid phase and the interphase are approximately equal.

$$\frac{\partial}{\partial t}(\alpha \rho_G h_G) + \nabla \cdot (\alpha \rho_G h_G \vec{U}_G) = -\nabla \cdot [\alpha(\vec{q}_G'' + \vec{q}_{G,\text{turb}}'')] + \alpha\left(\frac{\partial}{\partial t} + \vec{U}_G \cdot \nabla\right) P_G$$
$$+ h_{GI}\Gamma - a_I'' q_{GI}'' + (\mu\Phi)_G \tag{7.91}$$

$$\frac{\partial}{\partial t}[(1-\alpha)\rho_L h_L] + \nabla \cdot [(1-\alpha)\rho_L h_L \vec{U}_L] = -\nabla \cdot [(1-\alpha)(\vec{q}_L'' + \vec{q}_{L,\text{turb}}'')]$$
$$+ (1-\alpha)\left(\frac{\partial}{\partial t} + \vec{U}_L \cdot \nabla\right) P_L$$
$$- h_{LI}\Gamma + a_I'' q_{LI}'' + (\mu\Phi)_L. \tag{7.92}$$

Note that, consistent with the analysis in Section 5.7 (see Fig. 5.9), q_{GI}'' is positive when it represents heat flow from gas bulk towards the interphase, while positive q_{LI}'' is positive when it represents heat flow from the interphase towards the liquid bulk. Furthermore, average gas and liquid-phase parameters are defined as

$$\vec{q}_G'' = \frac{\alpha_1 \vec{q}_{G,1}'' + \alpha_2 \vec{q}_{G,2}''}{\alpha} \tag{7.93}$$

$$\vec{q}_{G,\text{turb}}'' = \frac{\alpha_1 \vec{q}_{G,\text{turb},1}'' + \alpha_2 \vec{q}_{G,\text{turb},2}''}{\alpha} \tag{7.94}$$

$$\vec{q}_{GI}'' = \frac{a_{I,1}'' \vec{q}_{GI,1}'' + a_{I,2}'' \vec{q}_{GI,2}''}{a_I''} \tag{7.95}$$

$$\vec{q}_{LI}'' = \frac{a_{I,1}'' \vec{q}_{LI,1}'' + a_{I,2}'' \vec{q}_{LI,2}''}{a_I''} \tag{7.96}$$

$$a_I'' = a_{I,1}'' + a_{I,2}''. \tag{7.97}$$

The viscous dissipation terms $(\mu\Phi)_G$ and $(\mu\Phi)_L$ are defined similarly to Eq. (5.106a).

The energy balance at the interphase can be formulated following Section 5.7. Neglecting the kinetic energy changed during phase change, one can derive

$$\Gamma_1 = \frac{a_{I,1}''(q_{GI,1}'' - q_{LI,1}'')}{h_{fg}} \tag{7.98}$$

$$\Gamma_2 = \frac{a_{I,2}''(q_{GI,2}'' - q_{LI,2}'')}{h_{fg}} \tag{7.99}$$

$$q_{GI,1}'' = \dot{H}_{GI,1}(T_G - T_{I,1}) \tag{7.100}$$

$$q_{GI,2}'' = \dot{H}_{GI,2}(T_G - T_{I,2}) \tag{7.101}$$

$$q_{LI,1}'' = \dot{H}_{LI,1}(T_{I,1} - T_L) \tag{7.102}$$

$$q_{LI,2}'' = \dot{H}_{LI,2}(T_{I,2} - T_L). \tag{7.103}$$

A multitude of closure relations are evidently needed, the most complicated among them being the bubble–bubble interactions. Detailed discussions about such hydrodynamic interactions can be found in Sun *et al.* (2004a, b), Ishii and Hibiki (2011), Smith *et al.* (2012), and Yang *et al.* (2013).

PROBLEMS

7.1 Prepare a flow chart that can be used for determining the flow regime in vertical upward pipe flow based on the flow regime transition models of Taitel *et al.* (1980).

7.2 A two-phase mixture of saturated water and steam at a pressure of 2 bars flows with a mass flux of $G = 600$ kg/m^2·s through a round vertical tube that has an inner diameter of 2.5 cm.

(a) Using the flow regime transition models of Taitel *et al.* (1980), estimate the qualities at which transitions from bubbly to slug, and from churn to annular, take place.

(b) Repeat part (a), this time using the flow regime map of Hewitt and Roberts (1969).

7.3 Solve Problem 4.1 using the flow regime transition models of Taitel *et al.* (1980).

7.4 Solve Problem 4.3 using the flow regime transition models of Taitel and Dukler (1976a, b).

7.5 A circular channel with uniform cross section is inclined with respect to the horizontal plane by an angle θ. The channel supports a stratified, steady countercurrent flow. Liquid and gas volumetric flow rates have equal absolute values. Wall–liquid, wall–gas, and gas–liquid frictional forces can be represented according to

$$F_{wL} = \frac{p_L}{A}\tau_{wL} = \frac{p_L}{A}f_L\frac{1}{2}\rho_L|U_L|U_L,$$

$$F_{wG} = \frac{p_G}{A}\tau_{wG} = \frac{p_G}{A}f_G\frac{1}{2}\rho_G|U_G|U_G,$$

$$F_I = \frac{p_I}{A}\tau_I = \frac{p_I}{A}f_I\frac{1}{2}\rho_G|U_G - U_L|(U_G - U_L).$$

Prove that the axial variation of the void fraction α follows

$$\frac{d\alpha}{dz} = -\frac{a}{c} + \frac{b + a\frac{d}{c}}{d - cj_L^2}.$$

Derive relations for a, b, c, and d.

7.6 A horizontal tube supports an adiabatic annular two-phase flow, under steady, equilibrium conditions. The wall–liquid and gas–liquid interfacial shear stresses can be represented, respectively, by

$$\tau_{wL} = f_{wL}\frac{1}{2}\rho_L|U_L|U_L,$$

$$\tau_{wG} = f_G\frac{1}{2}\rho_G|U_G|U_G.$$

(a) Derive a relation among α, j_L, and j_G.

(b) Prove that, when $U_G \gg U_L$, the relation derived in part (a) reduces to

$$\frac{\alpha^{5/2}}{(1-\alpha)^2} = \frac{f_I \rho_G}{f_{wL} \rho_L} \left(\frac{j_G}{j_L}\right)^2.$$

7.7 The subchannels in a once-through steam generator can be idealized as vertical tubes 3.7 m long with $D = 1.25$ cm.

(a) For the two-phase flow of saturated water and steam at 71 bar pressure, plot the flow regime map based on the transition models of Taitel *et al.* (1980).

(b) Repeat part (a), this time assuming that an atmospheric, room-temperature air–water mixture constitutes the two-phase flow. Compare the two flow regime maps, and discuss the similarities and differences.

7.8 Consider the concurrent flow of a two-phase liquefied petroleum gas (LPG) (assumed to be pure propane, for simplicity) in an insulated horizontal tube at a pressure of 8 bars. The tube diameter is 5 cm. The mixture mass flux is 3.0×10^3 kg/m^2·s. For qualities $x = 0.05$ and 0.10, do the following.

(a) Find h_L, the height of the liquid level if the flow regime is an equilibrium stratified.

(b) Find the flow regime using the flow regime transition models of Taitel and Dukler (1976a, b).

Hint: If you need the void fraction, you may use the correlation of Bhagwat and Ghajar (2014), discussed in Section 6.3.

7.9 A unit cell representing a developed slug flow in a vertical tube is shown in Fig. P7.9. In accordance with the definition of the mixture volumetric flux, it can be argued that the unit cell moves upward with a velocity equal to j. The mean velocity of the liquid slug, furthermore, is equal to j. Prove that the following relations also apply:

$$U_F = \frac{U_B(1-\xi)^2 - j}{2\xi - \xi^2},$$

$$U_B = \frac{j + U_F(2\xi - \xi^2)}{(1-\xi)^2},$$

where $\xi = 2\delta_F/D$. Also, show that the assumption that the liquid film thickness is similar to a turbulent falling film and the application of the correlation of Brötz (1954), Eq. (3.106), lead to

$$j_F = 9.916(1-\alpha_B)\sqrt{\frac{gD\Delta\rho(1-\sqrt{\alpha_B})}{\rho_L}},$$

where $\alpha_B = 1 - 4\delta_F/D$.

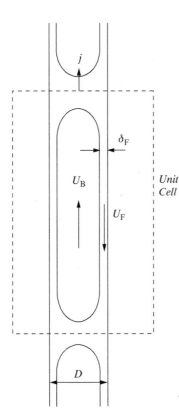

Figure P7.9. Unit cell for ideal slug flow, for Problem 7.9.

7.10 The experimental data in Table P7.10 represent slug-to-churn transition in tests with a room-temperature air–water mixture at 2.4 bars in a vertical tube with $D =$ 3.18 cm. Compare these data with the predictions of the model of Mishima and Ishii (1984).

Table P7.10. *Table for Problem 7.10.*

Liquid mass flux, G_L(kg/m^2·s)	Gas mass flux, G_G (kg/m^2·s)
5.3	6.5–8.0
10.5	7.1–9.0
111.8	9.6–11.3
297	9.0–14.3

7.11 Weisman *et al.* (1979) have proposed the following regime transition correlations for co-current gas–liquid flow in horizontal pipes:

$$\left[\frac{\sigma}{gD^2\Delta\rho}\right]^{0.2}\left[\frac{\rho_G\, j_G D}{\mu_G}\right]^{0.45} = 8(j_G/j_L)^{0.16} \quad \text{(stratified to wavy)},$$

$$\frac{j_G}{\sqrt{gD}} = 0.25(j_G/j_L)^{1.1} \quad \text{(stratified to intermittent)},$$

$$\left[\frac{(-dP/dz)_{\text{fr,L}}}{\Delta\rho g}\right]^{1/2}\left(\frac{\sigma}{\Delta\rho gD^2}\right)^{-0.25} = 9.7 \quad \text{(intermittent to dispersed bubbly)},$$

$$1.9(j_{\text{G}}/j_{\text{L}})^{1/8} = \left[\frac{j_{\text{G}}\sqrt{\rho_{\text{G}}}}{(\sigma g\Delta\rho)^{1/4}}\right]^{0.2}\left[\frac{j_{\text{G}}^2}{gD}\right]^{0.18} \quad \text{(transition to annular)}.$$

Assume atmospheric air–water flow in a 5-cm-diameter horizontal pipe.

(a) Calculate and plot the stratified-to-intermittent regime transition line on the Mandhane (j_{L}, j_{G}) coordinates and compare it with the relevant transition line(s) in the flow regime map of Mandhane *et al.* (1974).

(b) Calculate and plot the transition to the annular flow regime lines and compare them with the relevant transition line(s) of Baker (1954).

(c) For $j_{\text{G}} = 1$ and 5 m/s, calculate the j_{L} values for transition to dispersed bubbly flow, and compare with the relevant values according to the flow regime map of Mandhane *et al.* (1974).

7.12 A liquid–gas mixture with properties similar to water and air at room temperature and atmospheric pressure flows upward through a 5-cm-diameter tube that is inclined with respect to the horizontal plane. The superficial velocity of gas is 0.85 m/s.

(a) Calculate the minimum liquid superficial velocities needed for the finely dispersed bubbly flow regime for angles of inclination $\theta = 10°, 30°$, and $90°$.

(b) For $\theta = 30°$ and $60°$, calculate the liquid superficial velocity that would represent the bubbly–slug flow regime transition.

7.13 The interfacial surface area concentration a_I'' in two-phase pipe flow has been measured and correlated by many investigators. The following correlation has been developed for the bubbly flow regime by Delhaye and Bricard (1994);

$$a_I''\sqrt{\frac{\sigma}{g\Delta\rho}} = 10^{-3}\left[7.23 - \frac{6.82\,\text{Re}_{\text{L}}}{\text{Re}_{\text{L}} + 3240}\right]\text{Re}_{\text{G}}.$$

Consider an air–water mixture at room temperature and atmospheric pressure flowing in a vertical tube that is 1.25 cm in diameter, with $j_{\text{L}} = 6.5$ m/s. Using the Delhaye and Bricard correlation, and an appropriate method for estimation of the void fraction, calculate a_I'', and estimate the average bubble diameter and number density for $j_{\text{G}} = 0.75$ and 1.25 m/s. For the latter calculations, assume uniform-size bubbles.

8 Pressure Drop in Two-Phase Flow

8.1 Introduction

Consider the channel shown schematically in Fig. 8.1. The cross-section-averaged two-phase mixture momentum equation, Eq. (5.72), can be written as

$$\left(-\frac{\partial P}{\partial z}\right) = \left(-\frac{\partial P}{\partial z}\right)_{ta} + \left(-\frac{\partial P}{\partial z}\right)_{sa} + \left(-\frac{\partial P}{\partial z}\right)_{g} + \left(-\frac{\partial P}{\partial z}\right)_{fr}, \tag{8.1}$$

where

$\left(-\frac{\partial P}{\partial z}\right)$ = channel total pressure gradient,

$\left(-\frac{\partial P}{\partial z}\right)_{ta} = \frac{\partial G}{\partial t}$ = temporal mixture acceleration,

$\left(-\frac{\partial P}{\partial z}\right)_{sa} = \frac{1}{A}\frac{\partial}{\partial z}\left(A\frac{G^2}{\rho'}\right)$ = spatial mixture acceleration,

$\left(-\frac{\partial P}{\partial z}\right)_{g} = \rho g \sin\theta$ = hydrostatic pressure gradient,

$\left(-\frac{\partial P}{\partial z}\right)_{fr} = \tau_w P_f/A$ = frictional pressure gradient.

The acceleration terms are often important in two-phase flows with phase change. In steady-state boiling or condensing flows, for example, the magnitude of the spatial acceleration term is often larger than the frictional pressure gradient.

When Eq. (8.1) is integrated along a pipe system, other terms appear that cannot be included in the differential one-dimensional model equations. These terms result from the *form (minor) pressure drops* and are caused by abrupt changes in flow area or flow path, as well as various control and regulation devices (e.g., valves, orifices, bends, and perforated plates). These pressure drops (which, as will be shown later, can be positive or negative) result from complicated multi-dimensional hydrodynamic processes. Integration of Eq. (8.1) for Fig. 8.1 thus leads to

$$P_1 - P_O = \int_{z_1}^{z_0} \left\{\left(-\frac{\partial P}{\partial z}\right)_{ta} + \left(-\frac{\partial P}{\partial z}\right)_{sa} + \left(-\frac{\partial P}{\partial z}\right)_{g} + \left(-\frac{\partial P}{\partial z}\right)_{fr}\right\} dz + \sum_{i=1}^{N} \Delta P_i, \tag{8.2}$$

where ΔP_i is the *total* pressure drop due to flow disturbance i, and N is the total number of flow disturbances. In writing Eq. (8.2), the pressure drops associated with flow disturbances have been treated as if each one of them occurs at a single point. This of course is not physically true and would introduce discontinuity into the differential equations and cause difficulties for their numerical solution. In practice, when numerical solutions of conservation equations are sought, these pressure drops are often assumed to occur over a finite length of the piping system.

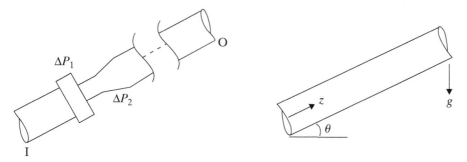

Figure 8.1. Schematic of a flow channel and a one-dimensional flow system.

In the forthcoming sections we first discuss frictional pressure drop. The minor pressure drops will then be reviewed. The discussion in this chapter will be focused on general-purpose predictive methods. Models and correlations that have been developed specifically for small flow passages (mini- and microchannels) will be discussed in Chapter 10. Predictive methods that are specifically derived for condensing internal flows will be discussed in Chapter 16.

8.2 Two-Phase Frictional Pressure Drop in Homogeneous Flow and the Concept of a Two-Phase Multiplier

In the homogeneous mixture (HM) model the two phases are assumed to remain well mixed and move with identical velocities everywhere. A homogeneous mixture thus acts essentially as a single-phase fluid that is compressible and has variable properties. A simple method for calculating the HM two-phase pressure drop can therefore be developed by analogy with single-phase flow.

Let us consider an HM flow along a one-dimensional conduit with s representing the axial coordinate along the conduit. Recall that, for a turbulent single-phase flow,

$$\left(-\frac{\partial P}{\partial z}\right)_{\text{fr}} = 4f\frac{1}{D_{\text{H}}}\frac{G^2}{2\rho}. \tag{8.3}$$

Let us use Blasius's correlation for the Fanning friction factor, f

$$f = 0.079\text{Re}^{-0.25}, \tag{8.4}$$

where $\text{Re} = GD/\mu$. Similarly, let us write for the homogeneous two-phase flow

$$\left(-\frac{\partial P}{\partial z}\right)_{\text{fr}} = 4f_{\text{TP}}\frac{1}{D_{\text{H}}}\frac{G^2}{2\rho_{\text{TP}}}, \tag{8.5}$$

$$f_{\text{TP}} = 0.079\text{Re}_{\text{TP}}^{-0.25}, \tag{8.6}$$

$$\rho_{\text{TP}} = \rho_{\text{h}} = \left(\frac{x}{\rho_{\text{G}}} + \frac{1-x}{\rho_{\text{L}}}\right)^{-1}, \tag{8.7}$$

$$\text{Re}_{\text{TP}} = \frac{GD_{\text{H}}}{\mu_{\text{TP}}}, \tag{8.8a}$$

where all parameters with subscript TP represent two-phase flow. An appropriate estimate for μ_{TP} is obviously needed. A widely used and simple correlation for the viscosity of a homogeneous gas–liquid two-phase mixture is (McAdams *et al.*, 1942)

$$\mu_{TP} = \left(\frac{x}{\mu_G} + \frac{1-x}{\mu_L} \right)^{-1}. \tag{8.8b}$$

Substitution of Eqs. (8.6) and (8.7) in Eq. (8.5) provides the pressure-drop calculation method we have been seeking. The resulting expression can be presented in four different but equivalent forms:

$$\left(-\frac{\partial P}{\partial z} \right)_{fr} = \Phi_{L0}^2 \left(-\frac{\partial P}{\partial z} \right)_{fr,L0}, \tag{8.9}$$

$$\left(-\frac{\partial P}{\partial z} \right)_{fr} = \Phi_{G0}^2 \left(-\frac{\partial P}{\partial z} \right)_{fr,G0}, \tag{8.10}$$

$$\left(-\frac{\partial P}{\partial z} \right)_{fr} = \Phi_{L}^2 \left(-\frac{\partial P}{\partial z} \right)_{fr,L}, \tag{8.11}$$

$$\left(-\frac{\partial P}{\partial z} \right)_{fr} = \Phi_{G}^2 \left(-\frac{\partial P}{\partial z} \right)_{fr,G}. \tag{8.12}$$

The right-hand-side pressure gradient terms are all single-phase-flow based. The terms with subscripts L0 and G0 correspond to frictional pressure gradients when all the mixture is liquid and gas, respectively, the term with subscript L is the frictional pressure gradient when only pure liquid at a mass flux $G(1-x)$ flows in the channel, and subscript G represents the case when pure gas at mass flux Gx flows in the channel. The parameters Φ_{L0}^2, Φ_{G0}^2, Φ_{L}^2, and Φ_{G}^2 are *two-phase multipliers*. When Eq. (8.9) is used, for example, we have

$$\left(-\frac{\partial P}{\partial z} \right)_{fr,L0} = f_{L0} 4 \frac{1}{D_H} \frac{G^2}{2\rho_L}, \tag{8.13}$$

$$f_{L0} = 0.079 \left(\frac{GD_H}{\mu_L} \right)^{-0.25}, \tag{8.14}$$

$$\Phi_{L0}^2 = \left[1 + x \frac{\mu_L - \mu_G}{\mu_G} \right]^{-1/4} [1 + x(\rho_L/\rho_G - 1)]. \tag{8.15}$$

When Eq. (8.12) is used, then

$$\left(-\frac{\partial P}{\partial z} \right)_{fr,G} = 4 f_G \frac{1}{D_H} \frac{(Gx)^2}{2\rho_G}, \tag{8.16}$$

$$f_G = 0.079 (GxD_H/\mu_G)^{-0.25}, \tag{8.17}$$

$$\Phi_{G}^2 = \left[1 + \frac{\rho_G}{\rho_L} (1-x) \right] x^{-7/4} \left[x + \frac{\mu_G}{\mu_L} (1-x) \right]^{-1/4}. \tag{8.18}$$

It can also be easily shown that

$$\Phi_{G0}^2 = \left[x + \frac{\rho_G}{\rho_L} (1-x) \right] \left[x + \frac{\mu_G}{\mu_L} (1-x) \right]^{-1/4}, \tag{8.19}$$

Table 8.1. *Correlations for two-phase frictional pressure drop based on homogeneous flow assumption.*

Author	Correlation	Comments
Akers *et al.* (1959)	$\mu_{TP} = \dfrac{\mu_L}{\left[(1-x)+x(\rho_L/\rho_G)^{0.5}\right]}$	Based on condensing two-phase flow data
McAdams *et al.* (1942)	Eq. (8.8b)	Effective mixture viscosity
Cicchitti *et al.* (1960)	$\mu_{TP} = x\mu_G + (1-x)\mu_L$	Effective mixture viscosity
Dukler *et al.* (1964)	$\mu_{TP} = \rho_h \left[\dfrac{x\mu_G}{\rho_G} + \dfrac{(1-x)\mu_L}{\rho_L}\right]$	Effective mixture viscosity using density-weighted averaging
Beattie and Whalley (1982)	Eqs. (8.21)–(8.23)	Effective mixture viscosity
Lin *et al.* (1991)	$\mu_{TP} = \dfrac{\mu_L\mu_G}{\mu_G+x^{1.4}(\mu_L-\mu_G)}$	Based on R-12 vaporization data
Awad and Muzychka (2008)	$\mu_{TP} = \mu_L \dfrac{2\mu_L+\mu_G-2x(\mu_L-\mu_G)}{2\mu_L+\mu_G+x(\mu_L-\mu_G)}$ (Definition 3) $\mu_{TP} = \mu_G \dfrac{2\mu_G+\mu_L-2(1-x)(\mu_G-\mu_L)}{2\mu_G+\mu_L+(1-x)(\mu_G-\mu_L)}$ (Definition 4) Arithmetic average of the above two equations (Definition 6)	Based on analogy with thermal conductivity of a porous medium

and

$$\Phi_L^2 = \Phi_{L0}^2(1-x)^{-7/4}. \tag{8.20}$$

This analysis also has introduced us to the concept of *two-phase multipliers*, which provides a good way for correlating two-phase frictional pressure losses. Historically, however, the idea of two-phase multipliers was developed based on an idealized annular flow (Lockhart and Martinelli, 1949).

Besides the correlation of McAdams, Eq. (8.8), several other correlations have been proposed for the homogeneous two-phase flow viscosity. Table 8.1 is a summary of some of these correlations. In these correlations $\beta = j_G/j$ is the volumetric quality, and is equal to α_h, the homogeneous-flow void fraction. The correlation of Akers *et al.* (1958) is based on condensing two-phase flow data. The correlation of Beattie and Whalley (1982) is part of their proposed method for calculating the frictional pressure drop in two-phase flow. The expressions derived by Awad and Muzychka (2008) are all based on the analogy between the viscosity of a gas–liquid two-phase mixture and the effective (volume-average) thermal conductivity of a two-phase porous medium. The expressions of Awad and Muzychka (2008) satisfy the limits of $\lim_{x\to 0}\mu_{TP} = \mu_L$ and $\lim_{x\to 1}\mu_{TP} = \mu_G$.

A simple correlation that appears to do well over a wide range of parameters, including some minichannels and narrow rectangular channels and annuli, is the correlation of Beattie and Whalley (1982). The correlation, which can be considered to be a modification of the homogeneous flow model, is in terms of a two-phase friction factor, f_{TP}:

$$\left(-\frac{\partial P}{\partial z}\right)_{fr} = 4f_{TP}\frac{1}{D_H}\frac{G^2}{2\rho_h}, \tag{8.21}$$

where ρ_h is the homogeneous density, Re_{TP} is defined in Eq. (8.8), and

$$\mu_{TP} = \alpha_h \mu_G + \mu_L(1 - \alpha_h)(1 + 2.5\alpha_h). \tag{8.22}$$

The two-phase friction factor is found from $f_{TP} = f'/4$, and f' is found from the Colebrook–White correlation with $\varepsilon_D = 0$:

$$\frac{1}{\sqrt{f'}} = 1.14 - 2\log_{10}\left[\frac{\varepsilon_D}{D} + \frac{9.35}{Re_{TP}\sqrt{f'}}\right], \tag{8.23}$$

where ε_D is the surface roughness and Re_{TP} is found using Eq. (8.8a).

8.3 Empirical Two-Phase Frictional Pressure Drop Methods

The HM model performs reasonably well when the two-phase flow pattern represents a well-mixed configuration (e.g., dispersed bubbly). It also appears to do well for well-mixed two-phase flow regimes in minichannels. In general, however, it deviates from experimental data. For flow patterns such as annular, slug, and stratified flows, some phenomenological models have been developed in the past, but available models are developmental, and are difficult to use because of the uncertainties associated with the flow regime transitions. Using empirical correlations remain the most widely applied method.

Most empirical correlations use the concept of two-phase flow multipliers and are applicable to all flow regimes (i.e., flow regime transition effects are implicitly included in them). The concept was originally proposed by Lockhart and Martinelli (1949) based on a simple separated-flow model. In general it indicates that

$$\Phi^2 = f(G, x, \text{fluid properties}). \tag{8.24}$$

Note that the HM model analysis in the previous section did not predict dependence on G.

The available empirical methods are numerous. Only a few of the most widely used will be reviewed here.

The *Lockhart–Martinelli* method is among the oldest techniques (Lockhart and Martinelli, 1949). More recent variations include correlations for non-Newtonian liquid–gas two-phase flows and two-phase flow in thin rectangular channels and microchannels. The method is based on a simple and inaccurate model, and it is therefore better to treat it as purely empirical. It assumes that the two-phase multipliers are functions of the *Martinelli parameter* (also referred to as the *Martinelli factor*) defined as

$$X^2 = \frac{\Phi_G^2}{\Phi_L^2} = \frac{\left(-\frac{\partial P}{\partial z}\right)_{fr,L}}{\left(-\frac{\partial P}{\partial z}\right)_{fr,G}}. \tag{8.25}$$

Lockhart and Martinelli (1949) graphically correlated Φ_G^2 and Φ_L^2 as functions of X. The phasic frictional pressure gradients evidently depend on the flow regimes of the phases (viscous or turbulent), when each is assumed to flow alone in the channel. The single-phase flow regimes depend on $Re_G = GxD_H/\mu_G$ and $Re_L = G(1-x)D_H/\mu_L$, and four different combinations could occur. When both Reynolds numbers correspond to turbulent flow (turbulent–turbulent flow), we can use Blasius's correlation (Eq. (8.4)) for the single-phase friction factor to easily show that

$$X_{tt}^2 = \left(\frac{\mu_L}{\mu_G}\right)^{0.25} \left(\frac{1-x}{x}\right)^{1.75} \frac{\rho_G}{\rho_L}, \tag{8.26}$$

where subscript tt is for turbulent–turbulent.

The *Martinelli parameter* contains flow quality x and the phasic properties that are important for most commonly encountered gas–liquid two-phase flows. It has been used in empirical and semi-analytical models dealing with many two-phase flow, boiling, and condensation problems. In such models, often the following approximate form is used:

$$X_{tt} = \left(\frac{\rho_G}{\rho_L}\right)^{0.5} \left(\frac{\mu_L}{\mu_G}\right)^{0.1} \left(\frac{1-x}{x}\right)^{0.9}. \tag{8.27}$$

Simpler algebraic correlations have been proposed based on the Lockhart–Martinelli approach. A widely used correlation is (Chisholm and Laird, 1958; Chisholm, 1967)

$$\Phi_L^2 = 1 + \frac{C}{X} + \frac{1}{X^2}. \tag{8.28}$$

Alternatively,

$$\Phi_G^2 = 1 + CX + X^2. \tag{8.29}$$

The values of coefficient C are (Chisholm, 1967) as follows:

Liquid	Gas	C
Turbulent	Turbulent	20
Viscous	Turbulent	12
Turbulent	Viscous	10
Viscous	Viscous	5

The experimental data of Lockhart and Martinelli (1949), based on which the aforementioned constant values of parameter C were proposed by Chisholm (1967), were obtained with water, oil, and hydrocarbons in flow passages with $D = 1.49$–25.8 mm. Equations (8.28) and (8.29) have been found to be a useful basis for correlating experimental data representing refrigerants as well as data representing mini- and microchannel two-phase flows (Mishima and Hibiki, 1996; Wang et al., 1997; Lee and Lee, 2001a; Hwang and Kim, 2006; Sun and Mishima, 2009; Lee et al., 2010; Zhang et al., 2010; Kim and Mudawar, 2012a, c). It has been found, however, that the parameter C is not a constant, and is sensitive to the size of the flow passage when the flow passage can be considered as a mini- or microchannel (i.e., when the channel size is smaller than the threshold whereby macro-scale models and correlations are no longer reliable, see Section 3.7.2). Table 8.2 is a summary of some of the recently proposed modifications for the Chisholm–Laird correlation. The parameter C also depends on fluid properties. Some of these correlations cover the mini- and

Table 8.2. *Some proposed methods for the calculation of the constant C in Chisholm–Laird correlation (Eq. (8.27)).*

Author	Correlation	Comments
Chisholm (1967)	See table below Eq. (8.28)	Based on experimental data of Lockhart and Martinelli (1949) with water, oil, and hydrocarbons
Mishima and Hibiki (1996)	$C = 21(1 - e^{-0.319 D_H})$, D_H in mm (non-circular channel) $C = 21(1 - e^{-0.333 D})$, D in mm (circular channel)	Based on minichannel data ($D_H = 1.4$ mm) with R-134a and R-236ea
Lee and Lee (2001a)	$C = A\left[\dfrac{\mu_L^2}{\rho_L \sigma D_H}\right]^q \left[\dfrac{\mu_L j}{\sigma}\right]^r Re_{L0}^s$ (Eq. (10.18)) See Table 10.1 for the values of coefficients	Based on 350 data from several sources and data in horizontal rectangular channels with 0.4 to 4 mm gap size
Lee and Mudawar (2005a)	$C = \begin{cases} 2.16 Re_{L0}^{0.047} We_{L0}^{0.6} \text{ viscous liquid and gas} \\ 1.45 Re_{L0}^{0.25} We_{L0}^{0.33} \text{ viscous liquid turbulent gas} \end{cases}$ $We_{L0} = (GD)/(\sigma \rho_L)$	Based on boiling data with R-134a in a 231 μm × 713 μm microchannel
Yue et al. (2007)	$C = 0.185 X^{-0.0942} Re_{L0}^{0.711}$	Horizontal 1000 μm × 500 μm rectangular channel, CO_2-aqueous salt solutions
Li and Wu (2010a)	$C = 11.9 Bd^{0.45}$ for $Bd \le 1.5$ $C = 109.4\left(Bd Re_{L0}^{0.5}\right)^{-0.56}$ for $1.5 < Bd \le 11$ $Bd = [g(\rho_L - \rho_G) D_H^2]/\sigma$	Based on circular and rectangular channels; with R-12, R-22, R-32, R-134a, R-245fa, R236ea, R-404a, R-410a, R-422d, liquid nitrogen, $0.22 \le D_H \le 3.25$ mm
Pamitran et al. (2010)	$C = 3 \times 10^{-3} We_{TP}^{-0.433} Re_{TP}^{1.23}$ $Re_{TP} = GD/\mu_{TP}$; $We_{TP} = (G^2 D)/(\bar\rho \sigma)$ $\bar\rho = \alpha \rho_G + (1 - \alpha) \rho_L$ Find μ_{TP} from Eq. (8.23) (Beattie and Whalley, 1982); find α from the DFM (Steiner, 1993), see Table 6.1	Based on data with R-22, R-134a, R-410A, R-290, and R-744 in horizontal heated tubes of 0.5, 1.5, and 3.0 mm diameter
Sun and Mishima (2009)	$C = 26\left(1 + \dfrac{Re_L}{1,000}\right)\left[1 - \exp\left(\dfrac{-0.153}{0.27 D_H + 0.8}\right)\right]$ for $Re_L < 2,000$, $Re_G < 2,000$ $C = 1.79 X^{-0.19}\left(\dfrac{Re_G}{Re_L}\right)^{0.4}\left(\dfrac{1-x}{x}\right)^{0.5}$ for $Re_L > 2,000$ or $Re_G > 2,000$ $X = $ Martinelli factor	Extensive data base with air–water, R-22, R-123, R-134a, R-236ea, R-245fa, R-404a, R-410a, R-407c, R-507 CO_2, circular, semi-triangular, and rectangular channels, $0.506 < D_H < 12$ mm
Zhang et al. (2010)	$C = \begin{cases} 21[1 - \exp(-0.358/D_H^*)] \text{ flow boiling} \\ 21[1 - \exp(-0.674/D_H^*)] \text{ adiabatic gas–liquid mixture} \\ 21[1 - \exp(-0.142/D_H^*)] \text{ adiabatic vapor–liquid mixture} \end{cases}$ $D_H^* = D_H/\sqrt{\dfrac{\sigma}{g(\rho_L - \rho_G)}}$	Data with water–air, water–N_2, ethanol–air, oil–air, ammonia; and liquid–vapor mixtures of ammonia, R-134a, R-22, R236ea, water–steam; $1.4 < D_H < 3.25$ mm, round and rectangular channels
Kim and Mudawar (2012c)	See Table 8.3	Based on 7115 data points with air/CO_2/N_2–water, N_2–ethanol; and liquid–vapor mixtures of water, R-12, R-22, R-134a, R-236ea, R-245fa, R-404a, R-410a, R-407c, propane, methane, ammonia, and CO_2; $0.0695 < D_H < 6.22$ mm, round and rectangular channels

Table 8.3. *The coefficients in the Chisholm–Laird correlation for two-phase pressure drop multiplier, as suggested by Kim and Mudawar (2012c).*

Liquid	Gas	C
Turbulent	Turbulent	$0.39 \mathrm{Re}_{L0}^{0.03} \mathrm{Su}_{G0}^{0.10} (\rho_L/\rho_G)^{0.35}$
Turbulent	Laminar	$8.7 \times 10^{-4} \mathrm{Re}_{L0}^{0.17} \mathrm{Su}_{G0}^{0.50} (\rho_L/\rho_G)^{0.14}$
Laminar	Turbulent	$1.5 \times 10^{-3} \mathrm{Re}_{L0}^{0.59} \mathrm{Su}_{G0}^{0.19} (\rho_L/\rho_G)^{0.36}$
Laminar	Laminar	$3.5 \times 10^{-5} \mathrm{Re}_{L0}^{0.44} \mathrm{Su}_{G0}^{0.50} (\rho_L/\rho_G)^{0.48}$

microchannel size ranges, and will be further discussed later in Chapters 10, 14, and 15 in relation to pressure drop in mini- and microchannels.

Kim and Mudawar (2012c) have developed a comprehensive method for two-phase frictional pressure drop, based on the aforementioned Chisholm–Laird method. The data base for the correlation includes 7115 adiabatic and condensing two-phase channel flow data points, and covers the following parameter range:

Working fluids: air/CO_2/N_2–water mixtures, N_2–ethanol mixture; and liquid–vapor mixtures of water, R-12, R-22, R-134a, R-236ea, R-245fa, R-404a, R-410a, R-407c, propane, methane, ammonia, and CO_2.

Channel geometry: circular, rectangular

$0.0695 < D_H < 6.22 \, \mathrm{mm}$
$4.0 < G < 8528 \, \mathrm{kg/m}^2$
$3.9 < \mathrm{Re}_L < 7.9 \times 10^4$
$0 < \mathrm{Re}_G < 2.5 \times 10^5$
$0 < x < 1$
$0.0052 < P_r < 0.91$

Kim and Mudawar (2012c) derived their correlations for the constant C in Chisholm's expression (Eq. (8.28)), as summarized in Table 8.3. The gas Suratman number is defined as

$$\mathrm{Su}_{G0} = \frac{\mathrm{Re}_{G0}^2}{\mathrm{We}_{G0}} = \mathrm{Su}_G = \frac{\mathrm{Re}_G^2}{\mathrm{We}_G} = \frac{\rho_G \sigma D_H}{\mu_G^2}. \tag{8.30}$$

Note that X, the Martinelli parameter, should be found from Eq. (8.25). Thus, the liquid-only and gas-only frictional pressure gradients, $\left(-\frac{\partial P}{\partial z}\right)_{\mathrm{fr,L}}$ and $\left(-\frac{\partial P}{\partial z}\right)_{\mathrm{fr,G}}$, respectively, need to be calculated. Kim and Mudawar recommend the following single-phase Fanning friction factors for the calculation of single-phase flow frictional pressure gradients.

Laminar flow, circular channels ($\mathrm{Re}_{D_H,k} < 2000$),

$$f_k = 16 \mathrm{Re}_k^{-1}. \tag{8.31}$$

Laminar flow, rectangular channels ($\mathrm{Re}_{D_H,k} < 2000$) (Shah and London, 1978),

$$f_k \mathrm{Re}_{D_H,k} = 24\left(1 - 1.3553\alpha^* + 1.9467\alpha^{*2} - 1.7012\alpha^{*3} + 0.9564\alpha^{*4} - 0.2537\alpha^{*5}\right). \tag{8.32}$$

Turbulent flow:

$$f_k = 0.079 \mathrm{Re}_k^{-0.25} \quad 2 \times 10^3 < \mathrm{Re}_{D_H,k} < 2 \times 10^4 \text{ (Blasius, 1913)}. \tag{8.33}$$

$$f_k = 0.046 \mathrm{Re}_k^{-0.2} \quad \mathrm{Re}_{D_H,k} \geq 2 \times 10^4 \text{ (Kays and London, 1984)}. \tag{8.34}$$

In the above expressions, $k = L$ or G for liquid or gas, respectively, and α^* represents the aspect ratio (smaller side divided by the longer side) of the rectangular cross section. The correlation of Kim and Mudawar predicts their entire data base with a maximum mean average error (which occurs for turbulent liquid and laminar gas conditions) of 26.8%.

EXAMPLE 8.1. For saturated water–steam flow at 11-MPa pressure with a mixture mass flux of $G = 1500$ kg/m^{2}s in a 1-cm-inner-diameter pipe, calculate and plot Φ_{L0}^2 using the HM model, and calculate and plot Φ_L^2 using the HM model and Chisholm's method for the range $0.01 < x < 0.97$.

SOLUTION. The important properties are $\rho_f = 672$ kg/m^3, $\rho_g = 62.56$ kg/m^3, $\mu_f = 7.92 \times 10^{-5}$ kg/m·s, and $\mu_g = 2.07 \times 10^{-5}$ kg/m·s. Equations (8.15), (8.20), (8.27), and (8.28) can now be applied. For $x \leq 0.97$, Re_g and Re_f are both larger than 2300, implying that Eq. (8.28) applies, and $C = 20$. The calculated Φ_L^2 and Φ_{L0}^2 are plotted in the figure below.

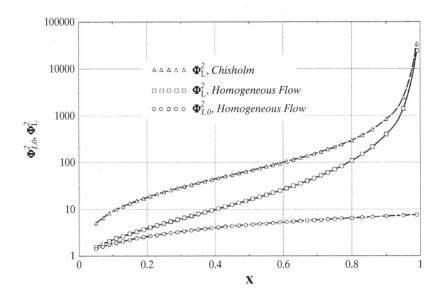

The trend of the curves in Example 8.1 is interesting. They confirm that the HM method disagrees significantly with Chisholm's method, except at very high qualities. Also, the HM model shows that Φ_{L0}^2 monotonically increases with increasing x at low and moderate values of x, but it becomes relatively insensitive to variations in x at high values of x. More accurate calculations that account for velocity slip between the two phases would indeed show that at very high qualities the trend is reversed, and Φ_{L0}^2 diminishes with increasing x. With increasing x, two opposite effects take place. On the one hand the mixture velocity increases, leading to higher pressure drop and therefore higher Φ_{L0}^2. On the other hand, increasing x implies lower mixture viscosity and therefore lower Φ_{L0}^2. The former effect is predominant at low x, but the latter takes over at high x.

The *Martinelli–Nelson method* (1948) was developed for the calculation of frictional pressure drop in boiling channels, assuming a saturated steam–water mixture everywhere. To be consistent with the notation in this book, we will therefore replace subscript L with f and G with g. The method applies to steam–water mixtures at all pressures between atmospheric (1 bar) and water's critical pressure (221 bars). The air–water data were assumed to represent atmospheric water–steam mixtures. At the critical pressure the distinction between the two phases disappears; therefore $\mu_f = \mu_g, \rho_f = \rho_g, \Phi_{f0}^2 = 1$, and $X_{tt}^2 = (\frac{1-x}{x})^{1.75}$. Having profiles of Φ_{f0}^2 as a function of X_{tt} for $P = 1$ bar and $P = P_{cr}$, one can calculate values for other pressures by interpolation and plot these graphically. Plots and tabulated values of Φ_{f0}^2 can be found in various places (including Collier and Thome (1994), Tong and Tang (1997), and Carey (2008)). For a uniformly heated boiling channel with uniform cross section, the total frictional pressure drop in the boiling part of the channel (where x varies from zero to x) can be calculated from

$$\Delta P_{fr} = \int_{0}^{Z_{out}} \left(-\frac{\partial P}{\partial z} \right)_{fr} dz. \tag{8.35}$$

This can be found by writing

$$\Delta P_{fr} = \left(-\frac{\partial P}{\partial z} \right)_{fr,f0} \int_{0}^{x} \Phi_{f0}^2(x)dz = \left(-\frac{\partial P}{\partial z} \right)_{fr,f0} \frac{L}{x} \int_{0}^{x} \Phi_{f0}^2(x)dx. \tag{8.36}$$

Martinelli and Nelson (1948) calculated Φ_{L0}^2 and plotted the quantity $\bar{\Phi}_{f0}^2 = \frac{1}{x} \int_{0}^{x} \Phi_{f0}^2(x)dx$. Note that the latter quantity only depends on the saturated steam and water properties and x. Once an average pressure for the boiling channel is assumed, the quantity $\bar{\Phi}_{f0}^2$ depends only on pressure and x. Using more experimental data, Thom (1964) calculated and tabulated $\bar{\Phi}_{f0}^2$ values at various pressures and qualities. Tabulated values of $\bar{\Phi}_{f0}^2$, along with void fraction and a parameter that represents the acceleration pressure change (see Problem 8.2), can also be found in Wallis (1969) and Collier and Thome (1994).

A useful approximation to Martinelli and Nelson's curves for the range $0 \leq X_{tt} \leq 1$ is the following correlation, proposed by Soliman *et al.* (1968):

$$\Phi_G = 1 + 2.85X_{tt}^{0.523}. \tag{8.37}$$

A widely used general-purpose correlation for two-phase frictional pressure drop, which appears to be the most accurate method available at this time, is the correlation of Friedel (1979). The correlation is based on a very extensive data bank. It is applicable to one- and two-component two-phase flows. For horizontal and vertical upward flow configurations, Friedel suggests

$$\Phi_{L0}^2 = A + 3.24x^{0.78}(1-x)^{0.24} \left(\frac{\rho_L}{\rho_L} \right)^{0.91} \left(\frac{\mu_G}{\mu_L} \right)^{0.19} \left(1 - \frac{\mu_G}{\mu_L} \right)^{0.7} Fr^{-0.0454}We^{-0.035}$$

$$\tag{8.38}$$

and for vertical, downward flow, Friedel's correlation gives

$$\Phi_{L0}^2 = A + 48.6x^{0.8}(1-x)^{0.29} \left(\frac{\rho_L}{\rho_G} \right)^{0.90} \left(\frac{\mu_G}{\mu_L} \right)^{0.73} \left(1 - \frac{\mu_G}{\mu_L} \right)^{7.4} Fr^{0.03}We^{-0.12}, \tag{8.39}$$

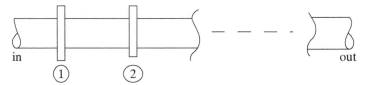

Figure 8.2. Schematic of a one-dimensional flow system.

where $A = (1-x)^2 + x^2 \rho_L f_{G0}(\rho_G f_{L0})^{-1}$, the Weber number is defined as $We = G^2 D/\rho_h \sigma$, and the Froude number is defined as $Fr = G^2/gD\rho_h^2$. The parameters f_{L0} and f_{G0} are single-phase friction factors, and are calculated by using Eq. (8.40) with $Re_{j0} = GD/\mu_j$ and $j = L$ or G. For turbulent flow ($Re_{j0} > 1055$), Friedel recommends

$$f_{j0} = 0.25[0.86859 \ln\{Re_{j0}/(1.964 \ln Re_{j0} - 3.8215)\}]^{-2}. \tag{8.40}$$

Liquid–vapor two-phase flow of various refrigerants as well as dielectric liquids in small flow passages is an important field of industry. Several studies have been reported where the widely referenced two-phase pressure drop correlations have been compared with extensive data banks that included experiments with refrigerants from different sources (Ould Didi *et al.*, 2002; Choi *et al.*, 2008; Sun and Mishima, 2009; Xu *et al.*, 2012). The correlation of Muller-Steinhagen and Heck (1986) has been found to be among the most accurate correlations (Tribbe and Muller-Steinhagen, 2000; Ould Didi *et al.*, 2002; Xu *et al.*, 2012). Using experimental data from 29 sources that covered hydraulic diameters in the 0.069–14 mm range, and 14 different working fluids including air–water, air–ethanol, R-134a, CO_2, R-410A, R-22, R-245fa, and ammonia, Xu *et al.* (2012) examined the overall accuracy of several correlations and noted that the correlation of Muller-Steinhagen and Heck (1986) performed the best.

The correlation of Muller-Steinhagen and Heck (1986) is:

$$\Phi_{L0}^2 = x^3 Y^2 + (1-x)^{1/3}[1 + 2x(Y^2 - 1)] \tag{8.41}$$

where Y is defined a

$$Y^2 = \left(-\frac{dP}{dz}\right)_{fr,G0} \bigg/ \left(-\frac{dP}{dz}\right)_{fr,L0}. \tag{8.42}$$

8.4 General Remarks about Local Pressure Drops

Flow disturbances such as bends, orifices, valves, and flow-area changes all cause changes in pressure. They also cause irreversible loss of fluid mechanical energy into heat. The flow and dissipation processes in most flow disturbances are complicated and multi-dimensional. In setting up one-dimensional conservation equations we often model them as local and sudden pressure drops. For a piping system such as the one shown in Fig. 8.2, for example, the total pressure drop can be obtained by integrating Eq. (8.1), and introducing the pressure drop terms from flow disruptions, which do not show up in the differential momentum equations,

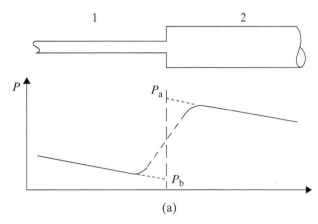

Figure 8.3. The definition of local
pressure drop in a sudden expansion
(a) and sudden contraction (b).

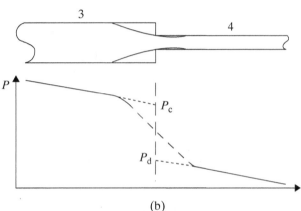

according to

$$P_{\text{in}} - P_{\text{out}} = \int \left[\left(-\frac{\partial P}{\partial z} \right)_{\text{ta}} + \left(-\frac{\partial P}{\partial z} \right)_{\text{sa}} + \left(-\frac{\partial P}{\partial z} \right)_{\text{fr}} + \left(-\frac{\partial P}{\partial z} \right)_{\text{g}} \right] dz + \sum_{i=1}^{N} \Delta P_i,$$

(8.43)

where ΔP_i is the *total* pressure drop across flow disturbance i. The flow phenomena
in the vicinity of a flow disturbance are complicated and multi-dimensional, as mentioned earlier. In interpreting experimental data, pressure drops at discontinuities
are defined such that they are consistent with their representation as local events.
For example, for the simple one-dimensional flow systems displayed in Figs. 8.3(a)
and (b), which are made of two straight channels and a sudden flow-area expansion
or contraction, Eq. (8.43) results in

$$P_1 - P_2 = \int_{z_1}^{z_i} \left(-\frac{\partial P}{\partial z} \right)_{\text{ta}} + \left(-\frac{\partial P}{\partial z} \right)_{\text{sa}} + \left(-\frac{\partial P}{\partial z} \right)_{\text{fr}} + \left(-\frac{\partial P}{\partial z} \right)_{\text{g}} dz + \Delta P_i$$

$$+ \int_{z_i}^{z_2} \left[\left(-\frac{\partial P}{\partial z} \right)_{\text{ta}} + \left(-\frac{\partial P}{\partial z} \right)_{\text{sa}} + \left(-\frac{\partial P}{\partial z} \right)_{\text{fr}} + \left(-\frac{\partial P}{\partial z} \right)_{\text{g}} \right] dz, \quad (8.44)$$

where $\Delta P_i = P_b - P_a$ (for expansion) or $P_c - P_d$ (for contraction). The pressures P_a and P_b, or P_c and P_d, are obtained in experiments by the extrapolation of the axial pressure profiles, as shown in Figs. 8.3(a) and (b).

In all flow disturbances, the total pressure drop ΔP_i (which, as mentioned, can be positive or negative) has two components, a reversible component and an irreversible one:

$$\Delta P_i = \Delta P_{i,R} + \Delta P_{i,I}. \tag{8.45}$$

The reversible component, $\Delta P_{i,R}$, can be positive (as in a sudden flow-area contraction) or negative (as in a flow-area expansion). The irreversible component, also referred to as the *pressure loss*, $\Delta P_{i,I}$, however, *is always positive*, as required by the second law of thermodynamics. It represents the transformation of mechanical energy into heat. An important point to remember is that the momentum equation always needs the total pressure drop across a flow disturbance, and not the pressure loss. The reversible pressure drop can be found from the integration of the mechanical energy equation (obtained by the dot product of $\vec{U}$ with the momentum equation), when all dissipation terms are neglected.

8.5 Single-Phase Flow Pressure Drops Caused by Flow Disturbances

Methods for the calculation of pressure drop in flow discontinuities are often based on the modification of single-phase flow correlations. Therefore, the fundamentals of local pressure changes in single-phase flow will be briefly discussed in this section. Methods for the prediction of two-phase flow pressure drop will be discussed in Section 8.6.

Consider a flow-area contraction followed by an expansion, shown in Fig. 8.4. Assume a horizontal configuration (so that any gravitational effect will be absent), incompressible flow, no frictional loss in the channels, one-dimensional flow, and flat velocity profiles in all three straight components of the system. Also, for clarity of discussion here, define flow-area ratios $\sigma'_A = A_2/A_1$ and $\sigma_A = A_2/A_3$, where A_1, A_2, and A_3 are flow cross-sectional areas of the three segments of the displayed piping system. Elsewhere in this chapter σ will always represent the ratio between smaller and larger flow areas.

For the flow-area contraction, mass continuity requires that $U_1/U_2 = \sigma'_A$. For ideal, reversible flow, where there is no loss in the vicinity of the flow-area change, the reversible mechanical energy equation (Bernoulli's equation in this case) then gives

$$P_1 + \frac{1}{2}\rho U_1^2 = P_{2'} + \frac{1}{2}\rho U_2^2, \tag{8.46}$$

where $P_{2'}$ is the pressure downstream from the flow-area contraction, had the flow been reversible. Elimination of U_1 using $U_1/U_2 = \sigma'$ then gives the reversible pressure drop across the flow-area contraction:

$$(P_1 - P_{2'}) = \Delta P_{R,con} = \frac{1}{2}\rho U_2^2\big(1 - \sigma'^2_A\big). \tag{8.47}$$

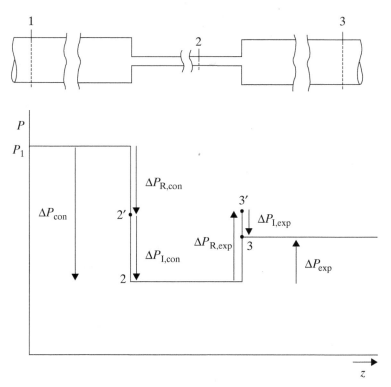

Figure 8.4. A one-dimensional flow system including a sudden expansion and a sudden contraction.

In practice, however, the measured value of P_2 is always lower than the ideal value $P_{2'}$, such that defining $\Delta P_{con} = P_1 - P_2$, we have

$$\Delta P_{con} = \Delta P_{R,con} + \Delta P_{l,con}, \tag{8.48}$$

$$\Delta P_{l,con} > 0.$$

Similarly, for the case of flow-area expansion, continuity requires that $U_3/U_2 = \sigma_A$, and Bernoulli's equation leads to

$$P_2 + \frac{1}{2}\rho U_2^2 = P_{3'} + \frac{1}{2}\rho U_3^2 \tag{8.49}$$

$$\Rightarrow P_2 - P_{3'} = \Delta P_{R,exp} = -\frac{1}{2}\rho U_2^2(1 - \sigma_A^2), \tag{8.50}$$

where $P_{3'}$ is the pressure downstream from the flow-area expansion, had the flow been without loss. Equation (8.50) thus suggests a recovery of pressure up to $P_{3'}$. In practice, the true recovered pressure P_3 is always lower than $P_{3'}$ because of irreversible losses; therefore, defining $\Delta P_{exp} = P_2 - P_3$, where

$$\Delta P_{exp} = \Delta P_{R,exp} + \Delta P_{l,exp}, \tag{8.51}$$

we get

$$\Delta P_{l,exp} = P_{3'} - P_3 > 0. \tag{8.52}$$

Note that the reversible pressure drop can be calculated from the reversible mechanical energy equation. A flow disturbance can therefore be characterized by knowing

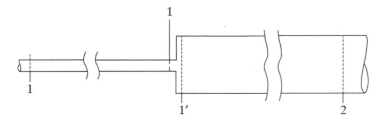

Figure 8.5. Flow-area expansion.

either the total pressure drop it causes or the irreversible pressure loss it causes. The irreversible pressure drop is often difficult to find from theory, and we often rely on empirical correlations for its calculation.

The discussion here holds true for flow disturbances other than simple flow-area expansions and contractions. The pressure drop across any disturbance is the summation of reversible and irreversible components; the reversible component can be found from theory, but the irreversible component often needs to be found from empirical methods.

8.5.1 Single-Phase Flow Pressure Drop across a Sudden Expansion

We now will discuss the magnitudes of the irreversible pressure loss for a simple expansion. This is a case where a simple theoretical analysis actually gives results that agree reasonably well with experimental data.

Assume that pressure just downstream from the expansion in Fig. 8.5 (point $1'$) is P_1, namely, equal to the pressure in the smaller channel. Conservation of momentum between points $1'$ and 2 then gives

$$P_1 A_2 - P_2 A_2 = \rho A_2 U_2 (U_2 - U_1) \tag{8.53}$$

$$\Rightarrow (P_1 - P_2) = \Delta P_{ex} = \rho U_1^2 \sigma_A (\sigma_A - 1). \tag{8.54}$$

The reversible mechanical energy equation gives

$$P_1 + \frac{1}{2}\rho U_1^2 = P_{2'} + \frac{1}{2}\rho U_2^2. \tag{8.55}$$

Therefore,

$$(P_1 - P_2)_R = \Delta P_{R,ex} = \frac{1}{2}\rho U_1^2 (\sigma_A^2 - 1). \tag{8.56}$$

Now, given that $\Delta P_{ex} = \Delta P_{ex,R} + \Delta P_{ex,I}$, we have

$$(P_1 - P_2)_I = \Delta P_{I,ex} = (1 - \sigma_A)^2 \frac{1}{2}\rho U_1^2. \tag{8.57}$$

We can now define a *loss coefficient*. For any flow disturbance, the loss coefficient K is defined as

$$\Delta P_1 = K \frac{1}{2}\rho U_{ref}^2, \tag{8.58}$$

where K is the loss coefficient for that particular flow disturbance and U_{ref} is the average velocity in a reference flow cross section.

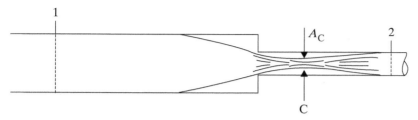

Figure 8.6. Flow-area contraction.

Using the average velocity in the smallest channel connected to the flow distur-bance as U_{ref}, we get for a sudden expansion (Borda–Carnot relation)

$$K_{\text{ex}} = (1 - \sigma_A)^2. \tag{8.59}$$

A very important issue must be pointed out here. The analysis presented thus far, and elsewhere in this chapter, assumes uniform velocity profiles in the channels con-nected to a discontinuity. Most of the tabulated and curvefitted values of loss coef-ficients are in fact based on assumed uniform velocity profiles. Flat velocity profiles are approximately true for fully developed turbulent flow. A similar analysis can be done for any known, nonuniform velocity profile, however, by using the following macroscopic conservation equation forms: for mass continuity,

$$A_1 \langle U_1 \rangle = A_2 \langle U_2 \rangle, \tag{8.60}$$

for momentum conservation,

$$P_1 - P_2 = \rho \left(\langle U^2 \rangle_2 - \langle U^2 \rangle_1 \right), \tag{8.61}$$

and for reversible mechanical energy,

$$P_1 + \frac{1}{2} \frac{\langle U^3 \rangle_1}{\langle U \rangle_1} = P_2 + \frac{1}{2} \frac{\langle U^3 \rangle_2}{\langle U \rangle_2}, \tag{8.62}$$

where we have used our usual cross-sectional averaging definition $\langle \xi \rangle_i = \frac{1}{A_i} \int_{A_i} \xi \, dA$.

8.5.2 Single-Phase Flow Pressure Drop across a Sudden Contraction

The hydrodynamics downstream from a flow-area contraction are different than for a flow-area expansion. When the smaller channel (channel 2 in Fig. 8.6) is suffi-ciently long, the flow undergoes a vena-contracta phenomenon. Irreversible losses associated with the sudden contraction occur primarily downstream from the vena-contracta point (point C in Fig. 8.6). Irreversible losses between points C and 2 in Fig. 8.6 can be modeled in the same way one would model flow across a sudden expansion, with A_C and A_2 representing the smaller and larger flow areas, respec-tively. The result will be

$$\Delta P_{\text{R,con}} = \frac{1}{2} \rho U_2^2 (1 - \sigma_A^2), \tag{8.63}$$

$$\Delta P_{\text{I,con}} = K_{\text{con}} \frac{1}{2} \rho U_2^2, \tag{8.64}$$

$$K_{\text{con}} = \left(\frac{1}{C_C} - 1 \right)^2, \tag{8.65}$$

where $\sigma_A = A_2/A_1$ is the ratio between smaller and larger flow areas, and $C_C = A_C/A_2$ is the *contraction coefficient*. Experimental data for C_C are available in handbooks. A useful expression for the estimation of C_C is (Geiger, 1966)

$$C_C = 1 - \frac{1 - \sigma_A}{2.08(1 - \sigma_A) + 0.5371}. \tag{8.66}$$

A simple curve fit to experimental data is (White, 1999)

$$K_{\text{con}} \approx \begin{cases} 0.42(1 - \sigma_A) & \text{for } \sigma_A \le 0.58, \\ (1 - \sigma_A)^2 & \text{for } \sigma_A > 0.58. \end{cases} \tag{8.67} \tag{8.68}$$

Note the similarity between Eqs. (8.59) and (8.68).

8.5.3 Pressure Change Caused by Other Flow Disturbances

In general, for any disturbance (valve, orifice, bend, partial blockage, etc.), one may write

$$\Delta P = \Delta P_R + \Delta P_I, \tag{8.69}$$

$$\Delta P_I = K \frac{1}{2} \rho U_{\text{ref}}^2, \tag{8.70}$$

where ΔP, ΔP_R, and ΔP_I all represent *pressure drops* (i.e., the pressure upstream from the discontinuity minus the pressure downstream from the discontinuity). The irreversible pressure loss will always be positive, with K often found from empirical correlations, tables, or charts. Table 8.4 lists loss coefficients for a number of common flow disturbances.

8.6 Two-Phase Flow Local Pressure Drops

In two-phase flow, usually only the total pressure drop across a flow disturbance (which, as in single-phase flow, can be positive or negative) is of interest. This is because the breakdown of the total pressure change into reversible and irreversible components is often not possible without making arbitrary assumptions, except for homogeneous flow, owing to the empirical nature of the void fraction correlations often used in the analysis of two-phase flow systems. There is also ambiguity about the exact definition of ideal, reversible flow conditions. The conditions downstream from the flow disturbance, in particular with respect to void fraction, can be different in reversible flow than in real situations.

Similar to the frictional pressure drop, often a *two-phase multiplier* is used, whereby

$$\Delta P = \Delta P_{L0} \Phi_{L0} = \Delta P_{G0} \Phi_{G0} = \Delta P_L \Phi_L = \Delta P_G \Phi_G, \tag{8.71}$$

where ΔP_{L0} is the total pressure drop when all the mixture is liquid and Φ_{L0} is the corresponding two-phase multiplier. ΔP_{G0} and Φ_{G0} are defined similarly, where all the mixture is gas. Likewise, ΔP_L represents the total pressure drop across the flow disturbance of interest when pure liquid at a mass flow rate of $GA(1 - x)$ flows through the inlet channel, and Φ_L is the corresponding two-phase multiplier; and ΔP_G and

Table 8.4. *Typical values of the loss coefficient* K *for various flow disturbances.*[a,b]

Flow disturbance	K
45° bend	0.35 to 0.45
90° bend	0.50 to 0.75
Regular 90° elbow	$K = 1.49\,\mathrm{Re}^{-0.145}$
45° standard elbow	0.17 to 0.45
180° return bend, flanged	0.2
180° return bend, threaded	1.5

Line flow, flanged tee	0.2
Line flow, threaded tee	0.9

Branch flow, flanged tee	1.0
Branch flow, threaded tee	2.0
Fully open gate valves	0.15
$\frac{1}{4}$-closed gate valve	0.26
Half-closed gate valve	2.1
$\frac{3}{4}$-closed gate valve	17
Open check valves	3.0
Fully open globe valve	6.4
Half-closed globe valve	9.5
Fully open ball valve	0.05
$\frac{1}{3}$-closed ball valve	5.5
$\frac{2}{3}$-closed ball valve	210

Entrance from a plenum into a pipe	
Sharp edged	0.5
Slightly rounded	0.23
Well-rounded	0.04
Projecting pipe	0.78
Exit from pipe into a plenum	1.0

[a] U_{ref} is the mean velocity in the pipe.
[b] From various sources, including White (1999) and Munson *et al.* (1998).

Φ_G are defined similarly. Note that, unlike the frictional pressure drop, the two-phase multipliers do not have a power of 2.

Let us analyze the two-phase pressure drop in a sudden expansion. Assume steady-state and uniform phasic velocities at any cross section. Mass continuity

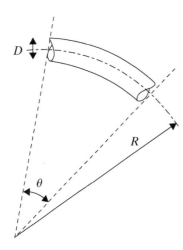

Figure 8.7. Configuration of a bend.

requires that $G_1 A_1 = G_2 A_2$. Also, similar to the case of single-phase flow, assume that the pressures on both sides of the flow-area expansion are equal ($P_1 = P_{1'}$ in Fig. 8.5). Momentum conservation between (1′) and (2) then gives

$$\Delta P_{ex} = P_{1'} - P_2 = P_1 - P_2 = G_2^2 \left(\frac{1}{\rho_2'} - \frac{1}{\sigma_A \rho_1'} \right), \qquad (8.72)$$

where the "momentum density" is defined as (see Chapter 5)

$$\rho' = \left[\frac{1 - x^2}{\rho_L (1 - \alpha)} + \frac{x^2}{\rho_G \alpha} \right]^{-1}. \qquad (8.73)$$

Assuming the conditions at the inlet (station 1) are known, we obviously need a model to predict α_2 and x_2. From intuition, strong mixing should be expected to take place downstream from a flow disturbance, which can have a significant impact, particularly in single-component liquid–vapor flows. If we assume that both phases are incompressible, $x_1 = x_2$, and $\alpha_1 = \alpha_2$ (assumptions that may be appropriate only for a two-component mixture), Eq. (8.72) would give

$$\Delta P_{ex} = \Delta P_{L0,ex} \Phi_{L0,ex}, \qquad (8.74)$$

$$\Delta P_{L0} = -\frac{G_1^2}{\rho_L} \sigma_A (1 - \sigma_A), \qquad (8.75)$$

where $\sigma_A = A_1 / A_2$ and

$$\Phi_{L0,ex} = \frac{\rho_L}{\rho'} = \left[\frac{(1 - x)^2}{(1 - \alpha)} + \frac{\rho_L}{\rho_G} \frac{x^2}{\alpha} \right]. \qquad (8.76)$$

For homogeneous flow we get

$$\Phi_{L0,ex} = (1 - x) + \left(\frac{\rho_L}{\rho_G} \right) x. \qquad (8.77)$$

We can derive expressions for reversible and irreversible pressure-drop components. (This is of course possible because we assumed known α and x on both sides of the flow-area expansion.) The conservation of mechanical energy for reversible flow gives

$$P_1\{[\alpha U_G + (1-\alpha)U_L]A\}_1 + \frac{1}{2}\{[\rho_L U_L^3(1-\alpha) + \rho_G U_G^3\alpha]A\}_1$$

$$= P_2\{[\alpha U_G + (1-\alpha)U_L]A\}_2 + \frac{1}{2}\{[\rho_L U_L^3(1-\alpha) + \rho_G U_G^3\alpha]A\}_2. \quad (8.78)$$

Using the identity expressions $\rho_L[U_L(1-\alpha)]_1 = G_1(1-x_1)$, $\rho_G(U_G\alpha)_1 = G_1 x_1$, $G_1 A_1 = G_2 A_2$, etc., we can recast Eq. (8.70) and solve for $\Delta P_{R,ex} = P_1 - P_2$ as

$$\Delta P_{R,ex} = -\frac{G_1^2}{2}(1-\sigma_A^2)\left[\frac{x^3}{\alpha^2\rho_G^2} + \frac{(1-x)^3}{(1-\alpha)^2\rho_L^2}\right]\left(\frac{x}{\rho_G} + \frac{1-x}{\rho_L}\right)^{-1}. \quad (8.79)$$

Using Eqs. (8.51) and (8.54) and assuming homogeneous flow then yields

$$\Delta P_{1,ex} = \frac{G_1^2}{2\rho_L}(1-\sigma_A^2)[1+x(\rho_L/\rho_G - 1)]. \quad (8.80)$$

Based on experimental data from several sources, Wang *et al.* (2010) have developed the following correlation for total pressure change across a sudden expansion:

$$\Delta P_{ex} = \Delta P_{ex,h}(1+\Omega_1 - \Omega_2)(1+\Omega_3) \quad (8.81)$$

where the homogeneous flow pressure drop $\Delta P_{ex,h}$ can be found from Eqs. (8.74) and (8.77), and

$$\Omega_1 = \left(\frac{\text{WeBd}}{\text{Re}_{L0}}\right)^2\left(\frac{1-x}{x}\right)^{0.3}\frac{1}{\text{Fr}^{0.8}} \quad (8.82)$$

$$\Omega_2 = 0.2\left(\frac{\mu_G}{\mu_L}\right)^{0.4} \quad (8.83)$$

$$\Omega_3 = 0.4\left(\frac{x}{1-x}\right)^{0.3} + 0.3e^{\frac{1.6}{\text{Re}_{L0}^{0.1}}} - 0.4\left(\frac{\rho_L}{\rho_G}\right)^{0.2}. \quad (8.84)$$

Here $\text{Bd} = \frac{(\rho_L - \rho_G)gD^2}{\sigma}$ is the Bond number, and Weber and Froude numbers are defined based on homogeneous flow density

$$\text{We} = \frac{G^2 D}{\sigma \rho_h} \quad (8.85)$$

$$\text{Fr} = \frac{G^2}{\rho_h^2 gD} \quad (8.86)$$

$$\rho_h = \left[\frac{x}{\rho_G} + \frac{1-x}{\rho_L}\right]^{-1}. \quad (8.87)$$

The range of parameters for the data base for this correlation is:

$506 < G < 5642\,\text{kg/m}^2\cdot\text{s}$; $0.002 < x < 0.99$; $0.057 < \sigma_A < 0.607$
$0.84 < D < 19\,\text{mm}$; $0.095 < \text{Bd} < 92$; $10.3 < \text{Fr} < 9.19 \times 10^5$
$100 < \text{We} < 8.3 \times 10^4$; $4.35 \times 10^2 < \text{Re}_{L0} < 4.95 \times 10^5$.

The data base evidently covers the minichannel size range.

A similar analysis can be performed for a two-phase pressure drop in a sudden flow-area contraction, if it is assumed that (a) the phases are both incompressible;

(b) α and x remain constant everywhere; (c) a vena-contracta occurs that has the same characteristics as those of the vena-contracta that happens in the system under single-phase flow conditions; and (d) all irreversibilities occur downstream from the vena-contracta point (between stations C and 2 in Fig. 8.6). The result will be

$$\Delta P_{\text{con}} = P_1 - P_2 = G_2^2 \left\{ \frac{\rho_h}{2\rho'''^2} \left(\frac{1}{C_C^2} - \sigma_A^2 \right) + \frac{1}{\rho'} (1 - C_C) \right\} \qquad (8.88)$$

$$\Delta P_{1,\text{con}} = \frac{G_2^2}{C_C^2} (1 - C_C) \left\{ -\frac{C_C}{\rho'} + \frac{1 + C_C}{2} \frac{\rho_h}{\rho'''^2} \right\}. \qquad (8.89)$$

where ρ' and ρ''' are defined in Eq. (5.73) and (5.89), respectively. However, experimental data indicate that strong mixing is caused by the contraction, which may justify the homogeneous flow assumption. The latter assumption leads to

$$\Delta P_{1,\text{con}} = \frac{1}{2} \frac{G_2^2}{\rho_L} \left(\frac{1}{C_C} - 1 \right)^2 [1 + x(\rho_L/\rho_G - 1)], \qquad (8.90)$$

$$\Delta P_{\text{con}} = P_1 - P_2 = \frac{1}{2} \frac{G_2^2}{\rho_L} \left[\left(\frac{1}{C_C} - 1 \right)^2 + (1 - \sigma_A^2) \right] [1 + x(\rho_L/\rho_G - 1)], \qquad (8.91)$$

where, again, $\sigma_A = A_2/A_1$ is the ratio between the smaller and larger flow areas. Equation (8.75) can be recast as

$$\Delta P_{\text{con}} = \Delta P_{L0,\text{con}} \Phi_{L0,\text{con}}, \qquad (8.91)$$

$$\Delta P_{L0,\text{con}} = \frac{1}{2} \frac{G^2}{\rho_L} \left[\left(\frac{1}{C_C} - 1 \right)^2 + (1 - \sigma_A^2) \right], \qquad (8.92)$$

$$\Phi_{L0} = 1 + x(\rho_L/\rho_G - 1). \qquad (8.93)$$

The homogeneous flow model has been found to do well in predicting experimental data (Guglielmini et al., 1986). More recently, Schmidt and Friedel (1997) performed careful experiments dealing with two-phase pressure drop across flow-area contractions using mixtures of air with water, Freon 12, an aqueous solution of glycerol, and an aqueous solution of calcium nitrate. The smaller tube diameters in their experiments varied in the $D \approx 17.2$ to 44.2 mm range, resulting in $\sigma_A \approx 0.057$ to 0.445. No vena-contracta was observed in these experiments.

Some useful and widely applied correlations for two-phase multipliers are now presented. All these correlations are for total pressure drop and it is assumed that the the pressure drop associated with single-phase flow is known.

For flow through *orifices*, Beattie (1973) proposed

$$\Phi_{L0} = [1 + x(\rho_L/\rho_G - 1)]^{0.8} \left[1 + x \left(\frac{\rho_L \mu_G}{\rho_G \mu_L} - 1 \right) \right]^{0.2}. \qquad (8.94)$$

The same author has proposed the following correlation for *spacer grids in rod bundles*:

$$\Phi_{L0} = [1 + x(\rho_L/\rho_G - 1)]^{0.8} \left[1 + x \left(\frac{3.5\rho_L}{\rho_G} - 1 \right) \right]^{0.2}. \qquad (8.95)$$

Return bends (180° bends) are of particular interest because of their widespread applications.

A correlation by Chisholm (1967, 1981) for two-phase pressure drop in a *bend* is

$$\Phi_{L0} = (1 - x^2) \left(1 + \frac{C}{X} + \frac{1}{X^2} \right), \tag{8.96}$$

where $X = [(\Delta P_L)/(\Delta P_G)]^{1/2}_{\text{bend}}$ is Martinelli's factor defined for the bend and

$$C = \left[1 + (C_2 - 1) \left(\frac{\rho_L - \rho_G}{\rho_L} \right)^{0.5} \right] \left[\sqrt{\frac{\rho_L}{\rho_G}} + \sqrt{\frac{\rho_G}{\rho_L}} \right], \tag{8.97}$$

with (see Fig. 8.7)

$$C_2 = 1 + \frac{2.2}{K_{L0}(2 + R/D)}, \tag{8.98}$$

where K_{L0} is the bend's single-phase loss coefficient for the conditions when all the mixture is pure liquid. A simpler alternative to Eq. (8.96) is (Chisholm, 1983)

$$\Phi_{L0} = 1 + \left(\frac{\rho_L}{\rho_G} - 1 \right) x \left[C_2 (1 - x) + x \right], \tag{8.99}$$

where C_2 is found from Eq. (8.98).

For a 180° bend K_{L0} can be found from (Idelchik, 1994)

$$K_{L0} = f_{L0} \frac{L_{\text{bend}}}{D} + 0.294 \left(\frac{R}{D} \right)^{1/2}, \tag{8.100}$$

where f_{L0} is the Fanning friction factor in the pipe, had the pipe been straight, and L_{bend} is the total linear length of the bend.

Based on data with R-22 and R-410A refrigerants, Domanski and Hermes (2008) have proposed

$$\Phi_{L0} = 6.5 \times 10^{-3} \left(\frac{GxD}{\mu_G} \right)^{0.54} \left(\frac{1}{x} - 1 \right)^{0.21} \left(\frac{\rho_L}{\rho_G} \right)^{0.34} \left(\frac{2R}{D} \right)^{-0.67}. \tag{8.101}$$

The range of parameters of the data base for this correlation is: $3.3 < D < 5.07$ mm; $3.91 < 2R/D < 8.15$.

Domanski and Hermes (2008) also recommend that the pressure drop in the bend, when it is assumed to be a straight tube, be found using the correlation of Muller-Steinhagen and Heck (1986)

$$\Delta P_{2P,\text{straight}} = \Delta P_{L0,\text{straight}} + 2x \left(\Delta P_{G0,\text{straight}} - \Delta P_{L0,\text{straight}} \right) (1 - x)^{1/3}$$
$$+ \Delta P_{G0,\text{straight}} x^{1/3}. \tag{8.102}$$

The pressure loss in the bend is then found from

$$\Delta P_{2P,\text{bend}} = \Phi_{L0} \Delta P_{2P,\text{straight}}. \tag{8.103}$$

Based on experimental data with air and water in a vertical tube, with bubbly and slug flow regimes, Alimonti (2014) noted that the homogeneous flow assumption (Eq. (8.93)) over predicted the two-phase pressure drop in gate and globe valves only slightly.

EXAMPLE 8.2. Calculate the total pressure drop in the system shown in Fig. 8.8 , for air–water mixture flow with the following specifications: pipe diameter $D = 3.7$ cm, liquid mass flux $G_L = 1500$ kg/m^2·s, gas mass flux $G_G = 130$ kg/m^2 s, temperature

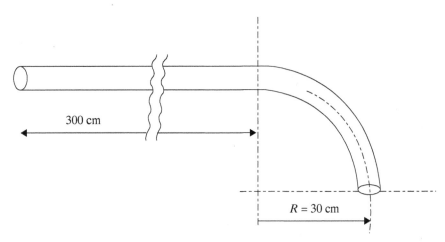

Figure 8.8. The piping system for Example 8.2.

$T = 25\,°C$, and average pressure $P = 10$ bars. Assume that the piping system lies in a horizontal plane.

SOLUTION. For properties we get $\rho_L = 997.5\,\text{kg/m}^3$, $\rho_G = 11.7\,\text{kg/m}^3$, $\mu_L = 8.93 \times 10^{-4}\,\text{kg/m·s}$, and $\mu_G = 1.85 \times 10^{-5}\,\text{kg/m·s}$.

For the bend, we can use the correlation of Chisholm, Eqs. (8.96)–(8.98). For the 90° bend, let us use $K_0 = 0.75$. Noting that $G_L = 1500\,\text{kg/m}^2\text{·s}$ and $G_G = 130\,\text{kg/m}^2\text{·s}$, we get

$$\Delta P_{L,\text{bend}} = K_0 \frac{1}{2} \frac{G_L^2}{\rho_L} = 845.9\,\text{N/m}^2,$$

$$\Delta P_{G,\text{bend}} = K_0 \frac{1}{2} \frac{G_G^2}{\rho_G} = 542.1\,\text{N/m}^2,$$

$$X = \left[\frac{\Delta P_{L,\text{bend}}}{\Delta P_{G,\text{bend}}} \right]^{1/2} = 1.249.$$

With $R = 0.3$ m and $D = 0.037$ m, Eq. (8.98) gives $C_2 = 1.29$. Equation (8.82) leads to $C = 12.04$. The flow quality is

$$x = \frac{G_G}{G_G + G_L} = 0.0797.$$

Equation (8.98) then gives $C_2 = 1.29$. Using this value, we can then solve Eq. (8.97), leading to $C = 12.04$. Equation (8.96) can now be applied to get $\Phi_{L0} = 11.2$. The total pressure drop in the bend will then be

$$\Delta P_{\text{bend}} = \Phi_{L0} \Delta P_{L0,\text{bend}} = \Phi_{L0} K_0 \frac{1}{2} \frac{(G_L + G_G)^2}{\rho_L} = 11\,196\,\text{N/m}^2.$$

We now need to calculate the pressure drop in the straight segment of the pipe. Let us use the method of Chisholm *et al.*, Eq. (8.28),

$$\text{Re}_G = \frac{G_G D}{\mu_G} = 2.6 \times 10^5,$$

$$\text{Re}_L = \frac{G_L D}{\mu_L} = 6.2 \times 10^4.$$

Clearly, both phases are turbulent; therefore $C = 20$ should be used in Eq. (8.28). The Martinelli parameter can be found from Eq. (8.27), leading to $X_{tt} = 1.44$. Application of Eq. (8.28) then gives $\Phi_L^2 = 15.35$. The pressure drop in the straight segment can be found by writing

$$f_L' = 0.316 \, \mathrm{Re}_L^{-0.25} = 0.020$$

and so

$$\Delta P_{\mathrm{straight}} = \Phi_L^2 \Delta P_L = \Phi_L^2 f_L' \frac{L}{D} \frac{G_L^2}{2\rho_L} = 2.81 \times 10^4 \, \mathrm{N/m^2}.$$

The total pressure drop will thus be

$$\Delta P_{\mathrm{tot}} = \Delta P_{\mathrm{bend}} + \Delta P_{\mathrm{straight}} = 3.93 \times 10^4 \, \mathrm{N/m^2}.$$

EXAMPLE 8.3. Ammonia with a mass flow rate of 35 g/s, a quality of 2% and a temperature of –25 °C flows in a horizontal tube with 6 mm inner diameter. The tube has a surface roughness of 1.5 μm. Find the frictional pressure gradient. Also, a gate valve is installed on the tube. The valve has a flow diameter that is equal to the inner diameter of the tube, and is half open. Calculate the pressure loss caused by the gate valve, using the HEM assumption.

SOLUTION. We deal with a saturated mixture of liquid and vapor ammonia at −25 °C. The saturation pressure of ammonia at this temperature is 1.504 bars. Other relevant properties are

$$\rho_f = 671.7 \, \mathrm{kg/m^2 \cdot s}$$
$$\rho_g = 1.287 \, \mathrm{kg/m^2 \cdot s}$$
$$\mu_f = 2.289 \times 10^{-4} \, \mathrm{kg/m \cdot s}$$
$$\mu_g = 8.295 \times 10^{-6} \, \mathrm{kg/m \cdot s}$$
$$h_{fg} = 1.345 \times 10^6 \, \mathrm{J/kg}.$$

We now calculate the mass flux and the homogeneous void fraction.

$$\frac{x}{1.0 - x} = \frac{\rho_g \alpha_h}{\rho_f (1 - \alpha_h)} \Rightarrow \alpha_h = 0.914$$
$$A = \pi D^2/4 = \pi (0.006 \, \mathrm{m})^2/4 = 2.827 \times 10^{-5} \, \mathrm{m^2}$$
$$G = \dot{m}/A = (35 \times 10^{-3} \, \mathrm{kg})/(2.827 \times 10^{-5} \, \mathrm{m^2}) = 1238 \, \mathrm{kg/m^2 \cdot s}.$$

To find the frictional pressure gradient, let us use the method of Beattie and Whalley (1982)

$$\rho_h = \frac{1}{x/\rho_g + \frac{1.-x}{\rho_f}} = 58.84 \, \mathrm{kg/m^2 \cdot s}$$

$$\mu_{TP} = \alpha_h \mu_g + \mu_f (1 - \alpha_h)(1 + 2.5\alpha_h) = 7.215 \times 10^{-5} \, \mathrm{kg/m \cdot s}$$

$$\mathrm{Re}_{TP} = \frac{G D_H}{\mu_{TP}} = 1.029 \times 10^5.$$

Following the recommendation of Beattie and Whalley (1982) we will apply the Colebrook equation (Eq. (8.23)), noting that

$$\varepsilon_D/D = (1.5 \times 10^{-6}\,\text{m})/(0.006\,\text{m}) = 2.5 \times 10^{-4}.$$

Iterative solution of Eq. (8.23) then gives

$$f' = 0.0159.$$

The frictional pressure gradient can now be found

$$\left(-\frac{\partial P}{\partial z}\right)_{\text{fr}} = f'\frac{1}{D}\frac{1}{2}\frac{G^2}{\rho_{\text{TP}}} = 0.0191\left(\frac{1}{0.006\,\text{m}}\right)\frac{1}{2}\frac{(1238\,\text{kg/m}^2\cdot\text{s})^2}{58.84\,\text{kg/m}^2\cdot\text{s}}$$
$$= 4.147 \times 10^4\,\text{N/m}^3.$$

We now calculate the pressure drop across the valve. The all-liquid velocity through the valve, when it is fully open, will be

$$U_{\text{f0}} = G/\rho_{\text{f}} = \frac{1238\,\text{kg/m}^2\cdot\text{s}}{671.7\,\text{kg/m}} = 1.843\,\text{m/s}.$$

This is the reference velocity that should be used for calculating the pressure drop across the valve. From Table 8.1 the loss coefficient for the half-open gate valve is 2.1. Assuming homogeneous flow, the pressure drop across the valve (which is the same as the pressure loss if we assume that phase density variations across the valve remain unchanged) can be found as

$$\Delta P_{\text{f0}} = K\frac{1}{2}\rho_{\text{f}}U_{\text{f0}}^2 = 2.1\left(\frac{1}{2}\right)(671.7\,\text{kg/m})(1.843\,\text{m/s})^2 = 2395\,\text{N/m}^2$$
$$\Phi_{\text{f0}} = 1.0 + x\left(\rho_{\text{f}}/\rho_{\text{g}} - 1.0\right) = 1.0 + (0.02)\left[(671.7/1.287) - 1\right] = 11.42$$
$$\Delta P_{\text{valve}} = \Phi_{\text{f0}}\Delta P_{\text{f0}} = 2.73 \times 10^3\,\text{N/m}^2.$$

8.7 Pressure Drop in Helical Flow Passages

Two-phase flow regimes in helically coiled flow passages were discussed earlier in Section 4.5, where definitions and terminology relevant to helically coiled flow passages were also discussed. In this section, methods for calculating the frictional pressure drop in helicoidally coiled flow passages will be discussed.

8.7.1 Hydrodynamics of Single-Phase Flow

Centrifugal forces that result from the flow path curvature influence the flow and heat transfer behavior of curved flow passages. These forces cause secondary flows that cause fluid mixing, move the location of maximum axial velocity towards the outer surface, and thus lead to complicated flow patterns. As a result, in comparison with flows in straight flow passages with similar cross section and boundary conditions, in curved flow passages the transition from laminar to turbulent flow regimes occurs at a higher Reynolds number when the cross section does not include sharp angles, the entrance lengths (both hydrodynamic and thermal) are shorter, and the circumferential-average friction factor and heat transfer coefficient both higher.

Friction factor and heat transfer coefficient have nonuniform distributions around the circumference, and their nonuniformity becomes stronger as the Dean number (to be defined shortly) is increased.

Consider a circular cross-section duct similar to the one shown in Fig. 4.19(a). The *Dean number* and *helical coil number*, respectively, are defined as

$$Dn = Re_{D_H}(R_i/R_{cl})^{1/2} \tag{8.104}$$

$$Hn = Re_{D_H}(R_i/R_c)^{1/2} \tag{8.105}$$

where R_i is the tube inner radius. The difference between these two dimensionless numbers, as noted, is that the coil radius (also referred to as pitch radius) R_{cl} is used in the definition of Dn, while the radius of curvature R_c is used in the definition of Hn. The Dean number, named after W. R. Dean who performed pivotal work on flow in curved flow passages (Dean, 1927), is used in models and correlations more often, and represents the significance of the influence of the aforementioned centrifugal forces that cause secondary flows and therefore enhance fluid mixing. (Similar arguments can of course be made using Hn.) As Dn increases, so do the strength of the secondary flows and the intensity of mixing of the fluid. As a result, increasing Dn leads to increasing circumferential average friction factor and heat transfer coefficient.

The entrance length is generally shorter in helically coiled tubes in comparison with straight ducts, by about 20% to 50% for laminar flow (Shah and Joshi, 1987). For most engineering applications, in particular when Dn >200, methods that are based on fully developed flow and thermally developed flow assumptions can be applied for design purposes.

Experimental and numerical studies dealing with laminar flow in curved tube passages are extensive. They show that under steady state, the deviation of the velocity profile from the velocity profile in a similar straight duct depends on the magnitude of Dn. For Dn < 20 this deviation is negligibly small and the velocity profile is similar to the velocity profile in a straight duct. At higher values of the Dean number the velocity and temperature profiles will not be symmetric with respect to the mid plane on which the flow path axis is located (the osculating plane described earlier). This asymmetry is caused by torsion (rotation). In a toroidal flow passage (i.e., a helical flow passage with zero pitch, for which the centerline of the flow passage is a circle), where there is no torsion, the velocity and temperature profiles would be symmetric with respect to the plane on which the flow axis is located. As a result of the asymmetry of the velocity profile in a helicoidal tube the radial location of the point where the axial velocity is maximum moves further towards the outer wall as Dn is increased. Experiment and simulation also suggest that the circumferentially-average friction factor and heat transfer coefficient are insensitive to coil pitch.

For a simple helicoidal tube, Schmidt (1967) has proposed the following correlation for the laminar–turbulent flow transition

$$Re_{D,cr} = 2300\left[1 + 8.6\left(\frac{R_i}{R_{cl}}\right)^{0.45}\right], \tag{8.106}$$

where, here and elsewhere in this section Re_D is defined based on the diameter of the tube. An alternative correlation for laminar-turbulent flow transition is

(Srinivasan *et al.*, 1970)

$$\text{Re}_{D,\text{cr}} = 2100 \left[1 + 12 \left(\frac{R_i}{R_{cl}} \right)^{0.5} \right]. \tag{8.107}$$

Friction Factor in Laminar Flow

For fully developed flow in a circular helicoidally coiled tube, when $7 < R_{cl}/R_i < 140$, the circumferentially averaged friction factor can be found from Srinivasan *et al..* (1970)

$$\frac{f_c}{f_s} = \begin{cases} 1 & \text{for} \quad \text{Dn} < 30 \\ 0.419 \, \text{Dn}^{0.275} & \text{for} \quad 30 < \text{Dn} < 300 \\ 0.1125 \, \text{Dn}^{0.5} & \text{for} \quad \text{Dn} > 300 \end{cases} \tag{8.108}$$

where f_c represents the Fanning friction factor in the curved pipe, and f_s is the Fanning friction factor in the pipe if it was straight. Based on an extensive experimental and numerical data base using regression analysis, Manlapaz and Churchill (1980) derived,

$$\frac{f_c}{f_s} = \left\{ \left[1.0 - \frac{0.18}{[1 + (35/\text{Dn})^2]^{0.5}} \right]^m + \left(1.0 + \frac{R_i/R_{cl}}{3} \right)^2 \left(\frac{\text{Dn}}{88.33} \right) \right\}^{0.5} \tag{8.109}$$

where

$$\begin{aligned} m &= 2 \quad \text{for} \quad \text{Dn} < 20 \\ m &= 1 \quad \text{for} \quad 20 < \text{Dn} < 40 \\ m &= 0 \quad \text{for} \quad \text{Dn} > 40. \end{aligned}$$

Evidently, the friction factor becomes insensitive to R_{cl}/R_i as R_{cl}/R_i increases, and at the limit of $R_{cl}/R_i \rightarrow \infty$ we get $f_c/f_s \rightarrow 1$, as expected.

The above correlations apply to isothermal flow. For non-isothermal situations a correction for the effect of temperature on fluid properties is needed. According to Kubair and Kuloor (1965)

$$f_c/f_{c,m} = 0.19(\mu_s/\mu_m)^{0.25} \tag{8.110}$$

where $f_{c,m}$ is the isothermal friction factor when the fluid average (mixed-cup) temperature is used for properties, and μ_s and μ_m are the fluid viscosities calculated at circumferentially averaged wall surface temperature and fluid average temperature, respectively.

For laminar or turbulent flow in a helicoidally coiled concentric annular flow passage, and based on experimental data with air and water, Xin *et al.* (1997) derived the following empirical correlation for Darcy friction factor

$$f_c' = 0.02985 + 75.89 \left[0.5 - \frac{\tan^{-1} \left(\frac{\text{Dn} - 39.88}{77.56} \right)}{\pi} \right] \Bigg/ \left(\frac{R_{cl}}{R_0 - R_i} \right)^{1.45} \tag{8.111}$$

where R_i and R_0 are the inner and outer radii of the annular flow-passage cross section and the Dean number is defined as $\text{Dn} = \text{Re}_{D_H} \left(\frac{R_0 - R_i}{R_{cl}} \right)^{1/2}$. The range of parameters for this correlation is

$$\text{Dn} = 35 \text{ to } 2 \times 10^4, \ R_0/R_i = 1.61 \text{ to } 1.67, \ \frac{R_{cl}}{R_0 - R_i} = 21 \text{ to } 32.$$

Friction Factor in Turbulent Flow

The hydrodynamic and transport phenomena for turbulent flow in curved ducts are strongly influenced by centrifugal forces that result from the curvature of the flow passage. The earlier discussion regarding the mechanisms that affect the flow and transport processes in laminar flow is at least qualitatively applicable to turbulent flow as well. Secondary flows develop during turbulent flow in curved ducts that cause mixing and lead to an increase in the circumferentially averaged friction factor and heat transfer coefficient in comparison with the same flow in a similar but straight duct. The circumferential distributions of the friction factor and heat transfer coefficient are nonuniform, but to a lesser extent than in laminar flow. The friction factor and heat transfer coefficient are both larger on the outer surface of the flow passage. The turbulent velocity profile in fully developed flow is also similar to the turbulent velocity profile in a similar but straight pipe, but the point of maximum axial velocity is shifted towards the outer surface.

It was mentioned earlier that the hydrodynamic and thermal entrance lengths for laminar flow in curved pipes are considerably shorter than the entrance lengths in straight tubes. (Note that fully developed conditions are not possible for all curved pipe types. They are possible for helicoidally coiled flow passages.) For turbulent flow the entrance lengths in curved pipes are even shorter than the entrance length in laminar flow. For turbulent flow in helicoidally coiled tubes, for example, the entrance length is typically about one-half round. As a result for typical helically coiled tubes models and correlations that are based on fully developed hydrodynamics and thermally developed conditions can be used for design purposes. Experiment and simulation also show that for turbulent flow in helicoidally coiled ducts the effect of coil pitch on the friction factor and heat transfer coefficient is small.

Turbulent flow through helically coiled ducts has been simulated using Reynolds-average Navier–Stokes (RANS)-type turbulence models by many investigators including Lin and Ebadian (1997), Kumar et al. (2006), Jayakumar et al. (2008), and Sleiti (2011). It has been found that commonly applied two-equation turbulence models including variations of the K–ε model lead to model predictions that agree with experimental data.

Some widely applied correlations for the friction factor are now presented. All correlations represent circumferentially averaged friction factors.

Equations (8.106) or (8.107) can be used for predicting the laminar–turbulent flow regime transition in fully developed flow in a helically coiled tube. As noted in the equations the flow passage curvature has a stabilizing effect on laminar flow for circular pipes and moves the critical Reynolds number for laminar–turbulent regime transition to a higher value in comparison with a straight tube. This trend, however, is likely to be true for channel cross sections that do not contain sharp angles. For helically coiled channels with square or rectangular cross sections the critical Reynolds number can be lower than the critical Reynolds number for a similar straight channel (Bolinder and Sundén, 1995).

For fully developed flow through circular helicoidally coiled tubes one of the most widely used correlations for the Darcy friction factor is (Ito, 1959)

$$f_c' = 0.304\mathrm{Re}_D^{-0.25} + 0.029\sqrt{\left(\frac{R_i}{R_{cl}}\right)}. \tag{8.112}$$

The range of applicability of this correlation is $0.034 < \mathrm{Re}_D(\frac{R_i}{R_{cl}})^2 < 300$.

Srinivasan *et al.* (1970) derived the following correlation based on extensive data

$$f'_c \left(\frac{R_{cl}}{R_i}\right)^{0.5} = 0.336 \left[\text{Re}_D \left(\frac{R_{cl}}{R_i}\right)^{-2} \right]^{-0.2}. \tag{8.113}$$

The range of applicability of this correlation is $\text{Re}_D(\frac{R_{cl}}{R_i})^{-2} < 700$ and $7 < \frac{R_{cl}}{R_i} < 104$. Hart *et al.* (1987) have derived the following correlation for helically coiled tubes

$$\frac{f'_c}{f'_s} = 1 + \left[\frac{0.090\text{Dn}^{3/2}}{70 + \text{Dn}}\right]. \tag{8.114}$$

Downing and Kojasoy (2002) compared the predictions of the above three correlations with their experimental data representing miniature helically coiled tubes with 234–592 μm inner diameters and 2.4–7.9 mm coil diameters in which R-134a was the working fluid, for $\text{Re}_D = 10^3$–10^4 and $3.3 < \frac{R_{cl}}{R_i} < 13.4$ range. The correlation of Hart *et al.* (Eq. (8.114)) performed best.

For turbulent flow in curved rectangular flow passages correlations that have been derived for circular flow passages, e.g., Eqs. (8.112) or (8.113), can be used by replacing the flow channel diameter with hydraulic diameter everywhere (Shah and Joshi, 1987) as long as $\text{Re}_{D_H} \gtrsim 8 \times 10^3$. For $450 < \text{Re}_{2b}(\frac{2b}{R_{cl}})^{1/2} < 7.5 \times 10^3$ and $25 \leq R_{cl}/2b \leq 164$, where $2b$ is the shorter side length of the cross section, Kadambi (1983) has proposed the following correlation:

$$\frac{f_c}{f_s} = 0.435 \times 10^{-3}\text{Re}_{2b}^{0.96}\left(\frac{R_{cl}}{2b}\right)^{0.22}. \tag{8.115}$$

8.7.2 Frictional Pressure Drop in Two-Phase Flow

For vertically oriented helicoidally coiled tubes with upward flow the method of Chisholm and Laird (1958), Eqs. (8.27) through (8.29) (often referred to as the Lockhart–Martinelli method), has been found to predict some experimental data reasonably well (Rippel *et al.*, 1966; Boyce *et al.*, 1969; Banerjee *et al.*, 1969; Kasturi and Stepanek, 1972a, b). In fact, the void-quality expression of Lockhart and Martinelli (1949) (see Eq. (6.60) and Table 6.1) could predict the measured void fractions in the experiments of these authors (Banerjee *et al.*, 1969; Kasturi and Stepanek, 1972a, b). More recent investigations have shown that the aforementioned correlations need to be adjusted. The Chisholm–Laird pressure drop correlation, in particular, needs to be adjusted to account for an additional mass flux effect that has been noticed in experiments. Modifications of the Lockhart–Martinelli method have been proposed.

Based on near-atmospheric air–water data in helicoidally coiled tubes with 12.7–38.1-mm inner diameter, coiled around the outside of cylinders with 305 and 609 mm diameters, Xin *et al.* (1996) derived

$$\Phi_L^2 \bigg/ \left(1 + \frac{20}{X_{tt}} + \frac{1}{X_{tt}^2}\right) = \left[1 + \frac{X_{tt}}{65.45F_d^{0.6}}\right]^2 \quad \text{for} \quad F_d \leq 0.1 \tag{8.116}$$

$$\Phi_L^2 \bigg/ \left(1 + \frac{20}{X_{tt}} + \frac{1}{X_{tt}^2}\right) = \left[1 + \frac{X_{tt}}{434.8F_d^{1.7}}\right]^2 \quad \text{for} \quad F_d > 0.1 \tag{8.117}$$

where

$$F_{\mathrm{d}} = \frac{j_{\mathrm{L0}}^2}{2R_{\mathrm{i}}g} \left(\frac{R_{\mathrm{i}}}{R_{\mathrm{cl}}}\right)^{1/2} (1 + \tan\theta)^{0.2}. \tag{8.118}$$

This and other correlations which are based on data representing flow in vertically oriented coils can be applied to horizontally oriented coiled tubes, provided that the mixture flow rate is high enough to render the buoyancy effect small, and more importantly, to prevent the occurrence of flow pulsations caused by phase separation and intermittent blockage of parts of the flow passage by liquid slugs. When such flow oscillations occur the pressure drop will evidently be intermittent. Awwad *et al.* (1995a, b) used test sections similar to those of Xin *et al.* (1996), with air and water, that were oriented horizontally. In experiments where flow intermittency did not occur their pressure drop data agreed with Chisholm and Laird (1958) (Lockhart–Martinelli correlation, Eq. (8.28)). When intermittency occurred, however, their data deviated from the Lockhart–Martinelli correlation. Awwad *et al.* applied nonlinear data regression and developed a correlation for their data.

As mentioned earlier, an important application of helically coiled tubes is in steam generators. Steam generators typically operate at high pressure. Correlations specifically developed for such steam generators have been proposed by several authors, a few of the most widely referenced of which are now presented.

Using experimental data with three coil curvatures over a parameter range relevant to Advanced Gas-cooled Reactors (AGRs), Ruffell (1974) derived the following correlation

$$\Phi_{\mathrm{f0}}^2 = (1 + F) \frac{\rho_{\mathrm{f}}}{\rho_{\mathrm{h}}}, \tag{8.119}$$

where ρ_{h} is the homogeneous density (see Eq. (7.7)) and

$$\begin{aligned} F = {} & \sin\left(\frac{1.16G}{10^3}\right) \\ & \times \left\{0.875 - 0.314y - \frac{0.74G}{10^3}(0.152 - 0.07y) - x\left(\frac{0.155G}{10^3} + 0.7 - 0.19y\right)\right\} \\ & \cdot [1 - 12(x - 0.3)(x - 0.4)(x - 0.5)(x - 0.6)] \end{aligned} \tag{8.120}$$

$$y = \frac{R_{\mathrm{cl}}}{100R_{\mathrm{i}}}. \tag{8.121}$$

The mass flux G should be in kg/m²·s. The range of experimental data used for the derivation of this correlation is

$$P = 6.0\text{–}18 \text{ MPa}; \ G = 300\text{–}1800 \text{ kg/m}^2 \cdot \text{s}; \ R_i = 5.35\text{–}9.3 \text{ mm};$$
$$R_{\mathrm{cl}}/R_{\mathrm{i}} = 6.25\text{–}185$$

Guo *et al.* (2001) have proposed

$$\Phi_{\mathrm{f0}}^2 = 142.2\left(\frac{P}{P_{\mathrm{cr}}}\right)^{0.62}\left(\frac{R_{\mathrm{i}}}{R_{\mathrm{cl}}}\right)^{1.04} \psi \left[1 + x\left(\frac{\rho_{\mathrm{f}}}{\rho_{\mathrm{g}}} - 1\right)\right] \tag{8.122}$$

where

$$\psi = 1 + \frac{x\,(1-x)\left[\left(10^3/G\right)-1\right](\rho_f/\rho_g)}{1+x\left[(\rho_f/\rho_g)-1\right]} \quad \text{for} \quad G \le 10^3\,\text{kg/m}^2\cdot\text{s} \quad (8.123)$$

$$\psi = 1 + \frac{x\,(1-x)\left[\left(10^3/G\right)-1\right](\rho_f/\rho_g)}{1+(1-x)\left[(\rho_f/\rho_g)-1\right]} \quad \text{for} \quad G > 10^3\,\text{kg/m}^2\cdot\text{s}. \quad (8.124)$$

The range of experimental data used for the derivation of this correlation is $P = 0.5$–3.5 MPa; $G = 150$–1760 kg/m^2·s; $R_i = 10$ mm; $R_{cl}/R_i = 13$–25; $x_{eq,exit} = 0$–1.2.

Other empirical correlations have been proposed for steam generator parameter ranges by Unal et al. (1981) and Zhao et al. (2003). The correlation of Unal et al. (1981) is based on data relevant to steam generators in Liquid Metal Fast Breeder Reactors (LMFBRs).

Based on experimental data with R-134a in miniature helically coiled tubes with 234–592 μm inner diameters and 2.4–7.9 mm coil diameters, Downing and Kojasoy (2002) found that Eq. (8.28) could predict their data well, provided that the constant C is found from

$$C = 3.598X^{-0.012}. \qquad (8.125)$$

Colorado et al. (2011) recently modeled boiling in a helically coiled steam generator using a one-dimensional model. Using the experimental data of Santini et al. (2008) they sought agreement between their model and experimental data with respect to pressure drop. Such agreement could only be obtained by modifying the correlation of Chisholm and Laird (1958) and Friedel (1979). Their modified Lockhart–Martinelli correlation is (compare with Eq. (8.28))

$$\Phi_L^2 = 1 + \frac{3.2789}{X_{tt}} + \frac{0.37}{X_{tt}^{2.0822}}. \qquad (8.126)$$

The data base for this correlation represents boiling water with 11–63 bars pressure, 192–811 kg/m^2·s mass flux, flowing in a helicoidally coiled tube with 12.53 mm inner diameter and 1000 mm coil diameter.

EXAMPLE 8.4. Saturated water–steam at a pressure of 70 bars and a quality of $x = 0.015$ flows in an adiabatic circular copper tube that has 6.0 mm inner diameter. The tube is helically coiled with $R_{cl} = 12$ cm and $l_{pch} = 4$ cm. The mixture mass flux is 340 kg/m^2·s. Find the local frictional pressure gradient.

SOLUTION. We are dealing with a high-pressure water–steam mixture, and the correlation of Ruffell (1974) will be appropriate.

First, we will find the relevant thermophysical and transport properties.

$$\rho_f = 739.9\,\text{kg/m}^3$$
$$\rho_g = 36.53\,\text{kg/m}^3$$
$$\mu_f = 9.129 \times 10^{-4}\,\text{kg/m·s}$$
$$h_{fg} = 1.345 \times 10^6\,\text{J/kg}$$
$$\sigma = 0.0176\,\text{N/m}$$

$$\rho_h = \frac{1.}{x/\rho_g + \frac{1.-x}{\rho_f}} = 574.1\,\text{kg/m}^3.$$

We now calculate the all-liquid frictional pressure gradient. First let us find the all-liquid flow regime.

$$Re_{f0} = (G \cdot D / \mu_f) = (340 \, \text{kg/m}^2 \cdot \text{s}) \, (0.006 \, \text{m}) / (9.129 \times 10^{-5} \, \text{kg/m} \cdot \text{s}) = 2.23 \times 10^4$$
$$f_{f0} = 0.079 \cdot Re_{f0}^{-0.25} = 0.006461.$$

The all-liquid flow regime is turbulent, and we can use the correlation of Ito (1959), Eq. (8.112).

$$f'_c = 0.304 Re_D^{-0.25} + 0.029 \sqrt{\left(\frac{R_i}{R_{cl}}\right)} = 0.304(2.23 \times 10^4)^{-0.25} + 0.029 \sqrt{\left(\frac{0.003 \, \text{m}}{0.12 \, \text{m}}\right)}$$
$$= 0.0294.$$

The all-liquid frictional pressure gradient would then be

$$\left(-\frac{\partial P}{\partial z}\right)_{\text{fr,f0}} = f'_c \frac{1}{D} \frac{1}{2} \frac{G^2}{\rho_{f0}} = 0.0294 \left(\frac{1}{0.006 \, \text{m}}\right) \frac{1}{2} \frac{(340 \, \text{kg/m}^2 \cdot \text{s})^2}{739.9 \, \text{kg/m}^3} = 383 \, \text{N/m}^3$$

The two-phase multiplier can now be found using Eqs. (8.119)–(8.121):

$$y = \frac{R_{cl}}{100 R_i} = 0.4$$

$$\sin\left(\frac{1.16 G}{10^3}\right) = \sin\left(\frac{1.16 \times 340 \, \text{kg/m}^2 \cdot \text{s}}{10^3 \, \text{kg/m}^2 \cdot \text{s}}\right) = 0.38.$$

Equation (8.120) then gives $F = 0.1704$, and from Eq. (8.119)

$$\Phi_{f0}^2 = (1 + F) \frac{\rho_f}{\rho_h} = (1 + 0.1704) \frac{739.9 \, \text{kg/m}^3}{574.1 \, \text{kg/m}^3} = 1.508$$

$$\Rightarrow \left(-\frac{\partial P}{\partial z}\right)_{\text{fr}} = \left(-\frac{\partial P}{\partial z}\right)_{\text{fr,f0}} \Phi_{f0}^2 = 578.4 \, \text{N/m}^3.$$

PROBLEMS

8.1 Using the separated-flow mixture momentum equation, Eq. (5.71) and Eq. (8.36), show that for a steady boiling flow in a straight channel with uniform cross section, the pressure drop between the point where pure saturated liquid is obtained (i.e., where $x_{eq} = 0$) and any arbitrary point is

$$\Delta P = \left(-\frac{\partial P}{\partial z}\right)_{\text{fr,f0}} \frac{L}{x_{eq}} \int_0^{x_{eq}} \Phi_{f0}^2(x) dx + \frac{G^2}{\rho_f} r(P, x_{eq}) + g \sin\theta \int_0^L [\rho_g \alpha + \rho_f(1 - \alpha)] dz,$$

(a)

where

$$r(P, x_{eq}) = \left[\frac{(1 - x_{eq})^2}{(1 - \alpha)} + \frac{\rho_f x_{eq}^2}{\rho_g \alpha} - 1\right].$$

(b)

8.2 Following Martinelli and Nelson (1948), Thom (1964) tabulated values of $r(P, x_{eq})$, $\overline{\Phi}_{L0}^2 = \frac{1}{x_{eq}} \int_0^{x_{eq}} \Phi_{L0}^2(x)dx$, and the void fraction for water. Tables P8.2a and P8.5b are summaries of $\overline{\Phi}_{L0}^2$ and $r(P, x_{eq})$ values.

Table P8.2a. *Selected values of $\overline{\Phi}_{L0}^2$ from Thom (1964).*

x_{eq} (%)	$P = 17.2$ bars (250 psia)	$P = 41$ bars (600 psia)	$P = 8.6$ MPa (1250 psia)	$P = 14.48$ MPa (2100 psia)	$P = 20.68$ MPa (3000 psia)
1	1.49	1.11	1.03	–	–
5	3.71	2.09	1.31	1.10	–
10	6.30	3.11	1.71	1.21	1.06
20	11.4	5.08	2.47	1.46	1.12
30	16.2	7.00	3.20	1.72	1.18
40	21.0	8.80	3.89	2.01	1.26
50	25.9	10.6	4.55	2.32	1.33
60	30.5	12.4	5.25	2.62	1.41
70	35.2	14.2	6.00	2.93	1.50
80	40.1	16.0	6.75	3.23	1.58
90	45.0	17.8	7.50	3.53	1.66
100	49.93	19.65	8.165	3.832	1.74

Table P8.2b. *Selected values of r (P, x_{eq}) from Thom (1964).*

x_{eq} (%)	$P = 17.2$ bars (250 psia)	$P = 41$ bars (600 psia)	$P = 8.6$ MPa (1250 psia)	$P = 14.48$ MPa (2100 psia)	$P = 20.68$ MPa (3000 psia)
1	0.4125	0.2007	0.0955	0.0431	0.0132
5	2.169	1.040	0.4892	0.2182	0.0657
10	4.62	2.165	1.001	0.4431	0.1319
20	10.39	4.678	2.100	0.9139	0.2676
30	17.30	7.539	3.292	1.412	0.4067
40	25.37	10.75	4.584	1.937	0.5495
50	34.58	14.30	5.958	2.490	0.6957
60	44.93	18.21	7.448	3.070	0.8455
70	56.44	22.46	9.030	3.678	0.9988
80	69.09	27.06	10.79	4.512	1.156
90	82.90	32.01	12.48	5.067	1.316
100	98.10	37.30	14.34	5.664	1.480

Saturated water enters a vertical and uniformly heated tube with $D = 2$ cm diameter and $L = 3.0$ m. The pressure at the inlet is 41 bars. The heat flux is such that for a mass flux of $G = 2000$ kg/m^2, $x_{eq} = 0.47$ is obtained at the exit. For mass fluxes in the $G = 25$–4200 kg/m^2·s range calculate and plot the frictional and total pressure drops in the tube using the method of Martinelli and Nelson. Note that, for the calculation of the last term on the right side of Eq. (a), an appropriate correlation for the void fraction is needed.

8.3. Repeat Problem 8.2, this time using the pressure-drop correlation of Friedel (1979) and the slip ratio correlation of Premoli *et al.* (1970) (Eqs. (6.54)–(6.59)).

8.4 A water evaporator consists of a vertical metallic tube that is 1.5 m long, with an inside diameter of 1.0 cm. A uniform heat flux of 1000 kW/m^2 is applied to the tube wall. Saturated liquid water enters the tube at a pressure of 2185 kPa. Using methods of your choice, calculate and plot the total pressure drop in the tube for flow rates between 30 and 800 g/s.

Note that, given the high pressure at the inlet, one can assume that the properties remain constant.

8.5 In Problem 8.4, select the range of flow rates that ensures that the flow regime at the exit of the boiler is either in the bubbly or slug flow regimes, but not churn or annular.

8.6 Calculate the total pressure drop around a 90°, 20-cm-radius bend in a horizontal 12-mm-diameter pipe for the flow of a steam–water mixture with qualities in the range 2.5%–45%. The system pressure is 10 bars and the mass flux is 850 kg/m^2·s.

8.7 Water flows upward in a tube with an inner diameter of 1.5 cm and a length of 1.75 m. The water enters the tube as saturated liquid at 1172 kPa. A heat flux of 1200 kW/m^2 (based on the inner tube surface) is applied uniformly to the system. For mass fluxes in the 100–600 kg/m^2 range do the following.

(a) Plot the variation of void fraction along the tube using the homogeneous equilibrium mixture model and the drift flux model.

(b) Determine the total and frictional pressure drops over the tube length, using the Lockhart–Martinelli–Chisholm approach (Eq. (8.28)).

8.8 In Problem 8.7, for a mass flux of 250 kg/m^2·s, determine the major two-phase flow regimes along the tube.

8.9 The correlation of Beattie (1973) for flow through orifices indicates that Φ_{L0} depends on flow quality and fluid properties.

(a) For saturated steam–water flow at 1, 10, and 50 bars pressures, calculate and plot Φ_{L0} as a function of x for the $x = 0.01$–0.90 range.

(b) Repeat part (a), this time for R-134a at 0 °C and 50 °C temperatures.

8.10 A correlation for interfacial area concentration in bubbly pipe flow, proposed by Hibiki and Ishii (2001), is

$$a_I'' \sqrt{\frac{\sigma}{g \Delta \rho}} = 0.5 \alpha^{0.847} \left\{ D_{\mathrm{H}} \left(\frac{\varepsilon}{v_{\mathrm{L}}^3} \right)^{1/4} \right\}^{0.283},$$

where ε, the turbulent dissipation rate, can be estimated by writing

$$\varepsilon = \frac{j}{\rho} \left(-\frac{\partial P}{\partial z} \right)_{\mathrm{fr}}.$$

Repeat Problem 7.12, this time using Hibiki and Ishii's correlation.

8.11 Consider the flow of a two-phase liquefied petroleum gas (LPG) (assumed to be pure propane, for simplicity) in an insulated horizontal tube at a pressure of 8 bars. The tube diameter is 5 cm. The mixture mass flux is 3.0×10^3 kg/m^2·s.

(a) For qualities $x = 0.05$ and 0.10, find the frictional pressure gradient using the correlation of Muller-Steinhagen and Heck (1986).

(b) Consider the segment of the tube where at the inlet you have pure saturated liquid and at the exit you have a flow quality of 10%. Assume that the tube segment is 2.5 m long and assume that quality varies linearly along the segment. Find the total pressure drop in the segment.

Hint: You need to calculate the acceleration pressure drop as well. If you need the void fraction, you may use the correlation of Bhagwat and Ghajar (2014). Knowing quality and void fraction you can find phase velocities.

8.12 For two-phase flow pressure drop in gate valves, Morris (1985) has proposed:

$$\Phi_{L0} = 1 + \left(x\frac{\rho_L}{\rho_G} + S_r(1-x)\right)\left\{x + \left(\frac{1-x}{S_r}\right)\left[1 + \frac{(S_r-1)^2}{\left(\frac{\rho_L}{\rho_G}\right)^{0.5}-1}\right]\right\}$$

where

$$S_r = \sqrt{\frac{\rho_L}{\rho_h}} = \left[x\frac{\rho_L}{\rho_G} + (1-x)\right]^{0.5}.$$

Consider air–water two-phase flow in a vertical pipe that is 2.5 cm in diameter, at atmospheric pressure and room temperature. The liquid mass flux is 250 kg/m^2·s. For gas mass fluxes of 50 and 100 kg/m^2·s, do the following.

(a) Find the flow regimes using the flow regime models of Taitel *et al.* (1980).

(b) Calculate the void fraction using the Drift Flux Model and choosing appropriate DFM parameters.

(c) Suppose a gate valve is installed on the tube, the inner diameter of which is equal to the inner diameter of the tube. Find the two-phase pressure drop that takes place in the gate valve when the gate valve is three-quaters closed. Compare the results with the prediction of the homogeneous flow model:

$$\Phi_{L0} = (1-x) + \left(\frac{\rho_L}{\rho_G}\right)x.$$

8.13 In a heat exchanger, a two-phase mixture of ammonia with a quality of 20% and temperature of $-20\,°C$ is to be completely evaporated into saturated vapor. The mass flow rate of ammonia is 16 g/s. The ammonia flows inside six parallel 8 mm inner diameter copper tubes that are 1.0 m long. The tubes have a surface roughness of 1.5 μm. For simplicity, assume that at the inlet to the tubes the mixture flow is homogeneous.

(a) Find the heat flux that is needed, assuming negligible contribution of kinetic energy to the energy conservation

(b) Repeat part (a), this time including the effect of kinetic energy

(c) Calculate the frictional pressure gradient at the middle of the tubes.

(d) Suppose a gate valve is located at the middle of the tube. The valve has a flow diameter of 2 mm, and is fully open. Find the pressure loss caused by the gate valve, using the HEM assumption.

Comment on the solution and validity of assumptions.

8.14 In a vertical, upward water boiling tube that is made of carbon steel the flow conditions are as follows:

Mass flow rate: 0.29 kg/s
Flow area: 1.5×10^{-4} m^2
Flow quality: 0.15
Pressure: 7.2 MPa.

(a) Calculate the frictional pressure gradient, using a model other than the homogeneous-equilibrium mixture model.

(b) Calculate the void fraction using the homogeneous-equilibrium mixture model.

(c) Calculate the void fraction using the drift flux model.

8.15 Refrigerant R-123 at a pressure of 5 bars flows in a horizontal circular tube that has 10.9 mm inner diameter.

(a) For a mass flux of 400 kg/m^2·s and for qualities of 0.01, 0.05, and 0.25 find the local frictional pressure gradient using the methods of Muller-Steinhagen and Heck (1986) and Friedel (1979).

(b) Repeat part (a), this time for a mass flux of 150 kg/m^2·s and a tube diameter of 6 mm, and compare the predictions of the correlations of Muller-Steinhagen and Heck (1986) and Kim and Mudawar (2012c).

8.16 Consider the system in the Problem 8.15. Assume that for a test with a mass flux of 900 kg/m^2·s, the quality at the inlet is 0.0 (i.e., saturated liquid at inlet), and increases to 0.25 over a tube length of 7 m. Calculate the total pressure drop caused by flow acceleration by

(a) assuming homogeneous flow and

(b) assuming separated flow and using a void-quality relation of your choice.

(c) What is the most likely flow regime near the exit?

8.17 Refrigerant R-134a at a temperature of 233 K flows in a horizontal circular tube that has a 0.51 mm inner diameter.

(a) For a mass flux of 300 kg/m^2·s and for qualities of 0.01, 0.04, and 0.08 find the local frictional pressure gradient using the methods of Muller-Steinhagen and Heck (1986) and Li and Wu (2010a).

(b) Repeat part (a), this time for a mass flux of 1100 kg/m^2·s, and compare the predictions of the correlations of Muller-Steinhagen and Heck (1986), Li and Wu (2010a), and Kim and Mudawar (2012c).

8.18 Liquefied petroleum gas (LPG) (assumed to be pure propane, for simplicity) flows in an insulated horizontal tube at an inlet pressure of 8 bars. The tube diameter is 5 cm. The mixture mass flux is 2.5×10^3 kg/m²·s. The tube has a 6 m long straight segment, followed by a return bend that has a radius of 25 cm. For flow qualities of 0.05 and 0.15 do the following.

(a) Find the total pressure drop over the entire tube, using appropriate correlations of your own choice.

(b) Find pressure loss over the bend using the method of Domanski and Hermes (2008).

8.19 Liquid nitrogen (LN₂) flows in a horizontal tube at a pressure of 8 bars. The tube diameter is 2.5 cm. The mixture mass flux is 2.0×10^3 kg/m²·s.

(a) For qualities $x = 0.05$ and 0.10, find the frictional pressure gradient using the homogeneous-equilibrium model, and the correlations of Muller-Steinhagen and Heck (1986). Which approach do you expect to be more accurate?

(b) Consider the segment of the tube where at the inlet you have pure saturated liquid and at the exit you have a flow quality of 25% and 100%. Assume that the tube segment is 2.0 m long and assume that quality varies linearly along the segment. Find the total pressure drop in the segment.

(c) Assume that, for case (b), a gate valve is installed midway along the tube. The gate valve has a flow area of 0.5 cm² when it is fully open. The gate valve is half open. Find the pressure loss caused by the gate valve.

8.20 Liquid R-134a flows in a horizontal tube at a pressure of 2 bars. The tube diameter is 0.8 mm. The mixture mass flux is 2.5×10^3 kg/m²·s.

(a) For qualities $x = 0.20$ and 0.45, find the frictional pressure gradient using the homogeneous-equilibrium model, and the correlations of Friedel (1979). Which approach do you expect to be more accurate?

(b) Assume that at the inlet you have pure saturated liquid and at the exit you have a flow quality of 60% and 80%. Assume that the tube segment is 2.0 m long and assume that quality varies linearly along the segment. Find the total pressure drop in the segment.

(c) Assume that, for the conditions of part (a), the tube is insulated and the horizontal tube consists of two straight segments each 40 cm long, and a U-bend in the middle, where the bend radius is 2 cm. Find the frictional and total pressure drop for qualities $x = 0.20$ and 0.45.

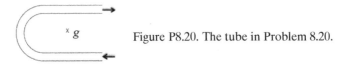

Figure P8.20. The tube in Problem 8.20.

8.21 Refrigerant R-123 at an inlet pressure of 5 bars flows in a horizontal circular tube that has 8 mm inner diameter. The tube has a uniformly heated segment that is 2.5 m long, followed by an adiabatic segment that is 1 m long.

(a) For a mass flux of 400 kg/m^2·s, calculate the wall heat fluxes that lead to qualities of 0.05 and 0.20 at the exit of the heated segment.

(b) Find the frictional and total pressure drops in the tube using the methods of Muller-Steinhagen and Heck (1986) and Friedel (1979).

8.22 Consider the two-phase flow of saturated methane at a pressure of 4 MPa in an insulated vertical pipe that is 2.0 cm in diameter. The mixture mass flux is 195.5 kg/m^2·s and the flow quality is 0.08.

(a) Find the flow regimes using an appropriate flow regime map.

(b) Calculate the void fraction using an appropriate DFM or slip ratio expression.

(c) Find the frictional pressure gradient.

(d) Suppose that a cryogenic valve is installed on the tube, the inner diameter of which is equal to the inner diameter of the tube. Find the two-phase pressure drop that takes place in the valve when the valve is half closed, assuming for simplicity that the valve is a globe valve. Compare the results with the prediction of the homogeneous flow model.

8.23 Refrigerant R-123 at a pressure of 4 bars flows in an adiabatic and helicoidally coiled circular copper tube that has 10.0 mm inner diameter. Assume that the coil's axis is vertical, and the total length of the tube is 7 m. Also, for the helically coiled tube $R_{cl} = 35$ cm and $l_{pch} = 3$ cm.

Assume that in a test a mass flux of 650 kg/m^2·s is underway. For inlet qualities of 0.01, 0.25, and 0.5 calculate the frictional and total pressure drops for the coil by

(a) assuming homogeneous flow and

(b) assuming separated flow and using an appropriate void-quality relation of your choice.

(c) How long would a straight tube with the same inner diameter have to be to cause the same frictional pressure drop as case (b)?

8.24 Liquefied petroleum gas (LPG) (assumed to be pure propane, for simplicity) is in an insulated horizontal tube at an inlet pressure of 7 bars. The tube diameter is 4 cm. The mixture mass flux is 2.5×10^3 kg/m^2·s. The tube has an 8-m-long straight segment. For flow qualities of 0.05 and 0.15 do the following.

(a) Find the total frictional pressure drop over the entire tube, using appropriate correlations of your own choice.

(b) Find the total (frictional and gravitational) pressure drop if the tube is coiled into a vertically oriented helical configuration with $R_{cl} = 25$ cm and $l_{pch} = 10$ cm.

8.25 Refrigerant R-123 at a pressure of 4 bars flows in a horizontal circular copper tube that has 10.0 mm inner diameter. The tube is adiabatic.

(a) For a mass flux of 400 kg/m^2·s and for qualities of 0.01, 0.25, and 0.5 find the local frictional pressure gradient using the methods of Muller-Steinhagen and Heck (1986).

(b) Repeat part (a) this time assuming that the flow tube is helically coiled with a vertical axis with $R_{cl} = 35$ cm and $l_{pch} = 3$ cm. The flow is upwards.

8.26 Refrigerant R-134a at a temperature of 233 K flows in a horizontal circular copper tube that has 4.0 mm inner diameter. The tube is adiabatic.

(a) For a mass flux of 300 kg/m^2·s and for qualities of 0.01, 0.4, and 0.8 find the local frictional pressure gradient using the methods of Muller-Steinhagen and Heck (1986).

(b) Repeat part (a) this time assuming that the tube is helically coiled with a vertical axis and upward flow, with $R_{cl} = 6$ cm and $l_{pch} = 1.5$ cm.

9 Countercurrent Flow Limitation

9.1 General Description

Countercurrent flow limitation (CCFL), or flooding, refers to an important class of gravity-induced hydrodynamic processes that impose a serious restriction on the operation of gas–liquid two-phase systems. Some examples in which CCFL is among the factors that determine what we can and cannot be done are the following:

(a) the emergency coolant injection into nuclear reactor cores following loss of coolant accidents,
(b) the "reflux" phenomenon in vertically oriented condenser channels with bottom-up vapor flow, and
(c) transport of gas–liquid fossil fuel mixtures in pipelines.

In the first example the coolant liquid attempts to penetrate the overheated system by gravity while vapor that results from evaporation attempts to rise, leading to a countercurrent flow configuration. The rising vapor can seriously reduce the rate of liquid penetration, or even completely block it. In the third example, the occurrence of CCFL causes a significant increase in the pressure drop and therefore the needed pumping power. CCFL represents a major issue that must be considered in the design and analysis of any system where a countercurrent of a gas and a liquid takes place.

To better understand the CCFL process, let us consider the simple experiment displayed in Fig. 9.1, where a large and open tank or plenum that contains a liquid is connected to a vertical pipe at its bottom. The vertical pipe itself is connected to a mixer before it drains into the atmosphere. Air can be injected into the mixer via the gas injection line. When there is no gas injection, liquid flows downward freely, and its flow rate is restricted by wall friction and other pressure losses. If in the pipe friction and end losses are assumed to be the dominant causes for pressure loss, one can write

$$(\rho_L - \rho_G)g(H + L) = \left(1 + f'\frac{L}{D} + K_{ent} + K_{ex}\right)\frac{G^2}{2\rho_L}, \tag{9.1}$$

where G is the mass flux. Now, let us consider an experiment using the same system, where a constant upward gas flow rate (equivalent to $j_G = $ const. in the pipe) is imposed while the tank is empty, and then liquid is injected into the tank at an increasing rate while the downward flow rate of liquid in the pipe is measured. The process line would look like the line ABC in Fig. 9.2. The rising gas in the pipe in this case imposes an interfacial force on the liquid that severely restricts the downward

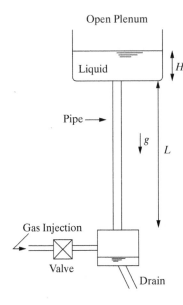

Open Plenum

Liquid

Pipe →

g

L

Gas Injection

Valve

Drain

H

Figure 9.1. Schematic of a vertical pipe subject to countercurrent gas and liquid flows.

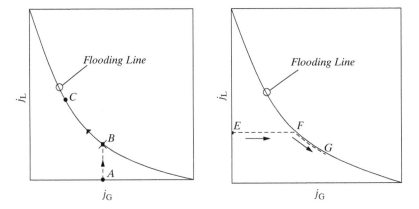

Figure 9.2. Process paths in flooding experiments.

penetration rate of liquid. (Note that everywhere in this chapter, unless otherwise stated, $j_L > 0$ for downward liquid flow, and $j_G > 0$ for upward gas flow.) Thus, with $j_G = $ const., when liquid injection is gradually increased from a small value, the liquid flow rate does not affect the upward flow rate of gas initially, and the process line is first AB. Beyond B, however, an increase in liquid delivery rate is only possible if j_G is reduced. At point B, the system is said to be *flooded. The* repetition of this simple experiment with different j_G values will result in a curve that is the locus of points B, similar to the one shown in Fig. 9.2.

A similar observation is made when the test system is modified so that a constant liquid flow rate (equivalent to $j_L = $ const.) can be maintained while j_G is increased. In this case the process path would follow the line EF first, where the increasing upward gas flow rate does not affect the downward flow rate of liquid. Beyond point F, however, increasing j_G is accompanied by a decreasing j_L. Once again, the pipe is *flooded* at point F.

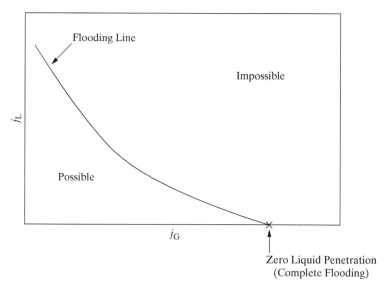

Figure 9.3. The flooding curve.

The experimentally obtained combinations of (j_L, j_G) define a *flooding line* or *curve* that divides the entire flow map into two regions: a physically allowable region and a physically impossible region, as shown in Fig. 9.3. The flooding curve is insensitive to both the channel length L and the depth of liquid in the plenum, H.

The mechanisms that cause flooding, and certain details of the flooding line, depend on the liquid and gas injection method. For the configuration shown in Fig. 9.1, the channel pressure gradient, void fraction, and qualitative flow patterns are displayed in Fig. 9.4 (Bharathan and Wallis, 1983), where τ_I and τ_w represent the gas–liquid interfacial shear stress and the shear stress at the liquid–wall interface, respectively, and

$$j_G^* = \sqrt{\frac{\rho_G}{\Delta\rho g D}}\, j_G, \tag{9.2}$$

$$j_L^* = \sqrt{\frac{\rho_L}{\Delta\rho g D}}\, j_L, \tag{9.3}$$

$$(-dP/dz)^* = \frac{-dP/dz}{\Delta\rho g}. \tag{9.4}$$

In region A, the flow pattern is annular and is composed of a falling liquid film on the wall and an upward gaseous core. In this regime, $\tau_I \ll \tau_w$, flow restriction happens at the top end of the channel, and the liquid film flows freely in the channel. In region B, the flow pattern is annular but comprises an agitated and wavy liquid–gas interphase (rough film). In this case $\tau_I \gg \tau_w$, and flow restriction happens at the bottom end. Region C represents transition conditions, with a discontinuity in film characteristics that may oscillate along the channel. Flow restriction occurs intermittently at the top or bottom, but not within the channel.

The most widely used arrangements in flooding experiments are depicted schematically in Fig. 9.5. For liquid, there are two basic injection modes, as shown in Fig. 9.5(a). In one liquid injection mode the liquid is injected into an upper plenum,

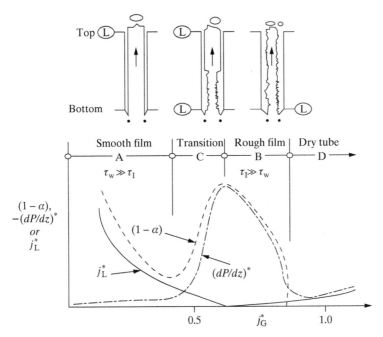

Figure 9.4. The flow patterns and axial variations of various parameters during flooding of a vertical pipe connected to a plenum at its top.

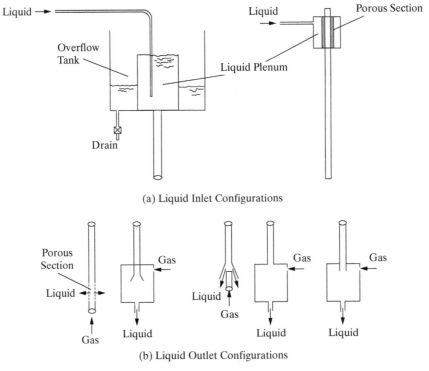

(a) Liquid Inlet Configurations

(b) Liquid Outlet Configurations

Figure 9.5. Schematics of typical gas and liquid injection arrangements in flooding experiments.

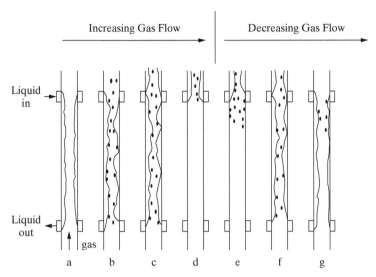

Figure 9.6. The flow patterns during flooding of a vertical pipe in which liquid injection takes place via a porous segment of the pipe. (From Bankoff and Lee, 1986.)

and from there it flows into the test channel. In the other major injection mode the liquid flows through a porous segment of the wall, or in some cases through a slot, leading to the immediate formation of a liquid film. When the plenum arrangement is used, the configuration of the channel inlet also affects the flooding behavior of the channel.

Figure 9.5(b) depicts some liquid outlet and gas injection arrangements. For flooding to take place, the bottom of the channel must lead to a volume that contains gas, so that the flow of the liquid leaving the channel is not restricted by the liquid collected there. In general, the more disturbance and turbulence generated at either liquid or gas injection locations, the more severe flooding will be (i.e., for the same flow rate of gas, a smaller flow rate of liquid is permitted), whereas injection arrangements where little flow disturbance, mixing, and turbulence is generated lead to less severe flooding limitations.

When the liquid injection is via a porous segment of the channel wall, a slot, or any other arrangement that leads immediately to the formation of a liquid film on the channel wall, the flow patterns have the qualitative appearances shown in Fig. 9.6 (Bankoff and Lee, 1983). The observed flow patterns are as follows.

(a) *Free fall*. All the injected liquid moves downward in this case.
(b) *Onset of flooding* (formation of large waves, entrainment of droplets). This is a point on the *flooding line*. A slight increase in gas flow rate will carry some of the liquid upward.
(c) *Partial delivery* of injected liquid to the exit. Falling and climbing film flows occur.
(d) *Zero liquid penetration*. This point corresponds to the minimum gas superficial velocity that would completely block the downward flow of liquid.
(e) *Flow reversal*. Because of a hysteresis phenomenon, downward liquid penetration starts at a lower gas flow rate than the zero liquid penetration point.
(f) *Partial delivery* of liquid (similar to (c)).

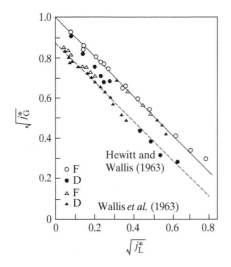

Figure 9.7. Comparison between flooding and deflooding curves.

(g) *Deflooding.* At gas flow rates smaller than this value, the downward delivery of injected liquid is complete.

The *flooding* and *deflooding* processes, depicted schematically in cases (b) and (g) in Fig. 9.6, are important thresholds. Flooding occurs in situations where j_L is maintained constant and j_G is increased. It represents the point where partial reduction of downward liquid delivery rate is initiated. Deflooding is the opposite process and occurs when gas flow rate is reduced starting from a very high value. For any particular liquid flow rate, it represents the point where complete, free, and unimpeded downward flow of liquid starts. In plenum-type liquid entry methods the flooding line represents the partial liquid delivery conditions, and flooding and deflooding points are essentially the same. However, when the method of liquid injection is meant to lead to an immediate formation of a liquid film on the flow passage wall (e.g., by injection through a sintered channel segment), flooding and deflooding lines are different. Figure 9.7 compares the flooding and deflooding experimental data obtained in such vertical channels (Hewitt and Wallis, 1963; Wallis *et al.*, 1963; Bankoff and Lee, 1986).

9.2 Flooding Correlations for Vertical Flow Passages

The correlation of Wallis (1961, 1969) is the most widely used and applies to small tubes for which

$$2 \leq D/\lambda_L \leq 40, \tag{9.5}$$

where $\lambda_L = \sqrt{\sigma/g\Delta\rho}$ is the Laplace length scale. Channels smaller in diameter than about $2\lambda_L$ do not support countercurrent gas–liquid flow. The correlation, which performs best with plenum-type liquid injection, can be written as

$$j_G^{*1/2} + m j_L^{*1/2} = C, \tag{9.6}$$

where j_G^* and j_L^* are defined in Eqs. (9.2) and (9.3), respectively.

Wallis's original expression, with $m = C = 1$, was based on a simple model that neglected the momentum interactions between the two phases (Wallis, 1961). It can,

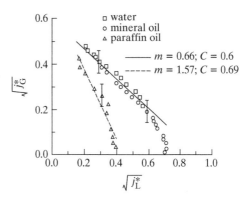

Figure 9.8. Correlation of some experimental data with Wallis's parameters (Wu, 1996; Ghiaasiaan et al., 1997a, b). Vertical channel of $D = 1.0$ cm, has air as the gas phase and a liquid inlet similar to Fig. 9.1.

however, be considered as an empirical correlation with values of parameters m and C that depend on inlet and exit conditions. These parameters vary approximately in the $m = 0.8$–1.0 and $C = 0.7$–1.0 ranges, and they depend mainly on geometry and flow passage end effects. Parameter C depends primarily on the liquid inlet condition; for example, $C = 0.9$ is recommended for a round-edged liquid inlet, and $C = 0.725$ for a sharp-edged liquid inlet (Bankoff and Lee, 1986). Figure 9.8 shows an example for the application of Wallis's correlation.

According to the correlation of Wallis the channel diameter has an important effect on the flooding superficial velocities. Experiments, however, have shown that with air and water-like fluid pairs the effect of channel size tends to disappear when the channel diameter exceeds about 5 cm.

The *Tien–Kutateladze flooding correlation* was proposed by Tien (1977), based on earlier theoretical analyses by Kutateladze (1972). Kutateladze analyzed gas–liquid interactions in countercurrent flow and introduced the following important dimensionless parameter (*Kutateladze number*):

$$K_G^* = j_G \frac{\rho_G^{1/2}}{(\sigma g \Delta \rho)^{1/4}}. \tag{9.7}$$

Equation (9.7) can in fact be obtained by replacing D in Eq. (10.1) with Laplace's length scale, λ_L. Earlier, Pushkina and Sorokin (1969) had noticed that the flow reversal point in flooding experiments could be correlated with

$$K_G^* = 3.2. \tag{9.8}$$

Tien (1977) proposed using λ_L, the Laplace length scale, instead of D in a Wallis-type correlation, leading to

$$K_G^{*1/2} + m_2 K_L^{*1/2} = C_2, \tag{9.9}$$

where

$$K_L^* = j_L \frac{\rho_L^{1/2}}{(\sigma g \Delta \rho)^{1/4}}. \tag{9.10}$$

Experiment shows that $m_2 \approx 1$ and $C_2 \approx 1.7$–2. The correlation is often used for large channels ($D/\lambda_L \geq 40$) and rod and tube bundles.

Using an extensive data bank dealing with flooding in vertical circular tubes, McQuillan and Whalley (1985a, b) assessed the accuracy of a large number of empirical and theoretical flooding correlations. Among the correlations examined was the following correlation of Alekseev *et al.* (1972):

$$K_G^* = 0.2576 \, Bd^{0.26} \, Fr^{*-0.22}, \tag{9.11}$$

where $Bd = D^2 \Delta \rho g / \sigma$ is the Bond number and Fr^* represents a modified Froude number defined as

$$Fr^* = \left(\frac{\Gamma_L}{\rho_L} \right) \frac{g^{0.25} \Delta \rho^{0.75}}{\sigma^{0.75}}, \tag{9.12}$$

where Γ_L is the liquid mass flow rate, per unit wetted perimeter (similar to the definition of Γ_F in Section 3.8). McQuillan and Whalley improved the correlation of Alekseev *et al.* according to

$$K_G^* = 0.286 \, Bd^{0.26} \, Fr^{*-0.22} \left\{ 1 + \frac{\mu_L}{\mu_W} \right\}^{-0.18}, \tag{9.13}$$

where $\mu_w = 0.001 \, \mathrm{N \cdot s/m^2}$ is the viscosity of water at room temperature.

For steam–water flow in a vertical tube in the parameter range $P = 0.3\text{–}8.0 \, \mathrm{MPa}$, $Bd = 64\text{–}1600$, and $L/D > 20$, Ilyukhin *et al.* (1999) recommend Eq. (9.9) with $m = 0.8$ and (Svetlov *et al.*, 1999)

$$C_2 = 1.2 Bd^{0.0625} \left(\frac{\rho_G}{\rho_L} \right)^{0.05}. \tag{9.13a}$$

EXAMPLE 9.1. Consider the system shown in Fig. 9.1. Assume that the vertical tube is 5 cm in diameter. The tank is partially full with water at room temperature and is open to the atmosphere. For upward air superficial velocities of 1.0, 3.0, and 12 m/s, estimate the drainage mass flux of water.

SOLUTION. For water and air, $\rho_L = 997 \, \mathrm{kg/m^3}$ and $\rho_G = 1.185 \, \mathrm{kg/m^3}$. For $j_G = 1 \, \mathrm{m/s}$,

$$j_G^* = j_G \sqrt{ \frac{\rho_G}{g D \Delta \rho} } = 0.0493.$$

We can now use the correlation of Wallis, Eq. (9.6), with $m = 1$ and $C = 0.725$, the latter corresponding to a sharp-edged liquid inlet. We thus get

$$j_L^* = \left[(C - \sqrt{j_G^*})/m \right]^2 = 0.253,$$

$$j_L = j_L^* \bigg/ \sqrt{ \frac{\rho_L}{g D \Delta \rho} } = 0.177 \, \mathrm{m/s}.$$

The drainage rate for water will be $\dot{m}_L = (\frac{\pi}{4} D^2) \rho_L j_L = 0.3466 \, \mathrm{kg/s}$.

A similar calculation, this time with $j_G = 3 \, \mathrm{m/s}$, leads to the following:

$$j_G^* = 0.148,$$
$$j_L^* = 0.116,$$
$$j_L = 0.081 \, \mathrm{m/s},$$
$$\dot{m}_L = 0.159 \, \mathrm{kg/s}.$$

With $j_G = 12$ m/s, however, no liquid drainage takes place. According to Eq. (9.6), the gas superficial velocity corresponding to zero liquid penetration can be found from

$$\sqrt{j^*_{G,min}} = C.$$

This leads to

$$j_{G,min} = C^2 \sqrt{\frac{gD\Delta\rho}{\rho_G}} = 10.67 \text{ m/s},$$

where $j_{G,min}$ is the minimum gas superficial velocity that would completely block the downward flow of liquid.

9.3 Flooding in Horizontal, Perforated Plates and Porous Media

Countercurrent flow limitation (CCFL) in a horizontal, perforated plate is of great interest, because of the wide applications of perforated plates as sieve trays and as the core upper tie plates in PWRs. As mentioned earlier, stable countercurrent gas–liquid flow is not sustainable in very small flow passages ($D/\lambda_L \leq 2$). Perforated plates with small holes can sustain a net countercurrent, however, because some holes carry downward-flowing liquid while others carry upward-flowing gas.

CCFL in horizontal perforated plates is more complicated than single channels, and can be affected by several parameters including flow passage size and geometric shape, perforation ratio (defined as the ratio between the total area of the holes divided by the total area of the plate), the size and geometric shape of the perforated plate, the depth of the liquid pool that may form over the perforated plate due to the accumulation of liquid, the method of liquid injection (i.e., the level of non-uniformity of the distribution of injected liquid over the plate), and even the size of the gas volume underneath the plate (the soft volume). A number of processes that do not occur in single flow passages are encountered in perforated plates. The CCFL phenomena in the flow passages will resemble the CCFL phenomena in a single flow passage only if the distributions of liquid and gas among the flow passages are uniform, in which case the aforementioned Wallis- or Tien–Kutateladze-type correlations would apply. Such uniform distribution does not occur, however.

For countercurrent flow in horizontal perforated plates a distinction should be made between the onset of liquid accumulation (OLA) and CCFL. Consider an experiment where liquid at a fixed flow rate is injected over a perforated plate, while an upward gas flow is imposed starting from a very low rate and is increased in small steps and after each step the gas flow rate is maintained constant until a quasi-steady state is reached. No liquid accumulation will be observed at very low gas flow rates. The gas flow rate at which liquid accumulation is first observed represents OLA. Increasing the gas flow rate by a small quantity beyond OLA level may not lead to CCFL, and may only lead to the development of a deeper swollen liquid pool above the plate (Lee *et al.*, 2007). As the gas flow rate is increased, CCFL occurs only when the depth of the swollen liquid pool above the plate keeps increasing without reaching a steady state. The depth of the swollen pool will influence the CCFL process, furthermore, as long as the pool is relatively shallow. In the air–water and steam–water experiments of Alekseev *et al.* (2000), for example, the liquid penetration rate increased with increasing pool height for pool heights up to about 0.5–0.6 m, and for deeper pools no noticeable effect occurred.

Depending on the method of introduction of liquid and gas, the liquid distribution is often uneven among the flow passages. As a result some undergo CCFL earlier. Hydrodynamic interaction occurs among neighboring flow passages, whereby the generated disturbances at the surface of the liquid collected above the perforated plate induce the neighboring flow passages to flood.

Under partially flooded conditions a liquid pool will form on the perforated plate, which will undergo oscillations. For small plates these oscillations do not cause a major change in the CCFL behaviour of the perforations. For large plates, however, the oscillations can lead to intermittency and "channeling", whereby some holes carry pure liquid in the downward direction, some do the opposite and carry pure gas upwards, while a third group undergo countercurrent flow (Sobajima, 1985; Kokkonen and Tuomisto, 1990). As a result of oscillations and channelling the CCFL line in the $K_L^{*1/2}$ versus $K_G^{*1/2}$ plot becomes strongly nonlinear at high gas velocities (i.e., at high K_G^*).

The effect of the soft volume (i.e., the gas volume underneath the perforated plate) is related to the compressibility of the gas and is noticeable at low gas superficial velocities when the gas leaves the flow channels undergoing a countercurrent flow condition as gas bubbles that form on the upper end of the flow passages. This effect depends on the size of the soft volume, and vanishes when the soft volume is large.

Increasing the perforation ratio or the flow passage size (hydraulic diameter) increases the the liquid penetration flow rate for a constant upward gas flow rate. The geometric shape of the flow passages appears to have little effect, however (Alekseev *et al.*, 2000). Thus, smaller spacing between neighboring flow passages promotes CCFL, evidently because it enhances the hydrodynamic interaction between flow passages. Significant hysteresis also occurs in perforated plates (Liu *et al.*, 1982).

Sobajima (1985) proposed the following "Wallis-type" correlation for perforated plates:

$$j_G^{*1/2} + m\, j_L^{*1/2} = C, \tag{9.14}$$

where j_L^* and j_G^* are defined similar to Eqs. (9.2) and (9.3), based on the hole hydraulic diameter D_H. For holes with $D = 10.5$ mm, the constant values were $m = 0.841$ and $C = 1.32$; and for holes with $D = 12$ mm, the constant values were $m = 1.0$ and $C = 1.25$.

A correlation proposed by Bankoff *et al.* (1981) is based on a length scale that is an interpolation among the Laplace length scale, hole hydraulic diameter, and the plate thickness:

$$K_G^{*1/2} + K_L^{*1/2} = C', \tag{9.15}$$

$$K_L^* = j_L \frac{\rho_L^{1/2}}{(g\Delta\rho\omega)^{1/2}}, \tag{9.16}$$

$$K_G^* = j_G \frac{\rho_L^{1/2}}{(g\Delta\rho\omega)^{1/2}}, \tag{9.17}$$

$$\omega = D_H^{1-\beta}\lambda_L^\beta, \tag{9.18}$$

$$\beta = \tanh\left(\frac{2\pi\gamma}{\Delta}D_H\right), \tag{9.19}$$

where γ, the perforation ratio, is the ratio between the total area of holes divided by the total plate area, and Δ represents the thickness of the perforated plate. Bankoff et al. (1981) performed experiments using perforated plates with 2–40 holes. For the constant C', Bankoff et al. (1981) derived the following correlation:

$$C' = \begin{cases} 1.07 + 4.33 \times 10^{-3} N_t \pi D_H/\lambda_L & \text{for } N_t D_H/\lambda_L \leq 200, & (9.20) \\ 2 & \text{for } N_t D_H/\lambda_L > 200, & (9.21) \end{cases}$$

where N_t is the total number of holes. The dependence of C' on N_t in Eq. (9.20) is evidently applicable to small perforated plates, and for large perforated plates $C' = 2$. The data base for the above correlation includes air–water and steam–water pairs obtained in experiments with 20-mm-thick rectangular plates with 71.5 mm × 42.9 mm dimensions. The plates had 2–40 holes with 3.8–28.6 mm diameters. The correlation also agreed with the steam–water experimental data of Naitoh et al. (1978) and Jones (1977). The idea of using a Tien–Kutateladze-type flooding correlation with the interpolated length given here has received attention for flooding in short passages.

The conditions leading to the onset of liquid accumulation (OLA) of a horizontal perforated plate were studied by Lee et al. (2007). Based on air–water data, including the data of Červenka and Kolář (1973), Takahashi et al. (1979), and Liu et al. (1982), which covered the parameter range $D_H/\Delta = 1 - 11$ and $n > 7$ they proposed a Wallis-type correlation (Eq. (9.14)) with $m = 1.01$ and $C = 0.67$. The experimental data of Červenka and Kolář (1973) and Takahashi et al. (1979) were obtained with sieve trays that had thicknesses of 0.2 mm and 1 mm, respectively, however.

CCFL phenomena in full-scale rod bundles and core upper tie plates of some PWR designs were studied by Kokkonen and Tuomisto (1990) using air and water. The perforated plates were 20 mm thick and were either circular or hexahedral. The holes were either circular with $D = 4$–12 mm or 4 mm × 102 mm rectangles, and the number of holes varied in the 33–889 range. Their experimental data show that the size of the perforated plate has a significant effect on CCFL phenomena, as discussed earlier.

Alekseev et al. (2000) have used the experimental data of Kokkonen and Tuomisto (1990), as well as their own saturated-steam–water experimental data for the development of the empirical correlation below that accounts for the size of the perforated plate. Their experiments were carried out at 0.6–4.0 MPa pressures, using circular plates with circular holes, with $D/\Delta = 1$. The holes were 2–14 mm in diameter, and numbered in the 4–100 range. The perforated plates were maintained under a deep swollen liquid pool during the experiments, and the soft vapor volume underneath their perforated plates was large enough to have a negligible effect on the CCFL process in the perforated plates. Their empirical correlation is

$$K_L^{*1/2} + 0.8 K_G^{*1/2} = 3.3(Bd)^{0.05} \varphi_0^{0.19} (\rho_G/\rho_L)^{0.05} \cdot \left[1 - \exp\left(-0.04\sqrt{Bd_{plate}}\right)\right]$$
$$\cdot \{1.03 \tanh[(\Delta/D)^{0.25}]\}^{0.5} \tag{9.21a}$$

where K_G^* and K_L^* are defined similar to Eqs. (9.7) and (9.10) based on the average gas and liquid superficial velocities in the holes, respectively, and φ_0 is the plate perforation ratio (the ratio between the total area of the holes divided by the total area of the plate). The Bond number Bd is defined based on the hole diameter

$$Bd = D^2 \Delta \rho g / \sigma.$$

The plate Bond number, Bd_{plate}, is defined similarly except that D is replaced with the effective plate diameter defined as four times the surface area of the plate divided by its perimeter (note the similarity with the hydraulic diameter of a flow passage). Alekseev et al. (2000) recommend the above correlation only for circular and hexahedral perforated plates, however. The correlation underpredicted the data of Bankoff et al. (1981), which were obtained with rectangular perforated plates.

CCFL in porous media is important during the emergency cooling of a nuclear reactor core following a severe accident. When coolant injection into a rubblized bed takes place, downward penetration of the coolant liquid can be seriously restricted by the rising steam. The CCFL phenomenon can make the cooling process in rubblized beds hydrodynamically controlled. A widely used empirical correlation for CCFL in beds composed of uniform-size spheres is (Schrock et al., 1984)

$$j_G^{*0.38} + 0.95 j_L^{*0.38} = 1.075, \tag{9.22}$$

where

$$j_L^* = j_L \left[\frac{6(1-\varepsilon)\rho_L}{\varepsilon^3 d_P g \Delta \rho}\right]^{1/2}, \tag{9.23}$$

$$j_G^* = j_G \left[\frac{6(1-\varepsilon)\rho_G}{\varepsilon^3 d_P g \Delta \rho}\right]^{1/2}, \tag{9.24}$$

and ε is the bed porosity and d_P is the diameter of particles forming the bed.

9.4 Flooding in Vertical Annular or Rectangular Passages

CCFL in vertical annular or rectangular passages is of interest because it represents the partial delivery of emergency coolant liquid into the annular downcomer of nuclear reactors and in thin rectangular heat pipes. The former application is particularly important with respect to the safety of nuclear reactors. Following a large-break loss of coolant accident (LB-LOCA), subcooled water is injected into an annular downcomer, as shown schematically in Fig. 9.9, and from there it flows upward into the bottom of the active core. The downward flow of the injected liquid is opposed by vapor flow originating from the lower plenum.

Experiments addressing CCFL in vertical, narrow channels using air and water were carried out by Osakabe and Kawasaki (1989), who correlated their data based on Wallis-type correlations. Osakabe and Kawasaki (1989) used air and water in rectangular channels with a length of $W = 100$ mm and widths of $\delta = 2.5$ and 10 mm

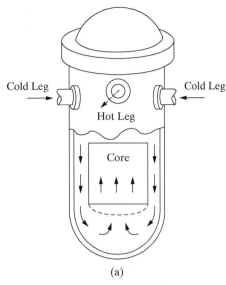

Figure 9.9. Schematic of a PWR reactor core: (a) normal operation; (b) cold leg emergency core coolant injection following a large break loss-of-coolant accident.

(a)

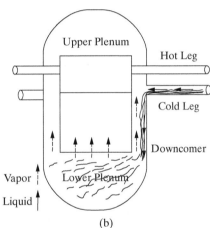

(b)

(see Fig. 9.10). Mishima (1984) also used air and water in rectangular channels with $W = 40$ mm and $\delta = 1.5, 2.4$, and 5 mm. Osakabe and Kawasaki (1989) correlated both data sets by the following Wallis-type correlation using W as the length scale:

$$j_G^{*1/2} + 0.8 j_L^{*1/2} = 0.58, \tag{9.25}$$

$$j_k^* = j_k \frac{\rho_k^{1/2}}{\sqrt{gW\,\Delta\rho}}. \tag{9.26}$$

Richter et al. (1979) conducted experiments in vertical annular test sections that had an outer diameter of 444.5 mm and inner diameters of 393.7 and 342.9 mm. They also noted that W, defined as the average of the circumferential lengths of an annular test section, was the proper length scale for the correlation of their data (Richter, 1981). Sudo et al. (1991) have proposed a Wallis-type correlation (Eq. (9.6)) for CCFL in vertical rectangular channels, with the coefficients defined as

$$m = 0.5 + 0.0015\,\mathrm{Bd}^{*1.3}, \tag{9.27}$$

$$C = 0.66(\delta/W)^{-0.25}, \tag{9.28}$$

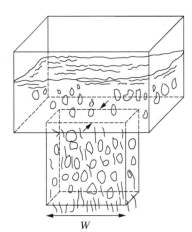

Figure 9.10. Experiments of Osakabe and Kawasaki (1989).

where δ is the gap spacing, and the modified Bond number is defined as

$$\mathrm{Bd}^* = W\delta\Delta\rho g/\sigma. \tag{9.29}$$

Experimental data show that strong multi-dimensional flow phenomena occur in complex systems, such as annular flow passages with asymmetric and nonuniform liquid injection arrangements. The most conspicuous multi-dimensional flow phenomenon is called *flow bypass*. Flow bypass occurs in annular and thin rectangular flow passages with asymmetric liquid injection. During the emergency core coolant injection into the downcomer of a PWR, for example, where liquid injection takes place through a cold leg, virtually undisturbed flow of the injected liquid may occur over a part of the annular flow passage, while vapor flows upward through the rest of the annulus. The liquid enters from one cold leg, and exits from the other, with little interaction taking place between the liquid and gas. With flow bypass and other complicating multi-dimensional phenomena, the flooding correlations quoted here, which are primarily based on simple channel experiments, do not apply. It has in fact been argued that, with complex system geometries, the system CCFL phenomena can be understood only in full-scale experiments, and no scaled-down experimental data or correlations should be trusted (Levy, 1999).

EXAMPLE 9.2. Water at room temperature and 10 bar pressure is injected over a large horizontal perforated plate. The holes are 10.5 mm in diameter and are arranged in a square lattice with a perforation ratio of 0.423. The plate is 20 mm thick. For upward superficial air velocities of 0.75 and 1.5 m/s, defined based on the total surface area of the plate, calculate the downward mass flux of water, also defined based on the total surface area of the plate.

SOLUTION. The fluid properties are $\rho_L = 997.5\,\mathrm{kg/m^3}$, $\rho_G = 11.69\,\mathrm{kg/m^3}$, $\mu_L = 8.93 \times 10^{-4}\,\mathrm{kg/m\cdot s}$, $\mu_G = 1.848 \times 10^{-5}\,\mathrm{kg/m\cdot s}$, and $\sigma = 0.07\,\mathrm{N/m}$. We can use the correlation of Bankoff *et al.* (1981). We have $\Delta = 0.02\,\mathrm{m}$, $\gamma = 0.423\,\mathrm{m}$, $D = D_H = 0.0105\,\mathrm{m}$, and $\lambda_L = \sqrt{\sigma/g\Delta\rho} = 0.00271\,\mathrm{m}$. Equations (9.19) and (9.18) then give, respectively,

$$\beta = 0.884$$

Figure 9.11. Schematic of a horizontal flow passage.

and

$$\omega = 0.00317 \text{ m}.$$

Also, since the plate is large, according to Eq. (9.21) we can assume $C' = 2$. With gas superficial velocity with respect to the total plate area of 0.75 m/s, we can find the gas superficial velocity based on the flow area as follows:

$$j_G = \frac{0.75 \text{ m/s}}{\gamma} = 1.773 \text{ m/s}.$$

We can then write

$$K_G^* = \sqrt{\frac{\rho_G}{g \Delta \rho W}} \, j_G = 1.095,$$

$$\sqrt{K_L^*} = C' - \sqrt{K_G^*} \Rightarrow K_L^* = 0.909.$$

$$j_L = K_L^* \Big/ \sqrt{\frac{\rho_L}{g \Delta \rho W}} = 0.1594 \text{ m/s}.$$

The downward liquid mass flux, with respect to the total plate area, will then be $\rho_L \gamma j_L = 67.25 \text{ kg/m}^2 \cdot \text{s}$.

Similar calculations, this time with a gas superficial velocity of 1.5 m/s with respect to the total plate area, give $j_G = 3.55$ m/s, $K_G^* = 2.19$, $K_L^* = 0.27$, $j_L = 0.0474$ m/s, and $\rho_L \gamma j_L = 20.0 \text{ kg/m}^2 \cdot \text{s}$.

9.5 Flooding Correlations for Horizontal and Inclined Flow Passages

CCFL in horizontal channels occurs during the emergency coolant injection into the cold leg of PWRs. Also, in a long horizontal pipeline, smooth stratified flow is the desirable flow pattern, since it requires low pumping power. A counterflow of gas can modify the flow pattern into wavy-stratified or slug.

The following correlation by Wallis (1969) for CCFL in horizontal flow passages is the outcome of an envelope method analysis:

$$j_L^{*1/2} + j_H^{*1/2} = 1, \tag{9.30}$$

$$j_H^* = j_H \left[\frac{\rho_H}{g \Delta \rho H} \right]^{1/2}, \tag{9.31}$$

$$j_L^* = j_L \left[\frac{\rho_L}{g \Delta \rho H} \right]^{1/2}, \tag{9.32}$$

where subscripts ρ_L and ρ_H are the densities of the light and heavy fluids, respectively, and H is the height of the cross section (see Fig. 9.11).

In horizontal and near-horizontal channels that support stratified flow the growth of interfacial waves can ultimately lead to the regime transition to slug flow and therefore to the formation of liquid slugs that block the gas flow. The growth of interfacial waves has in fact been identified as the primary mechanism responsible for flooding (Kordyban and Ranov, 1970; Mishima and Ishii, 1980; Ansari and Nariai, 1989). Accordingly, flow regime transition criteria for the disruption of stratified flow in favor of intermittent flow (e.g., Eq. (7.33), due to Mishima and Ishii (1980)) can be used for the prediction of flooding conditions.

9.6 Effect of Phase Change on CCFL

Evaporation or condensation can take place inside channels that support a counter-current flow. A good example is the countercurrent flow in the core upper tie plate of PWRs during emergency coolant injection into the upper plenum, where conden-sation can take place inside the holes of the perforated plate by subcooling of the emergency coolant water. Local evaporation promotes CCFL, whereas condensation has the opposite effect.

A common practice is to use the estimated local phasic superficial velocities in a correlation such as Bankoff's (Bankoff et al., 1981). Prediction of the local phasic mass fluxes is difficult, however, and requires a reasonable estimate of the extent of condensation or evaporation.

Block and Crowley (1976) have proposed that condensing countercurrent flow in a vertical channel can be modeled using Wallis's correlation (Eq. (9.6)), provided that the dimensionless vapor velocity is modified to

$$j_{G,\text{eff}}^* = j_G^* - f_{\text{cond}} \, \text{Ja}^* j_L^*, \tag{9.33}$$

where

$$\text{Ja}^* = \sqrt{\frac{\rho_f}{\rho_g}} \frac{C_{\text{PL}}(T_{\text{sat}} - T_L)}{h_{\text{fg}}}$$

is a modified Jakob number and f_{cond} is a condensation efficiency. A similar empirical framework for perforated plates is to recast Eq. (9.15) as (Bankoff et al., 1981)

$$\left[K_G^* - f_{\text{cond}} \, \text{Ja}^* \, K_L^* \right]^{1/2} + K_L^{*1/2} = C', \tag{9.34}$$

$$0.2 \leq f_{\text{cond}} \leq 0.8 \tag{9.35}$$

(Bankoff et al., 1981). The specification of f_{cond} is of course difficult.

Noncondensables reduce the effect of condensation on the breakdown of CCFL by reducing the condensation rate.

9.7 Modeling of CCFL Based on the Separated-Flow Momentum Equations

When CCFL occurs inside a long flow passage, it is possible to predict the conditions that cause CCFL based on the one-dimensional separated-flow momentum equa-tions. In these situations, CCFL is caused by the hydrodynamic interaction at the gas–liquid interphase in flow regimes such as annular (in vertical channels) or strati-fied (in horizontal and inclined channels). Before CCFL occurs, the base flow regime is maintained under counterflow conditions. CCFL takes place only when the hydro-dynamic interactions make the base flow regime impossible.

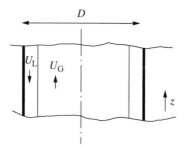

Figure 9.12. Countercurrent annular flow.

Consider countercurrent flow in a vertical pipe, as shown in Fig. 9.12. Assume that both phases are incompressible and that there is no CCFL. The equilibrium, steady-state momentum equations for the mixture and the gas phase will be, respectively,

$$-\frac{dP}{dz} + \frac{4}{D}\tau_w - [\rho_L(1-\alpha) + \rho_G\alpha]g = 0, \tag{9.36}$$

$$-\frac{dP}{dz}\alpha - \frac{4\tau_I\sqrt{\alpha}}{D} - \rho_G\alpha g = 0. \tag{9.37}$$

Elimination of the pressure gradient term between the two equations leads to

$$\frac{4\tau_w}{D} + \frac{4\tau_I}{D\sqrt{\alpha}} = (\rho_L - \rho_G)g(1-\alpha). \tag{9.38}$$

The wall and interfacial shear stresses can be represented as

$$\tau_w = \frac{1}{2}f_w\rho_L U_L^2 \approx \frac{1}{2}f_w\rho_L\frac{j_L^2}{(1-\alpha)^2} \tag{9.39}$$

and

$$\tau_I = \frac{1}{2}f_I\rho_G(U_G - U_L)^2 \approx \frac{1}{2}f_I\rho_G\frac{j_G^2}{\alpha^2}, \tag{9.40}$$

where f_w and f_I are skin friction coefficients. In Eq. (9.40), we have noted that at near-CCFL conditions $|U_L| \ll |U_G|$, and we have therefore neglected the liquid velocity. Combining Eqs. (9.38), (9.39), and (9.40), one derives

$$\frac{2f_I}{\alpha^{5/2}}j_G^{*2} + \frac{2f_w}{(1-\alpha)^2}j_L^{*2} - (1-\alpha) = 0. \tag{9.41}$$

For any specific value of the void fraction within the range applicable to the annular flow regime, Eq. (9.41) will provide a curve in the j_G^* versus j_L^* coordinate system, provided that appropriate correlations are used for the skin friction coefficients. Figure 9.13 shows qualitatively the j_G^* versus j_L^* curves. The envelope of the generated curves represents the CCFL line, since an equilibrium solution would be impossible for points located on the right side of the envelope. The equation defining the envelope of the curves can be found by eliminating a between the following two equations:

$$F(\alpha, j_G^*, j_L^*) = 0 \tag{9.42}$$

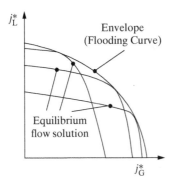

Figure 9.13. Equilibrium counterflow curves and their envelope.

and

$$G(a, j_G^*, j_L^*) = \frac{\partial}{\partial \alpha} F(\alpha, j_G^*, j_L^*) = 0, \tag{9.43}$$

where $F(\alpha, j_G^*, j_L^*)$ represents the left-hand side of Eq. (9.41). An analysis similar to this was performed by Bharathan *et al.* (1979). A similar approach can be used for CCFL in an otherwise stratified flow in a horizontal or inclined channel (Lee and Bankoff, 1983; Ohnuki, 1986).

The CCFL conditions can also be found simply by noting that it is impossible for an equilibrium separated flow to be sustained in the points to the right side of the flooding line in Fig. 9.13. Barnea *et al.* (1986) modeled CCFL in stratified flow in their inclined channel experiments by writing the one-dimensional momentum equations for the liquid and gas phases under equilibrium flow conditions, and eliminating the pressure gradient term between the two, to derive an equation similar to Eq. (7.22). The latter equation is to be solved along with Eqs. (7.23), (7.24), and (7.25) in Chapter 7. For any given j_L and j_G, this set of equations is closed and can be solved iteratively. Flooding can be assumed to occur when the set of equations cannot be satisfied for any α in the $0 < \alpha < 1$ range. However, Barnea *et al.* (1986) noted that this analysis represents only one possible flooding mechanism (namely, when gravity is overcome by pressure drop). Accordingly, they argued that flooding occurs when either an equilibrium solution to stratified flow becomes impossible or the conditions leading to regime transition from stratified to intermittent flow (e.g., Eq. (7.29)) are satisfied. Celata *et al.* (1991) carried out a similar analysis (i.e., based on the argument that the CCFL line represents conditions where an equilibrium separated countercurrent flow becomes impossible) for flooding in a vertical channel.

PROBLEMS

9.1 In Problem 3.8, suppose an upward flow of saturated vapor is under way in the heated tube. For $Re_F = 125$ and 1100 values, calculate the highest possible mass flow rate of vapor flow.

9.2 Figure P7.9, which is related to Problem 7.9, depicts a unit cell representing stable and developed slug flow in a vertical pipe. According to McQuillan and Whalley (1985a, b), the slug-to-churn flow regime transition occurs in a vertical, upward flow in a pipe when the combination of the upward-moving Taylor bubble and the falling liquid film surrounding it represent flooding conditions, in accordance with the flooding correlation of Wallis cast as

$$\sqrt{j_B^*} + \sqrt{j_F^*} = 1,$$

where

$$j_B^* = \sqrt{\frac{\rho_G}{\Delta\rho g D}}\, j_B,$$

$$j_F^* = \sqrt{\frac{\rho_L}{\Delta\rho g D}}\, j_F,$$

$$j_B = (1 - 4\delta_F/D)\left[1.2j + 0.35\sqrt{\Delta\rho g D/\rho_L}\right],$$

and

$$j_F = j_B - j.$$

The liquid film thickness in McQuillan and Whalley's model is to be calculated using the laminar and smooth falling film assumption (see Section 3.8).

(a) Interpret and comment on this model, using the expressions discussed in Problem 7.9.

(b) Show that Eq. (3.82) leads to

$$\delta_F \approx \left[\frac{3j_F D\mu_L}{4g\Delta\rho}\right]^{1/3}.$$

(c) Compare the experimental data given in Problem 7.10 (Table P7.10) with the predictions of the model of McQuillan and Whalley.

(d) Repeat part (c), this time assuming that the film thickness is found from the correlation of Brötz (1954), Eq. (3.106).

9.3 Discuss the physical meaning of the condensation efficiency f_{cond} used in Eqs. (9.33) and (9.35), and examine the limits on the magnitude of f_{cond} and their implication.

9.4 In horizontal channels the disruption of stratified flow and the establishment of intermittent flow leads to flooding. A horizontal pipeline is 25 cm in diameter and carries a mixture of kerosene ($\rho_L = 804\ \text{kg/m}^3$ and $\mu_L = 1.92 \times 10^{-3}\ \text{kg/m·s}$) and methane gas ($M = 16\ \text{kg/kmol}$ and $\mu_G = 1.34 \times 10^{-5}\ \text{kg/m·s}$) at 20 °C and 15 bars. For a mass flux of $G_L = 65\ \text{kg/m}^2\text{·s}$ calculate the gas mass flux that would flood the pipe, using the criterion of Mishima and Ishii (1980), Eq. (7.33).

9.5 The "hanging film phenomenon," according to which a liquid film is held at rest inside a vertical channel by rising gas, has been proposed as the mechanism responsible for complete flooding (zero downward liquid penetration) inside channels (Wallis and Makkenchery, 1974). Based on this phenomenology, Eichhorn (1980) proposed the following expression for the limit of zero liquid penetration:

$$K_G^*\sqrt{f/2}\sin\theta = 0.096\left\{1 + \frac{1}{3}\left[\frac{(\text{Bd}/8)^3 - 1}{(\text{Bd}/8)^3 + 1}\right]\right\},$$

where θ here is the contact angle, $Bd = D\sqrt{g\Delta\rho/\sigma}$ is the Bond number, and the skin friction coefficient can be found from an appropriate correlation, for example,

$$\sqrt{2/f} = 5.66 \log_{10}(Re_G\sqrt{f/2}) + 0.292,$$

where $Re_G = U_G D/\nu_G$.

For water flowing downward in a vertical pipe that is 15 cm in diameter while atmospheric air flows upward in the pipe, calculate the upward air superficial velocities corresponding to zero liquid penetration for contact angles in the $\theta = 30°$–$60°$ range, and compare the results with the predictions of the expression of Pushkina and Sorokin (1969).

9.6 Water vapor flows upward through a large perforated plate that has holes with $D = 10.5$ mm diameter. The holes are arranged in a square pitch with a perforation ratio of 0.423. The plate thickness is 20 mm. The system pressure is 14 bars. Subcooled water at 20° and 50° is injected onto the plate.

(a) For each subcooled water temperature calculate the minimum saturated water vapor flow rate, per hole, that would completely block the downward flow of water, assuming that $f_{cond} = 0.0$.

(b) Repeat part (a), assuming that the water vapor is saturated, for $f_{cond} = 0.5$ and 0.75.

(c) For parts (a) and (b), repeat the calculations assuming that the mass flow rate of steam leaving the perforated plate would be equal to the mass flow rate of water penetrating downward.

9.7. A large tank contains liquid methane at a pressure of 1 MPa and temperature of 147 K. The tank is connected to a drainage line at its bottom. The drainage line is a 2-m-long insulated vertical tube with an inner diameter of 1 cm, and drains into another tank that contains saturated methane at 1 MPa.

(a) Calculate the drainage mass flow rate assuming that no evaporation takes place.

(b) Because of evaporation in the bottom tank a countercurrent upward flow of vapor is possible in the drainage line. Find the flow rates of vapor that would

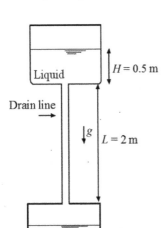

Figure P9.7. Figure for Problem 9.7.

result in the reduction of liquid drainage to 5% of the original rate. What is the minimum vapor flow rate that would completely stop the liquid drainage?

(c) Consider the tanks described above, when they are in a microgravity environment. Calculate the pressure difference between the two tanks that would cause a flow rate similar to part (a).

(d) Consider the tanks in microgravity, and assume that the mass flow rate of part (a) is needed. Assume that due to evaporation the fluid flowing between the two tanks has a flow quality of 0.04. Also, assume that a globe valve, which has an inner diameter that is equal to the pipe inner diameter when it is fully open, is installed on the drainage line. Calculate the total pressure drop when the valve is fully open, and when the valve is half closed.

9.8 Based on experiments with a vertical rectangular channel with 10 mm by 120 mm cross section, and with liquid injection through a porous segment of the channel wall, Droso *et al.* (2006) derived the following empirical CCFL correlation

$$\mathrm{Fr_G} = C\mathrm{Ka}^{0.4}\mathrm{Fr_L^{-0.15}}$$

$$\mathrm{Fr_G} = \frac{\rho_G j_G^2}{Sg\Delta\rho}; \quad \mathrm{Fr_L} = \frac{\rho_L j_L^2}{Sg\Delta\rho}; \quad \mathrm{Ka} = \frac{\sigma}{\mu_L}\left(\frac{\rho_L}{\mu_L g}\right)^{1/3}; \quad C = 0.0138,$$

where S is the channel width.

(a) Write the above correlation in terms of Wallis's dimensionless superficial velocities.

(b) Consider countercurrent flow conditions in a narrow channel with $S = 10$ mm, where water and air at room temperature and atmospheric pressure are used. For $\mathrm{Fr_G} = 0.1, 0.25$, and 0.7, calculate the downward superficial velocity of liquid from the aforementioned correlation, and compare your results with the predictions of the correlation of Sudo *et al.* (1991), given in Eqs. (9.27) through (9.29).

10 Two-Phase Flow in Small Flow Passages

The scale effect in two-phase flow and the classification of channel sizes were discussed in Section 3.7.2. The discussions in this chapter will primarily deal with channels with hydraulic diameters in the range $10 \ \mu m \lesssim D \lesssim 1 \ mm$, where the limits are understood to be approximate magnitudes. For convenience, however, channels with $10 \lesssim D_H < 100 \ \mu m$ will be referred to as microchannels, and channels with $100 \ \mu m \lesssim D_H \lesssim 1 \ mm$ will be referred to as minichannels. The two categories of channels will be discussed separately, furthermore, because, as will be seen, there are significant differences between them.

Single-phase and two-phase flows in minichannels have been of interest for decades. The occurrence of flashing two-phase flow in refrigerant restrictors formed the impetus for some of the early studies (Mikol, 1963; Marcy, 1949; Bolstad and Jordan, 1948; Hopkins, 1950). The number of investigations dealing with two-phase flow in minichannels is relatively large, but two-phase flow in microchannels is a more recent subject of interest.

In this chapter adiabatic two-phase flow in small flow channels is discussed. Flow regimes, heat transfer and other aspects of diabatic flow in snmall channels will be discussed in Chapters 14 (for flow boiling) and 16 (for internal flow condensation). The main difference between adiabatic and diabatic two-phase flows is that in adiabatic flow axial variations of flow parameters are often slow, whereas diabatic flows typically involve sharp axial variations in flow quality, void fraction, and flow pattern. These sharp axial variations are accompanied by significant flow acceleration (in boiling) or deceleration (in condensation). As a result, flow patterns, and consequently some of the other hydrodynamic characteristics of diabatic two-phase flow, can be different from adiabatic two-phase flows under seemingly similar hydrodynamic boundary conditions. The difference between the two types of flows becomes more significant as the intensity or rate of the phase change process increases. Thus, the relevance of the material that is presented in this chapter to diabatic two-phase flow is similar to the relevance of hydrodynamic aspects of diabatic two-phase flow to two-phase flow observed in internal boiling and condensation in large flow channels. Models, correlations, and other predictive techniques that are based on adiabatic flow experimental data can be comfortably applied to similar diabatic flow cases when the phase change processes are slow enough to render the axial variation of flow parameters slow. In general, however, the predictive techniques that are based on adiabatic two-phase flow experiments or simulation should be used for estimating the flow parameters in a diabatic flow situation only when predictive methods that are based on relevant diabatic experiments or simulations are not available.

10.1 Two-Phase Flow Regimes in Minichannels

Adaibatic two-phase flow regimes in minichannels under conditions where inertia is significant have been experimentally investigated rather extensively. Recent reviews of past investigations can be found in Cheng and Wu (2006), Shao *et al.* (2009), Rebrev (2010), and Choi *et al.* (2011), among others. Flow regime identification has been primarily by visual or photographic methods, and because of the subjective nature of these methods there is some disagreement with respect to the definition of the major flow regimes. However, experiments generally show that, with the exception of stratified flow, which does not occur when $D_H \lesssim 1$ mm with air/water-like fluid pairs (or, more generally, when the flow channel is smaller than the forthcoming macro–micro flow passage size threshold), all other major flow regimes (bubbly, slug, churn, annular, etc.) can occur in minichannels. The flow regimes and their parameter ranges are also similar for vertical and horizontal channels smaller than the macro–micro size threshold, and they are insensitive to channel orientation.

Bubbly, slug/plug, churn and annular are the primary flow regimes in adiabatic two-phase flow in mini- and microchannels. However, similar to the situation with large (macro) channels, many subtle and transitional flow patterns can be observed in miniature flow channels (Coleman and Garimella, 1999). Furthermore, with a non-wetting liquid–solid pair the rivulet flow regime is encountered under conditions that would lead to annular flow regime for a wetting liquid–solid pair. In heated channels where two-phase flow is the outcome of boiling, in addition to the aforementioned regimes, wispy annular and inverted annular flow regimes are also encountered.

The commonly observed flow patterns in minichannels are shown in Fig. 10.1 using the photographs of Triplett *et al.* (1999a). These experiments were performed using room-temperature air and water in horizontal round tubes with 1.1 and 1.45 mm inner diameters, and flow channels with semi-triangular cross sections (equilateral rectangle with two sharp corners and one corner smoothed) with 1.1 and 1.49 mm hydraulic diameters. The major flow regimes shown in these pictures are in good agreement with the observations of most other investigators, including Chung and Kawaji (2004) (for their room-temperature and atmospheric water–nitrogen tests with 250- and 526-μm-diameter test sections), although, as will be shown later, some flow patterns have been given different names by different authors.

Bubbly flow (Fig. 10.1(a)) is characterized by distinct and distorted (nonspherical) bubbles, typically considerably smaller in diameter than the channel. With increasing j_G while j_L remains constant (which leads to increasing void fraction), the flow field grows more crowded with bubbles, eventually leading to plug/slug flow (Fig. 10.1(b)), which is characterized by elongated cylindrical bubbles. This flow pattern has been called slug by some investigators (Suo and Griffith, 1964; Mishima and Hibiki, 1996), plug by others (Damianides and Westwater, 1988, Barajas and Panton, 1993), and Taylor flow by others (Hassan *et al.*, 2005; Shao *et al.*, 2009).

Figures 10.1(c) and (d) display the churn flow regime. Triplett *et al.* assumed two processes to characterize churn flow. In one process, the elongated bubbles become unstable as the gas flow rate is increased and their tails are disrupted into dispersed bubbles (Fig. 10.1(c)). This flow pattern has been referred to as pseudo-slug (Suo and Griffith, 1964), churn (Mishima and Hibiki, 1996), and frothy-slug (Zhao and Rezkallah, 1993). The second process that characterizes churn flow is the churning

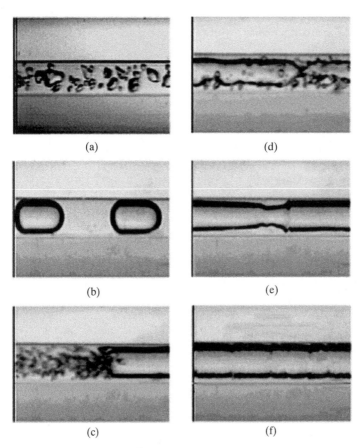

Figure 10.1. Photographs of flow regimes in the 1.1-mm-diameter test section of Triplett *et al.* (1999a): (a) bubbly ($j_L = 6$ m/s; $j_G = 0.396$ m/s); (b) plug ($j_L = 0.213$ m/s; $j_G = 0.154$ m/s); (c) churn ($j_L = 0.66$ m/s; $j_G = 6.18$ m/s); (d) churn ($j_L = 1.21$ m/s; $j_G = 4.63$ m/s); (e) slug-annular ($j_L = 0.043$ m/s; $j_G = 4.040$ m/s); (f) annular ($j_L = 0.082$ m/s; $j_G = 73.30$ m/s).

waves that periodically disrupt an otherwise wavy-annular flow pattern (Fig. 10.1(d)). This flow pattern has also been called frothy slug-annular (Zhao and Rezkallah, 1993). At relatively low liquid superficial velocities, increasing the mixture volumetric flux leads to the merging of long bubbles that characterize slug flow and to the development of the slug-annular flow regime displayed in Fig. 10.1(e). In this flow regime long segments of the channel support an essentially wavy-annular flow and are interrupted by large-amplitude solitary waves, which do not grow sufficiently to block the flow path. With further increase in the gas superficial velocity these large-amplitude solitary waves disappear and the annular flow pattern shown in Fig. 10.1(f) is established.

Experimental data indicate that, for air–water flow with $D_H \lesssim 1$ mm, the flow patterns depicted here and their morphology also apply to rectangular channels with small aspect ratios (i.e., near-square cross sections) (Coleman and Garimella, 1999, whose air–water experiments were conducted in a 1.3-mm-inner-diameter test section) and triangular channels (Zhao and Bi, 2001, who used air and water in experiments using equilateral cross-section vertical tubes with 0.87 and 1.44 mm hydraulic diameters). For smaller channels, however, the sharp corners can affect the flow

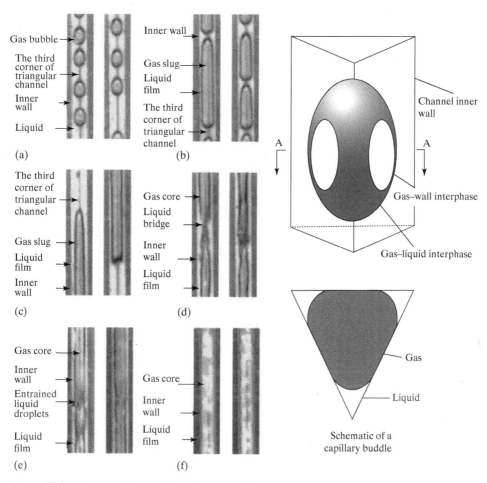

Figure 10.2. Air–water flow regimes in an equilateral triangular vertical channel with $D_H = 0.866$ mm: (a) capillary bubbly; (b) and (c) slug; (d) churn; (e) and (f) annular. (From Zhao and Bi, 2001.)

regimes, in particular at low gas and liquid flow rates. The sharp corners will retain liquid owing to the capillary effect. A bubble train flow pattern can thus be sustained with an essentially stagnant liquid. The geometry of the bubbles will depend on the capillary number, $Ca = \mu_L U_B / \sigma$ (Kolb and Cerro, 1993a, b). With very large Ca, the bubbles maintain a near-circular cross section, whereas the sharp corners maintain a considerable amount of liquid. For low Ca, however, the bubble will have nearly flat surfaces on the channel sides and will be curved in the corners. Figure 10.2 displays pictures from a vertical channel with an equilateral triangular cross section, with $D_h = 0.866$ mm (Zhao and Bi, 2001). The capillary bubbly flow in the figure, which displaces the dispersed bubbly flow observed in larger triangular channels, is composed of an approximately regularly spaced train of ellipsoidal bubbles that occupy most of the cross section. The two-phase flow regime data of Triplett *et al.* (1999a) are shown in Fig. 10.3, where the flow regimes representing the flow of air–water mixture in glass tubes reported by two other authors are also shown. The predominance of intermittent (slug, churn, and slug-annular) flow patterns can be noted. For

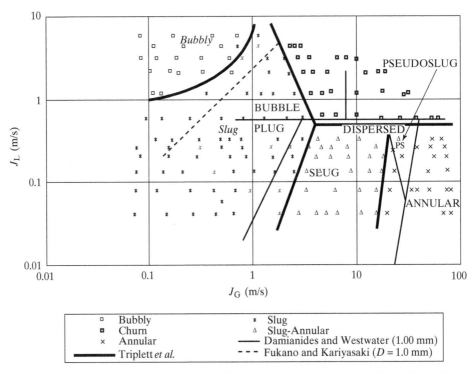

Figure 10.3. Comparison among air–water flow regime maps obtained in glass tubes with $D \cong$ 1 mm. Symbols represent the data of Triplett *et al.* (1999a), for their 1.09-mm-diameter tubular test section. The flow pattern names in capital and lower case letters represent those reported by Damianides and Westwater (1988) and Fukano and Kariyasaka (1993), respectively.

the air–water–Pyrex system, $\theta_0 \approx 34°$ (Smedley, 1990), implying a partially wetting liquid–solid pair.

The flow pattern identified as churn by Triplett *et al.* (Figs. 10.1(c) and (d)) appears to coincide with the flow pattern identified as dispersed by Damianides and Westwater. Furthermore, the slug and slug-annular regimes in Triplett's experiments (Figs. 10.1(e) and (f)) coincide with the plug and slug flow regimes in Damianides and Westwater (1988), respectively. These differences result from the subjective identification and naming of flow patterns, and the two experimental sets are otherwise in good overall agreement.

The data of Fukano and Kariyasaki (1993) are evidently in disagreement with the data of Triplett *et al.* (1999a) and Damianides and Westwater (1988), except for the intermittent-to-bubbly flow transition line where all three data sets are in good agreement.

Chung and Kawaji (2004) have investigated the hydrodynamic aspects of nitrogen–water two-phase flow in mini- and microchannels. In their minichannel experiments, performed with circular channels with $D = 250$ and 530 μm, they could observe bubbly, slug, churn, and slug-annular flow regimes. Their churn flow included a flow regime that they named the serpentine-like gas core. This regime was characterized by a deformed liquid film surrounding a serpentine-like gas core.

The experimental studies discussed so far all utilized materials that represented partial-wetting ($\theta_0 < 90°$) conditions. In view of the significance of surface

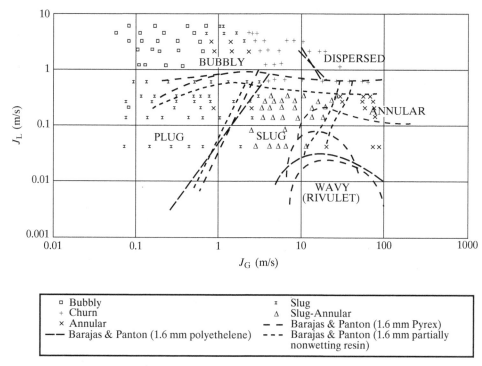

Figure 10.4. The effect of surface wettability on the air–water flow regimes. Symbols represent data of Triplett *et al.* (1999a), and flow regime names are from Barajas and Panton (1993).

tension in small flow passages, however, the surface wettability should impact the two-phase flow hydrodynamics. Experiments have confirmed the effect of surface wettability on two-phase flow regimes (Barajas and Panton, 1993; Cuband *et al.*, 2006). Barajas and Panton (1993) conducted experiments with air and water in 1.6-mm-inner-diameter tubes that were either horizontal or inclined at an angle of 34° with respect to the horizontal plane, using four different tube materials, including Pyrex ($\theta_0 = 34°$), polyethylene ($\theta_0 = 61°$), and polyurethane ($\theta_0 = 74°$) as partially wetting surfaces and the FEP fluoropolymer resin ($\theta_0 = 106°$) as a partially nonwetting combination. Figure 10.4 displays a summary of their flow regime maps, where the data of Triplett *et al.* (1999a) representing a 1.09-mm-diameter circular test section are also included for comparison. The data of Barajas and Panton (1993) indicate that with polyethylene and polyurethane the wavy flow pattern was replaced with a flow regime characterized by a single rivulet. A small multirivulet region also occurred on the flow regime map for the polyurethane test section. The flow regimes observed with the partially nonwetting channel FEP fluoropolymer were significantly different, however. In comparison with the partially wetting tubes, the ranges of occurrence of the rivulet and multirivulet flow patterns were now significantly wider. The flow channel in the experiments of Cuband *et al.* (2006) had a 525 μm × 525 μm square cross section. Air and water were used, and channel surface was made hydrophobic ($\theta_0 \approx 120°$) in some of the tests by using Teflon coating. Without coating the channel surface was hydrophylic ($\theta_0 \approx 9°$ for silicon and $\theta_0 \approx 25°$ for glass). Significantly different flow regimes occurred in hydrophylic and hydrophobic test sections. The flow regimes in the hydrophylic flow channel were in general similar to the flow regimes observed and reported by others, as described earlier in this section. In the hydophobic test sections the dominant flow regimes were isolated asymmetric bubbles at low void fractions and asymmetric slug flow in intermediate void fractions. At high void fractions the gas flowed through the center while droplets stuck to the wall and randomly moved, sometimes growing large enough to block the channel and form a plug.

The aforementioned flow regime maps were purely empirical and used liquid and gas superficial velocities as coordinates (also referred to as Mandhane's coordinates). This choice of coordinate in the flow regime map is simple and convenient, but does not consider the effect of fluid properties. Some authors have attempted to use dimensionless numbers that are relevant to the hydrodynamic behavior of the two-phase mixture as coordinates. Some of these investigations are now reviewed.

Akbar *et al.* (2003) argued that there is an important resemblance between two-phase flow in microchannels and in common large channels at microgravity. In both system types the surface tension, inertia, and the viscosity are important, while buoyancy is suppressed. They argued that consequently two-phase flow regime maps that have previously been developed for microgravity, or at least their underlying methodology, can be useful for microchannels. They compared the minichannel flow regime data from several sources, and developed a simple flow regime map based on analogy between minichannel flow and flow in large channels in microgravity, and using a methodology previously applied by Zhao and Rezkallah (1993) and Rezkallah (1996) to microgravity. They noted that available data are in reasonable agreement with respect to the major flow patterns in near-circular channels for $D_H \lesssim 1$ mm. However, the available data indicate that channels with sharp corners (e.g., rectangular) can support somewhat different regimes and transition boundaries.

In the Weber number-based two-phase flow regime map of Akbar *et al.* (2003) the entire flow regime map is divided into four zones, as depicted in Fig. 10.5. Their flow regime map can be represented as follows.

- Surface tension-dominated zone:
  - For $We_{LS} \leq 3.0$:

$$We_{GS} \leq 0.11 \, We_{LS}^{0.315} \tag{10.1}$$

  - For $We_{LS} > 3.0$:

$$We_{GS} \leq 1.0 \tag{10.2}$$

- Annular flow zone (inertia-dominated zone 1):

$$We_{GS} \geq 11.0 \, We_{LS}^{0.14} \tag{10.3}$$

$$We_{LS} \leq 3.0 \tag{10.4}$$

- Froth (dispersed) flow zone (inertia dominated zone 2):

$$We_{LS} > 3.0 \tag{10.5}$$

$$We_{GS} > 1.0 \tag{10.6}$$

where: $We_{LS} = j_L^2 D \rho_L / \sigma$ and $We_{GS} = j_G^2 D \rho_G / \sigma$.

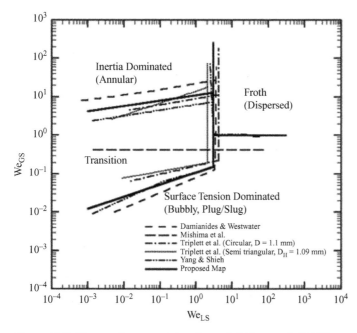

Figure 10.5. Flow regime transition lines of Akbar *et al.* (2003) compared with experimental data for circular and near-circular channels with $D_H \lesssim 1$ mm.

Figure 10.6. Comparison between the flow regime transition lines of Akbar *et al.* (2003) and the experimental transition lines of Yang and Shieh (2001) and Chung and Kawaji (2004). (From Shao *et al.*, 2009.) A = annular, B = bubbly, W = wavy, C = churn, T = Taylor (slug/plug), T-A Taylor-annular transition.

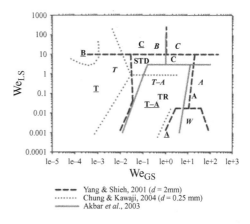

Figure 10.5 compares the flow regime map of Akbar *et al.* with data from several sources. Figure 10.6 compares the flow regime map of Akbar *et al.* with experimental transition lines of two investigations (Shao *et al.*, 2009). Note that in Figure 10.6 the coordinates have been reversed. The flow regime map of Akbar *et al.* predicts the experimental data well for air–water-like fluid pairs in channels with $D_H \lesssim 1$ mm channel size range. For air–water-like fluid pairs flowing in channels with $D_H > 1$ mm, however, the flow regime map is in fair agreement with available data, with the exception of the boundaries of the froth (dispersed) flow regime zone. This flow regime map agreed reasonably well with the data of Yue *et al.* (2007), for a horizontal 1000 μm × 500 μm rectangular channel, with CO_2–aquous salt solutions.

More recently Ong and Thome (2011a, b) performed experiments in circular heated channels with 1.03, 2.20, and 3.04 mm diameters, using R-134a, R-236fa, and R-245fa refrigerants as fluids. The major flow regimes they observed were bubbly, bubbly-slug (similar to plug), slug, slug-semi-annular, wavy annular, and smooth annular. Based on their experimental data they developed flow regime transition expressions based on dimensionless parameters. Their flow regime transition models will be described in Section 14.2.1 where flow regimes in small boiling channels are discussed. However, even though their flow regime transition models are primarily for diabatic (boiling) channels, with the exception of one regime transition (isolated bubbly to coalesced bubbly), the remainder of the expressions representing flow regime transitions depend only on hydrodynamic parameters. The flow regime transition models of Ong and Thome (2011a, b) may thus be useful for application to adiabatic flow conditions as well.

10.2 Void Fraction in Minichannels

Void fractions in microchannels have been measured by Kariyasaki *et al.* (1992), Mishima and Hibiki (1996), Bao *et al.* (1994), Triplett *et al.* (1999b), Kawahara *et al.* (2002), and Chung and Kawaji (2004). Fukano and Kariyasaki (1993) and Mishima and Hibiki (1996) also attempted to measure and correlate the velocity of large bubbles. Measurement of void fraction in mini- and microchannels is difficult. Most of the reported measurements have been based on image analysis. Simultaneous solenoid valves (Bao *et al.*, 1994) and neutron radiography and image processing (Mishima and Hibiki, 1996) have also been used.

It was mentioned earlier that bubbly, slug, and semi-annular flow regimes together occupy most of the entire flow regime map (see Fig. 10.3). Experimental data indicate that, in these flow regimes, there is little velocity slip between the two phases in minichannels.

Mishima and Hibiki (1996) correlated their void fraction data for upward flow in vertical channels, as well as the data of Kariyasaki *et al.* (1992), using the drift flux model. Since the buoyancy effect is suppressed by surface tension and viscous forces in mini- and microchannels, one would expect $V_{gj} \approx 0$. For bubbly and slug flow regimes, Mishima and Hibiki (1996) obtained $V_{gj} = 0$, and they correlated the distribution coefficient C_0 according to Eq. (6.48). More recently, however, Zhang *et al.* (2010) have refined the aforementioned correlation to

$$C_0 = 1.2 + 0.380 \exp(-1.39/D_H^*). \tag{6.49}$$

When the slip ratio $S_r = U_G/U_L$ is known, the void fraction can be calculated by using the fundamental void-quality relation in one-dimensional two-phase flow (see Eq. (3.39)). In homogeneous two-phase flow we have $S_r = 1$. Some slip ratio correlations were discussed in Section 6.6. Bao *et al.* (1994) measured the void fraction in tubes with 0.74–3.07 mm diameters for air and water mixed with various concentrations of glycerin. They compared their void fraction data with predictions of several correlations, all taken from the literature dealing with commonly used large channels, and based on the results they recommended the empirical correlations for the slip ratio proposed by the CISE group, Eq. (6.54) (Premoli *et al.*, 1970).

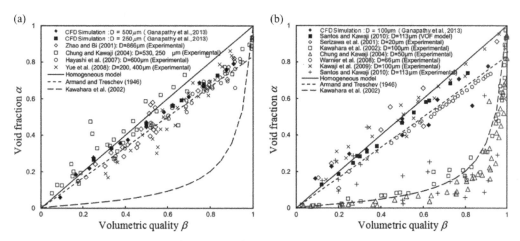

Figure 10.7. The relationship between void fraction and volumetric quality in small flow passages: (a) channels with characteristic dimension >150 μm; (b) channels with characteristic dimension <150 μm (from Ganapathy *et al.*, 2013a).

Triplett *et al.* (1999b) compared their void fraction data, estimated from photographs taken from their circular test sections, with predictions of several correlations. With the exception of the annular flow regime, where all the tested correlations overpredicted the data, the homogeneous model provided the best agreement with experiment.

Chung and Kawaji (2004) measured the time-averaged void fractions in circular channels with diameters of $D = 50, 100, 250$, and 530 μm and in a 96 μm square channel using image analysis. Their data, along with data from several other sources, are displayed in Figure 10.7. The homogeneous flow model agreed well with their 530-μm-diameter test data. Their data representing $D = 250$ μm deviated slightly from the homogeneous flow model, but they agreed well with the following Armand-type correlation that had been proposed earlier by Ali *et al.* (1993) for two-phase flow in narrow rectangular channels with $D_H \approx 1$ mm:

$$\alpha = C\beta, \quad C = 0.8. \tag{10.9}$$

where $\beta = j_G / j$ is the flow volumetric quality. The data of Chung and Kawaji (2004) and Chung *et al.* (2004) representing their 96-μm-square channel and their 50- and 100-μm-diameter test sections showed completely different trends than those just discussed, demonstrating a nonlinear relation between α and β, as noted in Fig. 10.7(b). The broken curves in the latter figures correspond to

$$\alpha = \frac{C_1 \beta^{0.5}}{1 - (1 - C_1)\beta^{0.5}}. \tag{10.10}$$

The constants were sensitive to channel size: $C_1 = 0.02$ for the 50-μm-diameter channel and $C_1 = 0.03$ for the two larger channels. Fu *et al.* (2008) and Zhang and Fu (2009) found that a similar expression could be used to curvefit their void fraction data that were obtained in experiments with nitrogen liquid–vapor flow in vertical tubes with 0.531 and 1.042 mm inner diameters. Xiong and Chung (2006) measured the void fraction by imaging in experiments with water–N_2 mixtures using multiple parallel horizontal channels that had near-square cross sections and were made of silicon carbide; with $D_{\mathrm{H}} = 0.209, 0.412$, and 0.622 mm.

Xiong and Chung (2006) derived the following constants for Eq. (10.10), based on their own data, as well as the 100-μm-diameter data of Kawahara *et al.* (2002):

$$C_1 = \frac{0.266}{1 + 13.8 \exp(-6.88 D_{\mathrm{H}})}, \qquad (10.11)$$

where D_{H} is in millimeters.

However, experiments by Ide *et al.* (2012) have shown that the reported strongly nonlinear relationship between α and β for minichannels may be due to two-phase flow intermittency caused by instability. Flow instability and oscillations were in fact observed in the aforementioned experiments of Fu *et al.* (2008) and Zhang and Fu (2009). In the experiments of Ide *et al.* (2012) with 100-μm-inner-diameter horizontal microtubes, gas injection took place through a 2-m-long 250-μm-inner-diameter tube which acted as a compressible volume that caused flow intermittency. With a 146-mm-long test section flow intermittency was rampant and the time-average void fraction at 37 mm from the inlet agreed with Eq. (10.10). When the test section length was extended to 1500 mm, however, the flow intermittency was significantly reduced and the time-average void fraction showed an approximately linear dependence between α and β. Their data in fact agreed with the prediction of a homogeneous flow ($\alpha = \beta$) when a T-junction was used at their test section inlet, and agreed with Eq. (10.9) with $C = 0.833$ when a flow-reduction-type mixer was used at their test section inlet.

Equation (10.11) also disagreed with the experimental data of Choi *et al.* (2011), which were generated with water and nitrogen in rectangular glass channels with hydraulic diameters in the 143–490 mm range and cross-section aspect ratios in the 0.16–0.92 range. The data of Choi *et al.* conformed with Eq. (10.9), but the constant C depended on the aspect ratio, and varied between 0.82 (for an aspect ratio of 0.92) and 0.92 (for an aspect ratio of 0.16).

The linear relationship between α and β has been confirmed by CFD simulations for Taylor flow (Santos and Kawaji, 2010; Ganapathy *et al.*, 2013b) (see Figure 10.7). CFD simulations by Ganapathy *et al.* (2013b) show that for air–water flow in a planar channel larger in width than 250 μm, for Taylor flow, the time-average void fraction follows Eq. (10.8) with $C = 0.833$. For channels with widths smaller than 250 μm CFD simulation also show a linear dependence between α and β, in agreement with experimetal data of Warnier *et al.* (2008) and Kawaji *et al.* (2009).

10.3 Two-Phase Flow Regimes and Void Fraction in Microchannels

For air–water-like fluid mixtures the flow behavior in microchannels (i.e., channels with $10 \lesssim D_{\mathrm{H}} \lesssim 100$ μm) is different than in minichannels. In bubbly or plug flow, the pressure difference between the liquid and the gas is large owing to the very small interfacial radii of curvature. Because of the predominance of surface tension, furthermore, small bubbles remain nearly spherical, and significant distortion from spherical shape occurs only when the bubble volume is about $\pi D^3/6$ or larger,

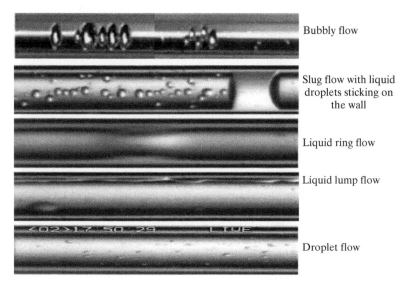

Bubbly flow

Slug flow with liquid droplets sticking on the wall

Liquid ring flow

Liquid lump flow

Droplet flow

Figure 10.8. Air–water two-phase flow regimes in a 100-μm-inner-diameter quartz tube. (From Serizawa *et al.*, 2002.)

making the spherical shape impossible. Bubble coalescence and breakup are rare, and consequently large bubbles that are generated remain large. Some flow regimes are encountered that are not seen in larger channels.

Two-phase flow in microchannels has been investigated by relatively few researchers, and there is disagreement among the few detailed investigations that have been published recently (Serizawa *et al.*, 2002; Kawahara *et al.*, 2002; Chung and Kawaji, 2004).

Figure 10.8 displays the air–water flow regimes in a quartz tube with $D = 100$ μm, recorded by Serizawa *et al.* (2002), who have observed that the slug flow regime in microchannels is primarily caused by entrance effects. In the slug flow regime, Serizawa *et al.* noted that dry zones develop in the liquid film separating the gas slug from the wall. At low velocities, liquid droplets were observed sticking to the dry areas. With increasing gas superficial velocity, the slug flow regime is replaced by liquid ring flow. The liquid ring flow regime itself is replaced with liquid lump flow with a further increase in gas superficial velocity. These flow regimes are not encountered in larger channels. The liquid ring appears to develop when, as a result of high gas velocity, the liquid slugs that separate the gas slugs from one another become unstable. The liquid ring flow regime has some resemblance to the slug-annular flow regime in the minichannel experiments of Triplett *et al.* (1999a) (see Fig. 10.1(e)). With increasing gas superficial velocity, the liquid rings are eventually transformed into liquid lumps, the motion of which resembles rivulets.

Surface wettability, as mentioned earlier, is an important parameter with respect to two-phase flow in mini- and microchannels. Figure 10.9 displays the air–water flow regimes in a quartz tube with $D = 100$ μm, recorded by Serizawa *et al.* (2002), when the tube surface was carefully cleaned. Interesting and important differences with flow regimes in Fig. 10.8 can be seen, including a dispersed bubbly flow pattern

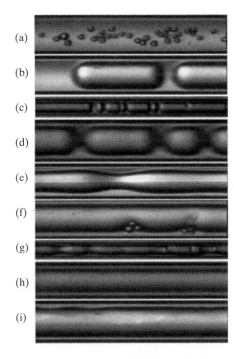

Figure 10.9. Air–water two-phase flow patterns in a 100-µm-inner-diameter cleaned quartz tube: (a) bubbly flow; (b) slug flow; (c) transition; (d) skewed flow (Yakitori flow); (e) liquid ring flow; (f) frothy annular flow; (g) transition; (h) annular flow; (i) rivulet flow. (From Serizawa *et al.*, 2002.)

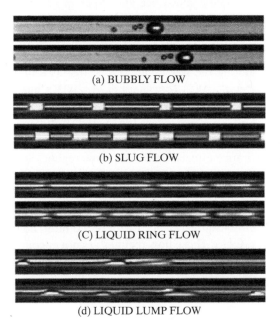

(a) BUBBLY FLOW

(b) SLUG FLOW

Figure 10.10. Air–water two-phase flow patterns in a 25-µm-inner-diameter silica tube. (From Serizawa *et al.*, 2002.)

(C) LIQUID RING FLOW

(d) LIQUID LUMP FLOW

(Fig. 10.9(a)) and the flow pattern in Fig. 10.9(d) where several bubbles are interconnected along the tube centerline. A stable annular flow regime was also observed (Fig. 10.9(h)).

Figure 10.10 displays air–water two-phase flow regimes in a 25-µm-diameter silica tube, reported by Serizawa *et al.* (2002). The surface tension is more predominant because of the smaller tube diameter. Moreover, rivulet and annular flow regimes were not observed.

Serizawa *et al.* (2002) compared their flow regime data representing a 20-μm-inner-diameter tube with the flow regime map of Mandhane *et al.* (1974) (discussed in Chapter 4). Flow stratification did not occur in the microchannel experiments; therefore stratified and wavy regions in the flow regime map of Mandhane *et al.* were irrelevant. They noticed that, provided that the liquid ring and liquid lump flow regime data were assumed to correspond to the annular flow regime in the map of Mandhane *et al.*, the agreement between data and the flow transition lines was actually reasonable.

Serizawa *et al.* (2002) measured the void fraction using video image analysis. For all bubbly and slug flow regimes, a linear correlation between α and β was obtained, leading to

$$\alpha = C\beta, \quad C = 0.833. \tag{10.12}$$

The good agreement between this result and the minichannel results of Chung and Kawaji (2004) (Eq. (10.9)) is noteworthy. The microchannel void fraction data of some authors, including Chung and Kawaji (2004) and Santos and Kawaji (2010), are in disagreement with the aforementioned data of Serizawa *et al.* (2002), however, and show a strongly nonlinear dependence between α and β (see Fig. 10.7(b)). However, data from several sources as well as CFD simulations for Taylor flow indicate that the above Armand-type correlation is valid for air–water-like two-phase mixtures.

Kawahara *et al.* (2002) and Chung and Kawaji (2004) conducted a detailed experimental study of nitrogen–water two-phase flow hydrodynamics in mini- and microchannels with diameters in the $D = 50$–526 μm range. Their minichannel results have already been discussed. Their microchannel data were obtained in circular fused silica capillaries with $D = 50$ and 100 μm. They did not observe bubbly flow, since their experiments did not cover a sufficiently low gas superficial velocity. Churn and slug-annular flow regimes were also completely absent in their experiments, however, and only variations of the slug flow pattern were observed. Furthermore, multiple flow patterns occurred at high liquid and gas flow rates in their 100-μm test section, including liquid-alone and gas slugs with various liquid film geometries. Unlike in minichannels, where significant gas–liquid agitation leads to strong momentum coupling between the two phases, little interphase agitation is observed in microchannels, leading to the conclusion that the liquid flow in microchannels is laminar (Kawahara *et al.*, 2002; Chung and Kawaji, 2004). The latter authors developed separate flow regime maps for their two microchannel test sections, where the major patterns are defined based on the probability of various specific flow regimes and the channel void fraction.

10.4 Two-Phase Flow and Void Fraction in Thin Rectangular Channels and Annuli

Two-phase flow in rectangular channels with $\delta \lesssim 1$ mm occurs in plate-type research nuclear reactors, in electronic components, and during critical flow through cracks that may occur in vessels containing pressurized fluids. Investigations have been reported by Lowry and Kawaji (1988), Wambsganss *et al.* (1991), Ali and Kawaji (1991), Ali *et al.* (1993), Mishima *et al.* (1993), Wilmarth and Ishii (1994), Fourar and

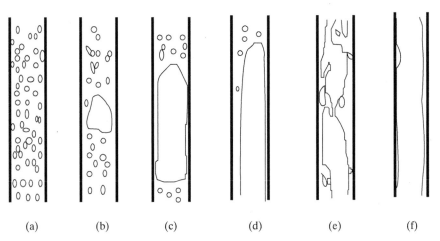

(a)   (b)   (c)   (d)   (e)   (f)

Figure 10.11. Flow regimes in vertical rectangular channels with $\delta \geq 0.5$ mm (medium-sized gaps): (a) bubbly; (b) cap-bubbly; (c) slug; (d) slug-churn; (e) churn-turbulent; (d) annular. (From Xu *et al.*, 1999.)

Bories (1995), Bonjour and Lallemand (1998), Xu *et al.* (1999), Hibiki and Mishima (2001), Warrier *et al.* (2002), Xiong and Chung (2007) and Choi *et al.* (2011).

Experiments in vertical narrow channels, using air and water, with consistent overall results with respect to the two-phase flow regime maps, have been reported by Kawaji and co-workers (Lowry and Kawaji, 1988; Ali and Kawaji, 1991; Ali *et al.*, 1993) (air–water; $\delta = 0.5$–2 mm), Mishima *et al.* (1993) (air–water; $\delta = 1.07$–5 mm), Wilmarth and Ishii (1994) (air–water; $\delta = 1, 2$ mm), and Xu *et al.* (1999) (air–water; $\delta = 0.3$–1 mm). Flow in horizontal channels has been studied by Fourar and Bories (1995) ($\delta > 0.5$; 1 mm; horizontal) and Ali *et al.* (1993) ($\delta = 0.78$; 1.46 mm; various orientations).

Four major flow regimes are often defined for vertical flow channels with $\delta > 0.5$ mm: bubbly, slug, churn-turbulent, and annular. Minor differences with respect to the description and identification of the flow patterns exist among these investigators, however. Figure 10.11 displays schematics of these flow regimes (Wilmarth and Ishii, 1994; Xu *et al.*, 1999). The cap-bubbly and slug-churn are transitional regimes that were defined by Xu *et al.* (1999).

In horizontal channels with $\delta > 0.5$ mm (i.e., flow between two parallel horizontal planes), the main flow regimes defined by Wilmarth and Ishii (1994) are displayed in Fig. 10.12. With $\delta < 0.5$ mm, however, significant changes occur in flow regimes, as will be discussed shortly.

10.4.1 Flow Regimes in Vertical and Inclined Channels

For vertical, upward flow, Mishima *et al.* (1993) identified four major flow regimes in their experiments: bubbly flow, characterized by crushed or pancake-shaped bubbles; slug flow, represented by crushed slug (elongated) bubbles; churn flow, in which the noses of the elongated bubbles were unstable and noticeably disturbed; and annular flow. The experimental data of Mishima *et al.* for their 1.07-mm gap, and the flow regimes of Wilmarth and Ishii (1994) for vertical, upward flow are compared

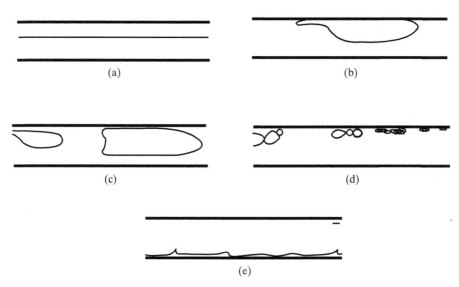

Figure 10.12. Flow regimes in flow between two horizontal planes: (a) stratified smooth; (b) plug flow; (c) slug flow; (d) dispersed bubbly; (e) wavy annular. (From Wilmarth and Ishii, 1994.)

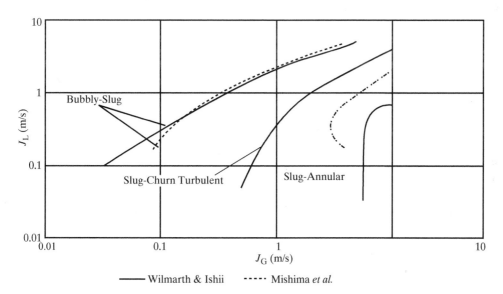

Figure 10.13. Flow patterns in the experiments of Mishima *et al.* (1993) and Wilmarth and Ishii (1994) (vertical, upflow) in test sections with 1-mm gap.

in Fig. 10.13, where bubbly and "cap-bubbly" flow patterns have been combined in the bubbly flow regime zone. Wilmarth and Ishii noted relatively good agreement between their data for the flow regime transition from bubbly to slug and the flow regime transition models of Taitel *et al.* (1980) (Eq. (7.3)) and Mishima and Ishii (1984) (Eq. (7.13)). Both models are based on maximum packing of bubbles. The regime transition model of Mishima and Ishii is based on the DFM, however, and Wilmarth and Ishii noted that a two-phase distribution coefficient C_0 for narrow channels is needed.

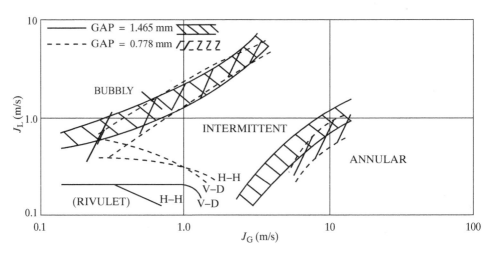

Figure 10.14. Flow patterns in the experiments of Ali *et al.* (1993) for all configurations except for horizontal flow between vertical plates (H–H = horizontal flow between horizontal plates; V–D = vertical downward flow; unmarked transition lines apply to all configurations).

Ali and Kawaji (1991) and Ali *et al.* (1993) performed an extensive experimental study using room-temperature and near-atmospheric air and water in rectangular narrow channels with six different configurations: vertical, co-current upward and downward flow; 45° inclined, co-current upward and downward flow; horizontal flow between horizontal plates; and horizontal flow between vertical plates. Their observed flow regimes and flow regime maps were similar for all configurations except for the last one and are displayed in Fig. 10.14. The rivulet flow pattern occurred at very low liquid superficial velocities. The flow regimes for horizontal flow between vertical plates included bubbly, intermittent, and stratified-wavy, and the flow regime maps for both $\delta = 0.778$ and 1.465 mm were similar to the flow regime maps observed in large pipes.

Hibiki and Mishima (2001) have modified the semi-analytical two-phase flow regime transition models of Mishima and Ishii (1984) for co-current upward flow in vertical tubes, described in Section 7.2.2, for application to thin rectangular vertical channels, using the experimental data of Mishima *et al.* (1993), Wilmarth and Ishii (1994), Xu *et al.* (1999), and others. The applicable data covered the range $s = 0.5$–17 mm.

10.4.2 Flow Regimes in Rectangular Channels and Annuli

For flow between two horizontal parallel plates, several experimental studies have been published. Differences with respect to the flow regime description and identification among various authors can be noted, however. The experimental flow regimes of Ali *et al.* (1993) were shown in Fig. 10.14. For the horizontal flow configuration, Wilmarth and Ishii (1994) could identify stratified, plug, slug, dispersed bubbly, and wavy annular flow patterns. In their experiments in similarly configured narrow channels, as mentioned before, Ali *et al.* (1993) identified bubbly, intermittent, and stratified-wavy flow regimes only. The two flow regime maps are compared

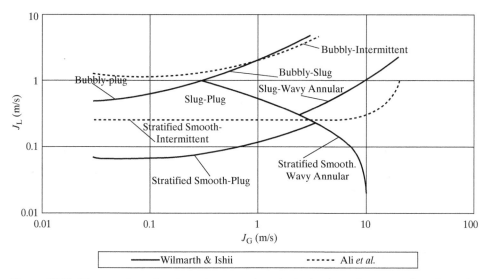

Figure 10.15. Flow regimes in air–water experiments in horizontal rectangular channels. (From Wilmarth and Ishii, 1994.)

in Fig. 10.15. The two sets of data are qualitatively in agreement with respect to the bubbly–plug/slug transition.

Two-Phase Flow in Rectangular Channels with $\delta \lesssim 0.5$ mm

The discussion of two-phase flow in narrow rectangular channels thus far has dealt primarily with $\delta \approx 1.0$ mm. Available data with low-viscosity (i.e., water-like) liquids show that with $\delta \leq 0.5$ mm the flow regimes are significantly different than those just discussed. However, the available experimental data dealing with such extremely narrow rectangular channels are few. Xu et al. (1999) performed experiments with a vertical channel with $W = 0.5$ m using air and water. Bubbly flow did not occur in their tests at all, and the main flow regimes were cap bubbly, slug-droplet, churn, and annular-droplet. In the slug-droplet flow, the flattened bubbles appeared to represent dry patches, with liquid droplets that were attached to the dry surface and were moved along the surface by the drag force. The liquid bridging between the flattened bubbles did not include entrained bubbles. The annular-droplet flow was similar to the annular-dispersed flow, where the gas core contained isolated entrained liquid droplets.

Two-Phase Flow in Thin Annuli

Ekberg et al. (1999) conducted experiments using two horizontal glass annuli with 1.02-mm spacing and studied the two-phase flow regimes, void fraction, and pressure drop. The two-phase flow patterns in vertical and horizontal large annular channels had earlier been studied by Kelessidis and Dukler (1989) and Osamusali and Chang (1988), respectively. Osamusali and Chang carried out experiments in three annuli, all with outer diameters $D_o = 4.08$ cm, and with inner to outer diameter ratios of $D_i/D_o = 0.375, 0.5$, and 0.625 ($\delta = 4.75, 6.35$, and 11.75 mm, respectively) and noted that the flow patterns and their transition lines were relatively insensitive to D_i/D_o. The experimental flow regime transition lines of Ekberg et al. (1999) are displayed in

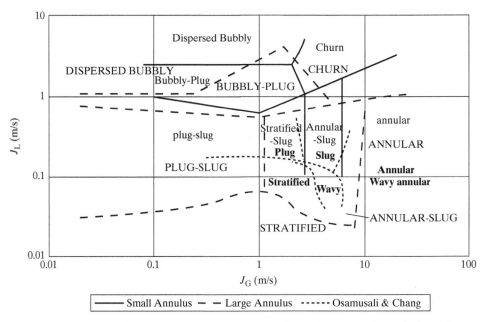

Figure 10.16. Flow regimes in narrow horizontal annuli. Regime names in capital and lower-case letters are for the small and large test sections of Ekberg *et al.* (1999), respectively. Regime names in bold letters are from Osamusali and Chang (1988).

Fig. 10.16, where they are compared with the experimental results of Osamusali and Chang (1988). These transition lines disagree with the flow regime map of Mandhane *et al.* (1974). Stratified flow occurred in the experiments of Ekberg *et al.* (1999).

Ekberg *et al.* (1999) compared their measured void fractions with the predictions of the homogeneous mixture model, the correlation of Lockhart and Martinelli (1949) as presented by Butterworth (1975) (Eq. (6.60) and Table 6.3), the correlations of Premoli *et al.* (1970) (Eqs. (6.54)–(6.59)), and the drift flux model with $C_0 = 1.25$ and $V_{gj} = 0$, following the results of Ali *et al.* (1993) for narrow channels. The Lockhart–Martinelli–Butterworth correlation best agreed with their data. Furthermore, in all flow regimes except annular, their test section void fraction closely agreed with the following correlation:

$$\alpha = 1 - \left[\frac{X}{1+X} \right]^2, \qquad (10.13)$$

where X represents Martinelli's factor.

10.5 Two-Phase Pressure Drop

The number of investigations dealing with two-phase pressure drop in minichannels (i.e., channels with $100\ \mu m \lesssim D_H \lesssim 1\ mm$) is relatively large, and useful summaries of these investigations have been provided by Kim and Mudawar (2012a) and Awad *et al.* (2014). Serious interest in two-phase flow in microchannels (i.e., channels with $10 \lesssim D_H < 100\ \mu m$) is more recent, however, and there is scarcity of experimental data and predictive models for pressure drop for these flow passages.

Measurement and correlation of frictional pressure drop in small channels are difficult for the following reasons.

(1) There is uncertainty with respect to channel geometry and wall roughness. Non-uniformities in channel cross-section geometry, and unknown roughness characteristics, can contribute to these uncertainties.

(2) There is uncertainty with respect to the magnitude of different pressure drop components, particularly acceleration. This is particularly true for experiments with evaporation and condensation where spatial acceleration is significant because of phase change. Only the total variations of pressure can be measured. The various pressure drop terms can then be calculated only with the help of a slip ratio or void-quality relation, or a model with phase-slip closure relations. The same is of course true for macro- and conventional channels. Because of the differences between the flow regimes in conventional and mini- and microchannels, however, the conventional slip ratio and void-quality relations may not always be applicable to mini- and microchannels.

(3) There is uncertainty with respect to entrance and exit pressure losses. Few experimental data are available on two-phase pressure losses caused by abrupt flow disturbances, including flow-area expansion and contraction (minor losses). The scant available data suggest that the minor pressure losses in mini- and microchannels are different than in large channels (see Section 10.8). In most of the published experimental studies, the test channels are connected to plenum(s) or larger flow passages at their inlet and outlet. Furthermore, virtually all studies have used macroscale models and methods for the calculation of test-section inlet and outlet pressure losses.

(4) Laminar flow in mini- and microchannels is likely to occur. Laminar flow is rare in large channels, and models and correlations that are based on large-channel data can be inadequate for laminar flow conditions.

(5) There are uncertainties and inconsistencies with respect to the single-phase flow frictional pressure losses. Some experimental data indicate that micro- and minichannels behave differently than conventional channels. Reported differences include transition to turbulent flow at a lower Reynolds number (Wu and Little, 1983; Choi et al., 1991; Peng et al., 1994a, b; Peng and Wang, 1998) and turbulent flow friction factors that are different than the predictions of well-established correlations (Yu et al., 1995; Peng and Wang, 1998; Hegab et al., 2002). However, other studies, including some recent careful experiments, have shown that laminar flow theory predicts mini- and microchannel data well (Mikol, 1963; Olson and Sunden, 1994; Kohl et al., 2005) and that the transition to turbulent flow occurs at a Reynolds number that is consistent with large channels (Kohl et al., 2005).

EXAMPLE 10.1. Saturated water enters a uniformly heated 10-cm-long tube that has a 1-mm inner diameter. At the exit, where the pressure is 18.7 bars, the equilibrium quality is 0.9. The mass flux is 1250 kg/m²s. Assuming homogeneous-equilibrium flow, estimate the acceleration and frictional pressure drops in the tube.

SOLUTION. Because an estimate of the pressure drop terms is sought, we will use conventional methods. For the frictional pressure drop, we will use the method of

Martinelli and Nelson (1948), Eq. (8.36). From Table P8.2a in Problem 8.2, for $x_{eq} = 0.9$ and an average pressure of 17.2 bars,

$$\bar{\Phi}_{f0}^2 = \frac{1}{x_{eq}} \int_0^{x_{eq}} \Phi_{f0}^2(x_{eq}) dx_{eq} = 45.$$

We will calculate the mean properties at 17.2 bars, whereby $\rho_f = 859$ kg/m^3, $\rho_g = 8.67$ kg/m^3, $\mu_f = 1.31 \times 10^{-4}$ kg/m·s, and $\mu_g = 1.59 \times 10^{-5}$ kg/m·s. To apply Eq. (8.36), we will proceed as follows:

$$Re_{f0} = GD/\mu_f = 9{,}543 \text{ (turbulent flow)},$$

$$f' = 0.316 Re_{f0}^{-0.25} = 0.032,$$

$$\left(-\frac{dP}{dz}\right)_{fr,f0} = f' \frac{1}{D} \frac{G^2}{2\rho_f} = 2.908 \times 10^4 \text{ Pa/m},$$

$$\Delta P_{fr} = L\left(-\frac{dP}{dz}\right)_{fr,f0} \bar{\Phi}_{f0}^2 = 1.309 \times 10^5 \text{ Pa}.$$

The acceleration pressure drop can be obtained by noting that (see Eq. (8.1))

$$\Delta P_{sa} = \int_0^L \left(-\frac{dP}{dz}\right)_{sa} dz = \left.\frac{G^2}{\rho'}\right|_0^L = \frac{G^2}{\rho_{h,ex}} - \frac{G^2}{\rho_{f,in}}.$$

For simplicity, let us use channel-average properties. For $x = 0.9$,

$$\rho_h = \left(\frac{x}{\rho_g} + \frac{1-x}{\rho_f}\right)^{-1} = 9.62 \text{ kg/m}^3.$$

As a result we will get

$$\Delta P_{sa} = 1.606 \times 10^5 \text{ Pa}.$$

The total pressure drop in the 10-cm-long capillary is thus about 2.915 bars.

As was the case for two-phase flow in conventional channels (see Chapter 8), the concept of a two-phase flow multiplier is often used in correlating mini- and microchannel data. Single-phase flow friction factors are therefore needed. The base single-phase flows are often laminar or transitional in mini- and microchannels. Furthermore, unlike in large channels where wall surface roughness is of little consequence in two-phase flow, the surface roughness can be important. The correlation of Churchill (1977) for single-phase friction factors in channels, which covers the laminar, transition, and turbulent regimes and accounts for the effect of surface roughness, has been chosen by some investigators (Lin et al., 1991; Zhao and Bi, 2001; Li and Wu, 2010a). The correlation of Churchill can be cast as

$$f' = 8\left[\left(\frac{C_1}{Re}\right)^{12} + \frac{1}{(A+B)^{3/2}}\right]^{1/12}, \tag{10.14}$$

where

$$A = \left\{ \frac{1}{\sqrt{C_t}} \ln \left[\frac{1}{\left(\frac{7}{Re}\right)^{0.9} + 0.27\frac{\varepsilon_D}{D}} \right] \right\}^{16},$$
(10.15)

$$B = \left(\frac{37530}{Re} \right)^{16},$$
(10.16)

where ε_D/D is the dimensionless surface roughness. For circular channels, $C_1 = 8$ and $1/\sqrt{C_t} = 2.457$.

Various widely used correlations for two-phase flow frictional pressure drop have been tested against minichannel data by many authors, often with unsatisfactory results. Modifications were therefore introduced into some correlations. The two most successful methods, however, appear to be the homogeneous flow model and the method of Chisholm (1967) (see Sections 8.2 and 8.3).

The homogeneous flow model had been popular in some early investigations for the interpretation of data involving phase change. A key element in the homogeneous flow model is the definition of mixture properties. Table 8.1 provides a summary of some widely used homogeneous flow models. Koizumi and Yokohama (1980) modeled the flow of R-12 in capillaries with $D = 1$ and 1.5 mm, using the HEM model with $\mu_{TP} = \rho_h v_L$, based on the argument that the flashing two-phase flow in their simulated refrigerant restrictor was predominantly bubbly. Using R-12 as the working fluid, and test sections with $D = 0.66$ mm ($\varepsilon_D = 2$ μm) and 1.17 mm ($\varepsilon_D = 3.5$ pm), Lin et al. (1991) measured the pressure drop with single-phase liquid flow, noting that their data could be well predicted by using the Churchill (1977) correlation. Based on an argument similar to that of Koizumi and Yokohama (1980), they applied the HEM model for two-phase pressure-drop calculations. For calculating the two-phase friction factor f_{Tp}, they used the Churchill correlation (Eq. (10.14)) by replacing Re with Re_{TP}, and over a quality range of $0 < x < 0.25$ they empirically correlated their two-phase mixture viscosity according to

$$\mu_{TP} = \frac{\mu_G \mu_L}{\mu_G + x^n(\mu_L - \mu_G)},$$
(10.17)

with $n = 1.4$ providing the best agreement between model and data. Bowers and Mudawar (1994) studied high-heat-flux boiling in channels with $D = 0.5$ and 2.54 mm. The HEM model with $f'_{TP} = 0.02$ could well predict their total experimental pressure drops.

It should be emphasized that because of the importance of the acceleration pressure drop in tests with significant phase change, the accuracy of the frictional model is often difficult to directly assess. The good agreement between model-predicted and measured total pressure drops when the homogeneous flow model is used, nevertheless may indicate that the homogeneous-equilibrium model in its entirety is adequate for such applications. Experimental studies that have supported the adequacy of the homogeneous model for application to adiabatic two-phase flow include those by Ungar and Cornwell (1992), Bao et al. (1994), and Triplett et al. (1999b). Ungar and Cornwell's data dealt with high-quality ammonium flow ($0.09 < x < 0.98$). Bao et al. (1994) performed an extensive experimental study using air and aqueous glycerin solutions with various concentrations and calculated the experimental friction factors using channel-average properties. They compared their data with various correlations. By implementing the simple modification given later into the correlation of Beattie and Whalley (1982), Eqs. (8.21) and (8.22), the latter correlation well predicted their entire data set. The correlation of Beattie and Whalley is based on the application of the homogeneous flow model and the Colebrook–White correlation (Eq. (8.23)) for the friction factor over the entire two-phase Reynolds number range. Bao *et al.* (1994) modified the correlation of Beattie and Whalley (1982) simply by using $f_{TP} - 16/Re_{TP}$ in the $Re_{TP} < 1000$ range. Triplett *et al.* (1999b) noted that, overall, the homogeneous mixture model applied with McAdams's correlation for mixture viscosity better predicted their data. However, the homogeneous flow model, as well as several other correlations, did poorly when applied to the annular flow regime data.

The experimental investigations of Kawahara *et al.* (2002) were discussed with respect to flow regimes and void fraction in microchannels earlier in the previous section. Kawahara *et al.* measured frictional pressure drop in their experiments as well, and noted that the homogeneous flow model agreed well with their data provided that the mixture viscosity was defined using the expression proposed by Dukler *et al.* (1964) (see Table 8.1). More recently, Zhang *et al.* (2010) compared the predictions of the homogeneous flow model with data from several sources, covering several fluid mixtures in minichannels with $1.4 < D_H < 3.25$ mm. The range of their data base is provided in Table 8.2. They observed that the homogeneous flow model performed best when applied with the mixture property definition of Dukler *et al.* (1964), followed by the homogeneous flow model with parameter defined by Beattie and Whalley (1982) (Eqs. (8.21) through (8.23)). Overall, they noted that the homogeneous mixture model with Dukler's mixture property definitions performed reasonably well for adiabatic two-phase flow in micro- and minichannels, but performed poorly for flow boiling conditions.

Numerous experimental investigations have been performed in the past decade with liquid–vapor two-phase flows of various refrigerants in minichannels, mostly involving diabatic flow conditions. The frictional pressure drop has been a key component of these investigations. Detailed reviews of past investigations dealing with two-phase flow pressure drop in small channels can be found in Sun and Mishima (2009), Kim and Mudawar (2012a, c), and Awad *et al.* (2014), among others. The review article by Kim and Mudawar (2012a) also contains a detailed list of references for available experimental data that deal with two-phase flow in miniature flow passages.

The method of Chisholm (1967), Eq. (8.28), has been modified for application to minichannels by many investigators. This approach appears to provide the best method at least for adiabatic two-phase flow. Table 8.2 is a summary of several recent modifications to the Chisholm–Laird method. The intention of most of these modifications has been to extend the applicability of the method to refrigerants. Some of the listed correlations have micro- and minichannel data included in their data bases. The correlation of Kim and Mudawar (2012b; 2013) (Table 8.3), for example, has extensive minichannel data in its data base. A few of those correlations have been specifically developed for miniature flow passages, however, and are therefore preferable for mini- and microchannel applications. These correlations include

Table 10.1. *Constants and exponents in the correlation of Lee and Lee (2001a).*

Liquid regime	Gas flow regime	A	q	r	s
Laminar	Laminar	6.833×10^{-8}	-1.317	0.719	0.577
Laminar	Turbulent	6.185×10^{-2}	0	0	0.726
Turbulent	Laminar	3.627	0	0	0.174
Turbulent	Turbulent	0.408	0	0	0.451

the channel hydraulic diameter and surface tension, as expected. Some of the most widely referenced and recent of these correlations re now reviewed.

Based on the observation that C depends on surface tension and hydraulic diameter, as well as phase mass fluxes, and using experimental data from several sources as well as their own data that covered channel gaps in the 0.4–4 mm range, Lee and Lee (2001a) derived the following correlation for C, for adiabatic flow in horizontal thin rectangular channels:

$$C = A\left[\frac{\mu_L^2}{\rho_L \sigma D_H}\right]^q \left[\frac{\mu_L j}{\sigma}\right]^r \mathrm{Re}_{L0}^s \tag{10.18}$$

where j represents the total mixture volumetric flux. The constants A, r, q, and s depend on the liquid and gas flow regimes (viscous or turbulent), and their values are listed in Table 10.1. For either phase, laminar (viscous) flow is assumed when $\mathrm{Re}_i = \rho_i j_i D_H / \mu_i < 2000$ ($i =$ L or G), and turbulent flow is assumed otherwise.

The correlation of Li and Wu (2010a) is based on circular and rectangular channels with $0.22 \leq D_H \leq 3.25$ mm, with R-12, R-22, R-32, R-134a, R-245fa, R236ea, R-404a, R-410a, R-422d, and liquid nitrogen as fluids. The correlation is remarkable as it accounts for all important fluid properties as well as phasic mass fluxes:

$$C = 11.9\,\mathrm{Bd}^{0.45} \quad \text{for Bd} \leq 1.5, \tag{10.19}$$

$$C = 109.4\left(\mathrm{Bd}\,\mathrm{Re}_{L0}^{0.5}\right)^{-0.56} \quad \text{for } 1.5 < \mathrm{Bd} \leq 11, \tag{10.20}$$

where the Bond number is defined as

$$\mathrm{Bd} = \left[g\left(\rho_L - \rho_G\right)D_H^2\right]/\sigma. \tag{10.21}$$

The correlations of Sun and Mishima (2009) (see Table 8.2) and Kim and Mudawar (2012c) (see Table 8.3) have extensive data bases and cover a wide range of channel sizes.

A number of correlations have also been developed based on condensing flows in mini- and microchannels. These correlations will be discussed in Chapter 16.

EXAMPLE 10.2. Consider a horizontal capillary tube with $D = 0.5$ mm, subject to air–water flow at 293 K. Assume a local pressure of 2 bars and a mass flux of $G = 500$ kg/m^2·s. For values of local quality in the $x = 0.2$–0.9 range, calculate the local frictional pressure gradient, using the correlations of Mishima and Hibiki (1996) and Lee and Lee (2001a).

SOLUTION. The relevant properties are $\rho_f = 998.3$ kg/m³, $\rho_g = 2.38$ kg/m³, $\mu_L = 1.0 \times 10^{-3}$ kg/m·s, and $\mu_G = 1.82 \times 10^{-5}$ kg/m·s. Let us proceed with the calculations for $x = 0.2$. Then

$$G_L = G(1 - x) = 400 \text{ kg/m}^2\cdot\text{s},$$
$$G_G = G_x = 100 \text{ kg/m}^2\cdot\text{s},$$
$$\text{Re}_L = G_L D/\mu_L = 198.9,$$

and

$$\text{Re}_G = G_G D/\mu_G = 2{,}740.$$

These and all other calculation results show that $42 < \text{Re}_L < 199$ and $2.7 \times 10^3 < \text{Re}_G < 1.39 \times 10^5$. The liquid and gas flows are thus viscous and turbulent, respectively, and

$$f'_L = 64/\text{Re}_L = 0.322,$$
$$f'_G = 0.316\text{Re}_G^{-0.25} = 0.0437,$$
$$\left(-\frac{dP}{dz}\right)_{\text{fr,L}} = f'_L \frac{G_L^2}{2D\rho_L} = 51{,}581 \text{ Pa/m},$$
$$\left(-\frac{dP}{dz}\right)_{\text{fr,G}} = f'_G \frac{G_G^2}{2D\rho_G} = 1.836 \times 10^5 \text{ Pa/m}.$$

The Martinelli factor is found from

$$X = \sqrt{\left(-\frac{dP}{dz}\right)_{\text{fr,L}} \Bigg/ \left(-\frac{dP}{dz}\right)_{\text{fr,G}}} = 0.53.$$

According to Mishima and Hibiki (1996),

$$C = 21\left[1 - \exp(-0.319 \times 0.5)\right] = 3.096,$$
$$\Phi_L^2 = 1 + \frac{C}{X} + \frac{1}{X^2} = 10.2,$$
$$\left(-\frac{dP}{dz}\right)_{\text{fr}} = \left(-\frac{dP}{dz}\right)_{\text{fr,L}} \Phi_L^2 = 5.37 \times 10^5 \text{ Pa/m}.$$

We now follow the correlation of Lee and Lee (2001a). Since we have viscous liquid and turbulent gas, according to Table 10.1,

$$A = 6.185 \times 10^{-2}, q = 0, r = 0, \text{ and } s = 0.726.$$

Therefore,

$$\text{Re}_{L0} = \frac{GD}{\mu_L} = 248.6$$

and

$$C = A\text{Re}_{L0}^s = 3.392.$$

Using these values of C and X in Eq. (8.28), we get $\Phi_L^2 = 10.96$. This leads to

$$\left(-\frac{dP}{dz}\right)_{\text{fr}} = 5.654 \times 10^5 \text{ Pa/m}.$$

The frictional pressure gradients for the $x = 0.2 - 0.9$ range are displayed in the figure below. Evidently the two methods provide very similar predictions for air–water mixtures.

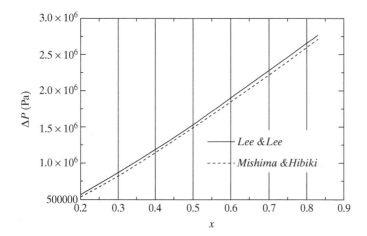

From laminar flow theory, the frictional pressure gradient associated with laminar liquid-only and gas-only flows in a thin rectangular channel with $W/\delta \to \infty$ can be found from

$$\left(-\frac{\partial P}{\partial z}\right)_{fr,i} = \frac{12\mu_i j_i}{\delta^2}, \tag{10.22}$$

where $i = $ L or G. The Martinelli parameter will then be

$$X = \left[\frac{\mu_L j_L}{\mu_G j_G}\right]^{1/2}. \tag{10.23}$$

Fourar and Bories (1995) conducted experiments using horizontal slits made by baked clay bricks with $\delta = 0.18, 0.40,$ and 0.54 mm and $W = 0.5$ m. Their two-phase frictional pressure drop data for all three δ values correlated well with

$$\Phi_G = 1 + X, \tag{10.24}$$

$$\Phi_L = \frac{1 + X}{X}. \tag{10.25}$$

These two expressions are of course equivalent.

10.6 Semitheoretical Models for Pressure Drop in the Intermittent Flow Regime

The ideal fully developed slug flow regime in minichannels is morphologically relatively simple and consists essentially of pure liquid slugs (liquid slugs without entrained microbubbles) and large gas bubbles (also sometimes referred to as gas slugs). The flow field in the slug flow regime can be idealized as an axisymmetric flow where the liquid and gas slugs are cylindrical, as in Fig. 10.17. The pressure drop can then be mechanistically modeled (Fukano et al., 1989; Garimella et al., 2002; Chung and Kawaji, 2004).

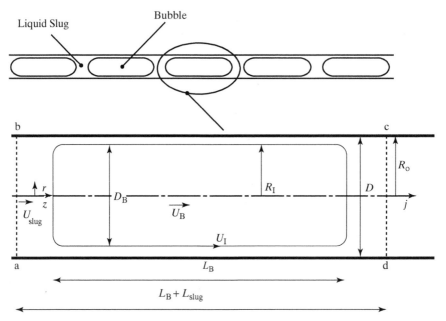

Figure 10.17. A unit cell in the slug flow regime.

Consider a unit cell abcd, as shown in Fig. 10.17. The unit cell moves with an apparent velocity of $j = j_G + j_L$ in the flow direction. This evidently implies that the liquid slug moves with a mean velocity equal to j (Govier and Aziz, 1972). This conclusion about the mean velocity in the liquid slug is general and is not limited to minichannels. The mean frictional pressure gradient in the unit cell can be represented as

$$\left(-\frac{dP}{dz}\right)_{fr} = \frac{1}{L_{slug} + L_B}\left\{\left(-\frac{dP}{dz}\right)_{fr,slug} L_{slug} + \left(-\frac{dP}{dz}\right)_{fr,B/F} L_B + \Delta P_{F/S}\right\}, \quad (10.26)$$

where $(-dP/dz)_{fr,slug}$ and $(-dP/dz)_{fr,F/B}$ are the mean frictional pressure gradients associated with the liquid slug and bubble/film regions in the unit cell, respectively, and $\Delta P_{F/S}$ is the pressure loss that results from the drainage of the liquid film into the liquid slug at the tail of the bubble.

The frictional pressure drop in the liquid slug can be estimated by assuming that the flow field in the slug is identical to fully developed single-phase liquid flow. In the ideal slug flow regime, $U_B = U_G$, and given that the average velocity of the liquid slug is equal to j, then

$$\left(-\frac{dP}{dz}\right)_{fr,slug} = f'_{slug}\frac{1}{D}\frac{1}{2}\rho_L j^2, \quad (10.27)$$

where f'_{slug} depends on whether the flow in the slug is laminar or turbulent. One can define $\mathrm{Re}_{slug} = \rho_L jD/\mu_L$, and use, for example (Chung and Kawaji, 2004),

$$f'_{slug} = \begin{cases} 64/\mathrm{Re}_{slug}, & \mathrm{Re}_{slug} < 2100, & (10.28) \\ 0.316\,\mathrm{Re}_{slug}^{-0.25}, & \mathrm{Re}_{slug} > 2100. & (10.29) \end{cases}$$

The frictional pressure drop in the bubble/film zone is likely to be small compared with the pressure drop in the slug in larger channels (Dukler and Hubbard, 1975), as well as in minichannels that carry air–water-like fluid mixtures (Fukano et al., 1989).

However, it can be significant for microchannels (Chung and Kawaji, 2004), and for minichannels that carry refrigerants with large ρ_G/ρ_L and μ_G/μ_L ratios (Garimella et al., 2002). The flow in the bubble/film zone can be treated as an ideal annular flow. Neglecting the effect of gravity (which is small in mini- and microchannels), we can express the momentum equation for laminar flow as

$$-\frac{dP}{dz} + \frac{\mu}{r}\frac{d}{dr}\left(r\frac{du}{dr}\right) = 0. \tag{10.30}$$

The general solution to Eq. (10.30) is

$$u = -\frac{1}{4\mu}\left(-\frac{dP}{dz}\right)r^2 + B\ln r + E. \tag{10.31}$$

Equations (10.30) and (10.31) in fact apply to either the liquid film or the gas core, as long as they are fully developed and laminar. The general solution, when both phases are laminar, can be derived by applying the no-slip condition at the wall and the continuity of velocity and shear stress at the liquid–gas interphase (see Eqs. (10.40)–(10.50)). The liquid film in most mini- and microchannel applications can be assumed to be laminar, but the gas phase can be turbulent in minichannel applications.

A simpler, semi-analytical solution can be formulated by first deriving the solution for the laminar liquid film by using the no-slip condition on the wall and the boundary condition $u_L = U_I$ at $r = R_I$. The result will be

$$u_L(r) = \frac{1}{4\mu_L}\left(-\frac{dP}{dz}\right)_{\text{fr,B/F}}\left[R_0^2 - r^2 - (R_0^2 - R_I^2)\frac{\ln(R_0/r)}{\ln(R_0/R_I)}\right] + U_I\frac{\ln(R_0/r)}{\ln(R_0/R_I)}. \tag{10.32}$$

The force balance on the gas core gives

$$\mu_L\left(\frac{dU}{dr}\right)_{r=R_I} = \frac{R_I}{2}\left(-\frac{dP}{dz}\right)_{\text{fr,B/F}}. \tag{10.33}$$

Using these two equations, it can be shown that the film–gas interphase velocity will be

$$U_I = \frac{1}{4\mu_L}\left(-\frac{dP}{dz}\right)_{\text{fr,B/F}}(R_0^2 - R_I^2). \tag{10.34}$$

For the gas phase one can also write

$$\left(-\frac{dP}{dz}\right)_{\text{fr,B/F}} = f_I'\frac{1}{2R_I}\frac{1}{2}\rho_G^2(U_G - U_I)^2. \tag{10.35}$$

The interphase friction factor f_I' can also be related to the gas-phase Reynolds number $\text{Re}_G = 2\rho_G R_I(U_B - U_I)/\mu_G$ by using laminar and turbulent channel flow correlations.

To solve these equations for $(-dP/dz)_{\text{fr}}$, expressions are needed for R_I/R_0, L_{slug}/L_B, and $\Delta P_{F/S}$. From the idealized flow field, one can think of the bubble as a cylinder and approximately write

$$\alpha = \frac{L_B R_I^2}{(L_B + L_{\text{slug}})R_0^2}. \tag{10.36}$$

Chung and Kawaji (2004), in modeling their experimental data dealing with flow in microchannels with $D = 50$ and $100\ \mu\text{m}$, assumed $R_I/R_0 = 0.9$ and $\Delta P_{F/S} = 0$ and

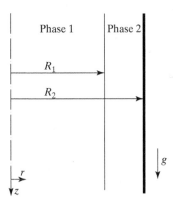

Figure 10.18. Ideal annular flow in a vertical tube.

used their measured void fractions. With the assumption $\Delta P_{F/S} = 0$ the L_{slug}/L_B ratio, rather than L_{slug} and L_B separately, is needed (see Eq. (10.26)) and that can be found from Eq. (10.36). Knowing j_G, j_L, and α, the equation set is then closed.

Earlier, Garimella *et al.* (2002) modeled the intermittent regime pressure drop in minichannels, when the liquid film is laminar and the gas core is turbulent. Rather than treating α as an input, however, they used the following two correlations:

$$\frac{L_{slug}}{L_B + L_{slug}} = [0.7228 + 0.4629 \exp(-0.9604 D_H)] \frac{j_L}{j_L + j_G}, \tag{10.37}$$

$$\frac{(L_{slug} + L_B)}{D_H} = \frac{1}{2.4369} Re_{slug}^{0.5601}, \tag{10.38}$$

where D_H is in millimeters. Garimella *et al.*, furthermore, used the following expression originally proposed by Dukler and Hubbard (1975):

$$\Delta P_{F/S} = \rho_L \left[1 - \frac{R_I^2}{R_0^2} \right] \frac{(U_{slug} - \overline{U}_F)(U_B - \overline{U}_F)}{2}, \tag{10.39}$$

where $\overline{U}_F = U_I/2$ is the mean liquid film velocity.

10.7 Ideal, Laminar Annular Flow

The general solution for ideal, annular flow where both phases are laminar is as follows (Hickox, 1971). Consider the flow field depicted in Fig. 10.18. The momentum equation for both phases will then be

$$\frac{\partial P}{\partial z} - \rho g = \frac{\mu}{r} \frac{\partial}{\partial r} \left(r \frac{\partial u}{\partial r} \right). \tag{10.40}$$

Let us define $K^* = \partial P/\partial z - \rho g$. The general solution for both phases is then

$$u = \frac{K^*}{4\mu} r^2 + B \ln r + E. \tag{10.41}$$

The no-slip boundary condition at the wall and equality of velocities and shear stresses at the interface between the two phases then give

$$u_1^* = a_1^* r^{*2} + 1, \tag{10.42}$$

$$u_2^* = a_2^* r^{*2} + b_2^* \ln r^* + e_2^*, \tag{10.43}$$

where

$$r^* = r/R_I, \, u_1^* = u_1/u_{1,\max}, \, u_2^* = u_2/u_{1,\max}, \, m^* = \mu_2/\mu_1,$$
$$k^* = K_2^*/K_1^*, \, d^* = R_0/R_I, \text{ and}$$

$$K_{1,2}^* = \left[\frac{\partial P}{\partial z} - \rho g \right]_{1,2}, \tag{10.44}$$

$$a_1^* = \frac{m^*}{D^* - m^*}, \tag{10.45}$$

$$a_2^* = \frac{k^*}{D^* - m^*}, \tag{10.46}$$

$$b_2^* = \frac{2(1 - k^*)}{D^* - m^*}, \tag{10.47}$$

$$e_2^* = \frac{D^* - k^*}{D^* - m^*}, \tag{10.48}$$

$$D^* = k^*(1 - d^{*2}) - 2(1 - k^*)\ln d^*, \tag{10.49}$$

and

$$u_{1,\max} = \frac{K_1^* R_I^2}{4\mu_2}(D^* - m^*). \tag{10.50}$$

10.8 The Bubble Train (Taylor Flow) Regime

10.8.1 General Remarks

The flow patterns in capillaries characterized by elongated capsule-like bubbles separated from one another by pure liquid slugs, and from the channel walls by a thin liquid film, are often referred to as the Taylor flow or bubble train regime. This flow pattern includes plug flow (Fig. 10.1(b)) and slug flow (Figs. 10.2(b) and 10.9(b)). The Taylor flow and bubble train regime terms are, however, applied to conditions where capillary effects are predominant (i.e., Eo ≲ 1, see Section 3.7.2), namely to mini- and microchannels under low- and moderate-flow-rate conditions where gas and liquid flows are both laminar, as opposed to the slug flow in conventional systems where inertia dominates. The Taylor flow (bubble train) regime is an important two-phase flow pattern in capillaries and minichannels that covers an extensive portion of their two-phase flow regime maps. Current interest in Taylor flow in capillaries is primarily due to their occurrence in a multitude of applications including chemical microprocess engineering, microfluidics and lab-on-a-chip, catalytic and multiphase reactors, blood flow in capillaries, cooling of multi-chip modules in supercomputers, miniature heat exchangers, fuel cells, and many others. Monolithic catalytic converter is a typical application. In non-boiling applications, for the same liquid flow rate, the Taylor flow can support a fluid–wall heat transfer coefficient that is larger by up to a factor of 2.5. The flow pattern provides a large interfacial area per unit volume, effective mixing within a bubble and in the liquid slug, and small wall–fluid thermal and mass transfer resistances. Monolithic converters made of arrays of parallel small

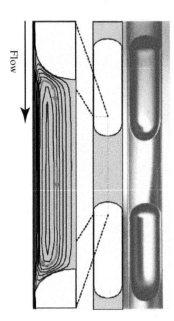

Figure 10.19. Taylor flow, and the CFD-predicted recirculation in the liquid slug. (From Nijhuis *et al.*, 2001.)

channels with diameters of about 1 mm provide high catalytic surface concentrations, highly efficient mass transfer, and low pressure drop (Heiszwolf *et al.*, 2001, Nijhuis *et al.*, 2001). The literature dealing with Taylor flow is extensive, and several reviews are available (Shao *et al.*, 2009; Gupta *et al.*, 2010b; Liu and Wang, 2011; Donaldson *et al.*, 2011; Talimi *et al.*, 2012; Howard and Walsh, 2013). Figure 10.19 displays the Taylor flow field and the flow recirculation in the liquid slug when such recirculation occurs (Nijhuis *et al.*, 2001). The liquid slug recirculation causes effective lateral mixing in the liquid, whereas the gas bubbles effectively block the axial dispersion of a transferred species from one liquid slug to another. The flow regime is thus ideal when liquid samples need to be maintained separate from one another.

The flow pattern in the liquid slug plays an important role with respect to the transport processes. With respect to wall–fluid heat transfer in the absence of boiling, for example, the heat transfer between the liquid slug and wall accounts for the bulk of such heat transfer, although the wall–liquid film heat transfer is not negligible. The contribution of the gas phase to wall–fluid heat transfer is negligibly small, however. The flow pattern in the liquid slug depends on the magnitude of the capillary number, defined as $Ca = \mu_L U_B / \sigma$. Based on an investigation dealing with the behavior of the liquid film that surrounds the bubble, Taylor (1961) speculated that three flow patterns can occur in the liquid slug that separated Taylor bubbles, as depicted in Figure 10.20 when the frame of reference is moving with the bubble. The flow patterns depend on the magnitude of $W = \frac{U_B - J_L}{U_B}$, and because this parameter is proportional to the square root of the capillary number (Thulasidas *et al.*, 1997), it depends on the magnitude of the capillary number. According to Taylor (1961), the central part of the liquid slug moves with the bubble For $W < 0.5$ one should expect two recirculation vortices in the liquid slug. For $W > 0.5$, however, no recirculation zones should be expected. This flow pattern is called the complete bypass. When

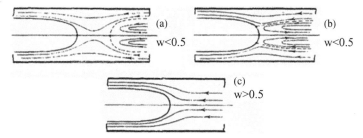

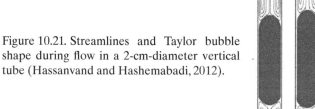

Figure 10.20. Flow patterns proposed by Taylor (1961) in the liquid slug in Taylor flow. (After Thulasidas *et al.*, 1997.)

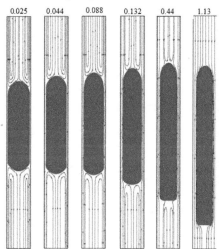

Figure 10.21. Streamlines and Taylor bubble shape during flow in a 2-cm-diameter vertical tube (Hassanvand and Hashemabadi, 2012).

recirculation zones occur, furthermore, two different conditions are possible. As shown in Fig. 10.20(a), there are two stagnation points, one at the bubble tip and one in the liquid slug. In the flow pattern shown in Fig. 10.20(b), however, there is a stagnation ring at the bubble nose. The occurrence of the flow patterns in Figs. 10.20(a) and (b) have been observed in experiments (Thulasidas *et al.*, 1997; Kreutzer *et al.*, 2005a). CFD simulations, furthermore, have confirmed the occurrence of all flow patterns shown (Taha and Cui, 2004; Hassanvand and Hashemabadi, 2012). Figure 10.21 displays the simulation results of Hassanvand and Hashemabadi (2012) for Taylor flow in a 2-cm-diameter tube. The transition to complete bypass flow occurs around $Ca = 0.5$. Thus for $Ca > 0.5$ a complete bypass flow should be expected.

The capillary number also determines the shape of the bubble and the behavior of the recirculation zones. At very low Ca, the bubble is approximately cylindrical with a hemispherical cap. With increasing Ca, the bubble nose becomes elongated and its tip becomes sharper. The centers of the recirculation vortices move closer to the centerline of the flow channel.

Two-phase flow through an array of parallel minichannels can lead to undesirable flow instability and oscillations, however. Research into the hydrodynamics of the bubble train regime is in response to the need for reliable models to develop strategies for avoiding these and other undesirable phenomena. Among the important hydrodynamic parameters are the pressure drop, slug length, and the velocity of Taylor bubbles. The average volumetric gas–liquid interfacial mass transfer coefficient and the average liquid–wall mass transfer coefficient are among the other needed properties of the Taylor flow regime.

When gas at a sufficient volumetric rate is released into the bottom of a vertical or inclined liquid-filled capillary, or the liquid-filled gap between two close parallel plates, the displacement of the liquid by the advancing gas finger or bubble leaves a stagnant thin film behind the advancing front. The thin liquid film separates the wall from the contiguous gas phase and is analogous to the liquid film that separates a Taylor bubble from the channel wall in which it flows. Fairbrother and Stubbs (1935) and Taylor (1961) were among the early experimental investigators of this phenomenon. Bretherton (1961) theoretically analyzed the advancement of gas into a liquid-filled channel and showed that the thickness of the liquid film, δ_{film}, deposited on the wall as the gas progresses through the channel depends on the capillary number defined as $\text{Ca} = \mu_L U_B/\sigma$, with U_B representing the propagation velocity of the gas into the liquid. For the limit of $\text{Ca} \to 0$, Bretherton derived

$$2\delta_F/D = 1.34 \, \text{Ca}^{2/3}. \tag{10.51}$$

For finite Ca values, this expression is valid if D is replaced with $D - 2\delta_F$. The above-mentioned experimental measurements by Taylor (1961) were performed by using various viscous oils. Aussillous and Quéré (2000) derived the following empirical curve fit to the data of Taylor:

$$\frac{2\delta_F}{D} = \frac{1.34 \, \text{Ca}^{2/3}}{1 + 2.5 \times 1.34 \, \text{Ca}^{2/3}}. \tag{10.52}$$

Equations (10.51) and (10.52) apply when inertial effects are negligible. Heil (2001) has shown that, at high values of Ca, inertia significantly influences the flow field and pressure distribution near the bubble tip and slightly modifies the behavior of the liquid film. Inertia thus tends to thicken the liquid film. By comparing the predictions of this correlation with experimental data Leung et al. (2012) noted that the correlation predicted the data very well for $\text{Re}_{\text{TP}}/\text{Ca} = 260$, where $\text{Re}_{\text{TP}} = (j_L + j_G)D/\mu_L$ (obtained with ethylene glycol) but underpredicted their data for $\text{Re}_{\text{TP}}/\text{Ca} \geq 8000$ (obtained with water and a water–ethylene glycol mixture). Also, as explained earlier, with increasing Ca the recirculation zones become smaller, but their period increases. Thus smaller and slower recirculation zones are to be expected when thickening of the liquid film takes place.

The above correlation does not consider inertia and should evidently be expected to be accurate for low-flow conditions. Han and Shikazono (2009) noted that for $\text{Ca} \gtrsim 0.02$ a transition into a visco-inertial flow pattern occurs, whereby the effect of flow velocity cannot be ignored. They developed the following correlation which is valid for Ca < 0.3:

$$
\frac{2\delta_{\mathrm{F}}}{D} =
\begin{cases}
\dfrac{1.34\,\mathrm{Ca}^{2/3}}{1+3.13\,\mathrm{Ca}^{2/3}+0.504\,\mathrm{Ca}^{0.672}\mathrm{Re}^{0.589}-0.352\,\mathrm{We}^{0.629}}, & (\mathrm{Re} < 2000) \quad (10.53)\\[4mm]
\dfrac{53\left(\frac{\mu_{\mathrm{L}}^2}{\rho_{\mathrm{L}}\sigma D}\right)^{2/3}}{1+497\left(\frac{\mu_{\mathrm{L}}^2}{\rho_{\mathrm{L}}\sigma D}\right)^{2/3}+7330\left(\frac{\mu_{\mathrm{L}}^2}{\rho_{\mathrm{L}}\sigma D}\right)^{0.672}-5000\left(\frac{\mu_{\mathrm{L}}^2}{\rho_{\mathrm{L}}\sigma D}\right)^{0.629}} & (\mathrm{Re} > 2000), \quad (10.54)
\end{cases}
$$

where $\mathrm{Re} = \rho_{\mathrm{L}} U_{\mathrm{B}} D/\mu_{\mathrm{L}}$ and $\mathrm{We} = \rho_{\mathrm{L}} U_{\mathrm{B}}^2 D/\sigma$. Equation (10.54), which is unlikely to apply to mini- and microchannels because of its high Reynolds number range, represents a turbulent liquid film and assumes that the film thickness remains independent of the bubble velocity and equal to the film thickness corresponding to $\mathrm{Re} = 2000$. Note that $\mathrm{We} = \mathrm{ReCa}$, and for $\mathrm{Re} = 2000$ we will have $\mathrm{We} = (2000)^2(\frac{\mu_{\mathrm{L}}^2}{\rho_{\mathrm{L}}\sigma D})$.

Taylor flow in noncircular flow passages, in particular square and rectangular channels, has been of interest because noncircular channels are used in electronic and other applications (Kolb and Cerro, 1993a; Fries *et al.*, 2008; Liu and Wang, 2008; Yue *et al.*, 2009). The published studies dealing with Taylor flow in noncircular channels are comparatively few, however (Talimi *et al.*, 2012). Brief discussions of the characteristics of Taylor flow in such channels can be found in Gupta *et al.* (2010a) and Talimi *et al.* (2012), among others. The flow and transport phenomena in these flow passages are affected by channel geometry. In noncircular channels the bubbles are axisymmetric only at high capillary numbers when viscosity dominates surface tension, and become more asymmetric as the capillary number decreases and the role of surface tension increases. For square channels, for example, the bubble cross section can remain symmetric and circular for Ca $\gtrsim$ 0.04, and flattens out against the walls at lower values of the capillary number. For this transition values of capillary number up to 0.1 have also been reported, however (Talimi *et al.*, 2012). Recirculation zones occur behind bubbles, and as the capillary number grows beyond a critical value, the aforementioned complete bypass flow pattern occurs and the recirculation zones completely disappear. For square channels, for example, Taha and Cui (2006b) reported a critical capillary number of 0.4 for the complete bypass flow pattern. For the same hydraulic diameter and bubble size, the liquid film in a channel with sharp corners is thinner than that at a circular channel, because in the former most of the liquid tends to collect in the corners. The liquid film can be extremely thin away from the corners, furthermore, and this may even lead to partial dryout in very small channels (Yu *et al.*, 2007).

Several investigators have performed detailed experimental studies of the Taylor bubble regime in capillaries. Thulasidas *et al.* (1995) performed experiments with circular and rectangular capillaries, using liquids that covered the range Ca = 10^{-3}–1.34. In a follow-up study, Thulasidas *et al.* (1997) used particle image velocimetry to elucidate the details of recirculation patterns in the liquid slugs. The liquid slug supported two counterflowing vortices. In a frame moving with a bubble, the vortices carry liquid ahead of the bubble near the channel centerline, while a counterflow occurs near the wall. The vortices completely disappeared in tests with Ca $\geq$ 0.52, however.

Taylor flow is morphologically simple and involves laminar flow. It is therefore relatively easy to simulate using CFD techniques. Such simulation of course needs

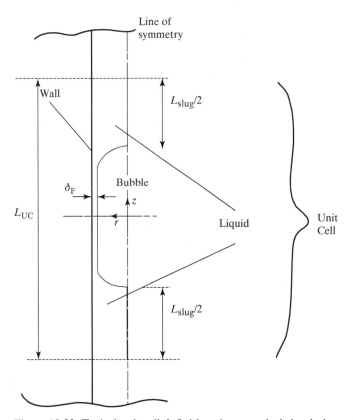

Figure 10.22. Typical unit cell definitions in numerical simulations of Taylor flow.

the resolution of the gas–liquid interphase. A number of methods for the numerical simulation of free surfaces and interfaces are available. A useful review of these methods can be found in Faghri and Zhang (2006). Several authors have performed CFD simulations of Taylor flow. The unit cell for idealized Taylor flow is shown in Fig. 10.22. The CFD simulations that model a single bubble apply the latter unit cell configuration, which is evidently more realistic in comparison with Fig. 10.17, where essentially the same flow regime was idealized for the mechanistic modeling of the pressure drop. Some invesigators have modeled the entire length of a capillary, however (Yang *et al.*, 2002; Qian and Lawal, 2006; Gupta *et al.*, 2009; Shao *et al.*, 2009). Among the pioneers, Edvinsson and Irandoust (1996) used the finite-element-based FIDAP code (Fluid Dynamics International, 1991) and modeled the Taylor bubble as essentially a void. They modeled the interphase by the spine method (Kistler and Scriven, 1984). In this method, two different types of elements, fixed and flexible, are defined, with the latter type used at the vicinity of the interphase. The nodes of the flexible elements can move to accommodate the motion of the interphase, but their motion is restricted to their corresponding prespecified spine lines. Several other investigators subsequently simulated Taylor bubble flow. These CFD simulations provide subtle details about the flow field, in agreement with experimental observations. Edvinsson and Irandoust (1996) predicted the occurrence of undulations near the bubble tail, in qualitative agreement with experimental observations. Giavedoni and Saita (1997, 1999) studied the geometric shape of the liquid meniscus

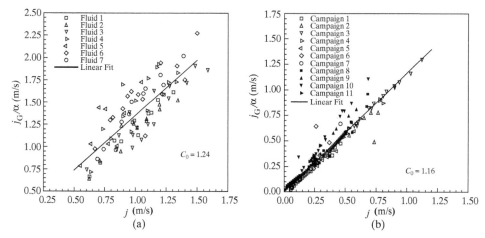

Figure 10.23. The DFM parameters based on (a) data of Laborie *et al.* (1999) for $D = 1$ mm and (b) all data of Liu *et al.* (2005). (From Akbar and Ghiaasiaan, 2006.)

that trails long Taylor bubbles, and Heil (2001) investigated the effect of inertia on the behavior of the liquid film surrounding a Taylor bubble. Kreutzer *et al.* (2005) performed both experimental and numerical investigations. Van Baten and Krishna (2004, 2005) have recently performed extensive CFD simulations of Taylor bubble flow in capillaries, focusing on the liquid-side mass transfer processes at the gas–liquid interphase and at the solid–liquid interphase, respectively. A common feature of all these simulations is that the bubble was essentially treated as a void. Recently, Taha and Cui (2006a, b) and Akbar and Ghiaasiaan (2006) performed simulations based on the volume-of-fluid (VOF) technique, using the CFD code Fluent (Fluent Inc., 2005). In the latter simulations, the gas phase is no longer treated as an inviscid fluid.

Numerous other simulations have been published more recently, including simulations based on the phase field method (a technique for modeling the liquid–gas interphase) (Ganapathy *et al.*, 2013a, b, c). Various aspects including heat transfer (Lakehal *et al.*, 2008; Asadolahi *et al.*, 2011; Leung *et al.*, 2012), mass transfer (Donaldson *et al.*, 2011; Hassanvand and Hashemabadi, 2012), and the effect of surfactants (Hayashi and Tomiyama, 2012) have been studied.

10.8.2 Some Useful Correlations

As mentioned earlier in Section 10.2, in applying the DFM to mini- and microchannels with $D_H \lesssim 1$ mm, one should expect $V_{gj} \approx 0$. Figure 10.23 displays the application of the DFM to some experimental data, where, as expected, $V_{gj} \approx 0$ and C_0 depends on D.

Liu *et al.* (2005) proposed the following correlation for the absolute bubble velocity:

$$\frac{U_B}{j} = \frac{1}{1 - 0.61 \, Ca_L^{0.33}}. \tag{10.55}$$

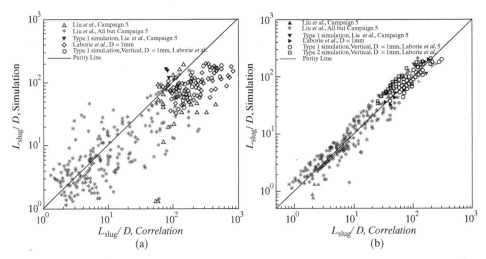

Figure 10.24. Comparison between the simulation results of Akbar and Ghiaasiaan (2006) for the liquid slug length in some reported experiments and (a) the correlation of Liu *et al.* (2005), Eq. (10.65); (b) the correlation proposed by Akbar and Ghiaasiaan (2006), Eq. (10.66).

The average frictional pressure gradient in a simulation for upward flow in a tube can be calculated from (see Fig. 10.24)

$$\left(-\frac{dP}{dz}\right)_{\text{fr}} = \frac{\Delta P_{\text{fr}}}{L_{\text{UC}}}, \tag{10.56}$$

$$\Delta P_{\text{fr}} = P_{z=-\frac{L_{\text{UC}}}{2}} - P_{z=+\frac{L_{\text{UC}}}{2}} - \frac{1}{\pi (D/2)^2} g \left[\rho_{\text{G}} V_{\text{B}} + \rho_{\text{L}} \left(\pi \frac{D^2}{4} L_{\text{UC}} - V_{\text{B}}\right)\right], \tag{10.57}$$

where V_{B} represents the volume of the Taylor bubble.

This pressure drop has two components: the frictional pressure drop in the liquid slugs, and pressure drop in bubbles. For a unit cell, assuming that the liquid slug is cylindrical and supports a fully developed laminar flow, one can write:

$$\Delta P_{\text{slug}} = \frac{64}{\text{Re}} \frac{1}{2} \rho_{\text{L}} j^2 \frac{L_{\text{slug}}}{D} \approx \frac{64}{\text{Re}} \frac{1}{2} \rho_{\text{L}} j^2 \frac{L_{\text{UC}}}{D} (1 - \beta), \tag{10.58}$$

where $\beta = j_{\text{G}}/j$ is the volumetric quality, and is approximately equal to the void fraction α. The pressure drop caused by a single bubble, according to the theoretical analysis of Bretherton (1961) for flow in a circular channel where film thickness is negligibly small compared with channel diameter, is

$$\Delta P_{\text{B}} = 7.16(3\text{Ca})^{\frac{2}{3}} \frac{\sigma}{D}. \tag{10.59}$$

In practice for flow conditions encountered in Taylor flow the bubble pressure drop depends on Re as well, and a better generic function would be (Kreutzer *et al.*, 2005)

$$\Delta P_{\text{B}} \approx \left(\frac{\text{Re}}{\text{Ca}}\right)^b \frac{1}{2} \rho_{\text{L}} j^2, \tag{10.60}$$

where experiments suggest that $b \approx 0.33$ (Kreutzer *et al.*, 2005; Walsh *et al.*, 2009). We can define a two-phase friction factor f_{TP} according to

$$\left(-\frac{dP}{dz}\right)_{fr} = 4f_{TP}\frac{1}{D}\frac{1}{2}\rho_L(1-\beta)j^2, \tag{10.61}$$

where $\beta = j_G/j$ is the volumetric flow quality. Summing Eqs. (10.58) and (10.60), and using the above definition we then get:

$$f_{TP} = \frac{16}{Re}\left[1 + a\frac{D}{L_{slug}}(Re/Ca)^{0.33}\right], \tag{10.62}$$

where $Re = \rho_L jD/\mu_L$ and $Ca = (j_L + j_G)\mu_L/\sigma$. The experimental data of Kreutzer *et al.* lead to $a = 0.17$.

Based on experimental data with air and several liquids covering the parameter range of $1.58 < Re < 1024; 0.0023 < Ca < 0.2$ and $6.4 \times 10^{-6} < Ca/Re < 0.019$, Walsh *et al.* (2009) proposed $a = 0.12$. Kreutzer *et al.* (2005)also performed extensive CFD-based simulations, noting that Re_{TP} affected the film thickness and the bubble shape. The simulations agreed with Eq. (10.61), provided that $a = 0.07$ was used in Eq. (10.62) when calculating f_{TP}. This discrepancy was attributed to the possibility of suppression of the gas–liquid interface motion in the experiments by surfactant contaminants. Walsh *et al.* (2009) have disputed this argument, however.

Akbar and Ghiaasiaan (2006) compared the data of Liu *et al.* (2005) with Kreutzer *et al.*'s correlation, as well as the correlations of Beattie and Whalley (1982) and Friedel (1979) described in Chapter 8. There was considerable data scatter in the very low Re_{TP} range, suggesting that these data should be treated with caution. The correlations of Kreutzer *et al.* and Friedel agreed with the data reasonably well, when the data for $Re_{TP} < 500$ were not considered.

The bubble and slug lengths are important parameters. When the unit cells are identical, the bubble and slug volume are related according to

$$\frac{V_B}{V_{slug} + V_B} = \alpha. \tag{10.63}$$

As an approximation, we can idealize the slugs and bubbles to be cylinders in which case the bubble and unit cell lengths are related according to

$$L_B/L_{slug} \approx \alpha/(1-\alpha). \tag{10.64}$$

The bubble length is a strong function of the process of bubble/slug formation, i.e., the type of mixing zone (Shao *et al.*, 2009, 2011). A variety of mixing zones are used, including T-, Y-, and M-junctions, annular zones, and gas injection into the liquid through a porous segment of the flow channel. Figure 10.25 depicts some common mixing techniques in microfludics. T-junctions are the most widely used. The process of formation, growth, and break-off of bubbles is affected by surface tension and viscous forces. During the injection of gas into the liquid in a T-junction, for $Ca \lesssim 0.01$, which is usually the case in microfluidics, the bubble generation process is a *squeezing regime*, and is dominated by surface tension. Long bubbles are then generated, which are sensitive to surface tension (Garstecki *et al.*, 2006; Shao *et al.*, 2011).

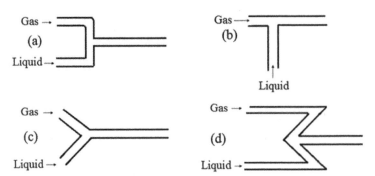

Figure 10.25. Some common mixing techniques in microfluidics: (a,b) T-junctions; (c) Y-junction; (d) M-junction.

Liu *et al.* (2005) proposed the following correlation for liquid slug length, based on their own experimental data obtained with a T-junction mixing arrangement:

$$\frac{j}{\sqrt{L_{\text{slug}}}} = 0.088 \text{Re}_G^{0.72} \text{Re}_L^{0.19}, \tag{10.65}$$

where $\text{Re}_G = (\rho_G j_G D)/\mu_G = GxD/\mu_G$ and $\text{Re}_L = (\rho_L j_L D)/\mu_L = G(1-x)D/\mu_L$, and L_{slug} is in meters. The data of Liu *et al.* and Laborie *et al.* (1999), as well as the simulation results of Akbar and Ghiaasiaan (2006), are plotted against this correlation in Fig 10.24.

Gas (air) was injected into the liquid (water, water/glycerol, ethyl alcohol, and water/ethyl alcohol) through a porous segment of the capillaries in the experiments of Laborie *et al.* Akbar and Ghiaasiaan (2006) developed the following correlation, in which L_{slug} and L_{UC} are in meters and U_{TP} is in meters per second:

$$\frac{j^{-0.33}}{\sqrt{L_{\text{slug}}}} = 142.6\alpha^{0.56}\left(\frac{D}{L_{\text{UC}}}\right)^{0.42} \text{Re}_G^{-0.252}. \tag{10.66}$$

This correlation is compared with data and simulation results in Fig. 10.24. The correlation predicts $j/\sqrt{L_{\text{slug}}}$ within a standard deviation of only 19.5%.

Bercic and Pintar (1997) experimentally measured and empirically correlated the liquid-side volumetric mass transfer coefficient and the volumetric solid–liquid mass transfer coefficient in capillaries with $D = 1.5, 2.5$, and 3.1 mm. Their experiments were on the dissolution of methane in water at 298 K and atmospheric pressure. They developed the following empirical correlation:

$$K_L a_I'' = \rho_L \frac{p_1 j^{p_2}}{[(1-\alpha)L_{\text{UC}}]^{p_3}}, \tag{10.67}$$

where $K_L a_I''$ is the liquid-side volumetric mass transfer coefficient (in kilograms per meter cubed per second), j and L_{UC} are in meters per second and meters, respectively, and

$$p_1 = 0.111 \pm 0.006,$$
$$p_2 = 1.19 \pm 0.02,$$
$$p_3 = 0.57 \pm 0.002.$$

Vandu *et al.* (2005), however, have indicated that some of the data of Bercic and Pintar (1997) were problematic since the liquid film in these experiments may have reached saturation with respect to the transferred chemical species. Based on their own experiments, in which the transfer of oxygen in Taylor flow was measured in circular channels with $D = 1, 2,$ and 3 mm and square channels with 1-, 2-, and 3-mm sides, Vandu *et al.* derived the correlation

$$K_L a_I'' = 4.5 \rho_L \frac{1}{D} \sqrt{\frac{D_{iL} j_G}{L_{UC}}}, \tag{10.68}$$

where $K_L a_I''$ is in kilograms per meter cubed per second, D_{iL} is the mass diffusivity of the transferred species in the liquid in meters squared per second, D, and L_{UC}, and j_G are in meters or meters per second. Vandu *et al.* recommend these correlations when $j_G/L_{UC} > 9\,\mathrm{s}^{-1}$.

Based on CFD simulations where Taylor bubbles were idealized as cylinders with hemispherical caps, Liu and Wang (2011) have proposed the following correlation for gas–liquid mass transfer coefficient in Taylor flow through circular capillaries,

$$\frac{K_L a_I'' D^2}{\rho_L D_{iL}} = 0.12 \left(\frac{L_{film}}{L_{slug}} \frac{D}{\delta_F} \right)^{0.44} \left(\frac{U_B D}{D_{iL}} \right)^{0.54}, \tag{10.69}$$

where the length of the liquid film can be found from

$$L_{film} = L_B - (D - 2\delta_F). \tag{10.70}$$

The liquid film thickness can be found from an appropriate correlation, e.g., the correlation of Aussillous and Quéré (2000) (Eq. (10.52)). The simulations of, Liu and Wang (2011) and the above correlation indicate that $K_L a''$ increases with increasing liquid-film length, liquid phase mass diffusivity, and bubble velocity, and it decreases with increasing liquid-slug length, capillary diameter, and liquid-film thickness.

For the bubble rise velocity in Taylor flow in a vertical upward tube, Rattner and Garimella (2015) have proposed the following,

$$U_B = C_0(j_G + j_L) + \Gamma \sqrt{gD}. \tag{10.71}$$

Note that, when the liquid slugs do not contain entrained bubbles, which is generally the case in micro- and minichannels, the left side can also be represented by

$$U_B = U_G = j_G/\alpha. \tag{10.72}$$

The distribution coefficient C_0 can be found from the following blended correlation:

$$C_0 = f_{LS} C_{0,LS} + (1 - f_{LS}) C_{0,Ca}. \tag{10.73}$$

The parameter $C_{0,Ca}$ represents the distribution coefficient for capillary scale flow passage and is to be found from (Liu *et al.*, 2005)

$$C_{0,Ca} = \frac{1}{1 - 0.61\,\mathrm{Ca}^{0.33}}, \tag{10.74}$$

where the capillary number is defined as $Ca = \mu_L j/\sigma$. The parameter $C_{0,LS}$ represents the distribution coefficient for large-scale flow passages and is correlated as

$$C_{0,LS} = 1.20 + \frac{1.09}{1 + \left(Re_j/805\right)^4}, \tag{10.75}$$

where $Re_j = \rho_L jD/\mu_L$. The blending parameter f_{LS} is to be found from

$$f_{LS} = \left[\frac{1}{1 + 4840 Re_j^{-0.163}}\right]^{0.816/Bd}. \tag{10.76}$$

The coefficient Γ is correlated as

$$\Gamma = \begin{cases} 0.344\left[1 - \exp\left(\frac{-0.01N_{\mu L}}{0.345}\right)\right]\left[1 + \frac{20}{Bd} - \frac{93.7}{Bd^2} - \frac{676.5}{Bd^3} + \frac{2706}{Bd^4}\right]^{1/2} \\ \quad \times \frac{1 - 0.96\exp\left(-0.0165 Bd\right)}{1 - 0.52\exp\left(-0.0165 Bd\right)} \qquad \text{for } Bd > 4.55 \\ 0 \qquad\qquad\qquad\qquad\qquad\qquad \text{for } Bd < 4.55 \end{cases}$$

where $N_{\mu L}$ is the viscosity number defined in Eq. (7.19). The condition $\Gamma = 0$, which applies when $Bd < 4.55$, is based on the observation that with $Bd < 4.55$ no bubble rise occurs without a net flow.

The above correlation of Rattner and Garimella (2015) has a very wide range of validity, and applies when $N_{\mu L} > 250$.

EXAMPLE 10.3. Air and water at room temperature and 4 bars flow through a 1.0-mm-inner-diameter tube that is 120 mm long. For liquid and gas superficial velocities $j_L = 0.5$ m/s and $j_G = 0.38$ m/s, calculate the frictional pressure gradient using the correlations of Beattie and Whalley (1982) and Kreutzer *et al.* (2005).

SOLUTION. The relevant properties are as follows: $\rho_L = 997.2$ kg/m³, $\rho_G = 4.68$ kg/m³, $\mu_L = 8.93 \times 10^{-4}$ kg/m·s, $\mu_G = 1.85 \times 10^{-5}$ kg/m·s, and $\sigma = 0.07$ N/m. First, consider the method of Beattie and Whalley (1982). The calculations proceed as follows (see Eqs. (8.21)–(8.23)):

$$G = \rho_G j_G + \rho_L j_L = 500.4 \text{ kg/m}^2\text{·s},$$

$$x = \rho_G j_G/(\rho_G j_G + \rho_L j_L) = 3.55 \times 10^{-3},$$

$$\frac{\rho_G \alpha_h}{\rho_G \alpha_h + \rho_L(1 - \alpha_h)} = \frac{x}{1 - x} \Rightarrow \alpha_h = 0.433,$$

$$\mu_{TP} = \alpha_h \mu_G + \mu_L(1 - \alpha_h)(1 + 2.5\alpha_h) = 1.06 \times 10^{-3} \text{ kg/m·s},$$

$$Re_{TP} = GD/\mu_{TP} = 470.7.$$

We note the small magnitude of Re_{TP}. Bao *et al.* (1994) performed pressure-drop experiments in minichannels and noted that the correlation of Beattie and Whalley performed well in predicting their data, provided that for $Re_{TP} < 1000$ the friction factor is obtained from a laminar flow correlation. Accordingly, we can write

$$f_{TP} = 16/\mathrm{Re}_{TP} = 0.034,$$

$$\left(-\frac{dP}{dz}\right)_{fr} = 4f_{TP}\left(\frac{1}{D}\right)\frac{G^2}{2\rho_h} = 2.99 \times 10^4 \text{ Pa/m}.$$

We will now apply the correlation of Kreutzer *et al.* (2005) to get

$$\mathrm{Re}_G = \rho_G j_G D/\mu_G = 96.2,$$
$$\mathrm{Re}_L = \rho_L j_L D/\mu_L = 558.1,$$
$$\beta = j_G/(j_G + j_L) = 0.432,$$
$$\mathrm{Ca} = \mu_L j/\sigma = 0.011.$$

We need the average slug length L_{slug}, which can be found from the correlation of Liu *et al.* (2005), Eq. (10.65):

$$L_{slug} = j^2/(0.088\mathrm{Re}_G^{0.72}\mathrm{Re}_L^{0.19})^2 = 0.0126 \text{ m}.$$

We can now continue by writing

$$\mathrm{Re} = \rho_L j D/\mu_L = 982.2,$$

$$f_{TP} = \frac{16}{\mathrm{Re}}\left[1 + 0.17\frac{D}{L_{slug}}\left(\frac{\mathrm{Re}}{\mathrm{Ca}}\right)^{0.33}\right] = 0.0257,$$

$$\left(-\frac{dP}{dz}\right)_{fr} = 4f_{TP}\frac{1}{D}\frac{1}{2}\rho_L(1 - \beta)j^2 = 2.257 \times 10^4 \text{ Pa/m}.$$

10.9 Pressure Drop Caused by Flow-Area Changes

Experimental data dealing with minor pressure drops in mini- and microchannel systems are scarce, and the common practice is to assume that conventional pressure-drop models and correlations are applicable. This assumption may not be justified, however, given the considerable differences between conventional and mini- and microscale systems with respect to velocity slip between gas and liquid phases. As noted earlier, uncertainties related to inlet and exit pressure drops may contribute to the data scatter and inconsistencies in mini- and microchannel thermal-hydraulics data.

The studies by Abdelall *et al.* (2005) and Chalfi and Ghiaasiaan (2008) are briefly reviewed here. In these investigations air–water pressure drops caused by abrupt area expansion and contraction were measured by using tubes with 0.84- and 1.6-mm diameters. Recall from Chapter 8 that when both phases are incompressible and there is no phase change, quality remains constant through the flow disturbance. Furthermore, by assuming that the void fraction also remains unchanged, the total pressure change across a sudden expansion can be found from Eqs. (8.74)–(8.76), and the total pressure drop across a sudden contraction can be found from Eqs. (8.91)–(8.93), provided that a void-quality (or slip ratio) relation is also used. The data of Abdelall *et al.* covered the range $\mathrm{Re}_{L0} \approx 1750$–3550, and the data of chalfi covered the range $\mathrm{Re}_{L0} \approx 430$–570. The homogeneous flow assumption, which has sometimes been used

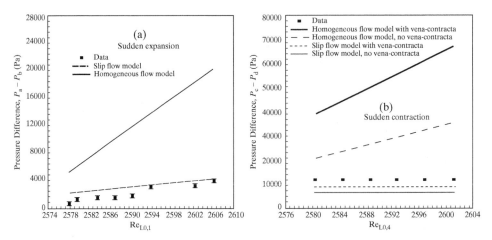

Figure 10.26. Two-phase pressure change caused by flow area expansion and contraction. (From Abdelall *et al.*, 2005.)

for the estimation of inlet and exit channel pressure drops in mini- and microchannel experimental investigations, was found to be inadequate and led to significant over-prediction of pressure changes. Figures 10.26(a) and (b) show typical results from Abdelall *et al.*, where the notation in the figures refer to Fig. 8.3. A slip flow model based on the slip ratio correlation of Zivi (1964) (Eq. (6.52)), however, agreed with the data well for both sudden expansion and contraction. Moreover, the assumption that no vena-contracta takes place in sudden contraction (i.e., $C_C = 1$) led to little change in the results.

For the flow area expansion data of Chalfi (2007) the slip flow model (i.e., Eqs. (8.74)–(8.76)) with Zivi's slip ratio expression slightly underpredicted the experimentally measured pressure changes, but good agreement between data and model was achieved by using

$$S_r = 0.7(\rho_L/\rho_G)^{1/3},\tag{10.71}$$

or

$$\alpha = 0.5\beta.\tag{10.72}$$

For their flow area contraction data, the slip flow model with Zivi's slip ratio expression predicted the data well provided that no vena–contracta was assumed. Thus, Eq. (8.88), with $C_C = 1$, along with Zivi's slip ratio for finding void fraction to be used for the calculation of ρ' and ρ''', agreed with the data well. (Note that Eq. (8.89) cannot be applied with $C_C = 1$, because the equation is based on the assumption that all irreversibilities occur due to vena–contracta.)

PROBLEMS

10.1 Using the flow regime transiton models of Akbar *et al.* (2003) construct the flow regime map in Mandhane coordinates (j_G, j_L) for saturated R-22 at 40 °C flowing in a circular channel with 1 mm inner diameter.

10.2 An air–water mixture at room temperature flows in a horizontal tube with $D = 1.1$ mm inner diameter. The tube is 120 mm long. The tube is connected to the atmosphere at its exit. For simplicity, the fluid is assumed to be isothermal everywhere. Consider the cases summarized in Table P10.2.

Table P10.2.

case number	j_L (m/s)	j_G (m/s)
1	0.104	0.1
2	0.104	1.0
3	0.104	10.0
4	0.841	0.1
5	0.841	1.0
6	0.841	10.
7	2.206	0.1
8	2.206	1.0
9	2.206	10.

(a) Determine the flow regimes based on the experimental flow regime map of Triplett *et al.* (1999a). compare the results with the predictions of the flow regime map of Mandhane *et al.* (1974) discussed in Chapter 4.

(b) Calculate and compare the mean void fractions based on the homogeneous flow model; using the correlation proposed by Ali, Kawaji, and co-workers (Eq. (10.9)); and using the drift flux model with parameters proposed by Zhang *et al.* (2010):

$$C_0 = 1.2 + 0.380 \exp\left(-1.39/D_H^*\right) ; V_{gj} \approx 0; D_H^* = \frac{D_H}{\sqrt{\sigma/[g\Delta\rho]}}.$$

10.3 Waelchli and von Rohr (2006) performed experiments with nitrogen as the gas phase, and de-ionized water, ethanol, and aqueous glycerol solutions as the liquid phase, all at room temperature. Their experimental test sections were horizontal, rectangular cross-section silicon microchannels with hydraulic diameters between 187.5 and 2181 μm. They divided their entire flow regime map into four major regimes: stratified (smooth and wavy), intermittent (slug and plug), annular (annular and annular-wavy), and bubbly. The stratified regime did not occur in their experiments, and for all their data obtained with channel hydraulic diameters of 187.5 and 210 μm they could develop the flow regime map shown in Figure P10.3. Note the dimensionless coordinates of the flow regime map, which include the effect of channel wall surface roughness.

For case numbers 3, 6, and 7 of Problem 10.2, find the flow regimes according to the flow regime map of Triplett *et al.*, and the aforementioned flow regime map of Waelchli and von Rohr (2006). Assume that the tube is made from free-machining copper (tellurium copper, C 14500), using the EDM machining technique, and has a mean surface roughness of 2 μm.

10.4 For the system described in Problem 10.2, and two cases of your choice in Table P10.2, do the following.

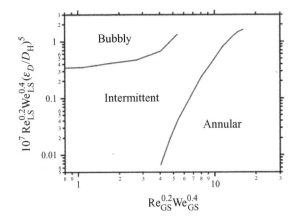

Figure P10.3. Flow regime map of Waelchli and von Rohr (2006).

(a) Find the pressure just upstream from the channel exit, by estimating the exit pressure drop based on common large-channel methods.

(b) Calculate the frictional pressure drop in the channel based on the frictional pressure gradient in the middle of the tube, using the homogeneous flow model and the DFM method of Zhang *et al.* (2010) for void fraction calculation:

$$C_0 = 1.2 + 0.380 \exp\left(-1.39/D_{\mathrm{H}}^*\right)\,;\, V_{gj} \approx 0.;\, D_{\mathrm{H}}^* = \frac{D_{\mathrm{H}}}{\sqrt{\sigma/[g\Delta\rho]}}.$$

(c) Estimate the pressure drop caused by acceleration resulting from the expansion of air caused by depressurization in the tube by assuming homogeneous flow in the channel.

(d) Repeat part (c), this time assuming that the correlation of Ali *et al.* (1993), Eq. (10.9), applies.

10.5 The calculations in Problem 10.4 were approximate. A more accurate calculation can be done by solving the relevant differential one-dimensional mixture momentum conservation equation using the methodology described in Section 5.10. Formulate the necessary system of differential equations, and explain the boundary conditions and the numerical solution procedure.

10.6 Refrigerant R-134a at a temperature of 233 K flows in a horizontal circular tube that has 0.48 mm inner diameter. The tube has a uniformly heated segment that is 6 cm long, followed by a 2-cm-long adiabatic segment.

(a) For a mass flux of 300 kg/m²·s, calculate the wall heat flux that would lead to a flow quality of 0.12 and 0.28 at the end of the heated segment.

(b) Find the frictional and total pressure drops in the flow channel using the methods of Muller-Steinhagen and Heck (1986) and Li and Wu (2010a).

(c) Repeat part (b), this time for a mass flux of 1100 kg/m²·s, and compare the predictions of Muller-Steinhagen and Heck (1986) and Kim and Mudawar (2012c).

10.7 Repeat Problems 10.4 and 10.5, this time assuming that the channel is a thin rectangular horizontal channel with $\delta = 0.75$ mm and with a much longer horizontal side. Use models and correlations that are appropriate to rectangular minichannels everywhere.

10.8 Prove the solution of Hickox (1971), Eqs. (10.40)–(10.50).

10.9 A 3-cm-long array of five identical parallel tubes, each $D = 0.8$ mm in diameter, carries a mixture of air and water at room temperature. The tubes are fed at their inlet from a large inlet plenum and are connected to another large plenum held at atmospheric pressure at their exit.

(a) For superficial velocities of $j_L = j_G = 1.5$ m/s at the exit, calculate the total liquid and gas mass flow rates, the average void fraction in the channels, and the pressure in the inlet plenum.

(b) suppose the tubes are to be replaced with an array of identical microtubes with $D = 50$ μm. It is required that, for the same inlet and exit plenum pressures, the microtubes carry the same total flow rates of water and air. How many microtubes are needed, and what will be the average void fraction in them?

For simplicity, assume that there is no nonuniformity with respect to flow distribution among the parallel tubes.

10.10 Table P10.10 contains simulation results for Taylor flow of air and water, under atmospheric pressure and room temperature (20 °C), in small tubes.

Table P10.10.

D (m)	j_G (m/s)	j_L (m/s)	$\left(\dfrac{dP}{dz}\right)_{fr}$ (k Pa/m)	L_{slug} (m)	L_{UC} (m)
1.0E-04	0.0733	0.1	121.51	3.422E-04	1.377E-03
1.0E-04	0.3758	0.5	625.34	3.056E-04	1.337E-03
1.0E-04	0.1857	0.25	317.76	3.229E-04	1.357E-03
1.0E-04	0.0749	0.1	156.83	1.421E-04	6.871E-04
1.0E-04	0.3836	0.5	203.73	1.165E-04	6.663E-04
1.0E-04	0.0896	0.1	66.34	1.443E-04	1.681E-03
1.0E-04	0.4507	0.5	106.01	1.129E-04	1.661E-03
2.5E-05	0.0744	0.1	1920.58	7.950E-05	3.352E-04
2.5E-05	0.3846	0.5	2853.08	7.014E-05	3.234E-04
2.5E-05	0.1900	0.25	2638.61	7.405E-05	3.252E-04
2.5E-05	0.0784	0.1	3562.33	3.248E-05	1.700E-04
2.5E-05	0.3939	0.5	4297.09	2.832E-05	1.654E-04
2.5E-05	0.0903	0.1	837.82	3.250E-05	4.168E-04
2.5E-05	0.4206	0.5	1796.31	2.679E-05	4.132E-04

(a) Compare the tabulated frictional pressure gradients with the predictions of Kreutzer *et al.* (2005) (Eqs. (10.61) and (10.62)), and Friedel (1979).

(b) Compare the tabulated simulation data with Eqs. (10.65) and (10.66).

(c) Calculate the thickness of the liquid film surrounding the Taylor bubbles, using Eq. (10.52).

10.11 For annular flow in common pipe flow conditions when the gas core is turbulent, Kocamustafaogullari *et al.* (1994) have derived the following semi-empirical correlations for the Sauter mean and maximum entrained droplet diameters, respectively:

$$\frac{d_{\mathrm{Sm}}}{D_{\mathrm{H}}} = 0.65K^*,$$

$$\frac{d_{\mathrm{max}}}{D_{\mathrm{H}}} = 2.609K^*,$$

where

$$K^* = C_{\mathrm{W}}^{-4/15}\mathrm{We}_{\mathrm{m}}^{-3/5}\left(\mathrm{Re}_{\mathrm{G}}^4/\mathrm{Re}_{\mathrm{L}}\right)^{1/15}\left(\frac{\rho_{\mathrm{G}}\mu_{\mathrm{G}}}{\rho_{\mathrm{L}}\mu_{\mathrm{L}}}\right)^{4/15},$$

$$C_{\mathrm{W}} = \begin{cases} \dfrac{1}{35.34}N_{\mu}^{4/5} & \text{when } N_{\mu} \leq 1/15, \\ 0.25 & \text{when } N_{\mu} > 1/15, \end{cases}$$

$$\mathrm{We}_{\mathrm{m}} = \frac{\rho_{\mathrm{G}}D_{\mathrm{H}}j_{\mathrm{G}}^2}{\sigma},$$

and

$$N_{\mu} = \frac{\mu_{\mathrm{L}}}{\left[\rho_{\mathrm{L}}\sigma\sqrt{\frac{\sigma}{g\Delta\rho}}\right]^{1/2}}.$$

The following correlations for the volume median diameter d_{vm} and maximum diameter of entrained droplets were also proposed earlier by Kataoka *et al.* (1983):

$$d_{\mathrm{vm}} = 0.01\frac{\sigma}{\rho_{\mathrm{G}}j_{\mathrm{G}}^2}\mathrm{Re}_{\mathrm{G}}^{2/3}\left(\frac{\rho_{\mathrm{G}}}{\rho_{\mathrm{L}}}\right)^{-1/3}\left(\frac{\mu_{\mathrm{G}}}{\mu_{\mathrm{L}}}\right)^{2/3},$$

$$d_{\mathrm{max}} = 0.031\frac{\sigma}{\rho_{\mathrm{G}}j_{\mathrm{G}}^2}\mathrm{Re}_{\mathrm{G}}^{2/3}\left(\frac{\rho_{\mathrm{G}}}{\rho_{\mathrm{L}}}\right)^{-1/3}\left(\frac{\mu_{\mathrm{G}}}{\mu_{\mathrm{L}}}\right)^{2/3}.$$

Apply these correlations for the cases of saturated mixtures of R-22 at 30 °C and R-123 at 70 °C, flowing in 1- and 1.5-mm-diameter tubes, where $j_{\mathrm{f}} = 4$ m/s and $j_{\mathrm{g}} = 11$ m/s. Based on the results, discuss the relevance of the correlations to mini- and microchannels.

10.12 In an experiment, mixtures of ethanol and air at 21 °C and atmospheric temperature are used in vertical capillaries to study Taylor flow phenomena.

Consider the flow in a capillary with a T-junction type inlet conditions. For the data points listed in the table below, do the following.

(a) Calculate the average bubble and slug lengths, the average void fraction, the liquid film thickness that separates the Taylor bubble from the wall, and the frictional pressure gradient based on the correlation of Kreutzer *et al.* (2005) (Eq. (10.62)).

(b) Compare the predicted frictional pressure gradient with the predictions of the correlation of Kim and Mudawar (2012c) (see Table 8.3).

(c) Using an appropriate flow regime map of your choice that is relevant to minichannels, examine whether the Taylor flow regime is consistent with the flow regime map:

D (mm)	j_L (m/s)	j_G (m/s)
0.715	0.185	0.1
0.715	0.145	0.14
0.715	0.4	0.055
0.715	0.3	0.155
0.487	0.185	0.1
0.487	0.145	0.14
0.487	0.4	0.055
0.487	0.3	0.155

10.13 In an experiment, mixtures of water and air at room temperature and atmospheric pressure are used in vertical capillaries to study Taylor flow phenomena. For the data points listed in the following table, assuming that a T-junction is used for mixing the gas and liquid at inlet, using the liquid superficial velocities, do the following.

(a) Using correlations of your choice, estimate the range of gas superficial velocity that leads to $1.8 < \frac{L_{slug}}{D} < 2.2$.

(b) Using a flow regime map of your choice which is relevant to minichannels, examine whether the aforementioned parameter range indeed leads to the Taylor flow regime.

(c) For the cases where $\frac{L_{slug}}{D} = 2.0$, find the frictional pressure gradient using the correlation of Kreutzer et al. (2005), and the correlation of Kim and Mudawar (2012c).

D (mm)	j_L (m/s)
1.0	0.49
1.0	0.74
1.0	0.88

10.14 Consider the flow of refrigerant R-134a at a mass flux of 1500 kg/m²s through a tube system, where there is a flow area expansion that leads from a 0.7 mm inner diameter tube to a tube that is of 1.9 mm inner diameter. The local pressure upstream from the flow area expansion is 1 MPa. For local qualities of 0.05 and 0.10, do the following.

(a) Estimate the pressure change associated with the flow are expansion assuming that the flow is homogeneous.

(b) Estimate the pressure change associated with the flow area expansion using the recommendations of Chalfi and Ghiaasiaan (2008).

(c) Examine the validity of the assumption of constant quality across the flow area expansion.

10.15 Consider the flow of ethanol at a mass flux of 450 kg/m²·s through a tube system, where there is a flow area contraction that leads from a 1.6-mm-inner-diameter tube to a tube that is of 0.65 mm inner diameter. The local pressure upstream from the flow area expansion is 3 bars. For local qualities of 0.04 and 0.09, do the following.

(a) Estimate the pressure change associated with the flow area contraction assuming that the flow is homogeneous.

(b) Estimate the pressure change associated with the flow area expansion using the recommendations of Chalfi (2007).

(c) Examine the validity of the assumption of constant quality across the flow area expansion.

PART TWO

BOILING AND CONDENSATION

11 Pool Boiling

11.1 The Pool Boiling Curve

Boiling is a process in which heat transfer causes evaporation. Pool boiling refers to boiling processes without an imposed forced flow, where fluid flow is caused by natural convective phenomena only.

A discussion of pool boiling should start with the *pool boiling curve and Nukiyama's experiment* (1934). Consider an experiment where an electrically heated wire is submerged in a saturated, quiescent liquid pool, where the wire temperature is measured, and the heat flux at the wire surface can be calculated from the supplied electric power. When heat flux, q_w'', is plotted as a function of the wall superheat, $\Delta T_w = T_w - T_{sat}$, the "boiling curve" displayed in Fig. 11.1 results.

The following important observations can be made about the boiling curve.

(a) The curve suggests at least three different boiling regimes, represented by the three segments of the curve. The three major regimes, depicted in Fig. 11.2 along with additional subregions, are nucleate boiling, transition boiling, and film boiling.
(b) The process paths for increasing and decreasing electric power (heat flux) are different. For increasing heat flux, the process path would follow the rightward-oriented arrows, and for decreasing power it would follow the leftward-oriented arrows, and the dashed part of the boiling curve is completely bypassed. Based on his experimental data, Nukiyama correctly conjectured that the dashed part of the curve (transition boiling) must be producible when $T_w - T_{sat}$, rather than q_w'', is controlled.

Figure 11.2 displays the boiling curve, with schematics of the flow field in the vicinity of the heated surface. For points situated on the left of point A, heat transfer is by natural convection and no boiling takes place. Boiling starts at the *onset of nucleate boiling point* (ONB), point A. Initiation of boiling is usually accompanied by a wall superheat excursion. This excursion is caused by the delay in the first-time nucleation of bubbles on wall crevices. This temperature excursion depends on the surface wettability by the liquid and is significant for wetting dielectric fluids. For the refrigerant R-113 on a platinum thin-film heater, for example, You *et al.* (1990) could measure wall superheat excursions as large as 73 °C. In the *partial boiling region* (AB), natural convection and boiling both contribute to heat transfer. The increasing slope, as $T_w - T_{sat}$ is increased, is due to the increasing contribution of boiling.

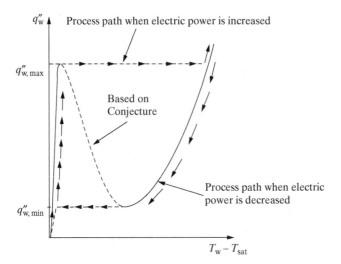

Figure 11.1. The pool boiling curve.

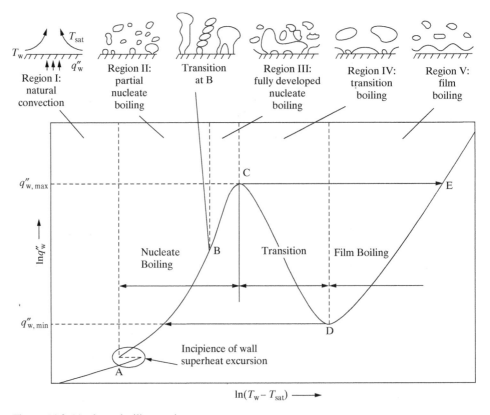

Figure 11.2. Nucleate boiling regimes.

In the *fully developed boiling region*, the contribution of natural convection heat transfer is negligible. The *critical heat flux point* (point C) represents the end of uninhibited macroscopic contact between liquid and the heated surface. When $q''_w > q''_{CHF}$ is imposed, hydrodynamic processes no longer allow for uninhibited contact between

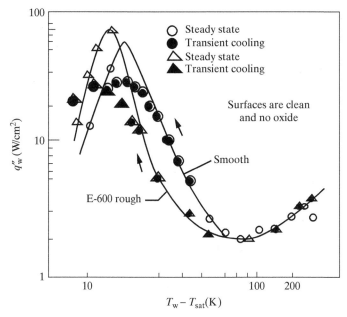

Figure 11.3. The effect of surface roughness on nucleate and transition boiling on a vertical copper surface (Bui and Dhir, 1985; Dhir, 1991).

solid and liquid. Depending on the magnitude of q_w'', partial or complete drying of the surface will occur.

In the *transition boiling regime*, the heated surface is intermittently dry or in macroscopic contact with liquid. Within the transition boiling region, the dry fraction of the heated surface increases as $T_w - T_{sat}$ is increased.

At and beyond the *minimum film boiling point, MFB* (point D), direct macroscopic contact between liquid and solid surface does not occur at all. The heated surface instead is covered by a vapor film.

Some important parametric effects on the pool boiling curve are now described. Increased surface wettability (reduction in contact angle) shifts the nucleate boiling line toward the right. Thus, with increased surface wettability, decreasing nucleate boiling heat transfer coefficients (for the same $T_w - T_{sat}$) are obtained. Increased surface wettability also increases the maximum heat flux (Liaw and Dhir, 1986). Increased surface roughness tends to move the nucleate and transition lines to the left, implying improvement in the nucleate boiling heat transfer characteristics. Figure 11.3 depicts the data of Bui and Dhir (1985). Surface contamination (deposition and oxidation) and improved surface wettability both have an effect similar to surface roughness. Liquid pool subcooling improves heat transfer in all boiling regimes, as shown in Fig. 11.4, except for the fully developed nucleate boiling region, where its effect is small. Also, as noted in Fig. 11.5, the surface orientation with respect to gravity has a strong effect on partial boiling and film boiling and little effect on fully developed nucleate boiling.

Nucleate boiling is the preferred mode of heat transfer for many thermal cooling systems, since it can sustain large heat fluxes with low heated surface temperatures. Important characteristics and parametric trends in nucleate pool boiling are as follows.

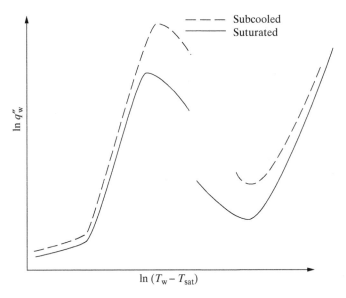

Figure 11.4. The effect of liquid pool subcooling on the boiling curve.

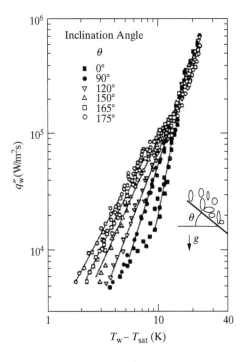

Figure 11.5. The effect of surface orientation with respect to gravity on nucleate boiling (after Nishikawa *et al.*, 1984).

In the fully developed nucleate boiling zone (the slugs and jets zone), as mentioned earlier, the heat transfer coefficient is insensitive to surface orientation. In the partial boiling zone, however, heat transfer is affected by orientation. Two effects contribute to the improvement of heat transfer on inclined and/or downward-facing surfaces. First, there is the effect of bubble rolling on inclined surfaces. Bubbles that are released from horizontal and upward-facing surfaces move primarily in the vertical direction, and the extent of the disruption of the thermal boundary layer caused by them is rather limited. The bubbles that are released from an inclined or downward-facing surface, however, roll on the surface for some distance before leaving the surface, thereby disrupting the thermal boundary layer over a rather significant part of the surface. Second, the effect of the thermal boundary layer on bubble nucleation should be considered. The natural convection boundary layer is thicker in downward-facing heated surfaces, thus promoting nucleation.

11.2 Heterogeneous Bubble Nucleation and Ebullition

Nucleate boiling under low-heat-flux conditions (i.e., in partial boiling) is characterized by heterogeneous bubble nucleation on the defects of the heated surfaces and the ensuing bubble ebullition phenomena. At higher-heat-flux conditions (corresponding to fully developed nucleate boiling) vapor jets and mushrooms predominate. These processes have been investigated for decades. We have a reasonable qualitative understanding of these processes. Theoretical models are capable of correct prediction of experimental data only for well-controlled experiments, however, owing to the complexity of the processes involved and the multitude of sources of uncertainty. In this section we provide a brief review of the theory of heterogeneous bubble nucleation and ebullition and describe the basic mechanisms that are at work during nucleate boiling. A detailed review of classical theory can be found in the monograph by Hsu and Graham (1986). Reviews of more recent research can be found in Dhir (1991, 1998) and Shoji (2004).

11.2.1 Heterogeneous Bubble Nucleation and Active Nucleation Sites

Solid surfaces are typically characterized by microscopic cavities and crevices. The number density, size range, and geometric shapes of these crevices depend on the surface material, finishing, and level of oxidation or contamination. Minute pockets of air are usually trapped in the crevices when the surface is submerged in a liquid, resulting in a preexisting gas–liquid interfacial area that can act as an embryo for bubble growth. With these preexisting interfacial areas, macroscopic liquid–vapor phase-change no longer needs homogeneous nucleation. Consequently, unlike in homogeneous boiling, where very large liquid superheats are needed for the initiation of the phase-change process, in heterogeneous boiling bubble nucleation needs only a relatively small wall superheat ($T_w - T_{sat}$ values of a few to several degrees Celsius for water). Nucleation on wall crevices in fact can take place within a thin, moderately superheated liquid layer adjacent to a heated surface even when the liquid bulk is subcooled. Nucleation on a crevice takes place when a microbubble residing inside or over the crevice can grow from evaporation. This can be understood by considering a conical crevice (see Fig. 11.6), bearing in mind that most cross-sectional profiles of surface crevices in metals are somewhat conical (Hsu and Graham, 1986). The embryonic bubble starts from a radius R_1 and grows until it extends outside the cavity. The largest curvature (corresponding to the smallest radius of curvature) occurs when the bubble forms a hemisphere with $R_B = R_C$, with R_C representing the radius of the cavity mouth. The condition $R_B = R_C$ represents the largest excess pressure that is needed for the bubble to remain at equilibrium.

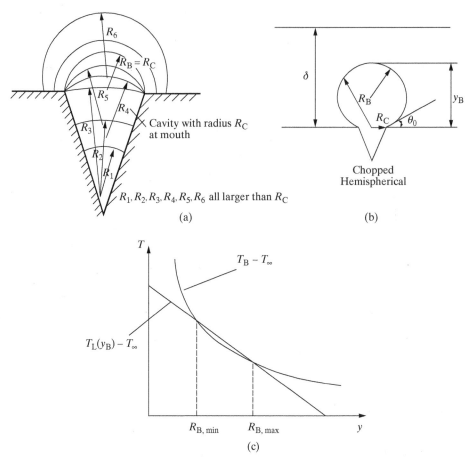

Figure 11.6. Bubble activation: (a) change of radius of curvature of bubble; (b) a chopped spherical bubble; and (c) criterion for the activation of the ebullition site.

For a bubble surrounded by a uniform-temperature liquid, therefore,

$$P_B - P_L = \frac{2\sigma}{R_C} \tag{11.1}$$

and

$$T_L - T_{sat} \approx \frac{T_{sat}}{\rho_v h_{fg}}(P_B - P_L) = \frac{2T_{sat}\sigma}{\rho_v h_{fg} R_C}, \tag{11.2}$$

where T_{sat} represents the saturation temperature at P_L and the Clausius–Clapeyron relation has been used.

Equation (11.2) is valid for a uniformly heated liquid and is applicable when liquid superheat occurs as a result of depressurization, for example. In practice, the liquid temperature adjacent to a heated surface can be nonuniform, and bubble formation on a heated wall requires that the wall superheat $T_w - T_{sat}$ be larger than what Eq. (11.2) predicts (Griffith and Wallis, 1960).

The bubble nucleation criterion of Hsu (1962) was developed based on the earlier experimental observations of Hsu and Graham (1961). According to Hsu (1962), for a bubble to grow, a bubble embryo must be surrounded by a liquid layer that is everywhere warmer than the bubble interior. Furthermore, experimental observations have indicated that growing bubbles are not hemispherical, but elongated. For a chopped-sphere bubble residing on a conical crevice (see Fig. 11.6(b)),

$$y_B = C_1 R_C, \tag{11.3}$$

$$R_B = C_2 R_C, \tag{11.4}$$

$$T_B = T_{sat} + \frac{2\sigma T_{sat}}{C_2 R_C \rho_v h_{fg}}, \tag{11.5}$$

where

$$C_1 = (1 + \cos\theta)/\sin\theta, \tag{11.6a}$$

$$C_2 = 1/\sin\theta. \tag{11.6b}$$

Hsu's aforementioned criterion requires that $T_L|_{y=y_B} \geq T_B$. The temperature profile in the liquid is thus needed. If it is assumed that a thermal boundary layer with thickness δ and with a linear temperature profile resides on the heated wall, then

$$\frac{T_L(y) - T_\infty}{T_w - T_\infty} = 1 - y/\delta. \tag{11.7}$$

Equations (11.3) and (11.7), and the requirement that $T_L(y_B) = T_B$, result in

$$\frac{T_B - T_\infty}{T_w - T_\infty} = 1 - C_1 \frac{R_C}{\delta}. \tag{11.8}$$

When the values of $T_B - T_\infty$ from Eq. (11.5) and the values of $T_L(y) - T_\infty$ predicted by Eq. (11.7) are plotted on the same graph for a surface that is supporting nucleate boiling, Fig. 11.6(c) is obtained. The two points of intersection represent the critical crevice sizes that lead to $T_L(y_B) = T_B$. Using Eqs. (11.5) and (11.7) one can show that

$$R_{C,min}, R_{C,max} = \frac{\delta(T_w - T_{sat})}{2C_1(T_w - T_\infty)} \left[1 \mp \sqrt{1 - \frac{8C_1}{C_2} \frac{(T_w - T_\infty)T_{sat}\sigma}{(T_w - T_{sat})^2 \delta \rho_v h_{fg}}} \right]. \tag{11.9}$$

For given T_w and T_∞, or equivalently for given q_w'' and T_∞ [since $q_w'' \approx H(T_w - T_\infty)$, with H representing the liquid single-phase convection heat transfer coefficient], only crevices in the range $R_{C,min} \leq R_C \leq R_{C,max}$ become activated. To apply Hsu's criterion, the superheated liquid film thickness δ is needed, and that can be estimated from $\delta = k_L/H$, with H representing the aforementioned convection heat transfer coefficient.

The expressions for C_1 and C_2 quoted here are based on a sharp cavity mouth. When the cavity mouth has a slope of θ_m, as shown in Fig. 11.7, the coefficients C_1 and C_2 can be modified simply by replacing θ with $\theta + \theta_m$ everywhere.

Several improvements have been introduced into Hsu's criterion. Howell and Siegel (1967) noted that the criterion was conservative (i.e., requires $T_w - T_{sat}$ values larger than measured values), and therefore they argued that the requirement of the bubble being surrounded everywhere by liquid warmer than the bubble should be replaced with the requirement that the net heat exchange rate between the bubble and surrounding liquid should be in favor of bubble growth. Based on this argument,

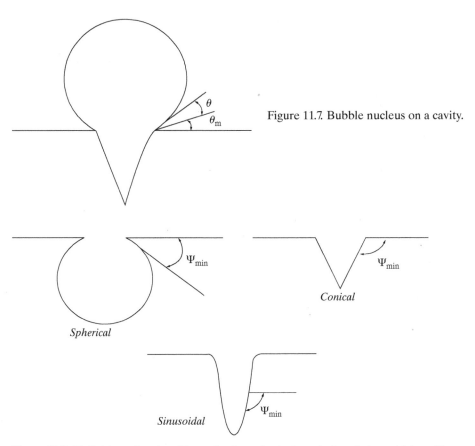

Figure 11.7. Bubble nucleus on a cavity.

Figure 11.8. Definition of cavity-side angles for spherical, conical, and sinusoidal cavities.

they derived the following criteria for bubble growth:

$$T_{\mathrm{w}} - T_{\mathrm{sat}} \geq \begin{cases} \dfrac{4\sigma T_{\mathrm{sat}}}{h_{\mathrm{fg}}\rho_{\mathrm{v}}\delta} & \text{for } R_{\mathrm{B}} > \delta, \quad (11.10) \\[2ex] \left(\dfrac{2\sigma T_{\mathrm{sat}}}{h_{\mathrm{fg}}\rho_{\mathrm{v}}R_{\mathrm{C}}}\right)\dfrac{1}{1 - \frac{R_{\mathrm{C}}}{2\delta}} & \text{for } R_{\mathrm{B}} < \delta. \quad (11.11) \end{cases}$$

Hsu's nucleation criterion deviates significantly from experimental data for highly wetting liquids. Such liquids tend to flood the cavities. Mizukami (1977) has argued that a vapor embryo in a cavity remains stable as long as the curvature of the interface is increased as a result of increasing vapor volume. Wang and Dhir examined the conditions that are sufficient for the occurrence of vapor/gas entrapment in a cavity, based on the minimization of the Helmholtz free energy of a system containing a gas–liquid interphase (Wang, 1992; Dhir, 1993). Their criterion for the entrapment of vapor/gas in a cavity (which is equivalent to the occurrence of a minimum of excess Helmholtz free energy on or in the cavity) is

$$\theta > \psi_{\min}, \quad (11.12)$$

where $\Psi_{\min}$ is the minimum cavity-side angle for spherical, conical, and sinusoidal cavities, as shown in Fig. 11.8. For a spherical cavity, based on the stability of the vapor–liquid interphase, Wang and Dhir derived the following inception criterion:

$$T_w - T_{sat} = \frac{2\sigma T_{sat}}{\rho_g h_{fg} R_C} K_{max},$$ (11.13)

where

$$K_{max} = \begin{cases} 1 & \text{for } \theta \leq \dfrac{\pi}{2}, \\ \sin\theta & \text{for } \theta > \dfrac{\pi}{2}. \end{cases}$$

EXAMPLE 11.1. A horizontal circular disk that is 10 cm in diameter is submerged in a shallow pool of quiescent water that is at 95 °C. Calculate the size range of active nucleation sites for $T_w = 109$ °C, assuming that the contact angle is 50°.

SOLUTION. The thermophysical properties of water at the film temperature $T_{film} = \frac{1}{2}(T_w + \overline{T}_L) = \frac{1}{2}(382 + 368) = 375$ K are $\rho_L = 957$ kg/m^3, $k_L = 0.666$ W/m·k, $\alpha_L = 1.648 \times 10^{-7}$ m^2/s, $\nu_L = 2.89 \times 10^{-7}$ m^2/s, $\sigma = 0.059$ N/m, and $Pr_L = 1.75$. Other properties are $T_{sat} = 373.1$ K, $\rho_g = 0.597$ kg/m^3, and $h_{fg} = 2.257 \times 10^6$ J/kg.

We need to estimate δ, the thickness of the thermal boundary layer, and for that we need to calculate the convection heat transfer coefficient. We can use a natural convection correlation. For an upward-facing, heated horizontal surface (Ghiaasiaan, 2011),

$$l_c = A/p,$$

where A and p are the surface area and perimeter, respectively, and l_c is the characteristic length of the surface. We thus get

$$l_c = D/4 = 0.025 \text{ m}.$$

The calculations then continue as follows. The thermal expansion coefficient is $\beta = 7.49 \times 10^{-4}$ K^{-1}. The Rayleigh number is therefore

$$Ra = \frac{g\beta(T_w - T_\infty)l_c^3}{\nu_L \alpha_L} = 3.377 \times 10^7.$$

The average Nusselt number is

$$Nu_{lc} = 0.15\, Ra^{1/3} = 48.2.$$

The average heat transfer coefficient is

$$H = Nu_{lc} k_L / l_c = 1{,}283 \text{ W/m}^2 \cdot \text{K},$$

with

$$\delta = \frac{k_L}{H} = 5.19 \times 10^{-4} \text{ m}.$$

The minimum and maximum crevice radii can now be found from Eq. (11.9), and that leads to

$$R_{C,min} = 2.87 \times 10^{-6} \text{ m} \approx 2.9 \text{ μm},$$
$$R_{C,max} = 1.50 \times 10^{-4} \text{ m} \approx 150 \text{ μm}.$$

Active nucleation sites represent perhaps the most difficult problem with respect to the mechanistic modeling of nucleate boiling. The number density and other characteristics of wall crevices depend on surface material, surface finishing, oxidation, and contamination. Furthermore, experiments show that the number of active nucleation sites increases as the wall heat flux (q_w'') or wall superheat ($T_w - T_{sat}$) is increased, and in general

$$N \sim (T_w - T_{sat})^m, \tag{11.14}$$

where m varies in the 4–6 range. The proportionality constant in this equation, as well as m, are likely to depend on the shape and size of the cavities. The thermal and hydrodynamic interaction among neighboring nucleation sites further complicates the problem. The conditions in the vicinity of an active nucleation site can activate or deactivate a neighboring active site (Kenning, 1989). The interaction between neighboring sites depends on the distance between them. According to the observations of Judd and Chopra (1993), when the distance is larger than about three times the diameter of departing bubbles, the two sites operate independently. When the distance is between one and three times the bubble diameter at departure, the formation of a bubble at one site inhibits the formation of a bubble at the other. For smaller distances, the formation of a bubble at one site promotes bubble formation at the other.

Kocamustafaogullari and Ishii (1983) have developed the following correlation for the cumulative number density of nucleation sites for boiling of water. The correlation has been utilized widely, even though its accuracy is only within about an order of magnitude. According to their correlation,

$$N^* = \left\{ \left[\frac{2R_C}{d_{Bd}} \right]^{-4.4} [2.157 \times 10^{-7} \rho^{*-3.2} (1 + 0.0049\rho^*)^{4.13}] \right\}^{1/4.4}, \tag{11.15}$$

$$N^* = N d_{Bd}^2, \tag{11.16}$$

$$R_C = \frac{2\sigma[1 + \rho_L/\rho_v]}{P_L} \left\{ \exp\left[\frac{h_{fg}(T_v - T_{sat})}{\frac{R_u}{M} T_v T_{sat}} \right] - 1 \right\}, \tag{11.17}$$

$$\rho^* = \Delta\rho/\rho_v, \tag{11.18}$$

$$d_{Bd} = 0.0012 \rho^{*0.9} \left[0.0208\,\theta \sqrt{\frac{\sigma}{g\Delta\rho}} \right], \tag{11.19}$$

where θ is the contact angle in degrees. The bracketed term on the right side of Eq. (11.19) represents the bubble departure diameter according to Fritz (1935); this will be discussed later.

11.2.2 Bubble Ebullition

In the isolated bubble regime in nucleate boiling, the activated nucleation sites undergo the following near-periodic processes. Following inception, a bubble grows to a critical size during the growth period t_{gr} and departs from the heated surface. The departing bubble leaves a small pocket of gas–vapor mixture behind. The departing bubble also disrupts the thermal boundary layer, and fresh and cool liquid from the ambient surroundings rushes in and replenishes the displaced superheated boundary layer. A new thermal boundary layer is then formed and grows in thickness during the waiting period t_{wt}, until the embryonic gas pocket left behind by the previous bubble starts to grow. The time period associated with the generation and release of a bubble is $t_{gr} + t_{wt}$, and the frequency of bubble release is $f_B = 1/(t_{gr} + t_{wt})$. Thus, should the active nucleation site density N, the departure diameter d_{Bd}, and the bubble release frequency f_B be known, the nucleate boiling component of heat transfer in the isolated bubble regime can be found from

$$q''_{NB} = N f_B \rho_g h_{fg} \frac{\pi}{6} d_{Bd}^3. \tag{11.20}$$

Attempts at the measurement and/or correlation of the various parameters affecting the bubble ebullition cycle have been under way for decades. A brief discussion follows.

Growth Period

Two different lines of thought have been followed with respect to bubble growth in nucleate boiling. In one line of thought, bubble growth is assumed to be caused by evaporation around the bubble while it is surrounded by superheated liquid. The aforementioned solutions of Plesset and Zwick (1954) and Forster and Zuber (1954) (Section 2.13) apply when a bubble is surrounded by a superheated liquid with uniform temperature. Refinements to these solutions have been made by Birkhoff et al. (1958), Scriven (1959), and Bankoff (1963). An analytical solution that accounts for the nonspherical shape of the bubble has also been derived by Mikic et al. (1970). These and other similar mathematical solutions may not realistically represent the growth of a bubble that is attached to a surface, however.

The second, and more realistic, line of thought is that a bubble that is growing while attached to a heated surface is separated from the heated surface by a thin liquid layer (the microlayer). Much of the the evaporation occurs in the microlayer (Snyder and Edwards, 1956; Moore and Mesler, 1961; Cooper and Lloyd, 1969). The average thickness of the microlayer can be estimated from (Cooper and Lloyd, 1969)

$$\delta_m = C(\nu_f t_{gr})^{1/2}, \tag{11.21}$$

where $C \approx 0.3$–1.3 (Cooper and Lloyd, 1969), with a preferred value of $C \approx 1$ (Lee and Nydahl, 1989). The thickness of the microlayer is nonuniform, however, and may be of the order of the molecular length near the center of the bubble base (Dhir, 1991).

Bubble Departure

One of the oldest and most widely used correlations is due to Fritz (1935):

$$d_{Bd} = 0.0208\,\theta \sqrt{\frac{\sigma}{g\Delta\rho}}, \tag{11.22}$$

where the contact angle θ must be in degrees. Fritz's correlation evidently considers only buoyancy and surface tension as forces determining the bubble departure. Phenomenologically, bubble departure occurs when forces tending to dislocate the bubble (buoyancy, wake caused by the preceding bubble, etc.) overcome the forces that resist bubble detachment (surface tension, drag, and inertia). The surface tension force is generally resistive, although it may also act in favor of bubble departure by making the bubble shape spherical (Cooper et al., 1978). Numerous models and correlations for bubble departure diameter have been proposed by attempting to include the effects of various forces (Staniszewski, 1959; Cole and Shulman, 1966; Zeng et al., 1993a, b; Chen et al., 1995). (See Hsu and Graham (1986) and Carey (2008) for reviews.) Application of these models and correlations often needs information about the bubble growth rate. The correlation of Cole and Shulman (1966), for example, is a simple modification of Fritz's correlation:

$$d_{Bd} = 0.0208\,\theta\sqrt{\frac{\sigma}{g\Delta\rho}}[1 + 0.0025(dd_B/dt)^{3/2}], \tag{11.23}$$

where dd_B/dt should be in millimeters per second. Based on an earlier correlation by van Stralen for thermally controlled bubble growth (which is predominant at high relative pressure P_r, because surface tension is weak), Gorenflo et al. (1986) have proposed

$$d_{Bd} = C\left[\frac{Ja^4\alpha_f^2}{g}\right]^{1/3}\left[1 + \left(1 + \frac{2\pi}{3Ja}\right)^{1/2}\right]^{4/3}, \tag{11.24}$$

where α_f is the thermal diffusivity of the liquid, $Ja = \rho_f C_{Pf}(T_w - T_{sat})/\rho_g h_{fg}$, and $C = 14.7$ for refrigerant R-12, 16.0 for refrigerant R-22, and 2.78 for propane.

There is considerable scatter in the bubble departure experimental data, and the existing correlations have limited accuracy. The physical processes leading to bubble departure are complicated, and there is coupling between momentum and energy exchange processes. Bubble departure as a consequence is a stochastic process even in well-controlled experiments (Klausner et al., 1997). Based on experimental data from several sources, Zeng et al. (1993a, b) have developed models for bubble detachment in pool and flow boiling where several forces are considered. The bubble growth rate is needed in these models. A simplified model for the prediction of vapor bubble growth has been developed by Chen et al. (1995).

The Waiting Period

A departing bubble disrupts the liquid thermal boundary layer over an area about four times the cross section of the departing bubble. Hsu and Graham (1961) modeled the development of the liquid thermal field as one-dimensional transient conduction in a slab with a known thickness δ. An improvement was made by Han and Griffith (1965), according to whom the waiting period can be estimated by using the solution to one-dimensional transient heat conduction into a semi-infinite medium that is initially at the liquid bulk temperature, and its surface temperature is suddenly raised to T_w as the waiting period starts. The time-dependent thermal layer thickness $\delta = \sqrt{\pi\alpha_L t}$ then follows, leading to the following expression for t_{wt}, the waiting period (Hsu and Graham, 1986):

$$t_{wt} = \frac{9}{4\pi\alpha_L}\left[\frac{(T_w - T_\infty)R_C}{T_w - T_{sat}\left(1 - \frac{2\sigma}{R_C\rho_g h_{fg}}\right)}\right]^2, \tag{11.25}$$

where α_L is the thermal diffusivity of the liquid.

These waiting period models disregard the local cooling that may develop in the solid surface as a result of thermal interaction with the liquid. The local cooling of the solid is negligible when the thermal capacity of the solid surface is infinitely large. Hatton and Hall (1966) have developed a model that accounts for the thermal response of the solid surface.

11.2.3 Heat Transfer Mechanisms in Nucleate Boiling

The phenomenology and related models and correlations described in the previous section dealt with the isolated bubble zone of the partial boiling regime. Recent direct simulations have shown that a mechanistic bubble ebullition model based on microlayer evaporation predicts well the experimental data obtained with a polished surface and a silicon wafer with a well-characterized artificial cavity (Dhir *et al.*, 2013). Heated surfaces have unknown cavity characteristics, however, and mechanistic models have limited practical and design application.

In the isolated bubble partial boiling regime, nucleate boiling and natural convection both contribute to heat transfer. The contribution of convection diminishes as the heat flux is increased, however. With increasing heat flux, furthermore, bubble frequency and the number of active nucleation sites both increase. Consequently, with increasing q''_w bubbles interact in the lateral direction, leading to the formation of vapor mushrooms (see Fig. 11.9). The transition from isolated bubbles to columns and mushrooms in fact represents a transition from partial to fully developed nucleate boiling, and a correlation for this transition is (Moissis and Berenson, 1963)

$$q''_w = 0.11\sqrt{\theta}\rho_g h_{fg}\left(\frac{\sigma g}{\Delta\rho}\right)^{1/4}, \tag{11.26}$$

where θ must be in degrees.

In the fully developed boiling regime, evaporation appears to occur primarily at the periphery of the vapor stems in the liquid macrolayer, which refers to the liquid film separating the heated surface from the base of the vapor mushroom.

As mentioned earlier, mechanistic or phenomenological models for nucleate boiling generally have limited use, except when they are applied to well-controlled experiments with well-characterized artificial cavities. In reality, nucleate boiling is more complex than the basic assumptions that are often made for the development of models. The main difficulty, besides the uncertainty associated with the characteristics of the nucleation sites (which has long been considered as the single most important impediment to successful mechanistic modeling of nucleate boiling), is the nonlinear and conjugate nature of a multitude of subprocesses in the vapor, liquid, and solid (heated surface) that participate in the bubble ebullition. Accordingly, the basic assumptions such as constant wall temperature or heat flux are flawed owing to the prevalence of temporal and spatial fluctuations, and the modeling of bubble behavior based on essentially static force balance considerations is invalid because of the dynamic and nonlinear phenomena involved. Nonlinear chaos dynamics has been proposed as an alternative methodology to mechanistic modeling. This is an emerging research field, however, and much more is needed in terms of measurement of local-instantaneous parameters as well as detailed simulations. Furthermore,

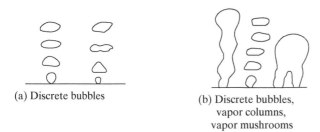

(a) Discrete bubbles

(b) Discrete bubbles,
vapor columns,
vapor mushrooms

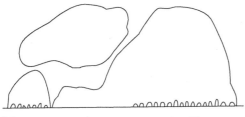

(c) Vapor columns, large vapor mushrooms

Figure 11.9. Vapor structures in pool boiling. (After Gaertner, 1965.)

(d) Large vapor mushrooms, vapor patches(?)

(a) Discrete bubble region
(b) First transition region
(c) Vapor mushroom region
(d) Second transition region

although nonlinear models are useful tools for a better understanding of boiling, they are not useful for prediction purposes (Shoji, 2004).

11.3 Nucleate Boiling Correlations

Mechanistic models based on bubble ebullition phenomena, however, are not yet able to predict the nucleate boiling heat transfer coefficients in general, because of the uncertainties related to surface characteristics. Empirical correlations are therefore often used. Numerous empirical correlations have been proposed for heat transfer in nucleate boiling in the past. The following correlations are among the most widely used.

The Correlation of Rohsenow (1952). This correlation is among the oldest and most widely used. The correlation uses the general form of the forced convection heat transfer:

$$\text{Nu} = \frac{h\lambda_\text{L}}{k_\text{f}} = \frac{1}{C_\text{sf}}\text{Re}^{1-n}\,\text{Pr}_\text{f}^m. \tag{11.27}$$

Table 11.1. *Values of the constants in the correlation of Rohsenow (Rohsenow, 1973; Thome, 2003).*

Surface combination	C_{sf}	$m+1$
Water–nickel	0.006	1.0
Water–platinum	0.013	1.0
Water–emery polished copper	0.0128	1.0
Water–brass	0.006	1.0
Water–ground and polished stainless steel	0.008	1.0
Water–Teflon pitted stainless steel	0.0058	1.0
Water–chemically etched stainless steel	0.0133	1.0
Water–mechanically polished stainless steel	0.0132	1.0
Water–emery polished, paraffin treated copper	0.0147	1.0
CCl_4–emery polished copper	0.007	1.7
Benzene–chromium	0.01	1.7
n-Pentane–chromium	0.015	1.7
n-Pentane–emery polished copper	0.0154	1.7
n-Pentane–emery polished nickel	0.0127	1.7
Ethyl alcohol–chromium	0.0027	1.7
Isopropyl alcohol–copper	0.0025	1.7
35% K_2CO_3–copper	0.0054	1.7
50% K_2CO_3–copper	0.0027	1.7
n-Butyl alcohol–copper	0.0030	1.7

Experimental data and the phenomenology of bubble ebullition on heated surfaces indicate that the physical scale of the surface has no effect on heat transfer. (This is of course not true for microscale objects, but our discussion here is about commonly used systems.) They also indicate that the effect of liquid pool temperature (subcooling) on fully developed nucleate boiling should be small (see Fig. 11.4). Rohsenow therefore used the Laplace length scale,

$$\lambda_L = \sqrt{\frac{\sigma}{g(\rho_g - \rho_f)}},$$

and $q_w''/\rho_f h_{fg}$ as the velocity U. Furthermore, he defined the heat transfer coefficient as

$$H = \frac{q_w''}{T_w - T_{sat}}. \tag{11.28}$$

Substitution of these parameters in Eq. (11.27) leads to

$$C_{Pf} \frac{T_w - T_{sat}}{h_{fg}} = C_{sf} \left[\frac{q_w''}{\mu_f h_{fg}} \sqrt{\frac{\sigma}{g \Delta \rho}} \right]^n \left(\frac{\mu C_p}{k} \right)_f^{m+1}, \tag{11.29}$$

where $n = 0.33$, $m = 0$ for water and $m = 0.7$ for other fluids. The value of the parameter C_{sf} depends on the solid–fluid combination, with some recommended values listed in Table 11.1. Its recommended value for unknown pairs is 0.013. As noted, this parameter varies in the relatively wide range $0.003 < C_{sf} < 0.0154$.

Despite its simplicity, the correlation of Rohsenow predicts the pool boiling data reasonably well. Its typical error in calculating q_w'' when T_w is known is about 100%, and in calculating $T_w - T_{sat}$ when q_w'' is known the error is about 25% (Lienhard and Lienhard, 2005). By using the fluid–surface pair constant C_{sf}, the correlation in fact accounts for the effects of surface characteristics and wettability, and this may be a main reason for its relative success (Dhir, 1991).

The Correlation of Forster and Zuber (1954). This correlation uses a generic expression similar to Eq. (11.27) and defines the length and velocity scales based on the growth process of microbubbles suspended in a superheated liquid. As described in Section (3.5), when a spherical vapor bubble is surrounded by an infinite, superheated liquid, its asymptotic growth rate from evaporation can be formulated by noting that the evaporation mass flux at the bubble surface is approximately equal to $m'' = k_f(T_L - T_{sat})/\delta h_{fg}$, with $\delta = \sqrt{\pi \alpha_f t}$. The asymptotic rate of bubble growth is related to the mass flux according to

$$\frac{d}{dt}\left(\frac{4}{3}\pi R^3 \rho_g\right) = 4\pi R^2 m''. \tag{11.30}$$

A more accurate solution is (see Eq. (2.225))

$$\dot{R} = \sqrt{\frac{\pi}{2}} \frac{k_f}{\rho_g h_{fg}} \frac{\Delta T_{sat}}{\sqrt{\alpha_f t}}, \tag{11.31}$$

where $\dot{R} = dR/dt$. If one assumes that Eq. (11.31) is valid starting from $R = 0$ an integration gives

$$R = 2\sqrt{\frac{\pi}{2}} \frac{k_f}{\rho_g h_{fg}} \frac{\Delta T_{sat}}{\sqrt{\alpha_f}} \sqrt{t}, \tag{11.32}$$

where $\Delta T_{sat} = T_w - T_{sat}$. Forster and Zuber used the generic correlation $Nu = 0.0015 Re^{0.62} Pr_f^{0.33}$, with $Re \sim \rho_f R \dot{R}/\mu_f$, and

$$Nu = \frac{q_w''}{\Delta T_{sat}} \frac{l}{k_f},$$

with the length scale l defined as

$$l = \frac{\Delta T_{sat} \rho_f C_{Pf} \sqrt{\pi \alpha_f}}{\rho_g h_{fg}} \sqrt{\frac{2\sigma}{\Delta P}} \left[\frac{\rho_f}{\Delta P}\right]^{1/4}, \tag{11.33}$$

where $\Delta P = P_{sat}(T_w) - P$. The final correlation of Forster and Zuber is

$$\frac{q_w''}{\rho_g h_{fg}}\left(\frac{\pi}{\alpha_f}\right)^{1/2}\left[\frac{\rho_f R^{*3}}{2\sigma}\right]^{1/4} = 0.0015\left\{\frac{\rho_f}{\mu_f}\left[\frac{(T_w - T_{sat})k_f}{\rho_g h_{fg}}\right]^2 \frac{\pi}{\alpha_f}\right\}^{5/8}(\mu C_P/k)_f^{1/3}, \tag{11.34}$$

where

$$R^* = \frac{2\sigma}{P_{sat}(T_w) - P}.$$

The correlation of Forster and Zuber (1955) thus does not account for the effect of surface properties.

Chen (1966) utilized the correlation of Forster and Zuber in his well-known and widely respected correlation for forced convection nucleate boiling (see Chapter 14).

The Correlations of Stephan and Abdelsalam (1980). These correlations have been found to have good accuracy. Define the Nusselt number as

$$Nu = H d_{Bd}/k_f, \tag{11.35}$$

where d_{Bd} is the bubble departure diameter according to the correlation of Fritz (1935) (Eq. (11.22)). For hydrocarbons, in the range $5.7 \times 10^{-3} \leq P/P_{cr} \leq 0.9$,

$$\text{Nu} = 0.0546 \left[\left(\frac{\rho_g}{\rho_f} \right)^{1/2} \left(\frac{q_w'' d_{Bd}}{k_f T_{sat}} \right) \right]^{0.67} \left(\frac{h_{fg} d_{Bd}^2}{\alpha_f^2} \right)^{0.248} \left(\frac{\Delta\rho}{\rho_f} \right)^{-4.33}, \qquad \theta_0 = 35°. \tag{11.36}$$

For water, in the range $10^{-4} \leq P/P_{cr} \leq 0.886$,

$$\text{Nu} = (0.246 \times 10^7) \left(\frac{q_w'' d_{Bd}}{k_f T_{sat}} \right)^{0.673} \left(\frac{h_{fg} d_{Bd}^2}{\alpha_f^2} \right)^{-1.58}$$
$$\times \left(C_{Pf} T_{sat} d_{Bd}^2 / \alpha_f^2 \right)^{1.26} \left(\frac{\Delta\rho}{\rho_f} \right)^{5.22}, \qquad \theta_0 = 45°. \tag{11.37}$$

For refrigerants (propane, n-butane, carbon dioxide, and several refrigerants including R-12, R-113, R-114, and RC-318), in the range $3 \times 10^{-3} \leq P/P_{cr} \leq 0.78$,

$$\text{Nu} = 207 \left(\frac{q_w'' d_{Bd}}{k_f T_{sat}} \right)^{0.745} \left(\frac{\rho_g}{\rho_f} \right)^{0.581} \text{Pr}_f^{0.533}, \qquad \theta_0 = 35°. \tag{11.38}$$

For cryogenic fluids, in the range $4 \times 10^{-3} \leq P/P_{cr} \leq 0.97$,

$$\text{Nu} = 4.82 \left(\frac{q_w'' d_{Bd}}{k_f T_{sat}} \right)^{0.624} \left[\frac{(\rho C_P k)_{cr}}{\rho_f C_{Pf} k_f} \right]^{0.117} \left(\frac{\rho_g}{\rho_f} \right)^{0.257} \left(\frac{C_{Pf} T_{sat} d_{Bd}^2}{\alpha_f^2} \right)^{0.374}$$
$$\times \left(\frac{h_{fg} d_{Bd}^2}{\alpha_f^2} \right)^{-0.329}, \qquad \theta_0 = 1°. \tag{11.39}$$

The Correlation of Cooper (1984). Based on an extensive data base, Cooper (1984) derived the following correlation, which is simple and general and applicable to various fluids:

$$\frac{q_w''^{1/3}}{T_w - T_{sat}} = 55.0 (P/P_{cr})^n \left(-\log_{10} \frac{P}{P_{cr}} \right)^{-0.55} M^{-0.5},$$
$$n = 0.12 - 0.21 \log_{10} R_P, \tag{11.40}$$

where q_w'' is the heat flux in watts per meter squared, $T_w - T_{sat}$ is in kelvins, M is the molecular mass number of the fluid, and R_P is the *roughness parameter*. Cooper has suggested some values for R_P. For cases where R_P is unspecified, Cooper recommends using $\log_{10} R_P = 0$. The range of applicability is $0.002 \leq P_r \leq 0.9, 2 \leq M \leq 200$. Thome (2003) has noted that the correlation without any correction gives accurate predictions for boiling of newer refrigerants on copper tubes. The correlation of Cooper has been used by some authors to represent the contribution of nucleate boiling to forced-flow boiling.

The Correlation of Gorenflo (1993). This widely respected correlation is fluid specific and has good accuracy when applied within its recommended ranges of parameters. The correlation is based on the modification of experimentally measured heat

Table 11.2. *Reference parameters for the correlation of Gorenflo (1993) for selected fluids.*

Fluid	P_{cr} (bar)	H_0 (W/m²·K)	Fluid	P_{cr} (bar)	H_0 (W/m²·K)
Water	220.6	5600	R-11	44.0	2800
Ammonia	113.0	7000	R-12	41.6	4000
Sulfur hexafluoride	37.6	3700	R-13	38.6	3900
Methane	46.0	7000	R-22	49.9	3900
Ethane	48.8	4500	R-23	48.7	4400
Propane	42.4	4000	R-113	34.1	2650
Benzene	48.9	2750	R-123	36.7	2600
n-Pentane	33.7	3400	R-134a	40.6	5040
i-Pentane	33.3	2500	R-152a	45.2	4000
Nitrogen (on Pt)	34.0	7000	RC-318	28.0	4200
Nitrogen (on Cu)	34.0	10 000	R-32	57.82	6550
Propane	42.48	5210	R-152a	45.17	5570
i-Butane	36.4	4320	R-143a	37.76	5410
Ethanol	63.8	4400	R-125	36.29	4940
Acetone	47.0	3950	R-227ea	29.80	4860

Note: Based in part on Thome (2003) and Gorenflo *et al.* (2004).

transfer coefficients obtained at standard conditions. The general form of the correlation is

$$\frac{H}{H_0} = F_{PR}(q''_w/q''_0)^n (R_P/R_{P0})^{0.133}, \tag{11.41}$$

where $q''_0 = 20\,000$ W/m², H_0 is the heat transfer coefficient corresponding to q''_0 obtained at the reference reduced pressure $P_{r0} = 0.1$, and the reference surface roughness parameter is $R_{P0} = 0.4$ µm. The pressure correction factor F_{PR} and the parameter n are to be calculated as follows. For water,

$$F_{PR} = 1.73 P_r^{0.27} + \left(6.1 + \frac{0.68}{1 - P_r}\right) P_r^2,$$

$$n = 0.9 - 0.3 P_r^{0.15}. \tag{11.42}$$

For other fluids included in the correlation's data base (excluding water and helium),

$$F_{PR} = 1.2 P_r^{0.27} + \left(2.5 + \frac{1}{1 - P_r}\right) P_r, \tag{11.43}$$

$$n = 0.9 - 0.3 P_r^{0.3}. \tag{11.44}$$

Values of H_0 and P_{cr} for some fluids are listed in Table 11.2. For fluids other than those listed in the correlation's data base, the experimental or estimated value of H_0 at the aforementioned reference conditions is needed. In the absence of such an experimental value, however, H_0 can be calculated by using other reliable correlations. The parameter range for the fluids listed in Table 11.2 is $0.0005 \le P_r \le 0.95$. When the roughness parameter is not known, $R_p = 0.4$ µm can be used. Gorenflo *et al.* (2004) have shown that the correlation does very well in predicting experimental data with newer refrigerants, and it correctly accounts for the effects of pressure and thermophysical properties on pool nucleate boiling heat transfer.

EXAMPLE 11.2. Using the correlations of Rohsenow (1952), Cooper (1984), and Gorenflo (1993), calculate the boiling heat transfer coefficient for a mechanically polished stainless-steel surface submerged in saturated water at a pressure of 17.9 bars. The wall is at $T_w = 490$ K. Assume a mean surface roughness of 2 μm.

SOLUTION. The relevant properties are $C_{Pf} = 4{,}524$ J/kg·K, $k_f = 0.647$ W/m·K, $\mu_f = 1.30 \times 10^{-4}$ kg/m·s, $\rho_f = 856.7$ kg/m³, $\rho_g = 9.0$ kg/m³, $T_{sat} = 480$ K, $h_{fg} = 1.913 \times 10^6$ J/kg, and $\sigma = 0.036$ N/m.

First, consider Rohsenow's correlation. From Table 11.1 we get $C_f = 0.0132$. We also have $m = 0$ and $n = 0.33$. Equation (11.29) can now be solved for q_w, resulting in

$$q''_{w,\text{Rohsenow}} = 9.147 \times 10^5 \text{ W/m}^2.$$

We now consider Cooper's correlation. We have

$$P_r = P/P_{cr} = 17.9 \text{ bars}/220.6 \text{ bars} = 0.0811,$$
$$M = 18,$$

and

$$R_P = 2,$$

and so

$$n = 0.12 - 0.21 \log_{10}(R_p) = 0.0568.$$

We can now solve Eq. (11.40) to get

$$q''_{W,\text{Cooper}} = 1.247 \times 10^6 \text{ W/m}^2.$$

Lastly, we consider the method of Gorenflo. From Table 11.2, we have $H_0 = 5600$ W/m². Furthermore, $q''_0 = 20{,}000$ W/m² and

$$n = 0.9 - 0.3P_r^{0.15} = 0.694,$$

$$F_{PR} = 1.73 \, \text{Pr}^{0.27} + \left[6.1 + \frac{0.68}{1 - \text{Pr}}\right] \text{Pr}^2 = 0.923.$$

We can now calculate the boiling heat transfer coefficient from Eq. (11.41), noting that

$$\frac{R_P}{R_{P0}} = \frac{2 \text{ μm}}{0.4 \text{ μm}} = 5.$$

Equation (11.41) must be solved simultaneously with the following equation:

$$q''_{w,\text{Gorenflo}} = H_{\text{Gorenflo}}(T_w - T_{sat}),$$

with $q''_{w,\text{Gorenflo}}$ and H_{Gorenflo} as the two unknowns. The result will be

$$q''_{w,\text{Gorenflo}} = 8.98 \times 10^5 \text{ W/m}^2.$$

11.4 The Hydrodynamic Theory of Boiling and Critical Heat Flux

The hydrodynamic theory of boiling is based on the argument that the vapor–liquid interfacial stability phenomena play a crucial role in processes such as the critical heat flux (CHF) and film boiling (Lienhard and Witte, 1985). The hydrodynamic limitations associated with the vapor–liquid interfacial stability and the transport of vapor near the heated surface thus determine the phenomenology of the afore-mentioned boiling regimes. Models that are based on the hydrodynamic theory of boiling have been relatively successful and extensively used, even though they are not always consistent with all data trends.

According to hydrodynamic theory, the CHF and minimum film boiling (to be discussed later) are Taylor instability-driven processes (Zuber, 1959; Zuber *et al.*, 1963). In CHF, vapor jets rise at the nodes of Taylor waves. The jets have the highest rise velocity that Helmholtz instability allows. In minimum film boiling, the surface is blanketed by vapor, and vapor bubbles are periodically released from the nodes of Taylor waves that develop at the liquid–vapor interface. In the transition boiling region, the surface partially supports rising jets and partially supports vapor bubbles.

The CHF also referred to as the peak heat flux, the boiling crisis, or burnout point (point C in Fig. 11.2), represents the maximum heat flux a heated surface can support without the loss of macroscopic physical contact between the liquid and the surface. Nucleate boiling is the heat transfer regime of choice for many industrial cooling systems, and the CHF represents the upper limit for the safe operation of these systems.

The CHF in pool boiling was modeled by Zuber *et al.* (1963) based on the postulation that it is a process controlled by hydrodynamic stability. The capability of the hydrodynamic system in preventing the development of large dry patches on the heated surface while transferring vapor from the vicinity of the surface is the controlling factor. Zuber's model assumes that rising vapor jets with radius R_j form on a square grid with a pitch equal to the fastest growing wavelength according to Taylor stability, $\lambda_d = 2\pi\sqrt{3}\sqrt{\sigma/g\Delta\rho}$, as displayed in Fig. 11.10. The rising jets are assumed to have the critical velocity dictated by the Helmholtz instability, $U_g = \sqrt{2\pi\sigma/(\rho_g\lambda_H)}$, where the neutral wavelength for the rising jets is assumed to be $\lambda_H = 2\pi R_j$. The critical heat flux will be equal to the rate of latent heat leaving by way of a single jet, divided by the area of a square grid, thereby

$$q''_{CHF} = \rho_g h_{fg} U_g \frac{\pi R_j^2}{\lambda_d^2}. \tag{11.45}$$

Zuber further assumed that $R_j = \lambda_d/4$. Substitution for λ_d, R_j, and U_g into Eq. (11.45) leads to

$$q''_{CHF,Z} \approx \frac{\pi}{24}\rho_g^{1/2}h_{fg}(\sigma g\Delta\rho)^{1/4}. \tag{11.46}$$

It is worth mentioning that essentially the same correlation, with a constant of 0.131 instead of $\pi/24$, had been derived earlier by Kutateladze and Borishansky based on dimensional analysis (Lienhard and Witte, 1985).

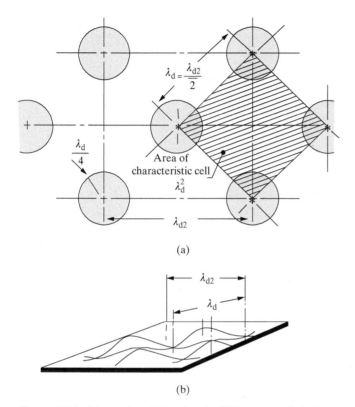

Figure 11.10. Schematic of rising jets in CHF on an infinitely large horizontal flat surface.

Sun and Lienhard (1970) noticed that Eq. (11.46) slightly underpredicted the experimental data. They noted, however, that an adequate adjustment in the expression can be obtained if for the neutral wavelength of the rising jets $\lambda_H = \lambda_{dl}$ is used, leading to

$$q''_{CHF} = 0.149 \rho_g^{1/2} h_{fg} (\sigma g \Delta \rho)^{1/4}. \tag{11.47}$$

Effect of Heated Surface Size and Geometry

Lienhard and co-workers also examined the effects of surface size and geometry on the CHF. The square pitch shown in Fig. 11.10 is a crucial element of the model, and the model should be expected to perform well only when the dimensions of the heated surface are much larger than λ_d. The analysis, furthermore, assumes a flat surface, which is an acceptable assumption as long as the principal radii of curvature of the surface are much larger than λ_d. Expressions (11.46) and (11.47) are thus valid when the surface is flat and large enough for its end effects to be unimportant. Otherwise, corrections are needed. Lienhard and Dhir (1973) developed a method for the required correction. Accordingly,

$$\frac{q''_{CHF}}{q''_{CHF,Z}} = f(l'), \tag{11.48}$$

where $l' = l / \sqrt{\sigma / g \Delta \rho}$, l is the characteristic length of the heated surface, q''_{CHF} is the average critical heat flux on the entire surface, and $q''_{CHF,Z}$ is the CHF predicted by

Table 11.3. *Size and shape corrections for CHF*

Situation	Characteristic length, l	Range	Correction factor, $f(l')$
Infinite flat heater	Heater width or diameter	$l' \geq 27$	1.14
Small flat heater	Heater width or diameter	$9 < l' < 20$	$1.14\lambda_d^2/A_{heat}$
Horizontal cylinder	Cylinder radius	$l' \geq 0.15$	$0.89 + 2.27\exp(-3.44\sqrt{l'})$
Large horizontal cylinder	Cylinder radius	$l' \geq 1.2$	0.90
Small horizontal cylinder	Cylinder radius	$0.15 \leq l' \leq 1.2$	$0.94l'^{-0.25}$
Large sphere	Sphere radius	$l' \geq 4.26$	0.84
Small sphere	Sphere radius	$0.15 \leq l' \leq 4.26$	$1.734/\sqrt{l'}$
Any large finite body	Characteristic length	Cannot specify generally; $l' \geq 4$	~ 0.90
Small horizontal ribbon oriented vertically			
Plain, both sides heated	Height of side	$0.15 \leq l' \leq 2.96$	$1.18/l'^{0.25}$
One-side insulated	Height of side	$0.15 \leq l' \leq 5.86$	$1.4/l'^{0.25}$
Small slender cylinder of any cross section	Transverse perimeter	$0.15 \leq l' \leq 5.86$	$1.4/l'^{0.25}$
Small bluff body	Characteristic length	Cannot specify generally; $l' \leq 4$	$const./\sqrt{l'}$

Note: Primed length parameters are all normalized with. $\sqrt{\sigma/g\Delta\rho}$.

Zuber's correlation (Eq. (11.46)). Table 11.3 gives a summary of the recommended values and empirical expressions for $f(l')$ (Sun and Lienhard, 1970; Ded and Lienhard, 1972; Lienhard and Dhir, 1973).

Other Parametric Effects

The CHF expressions quoted thus far are for horizontal surfaces and do not display any dependence on surface orientation, properties, etc. Experiments, however, indicate that certain surface properties have some effect on the CHF. Some important parametric effects on the CHF are now discussed, and relevant correlations are presented.

Surface wettability has been found to improve (increase) the CHF (Maracy and Winterton, 1988; Dhir and Liaw, 1989). An expression (based on a curve fit to the data of Dhir and Liaw (1989) and Maracy and Winterton (1988)) proposed by Haramura (1999) is

$$\frac{q''_{CHF}}{\rho_g^{1/2}h_{fg}(\sigma g\Delta\rho)^{1/4}} = 0.1\exp(-\theta_r/45°) + 0.055, \qquad (11.49)$$

where θ_r is the receding contact angle (in degrees).

Merte and Clark (1964) and Sun and Lienhard (1970) have reported that this hydrodynamic model of the CHF well predicts the effect of gravitational acceleration. Some more recent experiments have shown that the model is at least inaccurate in this respect, however. According to some experiments, for reduced gravity conditions the reduction in q''_{CHF} is significantly smaller than the predicted $g^{1/4}$. For example, at $10^{-5}g$, a reduction of only 60% in q''_{CHF} has been measured, whereas the correlation cited here predicts a 94% reduction (Abe *et al.*, 1994). Some experiments have shown an opposite trend, however (Shatto and Peterson, 1999). The model also does not consider the effects of surface conditions, for example the effect of surface wettability as discussed earlier. Experimental data also indicate that hydrodynamic theory deviates from data at very low pressures (Samokhin and Yagov, 1988). These and other shortcomings have in fact raised doubt about the fundamental assumptions underlying the hydrodynamic model of the CHF (Theofanous *et al.*, 2002a, b).

The effect of surface orientation is also important, and the CHF is lower on inclined
surfaces. A correlation proposed by Chang and You (1996), based on data with FC-72, is

$$\frac{q''_{\text{CHF}}}{q''_{\text{CHF,Horizontal}}} = 1 - 0.0012\theta \tan(0.414\theta) - 0.122 \sin(0.318\theta), \qquad (11.50)$$

where θ is the inclination angle with respect to the horizontal plane, in degrees.

Liquid subcooling increases the CHF. A correlation by Ivey and Morris (1962) is

$$\frac{q''_{\text{CHF}}}{q''_{\text{CHF,sat}}} = 1 + 0.1(\rho_{\text{f}}/\rho_{\text{g}})^{0.75} \frac{C_{\text{PL}}(T_{\text{sat}} - T_{\text{L}})}{h_{\text{fg}}}. \qquad (11.51)$$

A more recent study of the effect of liquid subcooling on the pool boiling CHF has been performed by Elkassabgi and Lienhard (1988), based on experimental data with isopropanol, methanol, R-113, and acetone. They used cylindrical electric resistance heaters with diameters of 0.8–1.54 mm. They noted three distinct regimes. For low subcooling conditions ($T_{\text{sat}} - T_{\text{L}}$ less than about 15 °C for isopropanol), they proposed

$$\frac{q''_{\text{CHF}}}{q''_{\text{CHF,sat}}} = 1 + 4.28 \frac{\rho_{\text{L}} C_{\text{PL}}(T_{\text{sat}} - T_{\text{L}})}{\rho_{\text{g}} h_{\text{fg}}} \left[\alpha_{\text{f}} \frac{[g\Delta\rho]^{1/4} \rho_{\text{g}}^{1/2}}{\sigma^{3/4}} \right]^{1/4}. \qquad (11.52)$$

Elkessabgi and Lienhard developed correlations for the effect of subcooling on the CHF for moderate and high subcooling regimes as well. The latter correlations include the radius of their cylindrical test section, however, and may therefore be limited to their range of geometric parameters (Dhir, 1991). A difficulty with respect to the application of the correlations of Elkassabgi and Lienhard is that in general it is not clear a priori which of the three regimes is applicable.

Surface roughness also increases the CHF, typically by 25%–35%.

11.5 Film Boiling

Let us postpone the discussion of the minimum film boiling point to after a discussion of film boiling. As noted earlier, hydrodynamic models with minor adjustments have done well in predicting the pool film boiling heat transfer in many situations. Film boiling models and correlations for some important heated surface configurations are now reviewed.

11.5.1 Film Boiling on a Horizontal, Flat Surface

Berenson (1961) has developed a well-known hydrodynamic model. According to Berenson's model, the surface is assumed to be covered by a contiguous vapor film

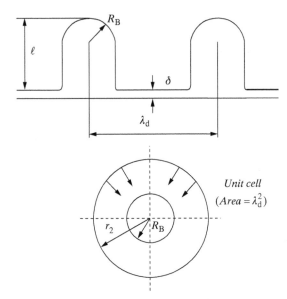

Figure 11.11. Film boiling on a horizontal surface, and schematic of Berenson's model.

(see Fig. 11.11). Standing Taylor waves with square λ_d pitch are assumed to occur at the liquid–vapor interphase. Vapor generated in a square unit cell with λ_d^2 area is assumed to flow toward each vapor dome. For simplicity of modeling, however, the square unit cell is replaced with a circle with radius $r_2 = \lambda_d/\sqrt{\pi}$. The vapor flow is assumed to be laminar, the thickness of the vapor disk is assumed to be constant, and inertia and kinetic energy of the vapor are neglected. The wall heat flux is assumed to be uniform over the unit cell. It is also assumed that (see Fig. 11.11)

$$R_B = 2.35\sqrt{\sigma/(g\Delta\rho)}, \tag{11.53}$$

$$\ell = 1.36 R_B = 3.2\sqrt{\sigma/(g\Delta\rho)}, \tag{11.54}$$

where $\Delta\rho = \rho_f - \rho_g$. The momentum equation for the vapor flow will be

$$\frac{dP}{dr} = C\frac{\mu_v \overline{U}_v}{\delta^2}, \tag{11.55}$$

where $\overline{U}_v$ is the average vapor velocity in the vapor film, $C = 12$ if the vapor velocity at the vapor–liquid interphase is assumed to be zero (i.e., no-slip condition), and $C = 3$ if zero shear stress at the interphase is assumed. Other equations resulting from these assumptions are

$$\dot{m}_v = 2\pi r \rho_v \delta \overline{U}_v, \tag{11.56}$$

$$\dot{m}_v h'_{fg} = \pi \left(r_2^2 - r^2\right) k_v \frac{\Delta T}{\delta}, \tag{11.57}$$

where $\Delta T = T_w - T_{sat}$ and

$$h'_{fg} = h_{fg}\left[1 + \frac{1}{2}C_{Pg}\frac{T_w - T_{sat}}{h_{fg}}\right]$$

is the latent heat of vaporization corrected for the effect of vapor superheating. This correction is needed because some of the heat lost by the heated surface is used up in superheating the vapor film. Equations (11.56) and (11.57) combined with $r_2 = \lambda_d/\sqrt{\pi}$ result in

$$\overline{U}_v = \frac{k_v \Delta T}{\rho_v h'_{fg} \delta^2} \left[\frac{(\lambda_d^2/2) - \pi r^2}{2\pi r} \right]. \tag{11.58}$$

Equations (11.55) and (11.58) can now be combined by eliminating $\overline{U}_v$ between them, and the resulting differential equation can be integrated between the limits R_B and r_2 to get

$$P_2 - P_1 = \frac{8C}{\pi} \frac{\mu_v k_v \Delta T}{\rho_v h'_{fg} \delta^4} \frac{\sigma}{g\Delta\rho}. \tag{11.59}$$

This pressure difference is assumed to be supplied by the hydrostatic and surface tension forces, so that

$$P_2 - P_1 = g\Delta\rho\ell - \frac{2\sigma}{R_B}. \tag{11.60}$$

Combining Eqs. (11.53), (11.54), (11.59), and (11.60), we find for the vapor film thickness

$$\delta = \left[1.09C \frac{\mu_v k_v \Delta T}{\rho_v g \Delta\rho h'_{fg}} \sqrt{\sigma/(g\Delta\rho)} \right]^{0.25}. \tag{11.61}$$

The heat transfer coefficient can now be found from $H = k_v/\delta$. The result, after the adjustment of the constant to match experimental data, is

$$H = 0.425 \left[\frac{k_v^3 \rho_v \Delta\rho g h'_{fg}}{\mu_v (T_w - T_{sat}) \sqrt{\sigma/(g\Delta\rho)}} \right]^{0.25}. \tag{11.62}$$

Properties with subscript v should be calculated at the mean vapor film temperature.

Berenson's modeling method has been successfully applied for modeling of other similar phase-change phenomena. An example is the melting of a miscible solid sublayer underneath a hot liquid pool. This phenomenon can occur during some severe nuclear reactor scenarios, where the fuel and structural material in the reactor core form a molten liquid pool with internal heat being generated by radioactive decay. The molten pool attacks its structural sublayer, gradually melting through it. This melting process has been modeled by using Berenson-type methods (Taghavi-Tafreshi et al., 1979).

EXAMPLE 11.3. Calculate the CHF, and estimate the wall temperature when the CHF occurs for the conditions of Example 11.2.

SOLUTION. All the properties that are needed were calculated in Example 11.2. Let us use the correlation of Zuber (1964), with the coefficient adjustment proposed by Lienhard and Dhir (1973), namely Eq. (11.47). The result will be

$$q''_{CHF} = 3.56 \times 10^6 \text{ W/m}^2.$$

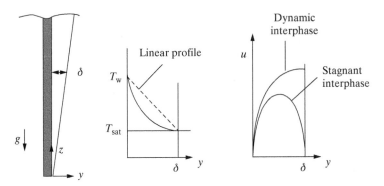

Figure 11.12. Laminar film boiling on a vertical surface.

To estimate the wall temperature at CHF conditions, we assume that the nucleate pool boiling correlations apply all the way to the CHF point. Thus, with Rohsenow's correlation, we will obtain $T_{w,CHF}$ by solving

$$C_{Pf}\frac{T_{w,CHF}-T_{sat}}{h_{fg}}=C_{sf}\left[\frac{q''_{CHF}}{\mu_f h_{fg}}\sqrt{\frac{\sigma}{g\Delta p}}\right]^n\left(\frac{\mu C_P}{k}\right)_f^{m+1},$$

where $n = 0.33$, $m = 0$, and $C_{sf} = 0.0132$. This gives $T_{w,CHF} = 495.7$ K.

Likewise, with Cooper's correlation, we need to find $T_{w,CHF}$ from

$$\frac{q''^{0.333}_{w,CHF}}{T_{w,CHF}-T_{sat}}=55P_r^n(-\log_{10}P_r)^{-0.55}M^{-0.5},$$

where $n = 0.0568$. The result is $T_{w,CHF} = 494.2$ K.

Finally, using Gorenflo's method, we have

$$\frac{q''_{w,CHF}}{T_{w,CHF}-T_{sat}}\cdot\frac{1}{H_0}=F_{PR}\left[\frac{q''_{w,CHF}}{q''_0}\right]^n\left(\frac{R_P}{R_{PO}}\right)^{0.133},$$

where $H_0 = 5600$ W/m^2, $q''_0 = 20\,000$ W/m^2, $R_P/R_{P0} = 5$, $F_{PR} = 0.923$, and $n = 0.694$. The result will be $T_{w,CHF} = 495.2$ K.

11.5.2 Film Boiling on a Vertical, Flat Surface

Analysis for a Coherent, Laminar Flow

The vapor film that forms on a vertical wall tends to rise because of buoyancy, much like the boundary layer that forms on vertical surfaces during free convection. In the simplest interpretation, the film can be assumed to remain laminar and coherent, without interfacial waves or instability. The vapor film can then be modeled by using the integral technique. For a contiguous laminar film rising in stagnant liquid in steady state, the vapor momentum conservation equation can be written as (see Fig. 11.12)

$$-\mu_v\frac{d^2U}{dy^2}-g(\rho_L-\rho_v)=0. \tag{11.63}$$

Assuming that the vapor layer has the thickness S, the boundary conditions for Eq. (11.63) are

$U = 0$ at $y = 0$,
$U = 0$ at $y = \delta$ for the stagnant interphase (i.e., no-slip),
$\frac{dU}{dy} = 0$ at $y = \delta$ for the dynamic interphase (i.e., zero shear stress).

The solution of Eq. (11.63) with these boundary conditions gives the velocity profile in the vapor film:

$$U(y) = \frac{g\Delta\rho}{2\mu_v}\left[C_1\delta y - y^2\right], \tag{11.64}$$

where $C_1 = 1$ for the stagnant interphase, and $C_1 = 2$ for the dynamic interphase. The vapor flow rate, per unit width of the vapor film, is related to the velocity profile according to

$$\Gamma_v = \rho_v \int_0^\delta U(y)dy. \tag{11.65}$$

It is now assumed that the temperature profile across the vapor film is linear. Energy balance for an infinitesimally thin slice of the film then gives

$$h_{fg}\frac{d\Gamma_v}{dz} = k_v\frac{T_w - T_{sat}}{\delta}. \tag{11.66}$$

One can now substitute Eq. (11.64) into Eq. (11.65), and then substitute the resulting expression for Γ_v into Eq. (11.66). A differential equation for δ is then obtained that, when solved, leads to

$$\delta = \left[\frac{8}{3(C_1/2 - 1/3)}\frac{k_v(T_w - T_{sat})\mu_v z}{\rho_v h_{fg}g\Delta\rho}\right]^{1/4}. \tag{11.67}$$

Knowing δ, one can calculate the local film boiling heat transfer coefficient from $H = k_v/\delta$. The average heat transfer coefficient for a vertical surface that of length L can then be found from

$$\overline{H}_L = \frac{1}{L}\int_0^L Hdz,$$

and integration yields (Bromley, 1950)

$$\overline{H}_{FB,L} = C\left[\frac{\rho_v h_{fg}g\Delta\rho k_v^3}{(T_w - T_{sat})\mu_v L}\right]^{1/4}, \tag{11.68}$$

where $C = 0.663$ for the stagnant interphase and $C = 0.943$ for the dynamic interphase. The stagnant interphase is evidently more appropriate for film boiling in a quiescent liquid pool.

The derivation thus far has neglected the occurrence of superheating in the vapor film. Some of the heat transferred from the wall to the flow field is evidently used up for the superheating of the vapor film. Equation (11.68) can be corrected for this effect simply by replacing h_{fg} with h'_{fg} where

$$h'_{fg} = H_{fg}\left[1 + 0.34\frac{C_{Pv}(T_w - T_{sat})}{h_{fg}}\right]. \qquad (11.69)$$

Improvements to the Simple Theory

Experiments have shown that Eq. (11.68) underpredicts experimental data when the length of the heated vertical surface is more than about one-half inch (Hsu and Graham, 1986). One reason could be the assumption of laminar film. Hsu and Westwater (1960) performed an analysis similar to Bromley's, but they assumed that the vapor film would become turbulent for $\delta^+ = \delta\sqrt{\frac{\tau_w}{\rho_v}}/\nu_v > 10$. The most serious shortcoming of Eq. (11.68), however, is that it does not account for the intermittency of the vapor film. Based on experimental observations, Bailey (1971) suggested that the vapor film supports a spatially intermittent structure. At the bottom of each spatial interval, the vapor film is initiated and grows, until it becomes unstable and eventually is dispersed by the time it reaches the top of the interval. Following its dispersal, a fresh film is initiated in the next interval. The vapor film remains laminar in the aforementioned intervals. The intermittency results from hydrodynamic instability, and the distance defining the intermittency, S (see Fig. 11.12), follows:

$$S \approx \lambda_{cr} = 2\pi\sqrt{\frac{\sigma}{g\Delta\rho}}. \qquad (11.70)$$

In view of the intermittency of the vapor film, Leonard et al. (1978) proposed that, for vertical surfaces, L in Eq. (11.68) should be replaced with λ_{cr}. With this substitution, Eq. (11.68) is often called the *modified Bromley correlation*. The correlation agrees well with inverted annular data in vertical tubes (Hsu, 1981).

Bui and Dhir (1985) studied saturated film boiling of water on a vertical surface. Their visual observations showed an intermittent, but considerably more complicated, vapor film behavior. Waves of small and large amplitude developed on the vapor–liquid interphase (Fig. 11.13(b)). The amplitude of the large waves was of the order of a few centimeters and grew with distance from the leading edge. The peaks of the waves evolved into bulges that resembled bubbles that were attached to the surface. The bulges acted as vapor sinks for the vapor flowing in the film and grew in size as they moved upward. The local heat transfer coefficient was highly transient as a result of intermittent exposure to vapor film and vapor bulges. Waves with small and large amplitudes, and intermittency with respect to film hydrodynamics as well as heat transfer, were also noted in experiments dealing with subcooled film boiling on vertical surfaces (Vijaykumar and Dhir, 1992a, b).

It should be noted that when the vertical surface is not flat, the analyses here apply as long as the vapor film thickness is much smaller than the principal radii of curvature of the surface. This condition is satisfied in many important applications (e.g., in the rod bundles of nuclear reactor cores and the tube bundles of their steam generator). Also, film boiling on moderately inclined flat surfaces can be treated by using vertical flat surface methods, provided that the gravitational constant g in the correlations is replaced with $g\sin\theta$, with θ representing the angle with the horizontal plane.

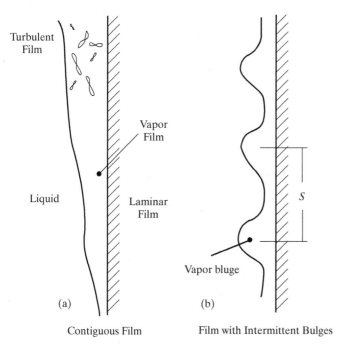

Figure 11.13. Film boiling on a long, vertical surface.

11.5.3 Film Boiling on Horizontal Tubes

This configuration is important for boilers and heat exchanges. A correlation by Breen and Westwater (1962) for film boiling on the outer surface of a horizontal cylinder with diameter D is

$$H = (0.59 + 0.069C)\left[\frac{g\Delta\rho\rho_v k_v^3 h'_{fg}}{\lambda_{cr}\mu_v(T_w - T_{sat})}\right]^{1/4}, \tag{11.71}$$

where

$$h'_{fg} = h_{fg}\left[1 + 0.34\frac{C_{Pv}(T_w - T_{sat})}{h_{fg}}\right]$$

and

$$C = \min(1, \lambda_{cr}/D).$$

11.5.4 The Effect of Thermal Radiation in Film Boiling

Thermal radiation becomes important only when very high heated surface temperatures are encountered. In that sense, film boiling is the only boiling heat transfer regime where radiation is significant. The following simple correction appears to do well in predicting experimental data:

$$H = H_{FB} + \frac{3}{4}H_{rad}, \tag{11.72}$$

where H_{FB} is the film boiling component of the heat transfer coefficient and should be predicted by using expressions similar to Eqs. (11.68), (11.71), etc., and H_{rad} is the

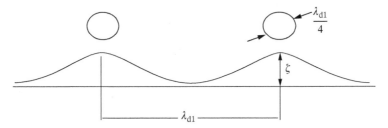

Figure 11.14. Hydrodynamics of MFB according to the model of Zuber (1959).

radiative component found from

$$H_{\mathrm{rad}} \approx \frac{\sigma \varepsilon_w \left(T_w^4 - T_{\mathrm{sat}}^4\right)}{T_w - T_{\mathrm{sat}}}, \tag{11.73}$$

where σ is the Stefan–Boltzmann constant and ε_w is the heated surface emissivity. This simple approximate correction for the effect of radiation is based on treating the heated surface as a small object surrounded by an infinitely large enclosure that has an isothermal surface at the temperature of the surrounding liquid. The correction factor 3/4 is empirical.

11.6 Minimum Film Boiling

The minimum film boiling (MFB) point is an important threshold. Models and correlations for the MFB temperature T_{MFB}, although many, are not very accurate or generally applicable. This is particularly true for transient boiling processes (quenching), where MFB may represent the position of the "quench front." A hydrodynamic model for MFB had been proposed by Zuber (1959) and improved upon by Berenson (1961). However, these models do not consider the effect of heated surface properties on the heat transfer process.

A phenomenon closely related to MFB is the Leidenfrost process, first reported in 1756. It refers to the dancing motion of a liquid droplet on a hot surface, which takes place because of the occurrence of film boiling and the formation of a vapor cushion between the droplet and the hot surface. If the surface temperature is gradually reduced, eventually the Leidenfrost temperature is reached, whereupon the droplet will partially wet the surface and stable film boiling is terminated.

Empirical correlations for MFB include a reduced-state Leidenfrost temperature correlation by Baumeister and Simon (1973):

$$T_{\mathrm{MFB}} = \frac{27}{32} T_{\mathrm{cr}} \left\{ 1 - \exp\left[-0.52 \left(\frac{10^4 (\rho_w / A_w)^{4/3}}{\sigma} \right)^{1/3} \right] \right\}, \tag{11.74}$$

where A_w and ρ_w are the atomic number and density (in grams per cubic centimeter) of the heated surface, T_{cr} is the critical temperature of the fluid, and σ is the liquid–vapor surface tension (in dynes per centimeter). The correlation evidently depends on the fluid–solid pair properties.

Zuber (1959) developed a model for MFB on a horizontal surface, the outline of which is as follows. The process is assumed to be driven by the Taylor instability, as depicted in Fig. 11.14. Bubbles are formed on a two-dimensional grid with a pitch that should be in the $\lambda_{cr} < \lambda < \lambda_d$ range. Let us proceed with $\lambda_d = 2\pi\sqrt{3}\sqrt{\sigma/g\Delta\rho}$ spacing. In each cycle a bubble grows and is released at every grid point (see Fig. 11.14). The bubbles grow as a result of the growth of Taylor waves, and the growth rate of the Taylor wave nodes corresponds to the fastest growing wavelength in the Taylor instability. It is also assumed that bubble release takes place when the peak rises to a height of $\lambda_d/2$, but the released bubble is a sphere with a radius of $\lambda_d/4$. The MFB heat flux is related to bubble release parameters according to

$$q''_{MFB} = f\frac{2}{\lambda_d^2}\frac{4\pi}{3}\left(\frac{\lambda_d}{4}\right)^3\rho_g h_{fg} = \frac{\pi}{24}\lambda_d\rho_g h_{fg}f. \tag{11.75}$$

The factor of 2 is based on the argument that in each complete cycle two bubbles are released from a unit cell (i.e., four one-quarter bubbles from the four corners of the unit cell in one-half cycle and one complete bubble from the center of the unit cell in the second half of the cycle). The bubble release frequency is found from

$$f = \overline{(d\zeta/dt)}/\lambda_d,$$

where the average growth rate of the wave displacement is represented as

$$\overline{(d\zeta/dt)} = \frac{1}{0.4\lambda_d}\int_0^{0.4\lambda_d}(d\zeta/dt)d\zeta.$$

The wave displacement follows

$$\zeta = \zeta_0\exp[i(\omega t - kx)]_d. \tag{11.76}$$

This would lead to $\overline{(d\zeta/dt)} = 0.2\omega_d\lambda_d$, and from there

$$f = 0.2\omega_d = 0.2\left[\frac{4(\Delta\rho)^3 g^3}{27\sigma(\rho_f + \rho_g)^2}\right]^{0.25},$$

where Eq. (2.178) in Chapter 2 has been used for ω_d. The analysis thus leads to the following expression:

$$q''_{MFB} = C_1\rho_v h_{fg}\left[\frac{\sigma g\Delta\rho}{(\rho_f + \rho_g)^2}\right]^{1/4}. \tag{11.77}$$

Zuber's analysis leads to $C_1 = 0.176$. This expression with the latter value for C_1 was found to overpredict experimental data, however. Based on experimental data, Berenson (1961) modified the coefficient to $C_1 = 0.091$ and replaced h_{fg} with h'_{fg} to account for the effect of vapor film superheating.

Note that by knowing q''_{MFB} and H_{MFB} (the latter from Eq. (11.62)), the surface temperature at MFB, namely, T_{MFB}, can be calculated from

$$(T_{MFB} - T_{sat})_{Berenson} = (q''_{MFB}/H_{FB})_{Berenson}. \tag{11.78}$$

Berenson performed the substitutions in Eq. (11.78); however, he adjusted the numerical coefficient in the resulting expression, making it applicable only with the English unit system. To avoid confusion, it is easier to directly use Eqs. (11.78) and (11.62).

In the analysis presented here it has evidently been assumed that surface properties have no effect on the MFB parameters. However, experimental data show that the thermophysical properties of the heated surface do affect T_{MFB}. A correlation that corrects Berenson's model for T_{MFB} for the effects of the solid surface thermophysical properties was proposed by Henry (1974). Accordingly,

$$\frac{T_{MFB} - T_{MFB}^*}{T_{MFB}^* - T_L} = 0.42 \left[\sqrt{\frac{(\rho C k)_f}{(\rho C k)_w}} \frac{h_{fg}}{C_w (T_{MFB}^* - T_{sat})} \right]^{0.6}, \tag{11.79}$$

where T_{MFB}^* is the MFB temperature predicted by Berenson's correlation.

EXAMPLE 11.4. Calculate the minimum film boiling heat flux and temperature for the conditions of Example 11.1. Assume that the disk is made of stainless steel.

SOLUTION. The saturation properties that are needed are $\rho_f = 958.4$ kg/m^3, $\rho_g = 0.597$ kg/m^3, $h_{fg} = 2.337 \times 10^6$ J/kg, $C_{Pg} = 1,987$ J/kg·K, $\sigma = 0.059$ N/m, $T_{sat} = 373$ K, $C_{Pf} = 4,217$ J/kg·K and $\alpha_f = 1.646 \times 10^{-7}$ m^2/s.

We need to estimate the mean vapor film properties. Let us use $T_{film} = T_{sat} + 40$ K as an estimation. The following vapor film properties accordingly represent superheated vapor at one atmosphere pressure and 413 K: $k_v = 0.0028$ W/m·K, $\mu_v = 1.38 \times 10^{-5}$ kg/m·s, $\rho_v = 0.537$ kg/m^3, and $h_{fg}' = 2.41 \times 10^6$ J/kg.

Equation (11.77) is now solved using $C_1 = 0.091$, resulting in

$$q_{MFB}'' = 17,679 \text{ W/m}^2.$$

The film boiling heat transfer coefficient is next calculated by using Berenson's correlation, Eq. (11.62), leading to

$$H_{Berenson} = 242.7 \text{ W/m}^2 \cdot \text{K}.$$

We can now write

$$T_{MFB,Berenson} - T_{sat} = q_{MFB}''/H_{Berenson} = 73 \text{ K} \Rightarrow T_{MFB,Berenson} = 446 \text{ K}.$$

We now apply the correlation of Henry (1974), Eq. (11.79), noting that $T_{MFB}^* = 446$ K, and $T_L = T_{sat}$. For the solid properties, let us use the properties of AISI 302 stainless steel at 446 K, whereby $\rho_w = 7,998$ kg/m^3, $C_w = 523$ J/kg·K, and $k_w = 17.9$ W/m·K. Equation (11.79) then gives $T_{MFB} = 579$ K.

11.7 Transition Boiling

In transition boiling, as mentioned earlier, the heated surface is partially in nucleate boiling and partially in film boiling. The transition boiling regime is poorly understood and has received relatively little research attention in the past. Industrial systems usually are not designed to operate in this regime. However, transition boiling is important in transient processes, particularly during the quenching of hot surfaces. Quenching of hot surfaces by liquid occurs during the reflood phase of a loss-of-coolant accident (LOCA), when the hot and partially dry fuel rods are subject to liquid supplied by the emergency cooling system.

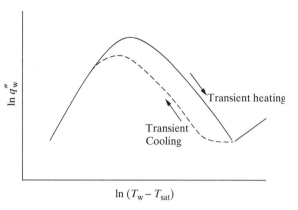

Figure 11.15. The transition boiling regime during heating and cooling transients.

Some important parametric trends in transition boiling are the following.

(a) Surface roughness moves the transition boiling line in the boiling curve toward the left.
(b) Improved wettability (lowering of contact angle) improves (increases) the transition boiling heat transfer coefficient.
(c) In transient tests, the transition boiling line obtained with transient heating (increasing T_w) is higher than with transient cooling (decreasing T_w), as shown in Fig. 11.15.
(d) Deposition of contaminants improves heat transfer in transient boiling.

Most of the widely used correlations for transition boiling are based on interpolations between CHF and MFB points. A few examples follow. The correlation of Bjonard and Griffith (1977) is

$$q''_{TB}(T_w) = Cq''_{CHF} + (1 - C)q''_{MFB}, \tag{11.80}$$

where

$$C = \left(\frac{T_{MFB} - T_w}{T_{MFB} - T_{CHF}}\right)^2. \tag{11.81}$$

Linear interpolation on a log–log scale is recommended by Haramura (1999):

$$\frac{\ln[q''_{TB}(T_w)/q''_{MFB}]}{\ln(q''_{CHF}/q''_{MFB})} = \frac{\ln[\Delta T_{MFB}/(T_w - T_{sat})]}{\ln(\Delta T_{MFB}/\Delta T_{CHF})}, \tag{11.82}$$

where $\Delta T_{MFB} = T_{MFB} - T_{sat}$ and $\Delta T_{CHF} = T_{CHF} - T_{sat}$.

11.8 Pool Boiling in Binary Liquid Mixtures

Boiling and condensation in miscible multicomponent liquids is common in petrochemical, refrigeration, and cryogenics engineering. The phenomenology of boiling and condensation in miscible multicomponent liquids has much in common with boiling and condensation in pure liquids. In pool boiling, for example, bubble nucleation and ebullition processes in multicomponent liquids are qualitatively similar to those in single-component liquids. Yet, important differences occur in the details of transport processes that, except for azeotropic mixtures, make single-component

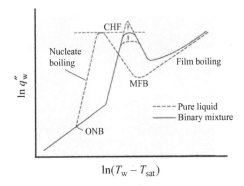

Figure 11.16. The pool boiling curves for a liquid substance and a binary mixture made of the same liquid mixed with another miscible liquid. (After Collier and Thome, 1994.)

predictive methods of little direct use for multicomponent mixtures. The most important difference between single-component liquids and multicomponent mixtures is that in the former transport processes are controlled by heat transfer, whereas in the latter we deal with combined heat and mass transfer processes. The mass transfer resistances that result from the uneven volatility of the components of the liquid mixture are often significant and deteriorate the wall–fluid heat transfer rate in comparison with a single-component fluid. The addition of a relatively small amount of one liquid to another results in considerable change in the boiling behavior of the liquid.

Figure 11.16 compares qualitatively the pool boiling curve of a pure substance with the pool boiling curve of the same substance when it is mixed with another miscible liquid. With the exception of the critical heat flux (CHF), the heat transfer coefficient is lower for the binary mixture. In the binary mixture the onset of nucleate boiling (ONB) occurs at a higher wall superheat, and in the nucleate boiling region the heat transfer coefficient is significantly lower. The effect on the CHF is complicated and the CHF in the binary mixture can be lower or higher than the CHF in single-component liquid.

In this section pool boiling in miscible binary liquid mixtures is discussed. The discussion will be limited to binary mixtures because: (a) binary mixtures are widely used, in particular in newer environmentally friendly refrigerants; and (b) there is a fairly well-developed literature dealing with such mixtures.

11.8.1 Nucleate Boiling Process

The behavior of miscible binary liquid mixtures when they are at equilibrium with their vapors was discussed in Section 1.4. In near-azeotropic mixtures, for which the dew point and bubble point temperatures are approximately equal, the mixture behaves similarly to a single-component fluid. Theory and practice associated with single-component boiling should then apply to such mixtures. The situation is different for zeotropic (non-azeotropic) mixtures, however. Intuition would suggest that the larger the difference between the dew and bubble temperatures, the more significant the deviation from the behavior of single-component liquids should be. Experience supports this intuition.

The role of mass transfer complicates the processes involved in multicomponent boiling and condensation. Boiling is accompanied by rapid evaporation at the liquid–vapor interphase, where the more volatile species evaporate faster.

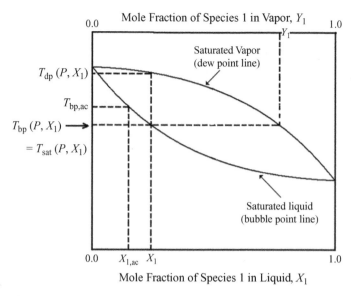

Figure 11.17. Constant-pressure phase diagram of a binary mixture during rapid evaporation.

Consider the phase diagram for a binary mixture shown in Fig. 11.17, where subscript 1 represents the more volatile species (i.e., the species with a lower boiling temperature), and X_1 represents the mole fraction of species 1 in the liquid bulk. Assume, for simplicity, that the liquid medium is infinitely large and is at the temperature T_{bp} (P, X_1), and assume that evaporation takes place at the interphase between this liquid medium and a vapor space. If evaporation at the gas–liquid interphase is very slow, then the mole fraction of species 1 in the generated vapor will be equal to Y_1, and the mole fraction of species 1 will remain uniform and equal to X_1 in the liquid. The situation will be different when rapid evaporation takes place, however. Preferential and stronger evaporation of the more volatile species will take place, and that leads to the partial depletion of species 1 in the liquid near the interphase. The mole fraction of species 1 in the liquid adjacent to the interphase will then move to, say, $X_{1,ac}$. (An opposite effect occurs during condensation, which will be discussed in Section 15.9.) This deteriorates the heat transfer in two ways. First, it causes the liquid–vapor interfacial temperature to increase from T_{bp} (P, X_1), which is often referred to as the saturation temperature of the mixture and is shown as T_{sat} (P, X_1) or simply T_{sat}, to $T_{bp,ac}$ (see Fig. 11.17), while the liquid bulk will remain at the temperature T_{bp} (P, X_1). The effect of increasing the liquid–vapor interphase temperature here is similar to increasing the saturation temperature in single-component boiling, which will lead to a reduction of the heat transfer rate when the wall surface temperature is constant. Second, because of the aforementioned preferential evaporation of the more volatile species the composition in the liquid phase will become non-uniform. The liquid adjacent to the interphase will have a higher concentration of the less volatile species (species 2) than the bulk liquid. Diffusive mass transfer sets up in the liquid phase, whereby species 1 diffuses towards the interphase while species 2 diffuses in the opposite direction and away from the interphase. The evaporation process in a binary liquid mixture is thus a combined heat and mass transfer process, and molecules of species 1 in the liquid have to overcome a mass transfer resistance before they can reach the interphase. This phenomenon is referred to as the diffusion resistance. The mass transfer resistance is often more important than the liquid-side thermal resistance.

In heterogeneous nucleate boiling on a heated surface a similar increase in the boiling temperature of the mixture occurs due to the following phenomenology. Suppose that the local mole fraction of species 1 is equal to X_1 in the liquid. The bubble that forms on the wall crevices will then contain species 1 at a mole fraction equal to Y_1. The formation, growth, and detachment of these bubbles, which contain species 1 at a concentration higher than the concentration of species 1 in the ambient liquid, will partially deplete the liquid in the vicinity of the heated surface from species 1, leading an increase of the local liquid saturation temperature from $T_{bp}(P, X_1)$ to $T_{bp,ac}$.

The significance of diffusion resistance was recognized early on (van Wijk *et al.*, 1956; Sternling and Tichacek, 1961), and its role has been demonstrated experimentally (see, for example, Wagner and Stephan, 2008). Attempts to include the mass transfer effect in heterogeneous bubble ebullition models has been made (van Stralen, 1966). For the asymptotic growth of a spherical vapor bubble surrounded by a superheated liquid, for example, Scriven (1959) was the first to develop a theoretical model, and van Stralen (1962) (see van Stralen and Cole, 1979) extended the analytical solution of Plesset and Zwick (1954) (see Eq. (2.267)) to a binary liquid. The solution of van Stralen (1962), for example, is

$$R = \sqrt{\frac{3}{\pi}} \frac{2k_L(T_\infty - T_{sat})}{\sqrt{\alpha_L}\rho_v h_{Lv}} \sqrt{t} N_{Sn}, \tag{11.83}$$

where the Scriven number is defined as (Thome and Shock, 1984):

$$N_{Sn} = \left\{ 1 - (Y_1 - X_1)\left(\frac{\alpha_L}{D_{12,L}}\right)^{1/2} \left(\frac{C_{PL}}{h_{Lv}}\right) \frac{dT_{bp}}{dX_1} \right\}^{-1}. \tag{11.84}$$

In this equation Y_1 is the mole fraction of the more volatile species in the vapor phase, which is assumed to be uniform, but X_1 is the mole fraction of the more volatile species in the liquid at the interphase. The term $\frac{dT_{bp}}{dX_1}$ represents the slope of the bubble point curve. Evidently, the factor N_{Sn} on the right side of Eq. (11.83) represents the slowing down of the growth of the bubble in comparison with the bubble growth in a pure liquid. The terms $(Y_1 - X_1)$ and $\frac{dT_{bp}}{X_1}$ always have opposite signs, and as a result $N_{Sn} < 1$ always.

11.8.2 Nucleate Boiling Heat Transfer Correlations

Boiling heat transfer for a binary fluid mixture is presented based on $T_w - T_{bp}(P, X_1)$ (or $T_w - T_{sat}(P, X_1)$) as the nominal temperature difference that drives heat transfer, for practical reasons. We can thus write

$$q''_w = H[T_w - T_{bp}(P, X_1)], \tag{11.85}$$

where H is the boiling heat transfer coefficient. Of course, as was mentioned before, $T_{bp}(P, X_1)$ is not equal to the real interfacial temperature due to the aforementioned mass transfer resistance and it would be more appropriate if the actual interfacial temperature were used instead of $T_{bp}(P, X_1)$. This is not practical, however, because the interfacial temperature is difficult to calculate.

To better understand the ideas behind the nucleate boiling heat transfer correlations, let us revisit the observations regarding the phenomenology of pool boiling in a non-azeotropic binary mixture. As explained earlier, the nucleate boiling heat transfer is slowed down because of the mass transfer resistance in the liquid at the vicinity of the interphase. Thus, in comparison with an ideal binary mixture with no mass diffusive resistance, the following two trends should be expected.

(1) When the wall heat flux is fixed, heat transfer coefficient is smaller (i.e. $H < H_{ideal}$), therefore a larger wall superheat will occur in the real system, i.e., $\Delta T_w > \Delta T_{w,ideal}$. In other words, $T_w > T_{w,ideal}$.
(2) When the wall temperature and coolant conditions (i.e., pressure, mass flux, quality, X_1, and therefore $T_{bp}(P, X_1)$) are fixed, then the heat flux will be lower, i.e., $q''_w < q''_{w,ideal}$.

Thus, for the development of a predictive model, two things are needed: a method to calculate the ideal heat transfer coefficient, H_{ideal}, and a method to adjust H_{ideal} in order to obtain H. Let us discuss H_{ideal} first.

Let us define H_1 and H_2 as the nucleate boiling heat transfer coefficients for pure species 1 and 2, respectively, at the same pressure and wall heat flux to which the binary mixture is subjected. We must evidently have

$$\lim_{X_1 \to 1} H_{ideal} = \lim_{m_{L,1} \to 1} H_{ideal} = H_1 \tag{11.86}$$

$$\lim_{X_1 \to 0} H_{ideal} = \lim_{m_{L,1} \to 0} H_{ideal} = H_2. \tag{11.87}$$

For an ideal binary liquid mixture the heat transfer coefficient can be estimated based on a mole fraction-weighted or mass faction-weighted heat transfer coefficient defined respectively as

$$H_{ideal} = H_1 X_1 + (1 - X_1) H_2 \tag{11.88}$$

$$H_{ideal} = H_1 m_1 + (1 - m_1) H_2. \tag{11.89}$$

Alternatively, we can write

$$H_{ideal} = \left[\frac{X_1}{H_1} + \frac{(1 - X_1)}{H_2} \right]^{-1} \tag{11.90}$$

$$H_{ideal} = \left[\frac{m_{L,1}}{H_1} + \frac{(1 - m_{L,1})}{H_2} \right]^{-1}. \tag{11.91}$$

These definitions all satisfy Eqs. (11.86) and (11.87).

Using these ideal binary mixture heat transfer coefficients, we can derive expressions for the ideal wall superheat. We note that for the ideal system we must have

$$q''_{w,ideal} = H_{ideal} \Delta T_{w,ideal}. \tag{11.92}$$

For the limiting conditions represented by Eqs. (11.86) and (11.87) the wall super-heats will respectively be $\Delta T_{w,1}$ and $\Delta T_{w,2}$, where:

$$\Delta T_{w,1} = q_w'' / H_1 \tag{11.93}$$

$$\Delta T_{w,2} = q_w'' / H_2. \tag{11.94}$$

Evidently, $\Delta T_{w,1}$ is the wall superheat needed if the fluid is pure species 1, when the same local pressure and heat flux are imposed; while $\Delta T_{w,2}$ is the wall superheat needed if the fluid is pure species 2, when the same local pressure and wall heat flux are imposed. It can be seen that the wall superheats for the ideal mixture, consistent with Eqs. (11.88) and (11.89) will then be, respectively:

$$\frac{1}{\Delta T_{w,ideal}} = \frac{X_1}{\Delta T_{w,1}} + \frac{1 - X_1}{\Delta T_{w,2}} \tag{11.95}$$

$$\frac{1}{\Delta T_{w,ideal}} = \frac{m_{L,1}}{\Delta T_{w,1}} + \frac{1 - m_{L,1}}{\Delta T_{w,2}}. \tag{11.96}$$

Likewise, it can be easily shown that the wall superheats for the ideal mixture, consistent with Eqs. (11.90) and (11.91) will then be, respectively:

$$\Delta T_{w,ideal} = \Delta T_{w,1} X_1 + \Delta T_{w,2}(1 - X_1) \tag{11.97}$$

$$\Delta T_{w,ideal} = \Delta T_{w,1} m_1 + \Delta T_{w,2}(1 - m_1). \tag{11.98}$$

Having derived expressions for the ideal binary mixture heat transfer coefficient, we now discuss methods that can be used to adjust H_{ideal} in order to obtain H, the real system heat transfer coefficient. We can define a correction factor F according to:

$$F = H_{ideal} / H, \tag{11.99}$$

where $F \geq 1$ is a correction factor which must satisfy the following limits:

$$\lim_{X_1 \to 0} F = \lim_{m_{L,1} \to 1} F = 1. \tag{11.100}$$

The deterioration in the heat transfer can be interpreted as a reduction in the heat transfer coefficient caused by the addition of mass transfer resistance to the evaporation process while the nominal wall superheat is used. One can write:

$$q_w'' = \frac{H_{ideal}}{F} [T_w - T_{bp}(P, X_1)]. \tag{11.101}$$

A number of empirical and semi-analytical correlations have been proposed for nucleate boiling heat transfer in non-azeotropic liquid mixtures. Some of the most widely applied correlations are now discussed.

Stephan and Körner (1969) used Eq. (11.90) for the definition of the ideal mixture heat transfer coefficient and argued that the increase in the liquid–vapor interphase temperature that results from preferential evaporation of the more volatile component must be proportional to $|Y_1 - X_1|$, and proposed

$$F = 1 + A_0(0.88 + 0.12P)|Y_1 - X_1|, \tag{11.102}$$

where P is in MPa, and the coefficient A_0 represents experimental measurements corresponding to the reference pressure 0.1 MPa. Table 11.4 is a summary of A_0 values for several binary mixtures.

Table 11.4. *Values of parameter A_0 in the correlation of Stephan and Körner (1969).*[a]

Mixture	A_0	Mixture	A_0
Acetone–ethanol	0.75	Methanol–benzene	1.08
Acetone–butanol	1.18	Methanol–amyl alcohol	0.80
Acetone–water	1.40	Methyl ethyl ketone–toluene	1.32
Ethanol–benzene	0.42	Methyl ethyl ketone–water	1.21
Ethanol–cyclohexane	1.31	Propanol–water	3.29
Ethanol–water	1.21	Water–glycol	1.47
Benzene–toluene	1.44	Water–glycerol	1.50
Heptane–methylcyclohexane	1.95	Water–pyridine	3.56
Isopropanol–water	2.04		

[a] From Celata *et al.* (1994c).

Calus and Leonidopoulos (1974) utilized the aforementioned analytical solution of Scriven (1959) for asymptotic growth of a spherical bubble in a binary liquid mixture, and consistent with Eqs. (11.91) proposed

$$F = N_{\text{Sn}}^{-1} = 1 - (m_{v,1} - m_{L,1})\left(\frac{\alpha_L}{D_{12,L}}\right)^{1/2}\left(\frac{C_{PL}}{h_{Lv}}\right)\frac{dT_{bp}}{dm_{L,1}}. \tag{11.103}$$

A widely referenced empirical correlation, derived by Schlünder (1982), is

$$F = 1 + \frac{H_{\text{ideal}}}{q_w''}[T_{\text{sat},2}(P) - T_{\text{sat},1}(P)](Y_1 - X_1)\left[1 - \exp\left(-\frac{B_0 q_w''}{\rho_L \beta_L h_{Lv}}\right)\right] \tag{11.104}$$

where H_{ideal} is found from Eq. (11.90), $B_0 \approx 1$ is an empirical coefficient, and β_L represents the mass transfer coefficient associated with the mass transfer near the interphase in the liquid phase that is caused by the nonuniformity of concentrations that result from preferential evaporation of the more volatile component. The typical values are in the 2×10^{-4} to 3×10^{-4} m/s range (Schlünder, 1982; Taboas *et al.* 2007; Wagner and Stephan, 2008). (Note that $\rho_L \beta_L$ is similar to the mass transfer coefficient K, defined and discussed in Chapter 2.)

Thome and Shakir (1987) modified the correlation of Schlünder (1982) according to

$$F = 1 + \frac{H_{\text{ideal}}}{q_w''}[T_{dp}(P, X_1) - T_{bp}(P, X_1)]\left[1 - \exp\left(-\frac{B_0 q_w''}{\rho_L \beta_L h_{Lv}}\right)\right] \tag{11.105}$$

where H_{ideal} is found from Eq. (11.90).

Using experimental data representing five different binary mixtures (methanol–water, ethanol–water, ethanol–*n*-butanol, methanol–ethanol, and methanol–benzene), and based on the argument that the correction factor F should depend on the ratio between the wall heat flux and the critical heat flux, q_w''/q_{CHF}'', Fujita and Tsutsui (1997) have proposed

$$F = 1 + \frac{H_{\text{ideal}}}{q_w''}[T_{dp}(P, X_1) - T_{bp}(P, X_1)]\left[1 - \exp\left(-\frac{60 q_w''}{q_{\text{CHF,ref}}''}\right)\right], \tag{11.106}$$

where H_{ideal} is found from Eq. (11.90), and the reference critical heat flux is defined as

$$q_{\text{CHF,ref}}'' = \rho_g^{1/2} h_{Lv}[\sigma g(\rho_L - \rho_g)]^{1/4}. \tag{11.107}$$

396 Pool Boiling

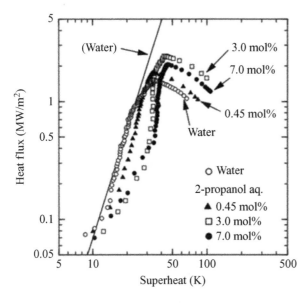

Figure 11.18. Pool boiling curves of water–propanol mixtures on a 12 mm horizontal disk under atmospheric pressure (after Sakashita *et al.*, 2010).

This heat flux is in fact the same as the critical heat flux according to Zuber's model (see Eq. (11.46)), without the multiplier constant in the latter correlation.

Kandlikar (1998b) has developed a predictive method based on theoretical considerations. The method of Kandlikar, although more complicated, has been shown to predict data representing benzene–methanol, R-23–R-13, R-22–R-12, water–ethylene glycol, and methane–water mixtures. Correlations that are accurate for specific widely encountered mixtures, such as ammonia–water, (Taboas *et al.*, 2007) and liquefied natural gas (Gong *et al.*, 2009) have also been recommended in the past.

11.8.3 Critical Heat Flux

Critical heat flux in binary liquids has been investigated for several decades (van Stralen, 1956; van Wijk *et al.*, 1956; Carne, 1964). It was observed early on that the addition of a small amount of alcohol (a few volume per cent) to water would increase the critical heat flux significantly in comparison with the critical heat flux of pure water at similar conditions. Similar observations have been made for other liquid pairs; the addition of a small amount of a more volatile liquid to another liquid increases the critical heat flux of the mixture in comparison with the critical heat flux of the less volatile liquid under similar conditions. Parametric dependence of the critical heat flux on the concentration make-up of a binary mixture is complicated, however, and the critical heat flux for a non-azeotropic binary liquid mixture in general cannot be predicted by simple weight-averaging of the critical heat fluxes associated with the components of the liquid at pure conditions. Figures 11.18, borrowed from Sakashita *et al.* (2010), is an example of the dependence of the critical heat flux on the composition of the binary liquid mixture.

Three mechanisms have been proposed for the complicated dependence of the critical heat flux on the composition of binary liquid mixtures, the first two of which contribute to the deterioration of nucleate boiling in binary liquids as well.

(1) The preferential evaporation of the more volatile species leads to the relative depletion of that species in the liquid in the vicinity of the interphase. The liquid near the interphase will then have a lower concentration of the more volatile species than the liquid bulk. Diffusion of the more volatile species will then take place from the liquid bulk towards the interphase, accompanied by the diffusion of the less volatile species in the opposite direction. The resulting mass transfer resistance slows the evaporation of the less volatile species and thereby slows the growth rate of bubbles during nucleate boiling. The outcome of the slower bubble growth is an increase of critical heat flux. This mechanism was proposed by van Stralen and co-workers (van Wijk *et al.*, 1956; van Stralen, 1967).

(2) The reduction in the concentration of the more volatile component in the liquid near the gas–liquid interphase causes the interphase temperature to be higher than $T_{bp}(P, X_1)$, the bubble point temperature of the bulk liquid. This phenomenon is called induced subcooling and causes an increase in critical heat flux (McEligot, 1964; Reddy and Lienhard, 1989).

(3) The above two mechanisms do not explain the magnitude of the difference in critical heat flux of binary mixtures in comparison with the critical heat flux in either pure component that is observed in many experiments. On the other hand, a clear correlation between the magnitude and sign of the change in critical heat flux and the Marangoni effect (i.e., the flow induced by the nonuniformity of the surface tension) can be observed in experiments. The third mechanism, originally suggested by Hovestreijdt (1963), is thus based on the Marangoni effect. The aforementioned preferential evaporation of the more volatile species occurs faster in areas with higher evaporation mass flux. The resulting nonuniformity in the mole fraction of the more volatile species causes nonuniformity in surface tension. The resulting gradient in surface tension causes the liquid underneath the interphase to move in the direction of the gradient of the surface tension (i.e., from a low-surface-tension area to a high-surface-tension area). (This can be shown by examination of Eq. (2.25) by considering a vapor–liquid mixture where the viscosity of the vapor phase is negligible.) When the more volatile species has a lower surface tension than the less volatile species, colder liquid from the bulk will move towards areas with high evaporation mass flux, in particular towards the thin liquid film (the macro layer) underneath bubbles growing on wall crevices, thereby postponing (hence increasing the magnitude of) the critical heat flux (Fujita *et al.*, 1995; McGillis and Carey, 1996; Fujita and Bai, 1998). This mechanism improves the critical heat flux only if $-\frac{\partial \sigma}{\partial X_1} > 0$, with X_1 representing the mole fraction of the more volatile species, however. In addition to increasing the critical heat flux by improving the replenishment of the liquid micro film underneath bubbles, this Marangoni effect also causes the average liquid layer underneath vapor pockets to be thicker near the center of a finite-sized heated surface (Sakashita *et al.*, 2010). A thicker average liquid film thickness evidently leads to a higher critical heat flux.

Many empirical and semi-empirical correlations have been proposed for the CHF of binary mixtures. The correlations discussed below are among the more recent, that are consistent with the aforementioned third mechanism.

Based on experimental data with methane–water, 2-propane–water, and ethylene glycol–water mixtures, McGillis and Carey (1996) proposed,

$$\frac{q''_{CHF}}{q''_{CHF,0}} = \left[1 + C_m \frac{1}{\sigma}\left|\left(\frac{\partial\sigma}{\partial X_1}\right)(Y_1 - X_1)\right|\right]^{1/4}, \tag{11.108}$$

where $q''_{CHF,0}$ is the CHF for a single-component liquid which has the properties of the liquid mixture, and $C_m \approx 1.14$.

Based on the argument that the Marangoni effect replenishes the liquid film (macro film) underneath an expanding bubble and therefore retards the expansion of the dry area beneath the bubble, Yagov (2004) proposed

$$\frac{q''_{CHF}}{q''_{CHF,0}} = 1 + C\frac{\sigma(T_{bp}, X_1) - \sigma(T_{dp}, X_1)}{\sigma(T_{bulk}, X_1)}; \qquad C \approx 1, \tag{11.109}$$

where T_{bulk} is the liquid mixture bulk temperature.

Sakashita *et al.* (2010) have found that the correlation of Yagov agreed well with their 2-propanol–water CHF data, while the aforementioned correlation of McGillis and Carey disagreed with their data.

Using experimental data covering the entire concentration range for seven binary liquid mixtures (methanol–water, ethanol–water, methanol–ethanol, methanol–n-butanol, methanol–benzene, benzene–heptane, water–ethylene glycol), Fujita and Bai (1998) derived

$$\frac{q''_{CHF}}{q''_{CHF,0}} = \left[1 - (1.83 \times 10^{-3})\frac{|Ma|^{1.43}}{Ma}\right]^{-1}, \tag{11.110}$$

where the Marangoni number is defined as

$$Ma = \frac{\sigma_D - \sigma_A}{\rho_L v_L^2}\left[\frac{\sigma}{g(\rho_L - \rho_v)}\right]^{1/2} Pr_L, \tag{11.111}$$

where subscripts A and D correspond to points A and D in Fig. 11.19. The single-component CHF is to be found from the correlation of Kutateladze and Borishanski (similar to the correlation of Zuber *et al.* (1963), see Eq. (11.46) and the discussion in Section 11.4), when adjustment is made for the finite size of the heated surface

$$q''_{CHF,0} = 0.131\{[f(l')]_1 X_1 + [f(l')]_2(1 - X_1)\}\rho_g^{1/2}h_{fg}[\sigma g\rho(\rho_L - \rho_v)]^{1/4}, \tag{11.112}$$

where $[f(l')]_1$ and $[f(l')]_2$ are the size and shape correction factors (see Eq. (11.48) and Table 11.3) calculated using pure liquid species 1 and 2, respectively.

EXAMPLE 11.5. A horizontal metallic heated surface is at 310 K temperature, and is cooled by a stagnant water–ammonia liquid mixture at 1 bar pressure. The mole fraction of ammonia in the liquid is 40%. Calculate the heat flux. Assume an equivalent surface roughness of 20 μm.

SOLUTION. We will use the correlation of Schlünder (1982). Let us use subscripts 1 and 2 to refer to ammonia and water, respectively. We first need to calculate the ideal mixture nucleate boiling heat transfer coefficient from Eq. (11.90). Let us use the correlation of Gorenflo, Eqs. (11.40)–(11.43). We need to calculate H_1 and H_2.

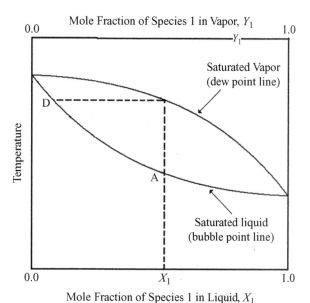

Figure 11.19. Definitions used in the correlation of Fujita and Bai (1998).

Therefore

$$q_0'' = 20{,}000 \text{ W/m}^2$$

$$H_{0,1} = 7{,}000 \text{ W/m}^2 \cdot \text{K}$$

$$T_{\text{sat},1}|_{P=1 \text{ bar}} = 239.6 \text{ K}$$

$$P_{\text{cr},1} = 130 \text{ bars}$$

$$P_{r,1} = \frac{1 \text{ bar}}{130 \text{ bars}} = 0.00885$$

$$n_1 = 0.9 - (0.3)(0.00885)^{0.3} = 0.828$$

$$F_{\text{PR},1} = 1.2(0.00885)^{0.27} + \left[2.5 + \frac{1}{1 - 0.00885}\right](0.00885) = 0.366$$

$$H_{0,2} = 5{,}600 \text{ W/m}^2 \cdot \text{K}$$

$$T_{\text{sat},2}|_{P=1 \text{ bar}} = 372.8 \text{ K}$$

$$P_{\text{cr},2} = 220.6 \text{ bars}$$

$$P_{r,2} = \frac{1 \text{ bars}}{220.6 \text{ bars}} = 0.004533$$

$$n_2 = 0.9 - (0.3)(0.004533)^{0.15} = 0.766$$

$$F_{\text{PR},2} = 1.73(0.004533)^{0.27} + \left[6.1 + \frac{0.68}{1 - 0.004533}\right](0.004533)^2 = 0.4031$$

$$\frac{H_1}{7{,}000 \text{ W/m}^2 \cdot \text{K}} = (0.3659)\left(\frac{q_w''}{2 \times 10^4 \text{ W/m}^2}\right)^{0.828}\left(\frac{2 \text{ μm}}{0.4 \text{ μm}}\right)^{0.133} \qquad \text{(a)}$$

$$\frac{H_2}{5{,}600 \text{ W/m}^2 \cdot \text{K}} = (0.3659)\left(\frac{q_w''}{2 \times 10^4 \text{ W/m}^2}\right)^{0.403}\left(\frac{2 \text{ μm}}{0.4 \text{ μm}}\right)^{0.133}. \qquad \text{(b)}$$

We will use the property routines in Engineering Equation Solver (EES) code. Accordingly, for the given pressure and ammonia mole fraction we get for liquid mixture properties:

$$T_{sat} = T_{bp} = 11.6\,°C = 284.7\ K$$
$$\rho_L = 867.3\ kg/m^3$$
$$\mu_L = 0.001332\ kg/ms$$
$$h_L = -176\ kJ/kg$$
$$h_v = 2183\ kJ/kg$$
$$h_{Lv} = h_v - h_L = 2359\ kJ/kg.$$

Also, from the phase diagram for water–ammonia mixture at 100 kPa pressure, we find that for $X_1 = 0.4, Y_1 = 0.995$. The ideal mixture heat transfer coefficient follows Eq. (11.90), we will therefore have

$$H_{ideal} = \left[\frac{X_1}{H_1} + \frac{(1 - X_1)}{H_2}\right]^{-1}. \tag{c}$$

We now need to apply the correlation of Schlünder (1982), Eq. (11.104). We assume $B_0 = 1$ and $\beta_L = 2 \times 10^{-4}$ m/s.

$$F = 1 + \frac{H_{ideal}}{q_w''}[372.8\ K - 239.6\ K](0.995 - 0.4)$$
$$\cdot \left[1 - \exp\left(-\frac{q_w''}{(867.3\ kg/m^3)(2 \times 10^{-4}\ m/s)(2.36 \times 10^6\ J/kg)}\right)\right], \tag{d}$$

$$q_w'' = H_{ideal}(T_w - T_{sat})/F$$
$$\Rightarrow q_w'' = H_{ideal}(310\ K - 284.7\ K)/F. \tag{e}$$

Equations (a) through (e) are now iteratively solved for H_1, H_2, H_{ideal}, F and q_w''. The solution gives

$$H_1 = 11.16\ kW/m^2{\cdot}K$$
$$H_2 = 8.964\ kW/m^2{\cdot}K$$
$$H_{ideal} = 9.729\ kW/m^2{\cdot}K$$
$$F = 2.687$$
$$q_w'' = 91.44\ kW/m^2$$

PROBLEMS

11.1 Using the boiling nucleation criteria of Hsu (1962), calculate the size ranges of sharp-edged wall crevices that can serve as active nucleation sites for a solid surface submerged in atmospheric saturated water, assuming contact angles of $\theta = 35°$ and $50°$.

11.2 Calculate and plot the bubble departure diameter as a function of pressure for refrigerant $R - 22$ in the 1- to 15-bar range, using the correlations of Fritz (1935) and Gorenflo *et al.* (1986), for a solid surface assuming $\theta = 45°$, and assuming a wall superheat of 5 °C.

11.3 A stainless-steel horizontal cylindrical heater 1 cm in diameter and 30 cm long is immersed in a pool of saturated water under atmospheric pressure conditions.

(a) Calculate the critical heat flux (CHF), and estimate the surface temperature associated with CHF.

(b) Calculate the minimum film boiling heat flux and surface temperature.

(c) Find the total heat transfer rate to the pool when the heater surface is at 108 °C, 115 °C, and 250 °C.

11.4 The heater in Problem 11.3 is immersed in a pool of saturated R-22 at 15-bar pressure. Repeat parts (a) and (b) of the calculations.

11.5 For the refrigerant R-134a, at 2-, 3-, and 10-bar pressures, calculate and compare the nucleate boiling heat flux from a stainless-steel surface with a wall superheat of 8 °C, using the correlations of Stephan and Abdelsalam (1980) for refrigerants, Cooper (1984), and Gorenflo (1993). Which correlation is likely to be the most accurate?

11.6 Using the method of Section 11.5.2, perform an analysis for film boiling over the surface of the conical object shown in Fig. P11.6.

Figure P11.6.

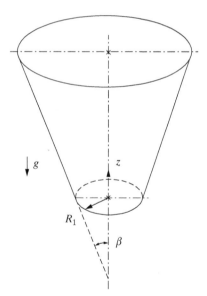

11.7 The bubble departure diameter has been suggested by some investigators as a threshold scale that distinguishes conventional and small channels. Using the correlations of Fritz (1935) (assuming $\theta = 50°$) and Gorenflo et al. (1986) (assuming $T_w - T_{sat} = 8\,°C$), calculate the bubble departure diameters for water at $P = 1, 10$, and 25 bars. Repeat the calculations for R-22 and R-134a at $T_{sat} = 30\,°C$ and 60 °C, using the correlation of Gorenflo with $T_w - T_{sat} = 8\,°C$. Compare the calculated departure diameters with the Laplace length scale. Discuss the adequacy of the two length scales for use as the aforementioned threshold scale.

11.8 For pool boiling on the outside of horizontal stainless-steel cylinders with $D = 5$ mm diameter, calculate q''_{CHF}, q''_{MFB}, and T_{MFB} when the coolant is saturated

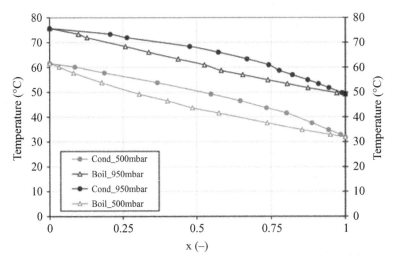

Figure P11.11. Equilibrium phase diagrams for FC-84/FC-3284 binary mixtures at 5 bars and 9.5 bars pressure (Wagner and Stephan, 2008).

R-134a at $T_{\text{sat}} = 40\,°C$ and $80\,°C$. Also, calculate the heat flux when $T_{\text{w}} = T_{\text{MFB}} + 200\,°C$.

11.9 A vessel contains a mixture of water and ammonia at a constant pressure of 3.8 bars. The initial mole fraction of ammonia in the liquid mixture is 40%. Pool boiling and evaporation occur in the vessel, until the mole fraction of ammonia in the liquid bulk reaches 25%. Nucleate boiling is under way on a heated surface that is at a temperature of 370 K.

(a) Find the mole fraction of ammonia in the vapor bulk.

(b) Calculate the heat flux using the correlation of Calus and Leonidopoulos (1974).

(c) Repeat part (b), this time using the correlation of Schlünder (1982).

11.10 In an experiment a liquid water–methanol mixture undergoes pool boiling on a heated flat surface that is at 360 K temperature. The pressure is one atmosphere and the mole fraction of ethanol in the liquid bulk is 30%.

(a) Calculate the wall heat flux using a pool boiling correlation of your choice.

(b) Using the method of Yagov (2004) estimate the wall heat flux that would lead to CHF.

(c) Repeat parts (a) and (b), this time assuming that the heated surface is at a temperature of 358 K, and that the mole fraction of methanol in the bulk liquid phase is 90%.

11.11 A heated surface is to be cooled via pool boiling. The coolant pressure is to be maintained at 5 bars. For a surface temperature that is at 10 °C above the saturation temperature (bubble point temperature in the case of a binary mixture), calculate the heat flux that can result from pool boiling when the coolant is:

(a) FC-84

(b) FC-3284

(c) 50% molar mixture of FC-84 and FC-3284.

For the equilibrium phase diagram, you may use the results of Wagner and Stephan (2008) in Fig. P11.11.

11.12 Consider pool boiling on a heated surface by a water–ammonia liquid mixture at 1 bar pressure. The surface temperature is to be kept at 310 K temperature.

(a) Using the correlations of Calus and Leonidopoulos (1974) and Schlünder (1982), calculate and compare the heat flux that can be dissipated from the surface if the mole fraction of ammonia in the liquid is 40%.

(b) Assuming that these correlations are valid up to the CHF point, what wall temperature would cause critical heat?

11.13 Kandlikar (2001) developed the forthcoming pool boiling CHF model by focusing on the behaviour of vapor bubbles on the heated surface, and assuming that CHF happens when the force resulting from momentum exchange of evaporating liquid overcomes the surface tension and buoyancy forces.

$$\frac{q''_{CHF}}{\rho_g^{1/2} h_{fg} (\sigma g \Delta \rho)^{1/4}} = \left(\frac{1 + \cos \theta_r}{16}\right) \left[\frac{2}{\pi} + \frac{\pi}{4} (1 + \cos \theta_r) \cos \theta\right]^{1/2},$$

where θ_r is the receding contact angle and θ represents the flow passage inclination angle with respect to the horizontal plane. The correlation applies to upward facing smooth heated surfaces, for $0 < \theta < \pi/2$. Consider atmospheric and saturated water boiling on a flat metallic surface.

(a) Using this correlation, find the critical heat flux on a horizontal upward-facing surface for receding contact angles of 30°, 40° and 60°. Compare the results with the predictions of the correlation of Zuber (1963), as corrected by Sun and Lienhard (1970).

(b) Repeat part (a) for a surface that is inclined by 45°.

11.14 The derivation of Eq. (11.46) was based on $\rho_g \leq \rho_f$. Without this assumption and related approximations the derivation would give (Dhir, 1998):

$$q''_{CHF, Z} \approx \frac{\pi}{24} \rho_g^{1/2} h_{fg} (\sigma g \Delta \rho)^{1/4} \left(\frac{\rho_f + \rho_g}{\rho_g}\right)^{1/2} \left[\frac{\rho_f (16 - \pi)}{\rho_f (16 - \pi) + \rho_g \pi}\right].$$

Compare the predictions of this expression with Eq. (11.46) when the coolant is water, for 1 bar, 10 bars and 50 bars pressures.

12 Flow Boiling

Flow boiling is considerably more complicated than pool boiling, owing to the coupling between hydrodynamics and boiling heat transfer processes. A sequence of two-phase and boiling heat transfer regimes takes place along the heated channels during flow boiling, as a result of the increasing quality. The two-phase flow regimes in a boiling channel are therefore "developing" everywhere and are morphologically different than their namesakes in adiabatic two-phase flows.

12.1 Forced-Flow Boiling Regimes

The preferred configuration for boiling channels is vertical upflow. In this configuration buoyancy helps the mixture flow, and the slip velocity between the two phases that is caused by their density difference actually improves the heat transfer. However, flow boiling in horizontal and even vertical channels with downflow is also of interest. Horizontal boiling channels are not uncommon, and flow boiling in a vertical, downward configuration may occur under accident conditions in systems that have otherwise been designed to operate in liquid forced convection heat transfer conditions.

Figure 12.1 displays schematically the heat transfer, two-phase flow, and boiling regimes that take place in a vertical tube with upward flow that operates in steady state and is subject to a uniform and moderate heat flux. The mass flow rate is assumed constant. When the fluid at the inlet is a highly subcooled liquid, at a very low heat flux, the flow field in the entire channel remains subcooled liquid (Fig. 12.1(a)). With increasing heat flux, boiling occurs in part of the channel, the flow regime at the exit depends on the heat flux (Figs. 12.1(b) and (c)), and with sufficiently high heat flux (or sufficiently low inlet subcooling), a complete sequence of boiling and related two-phase flow regimes takes place in the channel Fig. 12.1(c). Boiling starts at the *onset of nucleate boiling* (ONB) point. When the fluid at the inlet is saturated liquid, or a saturated liquid–vapor mixture, the boiling and two-phase flow patterns will be similar to those depicted in Fig. 12.1(d).

Figure 12.2 shows in more detail the flow and heat transfer regimes in a uniformly heated vertical channel with upward flow that is subject to a moderate heat flux, when the fluid at the inlet is subcooled liquid. The wall and fluid temperatures are also schematically displayed in the figure. Near the inlet where the liquid subcooling is too high to permit bubble nucleation, the flow regime is single-phase liquid, and the heat transfer regime is forced convection. Following the initiation of boiling, the sequence of flow regimes includes bubbly, slug, and annular, followed by dispersed

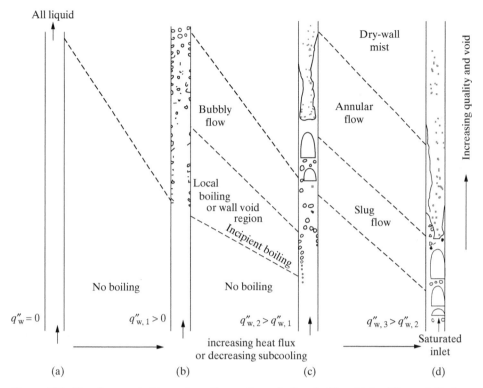

Figure 12.1. Development of two-phase flow patterns in flow boiling. (After Hsu and Graham, 1986.)

droplet flow, and eventually a single-phase pure vapor flow field. The two-phase flow regimes are evidently morphologically somewhat different than their namesakes in adiabatic two-phase flow.

Nucleate boiling is predominant in the bubbly and slug two-phase flow regimes and is followed by forced convective evaporation where the flow regime is predominantly annular. This is an extremely efficient heat transfer regime in which the heated wall is covered by a thin liquid film. The liquid film is cooled by evaporation at its surface, making it unable to sustain a sufficiently large superheat for bubble nucleation. Droplet entrainment can occur when the vapor flow rate is sufficiently high, leading to dispersed-droplet flow. Further downstream, the liquid film may eventually completely evaporate and lead to *dryout*. Sustained macroscopic contact between the heated surface and liquid does not occur downstream from the dryout point (the liquid-deficient region), although sporadic deposition of droplets onto the surface may take place. Further downstream, eventually the entrained droplets will completely evaporate, and a pure vapor single-phase flow field develops. The heat transfer coefficient in the liquid-deficient region is much lower than in the nucleate boiling or forced convective evaporation regimes. As a result, the occurrence of dryout is accompanied by a large temperature rise for the heated surface. The dryout phenomenon is thus similar to the critical heat flux previously discussed for pool boiling.

Figure 12.3 depicts the flow and heat transfer regimes in a vertical heated channel subject to a very high heat flux. The flow patterns are different than those described for Fig. 12.2. Because of the high wall heat flux, the ONB occurs in the channel while

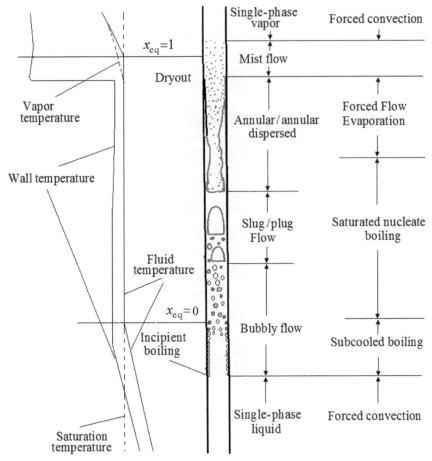

Figure 12.2. Two-phase flow and boiling regimes in a vertical pipe with a moderate wall heat flux.

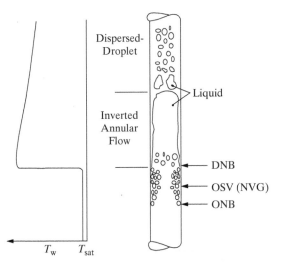

Figure 12.3. Two-phase flow and boiling regimes in a vertical pipe with a high wall heat flux.

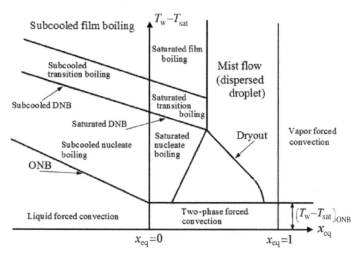

Figure 12.4. The flow boiling regimes in a tube with constant wall temperature, under $P = $ const. and $G = $ const. conditions.

the bulk liquid is still highly subcooled. Nucleate boiling takes place downstream from the ONB point, leading to increased voidage. A growing bubbly layer may form adjacent to the wall, and the bubbles may eventually crowd sufficiently to make a sustained macroscopic contact between the liquid and heated surface impossible. This leads to the *departure from nucleate boiling* (DNB), which is another mechanism similar to the critical heat flux in pool boiling. The heat transfer coefficient, which is very high in the subcooled boiling regime, deteriorates very significantly downstream from the DNB point, even though the bulk flow in the heated channel may still be highly subcooled.

For any particular uniformly heated vertical channel with upward flow, the various local heat transfer regimes constitute a surface in the mass-flux–heat-flux–equilibrium quality (G, q_w'', x_{eq}) coordinates. Likewise, the various local heat transfer regimes constitute a surface in the mass flux–wall superheat–equilibrium quality $(G, T_w - T_{sat}, x_{eq})$ coordinates for any particular vertical channel with upward flow and uniform wall temperature. It is easier to discuss these heat transfer regimes by investigating the intersection of these three-dimensional surfaces with a $G = $ const. plane, as in Figs. 12.4 and 12.5. A good thing about these figures is that they qualitatively show the evolution of the heat transfer regimes as one marches along a vertical channel with constant upward mass flux. A straight horizontal line in Fig. 12.4 shows the sequence of the heat transfer regimes for a fixed wall superheat (Carey, 2008). For the case of constant wall heat flux the sequence of heat transfer regimes depends strongly on the magnitude of heat flux, q_w'', as noted in Fig. 12.5. With a moderate heat flux such as $q_{w,2}''$, the sequence of regimes will follow the horizontal line designated with $q_{w,2}''$, consistent with Fig. 12.2, and will include, in order, liquid forced convection, subcooled boiling, saturated boiling, forced convective evaporation, dryout, and postdryout (post-CHF; liquid-deficient) heat transfer. When the heat flux is very high such as $q_{w,3}''$, however, the sequence of regimes may follow the line designated with $q_{w,3}''$, in agreement with Fig. 12.3. ONB occurs while the bulk liquid is highly subcooled, and instead of dryout, DNB takes place. At a yet much higher heat flux, such

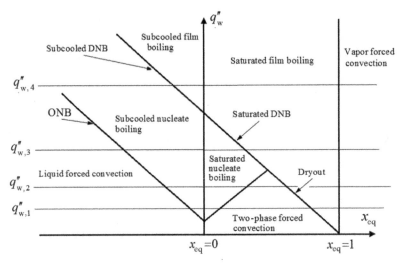

Figure 12.5. The boiling regimes map in two-dimensional heat flux–equilibrium quality coordinates, under $P = $ const. and $G = $ const. conditions.

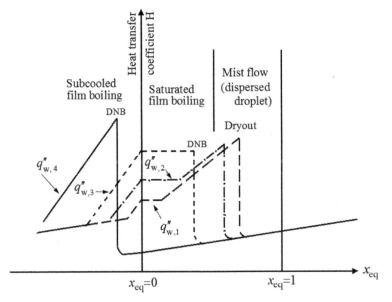

Figure 12.6. Variation of boiling heat transfer coefficient with quality, under $P = $ const. and $G = $ const. conditions. (Constant-heat-flux curves refer to Fig. 12.5.)

as $q''_{w,4}$ the sequence of regimes will follow the line designated with $q''_{w,4}$ in Fig. 12.5, where ONB and DNB both occur while the bulk fluid is highly subcooled.

The various heat transfer regimes lead to vastly different heat transfer coefficients. Figure 12.6 shows qualitatively the variation of the local heat transfer coefficients along a uniformly heated vertical channel. The designations $q''_{w,1}$ $q''_{w,2}$, etc., correspond to the lines shown in Fig. 12.5. As noted, the heat transfer coefficient is generally very high in nucleate boiling and forced-convective evaporation regimes, but it suffers a dramatic reduction once the critical heat flux (dryout of ONB) is

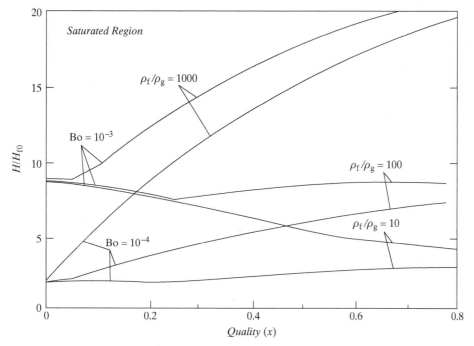

Figure 12.7. Saturated flow boiling map depicting the dependence of the boiling heat transfer coefficient on the equilibrium quality. (After Kandlikar, 1991.)

reached. It remains low in the post-CHF regime, in comparison with the nucleate boiling and convective evaporation regimes.

The map depicted in Fig. 12.6 suggests that, when the equilibrium quality (x_{eq}) is high, the heat transfer coefficient increases monotonically and slightly with increasing x_{eq}. This result is based on the assumption that heat transfer is essentially by nucleate boiling at low x_{eq} and by forced convective evaporation at high x_{eq}. Kandlikar (1991) has shown that at high x_{eq} the heat transfer coefficient may actually decrease with increasing x_{eq} in some circumstances, implying the significance of contributions from both nucleate boiling and forced convection (Kandlikar, 1991). The flow boiling map in Fig. 12.7 indicates that the trend in the variation of H/H_{L0} with x_{eq} depends on the boiling number, $q''_w/(Gh_{fg})$, as well as ρ_f/ρ_g. Water at high pressure exhibits a decreasing H/H_{L0} with x_{eq}, for example, whereas at low pressure the opposite trend is observed. Refrigerants that possess relatively low ρ_f/ρ_g ratios at normal refrigeration operating conditions also exhibit a decreasing H/H_{L0} with increasing x_{eq}.

Boiling and two-phase flow patterns for uniformly heated horizontal channels will now be discussed. In commonly applied channels (excluding mini- and microchannels) the tendency of the two phases to stratify affects the two-phase flow patterns, resulting in the occurrence of "early" dryout. When the coolant mass flux is very high, however, the flow and heat transfer patterns are insensitive to orientation.

Figure 12.8 displays schematically the boiling and heat transfer regimes in a uniformly heated horizontal pipe when the heat and mass fluxes are both moderate. The qualitative axial variations of the heat transfer coefficient are also shown in the figure. Although the main flow and heat transfer regimes are similar to those

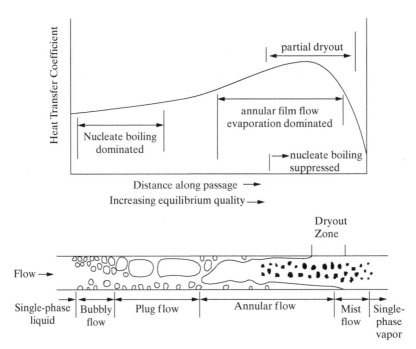

Figure 12.8. Flow and heat transfer regimes in a uniformly heated horizontal tube with moderate heat flux.

of Fig. 12.3, the effect of buoyancy can be important. Buoyancy tends to promote the stratification of the two phases. This effect becomes particularly important when the annular-dispersed flow regime (corresponding to the forced-convective evaporation regime) is reached. The liquid film tends to drain downward, often leading to *partial dryout*, where the liquid film breaks down near the top of the heated channel, while persisting in the lower parts of the channel perimeter. As a result of partial dryout, the CHF conditions in horizontal channels are generally reached at lower x_{eq} values than in vertical upflow channels. When the heat and mass fluxes are very high, flow and boiling regimes similar to those shown in Fig. 12.4 should be expected in horizontal channels because of the relatively small effect of gravity.

12.2 Flow Boiling Curves

Figure 12.9 displays the boiling curve for a vertical, upward pipe flow, for constant mass flux. The boiling curve in its entirety is unlikely to occur in a single heated pipe in steady state. An easy way to understand this is the following. Except for the subcooled liquid forced-convection and partial boiling regions, which evidently require variable quality, the remainder of the curve can represent measurements in a pipe when in repeated experiments the mass flux and local quality are maintained constant, while the heat flux is varied and the wall superheat is measured.

Figure 12.9 only applies to low quality conditions, however. The effects of mass flux G and equilibrium quality x_{eq} on the boiling curve are shown in Figs. 12.10 (a) and (b), respectively. The heat transfer coefficient is particularly sensitive to mass flux in single-phase liquid forced-convection, partial boiling, and post-CHF regimes, but it is insensitive to G in the fully developed nucleate boiling. (Detailed definitions

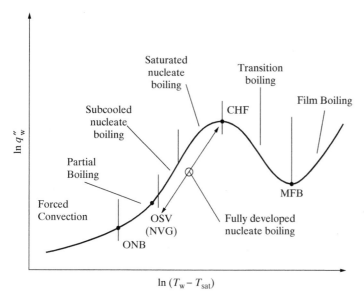

Figure 12.9. The flow boiling curve for low qualities.

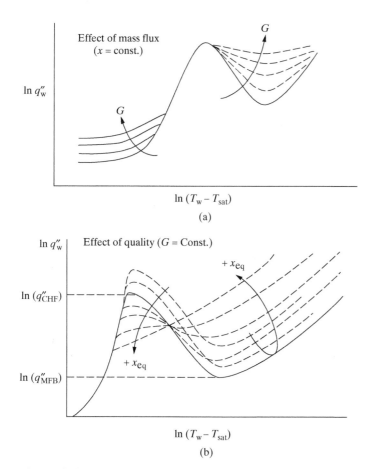

Figure 12.10. Effects of local quality and mass flux on the flow boiling curve.

Figure 12.11. Flow patterns and temperature variation in subcooled boiling.

for these heat transfer regimes will be given in the forthcoming sections.) The effect of local equilibrium quality is rather complicated. With increasing x_{eq}, the critical heat flux is decreased. (Note that in Fig. 12.10(b) the mass flux is assumed to remain constant.) When DNB-type CHF occurs, the post-CHF heat transfer coefficients also may decrease with increasing local x_{eq}. Following dryout, when x_{eq} is relatively high, however, increasing x_{eq} will lead to higher mixture velocity and therefore can actually increase the local heat flux (and therefore the local heat transfer coefficient).

12.3 Flow Patterns and Temperature Variation in Subcooled Boiling

The flow and heat transfer regimes associated with subcooled boiling (i.e., regions A and B of Fig. 12.2) are now discussed in some detail. The flow patterns are depicted in more detail in Fig. 12.11. The portion of the boiling curve representing subcooled boiling is also shown in Fig. 12.12. With respect to the main phenomenology shown in the figure, there is little difference between vertical and horizontal channels.

Forced convection to subcooled liquid occurs upstream of point B in Figs. 12.11 or 12.12. At point B (the ONB point) bubble nucleation starts, while the bulk liquid is still subcooled. With constant wall heat flux, the occurrence of ONB is often accompanied by a temperature undershoot, as shown in Fig. 12.11. The temperature undershoot is caused by a sudden increase in the local heat transfer coefficient resulting from the bubble nucleation process. Bubbles remain predominantly attached to the wall between points B and E. At and beyond E, where the bulk liquid is subcooled and the mixture mixed-cup enthalpy is still slightly below the saturated liquid

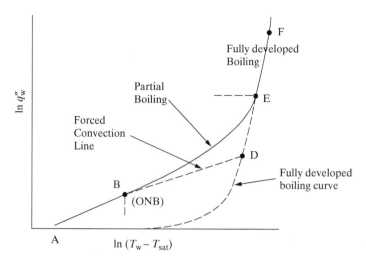

Figure 12.12. The boiling curve at the vicinity of subcooled boiling region.

enthalpy, bubbles departing from the wall can survive condensation. Point E is called the point of onset of significant void (OSV) or net vapor generation (NVG). At point F the mixed-cup fluid would be a saturated liquid. Thermodynamic nonequilibrium between the two phases persists, however. Thermodynamic nonequilibrium disappears at point G.

As noted earlier, the ONB point represents the point where boiling starts. Partial boiling takes place between points A and E. In partial boiling, forced convection and nucleate boiling mechanisms both contribute to heat transfer. Bulk boiling, in which the boiling mechanism is predominant and convection is insignificant, actually starts at a point slightly downstream from the OSV point. In bulk boiling, the contribution of convection to heat transfer can be small.

The OSB and OSV are important operational thresholds for boiling systems. These thresholds will be discussed in the following two sections.

12.4 Onset of Nucleate Boiling

The basic process for ONB is similar to heterogeneous nucleation on wall crevices in pool boiling. ONB occurs when some of the bubbles forming on crevices can survive. The main difference with pool boiling is that in forced-flow boiling the thickness of the thermal boundary layer can be assumed finite and stable.

The bubble nucleation criteria described in Chapter 11 are in principle valid for subcooled flow boiling, but improvements, primarily to account for the effect of flow, have been attempted. Kandlikar *et al.* (1997) performed numerical simulations of bubble nucleation in subcooled flow boiling and noted that for nucleation in the presence of liquid flow, flow stagnation occurred at $y = 1.1 R_B$ with y representing the distance from the wall. They modified the bubble nucleation model of Hsu (1962), described in Chapter 11, by using the liquid temperature at the stagnation point for the temperature of liquid at the bubble top, and derived

$$R_{C,min}, R_{C,max} = \frac{\delta \sin \theta}{2.2} \frac{(T_w - T_{sat})}{(T_w - \overline{T}_L)} \left[1 \mp \sqrt{1 - \frac{9.2(T_w - \overline{T}_L)T_{sat}\sigma}{(T_w - T_{sat})^2 \delta \rho_v h_{fg}}} \right]. \quad (12.1)$$

Mechanistic models based on the *tangency* concept have been successful and are widely applied. In the ONB models that are based on the tangency concept, it is assumed that bubbles attempt to grow on wall crevices that cover a wide range of sizes. ONB occurs when mechanically stable bubbles forming on any of the existing crevice size groups remain thermally stable.

The *ONB model of Bergles and Rohsenow* (1964) starts from Clausius's relation for vaporization:

$$\left(\frac{dP}{dT}\right)_{\text{sat}} = \frac{h_{\text{fg}}}{T v_{\text{fg}}} \approx \frac{h_{\text{fg}}}{T v_{\text{g}}} = \frac{h_{\text{fg}} P}{T^2 (R_{\text{u}}/M)}, \tag{12.2}$$

where the ideal gas law has been applied to the vapor. Equation (12.2) can be recast so that the variables (T and P) are separated:

$$\frac{dP}{P} = \frac{h_{\text{fg}}}{T^2 (R_{\text{u}}/M)} dT. \tag{12.3}$$

Integration of the two sides, using the saturation conditions associated with a flat interphase as the lower limit, and the conditions of the interior of a bubble as the upper limit, then gives

$$T_{\text{B}} - T_{\text{sat}} = \frac{(R_{\text{u}}/M) T_{\text{B}} T_{\text{sat}}}{h_{\text{fg}}} \ln\left(\frac{P_{\text{B}}}{P_{\infty}}\right), \tag{12.4}$$

where T_{B} is the bubble temperature, P_{∞} is the ambient pressure, P_{B} is the bubble pressure, and $T_{\text{sat}} = T_{\text{sat}}(P_{\infty})$. Now, mechanical equilibrium requires that

$$P_{\text{B}} - P_{\infty} = \frac{2\sigma}{R_{\text{B}}}, \tag{12.5}$$

where R_{B} is the bubble radius. The logarithmic term on the right side of Eq. (12.4) can now be written as $\ln[1 + (P_{\text{B}} - P_{\infty})/P_{\infty}]$ and combined with Eq. (12.5). For a bubble that is mechanically stable, therefore,

$$T_{\text{B}} - T_{\text{sat}} = \frac{R_{\text{u}} T_{\text{B}} T_{\text{sat}}}{h_{\text{fg}} M} \ln\left(1 + \frac{2\sigma}{P_{\infty} R_{\text{B}}}\right). \tag{12.6}$$

A bubble will not collapse from condensation if it is surrounded by liquid that is warmer than the content of the bubble (Hsu, 1962). The thermal boundary layer is modeled essentially as a stagnant film with thickness δ, with a linear temperature profile:

$$q_{\text{w}}'' = k_{\text{L}} \frac{T_{\text{w}} - T_{\text{L}}}{\delta} = k_{\text{L}} \frac{T_{\text{w}} - T(y)}{y}. \tag{12.7}$$

The film thickness is related to the local convection heat transfer coefficient H according to $H_{\text{L0}} = k_{\text{L}}/\delta$.

It is assumed that hemispherical bubbles form on the mouths of crevices of all sizes. ONB occurs, at $y = R_{\text{B}}$, when

$$T_{\text{B}}(R_{\text{B}}) = T(y) \tag{12.8}$$

and

$$\frac{dT_{\text{B}}}{dR_{\text{B}}} = \frac{dT(y)}{dy} \quad \text{(tangency condition)}. \tag{12.9}$$

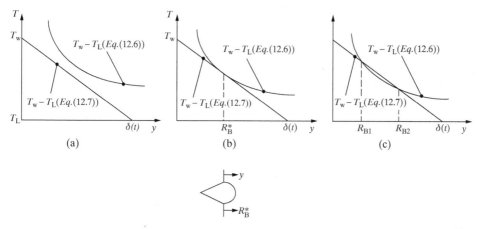

Figure 12.13. The bubble and its surrounding superheated liquid temperature in the ONB model of Bergles and Rohsenow (a) upstream of the ONB point, (b) at the ONB point, and (c) downstream of the ONB point.

Equations (12.6)–(12.9) include the unknowns q_w'', T_w, T, T_B, and $R^* = R_B$ (the critical bubble radius, or critical cavity mouth radius). Thus, in a uniformly heated channel with fixed q_w'' and mass flux, T_L (the bulk liquid temperature), T_w and $T(y)$ (for any given constant y) increase with distance from the inlet. Equations (12.6) and (12.7), when plotted together, will qualitatively appear as Fig. 12.13. (Note that, for consistency, predictions of Eq. (12.6) are displayed after writing $T_B - T_L = (T_B - T_{sat}) + (T_{sat} - T_L)$.) Upstream of the ONB point, the two equations do not intersect (Fig. 12.13(a)); tangency occurs at the ONB point (Fig. 12.13(b)); and downstream from the ONB point there is a range of bubble sizes (or equivalently crevice sizes), represented by the range between the two intersection points in Fig. 12.13(c) that support stable bubbles. [Note the similarity with the pool boiling nucleation criterion of Hsu (1962) described in Section 11.2.1.]

Since numerical solution of Eqs. (12.6)–(12.9) is rather tedious, Bergles and Rohsenow (1964) performed an extensive set of calculations for water and curve-fitted the predictions of the model described here for water according to

$$(T_w - T_{sat})_{ONB} = 0.556 \left[\frac{q_w''}{1082 P^{1.156}} \right]^n, \tag{12.10}$$

with

$$n = 0.463 P^{0.0234}, \tag{12.11}$$

where P is in bars, T is in kelvins, and q_w'' is in watts per meter squared. In English units, the correlation is

$$q_w'' = 15.60 P^{1.156} (T_w - T_{sat})_{ONB}^{n'}, \tag{12.12}$$

with

$$n' = 2.30 / P^{0.0234}, \tag{12.13}$$

where q_w'' is in British thermal units per foot squared per hour, the temperature difference is in degrees Fahrenheit, and P is in pounds per square inch absolute.

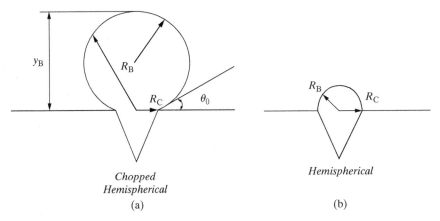

Figure 12.14. Geometry of a bubble on a cavity.

The *ONB model of Davis and Anderson* (1966) is based on an analytical solution of equations similar to those solved by Bergles and Rohsenow with some approximations. The assumption of a thermal boundary layer with a linear temperature profile is maintained, and the tangency of bubble and superheated liquid temperature profiles is applied. The model in its general form assumes a chopped-hemisphere bubble (Fig. 12.14(a)). For a mechanically stable bubble, Eq. (12.6) is accordingly cast as

$$T_B - T_{sat} = \frac{R_u T_B T_{sat}}{h_{fg} M} \ln \left(1 + \frac{2C_1 \sigma}{P_{sat} y_B} \right), \tag{12.14}$$

and, from there, one gets

$$\frac{dT_B}{dy_B} = -\frac{2C_1 (R_u/M) T_{sat}^2}{P_{sat} y_B^2 (1+\xi)} \left[1 - \frac{(R_u/M) T_{sat}}{h_{fg}} \ln(1+\xi) \right]^{-2}, \tag{12.15}$$

where y_B is the height of the bubble formed on a critical cavity, $C_1 = 1 + \cos\theta_0$, and $\xi = 2\sigma C_1/(P_{sat} y_B)$. Davis and Anderson argued that

$$\frac{(R_u/M) T_{sat}}{h_{fg}} \ln(1+\xi) \ll 1$$

for fluids at relatively high pressure or fluids that have low surface tensions. For such fluids, therefore, the second term in brackets on the right side of Eq. (12.15) can be neglected. They then applied conditions similar to Eqs. (12.7)–(12.9) and derived

$$(T_w - T_{sat})_{ONB}^2 = \frac{8(1 + \cos\theta_0)C}{k_L} q_w'', \tag{12.16a}$$

$$y_B = [2(1 + \cos\theta_0) k_L C/q_w'']^{1/2}, \tag{12.16b}$$

$$C = \sigma T_{sat}/(\rho_g h_{fg}), \tag{12.16c}$$

$$R_C^* = \left[\frac{2k_L(1 + \cos\theta_0)C}{q_w''} \right]^{1/2}, \tag{12.16d}$$

where R_C^* is the critical cavity mouth radius. For hemispherical bubbles residing on critical cavities [Fig. 12.14(b)] $\theta_0 = \pi/2$ and the model predicts:

$$(T_w - T_{sat})_{ONB} = \left(\frac{8\sigma q_w'' T_{sat}}{h_{fg} k_f \rho_g}\right)^{1/2}, \tag{12.17}$$

$$R_C^* = \left(\frac{2\sigma T_{sat} k_f}{h_{fg} q_w'' \rho_g}\right)_{ONB}^{1/2}. \tag{12.18}$$

The same equations were proposed by Sato and Matsumura (1963).

The *correlation of Marsh and Mudawar* (1989) is based on data dealing with the ONB phenomenon in subcooled turbulent liquid falling films, where commonly applied ONB methods do poorly. The ONB and other boiling thresholds in falling liquid films have been of interest because of the potential of falling film as a cooling method for microelectronics. Marsh and Mudawar (1989) attribute the inaccuracy of ONB methods to the assumed linear temperature profile of the liquid near the heated surfaces, which neglects the effect of turbulence. The correlation of Marsh and Mudawar (1989) is

$$q_{ONB}'' = \frac{1}{C} \frac{k_f h_{fg}}{8\sigma T_{sat} v_{fg}} (T_w - T_{sat})_{ONB}^2, \quad C = 3.5. \tag{12.19}$$

A method for the prediction of ONB in microchannels, based on the hypothesis that thermocapillary forces are crucial to the ONB process in microchannels, has been proposed by Ghiaasiaan and Chedester (2002). The method leads to Eq. (12.19), with constant C replaced by a correlation that is meant to account for the thermocapillary force that results from the nonuniformity of the bubble surface temperature. Qu and Mudawar (2002) have also proposed a bubble departure-type model for ONB in microchannels. These will be further discussed in Chapter 14.

None of the correlations considered here incorporate the effect of surface wettability on ONB. Experiments have shown that increased surface wettability (equivalent to smaller contact angle) leads to higher wall superheat for boiling incipience (You *et al.*, 1990). Basu *et al.* (2002) measured the boiling incipience of distilled water on flat unoxidized and oxidized copper blocks (with static contact angles $\theta_0 = 90°$ and $30 \pm 3°$, respectively) and a rod bundle with Zircalloy-4 cladding ($\theta_0 = 57°$). They noted that the tangency-based boiling incipience models discussed here did well for the data representing $\theta_0 = 57°$, but they underpredicted the boiling incipience superheat for the $\theta_0 = 30°$ data. Based on their own as well as others' data for water, R-113, R-11, and FC72 fluids heated on surfaces made from several metals and metallic alloys, Basu *et al.* (2002) developed the following empirical correlation:

$$(T_w - T_{sat})_{ONB} = \frac{2\sigma T_{sat}}{R_C^* F \rho_g h_{fg}}, \tag{12.20}$$

where R_C^* is found from Eq. (12.18), and the correction factor F is a function of the static contact angle θ_0 (in degrees) according to

$$F = 1 - \exp\left[-\left(\frac{\pi\theta_0}{180}\right)^3 - 0.5\left(\frac{\pi\theta_0}{180}\right)\right]. \tag{12.21}$$

The data of Basu *et al.* covered a range of static contact angles of $\theta_0 \approx 1°\text{--}85°$. Their correlation is valid for low-heat-flux conditions, so that $(T_w - T_{sat})_{ONB} \approx (T_B - T_{sat})_{ONB}$.

EXAMPLE 12.1. Water at 70-bar pressure and with an inlet subcooling of 15 °C flows into a uniformly heated vertical tube that is 1.5 cm in diameter and receives a heat flux of 1.8×10^5 W/m². The velocity of water at the inlet is 2 m/s. Find the location and the wall temperature where ONB occurs.

SOLUTION. We will use the correlation of Bergles and Rohsenow, Eq. (12.10). Since the inlet pressure is high, we can neglect the effect of pressure drop between the inlet and the ONB point on properties. Therefore, $T_{sat} = 559$ K, $\rho_g = 36.53$ kg/m³, $k_f = 0.56$ W/m·K, $h_{fg} = 1.505 \times 10^6$ J/kg, and $\sigma = 0.0174$ N/m.

The density of water at inlet conditions is $\rho_{L,in} = 768$ kg/m³. The mass flux is thus

$$G = \rho_{L,in} U_{L,in} = 1{,}573 \text{ kg/m}^2\text{·s}.$$

For average liquid properties, we will use 551 K as the approximation to the bulk liquid temperature at the ONB point, where, $\rho_L = 754.6$ kg/m³, $\mu_L = 9.46 \times 10^{-5}$ kg/m·s, $k_L = 0.574$ W/m·K, $C_{PL} = 5{,}220$ J/kg·K, and $Pr_L = 0.861$. Also, we will get

$$Re_L = GD/\mu_L = 2.39 \times 10^5.$$

For the convection heat transfer coefficient, let us use the correlation of Dittus and Boelter:

$$Nu_D = H_{L0}D/k_L = 0.023 Re_L^{0.8} Pr_L^{0.4} = 435.5 \Rightarrow H_{L0} = 16{\,}670 \text{ W/m}^2\text{·K}.$$

In accordance with Eqs. (12.10) and (12.11), furthermore,

$$n = 0.463(70)^{0.0234} = 0.511,$$

$$(T_w - T_{sat})_{ONB} = 0.556 \left[\frac{1.8 \times 10^5}{1082(70)^{1.156}} \right]^{0.511} = 0.62 °C.$$

This gives

$$T_{w,ONB} = 559.6 \text{ K}.$$

The local bulk liquid temperature can now be found:

$$q_w'' = H_{L0}(T_w - \overline{T}_L)_{ONB} \Rightarrow \overline{T}_{L,ONB} = 548.8 \text{ K}.$$

The location of the ONB point can now be found by performing the following energy balance:

$$\dot{m}_L C_{PL}(\overline{T}_{L,ONB} - T_{L,in}) = \pi D q_w'' Z_{ONB},$$

where

$$\dot{m}_L = G(\pi D^2/4) = 0.272 \text{ kg/s}.$$

The result will be $Z_{ONB} = 0.805$ m. We can double check the calculations by testing the correlation of Davis and Anderson, Eq. (12.17). This correlation will give $(T_w - T_{sat})_{ONB} = 0.69$ K.

12.5 Empirical Correlations for the Onset of Significant Void

The empirical correlations of Saha and Zuber (1974) are the most widely used for the specification of the OSV point. Saha and Zuber define two OSV regimes: the thermally controlled regime, which occurs when $Pe_L < 70{,}000$, and the hydrodynamically controlled regime, for which $Pe_L > 70{,}000$, where $Pe_L = GD_H C_{PL}/k_L$ is the Péclet number. In the thermally controlled regime, OSV occurs when either of the following equivalent criteria is met:

$$(h_f - h_L) \leq 0.0022 q_w'' D_H C_{PL}/k_L \tag{12.22}$$

or, equivalently,

$$\mathrm{Nu} = q_w'' D_H / \left[k_L (T_{sat} - \overline{T}_L) \right] \geq 455. \tag{12.23}$$

In the hydrodynamically controlled regime, OSV occurs when either of the following applies:

$$(h_f - h_L) \leq 154 q_w''/G, \tag{12.24}$$

or, equivalently,

$$\mathrm{St} = \frac{q_w''}{G C_{PL}(T_{sat} - \overline{T}_L)} \geq 0.0065. \tag{12.25}$$

Thus, in the thermally controlled regime, upstream of the OSV point, $(h_f - h_L)$ is larger than the right side of Eq. (12.22), and OSV occurs as soon as the equality represented by Eq. (12.22) is satisfied. Likewise, upstream of the OSV point $\mathrm{Nu} = q_w'' D_H / [k_L(T_{sat} - \overline{T}_L)] < 455$ applies, and OSV takes place once the two sides of this expression become equal. Equations (12.24) and (12.25) should be interpreted similarly.

The OSV correlation of Unal (1975) is among the simplest available methods, and its data base includes tests with $Pe_L \geq 12{,}000$ for water and R-22. It thus covers much of the aforementioned thermally controlled regime. The correlation is

$$\frac{H_{L0}(T_{sat} - \overline{T}_L)}{q_w''} = a, \tag{12.26}$$

where H_{L0} is the forced convection heat transfer coefficient. For water, $a = 0.11$ for $\overline{U}_L < 0.45\,\mathrm{m/s}$ and $a = 0.24$ for $\overline{U}_L \geq 0.45\,\mathrm{m/s}$ with $\overline{U}_L$ representing the bulk mean velocity. The threshold velocity 0.45 m/s is close to the velocity at which the effect of forced convection on bubble growth during subcooled boiling vanishes.

12.6 Mechanistic Models for Hydrodynamically Controlled Onset of Significant Void

These OSV models are based on a force balance on bubbles that have nucleated on wall crevices and have grown to the largest size thermally possible. OSV occurs at a location in the heated channel where the forces that tend to separate the bubble from the heated surface just overcome the surface tension force. These models evidently apply to hydrodynamically controlled OSV (i.e., $Pe_L > 70\,000$ according to Saha and Zuber (1974)). Models have been proposed by Levy (1967),

Staub (1968), and more recently by Rogers *et al.* (1987). Improvements of the latter model were published by Rogers and Li (1992). All the models are similar in their treatment of the crucial processes. The models of Levy (1967) and Staub (1968) have primarily been based on high-pressure data with water. The model of Rogers *et al.* is meant to represent water at low pressure. The models of Levy (1967) and Rogers *et al.* (1987) will be reviewed in the following.

The *OSV model of Levy* (1967) is probably the most widely used model of its kind. The model is based on the following assumptions.

Fully developed turbulent velocity and temperature boundary layers exist.
OSV happens when the largest thermally stable, heterogeneously generated bubbles are detached from the wall.
The detaching bubble contains saturated vapor corresponding to the local pressure.
The liquid temperature at the top of the largest thermally stable bubbles is at saturation.
The drag force on a bubble can be estimated by using the turbulent wall friction in a fully rough pipe.

For a vertical heated surface with upward flow (the predominant configuration of boiling systems), bubble departure occurs when

$$C_B g \Delta \rho R_B^3 + C_F \frac{\tau_w}{D_H} R_B^3 - C_s R_B \sigma = 0, \qquad (12.27)$$

where the three terms on the left side represent forces on a bubble from buoyancy, drag, and surface tension. This equation can be solved for R_B. The buoyancy term can be neglected when high-velocity and high-pressure data are of interest. Furthermore, the distance from the wall to the top of a just-departing bubble, y_B, is proportional to R_B. These lead to

$$y_B^+ = y_B \frac{U_\tau}{\nu_L} = c \frac{\sqrt{\sigma D_H \rho_L}}{\mu_L}, \qquad (12.28)$$

where

$$U_\tau = \sqrt{\tau_w / \rho_L}. \qquad (12.29)$$

The wall shear stress which can be induced by the presence of bubbles on the wall, is calculated by using a fully rough wall friction factor correlation,

$$\tau_w = \frac{f_{L0}'}{4} \frac{G^2}{2\rho_L}, \qquad (12.30)$$

$$f_{L0}' = 0.0055 \left\{ 1 + \left[2 \times 10^4 \frac{\varepsilon_D}{D_H} + \frac{10^6}{Re_{L0}} \right]^{1/3} \right\}, \quad \frac{\varepsilon_D}{D_H} = 10^{-4}, \qquad (12.31)$$

where $c = 0.015$ (and is empirically adjusted). As mentioned before, bubble thermal stability conditions require that the temperature of the liquid in contact with the point on the bubble surface that is the most distant from the heated surface be at T_{sat}. There is no need to adjust T_{sat} for the bubble interior curvature and Kelvin effect because the departing bubbles are typically relatively large. The temperature distribution in the liquid thermal boundary layer is evidently needed now. Levy's model uses the turbulent boundary layer temperature law of the wall derived by

Martinelli (1947). The turbulent boundary layer temperature profile, the derivation of which is based on the analogy between heat and momentum transfer processes, is discussed in Section 1.7. Accordingly,

$$T_w - T_L(y^+) = Q f(y^+, Pr_L),\qquad(12.32)$$

where

$$Q = \frac{q''_w}{\rho_L C_{PL} U_\tau},\qquad(12.33)$$

$$f(y^+, Pr_L) = \begin{cases} Pr_L\, y^+, & 0 \le y^+ \le 5, \\ 5\left\{Pr_L + \ln\left[1 + Pr_L\left(\dfrac{y^+}{5} - 1\right)\right]\right\}, & 5 < y^+ \le 30, \\ 5\{Pr_L + \ln[1 + 5\,Pr_L] + 0.5\ln(y^+/30)\}, & 30 < y^+. \end{cases}\qquad(12.34)$$

Using Eq. (12.32), along with $T_{sat} - \overline{T}_L = (T_w - \overline{T}_L) - (T_w - T_{sat})$ and $T_w - \overline{T}_L = q''_w/H_{L0}$, one gets

$$(T_{sat} - \overline{T}_L)_{OSV} = \frac{q''_{w,OSV}}{H_{L0}} - Q f(y_B^+, Pr_L).\qquad(12.35)$$

The *OSV model of Rogers et al.* (1987) is a modification of the OSV model of Staub (1968), which itself is similar to the model of Levy in many aspects. Staub's model accounts for the buoyancy effect, and in it the drag force on a bubble from shear stress is found by assuming a wall roughness equal to half the bubble departure diameter. The OSV model of Rogers *et al.* (1987) is meant to apply to low-pressure and low-flow-rate vertical heated channels. The basic assumptions are similar to those for Levy's model. However, bubbles are assumed to be chopped, distorted spheres, behaving as reported by Al-Hayes and Winterton (1981). In this respect, they are consistent with the experimental observations indicating that, at low pressure, bubbles departing from the wall slide before detaching (Bibeau and Salcudean, 1994a, b). Figure 12.15 depicts the bubble configuration. At bubble departure the forces that act on the bubble are the buoyancy force, drag force, and surface tension force, respectively:

$$F_B = \Delta\rho g \frac{\pi R_B^3}{3}\left[2 + 3\cos\theta_0 - \cos^3\theta_0\right],\qquad(12.36)$$

$$F_D = C_D \rho_f \frac{U_r^2}{2} R_B^2\left[\pi - \theta_0 + \cos\theta_0\sin\theta_0\right],\qquad(12.37)$$

$$F_\sigma = C_s \frac{\pi}{2} R_B \sigma \sin\theta_0(\cos\theta_r - \sin\theta_a)\qquad(12.38)$$

where θ_0, θ_r, and θ_a represent the static, receding, and advancing contact angles, respectively. The bubble height is related to its radius under static conditions according to

$$y_B^+ = R_B \frac{U_\tau}{\nu_f}(1 + \cos\theta_0).\qquad(12.39)$$

Bubble departure occurs when $F_D + F_B = F_\sigma$, and that leads to

$$R_B = \frac{3}{4\pi}\frac{C_2}{C_1} C_D \frac{U_r^2}{g}\left\{\left[1 + \frac{8\pi^2}{3}\frac{C_1 C_3}{C_2^2}\frac{C_s}{C_D^2}\frac{g\sigma}{\rho_f U_r^4}\right]^{1/2} - 1\right\},\qquad(12.40)$$

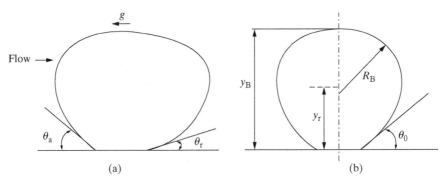

Figure 12.15. Bubble configuration in the OSV model of Rogers *et al.* (1987): (a) departing bubble; (b) bubble at equilibrium.

where

$$C_1 = 2 + 3\cos\theta_0 - \cos^3\theta_0, \tag{12.41a}$$

$$C_2 = \pi - \theta_0 + \cos\theta_0\sin\theta_0, \tag{12.41b}$$

$$C_3 = \sin\theta_0(\cos\theta_r - \cos\theta_a), \tag{12.41c}$$

$$C_D = \begin{cases} 1.22 & \text{for } 20 < \text{Re}_B < 400, \tag{12.41d} \\ 24/\text{Re}_B & \text{for } 4 < \text{Re}_B < 20, \tag{12.41e} \end{cases}$$

$$\text{Re}_B = \frac{2\rho_f U_r R_B}{\mu_f}, \tag{12.41f}$$

$$C_s = \frac{58}{\theta_0 + 5} + 0.14. \tag{12.41g}$$

In the last expression, θ_0 should be in degrees. Also, $\theta_r \approx \theta_0 - 10°$ and $\theta_a \approx \theta_0 + 10°$ are recommended. The quantity U_r represents the liquid time-averaged velocity, predicted by the turbulent boundary layer universal velocity profile, at the distance of $y_B/2$ from the wall.

These expressions account for the mechanical requirements for bubble departure. Bubbles large enough for departure would exist only if thermal conditions would allow, and that requires that the liquid temperature at y_B must be saturated. The liquid temperature distribution is found from the aforementioned turbulent temperature law of the wall.

EXAMPLE 12.2. For Example 12.1, estimate the location and the wall temperature where OSV occurs.

SOLUTION. All the needed properties were calculated in Example 12.1. We will proceed by calculating the Péclet number:

$$\text{Pe} = GDC_{PL}/k_L = 2.095 \times 10^5.$$

We deal with hyrdrodynamically controlled OSV, because Pe > 70000. Based on Eq. (12.24) we can write

$$C_{PL}(T_{sat} - \overline{T}_{L,OSV}) = 154 q_w''/G.$$

The solution of the equation gives $\overline{T}_{L,OSV} = 555.6$ K. The location of the OSV point is now found by an energy balance:

$$m_L C_{PL}(\overline{T}_{L,OSV} - T_{L,in}) = \pi D q_w'' Z_{OSV} \Rightarrow Z_{OSV} = 1.93\,\text{m}.$$

12.7 Transition from Partial Boiling to Fully Developed Subcooled Boiling

The point of onset of fully developed boiling (the vicinity of point E in Fig. 12.11 or 12.12) is often specified in experiments from the shape of the boiling curve. It represents a significant change in the gradient of the curve. According to Bowring (1962), the conditions of the onset of the fully developed boiling point (point E) can be found from

$$q_E'' = 1.4 q_D'', \tag{12.42}$$

where q_D'' is the heat flux at point D in Fig. 12.12; it represents the intersection of forced convection and fully developed boiling lines, when the lines are extended beyond their range of applicability. Point D must thus be obtained by intersecting appropriate correlations for single-phase liquid forced convection and fully developed boiling, and an iterative solution is often needed.

Shah (1977) has proposed

$$\left(\frac{T_{sat} - \overline{T}_L}{T_w - T_{sat}}\right)_E = 2. \tag{12.43}$$

Alternatively, since experiments show that the onset of significant void and onset of fully developed boiling are often very close to each other (Griffith *et al.*, 1958; Lahey and Moody, 1993), the OSV point, can be assumed to represent the beginning of fully developed subcooled boiling.

It should be noted that, in many design situations and even applications dealing with safety of boiling systems, there is no need for an accurate calculation of the heat transfer coefficient in the partial boiling regime or for the accurate location of point D. After all, in the partial boiling regime the heat transfer is efficient, the heated surface temperature is low, there is little voidage, and therefore the boiling process is virtually free from risk of burnout.

Bibeau and Sacudean (1990, 1994a, b) and Prodanovic *et al.* (2002a, b) have experimentally studied the bubble dynamics and voidage in subcooled boiling, for water under low-flow and low-pressure conditions. They did not observe a region of attached void. Bubble departure occurred downstream from the ONB point, and the bubble detachment mechanism did not appear to explain the OSV phenomenon. The latter observation is of course consistent with thermally controlled OSV for low-flow conditions. Departing bubbles slide and detach. Furthermore, unlike the high-pressure observation that OSV approximately coincides with the initiation of fully developed nucleate boiling (Griffith *et al.*, 1958), no correlation between the OSV and fully developed boiling initiation was observed. Prodanovic *et al.* (2002a) studied the behavior of bubbles in experiments in the 2- to 3-bar pressure range and noted that, during their growth, bubbles transform from a flattened to an elongated shape. The same authors (Prodanovic *et al.*, 2002b) studied and empirically correlated the transition from partial to fully developed nucleate boiling for the aforementioned tests with water under low-pressure and low-flow conditions.

12.8 Hydrodynamics of Subcooled Flow Boiling

Prediction of the void fraction profile during subcooled forced-flow boiling is important with respect to stability considerations in boiling channels and for neutron moderation in nuclear reactors. For boiling systems that operate at high pressure, the void fraction remains small upstream of the OSV point and typically is not more than a few percent. In these systems $a = 0$ is often assumed upstream of the OSV point. The error resulting from this assumption can be nontrivial for low-pressure boiling systems where a can be 2%–9% at the OSV point.

It is possible, in principle, to calculate the subcooled boiling local void fraction in a boiling channel by using mechanistic models that are consistent with the 2FM or DFM. This approach is consistent with the way some thermal-hydraulics computer codes solve the two-phase flow conservation equations. However, these mechanistic models require many constitutive relations that are not well understood. Simpler, semi-empirical methods are often used.

In this section the two-phase conservation equations in subcooled flow boiling and their closure requirements are first discussed. The subcooled flow boiling void-quality relations are then discussed.

In forced subcooled boiling the vapor phase is expected to remain saturated with respect to the local pressure. The one-dimensional, steady state, 2FM conservation equations for subcooled boiling in a channel with uniform cross section can be written as follows.

For the vapor mass,

$$\frac{d}{dz}(\rho_g U_g \alpha) = \Gamma. \tag{12.44}$$

For the mixture mass,

$$\frac{d}{dz}[\rho_L U_L (1 - \alpha) + \rho_g U_g \alpha] = 0. \tag{12.45}$$

For the vapor momentum,

$$\frac{d}{dz}\left[\rho_g \alpha U_g^2\right] - \Gamma U_I = -\alpha \frac{dP}{dz} - \rho_g g \alpha \sin\theta - F_I + F_{VM}. \tag{12.46}$$

For the mixture momentum,

$$\frac{d}{dz}\left[\rho_L (1 - \alpha) U_L^2 + \rho_g \alpha U_g^2\right] = -\frac{dP}{dz} - \bar{\rho} g \sin\theta - \frac{\tau_w p_f}{A}. \tag{12.47}$$

For the mixture energy

$$G\frac{d}{dz}[xh_g + (1 - x)h_L] + \frac{G^3}{2}\frac{d}{dz}\left[\frac{(1 - x)^3}{\rho_L^2 (1 - \alpha)^2} + \frac{x^3}{\rho_g^2 \alpha^2}\right] + Gg\sin\theta = p_{heat}q_w''/A. \tag{12.48}$$

In these equations θ is the angle of inclination with the horizontal plane, Γ is the vapor generation rate per unit mixture volume, subscript L stands for subcooled liquid, subscript g stands for saturated vapor, F_I is the interfacial drag and friction force, and F_{VM} is the virtual mass force.

The unknowns in these equations are U_L, U_g, P, x, h_L, and Γ. (Note that x can be replaced with α as an unknown.) The equation set is therefore not closed, because the unknowns outnumber the equations by one. The additional equation can be provided in two ways. The easier way is to seek an expression that relates quality x to h_L, or equivalently (see Eq. (12.51) to follow) an expression of the form $x = f(x_{eq})$. Alternatively, the equation set can be closed by modeling the volumetric evaporation rate, Γ. Methods based on the latter approach represent *mechanistic models*. However, Γ is determined by the wall heat flux, and the manner that heat flux is partitioned between absorption by the subcooled liquid and the evaporation processes should be determined. Calculation of Γ is thus difficult.

When changes in properties and mechanical energy terms are negligible, Eq. (12.48) can be integrated to obtain

$$[x(h_g - h_L) + h_L]_z - h_{in} = \frac{p_{heat}}{AG} \int_0^z q_w'' dz. \tag{12.49}$$

The local (flow-area-averaged) equilibrium quality can be represented as

$$x_{eq} = \frac{h - h_f}{h_{fg}} = \{[x(h_g - h_L) + h_L]_z - h_f\}/h_{fg}. \tag{12.50}$$

Note that Eq. (12.49) cannot be solved since it contains two unknowns: x and h_L. Again, an expression of the form $x = f(x_{eq})$ is needed.

The *quality profile fit* of Ahmad (1970) is a widely used empirical correlation between flow quality x and equilibrium quality in subcooled boiling. The correlation is

$$x = \frac{x_{eq} - x_{eq,OSV} \exp\left(\frac{x_{eq}}{x_{eq,OSV}} - 1\right)}{1 - x_{eq,OSV} \exp\left(\frac{x_e}{x_{eq,OSV}} - 1\right)}, \tag{12.51}$$

where $x_{eq,OSV}$ is the equilibrium quality at the OSV point. This expression obviously applies for $x_{eq} > x_{eq,OSV}$.

As mentioned earlier, the key to mechanistic modeling, namely the solution of Eqs. (12.44)–(12.48), is the specification of Γ in addition to the interfacial forces that determine the velocity slip between the two phases. Mechanistic modeling of these parameters requires detailed knowledge of bubble-related phenomena. When the DFM or some other diffusion model is applied, there is no need for specification of interfacial force terms. For example, using the DFM, we can write

$$U_g - j = (U_g - U_L)(1 - \alpha) = V_{gj}, \tag{12.52}$$

$$\alpha = \frac{x}{C_0 \left[x + \frac{\rho_g}{\rho_L}(1 - x)\right] + \frac{\rho_g V_{gj}}{G}}. \tag{12.53}$$

(Note that all parameters are cross-section averaged.) However, the need for modeling Γ still remains.

Let us discuss the phenomenology of void generation in subcooled boiling, which is needed for modeling the evaporation process. In subcooled flow boiling, a super-heated liquid layer is formed on the heated surface, while the bulk liquid is subcooled. Bubble nucleation takes place on wall crevices. Evaporation at the base of a bubble as it grows on the heated surface may be accompanied by condensation near its top. The wall heat flux is partially absorbed by the liquid phase, and the remainder goes to evaporation. Bubbles departing from the heated surface disrupt the thermal boundary layer, enhancing heat transfer between the wall and the liquid phase by their "pumping effect." Bubbles released from the wall may interact with other bubbles and will undergo partial condensation.

According to Bowring (1962), when the effect of recondensation is negligible, one can argue that

$$q_w'' = q_L'' + q_V''. \tag{12.54}$$

The sensible component of heat flux is due to convection as well as the pumping effect of the departing bubbles, that is,

$$q_L'' = c_L H_{L0}(T_w - \overline{T}_L) + q_P'', \tag{12.55}$$

where

q_L'' is the heat flux absorbed by subcooled liquid,

q_V'' is the heat flux associated with evaporation,

q_P'' is the heat flux associated with the pumping effect of bubbles departing from the wall,

c_L is the fraction of wall covered by liquid (≈ 1), and

H_{L0} is the single-phase-flow heat transfer coefficient (to be found from an appropriate correlation, e.g., the Dittus–Boelter correlation).

A pumping factor ε (Bowring's pumping parameter) can be defined, such that

$$\varepsilon = \frac{q_P''}{q_V''} = \frac{q_P''}{q_w'' - c_L H_{L0}(T_w - \overline{T}_L) - q_P''}. \tag{12.56}$$

The pumping factor can be found from (Rouhani and Axelsson, 1970)

$$\varepsilon = \frac{\rho_L(h_f - h_L)}{\rho_g h_{fg}}. \tag{12.57}$$

Thus, provided that condensation is neglected, and knowing H_{L0}, we can solve Eqs. (12.54)–(12.57) for ε, q_V'', q_P'', and q_L'', and the volumetric evaporation rate will be

$$\Gamma = \frac{p_{heat} q_V''}{A h_{fg}}. \tag{12.58}$$

In fact, in the absence of condensation, Eq. (12.44) can be recast and integrated to directly calculate x when the fluid properties and ε are assumed to remain constant:

$$G\frac{dx}{dz} = \Gamma \tag{12.59}$$

$$\Rightarrow x = \frac{p_{heat}}{AGh_{fg}} \int_{z_{OSV}}^{z} q_V'' dz. \tag{12.60}$$

The *model of Lahey and Moody* (1993) is a semi-empirical adjustment to Bowring's model and accounts for the effect of condensation. Equation (12.58) is recast as

$$\Gamma = p_{\text{heat}}(q_V'' - q_{\text{cond}}'')/(Ah_{\text{fg}}) = p_{\text{heat}}\left(\frac{q_V'' + q_P''}{1 + \varepsilon} - q_{\text{cond}}''\right) \bigg/ (Ah_{\text{fg}}). \quad (12.61)$$

Equation (12.59) will then lead to

$$x = \frac{p_{\text{heat}}}{AGh_{\text{fg}}}\left[\int_{z_{\text{OSV}}}^{z} \frac{q_V'' + q_P''}{1 + \varepsilon}dz - \int_{z_{\text{OSV}}}^{z} q_{\text{cond}}''dz\right]. \quad (12.62)$$

The integrands in Eq. (12.62) can be found from

$$q_V'' + q_P'' = \begin{cases} 0.0 & \text{for } z < z_{\text{OSV}}, \\ q_w''\left[1 - \dfrac{(h_f - h_L)z}{(h_f - h_L)_{\text{OSV}}}\right] & \text{for } z > z_{\text{OSV}}, \end{cases} \quad (12.63)$$

$$p_{\text{heat}}q_{\text{cond}}''/A = C\frac{h_{\text{fg}}}{v_{\text{fg}}}\alpha(T_{\text{sat}} - \overline{T}_L), \quad (12.64)$$

with

$$C = 150(\text{hr}\cdot{}^\circ\text{F})^{-1} = 0.075(\text{s}\cdot\text{K})^{-1}. \quad (12.65)$$

Models with more phenomenological detail also have been published (Hu and Pan, 1995; Hainoun et al., 1996; Zeitoun and Shoukri, 1997; Tu and Yeoh, 2002; Xu et al., 2006). The method of Zeitoun and Shoukri (1997) is based on the solution of one-dimensional conservation equations using the DFM. Hu and Pan (1995), Tu and Yeoh (2002), and Xu et al. (2006) have used the 2FM equations. In the model of Hu and Pan (1995), net vapor generation starts at the OSV point, which is modeled based on the correlation of Saha and Zuber (1974). Evaporation downstream from the OSV point is calculated by using the method of Moody and Lahey described earlier. In the model of Xu et al., evaporation starts at the ONB point, which is predicted using the correlation of Bergles and Rohsenow (1964).

Zeitoun and Shoukri (1996) developed the following empirical correlation for the Sauter mean diameter of bubbles during subcooled boiling:

$$\frac{d_{\text{Sm}}}{\sqrt{\frac{\sigma}{g\Delta\rho}}} = \frac{0.0683(\rho_L/\rho_g)^{1.326}}{\text{Re}^{0.324}\left[\text{Ja} + \frac{149.2(\rho_L/\rho_g)^{1.326}}{\text{Bo}^{0.487}\text{Re}^{1.6}}\right]}, \quad (12.66)$$

where $\text{Ja} = \rho_L C_{\text{PL}}(T_{\text{sat}} - \overline{T}_L)/\rho_g h_{\text{fg}}$ is the Jacob number, $\text{Bo} = q_w''/Gh_{\text{fg}}$ represents the boiling number, and $\text{Re} = GD_H/\mu_L$. The model of Zeitoun and Shoukri (1997) solves the one-dimensional, steady-state conservation equations, by using the DFM. Equations (12.54) and (12.55) are used with $c_L \approx 1$. Zeitoun and Shoukri argued that Bowring's method overpredicts q_P'' by assuming that the bubbles forming on the wall are surrounded by saturated liquid. The pumping factor should instead be defined as

$$\varepsilon = \frac{3}{4}\frac{\rho_L C_{\text{PL}}(T_w - \overline{T}_L)\delta}{\rho_g h_{\text{fg}}d_{\text{Sm}}}, \quad (12.67)$$

where d_{Sm} is found from Eq. (12.66), and $\delta = k_L(T_w - \overline{T}_L)/q''_w$ is the thickness of the thermal boundary layer. Zeitoun and Shoukri modeled condensation as

$$\frac{p_{heat}q''_{cond}}{A} = c_s a''_I H_I(T_{sat} - \overline{T}_L), \tag{12.68}$$

where c_s represents the fraction of bubble exposed to condensation (≈ 0.5 according to Zeitoun and Shoukri 1996). They obtained the condensation heat transfer coefficient at the interphase, H_I, from the following correlation (Zeitoun *et al.*, 1995):

$$\mathrm{Nu}_I = \frac{H_I d_{Sm}}{k_L} = 2.04 \mathrm{Re}_B^{0.61} \alpha^{0.328} \mathrm{Ja}^{-0.308}, \tag{12.69}$$

where

$$\mathrm{Re}_B = \rho_L U_B d_{Sm}/\mu_L$$

and

$$U_B = \frac{1.53}{1-\alpha}[\sigma g \Delta\rho/\rho_L^2]^{0.25}.$$

There is no need to define ONB or OSV points in the model of Zeitoun and Shoukri (1996). To start the calculations, one assumes a small finite α at the inlet.

The model of Tu and Yeoh (2002) is consistent with a six-equation model, allowing for subcooled/superheated liquid and subcooled/superheated vapor. It can thus account for condensation or evaporation in the bulk flow field. Equation (12.66) is used for bubble mean diameter in the bulk flow. The heat transfer rate associated with the volumetric phase change (defined to be positive for condensation) is found from

$$\frac{p_{heat}q''_{cond}}{A} = a''_I H_I(T_{sat} - \overline{T}_L), \tag{12.70}$$

where $a''_I = 6\alpha/d_{Sm}$ is the interfacial area concentration, and the interfacial heat transfer coefficient is found from the Ranz and Marshall (1952) correlation:

$$\frac{H_I d_{Sm}}{k_L} = 2 + 0.6\,\mathrm{Re}_B^{0.5}\,\mathrm{Pr}_L^{0.3}, \tag{12.71}$$

where $\mathrm{Re}_B = \rho_L d_{Sm}|U_g - U_L|/\mu_L$ is the bubble Reynolds number.

Xu *et al.* (2006) applied a five-equation 2FM (in which the vapor phase is assumed to remain saturated). The subcooled boiling phenomenology is based on Eqs. (12.54) and (12.55). In applying Eq. (12.55), they utilized the following correlation for c_L (Hainoun *et al.*, 1996):

$$c_L = \begin{cases} 1 - \frac{\pi}{16}\frac{\alpha}{\alpha_{OSV}} & \text{for } \alpha \leq 16\alpha_{OSV}/\pi, \\ 0 & \text{for } \alpha > 16\alpha_{OSV}/\pi. \end{cases}$$

Xu *et al.* examined the proposed pumping parameter expression of Zeitoun and Shoukri, Eq. (12.67), along with the following correlation that has been proposed by Hainoun *et al.* (1996):

$$\frac{1}{1+\varepsilon} = 2C_{ev}\left(\frac{T_w - T_{sat}}{T_w - \overline{T}_L}\right), C_{ev} \approx 0.5.$$

This correlation led to better agreement with experimental data.

One can observe that current mechanistic models need numerous closure relations that are sometimes poorly understood.

12.9 Pressure Drop in Subcooled Flow Boiling

The discussion in the previous section should have shown that the flow field downstream from the ONB point is a complicated, evolving two-phase mixture, with strong spatial acceleration. The two-phase flow pressure drop methods described in Chapter 8 can be used for estimating the frictional pressure drop. Those methods are primarily based on steady-state and developed-flow data, however, and may not be very accurate for subcooled boiling. Some empirical correlations have been specifically developed for subcooled boiling. These correlations often provide for the calculation of the total pressure drop over the subcooled boiling length of a heated channel, but they do not separate the frictional and acceleration pressure-drop terms.

The correlation of Owens and Schrock (1960) is

$$AC_{dw} = j + V_{gj} + \alpha \frac{\partial V_{gj}}{\partial \alpha}$$

$$\Phi_{L0}^2 = 0.97 + 0.028e^{6.13Y} \tag{12.72}$$

$$Y = 1 - \frac{T_w - T_{sat}}{(T_w - T_{sat})_{OSV}}. \tag{12.73}$$

The correlation of Tarasova *et al.* (1966) is

$$\Phi_{L0}^2 = 1 + \left(\frac{q_w'' \rho_L}{\rho_g G h_{fg}}\right)^{0.7} \left(\frac{\rho_L}{\rho_g}\right)^{0.08} \left[\frac{2.63}{Y} \ln\left(\frac{1.315}{1.315 - Y}\right) - 20\right]. \tag{12.74}$$

A correlation proposed by Ueda, and discussed by Kandlikar and Nariai (1999), is

$$\Phi_L^2 = 1 + 1.2x^n \left[\left(\frac{\rho_L}{\rho_g}\right)^{0.8} - 1\right], n = 0.75(1 + 0.01\sqrt{\rho_L/\rho_g}). \tag{12.75}$$

where x is the local quality. The correlation of Ueda has been found to perform best in predicting the experimental data of Nariai and Inasaka (1992).

12.10 Partial Flow Boiling

In the partial boiling regime, represented by the zone between points B and E in Figs. 12.11 or 12.12, single-phase liquid forced-convection and nucleate boiling both significantly contribute to heat transfer. The contribution of nucleate boiling increases as T_w is increased. The contribution of forced convection becomes small when the fully developed boiling (point E in Fig. 12.12) is reached. Heat transfer coefficient calculation methods for the partial boiling regime are mostly empirical curve fits done using interpolation.

The following correlation for the heat flux in partial boiling was proposed by Bergles and Rohsenow (1964):

$$q_w'' = q_{FC}'' \left[1 + \left\{\frac{q_{SB}''}{q_{FC}''} - \frac{q_{ONB}''}{q_{FC}''}\right\}^2\right]^{1/2}, \tag{12.76}$$

where q_{FC}'' is the forced convection heat flux, found from $q_{FC}'' = H_{L0}(T_w - \overline{T}_L)$, with H_{L0} representing the single-phase liquid forced-convection heat transfer coefficient. The parameter q_{SB}'' is the subcooled boiling heat flux, calculated by applying an appropriate fully developed subcooled boiling correlation. Equation (12.76) is evidently simple. More importantly, it does not include q_E'', the heat flux at the partial boiling–fully developed boiling transition point. This is important because when conservation equations are numerically solved in a boiling channel, q_E'' is not known *a priori*. Other empirical correlations that take into account this issue include those of Pokhalov *et al.* (1966) and Shah (1977).

The following interpolation method proposed by Kandlikar (1997, 1998a) is meant to provide smooth transition from the single-phase forced convection region to partial boiling, and from partial boiling to fully developed subcooled boiling:

$$q_w'' = \left[q_{ONB}'' - b(T_w - T_{sat})_{ONB}^m \right] + b(T_w - T_{sat})^m, \tag{12.77}$$

where

$$b = (q_E'' - q_{ONB}'')/\left[(T_w - T_{sat})_E^m - (T_w - T_{sat})_{ONB}^m \right], \tag{12.78}$$

$$m = n + pq_w'', \tag{12.79}$$

$$p = (1/0.3 - 1)/(q_E'' - q_{ONB}''), \tag{12.80}$$

$$n = 1 - pq_{ONB}''. \tag{12.81}$$

This correlation requires a priori knowledge of q_E'', however.

12.11 Fully Developed Subcooled Flow Boiling Heat Transfer Correlations

In this regime, because of the predominance of nucleate boiling, there is relatively little effect of coolant mass flux or coolant bulk temperature. As a result, the empirical correlations for water are very simple. The correlation of McAdams *et al.* (1949), for example, is

$$q_w'' = 2.26(T_w - T_{sat})^{3.86}. \tag{12.82}$$

The correlation is purely empirical and dimensional. The constant in the expression depends on the unit system. In the form presented here, the unit system must be as follows: q_w'' must be in watts per meter squared and $T_w - T_{sat}$ must be in kelvins. The range of applicability of Eq. (12.82) is $30 < P < 90$ psia.

The correlation of Thom *et al.* (1965) is for high-pressure water. In SI units, the correlation is

$$T_w - T_{sat} = 22.65q_w''^{0.5} \exp(-P/87), \tag{12.83}$$

where now q_w'' is in megawatts per meter squared, P is in bars, and $T_w - T_{sat}$ is in kelvins. The correlation's range of validity is $750 \leq P \leq 2000$ psia (51–136 bars).

A correlation that applies to moderately high pressures is due to Jens and Lottes (1951):

$$T_w - T_{sat} = 25q_w''^{0.25} \exp(-P/62), \tag{12.84}$$

where the units are the same as those for Eq. (12.83). The range of validity of this correlation is $7 \leq P \leq 172$ bars.

Table 12.1. *Values of the fluid-surface parameter F_{fl} in the correlation of Kandlikar (Kandlikar, 1997, 1998a; Kandlikar and Steinke, 2003).*

Fluid	F_{fl}	Fluid	F_{fl}
Water	1.00	R-32/R-1132	3.30
R-11	1.30	R-124	1.00
R-12	1.50	R-141b	1.80
R-13 BI	1.31	R-134a	1.63
R-22	2.20	R-152a	1.10
R-113	1.30	Kerosene	0.488
R-114	1.24	Nitrogen	4.70
		Neon	3.50

Note: Use 1.0 for any fluid with a stainless-steel tube.

A correlation by Kandlikar (1997, 1998a) is

$$q_w'' = \left[1058(Gh_{fg})^{-0.7} F_{fl} H_{L0} (T_w - T_{sat}) \right]^{3.33}, \tag{12.85}$$

where H_{L0} is to be calculated by using the correlation of Gnielinski (1976) or Petukhov and Popov (1963), and F_{fl} is the *fluid-surface parameter*. Values for this parameter are given in Table 12.1.

The correlation of Gnielinski (1976) for single-phase forced convection in tubes is applicable over the range $0.5 \le Pr_L \le 2000$ and $2300 \le Re_{L0} \le 10^4$:

$$Nu_{L0}^* = \frac{(Re_{L0} - 1,000)(f/2)Pr_L}{\left[1 + 12.7 \left(Pr_L^{2/3} - 1 \right)(f/2)^{0.5} \right]}. \tag{12.86}$$

The correlation of Petukhov and Popov (1963) is for the range $0.5 \le Pr_L \le 2000$ and $10^4 \le Re_{L0} \le 5 \times 10^6$:

$$Nu_{L0}^* = \frac{Re_{L0} Pr_L (f/2)}{\left[1.07 + 12.7 \left(Pr_L^{2/3} - 1 \right)(f/2)^{0.5} \right]}. \tag{12.87}$$

Equation (12.86) should be applied by using fluid properties calculated at mean fluid temperature. The value of Nu_{L0} calculated from Eq. (12.87) is thus a constant-property Nusselt number. It can be corrected for the effect of fluid property variation across the flow channel according to (Petukhov, 1970)

$$Nu_{L0} = Nu_{L0}^* \left(\frac{\mu_L}{\mu_w} \right)^{0.11}, \tag{12.88}$$

$$f = [1.58 \ln(Re_{L0}) - 3.28]^{-2}, \tag{12.89}$$

where μ_w is the liquid viscosity corresponding to the wall temperature.

In addition to these correlations, some of the recent saturated flow boiling correlations are also applicable to subcooled flow boiling, with minor modifications. These correlations will be discussed in Section 12.13.

12.12 Characteristics of Saturated Flow Boiling

Saturated, forced-flow boiling refers to the entire region between the point where $x_{eq} = 0$ and the critical heat flux point. A sequence of complicated two-phase flow patterns, including bubbly, churn, slug, and annular-dispersed, can take place, as noted in Fig. 12.2. The two-phase flow regimes cover a quality range of a few per cent, up to very high values characteristic of annular flow regime (sometimes approaching 100%). Nucleate boiling is predominant where quality is low (a few per cent), forced convective evaporation is predominant at high qualities representing annular flow, and elsewhere both mechanisms can be important. The relative contribution of forced convection increases as quality increases. In the two-phase forced convection region, bubble nucleation does not occur. Heat transfer occurs by evaporation at the liquid–vapor interface. In the annular-dispersed flow regime at very high qualities, the liquid film becomes so thin that nucleation is completely suppressed, and evaporation at the film surface provides an extremely efficient heat transfer process. As will be shown, most of the successful empirical correlations take this phenomenology into consideration.

As discussed earlier, the boiling heat transfer regimes in a boiling channel depend on the heat flux q_w'', mass flux G, and equilibrium quality x_{eq}. The horizontal lines in Fig. 12.5 show qualitatively the boiling regimes along a uniformly heated steady-state channel. At moderate heat fluxes, such as lines representing $q_{w,1}''$ and $q_{w,2}''$ in Fig. 12.5, the boiling regimes include subcooled nucleate boiling, saturated nucleate boiling, two-phase forced convection, dryout, and postdryout heat transfer.

The phenomenology of saturated nucleate boiling is essentially the same as that of subcooled nucleate boiling. The heat transfer coefficient is insensitive to mass flux, and all fully developed subcooled nucleate boiling predictive methods should in principle apply.

The nucleate boiling regime can be completely bypassed when the heat flux in the channel is very low. Suppression of nucleate boiling by forced convection is in principle possible in any two-phase flow regime. In practice, it occurs predominantly in the annular flow regime. The location along the boiling channel where suppression first occurs can be estimated by equating q_{NB}'' (obtained from an appropriate correlation) with $q_{FC}'' = H_{FC}(T_w - T_{sat})$, where H_{FC} is the heat transfer coefficient representing evaporative forced convection. The occurrence of a transition zone complicates the situation, however.

12.13 Saturated Flow Boiling Heat Transfer Correlations

Nucleate boiling and forced-convective evaporation both contribute to the heat transfer in saturated flow boiling. At low x_{eq}, the contribution of the nucleate boiling mechanism dominates, but the contribution of convection increases as x_{eq} is increased. Once the annular-dispersed flow regime is achieved, the contribution of convective evaporation becomes predominant. Forced-flow boiling correlations should thus take into account the composite nature of the boiling heat transfer mechanism.

Forced-flow boiling correlations can generally be divided into three groups. The first group uses the summation rule of Chen (Chen, 1966), who proposed one of the earliest and most successful correlations of this type, whereby $H = H_{NB} + H_{FC}$ is assumed. The second group uses the asymptotic model, whereby $H^n = H_{NB}^n + H_{FC}^n$. With $n > 1$, H asymptotically approaches H_{NB} or H_{FC} as $(H_{NB}/H_{FC}) \to \infty$ and vice versa, and $n \to \infty$ leads to the selection of the larger of the two. The third group constitutes the flow-pattern-dependent correlations.

In the forthcoming discussions, x and x_{eq} will be interchangeable since thermodynamic equilibrium prevails.

First, let us discuss the forced convective evaporation correlations. Many of these correlations are based on the two-phase multiplier concept, according to which $H = H_{f0} \cdot f(G, x, \ldots)$, where H_{f0} is the convection heat transfer coefficient when all of the mass is saturated liquid, and the function $f(G, x, \ldots)$ is a two-phase multiplier. The concept is thus similar in principle to the concept of the two-phase pressure-drop multiplier of Lockhart and Martinelli (1949), as discussed in Chapter 8. Some of the most widely used correlations are of the form

$$\frac{H}{H_{f0}} = C(1/X_{tt})^n, \tag{12.90}$$

where X_{tt} is the turbulent–turbulent Martinelli parameter (see Eq. (8.26)), given by

$$X_{tt} = \left(\frac{\rho_g}{\rho_f}\right)^{0.5} \left(\frac{\mu_f}{\mu_g}\right)^{0.1} \left(\frac{1-x}{x}\right)^{0.9}. \tag{12.91}$$

Dengler and Addoms (1956) suggest $C = 3.5$ and $n = 0.5$. Bennett et al. (1961) suggest $C = 2.9$ and $n = 0.66$.

Correlations for saturated flow boiling are now reviewed.

The Correlation of Chen (1966). Chen's correlation,

$$H = H_{NB}S + H_{FC}F, \tag{12.92}$$

is among the oldest and most successful and widely used correlations for saturated boiling. It works well for water at relatively low pressure and has been applied to a variety of fluids. It deviates from measured data for refrigerants, however. The forced convection component is found from

$$H_{FC}D_H/k_f = 0.023\mathrm{Re}_f^{0.8}\,\mathrm{Pr}_f^{0.4}, \tag{12.93}$$

where

$$\mathrm{Re}_f = G(1-x)D_H/\mu_f, \tag{12.94}$$

$$\mathrm{Pr}_f = (\mu C_P/k)_f. \tag{12.95}$$

The factor F is meant to represent $(\mathrm{Re}_{TP}/\mathrm{Re}_f)^{0.8}$ and was correlated by Chen empirically in a graphical form. A curve fit to the graphical correlation is (Collier, 1981)

$$F = \begin{cases} 1 & \text{for } \dfrac{1}{X_{tt}} < 0.1, \tag{12.96} \\[2ex] 2.35\left(0.213 + \dfrac{1}{X_{tt}}\right)^{0.736} & \text{for } \dfrac{1}{X_{tt}} > 0.1. \tag{12.97} \end{cases}$$

Another correlation for F is due to Bennett and Chen (1980):

$$F = \left(\frac{\mathrm{Pr}_f + 1}{2}\right)^{0.444} \left(\Phi_{f,tt}^2\right)^{0.444}. \tag{12.98}$$

The nucleate boiling component is based on the correlation of Forster and Zuber (1955) (see Section 11.3), modified to account for the reduced average superheat in the thermal boundary layer on bubble nucleation on wall cavities:

$$H_{NB} = 0.00122 \left\{ \frac{k_f^{0.79} C_{Pf}^{0.45} \rho_f^{0.49} \{g_c\}^{0.25}}{\sigma^{0.5} \mu_f^{0.29} h_{fg}^{0.24} \rho_g^{0.24}} \right\} \Delta T_{sat}^{0.24} \Delta P_{sat}^{0.75} S, \quad (12.99)$$

where $\Delta T_{sat} = T_w - T_{sat}$ and $\Delta P_{sat} = P_{sat}(T_w) - P$. Note that g_c is needed for English units only. The parameter S is *Chen's suppression factor* and is meant to represent $S = (\Delta T_{eff}/\Delta T_{sat})^{0.99}$, where ΔT_{eff} is the effective liquid superheat in the thermal boundary layer. S was also correlated graphically. An empirical curve fit to Chen's graphical correlation is (Collier, 1981)

$$S = \left[1 + (2.56 \times 10^{-6})(Re_f F^{1.25})^{1.17} \right]^{-1}. \quad (12.100)$$

Alternatively, according to Bennett and Chen (1980) (see Lahey and Moody, 1993),

$$S = 0.9622 - 0.5822 \tan^{-1} \left(\frac{Re_f F^{1.25}}{6.18 \times 10^4} \right). \quad (12.101)$$

The Correlation of Kandlikar (1990, 1991). Kandlikar's correlation is based on 10 000 data points covering water, refrigerants, and cryogenic fluids:

$$H = \max(H_{NBD}, H_{CBD}), \quad (12.102)$$

$$H_{NBD} = \left\{ 0.6683 Co^{-0.2}(1-x)^{0.8} f_2(Fr_{f0}) + 1058.0 Bo^{0.7}(1-x)^{0.8} F_{fl} \right\} H_{f0}, \quad (12.103)$$

$$H_{CBD} = \left\{ 1.136 Co^{-0.9}(1-x)^{0.8} f_2(Fr_{f0}) + 667.2 Bo^{0.7}(1-x)^{0.8} F_{fl} \right\} H_{f0}, \quad (12.104)$$

where, for the calculation of H_{f0}, the aforementioned correlation of Gnielinski (1976) (Eq. (12.86)) is recommended. Other parameters in Kandlikar's correlation are the convection number Co, the boiling number Bo, and the Froude number when all mixture is saturated liquid, Fr_{f0}, defined, respectively, as

$$Co = (\rho_g/\rho_f)^{0.5}[(1-x)/x]^{0.8}, \quad (12.105)$$

$$Bo = q_w''/(G h_{fg}), \quad (12.106)$$

and

$$Fr_{f0} = G^2/(\rho_f^2 g D). \quad (12.107)$$

The parameter F_{fl} is the aforementioned fluid-surface parameter (see Table 12.1). Finally,

$$f_2(Fr_{f0}) = 1 \quad (12.108)$$

for vertical tubes and for horizontal tubes with $Fr_{f0} \geq 0.04$, and

$$f_2(Fr_{f0}) = (25 Fr_{f0})^{0.3} \quad (12.109)$$

for $Fr_{f0} < 0.04$ in horizontal tubes.

The correlation of Gungor and Winterton (1986, 1987). Gungor and Winterton's correlation is based on 3,700 data points for water, refrigerants, and ethylene glycol. The original correlation (Gungor and Winterton, 1986) was subsequently simplified by the authors (Gungor and Winterton, 1987) to the following easy-to-use correlation:

$$H = H_f \left\{ 1 + 3{,}000 Bo^{0.86} + 1.12[x/(1-x)]^{0.75}[\rho_f/\rho_g]^{0.41} \right\} E_2. \quad (12.110)$$

For horizontal tubes with $Fr_{f0} < 0.05$,

$$E_2 = Fr_{f0}^{(0.1-2Fr_{f0})}, \tag{12.111}$$

Otherwise,

$$E_2 = 1. \tag{12.112}$$

The Correlation of Liu and Winterton (1991). Liu and Winterton's is a further improvement over the earlier correlation proposed by Gungor and Winterton (1986). This newer correlation is based on more than 4200 data points for saturated boiling and more than 990 data points for subcooled boiling. The fluids include water, refrigerants, and hydrocarbons. The form of the correlation is similar to a form suggested by Kutateladze (1961):

$$H = \left[(E_2 E H_{f0})^2 + (S_2 S H_{NB})^2\right]^{1/2}, \tag{12.113}$$

$$E = \left[1 + x Pr_f \left(\frac{\rho_f}{\rho_g} - 1\right)\right]^{0.35}, \tag{12.114}$$

$$S = \frac{1}{1 + 0.055 E^{0.1} Re_{f0}^{0.16}}. \tag{12.115}$$

The heat transfer coefficient H_{f0} is based on the Dittus–Boelter correlation

$$\frac{H_{f0} D_H}{k_f} = 0.023 Re_{f0}^{0.8} Pr_f^{0.4}. \tag{12.116}$$

The nucleate boiling heat transfer coefficient H_{NB} is to be calculated by using the pool boiling correlation of Cooper (1984):

$$H_{NB} = 55(P/P_{cr})^{0.12} \left(-\log_{10} \frac{P}{P_{cr}}\right)^{-0.55} M^{-0.5} q_w''^{2/3}. \tag{12.117}$$

When the channel is horizontal and $Fr_{f0} \leq 0.05$,

$$E_2 = Fr_{f0}^{(0.1-2Fr_{f0})}, \tag{12.118}$$

$$S_2 = \sqrt{Fr_{f0}}. \tag{12.119}$$

For vertical channels, and for horizontal channels for which $Fr_0 > 0.05$,

$$E_2 = S_2 = 1. \tag{12.120}$$

The correlation of Liu and Winterton (1991) can be applied to subcooled boiling as well, provided that $T_w - \overline{T}_L$ and $T_w - T_{sat}$ are used as temperature differences for forced convection and nucleate boiling components of the heat flux, respectively, thereby,

$$q_w'' = \sqrt{\left[S_2 S H_{NB}(T_w - T_{sat})\right]^2 + \left[E_2 E H_{f0}(T_w - \overline{T}_L)\right]^2}. \tag{12.121}$$

The Correlation of Steiner and Taborek (1992). The correlation of Steiner and Taborek is among the most accurate for nucleate boiling in vertical tubes. Accordingly,

$$H = \left[(H_{f0} F_{FC})^3 + (H_{NB,0} F_{NB})^3\right]^{1/3}, \tag{12.122}$$

where H_{f0} is to be found from a forced convection correlation, for example, the correlation of Gnielinski (1976) (see Eq. (12.86)). When $x \lesssim 0.6$ and the heat flux is high enough for nucleate boiling to occur, the correction factor for forced convection over the range $3.75 \le \rho_f/\rho_g \le 5000$ is found from

$$F_{FC} = \left[(1-x)^{1.5} + 1.9x^{0.6} \left(\frac{\rho_f}{\rho_g} \right)^{0.35} \right]^{1.1}. \tag{12.123}$$

It is assumed that nucleate boiling will not occur at all, and forced convective evaporation will be responsible for heat transfer when

$$q_w'' < q_{ONB}'' = \frac{2\sigma T_{sat} H_{f0}}{\rho_g h_{fg} R_C}$$

for an assumed nucleation site radius of $R_C = 0.3 \times 10^{-6}$ m. If nucleate boiling does not occur, for the range $3.75 \le \rho_f/\rho_g \le 017$, F_{FC} should be found from

$$F_{FC} = \left\{ \left[(1-x)^{1.5} + 1.9x^{0.6}(1-x)^{0.001} \left(\frac{\rho_f}{\rho_g} \right)^{0.35} \right]^{-2.2} \right.$$
$$\left. + \left[\frac{H_{g0}}{H_{f0}} x^{0.01} (1 + 8(1-x)^{0.7}) \left(\frac{\rho_f}{\rho_g} \right)^{0.67} \right]^{-2} \right\}^{-0.5}. \tag{12.124}$$

The parameter $H_{NB,0}$ in the correlation of Steiner and Taborek is a standard nucleate flow boiling heat transfer coefficient, and it should represent conditions where $P_r = 0.1$, the mean surface roughness is 1 μm, and a standard heat flux q_{NBO}'' is specified for each fluid. Table 12.2 summarizes values of q_{NBO}'' and other relevant parameters discussed in the following for various fluids. The nucleate boiling correction factor is found from

$$F_{NB} = F_{PR} \left(\frac{q_w''}{q_{NBO}''} \right)^n (D/D_0)^{-0.4} (R_P/R_{P0})^{0.133} F(M). \tag{12.125}$$

The standard diameter and roughness are $D_0 = 0.01$ m and $R_{P0} = 1$ μm, respectively. The parameters F_{PR} and n are defined similar to the correlation of Gorenflo (1993) (see Section 11.3), only with different coefficients:

$$F_{PR} = 2.816 P_r^{0.45} + \left(3.4 + \frac{1.7}{1 - P_r^7} \right) P_r^{3.7}. \tag{12.125}$$

For all fluids other than cryogens,

$$n = 0.8 - 0.1 \exp(1.75 P_r). \tag{12.127}$$

For cryogens,

$$n = 0.7 - 0.13 \exp(1.105 P_r). \tag{12.128}$$

The correction factor $F(M)$ is a function of the coolant molecular mass. $F(M) = 0.35$ and 0.86 for H_2 and He, respectively, and for $10 < M < 187$,

$$F(M) = 0.377 + 0.199 \ln M + 2.8427 \times 10^{-5} M^2, \quad F(M) \le 2.5 \tag{12.129}$$

Other widely referenced correlations include the correlations of Bjorge *et al.* (1982) and Klimenko (1988, 1990).

Table 12.2. *Reference nucleate boiling heat fluxes and heat transfer coefficients (Steiner and Taborek, 1992; Thome, 2007).*

Fluid	Formula	P_{cr} (bar)	M (kg/kmol)	q''_{NBO} (W/m^2)	H_{NB0} (W/m^2·K)
Methane	CH_4	46.0	16.04	20 000	8060
Ethane	C_2H_6	48.8	30.07	20 000	5210
Propane	C_3H_8	42.4	44.10	20 000	4000
n-Butane	C_4H_{10}	38.0	58.12	20 000	3300
n-Pentane	C_5H_{12}	33.7	72.15	20 000	3070
Isopentane	C_5H_{12}	33.3	72.15	20 000	2940
n-Hexane	C_6H_{14}	29.7	86.18	20 000	2840
n-Heptane	C_7H_{16}	27.3	100.20	20 000	2420
Cyclohexane	C_6H_{12}	40.8	84.16	20 000	2420
Benzene	C_6H_6	48.9	78.11	20 000	2730
Toluene	C_7H_8	41.1	92.14	20 000	2910
Diphenyl	$C_{12}H_{10}$	38.5	154.21	20 000	2030
Methanol	CH_4O	81.0	32.04	20 000	2770
Ethanol	C_2H_6O	63.8	46.07	20 000	3690
n-Propanol	C_3H_8O	51.7	60.10	20 000	3170
Isopropanol	C_3H_8O	47.6	60.10	20 000	2920
n-Butanol	$C_4H_{10}O$	49.6	74.12	20 000	2750
Isobutanol	$C_4H_{10}O$	43.0	74.12	20 000	2940
Acetone	C_3H_6O	47.0	58.08	20 000	3270
R-11 (fluorotrichloromethane)	$CFCl_3$	44.0	137.37	20 000	2690
R-12 (difluorodichloromethane)	CF_2Cl_2	41.6	120.91	20 000	3290
R-13 (trifluorochloromethane)	CF_3Cl	38.6	104.47	20 000	3910
R-13B1 (trifluorobromomethane)	CF_3Br	39.8	148.93	30 000	3380
R-22 (difluorochloromethane)	CHF_2Cl	49.9	86.47	20 000	3930
R-23 (trifluoromethane)	CHF_3	48.7	70.02	20 000	4870
R-113 (trifluorotrichloroethane)	$C_2F_3Cl_3$	34.1	187.38	20 000	2180
R-114 (tetrafluorodichloroethane)	$C_2F_4Cl_2$	32.6	170.92	20 000	2460
R-115 (pentafluorochloroethane)	C_2F_5Cl	31.3	154.47	20 000	2890
R-123 (1,1-dichloro-2,2,2-trifluoroethane)	$C_2HCl_2F_3$	36.7	152.93	20 000	2600
R-134a (1,1,1,2-tetrafluoroethane)	$C_2H_2F_4$	40.6	102.03	20 000	3500
R-152a (1,1-difluoroethane)	$C_2H_4F_2$	45.2	66.05	20 000	4000
R-226 (hexafluorochloropropane)	C_3HF_6Cl	30.6	186.48	20 000	3700
R-227 (heptafluoropropane)	C_3HF_7	29.3	170.03	20 000	3800
R-C318 (cyclooctafluorobutane)	C_4F_8	28.0	200.03	20 000	2710
R-502 (R-22 and R-115 mixture)	CHF_2Cl/C_2F_5Cl	40.8	111.6	20 000	2900
Chloromethane	CH_3Cl	66.8	50.49	20 000	4790
Tetrachloromethane	CCl_4	45.6	153.82	20 000	2320
Tetrafluoromethane	CF_4	37.4	88.0	20 000	4500
Helium I	He	2.275	4.0	1000	1990
Hydrogen (para)	H_2	12.97	2.02	10 000	12 220
Neon	Ne	26.5	20.18	10 000	8920
Nitrogen	N_2	34.0	28.02	10 000	4380
Argon	Ar	49.0	39.95	10 000	3870
Oxygen	O_2	50.8	32.0	10 000	4120
Water	H_2O	220.64	18.02	150 000	25 580
Ammonia	NH_3	113.0	17.03	150 000	36 640
Carbon dioxide	CO_2	73.8	44.01	150 000	18 890
Sulfur hexafluoride	SF_6	37.6	146.05	150 000	12 230

EXAMPLE 12.3. Water at 7 bars flows into a uniformly heated vertical tube that is 1.1 cm in diameter. The velocity of water at the inlet, where the temperature is 427.1 K, is 2.5 m/s. Using the correlation of Chen (1966), calculate the wall heat flux for $x_{eq} = 0.01, 0.05$, and 0.1. Assume that the wall temperature is 446 K.

SOLUTION. At inlet conditions, $\rho_L = 913.4$ kg/m^3. The mass flow rate will then be $\dot{m}_L = \rho_L(\pi D^2/4)\overline{U}_{L,in} = 0.217$ kg/s.

The saturation properties are $\rho_f = 902.6$ kg/m^3, $\rho_g = 3.66$ kg/m^3, $k_f = 0.665$ W/m·k, $C_{Pf} = 4{,}353$ J/kg·K, $\mu_f = 1.65 \times 10^{-4}$ kg/m·s, $h_{fg} = 2.066 \times 10^6$ J/kg, $Pr_f = 1.079$, and $\sigma = 0.046$ N/m.

First consider the $x_{eq} = 0.01$ case. The calculations will then proceed as follows:

$$Re_g = GxD/\mu_g = 1.731 \times 10^5,$$

$$Re_f = G(1-x)D/\mu_f = 1.509 \times 10^5,$$

$$Re_{f0} = GD/\mu_f = 1.524 \times 10^5.$$

From Eq. (12.91), we get $X_{tt} = 5.08$. Equation (12.98) then gives $F = 1.956$. Equation (12.101) can now be applied to get $S = 0.150$.

Also, for $T_w = 446$ K, we have $P_{sat}\big|_{T_w} = 849\,784$ Pa. Therefore,

$$\Delta P_{sat} = P_{sat}\big|_{T_w} - P = 149\,784\ \text{Pa}.$$

Equation (12.99) then gives $H_{NB} = 2{,}630$ W/m^2·K. Next, we will calculate the forced convection heat transfer coefficient from Eq. (12.93), and that gives $H_{FC} = 38{,}960$ W/m^2·K. The heat transfer coefficient is thus

$$H = H_{NB} + H_{FC} = 41\,590\ \text{W/m}^2 \cdot \text{K}.$$

Similar calculations for $x_{eq} = 0.05$ and 0.1 lead to the results summarized in the following table.

	$x_{eq} = 0.05$	$x_{eq} = 0.1$
X_{tt}	1.15	0.587
Re_f	144 780	137 170
Re_g	86 570	173 150
Re_{f0}	152 400	152 400
F	3.287	4.50
S	0.1037	0.088
H_{NB} (W/m^2·K)	1820	1540
H_{FC} (W/m^2·K)	63 350	83 040
H (W/m^2·K)	65 170	84 580

EXAMPLE 12.4. Water at 70-bar pressure and with an inlet subcooling of 5 °C flows into a uniformly heated vertical tube that is 1.5 cm in diameter and receives a heat flux of 4.2×10^5 W/m^2. The average velocity of water at the inlet is 2 m/s. Assuming that the heated pipe is made from stainless steel, find the locations where $x_{eq} = 0.02$ and 0.06, and calculate the heat transfer coefficient at these points using the correlations of Kandlikar (1990, 1991).

SOLUTION. The relevant properties are $\rho_f = 739.9$ kg/m^3, $k_f = 0.56$ W/m·K, $\mu_f = 9.13 \times 10^{-5}$ kg/m·s, $\sigma = 0.0174$ N/m, $C_{Pf} = 5{,}394$ J/kg, $Pr_f = 0.88$, $T_{sat} = 559$ K, and $h_{fg} = 1.505 \times 10^6$ J/kg.

At the inlet, the density is $\rho_L = 750$ kg/m^3; therefore,

$$G = \rho_L \overline{U}_{L,in} = 1{,}500\,\text{kg/m}^2,$$

$$\dot{m} = G\pi D^2/4 = 0.265\,\text{kg/s}.$$

Let us now consider the case where $x_{eq} = 0.02$. The location where x_{eq} occurs is found from

$$\dot{m}\left[C_{Pf}(T_{sat} - T_{L,in}) + x_{eq}h_{fg}\right] \approx \pi D q_w'' Z_{0.02} \Rightarrow Z_{0.02} \approx 0.764\,\text{m}.$$

Also, from Table 12.1, $F_{fl} = 1.0$. Furthermore,

$$\text{Re}_{f0} = GD/\mu_f = 2.464 \times 10^5.$$

The Fanning friction factor, found from Eq. (12.89), is $f = 0.00375$. The correlation of Petukhov and Popov, Eq. (12.87), then gives

$$H_{f0} = 14\,785\,\text{W/m}^2\cdot\text{K}.$$

The calculations proceed as follows:

$$\text{Co} = \sqrt{\frac{\rho_g}{\rho_f}\left(\frac{(1-x)}{x}\right)^{0.8}} = 5.0,$$

$$\text{Bo} = \frac{q_w''}{Gh_{fg}} = 1.861 \times 10^{-4},$$

$$\text{Fr}_{f0} = \frac{G^2}{g\rho_f^2 D} = 27.95,$$

$$f_2 = 1.$$

Equations (12.103) and (12.104) then give

$$H_{NBD} = 44\,730\,\text{W/m}^2\cdot\text{K},$$

$$H_{CBD} = 27\,647\,\text{W/m}^2\cdot\text{K}.$$

Thus,

$$H = 44\,730\,\text{W/m}^2\cdot\text{K}.$$

Similar calculations for $x_{eq} = 0.06$ lead to

$$Z_{0.06} \approx 1.57\,\text{m},$$

$$\text{Co} = 2.01,$$

$$H_{NBD} = 44\,630\,\text{W/m}^2\cdot\text{K},$$

$$H_{CBD} = 31\,520\,\text{W/m}^2\cdot\text{K},$$

$$H = 44{,}630\,\text{W/m}^2\cdot\text{K}.$$

EXAMPLE 12.5. Repeat the solution of Example 12.4, using the correlation of Liu and Winterton (1991).

SOLUTION. The relevant properties were all calculated in the previous example.

Let us start with the case where $x_{eq} = 0.02$. Equation (12.116) gives $H_{f0} = 16\,770$ W/m²·K. Also, from Eqs. (12.114) and (12.115) we find $E = 1.108$, $S = 0.712$, and $E_2 = S_2 = 1$.

With $P = 70$ bars, $P_{cr} = 220.6$ bars, $M = 18$ kg/kmol, and $q_w'' = 4.2 \times 10^5$ W/m², Eq. (12.117) gives

$$H_{NB} = 92\,090 \text{ W/m}^2\text{·K}.$$

Equation (12.113) then gives

$$H = 67\,560 \text{ W/m}^2\text{·K}.$$

Similar calculations for $x_{eq} = 0.06$ give $E = 1.278$, $S = 0.709$, and $E_2 = S_2 = 1$. The parameters H_{f0} and H_{NB} have the same values as before, and Eq. (12.113) leads to

$$H = 67\,955 \text{ W/m}^2\text{·K}.$$

12.14 Flow-Regime-Dependent Correlations for Saturated Boiling in Horizontal Channels

As noted earlier, there is strong interplay between hydrodynamics and heat transfer in boiling in horizontal channels (see Fig. 12.8). Flow stratification, in particular, can lead to early dryout. Kattan *et al.* (1998a, b, c) have developed a flow-regime-dependent method for saturated boiling in horizontal pipes. Further improvements of the technique have been made by Zurcher *et al.* (1999) and Wojtan *et al.* (2005a, b). The methodology consists of a flow regime map and regime-specific models and correlations for heat transfer.

The flow regime map associated with the saturated boiling of R-22 in a 13.84-mm-diameter pipe, according to the version of the aforementioned technique described by Wojtan *et al.* (2005a), is displayed in Fig. 12.16. The corresponding flow regime transition models are now briefly explained. The flow regimes can be determined for a heated pipe of uniform cross section by knowing the local pressure, mass flux, equilibrium quality, and wall heat flux.

First consider the conditions that lead to dryout. Dryout in a horizontal flow passage starts at the top of the passage where the liquid film is thinnest and expands until eventually it covers the entire channel perimeter. When the entire channel perimeter is dry, mist flow is established. The initiation of dryout, namely the disruption of the liquid film at the top of the heated flow passage, occurs under high quality conditions when

$$G_{dryout} = \left\{ 4.255[\ln(0.58/x) + 0.52](\rho_g \sigma/D)^{0.17} (gD\rho_g \Delta\rho)^{0.37} \right. $$
$$\left. \times (\rho_g/\rho_f)^{-0.25} (q_w''/q_{CHF}'')^{-0.7} \right\}^{0.926}, \tag{12.130}$$

where $q_{CHF}'' = 0.131\rho_g^{0.5} h_{fg} (\sigma_g \Delta\rho)^{1/4}$ is the pool boiling critical heat flux correlation of Kutateladze (see Section 11.4). The transition line from dryout to the mist flow

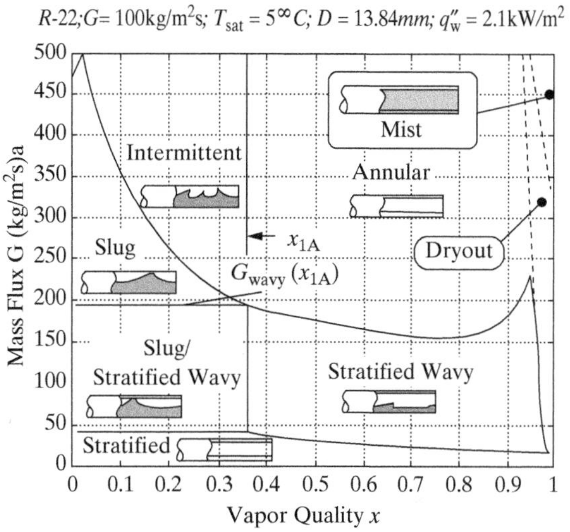

$R\text{-}22; G = 100\,\text{kg/m}^2\text{s}; T_{\text{sat}} = 5^\infty C; D = 13.84\,mm; q''_w = 2.1\,\text{kW/m}^2$

Figure 12.16. The flow regime map of Wojtan *et al.* (2005a) for saturated boiling of R-22 in a horizontal pipe.

regime can be represented by

$$G_{\text{mist}} = \left\{ \frac{1}{0.0058} \left[\ln \left(\frac{0.61}{x} \right) + 0.57 \right] (\rho_g \sigma / D)^{0.38} (gD\rho_g \Delta \rho)^{0.15} \right.$$
$$\left. \times \; (\rho_g / \rho_f)^{0.09} (q''_w / q''_{\text{CHF}})^{-0.27} \right\}^{0.943} . \qquad (12.131)$$

The extreme right portion of the regime map in Fig. 12.16 thus in fact represents partial dryout conditions. If conditions for the occurrence of dryout or mist flow are not met, the first step for the specification of the flow regime is to determine the geometric characteristics of stratified flow, should stratified flow have developed. With reference to Fig. 7.2, we need to calculate $\hat{h}_L = h_L/D$. This can be done by solving the equilibrium stratified flow momentum equations, as described in Chapter 7. The calculation would be iterative and tedious, however, and as an approximate alternative method Wojtan *et al.* used the DFM, with parameters borrowed from Rouhani and Axelsson (1970) (see Chapter 6), whereby

$$C_0 = 1 + 0.12(1 - x), \qquad (12.132)$$

$$V_{gj} = 1.18(1 - x) \left(\frac{\sigma g \Delta \rho}{\rho_f^2} \right)^{0.25} . \qquad (12.133)$$

The application of these parameters in the DFM void-quality expression leads to

$$\alpha = \frac{x}{\rho_g} \left\{ [1 + 0.12(1 - x)] \left(\frac{x}{\rho_g} + \frac{1 - x}{\rho_f} \right) + \frac{1.18(1 - x) \left[\frac{\sigma g \Delta \rho}{\rho_f^2} \right]^{0.25}}{G} \right\}^{-1} . \qquad (12.134)$$

The configuration of the stratified flow field in the pipe is shown for convenience in Fig. 12.17. Figure 12.17(a) is in fact similar to Fig. 7.2 in Chapter 7, bearing in mind that the angle γ in Fig. 7.2 is equivalent to $2\pi - \gamma_{\text{strat}}$ in Fig. 12.17(a). Knowing α, one can find $\hat{h}_L$ and other geometric parameters, including the angle γ_{dry} from Eqs. (7.26) and (7.27) in Chapter 7.

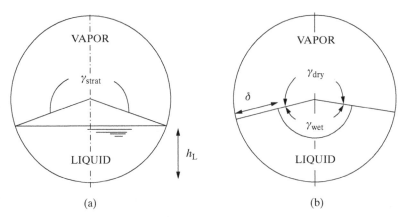

Figure 12.17. Definitions for the stratified flow regime.

The mass flux below which the stratified-wavy and slug flow regimes are possible is found from

$$G_{\text{wavy}} = \left\{ \frac{16\hat{A}_g^3 gD\rho_f\rho_g}{x^2\pi^2\sqrt{[1-(2\hat{h}_L-1)^2]}} \left[\frac{\pi^2}{25\hat{h}_L^2}\left(\frac{\text{Fr}}{\text{We}}\right)_{f0} + 1 \right] \right\}^{0.5} + 50, \quad (12.135)$$

where $\hat{A}_g = A\alpha/D^2$, $\text{Fr}_{f0} = G^2/(\rho_f^2 gD)$, and $\text{We}_{f0} = G^2 D/(\rho_f\sigma)$. Transition from stratified to stratified-wavy is represented by

$$G_{\text{strat}} = \left\{ \frac{5.12 \times 10^4 \hat{A}_f \hat{A}_g^2 \rho_g \Delta\rho\mu_f g}{x^2(1-x)\pi^3} \right\}^{1/3}, \quad (12.136)$$

where $\hat{A}_f = A(1-\alpha)/D^2$. When $G > G_{\text{wavy}}$, the transition from intermittent to annular flow occurs when $x > x_{1A}$, where

$$x_{1A} = \left\{ [0.291(\rho_g/\rho_f)^{-0.571}(\mu_g/\mu_f)^{1/7}] + 1 \right\}^{-1}. \quad (12.137)$$

The parameter $G_{\text{wavy}}(x_{1A})$, furthermore, is defined as the mass flux predicted by Eq. (12.135) when $x = x_{1A}$.

The flow regimes can now be specified as follows.

(a) Dryout is initiated when Eq. (12.130) is satisfied.
(b) Transition from dryout to mist flow occurs when Eq. (12.131) is met.

Under conditions where dryout has not occurred, the following apply.

(c) Intermittent flow occurs when $G < G_{\text{wavy}}$ and $x < x_{1A}$.
(d) Slug flow occurs when $G_{\text{strat}}(x_{1A}) < G < G_{\text{wavy}}$ and $x < x_{1A}$.
(e) The slug/stratified-wavy regime occurs when $G_{\text{strat}} < G < G_{\text{wavy}}(x_{1A})$ and $x > x_{1A}$, where $G_{\text{strat}}(x_{1A})$ is found from Eq. (12.136) with $x = x_{1A}$.
(f) Stratified-wavy flow occurs when $G < G_{\text{strat}}$ and $x > x_{1A}$.
(g) Transition from intermittent to annular flow occurs when $G > G_{\text{wavy}}$ and $x > x_{1A}$.
(h) At high qualities, the following substitutions, for an arbitrary local quality x_i, should be imposed:
  (1) if $G_{\text{stra}}(x_i) \geq G_{\text{dryout}}(x_i)$, then assume $G_{\text{dryout}}(x_i) \geq G_{\text{stra}}(x_i)$,
  (2) if $G_{\text{wavy}}(x_i) \geq G_{\text{dryout}}(x_i)$, then assume $G_{\text{dryout}}(x_i) \geq G_{\text{wavy}}(x_i)$,
  (3) if $G_{\text{dryout}}(x_i) \geq G_{\text{mist}}(x_i)$ then assume $G_{\text{dryout}}(x_i) \geq G_{\text{mist}}(x_i)$.

The saturated flow boiling heat transfer correlation of Kattan *et al.* (1998c), as modified by Wojtan *et al.* (2005b), is as follows. The circumferentially averaged heat transfer coefficient is found in general from

$$H = \frac{\gamma_{dry} H_g + (2\pi - \gamma_{dry}) H_{wet}}{2\pi}. \tag{12.138}$$

The vapor heat transfer coefficient is found from the correlation of Dittus and Boelter (1930) (see Eq. (12.117)), written based on vapor properties as

$$H_g = \frac{k_g}{D} 0.023 \mathrm{Re}_g^{0.8} \, \mathrm{Pr}_g^{0.4}, \tag{12.139}$$

where $\mathrm{Re}_g = GxD/(\mu_g \alpha)$. For all flow regimes, except for the dryout and mist regions, the heat transfer coefficient H_{wet} is found from the following asymptotic expression:

$$H_{wet} = \left[H_{FC}^3 + (0.8 H_{NB})^3 \right]^{1/3}, \tag{12.140}$$

where H_{NB} represents the nucleate boiling component and is to be calculated from the correlation of Cooper (1984) (Eq. (11.40)) in Section 11.3). The factor 0.8 is in fact a suppression factor and has been introduced by Wojtan *et al.* (2005b). The forced convection heat transfer coefficient is obtained from the following empirical correlation, which assumes a liquid film with uniform thickness δ over the wetted portion of the tube perimeter, assumed to cover an angle $\gamma_{wet} = 2\pi - \gamma_{dry}$ (see Fig. 12.17):

$$H_{FC} = 0.0133 \mathrm{Re}_\delta^{0.69} \, \mathrm{Pr}_f^{0.4} \frac{k_f}{D}, \tag{12.141}$$

$$\mathrm{Re}_\delta = \frac{4G(1-x)\delta}{(1-\alpha)\mu_f}, \tag{12.142}$$

$$\delta = \frac{D}{2} - \sqrt{\left(\frac{D}{2}\right)^2 - \frac{2A_f}{(2\pi - \gamma_{dry})}}, \tag{12.143}$$

with $\delta \leq \frac{D}{2}$ as the upper limit. The parameter γ_{dry} depends on the flow regime. For the intermittent, slug, and annular flow regimes, $\gamma_{dry} = 0$. In the stratified-wavy regime,

$$\gamma_{dry} = \left[\frac{G_{wavy} - G}{G_{wavy} - G_{strat}} \right]^{0.61} \gamma_{strat}. \tag{12.144}$$

For the slug/stratified-wavy regime,

$$\gamma_{dry} = \frac{x}{x_{1A}} \left[\frac{G_{wavy} - G}{G_{wavy} - G_{strat}} \right]^{0.61} \gamma_{strat}. \tag{12.145}$$

For the mist flow regime, Wojtan *et al.* developed the following correlation, by modifying the correlation of Groeneveld (1973), to be discussed in the next chapter:

$$H_{mist} = 0.0117 \mathrm{Re}_h^{0.79} \, \mathrm{Pr}_g^{1.06} Y^{-1.83} \frac{k_g}{D}, \tag{12.146}$$

where

$$\mathrm{Re}_h = \frac{GD}{\mu_g} \left(x + \frac{\rho_g}{\rho_f}(1-x) \right)$$

is the homogeneous Reynolds number and

$$Y = 1 - 0.1\left[\left(\frac{\rho_f}{\rho_g} - 1\right)(1 - x)\right]^{0.4}. \tag{12.147}$$

Finally, for the partial dryout regime (where part of the channel perimeter is covered with a liquid film, while another part of the perimeter is dry), Wojtan *et al.* developed the following interpolation:

$$H_{\text{dryout}} = H_{\text{TP}}(x_{\text{di}}) - \frac{x - x_{\text{di}}}{x_{\text{de}} - x_{\text{di}}}\left[H_{\text{TP}}(x_{\text{di}}) - H_{\text{mist}}(x_{\text{de}})\right], \tag{12.148}$$

where H_{TP} represents the two-phase heat transfer coefficient found from Eq. (12.138), and x_{di} and x_{de} represent equilibrium qualities at the beginning and end of the dispersed flow regimes. (Thus, x_{di} corresponds to the location where Eq. (12.131) is satisfied.) Wojtan *et al.* also suggest the following correlations:

$$x_{\text{di}} = 0.58 \exp\left[0.52 - 0.235 \text{We}_{\text{g0}}^{0.17} \text{Fr}_{\text{g0}}^{0.37} (\rho_g/\rho_f)^{0.25} (q''_w/q''_{\text{CHF}})^{0.7}\right]$$

$$x_{\text{de}} = 0.61 \exp\left[0.57 - 5.8 \times 10^{-3} \text{We}_{\text{g0}}^{0.38} \text{Fr}_{\text{g0}}^{0.15} (\rho_g/\rho_f)^{-0.09} (q''_w/q''_{\text{CHF}})^{0.27}\right],$$

where $\text{Fr}_{\text{g0}} = G^2/(\rho_g^2 gD)$, $\text{We}_{\text{g0}} = G^2 D/(\rho_g \sigma)$, and $q''_{\text{CHF}} = 0.131 \rho_g^{1/2} h_{\text{fg}} (\sigma g \Delta \rho)^{1/4}$.

The latter is in fact the correlation of Kutateladze and Borishansky for pool boiling critical heat flux (see Eq. (11.46) and its discussion).

Wojtan *et al.* (2005b) have shown that their flow-regime-dependent heat transfer models do well in comparison with data representing the flow boiling of refrigerants R-22, R-410, and R-407C.

12.15 Two-Phase Flow Instability

Distinction should be made between microscopic and macroscopic instabilities. Microscopic instabilities occur locally. The various interfacial hydrodynamic instabilities, discussed in Chapter 2, were examples of microscopic instability. Macroscopic instability deals with an entire two-phase flow system, or a portion thereof, and is the subject of discussion here.

Two-phase flow systems are susceptible to a number of instability and oscillation phenomena. For a fixed set of boundary conditions, there are often multiple solutions for the steady-state operation of a boiling/two-phase flow system, some of which are unstable. Small perturbations can cause a system that has multiple solutions for the given boundary conditions to move from one set of operating conditions to an entirely different set or to oscillate back and forth among two or more unstable operating conditions. Two-phase flow instability is of great concern for BWRs, steam generators and boilers, heat exchangers, and cryogenic equipment, among others, and has been extensively studied. A recent occasion of concern is the boiling instability in microchannel- and minichannel-based heat sinks (see Section 14.2). Only a brief review of two-phase flow instability will be provided in this section. More detailed discussions can be found in Bouré *et al.* (1973), Bergles (1978), Ishii (1976), Yadigaroglu (1981b), and Hsu and Graham (1986).

Two-phase flow instabilities can be divided into two groups: static and dynamic. Static instabilities represent discontinuities with respect to the steady-state operation

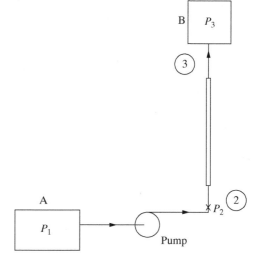

Figure 12.18. Schematic of a boiling system.

of a system, and these can be analyzed based on the system's steady-state conservation equations. Examples include flow regime transitions, flow excursions (Ledinegg instability), chugging and geysering, and burnout and quenching. Dynamic instabilities often lead to oscillations and can be analyzed by considering the transient dynamic and feedback characteristics of the system. Examples include density-wave oscillations, pressure-drop oscillations, and acoustic oscillations.

12.15.1 Static Instabilities

As mentioned, in a static instability a steady-state system becomes unstable under certain circumstances, and as a result of a perturbation it moves to an entirely different, steady-state operating condition.

Flow excursion, also referred to as *Ledinegg instability* (Ledinegg, 1938), is an important instability mode that results from the mass flux pressure-drop characteristics of boiling channels. Consider the system show in Fig. 12.18, where subcooled liquid is pumped from reservoir A and flows through the heated channel before entering reservoir B. First consider the heated channel alone and assume that the temperature and pressure at its inlet (point 2) and the total thermal load for the heated channel are all constant. The total pressure drop for the heated channel, $P_2 - P_3$, can be calculated by integrating the steady-state, one-dimensional mixture mass, momentum, and energy conservation equations (Eqs. (5.63), (5.71), and (5.86), for example), by using an appropriate correlation for the slip ratio and assuming thermodynamic equilibrium between vapor and liquid. When the total pressure drop for such a channel is plotted as a function of mass flux, often an S-shaped curve, similar to Fig. 12.19, is obtained. The curve is sometimes referred to as the demand curve, because the pressure difference $P_2 - P_3$ is needed for the flow to be established. Since the thermal load is constant, by reducing the mass flow rate the equilibrium quality at the exit, $x_{eq,exit}$, increases. With very high mass flow rates, the fluid throughout the channel remains in a subcooled liquid state, and $P_2 - P_3$ decreases with decreasing $\dot{m}$. Deviation from the single-phase liquid ΔP curve starts at the ONB point. With

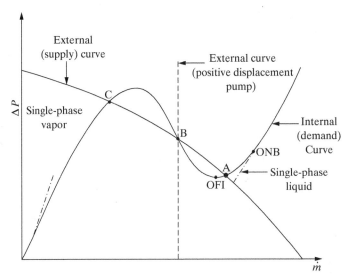

Figure 12.19. Internal (demand) and external (supply) pressure difference–flow rate characteristics.

the initiation of boiling, further reduction in $\dot{m}$ leads to an increase in flow quality at the exit and growth in the length of the channel where boiling is under way. The local minimum on the demand curve is referred to as the *onset of flow instability* (OFI) point. Beyond the OFI point, further reduction in $\dot{m}$ can lead to an increase in $P_2 - P_3$. The trend of the demand curve is changed for very low $\dot{m}$ values where the flow quality is large everywhere in the heated channel and $P_2 - P_3$ monotonically decreases as $\dot{m}$ is reduced.

The S-shaped channel characteristic curve indicates that for a range of $\dot{m}$ multiple solutions are possible. The portions of the channel's characteristic ΔP–$\dot{m}$ curve that have negative gradients can be unstable. This can be seen by plotting the typical characteristic ΔP–$\dot{m}$ curve of a pump, as shown in Fig. 12.19. Steady operation of course requires that the supply and demand ΔP values be the same, implying that only points A, B, and C are solutions. Points A and C are stable because a perturbation in $\dot{m}$ at these points causes an imbalance between the supply and demand values of ΔP that tends to bring the system back to the original steady state. Point B, however, is unstable. When the system operates at B, with a small positive perturbation in $\dot{m}$, the system moves all the way to the stable and steady condition A, whereas a small negative perturbation in $\dot{m}$ leads all the way to the steady and stable condition C.

By a simple analysis, it can be shown that the system is stable when

$$\frac{\partial \Delta P_{\rm P}}{\partial \dot{m}} < \frac{\partial \Delta P_{\rm C}}{\partial \dot{m}}, \tag{12.149}$$

where $\Delta P_{\rm P}$ and $\Delta P_{\rm C}$ represent the pump (supply) and channel (demand) pressure difference values. Evidently, any modification in the system that makes the slope of the demand curve more negative will be destabilizing, and a modification that leads to an opposite result is stabilizing. It can also be shown that increasing the channel exit pressure drop is destabilizing, whereas increasing the channel inlet pressure drop is stabilizing. The flow excursion instability can also be avoided if the pump characteristic curve is nearly a vertical line.

There are several other static instability modes. *Flow maldistribution instabilities* can occur in systems in which multiple parallel heated channels are connected at both ends to common inlet and outlet plenums, when the ΔP–$\dot{m}$ characteristic curve of the channels includes a negative-sloped portion.

Geysering and *chugging* are relaxation instabilities. Geysering takes place mostly in vertical heated channels of natural circulation loops. At the beginning of a cycle, liquid penetrates the channel. Evaporation follows and leads to the reduction of the hydrostatic pressure near the bottom of the channel. An explosive expulsion of vapor and liquid takes place when the pressure is reduced sufficiently to cause extensive evaporation. *Chugging* can occur in vertical heated flow channels when the reentry of liquid into the bottom of a heated channel leads to evaporation that causes a rapid increase in pressure. The high pressure pushes the liquid back, causing a reduction in vapor generation, reduction of pressure, and reentry of liquid into the channel's bottom. Chugging also takes place during venting of gas through vertical, submerged channels.

Relaxation instability can also be caused by thermodynamic nonequilibrium. An example is the flow boiling of a low-pressure liquid in a smooth heated tube, where because of poor nucleation the liquid may become considerably superheated before bubble nucleation takes place. Rapid evaporation happens once nucleation starts, and this may lead to the ejection of liquid from the heated channel.

Pressure drop–flow rate oscillation can take place as a result of delayed feedback between compressibility and inertia. This can happen, for example, when a compressible volume, such as a surge tank, is situated upstream of a heated channel.

12.15.2 Dynamic Instabilities

These instabilities, as mentioned, can be analyzed with consideration of the transient dynamic and feedback characteristics of the system, and they often lead to oscillations.

Density-wave oscillations are among the most common instabilities in boiling channels. These take place as a result of phase lag and feedback among flow rate, pressure drop, and phase-change processes. They mainly originate because waves resulting from perturbations in enthalpy or two-phase mixture density travel at speeds that are much lower than the speed of propagation of the pressure disturbances. Consider, for example, the heated channel shown in Fig. 12.20, where subcooled liquid enters the channel. The boiling boundary represents the OSV point, discussed in Sections 12.5 and 12.6. A periodic disturbance of the inlet mass flow rate will lead to the oscillation of the boiling boundary. The pressure drops in the liquid single-phase and two-phase regions will then oscillate. These, along with mass flux oscillations, will lead to perturbations in quality and void fraction, which travel downstream at velocities that are approximately equal to the two-phase mixture velocity. The pressure perturbation, however, travels much faster, at the velocity of sound. Phase lag will thus occur among the oscillating parameters, and these can lead to the enforcement of oscillations at the inlet flow rate.

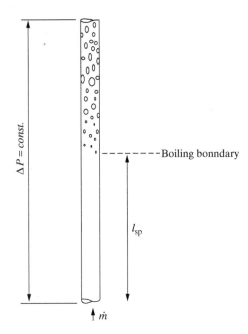

Figure 12.20. Schematic of a boiling channel.

The propagation velocity of density waves can be estimated based on the drift flux model (Zuber *et al.*, 1967). Density or concentration waves are kinematic waves that occur in systems where a functional relationship exists between the concentration and flux. Kinematic waves can be analyzed by considering only mass and energy conservation (whereas, for dynamic waves, momentum conservation is also needed). By assuming that the two phases are incompressible, and assuming that the vapor drift velocity is only a function of the total volumetric flux and void fraction, one can derive the following expression for a uniform-cross-section channel (see Problem 12.12):

$$\frac{\partial \alpha}{\partial t} + C_{\mathrm{dw}} \frac{\partial \alpha}{\partial z} = \frac{\Gamma}{\rho_{\mathrm{g}}} \left[1 - \alpha \left(1 + \frac{\partial V_{\mathrm{g}j}}{\partial j} \right) \frac{\Delta \rho}{\rho_{\mathrm{f}}} \right], \qquad (12.150)$$

where Γ is the vapor generation rate per unit mixture volume, and the propagation velocity of the density waves is

$$C_{\mathrm{dw}} = j + V_{\mathrm{g}j} + \alpha \frac{\partial V_{\mathrm{g}j}}{\partial \alpha}. \qquad (12.151)$$

Evidently, C_{dw} is of the same order of magnitude as j, as mentioned earlier. Density-wave oscillations, as a result, have low frequencies. Furthermore, unlike pressure disturbances, which travel in all directions, density waves only travel downstream.

The density-wave instability, like all macroscopic instabilities, can be modeled by solving the multiphase conservation equations with appropriate boundary conditions and perturbations. A simple and conservative criterion for stability, derived by Ishii (1971, 1976), is

$$x_{\mathrm{eq,exit}} \leq \frac{K_{\mathrm{in}} + f' \frac{L_{\mathrm{heat}}}{D} + K_{\mathrm{exit}}}{1 + \frac{1}{2} \left(f' \frac{L_{\mathrm{heat}}}{2D} + K_{\mathrm{exit}} \right)} \frac{\rho_{\mathrm{g}}}{\Delta \rho}, \qquad (12.152)$$

where f' is the Darcy two-phase friction factor and K_{in} and K_{exit} are the pressure loss coefficients for the heated channel inlet and exit, respectively. The applicability range of this equation is

$$\frac{(h_f - h_L)_{in}\Delta\rho}{h_{fg}\rho_g} \leq \pi. \tag{12.153}$$

Increasing the system pressure, and increasing K_{in} in particular, are stabilizing; increasing K_{exit} and increasing the pressure loss in the two-phase flow region are destabilizing.

12.16 Flow Boiling in Binary Liquid Mixtures

Flow boiling of multicomponent liquids is encountered in the chemical and petro-chemical, food, and cryogenics industries, to name a few. The application of environment-friendly binary refrigerant mixtures is a particularly important impe-tus for recent rather intense interest in flow boiling of binary liquid mixtures. In addition to the synthetic refrigerant mixtures, published studies in the recent past have also addressed water–ammonia, ammonia–lithium nitrate, and water–lithium bromide mixtures due to their extensive applications in absorption cycles. Useful reviews of the subject can be found in Thome and Shock (1984), Celata *et al.* (1994c), and Cheng and Mews (2006).

Much of the discussion in Section 11.8 regarding the phenomena that render the pool nucleate boiling heat transfer coefficient to be lower in non-azeotropic mixtures than the heat transfer coefficient associated with a pure liquid made of either com-ponent applies here as well. For an azeotropic liquid mixture there is little difference with pure liquids with respect to the flow boiling phenomena, and, provided that mixture properties are correctly accounted for, models and correlations for boiling in pure liquids can be applied. For non-azeotropic mixtures, however, the flow boil-ing heat transfer coefficient is always lower than the boiling heat transfer coefficient that would be obtained with either of the fluid components in pure form. The mixture boiling heat transfer coefficient is also generally lower than the heat transfer coef-ficient if mixture properties are used in models and correlations that apply to pure fluids. Figure 12.21 (Jung *et al.*, 1989a) shows the typical trends in an experiment with R-22–R-114 mixture in a 9-mm-inner-diameter horizontal tube.

The mechanisms responsible for the deterioration of heat transfer in flow boil-ing of non-azeotropic mixtures are similar to those described earlier in Section 11.8 for pool boiling of such mixtures. Boiling is accompanied by fast evaporation at the liquid–vapor interphase, where the more volatile species evaporates faster. The preferential evaporation of the more volatile species causes the mole fraction of the more volatile species to be lower near the liquid–vapor interphase than the mole fraction of the more volatile species in the liquid bulk. As a result the temper-ature of the interphase increases, and a mass transfer resistance is created that slows the transfer of the more volatile species from the liquid bulk towards the interphase. Both phenomena reduce the boiling heat transfer rate. A strong effect of mass trans-fer is observed in nucleate boiling. In the forced convective evaporation regime that takes place in the annular flow regime at high flow qualities, however, the heat trans-fer process is dominated by convection and the effect of mass transfer is diminished.

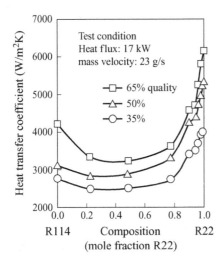

Figure 12.21. Flow boiling heat transfer coefficient for a R-22–R-114 binary mixture (Jung *et al.*, 1989a).

An observation directly related to the effect of mass transfer is the circumferential distribution of the heat transfer coefficient in horizontal channels in annular flow regime where, contrary to the situation observed in flow boiling of single-component liquids the heat transfer coefficient is actually smaller near the top than the bottom, while the liquid film near the top is thinner than the liquid layer near the bottom. The liquid film near the top, being thinner, is depleted of the more volatile species faster than the liquid near the bottom. As a result the liquid–vapor interphase near the top attains a higher temperature than the interphase near the bottom. Higher liquid–vapor interphase temperature leads to a lower heat transfer rate through the liquid layer.

A number of correlations have been proposed for flow boiling of non-azeotropic liquid mixtures. Only a few of the most widely referenced correlations will be reviewed here.

The correlation of Bennett and Chen (1980) is based on the modification of the correlation of Chen (1966) (Eqs. (12.92)–(12.101)) for fully developed flow boiling of single-component liquids. Its derivation is based on over 1000 data points for water, ethylene glycol and aqueous solutions of ethylene glycol in vertical and upward flow. Accordingly,

$$H = H_{\text{NB}} S_{\text{Binary}} + H_{\text{FC}}. \tag{12.154}$$

The forced convection component is found from (note that subscript L represents liquid mixture properties):

$$H_{\text{FC}} D_{\text{H}} / k_{\text{L}} = 0.023 \text{Re}_{\text{L}}^{0.8} \, \text{Pr}_{\text{L}}^{0.4} \, F_{\text{Binary}} \tag{12.155}$$

$$F_{\text{Binary}} = F \left(\frac{\text{Pr}_{\text{L}} + 1}{2} \right)^{0.444} \left[1 + \frac{m_{v,1} q_s''}{h_{\text{Lv}} K_{\text{FC}} \Delta T_{\text{sat}}} \left(\frac{dT_{\text{bp}}\,(P)}{dm_{\text{L},1}} \right) \right], \tag{12.156}$$

where

$$\text{Re}_{\text{L}} = G \, (1 - x) \, D_{\text{H}} / \mu_{\text{L}} \tag{12.157}$$

$$\text{Pr}_{\text{L}} = (\mu C_{\text{P}} / k)_{\text{L}}. \tag{12.158}$$

The factor F is meant to represent $(\mathrm{Re_{TP}/Re_f})^{0.8}$, and can be found from Eq. (12.98). The mass transfer coefficient K_{FC} is found from the correlation of Dittus and Boelter (1930), using heat and mass transfer analogy,

$$\frac{K_{FC}D_H}{\rho_L \boldsymbol{D}_{12,L}} = 0.023 \mathrm{Re}_L^{0.8} Sc_L^{0.4} F, \tag{12.159}$$

where $\boldsymbol{D}_{12,L}$ is the binary mass diffusivity of the two components of the liquid, and the Schmidt number is defined as

$$Sc_L \frac{\mu_L}{\rho_L \boldsymbol{D}_{12,L}}. \tag{12.160}$$

The nucleate boiling component is based on the correlation of Forster and Zuber (1955) (see Section 11.3), modified to account for the effect of reduced average superheat in the thermal boundary layer on bubble nucleation on wall cavities:

$$H_{NB} = 0.00122 \left\{ \frac{k_L^{0.79} C_{PL}^{0.45} \rho_L^{0.49} \{g_c\}^{0.43}}{\sigma^{0.5} \mu_L^{0.29} h_{Lv}^{0.24} \rho_v^{0.24}} \right\} \Delta T_{sat}^{0.24} \Delta P_{sat}^{0.75}, \tag{12.161}$$

where h_{Lv} is the mixture latent heat of vaporization, $\Delta T_{sat} = T_w - T_{sat} = T_w - T_{bp}(P, m_{L,1})$, and $\Delta P_{sat} = P_{bp}(T_w, m_{L,1}) - P$. The suppression factor S_{Binary} is meant to account for the suppression of bubble nucleation in forced convective evaporation as well as the effect of mass transfer resistance on nucleate boiling,

$$S_{Binary} = \left[1 + \frac{C_{PL}(m_{L,1} - m_{v,1})}{h_{Lv}} \left(\frac{dT_{bp}(P)}{dm_{L,1}} \right) \sqrt{\frac{\alpha_L}{\boldsymbol{D}_{12,L}}} \right]^{-1} S, \tag{12.162}$$

where α_L is thermal diffusivity of the liquid mixture. The parameter S is *Chen's suppression factor*, and is meant to represent $S = (\Delta T_{eff}/\Delta T_{sat})^{0.99}$, where ΔT_{eff} is the effective liquid superheat in the thermal boundary layer, and S can be found from Eq. (12.100) or (12.101).

Based on more than 3000 data points for R-22, R-12, R-152a, and R-114 mixtures for flow boiling in horizontal flow channels, Jung et al. (1989b) developed the following correlation.

$$H = \frac{S}{C_{UN}} H_{ideal} + C_{me} F H_L \tag{12.163}$$

$$F = 2.73 \left[0.29 + \frac{1}{X_{tt}} \right]^{0.85} \tag{12.164}$$

$$S = 4048 X_{tt}^{1.22} \mathrm{Bo}^{1.13}, \tag{12.165}$$

where H_{ideal} is defined according to Eq. (11.90), and X_{tt} is the turbulent–turbulent Martinelli parameter (see Eqs. (8.24)–(8.26)). The parameter F is identical to Chen's F parameter, and the above curve fit is meant to represent the convective evaporation heat transfer regime for pure fluids (note the similarity with Eq. (12.97)). The heat transfer coefficients associated with flow of pure components 1 and 2 (with mass flow rates equal to the total mixture mass flow rate), H_1 and H_2, are to be found from the pool boiling heat transfer correlations of Stephan and Abdelsalam (Eqs. (11.36) through (11.39)) for the relevant fluids. The correction factor C_{me} is found from:

$$C_{me} = 1 - 0.35|Y_1 - X_1|^{1.56}, \tag{12.166}$$

The correction factor C_{UN}, originally proposed by Unal (1986) in his proposed correlation for nucleate pool boiling in binary mixtures (Unal proposed $H_{NB} = H_{ideal}/C_{UN}$ as the nucleate boiling heat transfer coefficients), is found from:

$$C_{UN} = [1 + (b_2 + b_3)(1 + b_4)](1 + b_5),$$ (12.167)

where

$$b_2 = (1 - X_1)\ln\left(\frac{1.01 - X_1}{1.01 - Y_1}\right) + X_1\ln\left(\frac{X_1}{Y_1}\right) + |Y_1 - X_1|^{1.5}$$

$$b_3 = 0 \quad \text{if} \quad X_1 \geq 0.01$$

$$b_3 = \left(\frac{Y_1}{X_1}\right)^{0.1} - 1 \quad \text{if} \quad X_1 < 0.01$$

$$b_4 = 152(P/P_{1,cr})^{3.9}$$

$$b_5 = 0.92|Y_1 - X_1|^{0.001}(P/P_{1,cr})^{0.66}$$

$$X_1/Y_1 = 1 \quad \text{if} \quad X_1 = Y_1 = 0,$$

where $P_{1,cr}$ is the critical pressure of the more volatile component.

Using his pool boiling correlation as the starting point (Eqs. (12.102) to (12.109)), and including the effect of mass transfer on the liquid–vapor interphase temperature in nucleate boiling (where such effects are significant), Kandlikar (1998c) developed a correlation method for flow boiling of binary mixtures. First, define the *volatility parameter*:

$$V_1 = -\frac{C_{PL}(m_{G,1} - m_{L,1})}{h_{Lv}}\left(\frac{\alpha_L}{D_{12,L}}\right)^{0.5}\left(\frac{dT_{bp}(P)}{dm_{L,1}}\right).$$ (12.168)

Kandlikar defined three different boiling regimes, depending on the magnitude of the volatility parameter.

Regime I (near-azeotropic regime): $V_1 < 0.03$. In this regime the mixture acts as a near-azeotropic and the boiling heat transfer can be found using the methodology for flow boiling of single-component liquids modified for the effect of mixture properties. Thus, Eqs. (12.102) to (12.109) apply when appropriate mixture properties are used and the fluid-surface parameter for the mixture is calculated using the weighted-average values for the two components

$$F_{fi} = m_{L,1}F_{fi,1} + (1 - m_{L,1})F_{fi,2},$$ (12.169)

where $F_{fi,1}$ and $F_{fi,2}$ are fluid-surface parameters for components 1 and 2, respectively, and are to be found using Table 12.1. Properties of each of the two components are calculated at the phase equilibrium states.

Regime II (moderate diffusion-induced suppression regime): $0.03 < V_1 < 0.2$ and $Bo > 10^{-4}$. In this regime mass transfer effects tend to suppress bubble nucleation and convective evaporation dominates the heat transfer process. The heat transfer coefficient can be found from (see Eq. (12.104))

$$H = H_{CBD} = \left\{1.136Co^{-0.9}(1 - x)^{0.8}f_2(Fr_{f0}) + 667.2Bo^{0.7}(1 - x)^{0.8}F_{fi}\right\}H_{f0}.$$ (12.170)

Regime III (severe diffusion-induced suppression regime): $0.03 < V_1 < 0.2$ and Bo $<$ 10^{-4}, or $V_1 > 0.2$. In this regime the mass-transfer induced suppression of nucleation is severe, and the heat transfer coefficient can be found from

$$H = \left\{1.136\text{Co}^{-0.9}(1-x)^{0.8}\,f_2\,(\text{Fr}_{f0}) + 667.2\text{Bo}^{0.7}(1-x)^{0.8}F_{fi}F\right\}H_{f0}. \quad (12.171)$$

This is similar to Eq. (12.104), except for the correction factor F. The latter is to be obtained from

$$F = 0.678\left[1 - (m_{G,1} - m_{L,1})\left(\frac{\alpha_L}{D_{12,L}}\right)^{1/2}\frac{C_{PL}}{h_{Lv}}\frac{dT_{bp}}{dm_{L,1}}\right]^{-1}. \quad (12.172)$$

The above correlation agrees with more than 2500 data points within 8.3%–13.3% discrepancy, for R-12–R-22, R-22–R-114, R-22–R-152a, R-500, and R-132a–R-123 mixtures (Kandlikar, 1998c).

12.17 Flow Boiling in Helically Coiled Flow Passages

Helically coiled tubes are used extensively in boilers, and in the steam generators of gas cooled nuclear reactors, and liquid metal fast breeder reactors. They can provide very long flow passages in a compact geometry. They also generally support improved heat transfer in comparison with straight tube boilers because of the secondary flows that result from centrifugal forces. More recent applications deal with evaporation of refrigerants in compact tube-in-tube heat exchangers where helically coiled tubes with small diameters are used. Steam generator applications and the related investigations have been focused on helically coiled tubes with vertically oriented axes, with upward flow of steam at moderate and high pressures (Owhadi et al., 1968; Kozeki et al., 1970; Unal, 1978; Nariai et al., 1982; Zhao et al., 2003). More recent investigations with refrigerants typically utilize R-134a or some other refrigerant as the working fluid (Wongwises and Polsongkram, 2006a, b; Kim et al., 2006; Chen et al., 2011a, b; Balakrishnan et al., 2009; Aria et al., 2012; Elsayed et al., 2012).

Two-phase flow in helically coiled flow passages was discussed earlier in Section 4.5, where the important role of secondary flows that are caused by centrifugal forces was explained. These secondary flows play an important role during boiling and evaporation as well. The heat transfer coefficient is circumferentially nonuniform, and compared with flow in a vertical and straight tube the boiling heat transfer coefficient is generally higher in helically coiled tubes. The forced convective heat transfer in the annular flow regime occurs over a wider range of qualities, and the dryout CHF is postponed to a higher quality than in a straight tube because the secondary flows tend to spread the liquid film around. Dryout does not occur uniformly around the perimeter, however. It often starts at the inner surface of the tube, and is followed by a partially dried-out tube segment in which part of the perimeter is covered by a liquid film, while the remainder of the perimeter is dry. Total dryout eventually occurs. The quality at the beginning of dryout is typically 80% or higher, while total dryout typically occurs at qualities in excess of 90%. The partially dried-out segment can extend to much higher qualities, as much as 99%, however (Owhadi et al., 1968).

Single-Phase Forced Convection in Helically Coiled Flow Passages

Earlier in this chapter we noted that the single-phase flow convection heat transfer coefficient is often used in flow boiling correlations either as a scaling parameter, or as the basis for calculating the forced convection component of the total heat transfer coefficient. (See, for example, Eqs. (12.90) or (12.124) where H_{f0} represents the forced convection heat transfer coefficient when the flow is entirely liquid.) For helically coiled flow passages the single-phase forced convection heat transfer coefficient is sometimes utilized for the same purpose. Correlations for the laminar–turbulent flow regime transition, and correlations for the single-phase flow friction factor in smooth helically coiled flow passages were presented in Section 8.7.1. Some widely used single-phase forced convection heat transfer correlations are now reviewed.

The thermal entrance length is shorter for a curved duct than the thermal entrance length in a similar straight duct. As a result, for typical helicoid ducts with multiple rounds thermally developed methods are often adequate for design calculations. In what follows, some widely applied correlations for thermally developed flow in helicoid tubes will be reviewed.

For laminar, thermally developed flow in a helicoid tube with UWT, Manlapaz and Churchill (1980) derived the following correlation using regression

$$\mathrm{Nu}_{D,\mathrm{UWT}} = \left[\left(3.657 + \frac{4.343}{X_1} \right)^3 + 1.158 \left(\frac{\mathrm{Dn}}{X_2} \right)^{3/2} \right]^{1/3}, \tag{12.173}$$

where

$$X_1 = \left(1 + \frac{957}{\mathrm{Dn}^2\,\mathrm{Pr}} \right)^2 \tag{12.174}$$

$$X_2 = 1 + \frac{0.477}{\mathrm{Pr}}. \tag{12.175}$$

This correlation agrees well with experimental data and numerical simulations for $\mathrm{Pr} > 0.1$, but deviates from data for $\mathrm{Pr} < 0.1$ in the intermediate range of Dean number $50 \leq \mathrm{Dn} \leq 500$.

For thermally developed flow in a helicoid tube with $\widehat{H_1}$ boundary condition (i.e., axially constant wall heat flux and circumferentially constant wall temperature), Manlapaz and Churchill (1981) derived

$$\mathrm{Nu}_{D,\mathrm{H}_1} = \left[\left(4.364 + \frac{4.636}{X_3} \right)^3 + 1.816 \left(\frac{\mathrm{Dn}}{X_4} \right)^{3/2} \right]^{1/3} \tag{12.176}$$

$$X_3 = \left(1 + \frac{1342}{\mathrm{Dn}^2\,\mathrm{Pr}} \right)^2 \tag{12.177}$$

$$X_4 = 1 + \frac{1.15}{\mathrm{Pr}}. \tag{12.178}$$

However, many investigations have shown that the heat transfer coefficients for UWT and $\widehat{H_1}$ are approximately equal.

Manlapaz and Churchill (1981) have also computed and tabulated Nusselt numbers for thermally developed flow in a helicoid tube with UHF boundary condition.

Their tabulated results can be found in Shah and Joshi (1987) and Ebadian and Dong (1998).

The aforementioned correlations assume constant fluid thermophysical properties. They need to be corrected for the effect of temperature on thermophysical properties, should large temperature variations occur over the cross section of a tube. A correction method based on the proven correction factor in the correlation of Sieder and Tate (1936) is recommended for thermally developed flow in helicoid tubes (Shah and Joshi, 1987), whereby

$$\frac{\mathrm{Nu}_{D_\mathrm{H}}}{\mathrm{Nu}_{D_\mathrm{H},m}} = \left(\frac{\mu_\mathrm{m}}{\mu_\mathrm{s}}\right)^{0.14}, \tag{12.179}$$

where subscripts m and s represent the fluid bulk (mixed cup) and wall–fluid conditions.

For turbulent flow in a helically coiled tube the entrance length is only about one-half round. As a result, friction factor and heat transfer coefficients representing fully developed and thermally developed flow, respectively, are often quite adequate for design calculations. The majority of the past investigations have therefore been focused on hydrodynamic fully developed and thermally developed conditions.

Experiment and simulation show that for turbulent flow for $\mathrm{Pr} \gtrsim 0.7$ there is little dependence of the thermally developed Nusselt number on the type of thermal boundary conditions. This trend is similar to the trend observed in turbulent flow in straight flow passages. The correlations discussed below are thus circumferentially averaged, and they apply to UWT, UHF, and other convective heat transfer boundary conditions as long as $\mathrm{Pr} \gtrsim 0.7$.

A correlation for heat transfer in helically coiled tubes, with wide parameter range, is (Schmidt, 1967)

$$\frac{\mathrm{Nu}_{D,c}}{\mathrm{Nu}_{D,s}} = 1.0 + 3.6 \left[1 - \left(\frac{R_\mathrm{i}}{R_\mathrm{cl}}\right)\right] \left(\frac{R_\mathrm{i}}{R_\mathrm{cl}}\right)^{0.8}, \tag{12.180}$$

where $\mathrm{Nu}_{D,c}$ is the Nusselt number in the curve flow passage and $\mathrm{Nu}_{D,s}$ is the Nusselt number in a similar but straight tube subject to the same flow and boundary conditions. The range of validity of this correlation is $2 \times 10^4 < \mathrm{Re}_D < 1.5 \times 10^5$ and $5 < R_\mathrm{cl}/R_\mathrm{i} < 84$.

Based on data obtained with water and isopropyl alcohol, for $1.5 \times 10^3 < \mathrm{Re}_D < 2 \times 10^4$, Pratt (1947) has proposed for thermally developed flow in a helically coiled tube

$$\frac{\mathrm{Nu}_{D,c}}{\mathrm{Nu}_{D,s}} = 1.0 + 3.4 \left(\frac{R_\mathrm{i}}{R_\mathrm{cl}}\right). \tag{12.181}$$

To account for the effect of temperature-dependent properties, Orlov and Tselishchev (1964) proposed, for helically coiled tubes with $\frac{R_\mathrm{cl}}{R_\mathrm{i}} > 6$,

$$\frac{\mathrm{Nu}_{D,c}}{\mathrm{Nu}_{D,s}} = \left[1.0 + 3.54 \left(\frac{R_\mathrm{i}}{R_\mathrm{cl}}\right)\right] \left(\frac{\mathrm{Pr}_\mathrm{m}}{\mathrm{Pr}_\mathrm{s}}\right)^{0.25}. \tag{12.182}$$

Seban and McLaughlin (1963) derived the following correlation based on experimental data with water for $R_{cl}/R_i = 17$ to 104:

$$\text{Nu}_{D,c} = 0.023\text{Re}_D^{0.8}\text{Pr}^{0.4}\left[\text{Re}_D^{1/20}\left(\frac{R_i}{R_{cl}}\right)^{0.1}\right]. \tag{12.183}$$

The above correlations evidently do not include the effect of helical pitch, because experiment and analysis shows that the helix pitch has little effect of the heat transfer coefficient.

Flow Boiling Correlations for Helically Coiled Flow Passages

Several correlations have been proposed for boiling heat transfer coefficients in helically coiled tubes, and some of the most widely referenced correlations are reviewed below.

Recognizing the predominance of forced convective evaporation regime, some of the correlations dealing with boilers and steam generators are modifications to the correlation of Schrock and Grossman (1962). The latter correlation of Schrock and Grossman was developed based on flow boiling of water in straight and vertical tubes in the pressure range of 0.5 to 3.5 MPa, and with tube inner diameters of 2.9 mm to 11 mm

$$\text{Nu} = \frac{HD_H}{k_f} = \text{Re}_{f0}^{0.8}\,\text{Pr}_f^{1/3}\left\{170\left[\text{Bo} + 1.5\times10^{-4}X_{tt}^{-2/3}\right]\right\}, \tag{12.184}$$

where $\text{Bo} = q_w''/(Gh_{fg})$ is the boiling number, and the turbulent–turbulent Martinelli factor X_{tt} is to be found from Eq. (12.91). The correlation is applicable to saturated nucleate boiling as well as forced convective evaporation regime. This correlation predicted the data of Nariai et al. (1982) well for pressures up to 3.5 MPa, which were obtained with a coil diameter of 595 mm and tube inner diameter of 14.3 mm. The correlation under-predicted their data for 5 MPa pressure, however.

As mentioned, the correlation of Schrock and Grossman (1962) is based on boiling in straight flow channels, and does not include the effect of flow channel curvature. Owhadi et al. (1968) and Bell and Owhadi (1969) modified the correlation of Chen (1966) (see Eqs. (12.92) through (12.99)), based on atmospheric-pressure experimental data with water in helically coiled tubes with coil diameters of 25 cm and 52.2 cm, and 12.5 mm tube inside diameter. Their modification is simple. Accordingly, Eqs. (12.92) through (12.99) all apply, except that Eq. (12.93) is modified to include the effect of flow passage curvature:

$$H_{FC}D_H/k_f = 0.023\text{Re}_f^{0.8}\,\text{Pr}_f^{0.4}\left[\text{Re}_f^{1/20}\left(\frac{R_i}{R_{cl}}\right)^{0.1}\right]F. \tag{12.185}$$

Zhao et al. (2003) studied boiling heat transfer for water in a horizontally oriented heated coil, over a pressure range of 0.5–3.5 MPa, using a tube inner diameter of 9 mm and a coil diameter of 292 mm. The correlation of Schrock and Grossman (Eq. (12.184)) under-predicted their boiling heat transfer coefficient data only slightly, with a mean deviation of about 20%. Zhao et al. modified the correlation of Schrock and Grossman and derived

$$\frac{H}{H_{f0}} = 1.6X_{tt}^{-0.74} + 1.83\times10^5\text{Bo}^{1.46}, \tag{12.186}$$

where H_{f0} is found using the correlation of Seban and McLaughlin (1963), Eq. (12.183).

Recent investigations have been concerned with flow boiling of refrigerants. Wongwises and Polsongkram (2006a, b) and Aria *et al.* (2012) performed flow boiling experiments using R-134a as the working fluid in vertically oriented helical coils. Wongwises and Polsongkram used a 7.2 mm inner diameter tube and 305 mm coil diameter, and a coil pitch of 35 mm. The saturation pressure range in their experiments was 4.1–5.7 bars. The experiments of Aria *et al.* use a coil with 8.3 mm inner diameter, with a coil diameter of 35 mm and a coil pitch of 45 mm. Wongwises and Polsongkram (2006a, b) and Aria *et al.* (2012) have developed empirical correlations, which are only based on their own data.

Chen *et al.* (2011a) measured the boiling heat transfer coefficients of R-134a in a horizontally oriented helically coiled tube with 7.6 mm diameter and 300 mm coil diameter. Low pressures (0.2–0.75 MPa) and low flow rates (up to 260 kg/m²s) were used. Flow instability and intermittent flow consistent with the formation of liquid columns in near-vertical portions of the helical coil explained earlier in Section 4.5 were observed. Chen *et al.* (2011a) have also developed an empirical correlation based on their data.

EXAMPLE 12.6. Liquid nitrogen (LN$_2$) with a mass flux of 400 kg/m²·s and at 4 bars pressure flows through a copper tube that has an inner diameter of 5 mm. At inlet the temperature of LN$_2$ is 80 K. The wall heat flux (which is assumed to be uniform) is 30 kW/m². The tube is helicoidally coiled with a coil radius of 30 mm and a pitch of 1.5 cm.

(a) At what locations does the flow equilibrium quality reach 0.06?
(b) Calculate the heat transfer coefficient at the location of part (a), and compare it with the heat transfer coefficient if the tube were straight.

SOLUTION. Let us assume steady state, and neglect the contribution of potential energy to energy conversion. First, let us specify the relevant thermophysical and transport properties. We will use the property routines in Engineering Equation Solver (EES) software. We will find the saturation properties at 4 bars pressure, and assume that they remain constant (i.e., we neglect the effect of pressure drop in the tube on saturation properties). We then get

$$\rho_f = 738.6 \text{ kg/m}^3$$
$$\rho_g = 16.65 \text{ kg/m}^3$$
$$\mu_f = 0.000\,098\,87 \text{ kg/m·s}$$
$$\mu_g = 6.591 \times 10^{-6} \text{ kg/m·s}$$
$$k_f = 0.1174 \text{ W/m·K}$$
$$h_{fg} = 178\,354 \text{ J/kg}$$
$$\sigma = 0.005\,855 \text{ N/m}$$
$$\text{Pr}_f = 0.1178$$
$$T_{sat} = 91.23 \text{ K}$$

At the inlet the fluid is subcooled (compressed) liquid and its enthalpy is

$$h_{in} = -116\,402 \text{ J/kg}.$$

The mass flow rate through the tube is

$$\dot{m} = G\pi\frac{D^2}{4} = (400\,\text{kg/m}^2\text{·s})\pi\frac{(0.005\,\text{m})^2}{4} = 0.007854\,\text{kg/s}.$$

We must apply the energy conservation in order to solve part (a). Let us for simplicity assume that the two-phase mixture is homogeneous at the location where the quality of $x_2 = 0.06$ (referred to from now on as point 2) is achieved. We can then proceed as

$$\rho_{\text{in}} = 794.7\,\text{kg/m}^3$$

$$\rho_{\text{h},2} = \left[\frac{x_2}{\rho_{\text{g}}} + \frac{1-x_2}{\rho_{\text{f}}}\right]^{-1} = 205.1\,\text{kg/m}^3$$

$$h_2 = h_{\text{f}} + x_2 h_{\text{fg}} = -82\,149\,\text{J/kg}.$$

Equation (5.46), in steady-state form, should now be integrated between inlet and the point where $x = 0.06$ is achieved. Bearing in mind that the flow area of the pipe is constant, and noting that $A\rho_{\text{h}}j = \dot{m}$ and $p_{\text{heat}} = \pi D$ this equation can be cast as

$$\dot{m}\frac{d}{dz}\left(h + \frac{G^2}{2\rho_{\text{h}}}\right) = \pi D q_{\text{w}}''.$$

Integration between $z = 0$ (inlet) and $z = l$ (where $x = 0.02$), we get

$$\dot{m}\left[\left(h_2 + \frac{G^2}{2\rho_{\text{h},2}}\right) - \left(h_{\text{in}} + \frac{G^2}{2\rho_{\text{in}}}\right)\right] = \pi D q_{\text{w}}''l.$$

The only unknown in this equation is l, which gives

$$l = 0.1784\,\text{m}.$$

We now address part (b). We evidently deal with a flow boiling situation. We can use the correlation of Chen (1966), along with the modification proposed by Bell and Owhadi (1969):

$$X_{\text{tt}} = \sqrt{\rho_{\text{g}}/\rho_{\text{f}}} \cdot \left(\mu_{\text{f}}/\mu_{\text{g}}\right)^{0.1}\left(\frac{1-x_2}{x_2}\right)^{0.9} = 2.341$$

$$\text{Re}_{\text{f}} = G\,(1-x_2)\,D/\mu_{\text{f}} = \left(400\,\text{kg/m}^2\text{·s}\right)(1-0.06)\,(0.005\,\text{m})/(0.000\,098\,87\,\text{kg/m·s})$$
$$= 1.58 \times 10^4$$

$$\text{Re}_{\text{g}} = Gx_2D/\mu_{\text{g}} = (400\,\text{kg/m}^2\text{·s})\,(0.06)\,(0.005\,\text{m})/(6.591 \times 10^{-6}\,\text{kg/m·s})$$
$$= 3.03 \times 10^5$$

$$H_{\text{FC}} = \frac{k_{\text{f}}}{D}0.023\text{Re}_{\text{f}}^{0.8}\,\text{Pr}_{\text{f}}^{0.4} = \frac{0.1174\,\text{W/m·K}}{0.005\,\text{m}}\,(0.023)\left(1.58 \times 10^4\right)^{0.8}(0.1178)^{0.4}$$
$$= 1850\,\text{W/m}^2\text{·s}$$

$$F = 2.35(0.213 + 1/X_{\text{tt}})^{0.736} = 1.692.$$

For a straight tube, the following equations should now be solved iteratively to find H, H_{NB}, and T_w.

$$H_{NB} = 0.00122 k_f^{0.79} C_{Pf}^{0.45} \rho_f^{0.49} (T_w - T_{sat})^{0.24} \left[P_{sat}|_{T_w} - P \right]^{0.75} \frac{S}{\sqrt{\sigma} \mu_f^{0.29} h_{fg}^{0.24} \rho_g^{0.24}}$$

$$H = H_{NB} + F H_{FC}$$

$$q_w'' = H (T_w - T_{sat}).$$

Where S is calculated using Eq. (12.100). The result will be

$$H_{NB} = 5736 \, \text{W/m}^2 \cdot \text{K}$$
$$H = 8867 \, \text{W/m}^2 \cdot \text{K}$$
$$T_w = 94.62 \, \text{K}.$$

We now apply the method of Bell and Owhadi (1969). The equations to be iteratively solved are now,

$$H_{NB,c} = 0.00122 k_f^{0.79} C_{Pf}^{0.45} \rho_f^{0.49} (T_{w,c} - T_{sat})^{0.24} (P_{sat}|_{T_{w,c}} - P)^{.75} \frac{S}{\sqrt{\sigma} \cdot \mu_f^{0.29} \cdot h_{fg}^{0.24} \cdot \rho_g^{0.24}}$$

$$H_c = H_{NB,c} + F H_{FC} \left[Re_f^{1/20} \left(\frac{R_i}{R_{cl}} \right)^{0.1} \right]$$

$$q_w'' = H_c (T_{w,c} - T_{sat}),$$

where subscript c is meant to remind us that we deal with the curved tube. The factor S is found from Eq. (12.100). The result of the iterative solution will be

$$H_{NB,c} = 5410 \, \text{W/m}^2 \cdot \text{K}$$
$$H_c = 9369 \, \text{W/m}^2 \cdot \text{K}$$
$$T_{w,c} = 94.43 \, \text{K}.$$

It can be noted that the nucleate boiling component of the heat transfer coefficient is lower for the helicoidally coiled tube than the straight tube (i.e., $H_{NB,c} < H_{NB}$), while the convective component of the heat transfer, and the total heat transfer coefficient are both larger for the curved tube ($H_c > H$). The reduction in wall superheat for the helical tube due to its larger total heat transfer coefficient is the reason for $H_{NB,c} < H_{NB}$.

EXAMPLE 12.7. A water–ethanol liquid mixture at 3.8 bars pressure flows in a 1-cm-inner-diameter heated vertical tube with a mass flux of $G = 8800 \, \text{kg/m}^2 \cdot \text{s}$. At inlet where the temperature is 250 K the ethanol mole fraction is 45%. Calculate the heat transfer coefficient at a location where the mole fraction of ethanol in the liquid phase is 40%, assuming that the tube surface temperature is 10 K higher than the saturation (bubble point) temperature. For simplicity neglect the effect of pressure drop in the tube on properties.

SOLUTION. The binary phase diagram for the ethanol–water mixture at $P = 3.8$ bars pressure is depicted in Fig. 12.22, where point 2 represents the point-of-interest local conditions where $X_{1,in} = 0.45$ is reached. The temperature at the inlet is only $-23\,°\text{C}$,

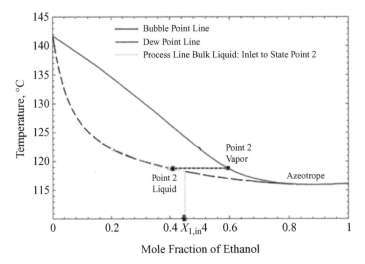

Figure 12.22. The binary phase diagram of ethanol–water at 3.8 bars pressure.

and evidently represents subcooled liquid. In the forthcoming analysis subscript 1 will represent ethanol, the more volatile component, and subscript 2 will represent water. At the location of interest where $X_1 = 0.40$ the phase equilibrium diagram shows that

$$Y_1 = 0.5953$$
$$T = T_{sat} = T_{bp}(X_1, P) = 118.8\,°C = 392\,K.$$

The mass fractions of ethanol the at inlet as well as the location of interest can be found from Eq. (1.44), whereby

$$m_{1,in} = 0.677$$
$$m_{1,L} = 0.63$$
$$m_{1,v} = 0.79.$$

The local quality x can be found by applying the principle of conservation of mass for either of the two components. When applied to ethanol, it leads to

$$Gm_{1,in} = (1-x)Gm_{1,L} + Gxm_{1,v}$$
$$\Rightarrow x = \frac{m_{1,in} - m_{1,L}}{m_{1,v} - m_{1,L}} = 0.29.$$

The properties of the ethanol–water liquid and vapor mixtures at the location of interest are found using ethanol–water liquid–vapor equilibrium (LVE):

$$\rho_L = 782.8\,kg/m^3$$
$$C_{PL} = 3993\,J/kg·K$$
$$\rho_v = 4.31\,kg/m^3$$
$$h_{LV} = 9.89 \times 10^5\,J/kg$$
$$\sigma = 0.038\,N/m$$
$$k_L = 220\,W/m·K$$
$$\mu_L = 2.5 \times 10^{-4}\,kg/m·s.$$

We need to estimate the ethanol–water binary mass diffusivity. The binary mass diffusivity of a dilute and non-dissociating species in a solute can be found from the correlation of Wilke and Chang (1954), Eq. (1.98). The correlation does not apply to high concentrations, however. Mass diffusion in dense liquid mixtures is rather complicated. A useful discussion of binary mass diffusivities in non-dilute liquid mixtures can be found in Cussler (2009). A simple and common approach for estimating the binary mass diffusivity in a dense liquid mixture is to use a geometric average defined as

$$D_{12,L} = [D_{12,L}|_{X_{1,L}\to 0}]^{(1-X_1)}[D_{21,L}|_{X_{2,L}\to 0}]^{X_1}\left[1 + \frac{\partial \ln \gamma_1}{\partial \ln X_1}\right], \quad (12.187)$$

where γ_1 is the activity coefficient of species 1 (see the discussion in Section 2.8), $D_{12,L}|_{X_{1,L}\to 0}$ is the ethanol–water binary diffusivity when species 1 (ethanol in our case) is dissolved at a trace level of concentration in species 2 (water in our case), and $D_{21,L}|_{X_{2,L}\to 0}$ is the ethanol–water binary diffusivity when species 2 (water in our case) is dissolved at a trace level of concentration in species 1 (ethanol). For an ideal liquid mixture the activity coefficient is equal to one and Eq. (12.187) becomes

$$D_{12,L} = [D_{12,L}|_{X_{1,L}\to 0}]^{(1-X_1)}[D_{21,L}|_{X_{2,L}\to 0}]^{X_1}. \quad (12.188)$$

This equation is used for the estimation of $D_{12,L}$ (see, for example, Kandlikar, 1998c), even though dense liquid solutions are often non-ideal. $D_{12,L}|_{X_{1,L}\to 0}$ and $D_{21,L}|_{X_{2,L}\to 0}$ can be found from the correlation of Wilke–Chang. (Note that $D_{12,L}|_{X_{1,L}\to 0} = D_{21,L}|_{X_{1,L}\to 0}$ and $D_{12,L}|_{X_{2,L}\to 0} = D_{21,L}|_{X_{2,L}\to 0}$.) For $D_{12,L}|_{X_{1,L}\to 0}$ we have

$$\phi_2 = 2.26$$
$$M_2 = 18\,\text{kg/kmol}$$
$$M_1 = 46\,\text{kg/kmol}$$
$$T = 392\,\text{K}$$
$$\mu \approx \mu_2 = 2.346 \times 10^{-4}\,\text{kg/m·s}$$
$$\tilde{V}_{b1} = \frac{M_1}{\rho_{f,1}\big|_{p=1.03\text{bars}}} = \frac{46\,\text{kg/kmol}}{736\,\text{kg/m}^3} = 0.0626\,\text{m}^3/\text{kmol}.$$

Equation (1.98) then gives (note that Eq. (1.98) is empirical and is not dimensionally consistent)

$$D_{12,L}|_{X_{1,L}\to 0} = 1.173 \times 10^{-16} \frac{(\Phi_2 M_2)^{1/2} T}{\mu \tilde{V}_{b1}^{0.6}}\,(\text{m}^2/\text{s}) = 6.595 \times 10^{-9}\,\text{m}^2/\text{s}.$$

For $D_{21,L}|_{X_{2,L}\to 0}$ we have

$$\phi_2 = 1.5$$
$$M_1 = 18\,\text{kg/kmol}$$
$$M_2 = 46\,\text{kg/kmol}$$
$$T = 392\,\text{K}$$
$$\mu \approx \mu_1 = 2.587 \times 10^{-4}\,\text{kg/m·s}$$
$$\tilde{V}_{b2} = \frac{M_2}{\rho_{f,2}\big|_{p=1.03\text{bars}}} = \frac{18\,\text{kg/kmol}}{958.1\,\text{kg/m}^3} = 0.0188\,\text{m}^3/\text{kmol}.$$

Equation (1.98) then gives

$$\boldsymbol{D}_{21,L}|_{X_{2,L} \to 0} = 1.173 \times 10^{-16} \frac{(\Phi_1 M_1)^{1/2} T}{\mu \tilde{V}_{b2}^{0.6}} \, (\text{m}^2/\text{s}) = 1.603 \times 10^{-8} \, \text{m}^2/\text{s}.$$

Therefore, from Eq. (12.187)

$$\boldsymbol{D}_{12,L} = [6.595 \times 10^{-9} \, \text{m}^2/\text{s}]^{0.6} \cdot [1.603 \times 10^{-8} \, \text{m}^2/\text{s}]^{0.4} = 9.41 \times 10^{-9} \, \text{m}^2/\text{s}.$$

We also need to calculate the viscosity of the vapor mixture, and for that we use Wilke's mixture rules (Eqs. (1.70) and (1.72)). The viscosities of saturated vapor of species 1 (ethanol) and 2 (water) at 392 K are, respectively,

$$\mu_{g,1} = 1.16 \times 10^{-5} \, \text{kg/m} \cdot \text{s}$$
$$\mu_{g,2} = 1.292 \times 10^{-5} \, \text{kg/m} \cdot \text{s}$$

$$\phi_{1,1} = \phi_{2,2} = 1$$
$$\phi_{1,2} = 0.574$$
$$\phi_{2,1} = 1.633.$$

Equation (1.70) then gives

$$\mu_g = \frac{Y_1 \mu_{g,1}}{Y_1 \phi_{11} + (1 - Y_1)\phi_{12}} + \frac{(1 - Y_1)\mu_{g,2}}{(1 - Y_1)\phi_{22} + Y\phi_{21}} = 1.242 \times 10^{-5} \, \text{kg/m} \cdot \text{s}.$$

We will use the correlation of Bennett and Chen (1980), Eqs. (12.154) through (12.162). First, let us calculate the additional properties that are needed:

$$\alpha_L = 7.04 \times 10^{-8} \, \text{m}^2/\text{s}$$
$$\text{Sc}_L = \frac{\mu_L}{\rho_L \boldsymbol{D}_{12,L}} = 51.74$$
$$\text{Re}_L = G(1 - x)D/\mu_L = 9.99 \times 10^{-5}$$
$$\text{Pr}_L = \frac{\mu_L}{\rho_L \alpha_L} = 4.54$$

The turbulent–turbulent Martinelli factor can be calculated using Eq. (8.26) to get

$$X_{tt} = 0.2719.$$

From the expressions given underneath Eq. (12.161) we get

$$\Delta P_{sat} = P_{bp}(T_w, X_1) - P = 5.1 \times 10^5 \, \text{Pa} - 3.8 \times 10^5 \, \text{Pa} = 1.31 \times 10^5 \, \text{Pa}$$
$$\Delta T_{sat} = T_w - T_{bp}(X_1, P) = 10 \, \text{K}.$$

We need to calculate $\frac{dT_{bl}}{dm_{1,L}}$. We can estimate this parameter numerically by writing

$$\frac{dT_{bl}}{dm_{1,L}} \approx \frac{\delta T}{m_{1,L}\left(P, T_{bP} + \frac{\delta T}{2}\right) - m_{1,L}\left(P, T_{bP} - \frac{\delta T}{2}\right)}$$

where $m_{1,L}(P, T)$ represents the equilibrium mass fraction of species 1 (ethanol) in the liquid phase at the specified pressure and temperature. With $T_{bP} = 284.7 \, \text{K}$ and $\delta T = 1 \, \text{K}$, we get

$$\frac{dT_{bl}}{dm_{1,L}} \approx 11.8 \, \text{K}.$$

We can now apply Eq. (12.161) to get (note that in this equation $\{g_c\} = 1$ when the SI unit system is used, and $\{g_c\} = 32.2\ \text{lb}_f/(\text{lb}_m \cdot \text{ft/s}^2))$ in Imperel writs

$$H_{NB} = 7.03 \times 10^3\ \text{W/m}^2 \cdot \text{K}.$$

From Eq. (12.97) we find

$$F = 6.387.$$

Next, we apply Eq. (12.159), which leads to

$$K_{FC} = 2.74\ \text{kg/m}^2 \cdot \text{s}.$$

Chen's suppression factor, S, can now be calculated using Eq. (12.100), and that gives

$$S = 0.0354.$$

The parameter S_{Binary} is calculated next from Eq. (12.162) whereby

$$S_{Binary} = 0.03402.$$

At this point Eqs. (12.155) and (12.156) must be solved simultaneously with the following two equations:

$$H = H_{NB}\, S_{Binary} + H_{FC}$$

(which is Eq. (12.154)) and

$$q_s'' = H[T_w - T_{bP}(X_1, P)].$$

The unknowns in these four equations are F_{Binary}, H_{FC}, H, and q_s''. This results in

$$F_{Binary} = 8.528$$
$$H_{FC} = 1.082 \times 10^5\ \text{W/m}^2 \cdot \text{K}$$
$$H = 1.084 \times 10^5\ \text{W/m}^2 \cdot \text{K}$$
$$q_s'' = 1.084 \times 10^6\ \text{W/m}^2.$$

One can observe that $H_{FC} \gg H_{NB}$, indicating that due to the very high flow quality the forced convective evaporation is the dominant heat transfer mechanism and the contribution of nucleate boiling is small.

PROBLEMS

12.1 In Problem 3.8, for $\text{Re}_F = 125$ and 1100 values, what local wall temperature would be needed to cause onset of nucleate boiling?

12.2 Water at 1 bar pressure and 95 °C temperature flows through a 5-mm-diameter heated tube. For Péclet numbers of 35 000 and 80 000, calculate the heat flux that would cause the onset of significant void. For the higher Pe case, perform the calculations using the correlation of Saha and Zuber (1974) and Unal (1975). Compare the results with $q_E'' = 1.4 q_D''$, the predictions of the correlation of Moissis and Berenson (1963), Eq. (11.26), for the onset of fully developed nucleate boiling in pool boiling.

12.3 Repeat Problem 12.2, for the Pe = 80 000 case, using the OSV model of Levy (1967). Compare the result with the prediction of the correlation of Forster and Greif (1959), according to whom, in reference to Fig. 12.12, $q_E'' = 1.4 q_D''$.

12.4 Water, initially at 70 °C, flows upward into a vertical heated tube with a mean velocity of 1.5 m/s. The tube is 4 cm in diameter and receives a uniform heat flux of 6×10^5 W/m².

(a) Find the location where OSV occurs.

(b) What are the local quality and void fraction 2 m downstream from the entrance of the tube?

(c) What would the two-phase flow regime be if the conditions calculated in part (b) represented an adiabatic flow?

12.5 The ONB model of Bergles and Rohsenow is based on the assumption of a quasi-steady superheated liquid film adjacent to the heated surface, with a linear temperature profile. How would you modify the model to use the turbulent boundary layer temperature law of the wall described in Section 1.7?

12.6 The fuel rods in a BWR are 1.14 cm in diameter and 3.66 m long. The rods are arranged in a square lattice (see Fig. P12.6), where the pitch is 1.65 cm, as shown in the figure. The core operates at 6.9 MPa, and the water temperature at the inlet is 544 K. Heat flux along one of the channels is assumed to be uniform and equal to 6.31×10^5 W/m². The flow is assumed to be one-dimensional and the equilibrium quality at the channel exit is 0.12.

(a) Calculate the coolant velocity at the inlet.

(b) Calculate the location of the OSV point.

(c) Using the homogeneous-equilibrium model, calculate the total and frictional pressure drops for the channel.

(d) Using the drift flux model, estimate the void fraction at the channel exit.

(e) Sketch and discuss the two-phase flow regimes along the channel.

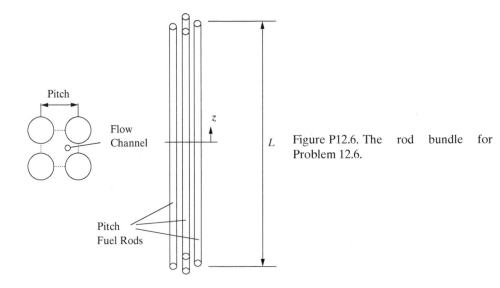

Figure P12.6. The rod bundle for Problem 12.6.

12.7 A vertical metallic tube with 5-cm inner diameter is heated at a rate of 28.9 kW/m. Saturated water at 200 °C and a mass flux of 306 kg/m²·s flows into the channel. The equilibrium quality at the exit is 5%.

(a) Calculate the channel length.

(b) Calculate the void fractions, and identify the two-phase flow regimes at locations where $z = 0.1\ L$, $0.25\ L$, $0.5\ L$, and $0.8\ L$, where z is the axial coordinate and L is the total length of the channel.

12.8 In an experiment liquid nitrogen (LN_2) with a mass flux of 500 kg/m²·s, at a pressure of 3 bars and a temperature of 75 K flows into a vertical stainless steel tube. The tube is 4.3 mm in diameter and 0.8 m long, and the wall heat flux is uniform. The tube is electrically heated at the rate of 1.6 kW/m of tube length.

(a) Find the location along the tube where ONB takes place.

(b) Find the equilibrium quality, the void fraction, and the local frictional pressure gradient at the middle and exit of the tube.

(c) Find the heat transfer coefficient at the location that is 15 cm downstream from the ONB point.

12.9 A PWR core is undergoing a slow transient, where the axial conduction in the fuel rod is negligible. The fuel rods are arranged as shown in the figure for Problem P12.6, and the flow channel hydraulic diameter is 1.5 cm. The fuel rod outer radius is 0.55 cm. The active fuel rods are $L = 3.66$ m long. The average power generation for the hottest channel in the core is 20 kW/m. The reactor is maintained at 14.6 MPa pressure, and the axial power distribution can be represented as

$$q' = q'_{max} \cos\left(\frac{\pi z}{1.2L}\right),$$

where q' is the power generation per unit length. The mass flow rate in the hottest flow channel is 0.1 kg/s, and the coolant inlet enthalpy is 1.174×10^3 kJ/kg.

(a) Calculate the locations of the ONB point and the point where fully developed boiling starts.

(b) Determine the heat transfer regime at the center of the channel.

(c) What are the most likely two-phase flow and heat transfer regimes at the exit of the channel?

12.10 In the composite correlations for flow boiling heat transfer discussed in Section 12.13, examine and assess the behavior of the convective terms for the limits $x_{eq} \to 0$ and $x_{eq} \to 1$. For each correlation, determine whether it meets the required physical conditions at the two limits.

12.11 The heated section of the apparatus shown in Fig. P12.11 is a tube that is 3.66 m long and 1.0 cm in inner diameter. It is uniformly heated at the rate of 1.4 kW/m. The downcomer is a large adiabatic vessel. The system is at 14.6 MPa, and water with 20 K subcooling is injected into the downcomer. The void fraction at the exit of the heated section is 0.25. The system operates at steady state. Assume that the two-phase flow in the heated section is in homogeneous equilibrium.

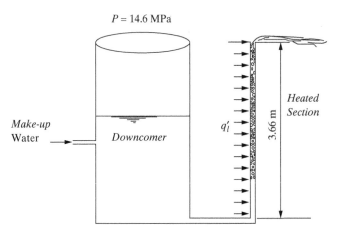

Figure P12.11. The schematic for Problem P12.11.

(a) Calculate the water mass flow rate in the system.

(b) Calculate the frictional pressure drop in the heated section.

(c) Calculate the water level height in the downcomer.

(d) Determine the heat transfer regimes at the center and exit of the heated section.

12.12 Repeat the solution of Example 12.4, this time using the correlation of Steiner and Taborek (1992).

12.13 Starting from the following one-dimensional mass conservation equations:

$$\frac{\partial(\alpha\rho_g)}{\partial t} + \frac{\partial}{\partial z}(\alpha\rho_g U_g) = \Gamma,$$

$$\frac{\partial[(1-\alpha)\rho_f]}{\partial t} + \frac{\partial}{\partial z}[(1-\alpha)\rho_f U_f] = -\Gamma,$$

and assuming that both phases are incompressible, derive Eqs. (12.150) and (12.151).

12.14 Using Eqs. (12.151) and appropriate DFM correlations for vertical boiling channels (see Chapter 6), derive expressions for the velocity of density waves in bubbly and slug flow regimes.

12.15 Two-component gas–liquid two-phase flow is common in some branches of industry. The flow of mixtures of oil and natural gas is a good example. Heat transfer in these mixtures often does not involve boiling. Based on an extensive experimental data base, Kim and Ghajar (2006) have derived the following correlation for heat transfer in horizontal tubes that carry a non-boiling gas–liquid two-phase mixture:

$$H_{TP} = F_P H_L \left\{ 1 + 0.7 \left[\left(\frac{x}{1-x}\right)^{0.08} \left(\frac{1-F_P}{F_P}\right)^{0.06} \left(\frac{Pr_G}{Pr_L}\right)^{0.03} \left(\frac{\mu_G}{\mu_L}\right)^{-0.14} \right] \right\}, \quad \text{(a)}$$

where F_P, a flow pattern factor, is to be found from

$$F_P = (1-\alpha) + \alpha F_s. \quad \text{(b)}$$

The parameter F_S is a shape factor and is defined as

$$F_s = \frac{2}{\pi}\tan^{-1}\left(\sqrt{\frac{\rho_G(U_G - U_L)^2}{gD(\rho_L - \rho_G)}}\right). \tag{c}$$

The liquid-only heat transfer coefficient H_L can be obtained from an appropriate correlation, by using the liquid phase mean velocity (rather than superficial velocity) to calculate the Reynolds number. Using the correlation of Sieder and Tate (1936), for example, one has

$$H_L = 0.027\mathrm{Re}_L^{0.8}\mathrm{Pr}_L^{0.33}\left(\frac{k_L}{D}\right)\left(\frac{\mu_L}{\mu_{L,w}}\right)^{0.14}, \tag{d}$$

where μ_L and $\mu_{L,w}$ represent the liquid viscosities at bulk and wall temperatures, respectively, and $\mathrm{Re}_L = \rho_L U_L D/\mu_L$. The parameter ranges of the data base used by Kim and Ghajar were $835 < j_L D/\nu_L < 25,900$, $1.16 \times 10^{-3} < x < 0.487$, $0.092 < \mathrm{Pr}_G/\mathrm{Pr}_L < 0.11$, and $0.016 < \mu_G/\mu_L < 0.02$. To apply the correlation, since phase velocities are involved, the void fraction is needed. Kim and Ghajar used the following correlation for calculation of the void fraction:

$$\alpha = \frac{1}{1 + \left(\frac{1-x}{x}\right)\left(\frac{\rho_G}{\rho_L}\right)\left(\frac{\rho_L}{\rho_h}\right)^{0.5}}, \tag{e}$$

where ρ_h is the homogeneous flow mixture density.

Calculate the heat transfer coefficient for an air–water mixture, at 2 bar pressure and room temperature, flowing in a horizontal pipe that has a diameter of 5 cm. The gas and liquid superficial velocities are 0.5 m/s and 3 m/s, respectively.

12.16 A horizontal pipeline with an inner diameter of 10.23 cm carries a mixture of petroleum and natural gas. The volumetric flow rate of the gas is one-third that of the liquid.

For petroleum at 5 bar pressure and 45 °C, assume the following properties:

$$\rho_L = 850\,\mathrm{kg/m^3}, \; \nu_L = 35 \times 10^{-6}\,\mathrm{m^2/s},$$

$$C_{PL} = 2.19\,\mathrm{kJ/kg \cdot K}, \quad \text{and} \quad k_L = 0.16\,\mathrm{W/m \cdot K}.$$

For natural gas at 5 bar pressure and 45 °C, assume

$$\rho_G = 8.9\,\mathrm{kg/m^3}, \; \mu_G = 8.97 \times 10^{-6}\,\mathrm{kg/m \cdot s}$$

and

$$C_{PG} = 1.86\,\mathrm{kJ/kg \cdot K}, \quad \text{and} \quad k_G = 0.0208\,\mathrm{W/m \cdot K}.$$

(a) Determine the total mass flow rates of petroleum and natural gas, when the gas superficial velocity is 0.75 m/s.

(b) Using the flow regime map of Mandhane *et al.* (1974), determine the two-phase flow regime.

(c) Examine the applicability of the correlation of Kim *et al.* (2000), described in the next problem, for the problem. Using the correlation, estimate the heat transfer coefficient between the mixture and the inner surface of the pipe, assuming that the inner surface is at a temperature of 65 °C.

12.17 Kim *et al.* (2000) have proposed the following correlation for heat transfer to a non-boiling gas–liquid mixture flowing in a vertical pipe:

$$H_{TP} = (1 - \alpha)H_L \left\{ 1 + 0.27 \left(\frac{x}{1-x}\right)^{-0.04} \left(\frac{\alpha}{1-\alpha}\right)^{1.21} \left(\frac{Pr_G}{Pr_L}\right)^{0.66} \left(\frac{\mu_G}{\mu_L}\right)^{-0.72} \right\}. \quad (f)$$

The parameter ranges of this correlation are

$$4000 < Re_L < 1.26 \times 10^5, 8.4 \times 10^{-6} < \frac{x}{1-x} < 0.77,$$

$$0.01 < \frac{\alpha}{1-\alpha} < 18.61, 1.18 \times 10^{-3} < \frac{Pr_G}{Pr_L} < 0.14.$$

Repeat Problem 12.15, this time assuming that the pipe is vertical. Use appropriate correlations of your own choice for the void fraction and the liquid single-phase heat transfer.

12.18 A vertical heated channel with UHF boundary condition is cooled by water at near-atmospheric pressure. The tube is made of copper and is 2 cm in diameter. The mass flow rate is such that $Re_{f0} = 6.3 \times 10^3$.

(a) At a location where a equilibrium quality is 2%, calculate the critical heat flux using the table look-up method.

(b) Assume that in a separate experiment the heat flux is kept uniform and constant and equal to one-half of the critical heat flux that was calculated in part (a). At a location where the equilibrium quality is 2%, calculate the heat transfer coefficient and wall temperature by using the correlation of Kandlikar (1990, 1991).

(c) Repeat part (b), this time assuming that the liquid at the inlet is a water–methanol mixture and the mole fraction of methanol is 30%.

12.19 A water–ammonia mixture with a 50% ammonia mole fraction enters a heated vertical tube that has an inner diameter of 2.5 cm. Near the inlet where the pressure is atmospheric and temperature is 250 K, the mixture flow rate gives $Re_{L0} = 9.5 \times 10^3$. The tube is heated with a uniform heat flux boundary condition. At a location where the mole fraction of ammonia in the liquid bulk is 30%, the wall temperature is 360 K.

(a) Calculate the heat transfer coefficient at the location described above, and the wall heat flux, using the correlation of Kandlikar (1998c).

(b) Repeat part (a), this time using the correlation of Jung *et al.* (1989b).

12.20 Using the correlation of Bennett and Chen (1980) calculate the heat transfer coefficient for a water–ethanol mixture flowing in a heated vertical tube that is 1 cm in diameter, for the following conditions.

(a) The mole fraction of ethanol in the liquid bulk is 40%.

(b) The mole fraction of ethanol in the liquid bulk is 20%.

Assume that the pressure is 3.8 bars, and at the inlet where the mixture temperature is 250 K the mole fraction of ethanol is 45%, we have $Re_{L0} = 11 \times 10^3$. For both parts assume that the local wall temperature is 10 K higher than the saturation (bubble point) temperature.

12.21 Liquid nitrogen at atmospheric pressure flows in a 0.6 cm-diameter copper tube, the inner surface of which is at 90 K. The mass flux is 450 kg/m²·s.

Assuming that the tube is straight and horizontal, do the following.

(a) Calculate the wall heat flux if the LN₂ bulk temperature is 74 K.

(b) Calculate the wall heat flux and local frictional pressure gradient assuming that the LN₂ is a saturated mixture with a quality of 0.05 and 0.1.

(c) Repeat part (b), this time assuming that the tube is helically coiled with vertical axis, where the coil radius is 6 cm and the coil pitch is 1.5 cm.

12.22 Liquid nitrogen (LN₂) with a mass flux of 350 kg/m²·s and at 3 bars pressure flows through a copper tube that has an inner diameter of 5 mm. At the inlet the temperature of LN₂ is 77 K. The wall heat flux (which is assumed to be uniform) is 15 kW/m².

(a) Calculate the distance from the inlet where the onset of nucleate boiling (ONB) occurs?

(b) At what locations does the flow equilibrium quality reach 0.08 and 0.15?

(c) Calculate the wall surface temperatures for part (b).

(d) For the local conditions of part (b), what local heat flux would cause CHF to occur?

(e) Repeat parts (b) and (c) this time assuming that the tubing is helically coiled with a coil radius of 30 mm and a pitch of 1.5 cm.

12.23 Wongwises and Polsongkram (2006a, b) and Aria *et al.* (2012) performed flow boiling experiments using R-134a as the working fluid in vertically oriented helical coils. Wongwises and Polsongkram used a 7.2-mm-inner-diameter tube and 305 mm coil diameter, and a coil pitch of 35 mm. The saturation pressure range in their experiments was 4.1–5.7 bars. The experiments of Aria *et al.* use a coil with 8.3 mm inner diameter, with a coil diameter of 35 mm and a coil pitch of 45 mm. The correlation of Wongwises and Polsongkram is

$$HD_{\mathrm{H}}/k_{\mathrm{f}} = 6895.98 \mathrm{Dn}_{\mathrm{Eq}}^{0.432} \, \mathrm{Pr}_{\mathrm{f}}^{-5.055} \, (\mathrm{Bo} \times 10^4)^{0.132} X_{\mathrm{tt}}^{-0.0238}.$$

The correlation of Aria *et al.* (2012) is simply an adjustment to the aforementioned correlation of Wongwises and Polsongkram:

$$HD_{\mathrm{H}}/k_{\mathrm{f}} = 7850 \mathrm{Dn}_{\mathrm{Eq}}^{0.43} \, \mathrm{Pr}_{\mathrm{f}}^{-5.055} \, (\mathrm{Bo} \times 10^4)^{0.125} X_{\mathrm{tt}}^{-0.036}.$$

In these equations the equivalent Dean number and phase Reynolds numbers are defined as

$$\mathrm{Dn}_{\mathrm{Eq}} = \left[\mathrm{Re}_{\mathrm{f}} + \mathrm{Re}_{\mathrm{g}} \left(\frac{\mu_{\mathrm{g}}}{\mu_{\mathrm{f}}} \right) \left(\frac{\rho_{\mathrm{f}}}{\rho_{\mathrm{g}}} \right)^{1/2} \right] \left(\frac{R_{\mathrm{i}}}{R_{\mathrm{cl}}} \right)^{1/2}$$

$$\mathrm{Re}_{\mathrm{f}} = \frac{G(1-x)D_{\mathrm{H}}}{\mu_{\mathrm{f}}}$$

$$\mathrm{Re}_{\mathrm{g}} = \frac{GxD_{\mathrm{H}}}{\mu_{\mathrm{g}}}.$$

We would like to examine the applicability of this correlation for liquid nitrogen (LN_2). LN_2 with a mass flux of 400 kg/m^2·s and at 3 bars pressure flows through a copper tube that has an inner diameter of 6 mm. At the inlet the temperature of LN_2 is 77 K. The wall heat flux (which is assumed to be uniform) is 180 kW/m^2. The copper tubing is helically coiled with a coil radius of 30 mm and a pitch of 1.5 cm.

(a) Calculate the distance from the inlet where saturation occurs.

(b) At what locations does the flow equilibrium quality reach 0.08 and 0.15?

(c) For the local conditions of part (b), calculate the local frictional pressure gradient and heat transfer coefficient, and compare them with the heat transfer coefficients that would be encountered if the tube was straight and horizontal.

12.24 Refrigerant R-123 at a pressure of 4 bars flows in a horizontal circular copper tube that has 10.0 mm inner diameter. The tube is helically coiled with vertical axis and $R_{cl} = 35$ cm and $l_{pch} = 3$ cm. Consider a mass flux of 400 kg/m^2·s, and qualities of 0.01, 0.25, and 0.5. Suppose the flow is upwards and the conditions represent the local conditions while boiling took place. For each flow quality calculate the boiling heat transfer coefficient. Assume a wall temperature of 82 °C.

12.25 Refrigerant R-134a at a temperature of 233 K flows in a circular copper tube that has 4.0 mm inner diameter. The tube is helically coiled with a vertical axis and upward flow with $R_{cl} = 6$ cm and $l_{pch} = 1.5$ cm. For a mass flux of 300 kg/m^2·s and for qualities of 0.01, 0.4, and 0.8 suppose the flow conditions represent the local conditions while boiling takes place. For each flow quality calculate the boiling heat transfer coefficient. Assume a wall temperature of 243 K.

12.26 Saturated liquid methane at a pressure of 4.8 bars flows into two parallel helicoidally coiled copper tubes with a vertically oriented axis, as shown in Figure P12.26. The tubes are identical and are 10.16 mm in inner diameter, the coil radius is 4 cm, and the coil pitch is 4.25 cm. The coiled tubes form ten rounds. The tubes are heated in a heat exchanger as shown in the figure, and for simplicity we assume that the heat

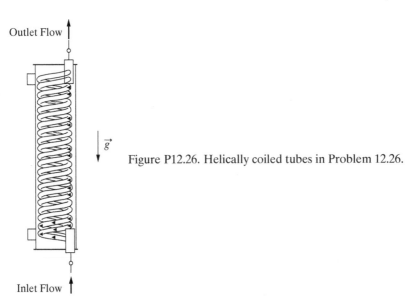

Figure P12.26. Helically coiled tubes in Problem 12.26.

flux imposed on the helicoidally coiled tubes is uniform. Experiment shows that for a total mass flow rate of 60 g/s complete evaporation of methane is achieved half way through the coil.

(a) Find the heat flux and the total and frictional pressure drops in the coil for the aforementioned 60 g/s mass flow rate

(b) Assuming that the heat flux remains constant, repeat part (a) for a mass flux of 150 g/s.

(c) Two modifications are considered for part (a): removing one of the parallel coils and keeping everything else unchanged; or reducing the total length of the coil by one half. Assuming that the heat flux remains constant, which design would be preferable from pumping power point of view?

(d) A final design option is to reduce the total length of the heat exchanger by one half, and replace the helix coil with a number of parallel straight tubes. Using the conditions of part (a) find the number of parallel tubes that are needed, assuming that the heat flux remains constant and that the total mass flow rate is divided equally among the parallel tubes, and discuss whether based on pumping power consideration this design is worth pursuing.

For simplicity you can assume homogeneous-equilibrium flow, and for pumping power only consider frictional losses inside the tubes.

13 Critical Heat Flux and Post-CHF Heat Transfer in Flow Boiling

13.1 Critical Heat Flux Mechanisms

The critical heat flux (CHF) is the most important threshold in forced-flow boiling. Forced-flow CHF is equivalent to peak heat flux in pool boiling and represents the upper limit for the safe operation of many cooling systems that rely on boiling heat transfer. The occurrence of CHF can cause a large temperature rise at the heated surface, potentially leading to its physical burnout. Moreover, the post-CHF heat transfer regimes are inefficient. Depending on circumstances, the CHF is also referred to as boiling crisis, departure from nucleate boiling, dryout heat flux, and burnout heat flux. Processes leading to forced-flow CHF are very complicated, involving the coupling of heat transfer, phase change, and two-phase flow hydrodynamics phenomena.

Consider the CHF line depicted in Fig. 13.1 which displays a portion of the boiling map previously shown in Figs. 12.4 and 12.5. Horizontal lines in this figure show qualitatively the sequence of heat transfer regimes encountered along a uniformly heated channel in steady state. Thus, moving along a horizontal line from left to right is similar to moving along a boiling channel. As seen in the figure, depending on the heat flux, CHF can occur under subcooled or saturated boiling conditions. When CHF takes place in subcooled boiling or saturated boiling at low flow qualities, the process is called departure from nucleate boiling (see Section 12.1), a title that is descriptive of the mechanism involved. The mechanism responsible for CHF at high quality boiling is the depletion of the liquid film in the annular flow regime, and the CHF is called dryout. Figure 13.1 also shows that the post-CHF heat transfer regime (i.e., the regime downstream from the CHF point) depends strongly on the type of CHF conditions.

CHF mechanisms are sensitive to orientation of the flow passage, except when the mass flux is very high. Because most boiling systems are vertical and operate under upward flow, this configuration will be emphasized in this chapter, but CHF in horizontal channels will also be discussed.

The phenomenology of CHF is strongly coupled with the two-phase flow regime. The physical processes responsible for CHF can be better understood by examining the phenomenological models for various CHF types.

DNB in subcooled and low-quality saturated flow has been studied extensively, and semi-empirical and mechanistic models with reasonable accuracy have been proposed. Basically, DNB occurs when the vapor generated on the wall is not removed from the vicinity of the wall fast enough, leading to the termination of macroscopic

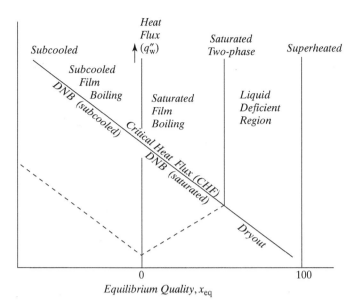

Figure 13.1. Qualitative depiction of the flow boiling map and critical heat flux.

contact between liquid and wall. The successful models are based on three different phenomenological arguments.

(1) *Critical liquid superheat.* Experiments show that in flow boiling at high flow rate and high pressure, a crowded bubble layer forms near the wall, flows parallel to the heated surface, and covers a layer of superheated liquid. In this model, the small bubbles generated at the wall are assumed to isolate the thin liquid film trapped underneath them. CHF is assumed to happen when this liquid film reaches some critical superheat (Tong *et al.*, 1965).

(2) *Coalescence of bubbles generated at the wall.* In this model, which is schematically displayed in Fig. 13.2, a thin bubbly layer forms adjacent to the wall. The void fraction in the layer is determined by the outward flow of vapor bubbles and the inward flow of liquid. The bubbly layer thus becomes more and more crowded as the near-wall turbulent eddies are unable to transport the bubbles away from the wall fast enough. CHF occurs when the void fraction in the bubbly layer exceeds a threshold above which the bubbles will be forced to coalesce (Weisman and Pei, 1983; Weisman, 1992).

(3) *Formation of a vapor blanket.* This is the best accepted model at present. In this model, vapor clots form near the heated wall as a result of the coalescence of small bubbles. Figure 13.3 shows a schematic of this mechanism. The vapor clots are separated from the wall by a thin liquid film. CHF occurs when, during the residence time of the liquid film beneath a vapor clot, the film evaporates completely (Lee and Mudawar, 1988; Galloway and Mudawar, 1993; Katto, 1992; Celata *et al.*, 1994b).

Because the accumulation of small bubbles near the heated surface is the primary cause of DNB, q''_{CHF} should be expected to depend on how the accumulation has proceeded. In other words, q''_{CHF} should depend on upstream conditions, and in particular on the axial profile of heat flux.

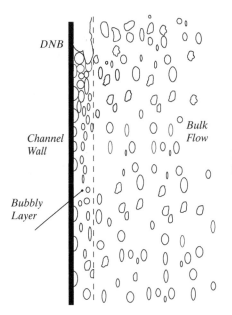

Figure 13.2. DNB caused by the coalescence of bubbles crowded near the heated surface.

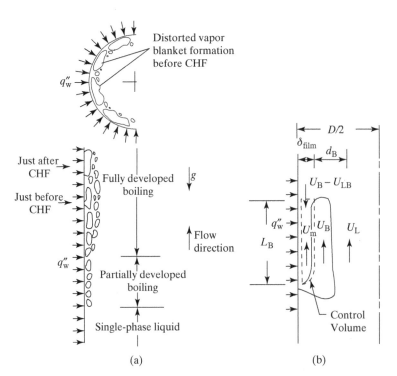

Figure 13.3. DNB caused by the formation of a vapor blanket: (a) subcooled CHF at high pressure and high mass flux; (b) onset of liquid sublayer dryout. (After Katto, 1990.)

DNB can occur in intermittent (slug or plug) flow regimes as well, as depicted in Fig. 13.4 (Weisman, 1992). In flow boiling at low flow rates and low pressure, large bubbles are generated, and an intermittent (slug or plug) flow pattern is often encountered. In this flow pattern CHF can occur when the liquid film separating a

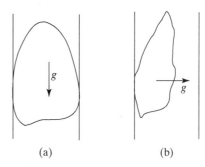

Figure 13.4. CHF in (a) slug flow and (b) plug flow in a horizontal channel.

(a)   (b)

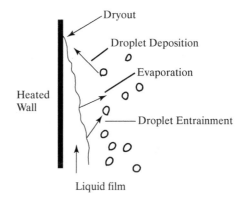

Figure 13.5. The dryout mechanism.

large vapor plug from the heated surface is sufficiently evaporated to create a dry patch before the vapor plug is passed over (Fiori and Bergles, 1970).

Dryout-type CHF is triggered by the breakdown of the contiguous liquid film in the annular-dispersed flow regime. A schematic of the dryout mechanism is shown in Fig. 13.5. Important processes affecting the liquid film include evaporation, entrainment of droplets from the film, and deposition of droplets. Droplet entrainment and evaporation tend to cause the breakdown of the liquid film, whereas droplet deposition replenishes the film and helps prevent dryout.

Dryout typically takes place a long distance from a heated channel inlet, and the film evaporation process is not strongly affected by upstream conditions. As a result, dryout CHF data typically have little dependence on inlet conditions.

13.2 Experiments and Parametric Trends

To understand the CHF data and their trends, it is important to see how CHF experiments are usually performed. A typical procedure for CHF experiments is as follows.

(a) A channel with fixed geometry and a well-defined thermal load (usually uniformly distributed heat flux) is used as the test section.

(b) Known (controlled) inlet and boundary conditions (coolant type, G, q_w'', $\Delta T_{\text{sub,in}}$, P_{exit}) are imposed.

(c) One inlet or exit parameter is changed while other controllable parameters are kept constant until a large wall temperature excursion is detected in the test section. Then, $q_w'' = q_{\text{CHF}}'$.

(d) In uniformly heated vertical channels, the temperature excursion usually occurs at the heated channel exit. The channel exit conditions at CHF for these cases thus also represent the local conditions that lead to CHF. In horizontal channels the boiling crisis is distributed over a finite length near the exit of the heated channel.

(e) In nonuniformly heated channels, CHF can occur upstream from the exit.

It should be mentioned that an upstream boiling crisis is sometimes observed in uniformly heated channels, where dry patches are generated at locations upstream from the channel exit but do not expand. Macroscopic physical contact between liquid and surface is thus reestablished downstream from the dry patches. The upstream boiling crisis has been observed in both vertical and horizontal heated channels (Becker, 1971; Merilo, 1977). In horizontal heated channels, however, the distributed boiling crisis (to be described shortly) is the most prominent observation.

Experimental data and their trends, as well as predictive correlations, can be presented in terms of inlet conditions only, local conditions (the same as exit conditions in most uniformly heated tests) only, or a combination of these conditions. In most cases, however, either inlet or local conditions are used in correlations. Although inlet condition trends and correlations are handy for the design calculations of boiling channels, local-condition predictive methods are more appropriate for use in thermal hydraulics codes.

Neglecting second-order effects, we can present the main parametric dependencies of CHF in a uniformly heated circular pipe, when inlet parameters are considered, in the following two equivalent generic forms:

$$q''_{CHF} = f[L_{heat}, D, (h_f - h)_{in}, G, P] \tag{13.1}$$

or

$$q''_{CHF} = f(L_{heat}, D, x_{eq,in}, G, P). \tag{13.2}$$

The major trends in CHF experiments can be summarized as follows (Hewitt, 1977; Yadigaroglu, 1981b).

(1) When all other parameters are kept constant, q''_{CHF} increases approximately linearly with inlet subcooling.

(2) When L_{heat}, D, and $(h_f - h)_{in}$ are maintained constant, q''_{CHF} increases monotonically with G. The effect of G is stronger in low-mass-flux conditions.

(3) When G, D, and $(h_f - h)_{in}$ are maintained constant, q''_{CHF} decreases with increasing L_{heat}; however, the total power needed to cause an actual burnout increases with increasing L_{heat}.

(4) When L_{heat}, $(h_f - h)_{in}$, and G are maintained constant, q''_{CHF} increases with increasing channel diameter D, and the effect is stronger for smaller channels.

The trend in item (2) is of particular interest, since it represents the difference between DNB-type and dryout-type CHF processes. Figure 13.6 qualitatively shows this trend.

When local conditions are considered, the generic form of the main parametric dependencies will be

$$q''_{CHF} = f(D, x_{eq}, G, P), \tag{13.3}$$

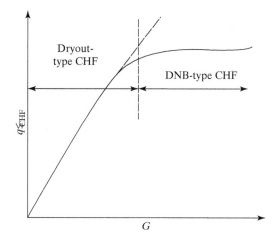

Figure 13.6. Effect of mass flux on q''_{CHF} in low- and high-flow regimes. (After El-Genk and Rao, 1991b.)

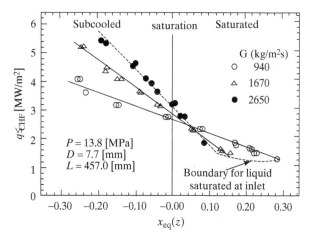

Figure 13.7. The effects of exit quality and mass flux on q''_{CHF} in a uniformly heated round tube.

where x_{eq} is the quality at the location where CHF has occurred. The parametric effects are more complicated here. For example, q''_{CHF} decreases with increasing x_{eq}, but it depends on G in a complex manner, as seen in Fig. 13.7. The figure depicts curve fits to experimental data of Weatherhead (1963), with water as the working fluid (Collier and Thome, 1994). At low qualities q''_{CHF} increases with increasing G, but at higher qualities a reverse trend is noted. Some investigators have proposed the $q''_{CHF}-x_{eq}$ dependence as shown in Fig. 13.8 (Doroschuk et al., 1975; Subbotin et al., 1982). DNB (zone III) and dryout (zone I) are the main patterns of CHF, and in between them occurs a zone II where q''_{CHF} is extremely sensitive to x_{eq}. In this zone, dryout is the CHF mechanism, and the sharp drop at q''_{CHF} occurs because the very strong evaporation from the film essentially blocks the deposition of entrained droplets onto the liquid film.

Figure 13.9 displays the effect of diameter D in subcooled CHF when x_{eq}, G, and P are maintained constant. Clearly, q''_{CHF} increases with decreasing D, and the effect is stronger for smaller channels. The contrast with the dependence of q''_{CHF} on D, when inlet and integral characteristics of the system were maintained constant, is worthy of notice. The important point to note is that, when D is changed while $L_{heat}, (h_f - h)_{in}$,

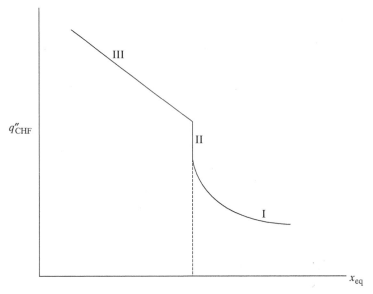

Figure 13.8. The effect of exit (local) quality on critical heat flux, when G, D, and P are maintained constant.

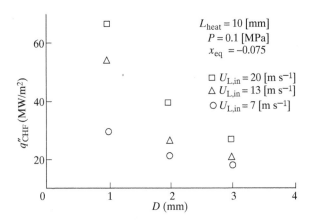

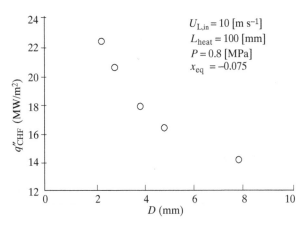

Figure 13.9. Effect of diameter on CHF in small channels. (Based on Celata et al., 1993c.)

and G are kept constant, the quality at the exit also changes. The increase in q''_{CHF} is thus a consequence of a change not only in D but also in the quality.

13.3 Correlations for Upward Flow in Vertical Channels

Empirical and semi-empirical correlations are often used for predicting CHF. A vast number of empirical correlations for CHF have been proposed in the past (Groeneveld and Snoek, 1986), but relatively few of these correlations have proven to be reasonably accurate. The empirical correlations are generally of three types.

(1) *Local-conditions correlations.* These are based on the assumption that local parameters control CHF. They are generically in the form

$$q''_{CHF} = f(G, D_H, \text{ fluid properties}, P, x_{eq}, \text{ power profile}). \qquad (13.4)$$

Although the local-conditions hypothesis has serious limitations and does not agree with all CHF data, correlations based on local conditions are widely used.

(2) *Inlet-conditions correlations.* The generic form for these correlations is

$$q''_{CHF} = f(G, D_H, \text{ fluid properties}, P_{in}, \Delta T_{sub,in}, \text{ power profile}, L_{heat}/D_H), \qquad (13.5)$$

where $\Delta T_{sub,in} = (T_{sat} - \overline{T}_L)_{in}$. The $\Delta T_{sub,in}$ term can of course be replaced with $x_{eq,in}$ the equilibrium quality at inlet, from

$$x_{eq,in} = [(h - h_f)/h_{fg}]_{in} = \{[C_{PL}(\overline{T}_L - T_{sat})]/h_{fg}\}_{in}. \qquad (13.6)$$

(3) *Global-conditions (critical quality–boiling length) correlations.* These correlations are meant to predict the occurrence of CHF based on the global characteristics of a boiling medium that can have a nonuniform power distribution and a complex geometry. The most successful among such correlations are the critical quality–boiling length correlations, which are based on the hypothesis that a unique relationship exists between the local quality at the CHF point, $x_{eq,cr}$, and the boiling length upstream from the CHF point. An example where such correlations have wide application is the rod bundles in a BWR core. These correlations are usually in the following generic form:

$$x_{eq,cr} = f(G, D_H, L_b, P, \ldots), \qquad (13.7)$$

where $x_{eq,cr}$ is the critical quality and L_b is the boiling length, that is, the distance between the the point where ONB has occurred (or, for simplicity, the location where $x_{eq} = 0.0$ has occurred) and the location where $x_{eq,cr}$ has been reached. When applied to a channel with constant wall heat flux, the left-hand side of Eq. (13.7) can be replaced with $(q''_{CHF} p_{heat} L_b)/(AGh_{fg})$, with p_{heat} representing the heated perimeter. The critical quality and critical boiling length are thus assumed to contain all the upstream effects. This hypothesis has been particularly successful for dryout-type CHF data.

Critical quality–boiling length correlations, as mentioned, are meant to predict the occurrence of dryout based on the global characteristics of a boiling medium. In fact, for a uniformly heated tube with long boiling length, simple droplet entrainment–deposition theory leads to an equation similar to Eq. (13.7) (Weisman, 1992).

We will now review some widely used methods and correlations.

Table Look-up Method. This is among the simplest and most reliable local-conditions methods for CHF in round, vertical, and uniformly heated pipes for water (Groeneveld *et al.*, 1996; Kirillov *et al.*, 1991a, b). It can be considered as a local-parameters correlation and provides the following relation in tabular form:

$$q''_{CHF} = f(P, G, x_{eq}). \tag{13.8}$$

Extensive tables are available for $D_{ref} = 8$ mm. For other diameters, the following correction must be made:

$$q''_{CHF} = q''_{CHF, D_{ref}} (D_{ref}/D)^{0.5}. \tag{13.9}$$

If the correction factor is smaller than 0.6, use 0.6. The most recent version (the 1995 look-up table) is valid for the following ranges (Groeneveld *et al.*, 1996):

$$0.1 < P < 20.0 \, \text{MPa},$$

$$0.0 \leq G \leq 8000 \, \text{kg/m}^2 \cdot \text{s},$$

$$-0.5 < x_{eq} < 1.0.$$

This look-up table, in its entirety, is available at the Internet site www.magma.ca/~thermal/.

The Correlation of Bowring (1972). This purely empirical local-conditions-type correlation for water was originally developed for the prediction of critical heat flux in rod bundles during blowdown transients. It is

$$q''_{CHF} = \frac{A_B - D_H G h_{fg} x_{eq}/4}{C_B}, \tag{13.10}$$

where q''_{CHF} is in watts per meter squared, D_H is in meters, P (which will be used shortly) is in megapascals, and G is in kilograms per meter squared per second, and where

$$A_B = \frac{2.317(h_{fg} D_H G/4) F_1}{1 + 0.0143 F_2 D_H^{1/2} G}, \tag{13.11}$$

$$C_B = \frac{0.077 F_3 D_H G}{1 + 0.347 F_4 (G/1356)^n}, \tag{13.12}$$

$$n = 2.0 - 0.5 P_R, \tag{13.13}$$

$$P_R = 0.145 P. \tag{13.14}$$

For $P_R < 1 \, \text{MPa}$

$$F_1 = \{ P_R^{18.492} \exp[20.89(1 - P_R)] + 0.917 \}/1.917, \tag{13.15a}$$

$$F_2 = 1.309 F_1 / \{ P_R^{1.316} \exp[2.444(1 - P_R)] + 0.309 \}, \tag{13.16a}$$

$$F_3 = \{ P_R^{17.023} \exp[16.658(1 - P_R)] + 0.667 \}/1.667, \tag{13.17a}$$

$$F_4 = F_3 P_R^{1.649}, \tag{13.18a}$$

For $P_R > 1\,\mathrm{MPa}$

$$F_1 = P_R^{-0.368} \exp[0.648(1 - P_R)], \tag{13.15b}$$

$$F_2 = F_1 / \{P_R^{-0.448} \exp[0.245(1 - P_R)]\}, \tag{13.16b}$$

$$F_3 = P_R^{0.219}, \tag{13.17b}$$

$$F_4 = F_3 P_R^{1.649}. \tag{13.18b}$$

The experimental data base for this correlation had the following ranges of parameters:

$$2 < P < 190\,\mathrm{bars},$$
$$136 \le G \le 18\,600\,\mathrm{kg/m^2 \cdot s},$$
$$2 < D_H < 45\,\mathrm{mm},$$
$$0.15 < L_{\mathrm{heat}} < 3.7\,\mathrm{m}.$$

The CISE-4 Correlation. This correlation is of the *critical quality–boiling length* type (Bertoletti *et al.*, 1965):

$$x_{\mathrm{eq,cr}} = \frac{p_{\mathrm{heat}}}{p_{\mathrm{tot}}} c_1 \frac{L_b}{L_b + c_2}, \tag{13.19}$$

$$c_1 = \frac{1 - (P/P_{\mathrm{cr}})}{(G/1000)^{1/3}}, \tag{13.20}$$

$$c_2 = 0.2 \left(\frac{P_{\mathrm{cr}}}{P} - 1 \right)^{0.4} D_H^{1.4} G \tag{13.21}$$

where p_{heat} and p_{tot} represent the heated and total wetted perimeters, respectively, and D_H is the hydraulic diameter. All parameters in these equations are in SI units. The correlation is applicable over the following ranges of parameters (Hsu and Graham, 1986): $P > 44\,\mathrm{bars}$, $G > 1000[(1 - P/P_{\mathrm{cr}})p_{\mathrm{heat}}/P_{\mathrm{tot}}]^3$, and $x_{\mathrm{eq,in}} < 0.5\frac{p_{\mathrm{heat}}}{p_{\mathrm{tot}}} c_1$.

The Correlation of Caira *et al.* (1995). This correlation is based on inlet conditions and is among the most accurate correlations for CHF. The correlation reads

$$q''_{\mathrm{CHF}} = \frac{c_1 + [0.25(h_f - h)_{\mathrm{in}}]^{y_3} c_2}{1 + c_3 L_{\mathrm{heat}}^{y_{10}}}, \tag{13.22}$$

$$c_1 = y_0 D^{y_1} G^{y_2}, \tag{13.23}$$

$$c_2 = y_4 D^{y_5} G^{y_6}, \tag{13.24}$$

$$c_3 = y_7 D^{y_8} G^{y_9}. \tag{13.25}$$

All parameters are in SI units, and

$$y_0 = 10\,829.55, \; y_1 = -0.0547, \; y_2 = 0.713,$$
$$y_3 = 0.978, \; y_4 = 0.188, \; y_5 = 0.486, \; y_6 = 0.462,$$
$$y_7 = 0.188, \; y_8 = 1.2, \; y_9 = 0.36, \; y_{10} = 0.911.$$

The parameter ranges of the experimental data base for the correlation are

$$0.1 < P < 8.4 \, \text{MPa},$$
$$0.3 < D < 25.4 \, \text{mm},$$
$$0.25 < L_{\text{heat}} < 61 \, \text{cm},$$
$$900 < G < 90\,000 \, \text{kg/m}^2\cdot\text{s},$$
$$0.3 < T_{\text{in}} < 242.7 \, ^\circ\text{C}.$$

The Correlation of Shah (1987). This correlation is based on a vast data pool and can be applied to various fluids. The correlation is in two different versions: the upstream-conditions correlation (UCC) (meaning the inlet conditions) and the local-conditions correlation (LCC).

For the UCC version,

$$\text{Bo} = q''_{\text{CHF}}/Gh_{\text{fg}} = 0.124(D/L_{\text{E}})^{0.89}(10^4/Y)^n(1 - x_{\text{iE}}). \tag{13.26}$$

When the inlet quality is negative ($x_{\text{eq,in}} \leq 0$), then L_{E} is the axial distance from the channel inlet, L, and $x_{\text{iE}} = x_{\text{eq,in}}$. However, when $x_{\text{eq,in}} > 0$, then L_{E} is the boiling length and $x_{\text{iE}} = 0$. The boiling length is found from

$$L_{\text{b}} = L + Dx_{\text{eq,in}}/(4\text{Bo})$$

For all fluids, $n = 0$ when $Y \leq 10^4$. For helium, when $Y > 10^4$, n should be found from

$$n = (D/L_{\text{E}})^{0.33}. \tag{13.27}$$

For other fluids, when $Y > 10^4$,

$$n = \begin{cases} (D/L_{\text{E}})^{0.54} & \text{for } Y \leq 10^6, \tag{13.28} \\ \dfrac{0.12}{(1-x_{\text{iE}})^{0.5}} & \text{for } Y > 10^6. \tag{13.29} \end{cases}$$

The parameter Y is defined as

$$Y = \frac{GDC_{\text{PL}}}{k_{\text{L}}}\left(\rho_{\text{L}}^2 gD/G^2\right)^{-0.4}(\mu_{\text{L}}/\mu_{\text{G}})^{0.6}. \tag{13.30}$$

For the LCC version,

$$\text{Bo} = q''_{\text{CHF}}/Gh_{\text{fg}} = F_{\text{E}}F_x \, \text{Bo}_0, \tag{13.31}$$

where L_C is the axial distance from the entrance and

$$F_{\text{E}} = 1.54 - 0.032(L_C/D). \tag{13.32}$$

However, it is required that $F_{\text{E}} \geq 1$; therefore if $F_{\text{E}} < 1$ use $F_{\text{E}} = 1$. Parameter Bo_0 has the highest value provided by the following three expressions:

$$\text{Bo}_0 = 15Y^{-0.612}, \tag{13.33}$$

$$\text{Bo}_0 = 0.082Y^{-0.3}\left[1 + 1.45P_{\text{r}}^{4.03}\right], \tag{13.34}$$

or

$$\text{Bo}_0 = 0.0024Y^{-0.105}\left[1 + 1.15P_{\text{r}}^{3.39}\right], \tag{13.35}$$

where $P_r = P/P_{cr}$ is the reduced pressure. If $x_{eq} \geq 0$ then

$$F_x = F_3\left[1 + \frac{(F_3^{-0.29} - 1)(P_r - 0.6)}{0.35}\right]^c \tag{13.36}$$

$$F_3 = \left(\frac{1.25 \times 10^5}{Y}\right)^{0.833x_{eq}}, \tag{13.37}$$

$$c = \begin{cases} 0 & \text{for } P_r \leq 0.6, \\ 1 & \text{for } P_r > 0.6. \end{cases} \tag{13.38} \\ \tag{13.39}$$

If $x_{eq} < 0$,

$$F_x = F_1\left[1 - \frac{(1 - F_2)(P_r - 0.6)}{0.35}\right]^b, \tag{13.40}$$

$$F_1 = 1 + 0.0052(-x_{eq})^{0.88}Y^{0.41}, \tag{13.41}$$

If $Y \geq 1.4 \times 10^7$, then $Y = 1.4 \times 10^7$ must be used in Eq. (13.41). Also

$$F_2 = \begin{cases} F_1^{-0.42} & \text{when } F_1 \leq 4, \\ 0.55 & \text{when } F_1 > 4, \end{cases} \tag{13.42} \\ \tag{13.43}$$

$$b = \begin{cases} 0 & \text{for } P_r \leq 0.6, \\ 1 & \text{for } P_r > 0.6. \end{cases} \tag{13.44} \\ \tag{13.45}$$

Shah (1987) recommends the following with regard to the choice between UCC and LCC correlations. For helium, always use UCC. For other fluids, use UCC when $Y \leq 10^6$ or $L_E > 160/P_r^{1.14}$. Otherwise, use the correlation that predicts a lower value for Bo. The ranges of data for Shah's correlation for water, R-11, R-12, R-21, R-22, R-113, R-114, ammonia, N_2O_4, helium, nitrogen, CO_2, hydrogen, acetone, benzene, diphenyl, ethanol, ethylene glycol, potassium, rubidium, and o-terphenyl are as follows:

$$0.32 < D < 37.8\,\text{mm},$$
$$0.0014 < P_r < 0.961,$$
$$4.0 < G < 2.9 \times 10^5\,\text{kg/m}^2\cdot\text{s},$$
$$0.11 < q_w'' < 4.5 \times 10^4\,\text{kW/m}^2,$$
$$1.3 < L_C/D < 940,$$
$$-4.0 < x_{eq,in} < 0.81,$$
$$-2.6 < x_{eq,CHF} < 1.0.$$

Katto (1994) has indicated that the strong dependence of the parameter Y in Shah's CHF correlation on g for high-mass-flux forced flow may be physically questionable.

EXAMPLE 13.1. In an experiment, a vertical rod bundle is used for CHF measurement. The rods are 9.5 mm in diameter and are arranged on a square lattice with a pitch of 12.6 mm. The heated length of the rods is 4.27 m. Water at 14.48 MPa pressure and

309 °C temperature, with a mass flux of $G = 3428$ kg/m²·s, flows into the rod bundle at its bottom.

(a) In one test, a heat flux of 0.85 MW/m² is to be imposed on the rod bundle. Should we expect to see CHF occurring at any location below 3.1 m from the inlet?
(b) Calculate the heat flux that would cause CHF conditions at a location that is 3.1 m above the entrance.

SOLUTION.

(a) We will use the correlation of Bowring (1972). The properties needed are as follows: $T_{sat} = 612.5$ K, $h_f = 1.589 \times 10^6$ J/kg, $h_{fg} = 1.035 \times 10^6$ J/kg, and $h_{in} = 1.388 \times 10^6$ J/kg. Also,

$$A_c = p^2 - \pi D^2/4 = 9.041 \times 10^{-5} \, m^2,$$

$$D_H = \frac{4A_c}{\pi D} = 0.0121 \, m,$$

where A_c is the subchannel flow area.

The local quality can be estimated by performing an energy balance between the inlet and the point where $z = 3.1$ m. Neglecting the kinetic and potential energy changes, and assuming constant h_{fg}, we can write

$$A_c G[(h_f + x_{eq} h_{fg}) - h_{in}] = p q''_w z. \tag{a}$$

This gives $x_{eq} = 0.0502$ for $z = 3.1$ m. The calculations proceed as follows:

$$P_R = (0.145)(14.48) = 2.1 \, MPa,$$

$$n = 2.0 - 0.5 P_R = 0.9502.$$

Equations (13.15b)–(13.18b) give

$$F_1 = 0.3733,$$
$$F_2 = 0.6813,$$
$$F_3 = 1.176,$$
$$F_4 = 3.997.$$

Equations (13.11) and (13.12) can now be solved to get $A_B = 1.988 \times 10^6$ and $C_B = 0.8653$. Equation (13.10) is used next, leading to $q''_{CHF} = 1.675$ MW/m². This is the local heat flux that would cause CHF conditions. The local heat flux is 0.85 MW/m², however. We therefore should not expect CHF to occur at or below $z = 3.1$ m.

(b) For this part, all the correlation coefficients calculated in part (a) are valid. We should, however, simultaneously solve Eq. (13.10) and Eq. (a) with $z = 3.1$m, bearing in mind that here $q''_w = q''_{CHF}$. The unknowns in the two equations are thus q''_{CHF} and x_{eq}. The iterative solution leads to

$$x_{eq} = 0.102,$$
$$q''_w = q''_{CHF} = 1.03 \, MW/m^2.$$

EXAMPLE 13.2. For the experimental rod bundle of Example 13.1, suppose in an experiment the mass flux is $G = 4500$ kg/m^2·s, and the pressure near the exit of the rod bundle is 15 MPa. Estimate the heat flux that would lead to CHF conditions at 3.22 m above the inlet. Use the information in the following table, which has been extracted from the 1995 CHF look-up table of Groeneveld et al. (1996).

x_{eq}	**−0.4**	**−0.3**	**−0.2**	**− 0.15**	**−0.1**	**−0.05**
q''_{CHF} (kW/m^2)	8512	6767	6565	6179	5561	4808
x_{eq}	**0.0**	**0.05**	**0.1**	**0.15**	**0.2**	**0.25**
q''_{CHF} (kW/m^2)	3552	3057	2953	2472	1951	1607
x_{eq}	**0.3**	**0.35**	**0.4**	**0.45**	**0.5**	**0.55**
q''_{CHF} (kW/m^2)	1355	1103	934	841	676	455
x_{eq}	**0.6**	**0.7**	**0.8**	**0.9**	**1.0**	
q''_{CHF} (kW/m^2)	291	164	88	43	0	

SOLUTION. The table represents the typical information that can be found in the CHF look-up table. Let us assume that the local pressure at $z = 3.22$ m is 15 MPa, and for simplicity use the properties at 15 MPa. The local quality and the imposed heat flux are then related according to

$$G[h_f + x_{eq}h_{fg} - h_{in}] = q''_w p_{heat} z/A,$$
$$q''_w = q''_{CHF}\sqrt{(0.008/D_H)},$$

where $z = 3.22$ m, $h_{in} = 1.387 \times 10^6$ J/kg, $h_{fg} = 1.00 \times 10^6$ J/kg, and q''_{CHF} comes from the given table. These equations, with data from the table, must be solved iteratively to specify x_{eq} and q''_w. The iterative solution gives $x_{eq} = 0.20$ and $q''_{CHF} = 1.951$ MW/m^2. The heat flux that causes CHF conditions at $z = 3.22$ m will then be $q''_w = 1.79$ MW/m^2.

Empirical CHF Correlations for Nuclear Reactor Design. These purely empirical correlations are based on steady-state, vertical, upward flow data with water and cover the parameter range of interest for the specific type of reactors they represent. The correlations are often dimensional and have little phenomenological basis; they are not recommended outside their data-base parameter range. An example is the W-3 correlations for DNB in PWRs (Tong, 1967, 1972):

$$q''_{CHF}/10^6 = \{(2.022 - 0.000\,4302P) + (0.1722 - 0.000\,0984P)$$
$$\times \exp[(18.177 - 0.004\,129P)x_{eq}]\}[(0.1484 - 1.596x_{eq} + 0.1729x_{eq}|x_{eq}|)$$
$$\times G/10^6 + 1.037](1.157 - 0.869x_{eq})[0.2664 + 0.8357\exp(-3.151D_H)]$$
$$\times [0.8258 + 0.000\,794(h_f - h_{in})].$$
(13.46)

The ranges of parameters are as follows:

$$1{,}000 < P < 2{,}300\,\text{psia},$$
$$10^6 < G < 5 \times 10^6\,\text{lb/hr·ft}^2,$$
$$0.2 < D_H < 0.7\,\text{in.}\quad\text{(equivalent heated diameter)},$$
$$-0.15 < x_{eq} < +0.15,$$
$$h_{in} \geq 400\,\text{Btu/lb},$$
$$10 < L_{heat} < 144\,\text{in.},$$

where q''_{CHF} is in British thermal units per hour per square foot. Similar correlations have been proposed by various reactor vendors (Todreas and Kazimi, 2011).

The reactor design correlations for BWRs are usually of the critical quality–boiling length type. A useful discussion can be found in Lahey and Moody (1993).

EXAMPLE 13.3. In Example 13.1, assume that the heat flux is uniform and equal to 0.7 MW/m². At the center of the rod compare the local heat flux with the heat flux that would cause CHF conditions to occur.

SOLUTION. Let us use the aforementioned W-3 correlation. The properties and parameters that are needed for the correlation are

$$G = 3{,}428 \, \text{kg/m}^2 \cdot \text{s} = 2.527 \times 10^6 \, \text{lb/hr.ft}^2,$$
$$P = 14.48 \, \text{MPa} = 2{,}101 \, \text{psia},$$
$$h_{in} = 1.388 \times 10^6 \, \text{J/kg} = 596.5 \, \text{Btu/lb},$$
$$h_f = 1.589 \times 10^6 \, \text{J/kg} = 683.3 \, \text{Btu/lb}.$$

We also need the local quality at $z = 2.135$ m, and that can be found from Eq. (a) in Example 13.1, leading to $x_{eq} = -0.056$. We can now use Eq. (13.46), to find

$$q''_{CHF} = 876\,746 \, \text{Btu/hr} \cdot \text{ft}^2 = 2.766 \, \text{MW/m}^2.$$

We can calculate the local departure from nucleate boiling ratio (DNBR) as

$$\text{DNBR} = q''_{CHF}/q''_w = 2.766/0.7 = 3.95.$$

Clearly, there is little danger of reaching CHF conditions at that particular location.

The correlation used here and many other similar reactor design correlations are based on uniformly heated flow channel data. However, the heat generation along fuel rods in nuclear reactors is axially nonuniform, and the correlations are inaccurate when applied to cases involving strongly nonuniform power distribution. A simple method proposed by Tong (1975) can account for the effect of power nonuniformity, when CHF correlations for nuclear reactor design are used. The method is consistent with the assumption that DNB is caused by the occurrence of a critical liquid superheat described earlier in Section 13.1 (Tong and Tang, 1997). Accordingly, q''_{CHF} is found from

$$q''_{CHF}(z) = q''_{CHF,u}(z)/F, \tag{13.47}$$

where

$$F = \frac{c}{q''_w(l_{DNB})[1 - \exp(-c l_{DNB})]} \int_0^{l_{DNB}} q''_w(z') \exp[-c(l_{DNB} - z')]dz', \tag{13.48}$$

$$c = 1.8 \frac{(1 - x_{eq})^{4.31}}{(G/10^6)^{0.478}} \, \text{ft}^{-1}, \tag{13.49}$$

G is the mass flux (in lb/ft²·hr),
x_{eq} is the local equilibrium quality,

$q''_{\text{CHF}}(z)$ is the local heat flux that would cause DNB,

$q''_{\text{CHF,u}}(z)$ is the local CHF, as predicted by design correlations that are based on uniform heat flux data,

l_{DNB} is the distance of the DNB point from the point where boiling starts, and z' is a dummy variable, representing distance from the point where boiling starts (chosen for simplicity to be the point where $x_{\text{eq}} = 0$).

Tight Lattice Rod Bundles

The aforementioned W-3 and other similar reactor design correlations are based on high-pressure and high-flow conditions and are meant to apply to rod bundles that are common in light water reactors. These rod bundles have a square lattice, with rods that are typically about 1 cm in diameter, and have a pitch-to-diameter ratio of approximately 1.30. Tightly latticed PWR cores operating under low-pressure and/or low-flow conditions have been proposed, however, for improved fuel utilization, or higher fuel conversion. Tight, hexagonal rod bundles in which the rods have a triangular pitch are used in these reactor designs. CHF experiments with tight, rectangular-pitched rod bundles have been performed by Zeggel *et al.* (1990), Yoshimoto *et al.* (1993), and Iwamura *et al.* (1994).

The aforementioned reactor design correlations generally do poorly when they are applied to data obtained with tight-latticed rod bundles under low-flow conditions (El-Genk and Rao, 1991b; Iwamura *et al.*, 1994). The experimental data of Iwamura *et al.* were obtained in a seven-rod rectangular-pitch bundle, at 15.8 MPa, with bundle exit qualities in the –0.01 to –0.19 range, and mass fluxes in the range 820–3100 kg/m²·s. The rods were 9.5 mm in diameter, and the spacing between the adjacent rods was 2.2 mm. Iwamura *et al.* carried out critical heat flux experiments under steady-state as well as transient conditions. They compared their data with several empirical correlations as well as with the predictions of the mechanistic DNB models of Lee and Mudawar (1988), Katto (1990), and Weisman and Pei (1983). All models performed rather poorly. The *KfK correlation* (Dalle Donne and Hame, 1985), which is in fact a modification of the WSC-2 correlation (Bowring, 1979), could predict their experimental results with reasonable accuracy. The correlation could in fact predict both the steady-state and transient data with similar accuracy. The WSC-2 is a flexible correlation developed for subchannel analysis, and it has been optimized for triangular-pitched and square-pitched subchannels separately. The correlation also accounts for the nonuniformity of heat flux in a rod bundle.

The relevance of data obtained with a small rod bundle to conditions in much larger rod bundles is doubtful because of the effect of cold bundle walls on CHF. Cheng (2005) reported on experiments in a vertical 37-rod, hexagonal bundle. The triangular-pitched rods were 9.0 mm in diameter and had a pitch-to-diameter ratio of 1.178. The test fluid was Freon-12, the system pressure was varied in the 1.0–2.7-MPa range, and the exit quality varied in the –0.4 to –0.2 range. The parametric dependencies in the data were similar to the parametric dependencies typically observed in heated tubes. Cheng compared his experimental data with the EPRI-1 (Reddy and Fighetti, 1983) correlation, the correlation of Courtaud *et al.* (1988), and the aforementioned correlation of Dalle Donne and Hame (1985). All three correlations underpredicted the experimental data.

El-Genk *et al.* (1988) conducted CHF experiments with water in uniformly heated vertical annuli under low-flow ($G \leq 250$ kg/m^2 s) and low-pressure ($P = 1.18$ bars) conditions and empirically correlated their CHF data. Their data occurred at churn–annular and annular–annular mist flow regimes, and separate correlations were developed for each regime. El-Genk and Rao (1991b) examined the applicability of their correlations to low-flow CHF data in tight rod bundles. They showed that their annular channel correlations could predict the low-flow rod bundle data well, provided that the CHF data corresponded to the same two-phase flow regime as the correlation.

13.4 Correlations for Subcooled Upward Flow of Water in Vertical Channels

Cooling by a highly subcooled liquid flow is very efficient. CHF under subcooled liquid flow conditions is thus of particular interest, since it represents the threshold for safe operation when forced subcooled boiling is the cooling mechanism. Subcooled CHF is also of great interest in the safety analysis of pressurized water nuclear reactors. These reactors are designed to operate such that their primary coolant systems contain pressurized and subcooled water everywhere and at all times, and rules for safe normal operation require that the CHF conditions never be approached anywhere in the reactor core. The criterion is represented in terms of a maximum DNBR, according to $\mathrm{DNBR} = q''_{\mathrm{CHF}}/q''_{\mathrm{w}} > \mathrm{DNBR}_{\mathrm{min}}$, where $\mathrm{DNBR}_{\mathrm{min}} > 1$ should apply everywhere in the core. Some empirical correlations for CHF are reviewed in the following.

The correlation of Tong (1969) is among the oldest:

$$\frac{q''_{\mathrm{CHF}}}{h_{\mathrm{fg}}} = C\frac{G^{0.4}\mu_{\mathrm{f}}^{0.6}}{D^{0.6}}. \tag{13.50}$$

This is equivalent to

$$\mathrm{Bo} = C/\mathrm{Re}^{0.6}. \tag{13.51}$$

Tong suggested (1969)

$$C = 1.76 - 7.433x_{\mathrm{eq}} + 12.222x_{\mathrm{eq}}^2. \tag{13.52}$$

The ranges of validity of data for Tong's correlation are

$$0.1 < P \leq 5.5\,\mathrm{MPa},$$
$$2.2 < G < 40\,\mathrm{Mg/m^2 \cdot s},$$
$$15 < \Delta T_{\mathrm{sub,exit}} < 190\,\mathrm{K},$$
$$2.5 < D < 8.0\,\mathrm{mm},$$
$$12 < L_{\mathrm{heat}}/D < 40,$$
$$4.0 < q''_{\mathrm{CHF}} < 60.6\,\mathrm{MW/m^2}.$$

Celata *et al.* (1994a) improved the accuracy of the aforementioned correlation of Tong (1969) by proposing

$$\text{Bo} = C/\text{Re}^{0.5}, \tag{13.53}$$

$$C = C_1(0.216 + 4.74 \times 10^{-2}P), \tag{13.54}$$

$$C_1 = \begin{cases} 0.825 + 0.987x_{\text{eq}} & \text{for } -0.1 < x_{\text{eq}} < 0, & (13.55) \\ 1 & \text{for } x_{\text{eq}} < -0.1, & (13.56) \\ 1/(2 + 30x_{\text{eq}}) & \text{for } x_{\text{eq}} > 0, & (13.57) \end{cases}$$

where P in Eq. (13.54) is in megapascals. The parameter ranges for this correlation are

$$0.1 < P < 8.4 \text{ MPa},$$
$$2 \times 10^3 < G < 90.0 \times 10^3 \text{ kg/m}^2 \cdot \text{s},$$
$$0.3 < D < 25.4 \text{ mm},$$
$$0.1 < L_{\text{heat}} < 0.61 \text{ m},$$
$$90 < \Delta T_{\text{sub,in}} < 230 \text{ K}.$$

The correlations of Hall and Mudawar (2000a, b) are for steady-state subcooled water flow in uniformly heated round vertical tubes. The correlations are based on the PU-BTPFL data base, which includes a massive number of qualified CHF data points for water, covering a very wide range of parameters. Hall and Mudawar proposed two separate correlations, one based on inlet conditions and the other based on exit (local) conditions. Their inlet-conditions correlation is

$$\text{Bo} = \frac{c_1 \text{We}_D^{c_2}(\rho_f/\rho_g)^{c_3}[1 - c_4(\rho_f/\rho_g)^{c_5}x_{\text{in}}^*]}{1 + 4c_1c_4\text{We}_D^{c_2}(\rho_f/\rho_g)^{c_3+c_5}(L_{\text{heat}}/D)}. \tag{13.58}$$

Their exit-conditions-based (local-conditions-based) correlation is

$$\text{Bo} = c_1 \text{We}_D^{c_2}(\rho_f/\rho_g)^{c_3}[1 - c_4(\rho_f/\rho_g)^{c_5}x_{\text{eq,out}}], \tag{13.59}$$

where

$$\text{We}_D = G^2D/\rho_f\sigma,$$
$$x_{\text{in}}^* = (h_{\text{in}} - h_{\text{f,out}})/h_{\text{fg,out}},$$

$h_{\text{f,out}}$ and $h_{\text{fg,out}}$ are the properties at exit, and

$$c_1 = 0.0722, c_2 = -0.132, c_3 = -0.644,$$
$$c_4 = 0.900, c_5 = 0.724.$$

The parameter ranges of the data base for both correlations are

$$0.25 \text{ mm} < D < 1.5 \text{ cm}, 300 \le G < 30\,000 \text{ kg/m}^2 \cdot \text{s}, 1 \le P \le 200 \text{ bars}.$$

For their inlet-conditions correlation, furthermore,

$$-2 < x_{\text{eq,in}} < 0.0, -1 < x_{\text{eq,out}} < 0.0, 2 \le L_{\text{heat}}/D \le 200.$$

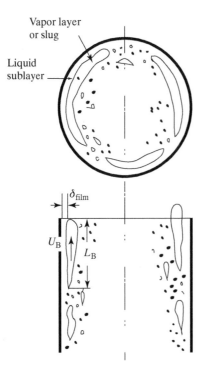

Figure 13.10. Schematic of the flow field in the vicinity of the CHF point in subcooled boiling. (After Katto, 1990.)

For their exit-conditions correlation, however, $-1 < x_{eq,out} < 0.05$. (Note that parameters $x_{eq,in}$ and L_{heat}/D are not needed for exit-conditions correlations.)

13.5 Mechanistic Models for DNB

The phenomenology of DNB in subcooled or low-quality saturated boiling is relatively well understood. The same can be said about dryout. Accordingly, mechanistic models with good accuracy have been developed for these processes. Some recent models are discussed in this section.

The DNB Model of Katto (1992)

This DNB model for vertical upward flow in pipes is based on the concept of liquid film dryout caused by an overlying vapor clot (mechanism (3) described in Section 13.1), as proposed and modeled earlier by Lee and Mudawar (1988). Further improvement on the same model has been also proposed by Celata *et al.* (1994b) to extend its range of applicability to local void fractions of more than 70%. Katto's model is based on data with water for $2.5 \leq D \leq 12.9$ mm and $P = 0.1$–19.6 MPa and data with the following fluids: water, R-11, R-12, R-113, helium, and nitrogen. The data are also limited to $\alpha < 0.70$, and the data with helium include $D = 1$ mm. The outline of the model is as follows.

Figure 13.10 schematically shows the phenomenology assumed in the model. The vapor clots generated by the coalescence of microbubbles near the wall are separated from the wall by a liquid sublayer whose initial thickness (namely, the thickness at the front end of the bubble) is (Haramura and Katto, 1983)

$$\delta_{\text{film}} = 1.705 \times 10^{-3} \pi \left(\frac{\rho_v}{\rho_L}\right)^{0.4} \left(1 + \frac{\rho_v}{\rho_L}\right) \frac{\sigma}{\rho_v} \left(\frac{\rho_v h_{\text{fg}}}{q_b''}\right)^2. \tag{13.60}$$

The liquid film undergoes evaporation while the vapor clot moves over it. Only part of the wall heat flux is used up for boiling, however, and the remainder is convected into the subcooled liquid. Following Shah (1977) the component of heat flux that is used for boiling is found from

$$q_b'' = q_w'' - H_{\text{FC}}(T_w - \overline{T}_L), \tag{13.61}$$

where H_{FC} is found from the Dittus–Boelter correlation (see Eq. (12.116) for turbulent forced convection in pipes), and

$$T_w - \overline{T}_L = \frac{(C-1)(T_{\text{sat}} - \overline{T}_L) + q_w''/H_{\text{FC}}}{C}, \tag{13.62}$$

$$C = 230(q_w''/Gh_{\text{fg}})^{0.5}. \tag{13.63}$$

CHF occurs when the liquid film is completely evaporated during the residence time of a vapor clot over it, namely when

$$q_b'' = \rho_L \delta_{\text{film}} h_{\text{fg}}/t_{\text{res}}. \tag{13.64}$$

The residence time of the bubble clot over the liquid sublayer depends on the length of the bubble clot, which is found from Helmholtz stability theory, and the velocity difference between the vapor clot and the liquid sublayer:

$$t_{\text{res}} = \frac{L_B}{U_B - U_{\text{LB}}} = \frac{2\pi\sigma(\rho_L + \rho_v)}{\rho_L \rho_v (U_B - U_{\text{LB}})^3}. \tag{13.65}$$

The relative velocity $U_B - U_{\text{LB}}$ is evidently needed. In a fully turbulent flow in the channel, the velocity of the vapor clot and liquid sublayer should both depend on the turbulent velocity profile near the wall. However, the velocity profile will be affected by the presence of the bubbles. It is therefore assumed that

$$U_B - U_{\text{LB}} = KU_{L,\delta}, \tag{13.66}$$

where $U_{L,\delta}$ is the velocity in the turbulent boundary layer at a distance of δ_{film} from the wall, found from the universal turbulent boundary layer velocity profile (see Section 1.7), using the homogeneous flow density $\rho = [\frac{x}{\rho_v} + \frac{1-x}{\rho_L}]^{-1}$ and mixture viscosity defined as $\mu = \mu_v \alpha + \mu_L(1 - \alpha)(1 + 2.5\alpha)$ as the fluid properties. The parameter K is empirically correlated as follows. For $(\rho_g/\rho_f) > (\rho_g/\rho_f)_B$,

$$K = \frac{242[1 + K_1(0.355 - \alpha)][1 + K_2(0.1 - \alpha)]}{[0.0197 + (\rho_g/\rho_f)^{0.733}][1 + 90.3(\rho_g/\rho_f)^{3.68}]} \text{Re}^{-0.8},$$

$$K_1 = \begin{cases} 0 & \text{for } \alpha > 0.355, \\ 3.76 & \text{for } \alpha < 0.355, \end{cases} \tag{13.67}$$

$$K_2 = \begin{cases} 0 & \text{for } \alpha > 0.1, \\ 2.62 & \text{for } \alpha < 0.1. \end{cases}$$

For $(\rho_g/\rho_f) < (\rho_g/\rho_f)_B$,

$$K = \frac{22.4[1 + K_3(0.355 - \alpha)]}{(\rho_g/\rho_f)^{1.28}} Re^{-0.8}, \tag{13.68}$$

$$K_3 = \begin{cases} 0 & \text{for } \alpha > 0.355, & (13.69) \\ 1.33 & \text{for } \alpha < 0.355. & (13.70) \end{cases}$$

The threshold density ratio $(\rho_g/\rho_f)_B$ itself is found by intersecting these two K equations.

To apply this model, one evidently needs to calculate the local void fraction. According to Katto's model, the OSV correlations of Saha and Zuber (1974) (Eqs. (12.22)–(12.25)) are to be used for locating the OSV point. The local quality downstream from the OSV point is found by using the profile-fit method of Ahmad (1970) (Eq. (12.51)), and the local void fraction is obtained by assuming homogeneous flow.

An interesting feature of Katto's model is its lack of explicit dependence on flow channel orientation.

The DNB Model of Celata *et al.* (1994b)

A modified version of Katto's model has been proposed by Celata *et al.* (1994b), with the goal of extending its range of applicability. Celata *et al.* (1994b) pointed out that Katto's model is unsuitable for calculating the CHF when the void fraction is larger than 70%. The differences between this model and Katto's method are as follows.

(1) The initial liquid film thickness is found from

$$\delta_{film} = y^* - d_B, \tag{13.71}$$

where y^*, the thickness of the superheated liquid layer next to the wall, is found from the turbulent boundary layer temperature law-of-the-wall profile (Eqs. (1.112)–(1.114)). (In other words, in Eqs. (1.112)–(1.114), we look for a value of y that corresponds to $T(y) = T_{sat}$.) The parameter d_B is the vapor clot equivalent diameter, and it is assumed to be equal to the diameter of bubbles departing from the wall at the OSV conditions (Staub, 1968):

$$d_B = \frac{32}{f'} \frac{\sigma F(\theta)\rho_L}{G^2},$$
$$F(\theta) = 0.03, \tag{13.72}$$

where f' is the friction factor, found from the Colebrook correlation (Eq. (8.34)) by assuming an effective wall roughness of $0.75 d_B$.

(2) The vapor clot velocity relative to the liquid sublayer is found from the following force balance on a vapor clot:

$$\frac{\pi}{4} d_B^2 L_B g \Delta\rho = \frac{1}{2}\rho_L C_D (U_B - U_{BL})^2 \frac{\pi d_B^2}{4}$$

$$\Rightarrow U_B - U_{LB} = \left(\frac{2 L_B g \Delta\rho}{\rho_L C_D}\right)^{1/2}. \tag{13.73}$$

The drag coefficient for a deformed bubble is found from (Harmathy, 1960)

$$C_D = \frac{2}{3} \frac{d_B}{\left(\frac{\sigma}{g \Delta \rho}\right)^{0.5}}.$$ (13.74)

(3) The liquid film velocity U_{LB} is found from the universal velocity profile, at a distance of $y^* - 0.5 d_B$ from the wall, by using the properties of the pure liquid.
(4) CHF happens when

$$q''_{CHF} = \frac{\rho_L \delta_{film} h_{fg}}{L_B} U_B.$$ (13.75)

The ranges of the data used for validation of the correlation of Celata *et al.* (1994b) are

$$0.1 \le P \le 8.4\,\text{MPa},$$
$$0.3 \le D \le 25.4\,\text{mm},$$
$$0.0025 \le L_{heat} \le 0.61\,\text{m},$$
$$10^3 \le G \le 90 \times 10^3\,\text{kg/m}^2\text{·s},$$
$$25 \le \Delta T_{sub,in} \le 255\,\text{K}.$$

The data base evidently includes the minichannel-size range. One can also make the following observations.
(1) Compared with Katto's model, the model of Celata *et al.* is strictly for vertical channels.
(2) The correlation used for drag coefficient (Eq. (13.74)) is unlikely to be applicable to small vapor clots moving near a solid wall. This is particularly true when the channel diameter is very small.

13.6 Mechanistic Models for Dryout

Mechanistic models for dryout are probably the most successful and accurate among all the mechanistic models dealing with various CHF types. Models for dryout have been developed by several authors and research groups (Whalley, 1977; Saito *et al.*, 1978; Levy *et al.*, 1981; Sugawara, 1990a, b; Hewitt and Govan, 1990; Celata *et al.*, 2001).

Consider a heated channel undergoing dryout, as depicted in Fig. 13.5, and for simplicity assume steady state. Also, let us start our discussion from the axial location where the annular flow regime starts. Note that an appropriate set of conservation equations (e.g., based on separated flow, or a slip flow model) can be set up for the channel, and these can be solved by starting from the channel inlet, up to the point where the annular-dispersed flow regime starts. The annular-dispersed flow regime can be assumed to start when Eq. (7.16) applies (Mishima and Ishii, 1984; Celata *et al.*, 2001). The fraction of the liquid that is in the dispersed phase at the point where the annular-dispersed flow regime starts can vary typically in the 90%–99% range (Whalley *et al.*, 1974) and should also be specified. All three fluids (the liquid film, the droplets, and the vapor) remain saturated up to the dryout point. Hewitt and Govan, for example, assumed 99% entrainment at the point where the annular-dispersed flow pattern started when the local quality was 1%.

The conservation equations for the annular-dispersed flow regime can now be set up and numerically solved along the channel, until the dryout condition is reached. Dryout can occur when the liquid film flow rate approaches zero or when the film thickens diminishes below some critical value ($4\delta_F/D \le 10^{-5}$, according to Sugawara (1990a, b)). Saito *et al.* (1978) and Sugawara (1990a, b) applied a three-fluid model, whereby separate mass and momentum conservation equations were solved for each of the three fluids, namely, for the liquid film, the dispersed droplets, and the vapor. (Because all three fluids remain saturated, only one energy equation is needed.) Extensive closure relations are needed for this type of modeling. Processes in need of closure relations include rates of dispersed droplet entrainment and deposition, effective droplet size, film evaporation rate, film–wall and film–vapor interfacial shear stresses, and the vapor–droplet interfacial force. Sugawara (1990a, b) also included a model for the suppression of droplet deposition by the flow of vapor from the evaporating film.

Droplet entrainment and deposition are probably the most important among the closure relations. Let use define $\dot{m}_F$, $\dot{m}_d$, and $\dot{m}_g$, as the mass flow rates of the liquid film, the dispersed droplets, and the vapor, respectively. The liquid film mass conservation equation can then be written as

$$\frac{d\dot{m}''_F}{dz} = \frac{4}{D}\left[\dot{D} - \dot{E} - \frac{q''_w}{h_{fg}}\right],\tag{13.76}$$

where $\dot{m}''_F = 4\dot{m}_F/\pi D^2$ is the film mass flow rate per unit cross-sectional area. Evaporation resulting from the pressure drop has been neglected. Parameters $\dot{D}$ and $\dot{E}$ are, respectively, the deposition and entrainment mass fluxes per unit flow area. According to Hewitt and Govan (1990),

$$\dot{D} = k\mathbf{C},\tag{13.77}$$

where $\mathbf{C}$ is the concentration of droplets in the vapor-dispersed droplet mixture (in kilograms per meter cubed in SI units) and k is a deposition mass transfer coefficient (in meters per second in SI units), found from

$$k\sqrt{\frac{\rho_g D}{\sigma}} = \begin{cases} 0.18 & \text{for } \mathbf{C}/\rho_g < 0.3, \tag{13.78} \\ 0.083\left(\frac{\mathbf{C}}{\rho_g}\right)^{-0.65} & \text{for } \mathbf{C}/\rho_g > 0.3. \tag{13.79} \end{cases}$$

Entrainment occurs only when $\dot{m}''_F > \dot{m}''_{FC}$, where $\dot{m}''_{FC}$ is the critical film flow rate for the initiation of entrainment and is found from

$$\frac{\dot{m}''_{FC}D}{\mu_f} = \exp\left[5.8504 + 0.4249\frac{\mu_g}{\mu_f}\sqrt{\frac{\rho_f}{\rho_g}}\right].\tag{13.80}$$

When $\dot{m}''_F > \dot{m}''_{FC}$, then

$$\dot{E} = 5.75 \times 10^{-5}\rho_g j_g\left[(\dot{m}''_F - \dot{m}''_{FC})^2\frac{\rho_f D}{\sigma \rho_g^2}\right]^{0.316}.\tag{13.81}$$

An alternative set of correlations for droplet entrainment and deposition rates are provided by Kataoka and Ishii (1983), and have been used by Celata *et al.* (2001).

By assuming an ideal liquid film with a velocity profile similar to the universal turbulent boundary layer velocity profile, and using the fundamental void-quality relation (Eq. (3.39)) along with the slip ratio correlation of Premoli *et al.* (1971) (Eq. (6.40)), Celata *et al.* essentially decoupled the liquid film momentum equation from the momentum equations of vapor and droplets.

When dryout in large, open-lattice rod bundles is modeled, the effect of turbulent mixing between adjacent subchannels must also be considered (Whalley, 1977). Furthermore, the modeling elements described here apply to transient dryout and rewetting processes as well. The main difference between steady-state and transient dryout is that during a transient rewetting or dryout process the location of the quench front varies with time with a velocity that depends on heat conduction in the solid (Catton *et al.*, 1988).

13.7 CHF in Inclined and Horizontal Systems

Horizontal and inclined boiling flow passages are used in boilers, and horizontally oriented rod bundles are used in some nuclear reactor designs. Gravity has an important effect on the boiling two-phase flow patterns in inclined and horizontal flow passages (see Fig. 12.8).

Investigations dealing with CHF in horizontal and inclined large channels (excluding mini- and microchannels) are relatively few (Becker, 1971; Merilo, 1977, 1979; Fisher *et al.*, 1978; Kefer *et al.*, 1989; Wong *et al.*, 1990). The published investigations, however, indicate that the impact of gravity on CHF is particularly important, and in general, with all parameters identical, q''_{CHF} in a horizontal heated flow passage is always smaller than in a vertical flow passage. The effect of orientation is diminished as the mass flux is increased, however, and for extremely high mass fluxes the orientation effect essentially vanishes.

In an inclined or horizontal heated pipe, depending on the heat and mass fluxes, CHF can occur over a wide range of equilibrium qualities. The mechanisms that cause CHF, and the effect of gravity on them, are as follows (Fisher *et al.*, 1978; Wong *et al.*, 1990). In highly subcooled flow, bubbles that form on the top surface are forced against the wall by buoyancy, and their departure is postponed, leading to earlier CHF in comparison with vertical flow. When CHF takes place at very low x_{eq}, flat, ribbon-like bubbles form near the wall and are separated from the heated wall by a thin liquid film. Depletion of the liquid film by evaporation causes CHF. At low and intermediate x_{eq}, the flow field is characterized by large splashing waves, or surges, and little droplet entrainment. The liquid film on the top surface is not effectively replenished by droplet impingement, while it loses liquid to evaporation and drainage. The outcome is an earlier CHF in comparison with vertical channels. At high x_{eq} values the most likely flow regime is annular. The annular film on the top is always thinner than the film near the bottom owing to gravity-induced drainage. Furthermore, although large-amplitude waves and entrainment take place at the channel bottom, little of either process occurs at the top. Consequently, the film on the top surface is depleted faster, leading to early CHF.

Figure 13.11 shows a typical set of wall temperature profiles that are observed in near-horizontal boiling channels. A transition region, representing a partially wetted

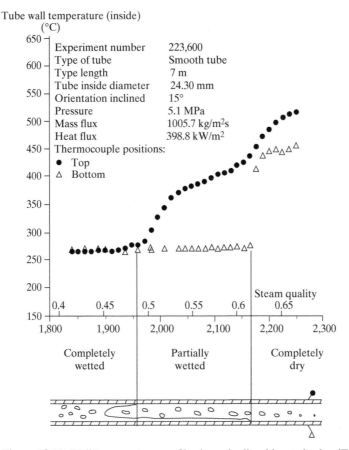

Figure 13.11. Wall temperature profiles in an inclined heated tube. (From Kefer *et al.*, 1989.)

channel perimeter (also referred to as the distributed boiling crisis region), separates the fully wetted heated surface region from the completely dry heated surface region. The distributed boiling crisis region thus starts at the point where a dry patch is developed at the top surface of the heated channel.

Correlations for CHF in Horizontal Pipes. As mentioned earlier, at very high flow rates there is little difference between CHF in horizontal and vertical channels (Merilo, 1977). Since stratification is the main cause for early CHF in inclined channels, one can argue that vertical-channel CHF correlations can be used for CHF in inclined or horizontal channels as long as the threshold represented by Eq. (7.32) is not approached (Wong *et al.*, 1990). Equation (7.32), as discussed in Section 7.3, is a curve fit to the model of Taitel and Dukler (1976a, b) (Eq. (7.29)) for transition to stratified flow.

Based on experimental data for water and Freon-12, and using the method of compensated distortions for fluid-to-fluid modeling of CHF (Ahmand, 1973), Merilo (1979) developed the following correlation:

$$\left(\frac{q_w''}{Gh_{fg}}\right)_{CHF} = 575 \mathrm{Re}_{f0}^{-0.34} [Z^3 \mathrm{Bd}]^{0.358} (\mu_f/\mu_g)^{-2.18} (L_{heat}/D)^{-0.511}$$

$$\times [(\rho_f/\rho_g) - 1]^{1.27} (1 - x_{eq,in})^{1.64}, \tag{13.82}$$

where

$$Z = \frac{\mu_f}{\sqrt{\sigma D \rho_f}},$$ (13.83)

$$\text{Bd} = (\rho_f - \rho_g)gD^2/\sigma.$$ (13.84)

The valid ranges of experimental parameters for this correlation are

$$5.3 \leq D \leq 19.1 \, \text{mm}, \, 112 \leq L_{\text{heat}}/D \leq 571, \, 13 \leq \rho_f/\rho_g \leq 20.5,$$

$$700 \leq G \leq 5400 \, \text{kg/m}^2 \cdot \text{s}, \, -0.35 \leq x_{\text{eq,in}} \leq 0.0.$$

Wong *et al.* (1990) compared this correlation with several sets of data. The correlation was in reasonable agreement with some data but overpredicted others.

A method for empirically correlating CHF in horizontal channels, suggested by Groeneveld (1986), is to write

$$q''_{\text{CHF,hor}} = K_{\text{hor}} q''_{\text{CHF,ver}},$$ (13.85)

where $q''_{\text{CHF,hor}}$ and $q''_{\text{CHF,ver}}$ are critical heat fluxes for horizontal and vertical channels that are otherwise identical, and K_{hor} is a correction factor. Wong *et al.* (1990) derived expressions for K_{hor} based on several phenomenological arguments. Among them, the one that provided the best agreement with experimental data was based on the balance between buoyancy and turbulent forces, leading to

$$K_{\text{hor}} = 1 - \exp\left[-\sqrt{0.0153 \text{Re}_{f0}^{-0.2} \left(\frac{1-x}{1-\alpha}\right)^2 \frac{G^2}{gD\rho_f(\rho_f - \rho_g)\sqrt{\alpha}}} \right].$$ (13.86)

EXAMPLE 13.4. Water at a mass flux of 1,000 kg/m^2·s, with a local pressure of 70 bars and a local equilibrium quality of $x_{\text{eq}} = 0.45$, flows through a uniformly heated vertical tube with 0.95-cm inner diameter. Using the information in the table that follows, which has been taken from the 1995 CHF look-up table (Groeneveld *et al.*, 1996), calculate the heat flux that would cause CHF to occur at that location. Can you determine the type of the CHF?

x_{eq}	-0.4	-0.3	-0.2	-0.15	-0.1	-0.05
q''_{CHF} (kW/m^2)	6930	6386	6216	6135	5799	5604
x_{eq}	0.0	0.05	0.1	0.15	0.2	0.25
q''_{CHF} (kW/m^2)	5505	5318	5070	4472	3892	3626
x_{eq}	0.3	0.35	0.4	0.45	0.5	0.6
q''_{CHF} (kW/m^2)	3347	3136	3031	3028	2838	1774
x_{eq}	0.7	0.8	0.9	1.0		
q''_{CHF} (kW/m^2)	1121	735	613	0		

SOLUTION. The table indicates that $q''_{\text{CHF}} = 3.028 \, \text{MW/m}^2$ for an 8-mm-diameter tube. A correction for the tube diameter is needed. Equation (13.9) then gives

$$q''_w = (3.028 \, \text{MW/m}^2)\sqrt{0.008/0.0095} = 2.779 \, \text{MW/m}^2.$$

We can estimate the void fraction using the slip ratio correlation of Chisholm (1973), Eq. (6.39). That results in $S_r = 3.11$. We can then use the fundamental void-quality

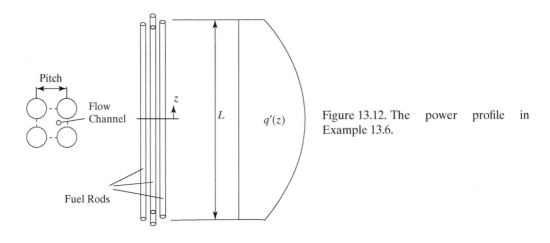

Figure 13.12. The power profile in Example 13.6.

relation, Eq. (6.39), which results in $\alpha \approx 0.84$. The flow regime is thus likely to be churn or annular-dispersed.

EXAMPLE 13.5. In Example 13.4, assume that the tube is rotated and made horizontal, while the heat flux is maintained. Estimate the CHF at that point.

SOLUTION. We calculated the critical heat flux for the vertical and upward configuration in the tube. We can now use Eqs. (13.85) and (13.86), with $x = 0.45$ and $\alpha \approx 0.84$. Using the saturation properties of water and steam at 70 bars, we will get $\mathrm{Re}_{f0} = 1.041 \times 10^5$, Eq. (13.86) then gives $K_{\mathrm{hor}} = 0.474$, and Eq. (13.85) then gives $q''_w \approx 1.32 \, \mathrm{MW/m^2}$.

EXAMPLE 13.6. A PWR operates at 15 MPa pressure. The fuel rods are 1.07 cm in diameter and are arranged as shown in Fig. 13.12 . The pitch is $p = 1.42$ cm. The water temperature at the inlet is 280 °C, and the coolant mass flow rate is 0.2 kg/s per rod. The axial power distribution in the hot channel of the core can be represented as

$$q' = q'_{\max} \cos\left(\frac{\pi z}{1.2L}\right),$$

where q' is the power generation per unit length, z is the coordinate defined in Fig. 13.12, and $q'_{\max} = 33.6$ kW/m. Calculate the local critical heat flux at the location 1.4 m above the center.

SOLUTION. Assume for simplicity that the flow in the subchannel representing the unit cell composed of a single tube and a 1.42 cm × 1.42 cm square surrounding it is one-dimensional. We will also assume that the properties of saturated water and steam correspond to the inlet pressure. The property tables then give $h_{\mathrm{in}} = 1.231 \times 10^6$ J/kg = 529.4 Btu/lb, $h_f = 1.61 \times 10^6$ J/kg = 692.1 Btu/lb, $h_{\mathrm{fg}} = 1.00 \times 10^6$ J/kg, $T_{\mathrm{sat}} = 615.3$ K, and $\rho_f = 603.4$ kg/m^3.

For convenience, let us define $z^* = 1.4$ m as the coordinate of the location where CHF is to be calculated. We will calculate the equilibrium quality at z^* by performing an energy balance on the subchannel and assuming that the potential and kinetic energy changes are negligible. We can then write

$$\dot{m}[(h_f + x_{eq}h_{fg}) - h_{in}] = \int_{z=-L/2}^{z^*} q'_{max} \cos\left(\frac{\pi z}{1.2L}\right) dz. \tag{a}$$

The right side of this equation can be written as

$$q'_{max} \frac{1.2L}{\pi} \left[\sin\frac{\pi z^*}{1.2L} + \sin\frac{\pi}{2.4}\right]. \tag{b}$$

By substitution from Eq. (b) into Eq. (a), and plugging in numbers, we find $x_{eq} = 0.0462$.

Boiling starts in the subchannel at the ONB point. For simplicity, we assume that boiling starts approximately where $x_{eq} = 0$. We therefore calculate z_B, representing the coordinate of the point where boiling starts, from

$$\dot{m}(h_f - h_{in}) = q'_{max} \frac{1.2L}{\pi} \left[\sin\frac{\pi z_B}{1.2L} + \sin\frac{\pi}{2.4}\right]. \tag{c}$$

The solution of Eq. (c) gives $z_B = 0.98$ m.

We next calculate $q''_{CHF,u}(z^*)$, and this can be found from Eq. (13.46). To apply this equation, we note that

$$G = \dot{m}/(p^2 - \pi D^2/4) = 1{,}777 \text{ kg/m}^2\text{·s} = 1.31 \times 10^6 \text{ lb/ft}^2\text{·hr},$$
$$P = 15 \text{ MPa} = 2{,}176 \text{ psia},$$
$$D_H = 4(p^2 - \pi D^2/4)/\pi D = 0.0134 \text{ m} = 0.421 \text{ in.}$$

The solution of Eq. (13.46) then gives

$$q''_{CHF,u} = 6.036 \times 10^5 \text{ Btu/ft}^2\text{·hr} = 1.905 \times 10^6 \text{ W/m}^2\text{·s}.$$

We now need to find the correction factor F, using Eqs. (13.47)–(13.49). First, find l_{DNB} from

$$l_{DNB} = z^* - z_B = 0.419 \text{ m}.$$

Also,

$$q''_w(l_{DNB}) = \frac{1}{\pi D} q'_{max} \cos\left(\frac{\pi z^*}{1.2L}\right) = 5.39 \times 10^5 \text{ W/m}^2,$$
$$C = (0.4 \times 12)\frac{(1 - 0.0462)^{4.31}}{(1.31)^{0.478}} \text{ft}^{-1} \cdot \left(\frac{3.28\text{ft}}{\text{m}}\right) = 4.23 \text{ m}^{-1}.$$

We thus get

$$\frac{C}{q''_w(l_{DNB})[1 - \exp(-Cl_{DNB})]} = 0.946 \times 10^{-5} (\text{W/m}^2)^{-1}.$$

To calculate the integral on the right side of Eq. (13.48), let us change the variable in that integral from z' (which is measured from the z_B point) to z, by noting that $z' = z - z_B$. The integral will then be

$$\int_{z_B}^{z*} q''_{w,max} \cos\left(\frac{\pi z}{1.2L}\right) \exp[-C(z^* - z)]dz, \tag{d}$$

where

$$q''_{w,max} = q'_{max}/(\pi D) = 9.996 \times 10^5 \text{ W/m}^2.$$

We can use the following identity for the integration:

$$\int e^{ax} \cos(bx)dx = \frac{e^{ax}[a\cos(bx) + b\sin(bx)]}{a^2 + b^2}.$$

The integral in Eq. (d) then leads to

$$q''_{w,max}e^{-Cz*}\left\{\frac{C\cos\left(\frac{\pi z}{1.2L}\right) + \left(\frac{\pi}{1.2L}\right)\sin\left(\frac{\pi z}{1.2L}\right)}{C^2 + \left(\frac{\pi}{1.2L}\right)^2}e^{Cz}\right\}_{z=z_B}^{z=z^*} = 1.225 \times 10^6 \text{ W/m}^2.$$

Equation (13.48) then gives $F = 1.158$. The local heat flux that would have caused CHF to occur at z^* would thus be

$$q''_{CHF}(z^*) = q''_{CHF,u}/F = 1.644 \times 10^6 \text{ W/m}^2.$$

The local DNBR can now be calculated:

$$\text{DNBR}(z^*) = q''_{CHF}(z^*)/q''_w(z^*) = 1.644 \times 10^6/5.39 \times 10^5 = 3.05.$$

13.8 Post-Critical Heat Flux Heat Transfer

Post-CHF regimes include transition boiling, (possibly) stable film boiling, liquid-deficient boiling, and single-phase vapor-forced convection regimes (Figs. 12.2 and 12.3). Transition boiling is mostly encountered in transient processes, such as quenching of heated objects.

An area of application where transition boiling as well as all other post-CHF regimes can occur is the rewetting (quenching) of hot surfaces. In fact, one of the most important applications of MFB is that it represents the quenching temperature in the rewetting process of hot surfaces, and rewetting is a crucial process in the emergency cooling of nuclear fuel rods. During the early stages of the reflood phase of a LB-LOCA in most PWRs, for example, subcooled water from the emergency core cooling system (ECCS) is injected into one of the cold legs of the primary coolant system, from there it flows into the downcomer of the reactor, and subsequently it enters and fills the lower plenum. The ECCS water then enters the bottom of the core, leading to the formation of a swollen two-phase level and a quench front that advances upward along the hot fuel rods (Ghiaasiaan and Catton, 1983; Ghiaasiaan et al., 1985; Catton et al., 1988). The flow and heat transfer regimes in the vicinity of the quench front are similar to those shown in Fig. 13.13 (Sepold et al., 2001).

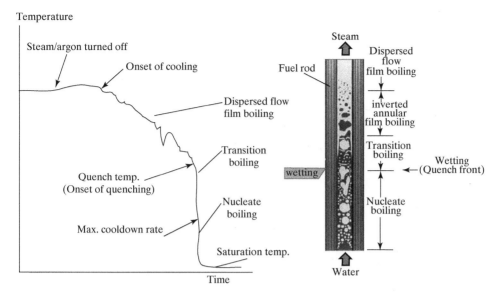

Figure 13.13. Flow and heat transfer regimes in the vicinity of the quench front during rewetting of a hot rod. (From Sepold *et al.*, 2001.)

Transition and nucleate boiling regimes take place upstream from the quench front, while stable film boiling, liquid deficient boiling, and cooling by a dispersed-droplet flow are all observed. The speed of the propagation of the quench front is the single most important parameter that determines the effectiveness of the emergency cooling system of the reactor. The propagation speed of the quench front itself is determined by conduction in the fuel rod and the convective heat transfer on both sides of the quench front. The convective cooling behind (upstream of) the quench front is very strong, being typically two or more orders of magnitude larger than the heat transfer coefficient ahead of (downstream from) the quench front. Consequently, inaccuracies in the heat transfer coefficient in transition boiling are relatively unimportant. Accuracy with respect to the heat transfer immediately ahead of the quench front is in practice much more important, but it is the conditions that lead to quenching that comprise the most important aspect in modeling the rewetting process. It is often assumed that the quench front is actually the point where the surface temperature has just dropped below the MFB temperature (Ghiaasiaan and Catton, 1983; Catton *et al.*, 1988).

Experimental data show that the quench front temperature is a complicated function of the solid and fluid properties as well as local hydrodynamic parameters. Various aspects of the process of rewetting of hot surfaces have been investigated extensively (Catton *et al.*, 1988). Forced-flow transition boiling and MFB are not well understood, however, despite extensive past research. The main difficulties that complicate these processes are the complex coupling between boiling and hydrodynamics and the fact that much of the available data deals with transient processes.

Minimum Film Boiling Point and Transition Boiling
For MFB, sometimes correlations for Leidenfrost (in pool boiling) are used (see Section 11.6).

Correlations for flow MFB are often based on quenching data over limited ranges of parameters. Unlike pool boiling, it is not practical to use a simple interpolation between the CHF and the MFB points, because the conditions at the MFB point are not unique and are often not known a priori. Some examples, all of which are for water, follow.

Bjonard and Griffith (1977) have proposed the following simple interpolation between pool boiling CHF and MFB heat fluxes:

$$q''_{TB} = q''_{CHF}\delta + q''_{MFB}(1 - \delta),   \text{(13.87)}$$

where

$$\delta = \left(\frac{T_{MFB} - T_w}{T_{MFB} - T_{CHF}} \right)^2.$$

The correlation of Ramu and Weisman (1974) is

$$H_{TB} = 500S\{\exp[-0.14(\Delta T - \Delta T_{CHF})] + \exp[-0.125(\Delta T - \Delta T_{CHF})]\},   \text{(13.88)}$$

where T is in kelvins, H is in watts per meter squared per kelvin, S is Chen's suppression factor (see Eqs. (12.100) or (12.101)) based on local mass flux and quality, and $\Delta T = T_w - T_{sat}$.

Based on experimental data from the Harwell Atomic Energy Research Establishment (AERE), Tong and Young (1974) derived

$$q''_{TB} = q''_{FB} + q''_{NB} \exp \left[-0.0394 \frac{x_{eq}^{2/3}}{(dx_{eq}/dz)} \left(\frac{T_w - T_{sat}}{55.6} \right)^n \right],   \text{(13.89)}$$

with

$$n = 1 + 0.002\,88(T_w - T_{sat}),   \text{(13.90)}$$

where temperatures are degrees Celsius or kelvins and q''_{FB} and q''_{NB} are the film and nucleate boiling heat fluxes calculated based on local conditions. The parameter ranges of the data points were as follows: $P = 3.5–9.7$ MPa, $G < 700 \sim 4000$ kg/m^2·s, and $x_{eq} = 0.15–1.10$.

Stable Film Boiling

Stable film boiling occurs when a liquid-dominated bulk flow exists in the vicinity of a dried-out surface. The film boiling process in this case should closely resemble the film boiling process in pool boiling. Stable film boiling can thus be assumed when

$$T_w > T_{MFB}   \text{(13.91)}$$

and

$$\alpha \leq 0.4,   \text{(13.92)}$$

where α is the local void fraction. When stable boiling is encountered, pool film boiling methods are recommended.

Table 13.1. *Numerical values of the constants in Groeneveld's correlation.*

Correlation	a	b	c	d	Pressure (MPa)	Mass flux (kg/m²·s)	Quality
Groeneveld 5.7	5.2×10^{-2}	0.688	1.26	−1.06	3.44–10.1	0.8–4.1	0.10–0.90
Goreneveld 5.9	3.27×10^{-3}	0.901	1.32	−1.50	3.44–21.8	0.70–5.30	0.10–0.90

Liquid-Deficient Heat Transfer Regime

In this regime, the two-phase flow pattern is primarily dispersed-droplet. Thermo-dynamic nonequilibrium (with saturated droplets and superheated vapor) is possible. Several hydrodynamic and thermal processes simultaneously contribute to heat transfer, including convection from wall to vapor, convection and radiation from wall to droplets, convection from vapor to droplets, evaporation from droplets, droplet impingement on the wall, and the enhancement of turbulence in the vapor phase by the droplets. The correlation of Groeneveld (1973) is among the most accurate, according to which

$$\frac{HD_{\mathrm{H}}}{k_{\mathrm{g}}} = a \left\{ \mathrm{Re}_{\mathrm{g}} \left[x_{\mathrm{eq}} + \frac{\rho_{\mathrm{g}}}{\rho_{\mathrm{f}}} (1 - x_{\mathrm{eq}}) \right] \right\}^{b} \mathrm{Pr}_{\mathrm{v,w}}^{c} Y^{d}, \tag{13.93}$$

$$Y = 1 - 0.1 \left(\frac{\rho_{\mathrm{f}}}{\rho_{\mathrm{g}}} - 1 \right)^{0.4} (1 - x_{\mathrm{eq}})^{0.4}, \tag{13.94}$$

where $\mathrm{Re}_{\mathrm{g}} = G x_{\mathrm{eq}} D / \mu_{g}$, $\mathrm{Pr}_{\mathrm{v,w}}$ is the vapor Prandtl number at wall temperature T_{w}, and the heat flux is related to the wall temperature according to

$$q_{\mathrm{w}}'' = H(T_{\mathrm{w}} - T_{\mathrm{sat}}). \tag{13.95}$$

The constants $a, b, c,$ and d depend on the flow channel geometry and are summarized in Table 13.1. The range of validity of the correlation is also summarized in Table 13.1.

Several other correlations have also been proposed. See Groeneveld and Snoek (1986) for a good review. Among them, the correlation of Dougall and Rohsenow (1963) has been widely applied in nuclear reactor licensing calculations:

$$\frac{HD_{\mathrm{H}}}{k_{\mathrm{g}}} = 0.023 \left\{ \frac{GD_{\mathrm{H}}}{\mu_{\mathrm{g}}} \left[x_{\mathrm{eq}} + \frac{\rho_{\mathrm{g}}}{\rho_{\mathrm{f}}} (1 - x_{\mathrm{eq}}) \right] \right\}^{0.8} \mathrm{Pr}_{\mathrm{g}}^{0.4}, \tag{13.96}$$

where the heat flux and wall temperature are related according to Eq. (13.95).

Yoder *et al.* (1982) compared the predictions of several correlations with their core uncovery experimental data. The data were obtained in a 17 × 17 rod bundle with typical geometric characteristics of the rod bundles in PWR cores, in the 6.01–13.7 MPa pressure range. Among the tested correlations the aforementioned correlations of Groeneveld both performed well. The correlation of Dougal and Rohsenow, however, overpredicted their data.

Table Look-up Method for Steam–Water Fully Developed Boiling

For fully developed flow boiling of water in vertical tubes, a direct table look-up method similar to the aforementioned table look-up method for CHF has been proposed as an alternative to the application of correlations (Kirillov *et al.*, 1996; Leung *et al.*, 1997; Groeneveld *et al.*, 2003). The argument in favor of this method is that the available experimental data are vast, and the existing empirical and semi-analytical correlations are generally valid over relatively limited parameter ranges. The most recent tables are based on more than 77 000 data points and cover the inverted annular and dispersed-flow boiling regimes (Groeneveld *et al.*, 2003). The experimental data base for these tables covers the following parameter ranges:

$$2.5 \leq D \leq 24.7 \, \text{mm}, \, 12 \leq G \leq 6995 \, \text{kg/m}^2 \cdot \text{s},$$

$$-0.1 \leq x_{eq} \leq 2.0, \, 0.1 \, \text{MPa} \leq P \leq 20 \, \text{MPa}.$$

The conditions with $x_{eq} > 1$ evidently imply thermodynamic nonequilibrium, where saturated droplets are entrained in superheated vapor. The look-up table of Groeneveld *et al.* (2003) is available in its entirety at www.magma.ca/~ thermal/. Accordingly, the film boiling coefficient should be found from

$$H = H_{D_{ref}}[P, G, x_{eq}, (T_w - T_{sat})](D_{ref}/D)^{0.2}, \tag{13.97}$$

where $D_{ref} = 8 \, \text{mm}$ and $H_{D_{ref}}[P, G, x_{eq}, (T_w - T_{sat})]$ is read from the tables.

13.9 Critical Heat Flux in Binary Liquid Mixtures

Dryout and DNB have been recognized as two different mechanisms that cause CHF in flow boiling of binary liquid mixtures.

In DNB-type CHF the phenomenology that leads to the disruption of sustained macroscopic contact between liquid coolant and the heated surface is similar to single-component flow boiling, as described earlier. The composition of the mixture affects the CHF, however. The heat and mass transfer processes near the liquid–vapor interphase that were described earlier in Section 11.8 (i.e., the rise in interfacial temperature due to the partial depletion of the more volatile species, and the occurrence of a mass transfer resistance resisting the diffusion of the more volatile species from the liquid bulk towards the interphase) are responsible for the effect of composition on CHF. The addition of a small amount of a volatile species to a liquid appears to lead to an increase in the CHF (Tolubinsky and Matorin, 1973; Celata *et al.* 1994d) and the critical heat flux varies approximately linearly with the concentration of the added species (or the less volatile species), although an opposite trend has also been reported (Bergles and Scarola, 1966). Bergles and Scarola reported a distinct reduction in CHF when they added 2.2% by weight of 1-pentanol to water, and explained the phenomenon as follows: with the added more volatile species the bubbles generated at the heated surface tend to be smaller. This is a disadvantage with respect to CHF under low subcooling and pressure because a larger bubble causes higher local velocity and a higher local velocity prevents the formation of a vapor blanket that leads to DNB (Collier and Thome, 1994).

For DNB-type CHF, one may write (Collier and Thome, 1994)

$$q''_{CHF} = q''_{CHF, ideal}(1 + \chi) \tag{13.98}$$

where the ideal mixture CHF is to be found from the following linear interpolation using the CHF values that would occur for pure species 1 ($q''_{CHF,1}$) and pure species 2 ($q''_{CHF,2}$)

$$q''_{CHF,\,ideal} = q''_{CHF,1} X_1 + q''_{CHF,2}(1 - X_1). \tag{13.99}$$

Sterman et al. (1968) observed that χ varied between zero and 0.8 and proposed

$$\chi = (3.2 \times 10^5)\frac{|Y_1 - X_1|^3}{\mathrm{Re}_{L0,2}} + 6.9\frac{|Y_1 - X_1|^{1.5}}{\mathrm{Re}_{L0,2}^{0.4}}\left(\frac{T_{sat,1}(P)}{T_{bp}(X_1, P) - T_{sat,1}(P)}\right) \tag{13.100}$$

where $\mathrm{Re}_{L0,2} = GD_H/\mu_{f,2}$ represents the Reynolds number using the saturation properties of the less volatile component, and is used because the less volatile component has a high concentration near the heated surfaces. Using experimental data with ethanol–water, ethanol–benzene, acetone–water, and ethylene glycol–water mixtures, Tolubinsky and Matorin (1973) have proposed

$$\chi = 1.5|Y_1 - X_1|^{1.8} + 6.8|Y_1 - X_1|\frac{T_{bp}(X_1, P) - T_{sat,1}}{T_{sat,1}}. \tag{13.101}$$

This correlation agreed well with the experimental data of Celata et al. (1994d) that were obtained with R-12–R-114 mixtures within $\mp 20\%$.

Dryout occurs in binary mixtures just as in single-component liquids. It occurs when the flow regime is annular-dispersed, and is caused by the breakdown or depletion of the liquid film on the heated surface. For binary mixtures the effect of mole fraction on dryout is small, and techniques that are successful in modeling the dryout in single-component fluids can be applied to binary mixtures as well (Celata et al., 1994d). The empirical correlations that have been derived based on single-component liquid dryout data, as well as the mechanistic technique described in Section 13.6, can be applied. Celata et al. (1994c) found good agreement between their experimental data that were obtained with R-12 refrigerant and the predictions of a modified version of the CISE correlation (Celata et al., 1986) (see Eqs. (13.22) to (13.25) for the original CISE correlation), as well as a correlation developed earlier by Katto and Ohno (1984).

PROBLEMS

13.1 The data points in Table P13.1, which are from Becker (1971), are included in the PU-BTPFL CHF Database (Hall and Mudawar, 1997). The test section is a uniformly heated vertical tube, with $D = 10$ mm and $L/D = 100$. The heat fluxes listed in the table have caused the CHF to occur in the test section. You have been asked to compare the predictions of the correlations of Bowring (1972), Caira et al. (1995), and Hall and Mudawar (2000a, b), where applicable, with these data. Assuming that CHF occurs at the exit of the test section for all the data points, compare the predictions of the aforementioned correlations with the data, and comment on the results.

Table P13.1. *Selected data from Becker (1971).*

Test number	$G(\text{kg/m}^2 \cdot \text{s})$	P_{exit} (bar)	$\Delta T_{\text{sub,in}}(^\circ\text{C})$	$x_{\text{eq,exit}}$	$q_w''(\text{MW/m}^2)$
3	1.328×10^3	80	191.5	0.153	3.646
6	2.007×10^3	80	194.7	0.016	4.589
9	2.784×10^3	80	192.3	−0.065	5.476
35	1.947×10^3	100	181.9	−0.002	4.17
70	0.391×10^3	140	236.6	0.374	1.506

In particular, discuss the difference among inlet- (global-) and local-conditions correlations. For simplicity, and in view of the high pressures in the tests, you can assume that for each data point the pressure in the test section is constant.

13.2 The departure from nucleate boiling ratio is an important parameter for pressurized water nuclear reactors. It is defined at any location in the reactor core according to

$$\text{DNBR} = q_{\text{CHF}}''/q_w'',$$

where q_w'' is the local heat flux at the fuel rod surface, and q_{CHF}'' is the local surface heat flux that would have caused CHF. Safe operation is ensured by maintaining $\text{DNBR} > (\text{DNBR})_{\text{min}}$, where $(\text{DNBR})_{\text{min}} > 1$.

A PWR operates at 15 MPa pressure. Water coolant enters the core bottom (inlet) at 280 °C, with a mass flux of 0.25 kg/s per fuel rod. The hottest fuel rod generates power that is nonuniformly distributed, according to (see Fig. 13.13)

$$q' = q_{\text{max}}' \cos\left(\frac{\pi z}{1.2L}\right),$$

where q' is the power generation per unit length, $L = 3.66$ m is the total active height of the fuel rods, and $q_{\text{max}}' = 42.0$ kW/m. The rods are 0.9 cm in diameter and the pitch-to-diameter ratio for the rod bundle is 1.33. Plot the axial variation of DNBR along the aforementioned hottest channel using the Westinghouse W-3 correlation.

13.3 For the previous problem, suppose that the minimum allowable value for DNBR is 2.0. Calculate the maximum allowable q_{max}'.

13.4 Consider test numbers 3 and 35 in Problem 13.1. Assuming a horizontal test section, at what distance from the inlet would CHF take place? For simplicity, assume homogeneous equilibrium flow.

13.5 In an experiment, a vertical, uniformly heated rod bundle is cooled by water. The rods are patterned on a square lattice (see Fig. P4.4, Problem 4.4). The diameter and pitch are 14.3 and 18.75 mm, respectively. Water at 6.9-MPa pressure and 262 °C temperature, with a mass flux of $G = 1380$ kg/m²·s, flows upward in the rod bundle. The total heated length of the bundle is 1.83 m. Experiment shows that CHF occurs at the exit when a uniform wall heat flux of 1.92 MW/m² is imposed. Use this data point to assess the performance of the correlation of Bowring (1972).

13.6 The following generalized subchannel CHF correlation (referred to as the EPRI correlation) has been developed for operating conditions of PWRs and BWRs, as well as postulated LOCAs (Reddy and Fighetti, 1983):

$$q''_{CHF} = \frac{A - x_{eq,in}}{C + \left[\frac{x_{eq} - x_{eq,in}}{q''_w}\right]}, \tag{13.98}$$

where $x_{eq,in}$ and x_{eq} represent the inlet and local equilibrium qualities, respectively, q''_w is the local heat flux, and

$$A = p_1 P_r^{p_2} G^{(p_5 + p_7^{P_r})},$$
$$C = p_3 P_r^{p_4} G^{(p_6 + p_8^{P_r})}.$$

The parameters and their units are

$q''_{CHF}, q''_w =$ CHF and local heat flux, respectively (MBtu/hr·ft^2),
$G =$ local mass flux (Mlb$_m$/hr·ft^2), and
$P_r = (P/P_{cr}) =$ reduced pressure.

The optimized constants are

$p_1 = 0.5328, p_2 = 0.1212,$
$p_3 = 1.6151, p_4 = 1.4066,$
$P_5 = -0.3040, P_6 = 0.4843,$
$P_7 = -0.3285, P_8 = -2.0749.$

This correlation is for a bare fuel rod. Empirical correction factors are also pro-posed for the effects of spacer grids and the effect of cold bundle walls. The valid data ranges for the correlation are $0.2 < G < 4.1$ Mlb/hr·ft^2, $200 < P < 2450$ psia, $-1.10 \leq x_{eq,in} \leq 0.0$, $-0.25 \leq x_{eq} \leq 0.75$, $1 \leq L \leq 5.6$ ft, $0.35 \leq D_H \leq 0.55$ in., and $0.38 \leq$ rod diameter ≤ 0.63 in.

Using this correlation, calculate the DNBR (defined in Problem 13.2) at the mid-height of an experimental fuel rod bundle that has geometric and flow conditions similar to Problem 13.2 but is subject to uniform heat generation at the rate of 31 kW/m per rod.

13.7 A uniformly heated vertical tube that is 8 mm in diameter and operates at 78.5 bars is cooled with water flowing at a mass flux of $G = 3.2 \times 10^3$ kg/m^2·s. At the inlet to the tube, the water has a temperature of $T_{in} = 157$ °C. A heat flux of 4.96 MW/m^2 is imposed on the tube. Experiment has shown that DNB occurs at a location where $x_{eq} = -0.031$. Calculate the local critical heat flux at the latter point using the DNB model of Katto (1992), and compare the result with the experimental measurement.

13.8 In Problem 13.7,

(a) determine the axial location where CHF has occurred in the reported experiment and

(b) determine the axial location where CHF occurs using the CHF correlation of Celata et al. (1994a), Eqs. (13.53)–(13.57), if the imposed heat flux is 4.96 MW/m^2.

13.9 A uniformly heated vertical tube that is 12.9 mm in diameter is cooled with water flowing at a mass flux of $G = 0.717 \times 10^3$ kg/m^2·s. The pressure is 71.2 bars. At the inlet to the tube, the water has a specific enthalpy of 509.2 kJ/kg. A heat flux of 0.44 MW/m^2 is imposed on the tube.

(a) Find the distance from inlet where $x_{eq} = 0.91$ is reached.

(b) Assuming that the flow regime is annular-dispersed, determine whether droplet entrainment occurs at the location where $x_{eq} = 0.91$ is reached, and calculate the droplet entrainment and deposition rates.

(c) Assuming that the heat transfer regime is post-CHF, calculate the local heat transfer coefficient and local tube surface temperature.

13.10 Water, at a mass flow rate of 1.67 kg/s, and at 10 bar pressure and 160 °C temperature enters a vertical tube which is 45 mm in diameter and 3 m long. The tube is uniformly heated at q_l(kW/m).

(a) Determine q_l that results in critical heat flux at the tube exit.

(b) Repeat part (a), this time assuming critical heat flux occurs at the mid-point of the tube.

(c) For both parts (a) and (b) determine the location of onset of nucleate boiling.

(d) For the conditions of part (a), determine the heat transfer regime, and the heat transfer coefficient, at the mid-point of the channel.

14 Flow Boiling and CHF in Small Passages

14.1 Mini- and Microchannel-Based Cooling Systems

Compact heat exchangers, refrigeration systems, the cooling systems for microelectonic devices, and the cooling systems for the first wall of fusion reactors are some examples for the applications of mini- and microchannel-based cooling systems. Compact heat exchangers and refrigeration systems in fact represent an important current application of minichannels. Figure 14.1 displays typical minichannel flow passages in compact heat exchangers. In this chapter, flow boiling and CHF in channels with $10 \ \mu\text{m} \lesssim D_H \lesssim 3 \ \text{mm}$ are discussed.

A distinction should be made between mini- and microchannel-based systems because they are different for several phenomenological and practical reasons. Some important differences between the two channel size categories with respect to the basic two-phase flow phenomena, in particular the flow patterns and the gas–liquid velocity slip, were discussed in Section 3.7 and Chapter 10. Other important differences between the two categories are as follows.

(1) For practical reasons microchannel cooling systems are typically designed as arrays of parallel channels connected at both ends to common inlet and outlet plena, manifolds, or headers. Multiple parallel channels with common inlet and outlet mixing volumes are susceptible to instability, flow maldistribution, and oscillations. Minichannel cooling systems, in contrast, are designed both as parallel arrays as well as individual channels with independent flow controls.

(2) Also for practical reasons, microchannels operate under low-flow conditions. Minichannel systems can operate over a wide range of coolant flow rates, however.

(3) It is usually feasible to measure or at least indirectly quantify the local flow and heat transfer process rates in experiments with minichannels. In microchannels, however, heat conduction in both axial and circumferential directions in the solid structures is usually very important and can cause significant temperature nonuniformity. The heat transfer in microchannel systems is consequently always a conjugate problem, and the correct interpretation of experimental data requires careful and detailed thermal analysis of the entire flow field and its confining solid structure.

(4) Bubble ebullition processes that follow bubble nucleation in microchannels are different than in minichannels.

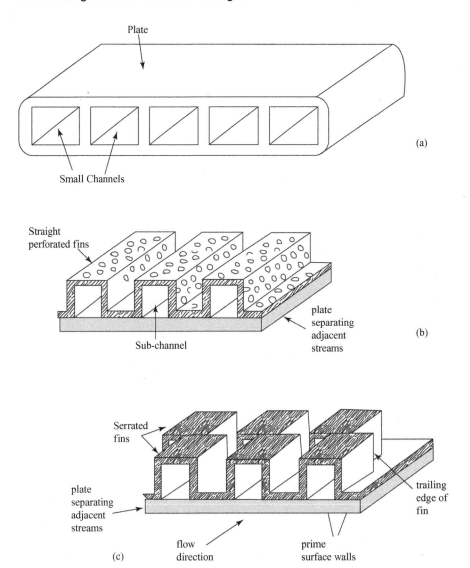

Figure 14.1. Schematics of minichannel passages: (a) straight parallel channels, (b) straight perforated passages, and (c) fin passages. (From Watel, 2003.)

Channels with $200\ \mu m \lesssim D_H \lesssim 3$ mm can be used in compact heat exchangers and refrigeration systems, and they are often included in the design of the first wall of fusion reactors. A wide variety of designs and coolant flow rates are possible for systems that utilize these channels. Thus, although most applications may use arrays of parallel channels connected to common inlet and outlet mixing volumes (see Fig. 14.2), applications where an individual heated channel is provided by an independent flow control system are also encountered. The coolant mass flux can be small, as in miniature evaporators, or it can be very large in applications where cooling by highly subcooled liquid forced convection is desired (e.g., in the cooling systems of fusion reactor first walls). In the latter application the channels are designed to operate in the single-phase liquid forced convection regime, and therefore the flow

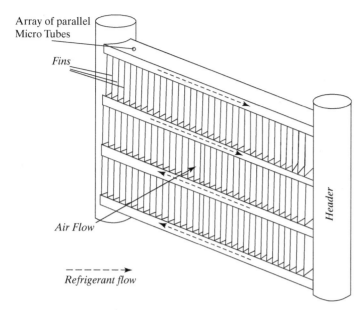

Array of parallel
Micro Tubes

Fins

Header

Air Flow

Refrigerant flow

Figure 14.2. Schematic of a minichannel-based cooling system consisting of an array of parallel channels connected to common inlet and outlet plena.

boiling thresholds such as ONB, OFI, and CHF are of concern for the safe operation of these systems.

Microchannels and minichannels (channels with 10 μm $\lesssim D_H \lesssim$ 3 mm) are particularly useful for volumetric cooling in heated blocks or heat sinks configured as plate substrates. Two-dimensional arrays of about 100 or more parallel microchannels connected to common inlet and outlet plena or manifolds represent probably the most practical design for the latter application. They are relatively easy to construct, install, and maintain. An array of microchannels is cut into a two-dimensional block of silicon or copper, and the block is used as a heat sink with heating imposed typically on only one side of the block. Boiling is a desirable heat transfer regime in microelectronic cooling systems because during saturated boiling the heated surface remains at a temperature that is only slightly higher than the saturation temperature, and so relatively uniform wall temperatures are obtained. If the pressure along the microchannel is maintained reasonably uniform, the heated surface temperature will also remain approximately uniform. The coolant flow rate needs to be low to avoid excessive pressure loss in microchannels. A large pressure drop in microchannels would not only require a powerful pump but also lead to a significant variation of saturation temperature along the microchannel. The available experimental data dealing with boiling in microchannels are primarily obtained in test sections made of parallel arrays of channels.

It is emphasized that, although the past studies dealing with boiling in minichannels are rather extensive, in comparison, experimental investigations addressing boiling in microchannels are scarce. In view of the difference between mini- and microchannels with respect to two-phase flow phenomena, as discussed in Chapter 10, the minichannel-scale experimental observations and data trends are unlikely to apply to microchannels; nor can minichannel data be extrapolated to microchannels with confidence.

14.2 Boiling Two-Phase Flow Patterns and Flow Instability

Table 14.1 summarizes some experimental investigations dealing with boiling in small flow passages. A summary of some experimental investigations dealing with critical heat flux in small flow passages is provided in Table 14.2. As mentioned earlier, a distinction should be made between parallel mini- and microchannels with common inlet and outlet plena, manifolds, or headers (i.e., systems that are susceptible to pressure drop and parallel channel instability) and minichannels that are subject to "hard" inlet conditions. We will use the phrase "hard inlet conditions" to refer to channels that are equipped with throttle valves and/or orifices upstream of their inlets, so that backflow does not occur in them. To ensure hard inlet conditions, one needs to use a separate flow control system for an individual channel, so that the channel can be provided with stable inlet flow conditions. In comparison, parallel channels with a common inlet mixing volume have "soft inlet conditions," meaning that the inlet flow rate will change in response to a change in the total pressure drop in the channels. The two types of channels behave differently.

14.2.1 Flow Regimes in Minichannels with Stable Flow Rates

In this section we are interested in two-phase flow regimes in channels that do not undergo any significant flow oscillations.

Experimental observations and physical arguments indicate that the basic phenomenology of flow boiling in minichannels is similar to that in large channels as long as there are defects on the heated surface that have characteristic sizes that are smaller than the flow channel cross-sectional dimensions. Bubbles nucleate on the heated wall crevices in such minichannels, leading to the onset of nucleate boiling, and further downstream the bubbles are released into the bulk flow and lead to the development of a two-phase flow field. The confinement resulting from the small size of the channel, however, can affect the bubble dynamics.

Cornwell and Kew (1992) have concluded that spatial confinement becomes important when

$$N_{con} = \sqrt{\sigma/g\Delta\rho}/D_H > 0.5, \qquad (14.1)$$

where N_{con} is the confinement number and is related to the Bond number $Bd = D_H^2/(\sigma/g\Delta\rho)$ according to $N_{con} = 1/\sqrt{Bd}$. This is evidently equivalent to $Bd < 1.4$. Cornwell and Kew have proposed that it is sufficient to define only three major flow regimes for flow boiling in minichannels. These flow regimes are shown schematically in Fig. 14.3. *Isolated bubble flow* is similar to bubbly flow in large conventional channels and is characterized by individual bubbles typically significantly smaller in size than the channel's smaller lateral dimension. *Confined bubble flow* is characterized by bubbles that span the entire smaller lateral dimension of the channel and are separated from the channel walls by thin evaporating liquid films. These confined bubbles can result from the growth of isolated bubbles or their coalescence. The confined bubble flow pattern thus resembles plug flow (Kandlikar, 2002). *Annular-slug flow* represents all the flow patterns that may occur when confined bubble flow is terminated. Confined bubble flow ends when the bubbles grow significantly in the axial direction and eventually leads to the collapse and dispersal of some

Author	Working fluid	Channel specifications	Heat flux (kW/m²)	Mass flux (kg/m²·s)	Vapor quality	Comments
Lazarek and Black (1982)	R-113	Vertical, circular stainless steel, D = 3.15 mm	14–380	125–750	−0.2 to 0.6	Single channel
Wambsganss et al. (1993)	R-113	Horizontal, circular, stainless steel, D = 2.92 mm	8.8–90.75	50–300	0.0 to 0.9	Single channel
Tran et al. (1993, 2000)	R-12, R-113, R-134a	Horizontal circular (D = 2.46, 2.92 mm) and rectangular (1.7 × 4.06 mm), stainless steel or brass	3.6–129	33–832	Up to 0.95	Single channel
Cornwell and Kew (1992)	R-113	Vertical rectangular (1.2 × 0.9, 3.25 × 1.1 mm²)	3–20	117–627		Multichannel
Kew and Cornwell (1997)	R-141b	Horizontal circular (D = 1.39–3.69 mm), stainless steel	9.7–90	188–1480	−0.05 to 0.9	Single channel
Yan and Lin (1998)	R-134a	Horizontal circular (D = 2 mm)	5–20	50–200	0.1 to 0.9	28 parallel channels
Oh et al. (1998)	R-134a	Horizontal circular (D = 0.75, 1, 2 mm), copper	10, 15, 20	240–720	0.1 to 1.0	Single channel
Bao et al. (2000)	R-11, R-123	Horizontal circular (D = 1.95 mm), copper	5–200	50–1800	−0.3 to 0.9	Single channel
Kuwahara et al. (2000)	R-134a	Horizontal circular (D = 0.84, 2 mm), stainless steel	1.16–46.8	100–600	0.01 to 0.84	Single channel
Lin et al. (2001a)	R-141b	Vertical circular (D = 1.0 mm)	10–1150	300–2000	−0.2 to 0.99	Single channel
Lin et al. (2001b)	R-141b	Vertical circular (D = 1.1–3.6 mm) and rectangular (2 × 2 mm)	1–300	50–3500	0 to 1.0	Single channel
Lee and Lee (2001b)	R-113	Horizontal rectangular (20 × 0.4, 20 × 1, 20 × 2 mm²), stainless steel	3–15	50–200	0.15 to 0.75	Single channel
Warrier et al. (2002)	FC-84	Horizontal rectangular (D_{H} = 0.75 mm), aluminum	0–59.9	557–1600	0.03 to 0.55	5 parallel channels
Yu et al. (2002)	Water	Horizontal circular (D = 2.98 mm), stainless steel	50–300	50–200	0 to 1.0	Single channel
Qu and Mudawar (2003b)	Water	Horizontal rectangular (0.231 × 0.713 mm²), copper	220–1300	135–402	0.01 to 0.17	21 parallel channels
Sumith et al. (2003)	Water	Vertical circular (D = 1.45 mm) stainless steel	10–715	23.4–152.7	0 to 0.8	Single channel
Yen et al. (2003)	HCFC-123, FC-72	Horizontal circular (D = 0.19–0.51 mm), stainless steel	1–13	50–300	0.01 to 0.27	Single channel
Pettersen (2004)	CO_2	Horizontal circular (D = 0.81 mm), aluminum	5–20	190–570	0.2 to 0.8	25 parallel channels
Saitoh et al. (2005)	R-134a	Horizontal circular (D = 0.51, 1.12, 3.1 mm), stainless steel	5–39	150–450	0.2 to 1.0	Single channel
Agostino and Bontemps (2005)	R-134a	Vertical rectangular (3.28 × 1.47 mm²), aluminum	6–31.6	90–295	Not given	11 parallel channels

(cont.)

Table 14.1. (cont.)

Author	Working fluid	Channel specifications	Heat flux (kW/m²)	Mass flux (kg/m²·s)	Vapor quality	Comments
Grohmann (2005)	Argon	Horizontal circular (D = 250, 500 μm)	Not given	Not given	Not given	Single channel
Lie et al. (2006)	R-134a, R-407C	Horizontal circular (D = 0.83, 2.0 mm), copper	5–15	200–1500	0.2 to 0.8 (inlet quality)	28 parallel channels
Yun et al. (2006)	R-410A	Horizontal rectangular (1.79 × 1.2 and 1.57 × 1.2 mm²)	10–20	200–400	0.05 to 0.85	7 and 8 parallel channels, respectively
Tran et al. (1996)*	R-12	Horizontal, circular brass, D = 2.46 mm	7–59	89–300	0.2–0.8	Single channel
Agostini et al. (2003)*	R-134a	Vertical, rectangular aluminum, D_H = 0.72, 2.01 mm	4.4–15	83–467	0.0–1.0	7 or 18 parallel channels
Chen and Garimella (2006)*	FC-77	Vertical, rectangular copper or silicon, D_H = 0.84, 0.39 mm	20–800	63.5–440	0.01–0.99	10 or 24 parallel channels
Qi et al. (2007)	N₂	Vertical, circular stainless steel, D = 0.53–1.93 mm	50–210	440–3000	0.0–0.90	Single channel
Agostini et al. (2008a, b),*	R-236fa, R-245fa	Horizontal, rectangular D_H = 0.34 mm	7–420	281–501	0.02–0.78	67 parallel channels
Bertsch et al. (2008, 2009a)*	R-134a, R-245fa,	Horizontal, rectangular copper, D_H = 0.54, 1.09 mm	50–220	20–350	0.0–0.95	17 or 33 parallel channels
Harirchian and Garimella (2008)*	FC-77	Horizontal, rectangular silicon, D_H = 0.16–0.57 mm	0–300	250–1600	0.0–1.0	2 to 60 parallel channels
Ong and Thome (2011b)	R-134a, R-236fa, R-245fa	Horizontal, circular stainless steel, D = 1.03, 2.20, 3.04 mm	9.6–221.6	100–1290	0.0–1.0	Single channel
Lee et al. (2011)	Water	Horizontal, square or rectangular, 50–100 μm wide, silicon bed and Pyrex cover		≤ 139.8		Single channel, focused on ONB
Vakili-Farahani et al. (2012, 2013)	R-245fa, R-1234ze(E), R-134a	Vertical, rectangular aluminum, D_H = 1.4 mm	3–107	50–400	0.0–1.0	7 parallel channels
Costa-Patry et al. (2011a)[a]	R-245fa and R-236fa	Horizontal, rectangular silicon (85 μm wide × 560 μm high)				135 parallel channels
Costa-Patry et al. (2011b)[a]	R-245fa, R-1234ze(E), and R-134a	Horizontal, rectangular copper, (163 μm wide × 1560 μm high)				52 parallel channels
Mosyak et al. (2012)	Water	Horizontal, rectangular aluminum, D_H = 0.297 mm		15.5–77.1		25 parallel microchannels, focused on ONB

* Data utilized by Bertsch et al. (2009b) for the validation of their boiling heat transfer correlation.

Table 14.2. *Summary of some investigations dealing with CHF in small channels.*

Source	Channel characteristics	Fluid	Pressure (MPa)	Mass flux (kg/m²·s)	Inlet conditions	Critical heat flux (MW/m²)
Ornatskiy (1960)[a]	$D = 0.5$ mm, $L = 14$ cm, vertical	Water	1.0–3.2	20 000–90 000	$T_{in} = 1.5$–154 °C	41.9–224.5
Ornatskiy and Kichigan (1962)[b]	$D = 2$ mm, $L = 56$ mm, vertical	Water	1.0–2.5	5000–30 000	$T_{in} = 2.7$–204.5 °C	6.4–64.6
Ornatskiy and Vinyarskiy (1964)[b]	$D = 0.4$–2.0 mm, $L = 11.2$–56 mm, vertical	Water	1.1–3.2	10 000–90 000	$T_{in} = 6.7$–155.6 °C	279–2279
Lowdermilk et al. (1958)	$D = 1.30$ mm, $L = 65$ cm	Water				
Weatherhead (1963)	$D = 1.14$ mm, $L = 114$ mm	Water				
Lezzi et al. (1994)	$D = 1$ mm, $L = 239$ mm	Water				
Loomsmore and Skinner (1965)[a]	$D = 0.6$–2.4 mm, $L = 6.3$–150 mm, vertical	Water	0.1–0.7	3000–25 000	$T_{in} = 3.2$–130.9 °C	6.7–44.8
Daleas and Bergles (1965)[b]	$D = 1.2$–2.4 mm, $L/D = 14.9$–26, vertical	Water	0.2	1520–3000		0.31–3.1
Subbotin et al. (1982)	$D = 1.63$ mm, $L = 180$ mm, vertical	Helium	0.1–0.2	80–320	$x_{in} \geq -0.25$	
Katto and Yokoya (1984)	$D = 1$ mm, $L/D = 25$–200, vertical	Liquid He	0.199	11–108	$h_f - h_{in} =$ -3.5–$+7.0$ kJ/kg	
Boyd (1988)	$D = 3$ mm, $L/D = 96.9$, horizontal	Water	0.77 at exit	4600–40 600	$T_{in} = 20$ °C	6.25–41.58
Nariai et al. (1987, 1989)	$D = 1, 2, 3$ mm; $L = 1.0$–100 mm, vertical	Water	0.1	6700–20 900	$T_{in} = 15.464$ °C	4.6–70
Oh and Englert (1993)	Rectangular, $\delta = 1.98$ mm, vertical	Water	0.02–0.085	30–80	$\Delta T_{sub,in} = 5$–72 °C	
Inasaka and Nariai (1993)	$D = 3$ mm, $L = 100$ mm, vertical	Water	0.3–1.1	4300–30 000	$T_{in} = 25$–78 °C	7.3–44.5
Hosaka et al. (1990)	$D = 0.5, 1, 3$ mm, $L/D = 50$, vertical	R-113	1.1–2.4	9300–32 000	$\Delta T_{sub,in} = 50$–80 °C	
Celata et al. (1993a)	$D = 2.5$ mm, $L = 100$ mm, vertical	Water	0.6–2.6	10 100–40 000	$T_{in} = 29.8$–70.5 °C	12.1–60.6
Vandervort et al. (1992, 1994)	$D = 0.3$–2.6 mm, $L = 2.5$–66 mm, vertical	Water	0.1–2.3	8400–42 700	$T_{in} = 6.4$–84.9 °C	18.7–123.8
Bowers and Mudawar (1994)	17 parallel, $D = 0.51$; 3 parallel, $D = 2.54$ mm, $L = 10$ mm; all horizontal	R-113	0.138 at inlet	31–150 for $D = 2.54$ mm; 120–480 for $D = 0.5$ mm	$\Delta T_{sub,in} = 10$–32 °C	
Roach et al. (1999a, b)	$D = 1.17, 1.45$ mm, circular; $D_H = 1.13$ mm, semitriangular; $L = 160$ mm; horizontal	Water	0.344–1.043 at exit	250–1000	$T_{in} = 49$–72.5 °C	0.86–3.7
Jiang et al. (1999)	10–58 parallel diamond-shaped channels with $D_H = 40, 80$ μm, horizontal	Water	–	–	–	–

(cont.)

Table 14.2. (cont.)

Source	Channel characteristics	Fluid	Pressure (MPa)	Mass flux (kg/m²·s)	Inlet conditions	Critical heat flux (MW/m²)
Qu and Mudawar (2004b)	21 parallel channels with 215 × 821 μm² cross section; horizontal, L = 44.7 mm	Water	0.113 at exit	86–368	T_{in} = 30 °C, 60 °C	0.269–0.542
Wojtan et al. (2006)	Circular, D = 0.509, 0.79 mm, L = 20–70.5 mm single channels	R-134a, R-245fa		400–1600	2–15 °C inlet subcooling	3.2–600
Kosar and Peles (2007)[c]	Horizontal, rectangular (200 μm × 264 μm) silicon, L/D_H = 44	R-123	0.227–0.520	292–1117		0.28–1.06
Agostini et al. (2008b)	Rectangular (223 μm × 680 μm), silicon, 67 parallel channels, L = 20 mm	R-236fa		276–992	0.4–15.3 °C inlet subcooling	219–522
Martin-Callizo et al. (2008)[c]	Circular, stainless steel, D = 0.64 mm, L/D = 481	R22, R-134a, R-245fa	0.112–0.114	180–535		0.025–0.069
Kosar et al. (2009a)[c]	Horizontal, circular, stainless steel, D = 0.127–0.254 mm, L/D = 88–352	Water	0.42–1.55	1.2×10^4–5.3×10^4		‑5.5–41.0
Kuan and Kandlikar (2008)[c]	Horizontal, rectangular copper, (1.05 mm × 0.157 μm, L/D = 220	Water, R-123	0.101 for water, 0.23–0.27 for R-123	50.4–231.7 for water, 410–534 for R-123		6 parallel channels
Roday and Jensen (2009)	Circular, D = 0.286, 0.427, 0.70 mm, L_{heat} = 21.66–90.84 mm, single channels	Water, R-123	0.025–0.225	320–1570	2–80 °C inlet subcooling	
Mauro et al. (2010)	Rectangular (199 μm × 756 μm), copper, 29 67 parallel channels, L = 30 mm	R-134a, R-236fa, R-245fa		250–1500	5–25 °C inlet subcooling	100–325
Park and Thome (2010)	Rectangular (1) Rectangular, copper (467 μm × 4052 μm), 20 parallel channels, L = 30 mm, L_{heat} = 20 mm (2) Rectangular, copper (199 μm × 756 μm), 29 parallel channels, L = 30 mm, L_{heat} = 20 mm	R-134a, R-236fa, R-245fa		100–4000	3–20	370–3420
Chen and Garimella (2012)	Rectangular (100 μm × 389 μm), silicon, 60 parallel channels, L = 12.7 mm	FC-77		253.7–1015	T_{in} = 71 °C (T_{sat} = 97 °C)	30 ~ 110

[a] From Celata et al. (1994a).
[b] From Boyd (1985a, b).
[c] Based on Kandlikar (2010).

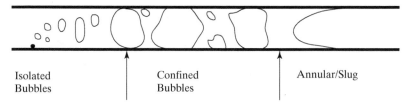

Isolated
Bubbles

Confined
Bubbles

Annular/Slug

Figure 14.3. The major flow boiling regimes in small passages according to Cornwell and Kew (1992). (After Kandlikar, 2002.)

Figure 14.4. Steady flow boiling regimes in a 0.5 mm × 4 mm minichannel. (From Brutin *et al.*, 2003.)

of the liquid slugs that separate neighboring bubbles. The resulting regime is predominantly annular-dispersed, with random irregular liquid slugs interspersed in the vapor.

Several other criteria that can be used to determine whether a channel can be considered a microchannel (i.e., when models and correlations representing commonly applied large flow channels do not apply) were discussed earlier in Section 3.7.2. Included among these criteria is the following expression (Harirchian and Garimella, 2010) (see Figure 3.4):

$$\text{Bd}^{0.5}\,\text{Re}_{\text{L0}} = \frac{1}{\mu_{\text{L}}}\left[\frac{g(\rho_{\text{L}} - \rho_{\text{G}})}{\sigma}\right]^{0.5} GD_{\text{H}}^2 = 160. \tag{3.82}$$

Figure 14.4 shows the steady boiling flow regimes in a rectangular minichannel with 0.5 mm × 4 mm cross section, with *n*-pentane as the coolant (Brutin *et al.*, 2003). The displayed picture represents a mass flux of $G = 240$ kg/m^2·s. As noted, the sequence of the two-phase flow patterns is generally consistent with the aforementioned discussion. Brutin *et al.* noted that, when confined bubbles dominated the flow field, small bubbles were also present. The smaller bubbles occurred near the channel sides, whereas the confined bubbles occupied the middle. The smaller bubbles, furthermore, moved considerably faster than the large bubbles.

The flow patterns just described occur for moderate heat and mass fluxes. Vandervort *et al.* (1992) investigated the heat transfer to highly subcooled water at very high mass fluxes (with up to $G = 42\,700$ kg·m^2·s) in tubes with 0.3–2.5 mm diameter. At a mass flux of $G = 5000$ kg/m^2·s, and with a wall heat flux that was about 70% of the heat flux that would lead to CHF near the exit of their heated tubes, the flow field at the exit of their test section was foggy, indicating the presence of a large number of microbubbles too small to be discernible individually. As the mass flux was increased, higher heat fluxes were required for fogging to occur.

There is a scarcity of information about boiling and the two-phase flow structures that may occur in microchannels subject to hard inlet conditions. It has been argued, however, that even in microchannels the onset of boiling must be caused by bubble nucleation on heated surface crevices and defects, as long as crevices and defects smaller in size than the channel lateral dimensions are present (Kendall *et al.*,

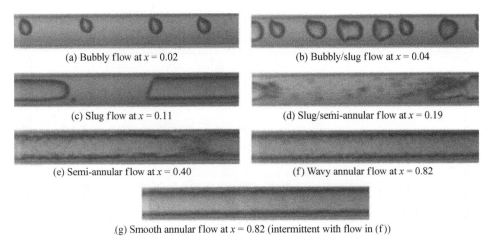

(a) Bubbly flow at $x = 0.02$ (b) Bubbly/slug flow at $x = 0.04$

(c) Slug flow at $x = 0.11$ (d) Slug/semi-annular flow at $x = 0.19$

(e) Semi-annular flow at $x = 0.40$ (f) Wavy annular flow at $x = 0.82$

(g) Smooth annular flow at $x = 0.82$ (intermittent with flow in (f))

Figure 14.5. Flow patterns and transitions observed by Revellin *et al.* (2006) for R-134a at the exit of a 0.509 mm heated glass channel with a mass flux of 500 kg/m²·s and saturation temperature of 30 °C.

2001; Zhang *et al.*, 2005). The heterogeneous nucleation phenomenology associated with wall crevices is thus expected to break down only when the heated channel lateral dimensions are comparable with the characteristic size of wall defects. Since the current methods for microchannel construction typically lead to smooth channel surfaces with micron-size defects (Zhang *et al.*, 2005), the heterogeneous bubble nucleation phenomenology should apply to the entire microchannel size range. Bubble ebullition and growth, and the subsequent two-phase flow regime evolution, however, are likely to be significantly different in microchannels in comparison with minichannels.

Two-phase flow regimes in boiling flow channels have been studied extensively in recent years (Revellin *et al.*, 2006; Revellin and Thome, 2007a, b; Bertsch *et al.*, 2009a, b; Singh *et al.*, 2009; Harirchian and Garimella, 2010; Ong and Thome, 2011a, b; Galvis and Culham, 2012; Thome *et al.*, 2013). The flow regimes in boiling microchannels are affected by mass flux, quality and channel size, as well as the wall heat flux. In general, however, bubbly, plug and slug flow regimes occur at low heat flux.

Revellin *et al.* (2006) carried out experiments with a 0.509 mm glass tube with refrigerant R-134a, and Ong and Thome (2011a) performed experiments in heated circular channels with 1.03, 2.20, and 3.04 mm diameters, using R-134a, R-236fa, and R-245fa refrigerants as fluids. The major flow regimes they observed were bubbly, bubbly-slug (similar to plug), slug, slug-semiannular, wavy annular, and smooth annular. Figure 14.5 displays the aforementioned flow regimes photographed by Revellin *et al.* (2006). As can be noted, the flow regimes resemble the flow regimes displayed for adiabatic two-phase flow (see Fig. 10.1). They did not include the churn flow regime among their observed regimes. These flow regimes are representative of what other researchers have reported, with minor system-specific differences in some cases. Figure 14.6, for example, displays the flow regimes observed by Galvis and Culham (2012) in experiments with water in a rectangular microchannel. Their reported flow regimes include churn. The flow regimes observed by Harirchian and Garimella

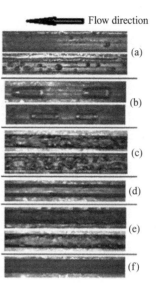

Figure 14.6. Computer enhanced flow patterns for boiling water flow in rectangular microchannels: (a) bubbly flow, (b) slug flow, (c) churn flow, (d) annular flow, (e) wavy annular (Galvis and Culham, 2012).

(2009), which were obtained with the dielectric fluid F-77 in rectangular channels with hydraulic diameters of 96–707 μm were similar to those shown in Fig. 14.6, and included the churn flow regime.

Based on their aforementioned experimental data, Ong and Thome (2011a, b) developed the following expressions for flow regime transitions in a boiling microchannel. According to Ong and Thome, regime transitions occur at critical flow qualities which themselves depend on mass flux and fluid properties.

Isolated bubbly (IB) to coalescing bubbly (B):

$$x_{IB/CB} = 0.36(N_{con})^{0.2}\left(\frac{\mu_g}{\mu_f}\right)^{0.65}\left(\frac{\rho_g}{\rho_f}\right)^{0.9}\mathrm{Re}_{g0}^{0.75}(\mathrm{Bo})^{0.25}\mathrm{We}_{f0}^{-0.91}. \qquad (14.2)$$

Coalescing bubbly (CB) to annular (A):

$$x_{CB/A} = 0.047(N_{con})^{0.05}\left(\frac{\mu_g}{\mu_f}\right)^{0.7}\left(\frac{\rho_g}{\rho_f}\right)^{0.60}\mathrm{Re}_{g0}^{0.8}\mathrm{We}_{f0}^{-0.91}. \qquad (14.3)$$

Slug/plug (S-P) to coalescing bubbly if $x_{S\text{-}P/CB} < x_{CB/A}$:

$$x_{S\text{-}P/CB} = 9.0(N_{con})^{0.20}\left(\frac{\rho_g}{\rho_f}\right)^{0.90}\mathrm{Re}_{f0}^{0.1}\mathrm{Fr}_{f0}^{-1.2}. \qquad (14.4)$$

Slug/plug to annular if

$$x_{S\text{-}P/CB} > x_{CB/A}: \qquad (14.5a)$$

Annular to dryout (DO):

$$x_{crit} = 1.748\left(\frac{\rho_g}{\rho_f}\right)^{0.073}\left(\frac{\rho_f^{0.24}\sigma^{0.24}}{G^{0.48}}\right)\frac{L_{heat}^{0.04}}{D_i^{0.28}}. \qquad (14.5b)$$

Equation (14.5b) has been developed based on experimental data of Wojtan et al. (2006) to complement the aforementioned flow regime model (Thome et al., 2013). In Eq. (14.5b) L_{heat} and D_i are the heated length and inner diameter of the flow passage,

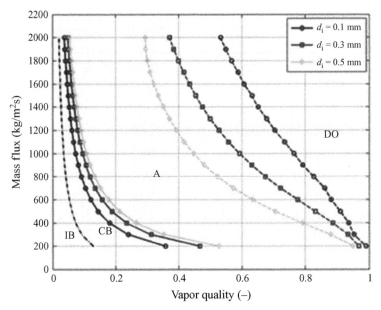

Figure 14.7. Flow regime transition lines for three tube inner diameters for R-134a at 30 °C saturation temperature ($q''_s = 60$ kW/m², $L_{\text{heat}} = 20$ cm, saturated liquid at inlet). (Thome *et al.*, 2013.)

respectively. Note that the boiling number, Bo $= \frac{q''_w}{Gh_{fg}}$, only shows up in Eq. (14.2) for bubbly to coalescing bubbly regime transition, and Eqs. (14.3) through (14.4) do not depend on the channel heat transfer conditions. The flow regime transition model may thus be applicable to adiabatic flow conditions as well. (Recall that dryout is rare in adiabatic flow.) The definitions and ranges of other dimensionless numbers used in the above expressions are:

$$0.27 < N_{\text{con}} = \frac{\sqrt{\sigma / [g(\rho_f - \rho_g)]}}{D_i} < 1.67$$

$$8.5 \times 10^3 < \text{Re}_{g0} = \frac{GD_i}{\mu_g} < 4.5 \times 10^5$$

$$1.57 \times 10^3 < \text{Re}_{f0} = \frac{GD_i}{\mu_f} < 3.1 \times 10^4$$

$$0.58 < \text{We}_{f0} = \frac{G^2 D_i}{\sigma \rho_f} < 1247$$

$$0.24 < \text{Fr}_{f0} = \frac{G^2}{\rho_f^2 g D_i} < 79.7.$$

Figure 14.7 displays a typical flow regime map based on the aforementioned regime transitions (Thome *et al.*, 2013).

A flow regime map based on dimensionless parameters has also been proposed by Harirchian and Garimella (2010), primarily based on their own flow boiling data with the perfluorinated dielectric fluid FC-77, representing rectangular cross-section heated channels that had hydraulic diameters in the 96–707 µm range. They derived a criterion for the macro–micro size threshold based on the confinement of bubbles

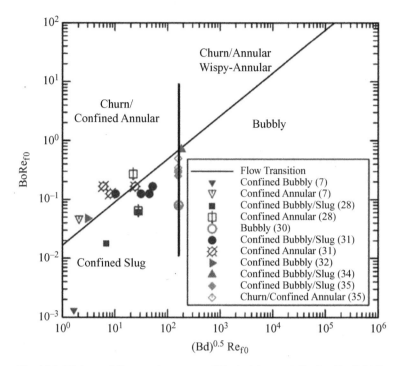

Fig. 14.8. Universal flow regime map of Harirchian and Garimella (2010).

that are generated and grow during nucleate boiling. Their macro-micro size criterion was presented in Section 3.7.2 (see Eq. (3.82)). Their flow regime map is depicted in Fig. 14.8, where its predictions are confirmed by comparison with the experimental data of the authors themselves which were obtained with FC-77 fluid, as well as data representing water, deionized water, R-134a, Vertel XF (a fluorocarbon solvent), and FC-72. The vertical transition line in the flow regime map represents the macro-micro channel size threshold, and represents Eq. (3.82). Mathematical expressions for the other flow regime transition boundaries are as follows.

Transition from confined slug to the alternating regime of churn/confined annular, or from bubbly to alternating churn/annular or wispy-annular can be represented by:

$$Bo\,Re_{f0} = 0.017(Bd)^{0.4}\,Re_{f0}^{0.7} \tag{14.6}$$

Alternatively, this equation can be represented as,

$$Bo = 0.017(Bd)^{0.4}\,Re_{f0}^{-0.3}. \tag{14.7}$$

According to this flow regime map, for $Bd^{0.5}Re_{f0} < 160$ vapor confinement occurs in slug and churn annular flow regimes, while there is no vapor confinement when $Bd^{0.5}Re_{f0} > 160$. For heat fluxes that are low enough to give $Bo < 0.017(Bd)^{0.4}Re_{f0}^{-0.3}$ the flow pattern will be slug when $Bd^{0.5}Re_{f0} < 160$, and it will be bubbly if $Bd^{0.5}Re_{f0} > 160$. For the latter case, when the heat flux is high enough so that $Bo > 0.017(Bd)^{0.4}Re_{f0}^{-0.3}$ vapor bubbles coalesce and lead to alternating churn/annular or churn/wispy annular flow patterns.

14.2.2 Flow Phenomena in Arrays of Parallel Channels

Figure 14.2 is a schematic of a minichannel cooling system in which parallel channels with common inlet and exit mixing volumes are used. Experimental investigations using this type of parallel channel-system have been reported by Peng and Wang (1994), Hetsroni *et al.* (2003), Qu and Mudawar (2002, 2003a, b), Kandlikar and Balasubramaman (2005), Wu and Cheng (2003, 2004), Kosar *et al.* (2006), Jiang *et al.* (1999, 2001), Zhang *et al.* (2005), and Tian *et al.* (2005), among others.

Two different types of flow instability can occur in these systems. The first, and by far the more disruptive, is the severe pressure-drop instability (Qu and Mudawar, 2002, 2004a), also referred to as the upstream compressible flow instability (Kosar *et al.*, 2006), or periodic boiling (Wu and Cheng, 2003, 2004). The second type of instability, referred to as the parallel-channel instability (Qu and Mudawar, 2002, 2004a), is often relatively mild and leads to channel-to-channel flow oscillations.

Severe Pressure-Drop Oscillations

This type of instability results from the coupling between the system of parallel channels and a compressible volume upstream from the inlet plenum or manifold. The compressible volume can be an entrained bubble, a flexible hose, or a large container of liquid in a slightly flexible container. The phenomenological sequence of events in an oscillation cycle is as follows. An oscillation cycle starts when liquid coolant flows into the channels, whereby boiling and vapor generation cause the flow resistance in the channel to increase. This leads to a reduction in the mass flow rate. The reduction in the mass flow rate in turn leads to a reduction in the pressure drop, and consequently the liquid flow into the channel is re-established, sometimes leading to the expulsion of vapor from the heated channels into their outlet header or plenum. A severe oscillatory flow is thus established that can become self-sustained under some circumstances. The flow field in each channel includes a liquid single-phase flow zone and a two-phase mixture zone, and the boundary between the two flow regime zones oscillates in relative unison in all of the channels (see Fig. 14.9(a)). Severe pressure-drop oscillations are not limited to systems with multiple channels. They can occur in a single-channel system as well, as long as the boiling channel is attached to a large compressible volume at its upstream end (Brutin *et al.*, 2003). When severe oscillations take place, unstable slug and annular-dispersed flows with a significant entrained droplet component appear to be the predominant flow patterns in the boiling two-phase zone (Jiang *et al.*, 2001; Qu and Mudawar, 2004a).

The oscillations are sometimes severe enough that they lead to the periodic injection of vapor into the inlet plenum or header. Near-stagnation conditions can occur in the heated channels during a portion of each cycle, and the parts of the heated channels that are covered with either vapor or a thin liquid film become susceptible to premature CHF.

Wu and Cheng (2003, 2004) studied the periodic boiling instability in detail, using parallel trapezoidal flow channels with 82.8- and 158.8-μm hydraulic diameters, with water as the working fluid. The amplitude and frequency of the oscillations and the sequence of flow patterns in the heated channels all depended strongly on heat and mass fluxes. At low heat flux and high mass flux, the flow patterns were liquid–two-phase alternating flow, and the behavior of the flow field was similar to that described

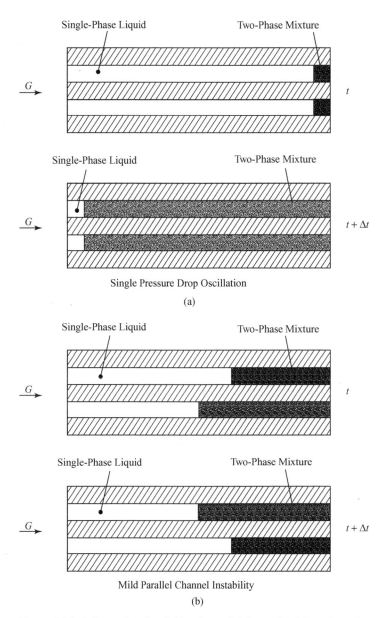

Figure 14.9. Schematic of neighboring minichannels: (a) undergoing severe pressure drop oscillations and (b) during parallel channel oscillations.

in the previous paragraph. At high heat flux and low mass flux, however, a liquid–two-phase–vapor alternating flow was observed. In each period the flow field in a channel varied from a liquid single-phase flow into the outlet manifold, to a boiling two-phase flow, to a pure vapor flow into the inlet manifold.

Severe pressure oscillations can evidently cause serious problems, including system vibration, and are therefore unlikely to be tolerated in practice. Furthermore, it is relatively simple to suppress them. They can be eliminated if the two-way coupling between the parallel-channel system and its upstream flow delivery system is broken.

A practical and simple way to mitigate them effectively is to install a throttle valve upstream of the inlet mixing volume for the microchannels and maintain the pressure upstream of the valve considerably higher than the pressure in the inlet mixing volume. The coupling between the two sides of the valve will then become one way, and pressure fluctuation in the parallel channels will have little effect (or no effect in the ideal case) on the mass flow rates (Qu and Mudawar, 2004a).

Parallel-Channel Instability

This is a flow excursion instability that occurs in parallel-channel systems when severe pressure oscillations are absent and leads to channel-to-channel oscillations. These oscillations take place when, for a given total pressure drop, the solution to the coupled quasi-steady thermal-hydraulics equations for a single channel is not unique. Consequently, the parallel channels can sustain different flow patterns, and that in turn leads to flow maldistribution among the channels. Neighboring heated channels that are seemingly identical with respect to their boundary conditions support oscillating and different flow regimes. The flow regimes are thus time-dependent and vary from channel to channel. Annular–vapor–two-phase–liquid and purely liquid single-phase flows can all take place simultaneously in different channels. The boundary between single-phase liquid and boiling two-phase flow regions oscillates in the heated channels, and the oscillations are out of phase among the channels.

Parallel-channel flow oscillations, like other types of instability and oscillations, are deleterious to the performance of boiling systems. A practical way for suppressing these oscillations is to increase the hydraulic flow resistance at the inlet to each channel and this can be done by installing an orifice at the entrance of every channel. The orifices cause a large pressure drop, and as a result the inlet plenum or header remains at a significantly higher pressure than that at the individual channels. With sufficiently large hydraulic resistance at the entrance to the channels, reverse flow from the channels into the inlet volume will no longer take place. Kandlikar (2005) and Kosar et al. (2006) have demonstrated that the parallel-channel oscillation can be suppressed in minichannel arrays effectively by the installation of proper orifices. This technique has been successfully applied to parallel microchannels as well. Szczukiewicz et al. (2013a, b) performed two-phase flow boiling experiments with a silicon micro-evaporator comprising 67 parallel channels, each 100 μm × 100 μm in cross section, using R245fa, R-236fa, and R-1234ze as working fluids. The parallel channels were connected to plena at their inlet and outlet, and heating was provided on one side of the micro-evaporator using serpentine micro-heaters. The temperature distribution in the micro-evaporator was recorded during experiments using an infrared camera. Without any flow restrictions at the inlet to the parallel channels the experiments indicated strongly nonuniform time-average heat transfer and flow distributions, as expected. However, by installing 100 μm long, 50 μm ×100 μm cross-section orifices at inlets to all channels the nonuniformity in the temperature of the micro-evaporator essentially disappeared. Such flow restrictions are particularly effective when the liquid flashes downstream from the flow restriction. Flashing occurs when the liquid upon depressurization finds itself superheated. As a result of flashing bubbles are generated and increase the mixture velocity. They may even cause choking (see Section 17.9), which will effectively prevent hydrodynamic signals from traveling upstream and will therefore decouple the flow channels from the flow

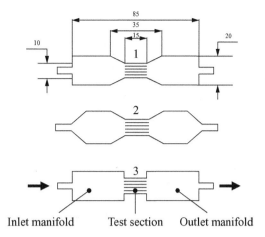

Figure 14.10. The manifold types investigated by Mosyak *et al.* (2012). (The depicted dimensions are in mm.) In the absence of boiling manifold type 1 resulted in uniform distribution of liquid among the microchannels.

field upstream of the flow restriction. Parallel channel oscillations in small channels have been studied by Hetsroni *et al.* (2003, 2005, 2006) and Qu and Mudawar (2004b). These investigations show that parallel-channel flow oscillations are relatively mild at low and moderate heat fluxes, but they can become severe as CHF conditions are approached. Qu and Mudawar (2004b) studied CHF in a test section consisting of 21 parallel channels, each 215 μm × 821 μm in cross section (see Table 14.2). A throttle valve was installed upstream of the inlet plenum to eliminate the severe pressure-drop oscillations. At near-CHF conditions the flow oscillations led to significant backflow of vapor into the inlet plenum. Similar observations were reported by Hetsroni *et al.* (2006), who studied various boiling phenomena in a test section consisting of 13, 21, and 26 parallel microchannels with identical triangular cross sections with hydraulic diameters of 220, 130, and 100 μm, respectively. Periodic and in-phase oscillations in pressure drop and exit manifold temperature could be observed, which became particularly intense as CHF was approached.

The absence of large cavities on the channel walls enhances the parallel-channel instability, by causing a delay in nucleation. This is particularly relevant to mini- and microchannels that have very smooth wall surfaces. The liquid that flows into a channel will need to reach a relatively high level of superheat before the activation and growth of bubbles on the existing wall cavities take place. With a high liquid superheat, the growth of the activated bubbles will be extremely fast, and that leads to the rapid flow blockage by the bubble and the axial growth of the bubble in both directions. Kandlikar *et al.* (2005) have studied the effect of artificial large wall cavities on the flow instability of parallel channels.

A recent investigation by Mosyak *et al.* (2012) has shown that the parallel channel oscillations can be also significantly mitigated by designing the inlet and outlet plena such that in the absence of boiling the liquid flow would be uniformly distributed among the parallel channels. Mosyak *et al.* investigated ONB in 25 parallel rectangular channels with 200 μm × 580 μm cross-sections. CFD simulations by Mosyak *et al.* showed that among the three types of manifold depicted in Fig. 14.10, the manifold type 1 provided the result and led to a uniform distribution of water flow among the channels. With boiling, however, periodic temperature oscillations that indicated flow oscillations occurred.

14.3 Onset of Nucleate Boiling and Onset of Significant Void

Because forced convection to subcooled liquid is the preferred heat transfer regime in many minichannel applications, phenomena that represent thresholds for the safe operation in the subcooled flow convection mode, including ONB, OSV, and CHF, have been investigated in the past.

14.3.1 ONB and OSV in Channels with Hard Inlet Conditions

Experimental data dealing with ONB and OSV in microchannels with hard inlet conditions are scarce. Experiments aimed at the measurement of conditions leading to either phenomenon in minichannels have been carried out by several authors, however, including Inasaka *et al.* (1989), Vandervort *et al.* (1992), Kennedy *et al.* (2000), Hapke *et al.* (2000), and Mosyak *et al.* (2012). The trends in these experimental data suggest that the basic phenomenology of ONB and OSV, as described in Chapter 12, should be similar to that in large channels. Some differences should be expected with respect to the details of bubble nucleation and ebullition between minichannels and large channels, however. The primary cause for these differences is that the temperature and velocity gradients in the vicinity of wall crevices are much larger in minichannels. As a result of extremely large temperature and velocity gradients, the relative magnitudes of forces that act on bubbles are different than the forces that act on bubbles in conventional flow channels. Some deviation from the correlations that are based on data in large channels should therefore be expected.

Inasaka *et al.* (1989) performed experiments in heated tubes with 1 mm and 3 mm diameters, under high-mass-flux conditions ($G = 7000$–$20\,000$ kg/m^2·s). Their working fluid was water. The correlation of Bergles and Rohsenow (1964) agreed with their data reasonably well. Vandervort *et al.* (1992) also studied subcooled boiling in heated tubes with $D = 0.3$–2.6 mm. The tubes were subject to high heat and mass fluxes ($G = 8400$–$42\,700$ kg/m^2·s). They have also reported that the correlation of Bergles and Rohsenow predicted their ONB data well. They also applied the model of Levy (1967) (see Section 12.6) for the estimation of the diameter of bubbles released at the OSV point and showed that the released bubbles should have been typically only a few micrometers in diameter in their tests. They then estimated the orders of magnitude of various forces that act on such microbubbles, showing that the thermocapillary (Marangoni) force and the lift force resulting from the ambient fluid velocity gradient are significant in mini- and microchannels. Kennedy *et al.* (2000) studied the ONB and OFI phenomena in water-cooled heated channels with 1.17 and 1.45 mm diameters. The mass flux in their experiments varied in the range $G = 800$–4500 kg/m^2·s. The correlation of Bergles and Rohsenow was found to be in agreement with the experimental data with respect to the major trends, but it underpredicted the heat flux that caused ONB at a given wall superheat. Based on the experimental data of Inasaka *et al.* (1989) and Kennedy *et al.* (2000), Ghiaasiaan and Chedester (2002) empirically modified the ONB correlation of Davis and Anderson (1966) (described in Section 12.4.2) to account for the effect of the thermocapillary force. The correlation of Davis and Anderson for hemispherical bubbles can be represented as (see Eq. (12.17)):

$$q''_{ONB} = \frac{k_f h_{fg}}{C(8\sigma T_{sat} v_{fg})} (T_w - T_{sat})^2_{ONB}, \tag{14.8}$$

where $C = 1$. The analysis leading to the correlation also gives Eq. (12.18) for R_C^*, the critical cavity mouth radius:

$$R_C^* = \left(\frac{2\sigma\, T_{sat} k_f}{h_{fg} q_{ONB}'' \rho_g} \right)^{1/2}. \tag{14.9}$$

Chedester and Ghiaasiaan correlated the parameter C in terms of σ_f and σ_w, which represent the values of the fluid surface tension at saturation and at the wall temperature. Hapke et al. (2000) conducted experiments with water in heated tubes with 1.5-mm inner diameter and observed similar trends to those of Kennedy et al. They also developed an empirical correlation. The correlations of Chedester and Ghiaasiaan and Hapke et al. are both valid over limited range of parameters, however. At this time, the well-established macroscale methods, such as those of Bergles and Rohsenow (1964) or Davis and Anderson (1966), appear to be adequate for estimation of ONB conditions.

EXAMPLE 14.1. Consider the flow of subcooled water at a pressure of 2 bars through a capillary tube. The tube is uniformly heated with a heat flux of $q_w'' = 3.0 \times 10^5$ W/m^2. The liquid mean velocity is $\overline{U}_L = 1.75$ m/s. Using the correlation of Davis and Anderson, calculate the wall and liquid mean temperatures where ONB occurs, for capillary diameters in the range 0.4–2 mm.

SOLUTION. The relevant saturation properties are $T_{sat} = 393.4$ K, $\rho_f = 943$ kg/m^3, $\rho_g = 1.128$ kg/m^3, $k_f = 0.669$ W/m·K, $h_{fg} = 2.202 \times 10^6$ J/kg, $\sigma_f = 0.0553$ N/m, and $C_{Pf} = 4,249$ J/kg·K.

Let us use 385.4 K (corresponding to about 8 K subcooling) as the temperature for calculating the bulk liquid properties. Then $\rho_L = 949.3$ kg/m^3, $k_L = 0.668$ W/m·K, $\mu_L = 2.494 \times 10^{-4}$ kg/m·s, and $Pr_L = 1.58$.

We will now go through the details for $D = 0.8$ mm. The calculations proceed as follows. For the Reynolds number, we get

$$Re_{L0} = \rho_L U_L D / \mu_L = 5328.$$

The heat transfer coefficient can be estimated by using the correlation of Dittus and Boelter,

$$H_{L0} D / k_L = 0.023\, Re_{L0}^{0.8} Pr_L^{0.4} \Rightarrow H_{L0} \approx 22\,100 \text{ W/m}^2\text{·K}.$$

Equation (14.8) with $C = 1$ can now be solved to obtain the wall temperature, and that leads to $T_{w,ONB} = 399$ K. The liquid bulk temperature can now be found from

$$\overline{T}_L = T_{w,ONB} - q_{w,ONB}'' / H_{L0} = 385.4 \text{ K}.$$

Similar calculations lead to the following table.

D (mm)	2	1.5	1.0	0.8	0.6	0.4
$\overline{T}_L$ (K)	382.7	383.6	384.8	385.4	386.2	387.2
T_w (K)	399	399	399	399	399	399

The foregoing discussion shows that for high flow rate, high heat flux, and highly subcooled coolants, microbubbles generated heterogeneously on the wall play the crucial role in the phenomenology of ONB in minichannels with hard inlet conditions. Experimental data in fact suggest that in this type of flow channel OSV is also controlled by heterogeneously generated microbubbles and the phenomenology of the occurrence of this important threshold is likely to be similar to that in large channels. There is little separation among the ONB, OSV, and DNB points for high-heat-flux conditions, however.

The OSV point is usually identified in experiments with large channels by measuring the void fraction profile along the channel and defining the OSV as the point downstream of which the slope of the void fraction profile is increased significantly (see Section 12.3). This technique is not practical in mini- and microchannels owing to the difficulty of accurately measuring the void fraction. However, as noted in Chapter 12, OSV typically occurs only slightly before the onset of the flow instability. The specification of conditions heading to OFI is rather straightforward, furthermore. Some researchers have measured the conditions leading to OFI in minichannels and have used them as estimation for conditions that lead to OSV (Inasaka *et al.*, 1989; Roach *et al.*, 1999a, b; Kennedy *et al.*, 2000).

For hydrodynamically controlled OSV conditions (i.e., when $Pe_L = GD_H C_{PL}/k_L \geq 70\,000$), the correlation of Levy (1967) underpredicted the heat flux that caused OSV in the experimental minichannel data of Inasaka *et al.* (1989) and Kennedy *et al.* (2000). The underprediction of the data was relatively slight, however (with a typical discrepancy of less than 40%). The correlation of Saha and Zuber (1974) also agreed with the hydrodynamically controlled OSV data of Inasaka *et al.* reasonably well. Chedester and Ghiaasiaan (2002) argued that the underprediction of the OSV heat flux in minichannels by bubble departure models such as Levy's (1967) can be due to the neglect of the thermocapillary effect in these models. They developed an analytical bubble-departure model that accounts for the effect of the thermocapillary force semi-empirically.

Experimental data dealing with low-flow OSV in minichannels with hard inlet conditions are scarce. Roach *et al.* (1999a, b) performed experiments with degassed water in circular and semitriangular flow passages representing the cross section of a subchannel defined by three parallel heated rods arranged in an equilateral triangular pitch, with $D_H = 1.13$–1.45 mm. Most of their OFI data occurred when the local equilibrium quality was positive, indicating that subcooled voidage was insignificant in the experiments. The current models and correlations for thermally controlled OSV thus appear to be inapplicable to minichannels. Roach *et al.* (1999a, b) also examined the effect of dissolved air in water on OFI. The effect was small.

14.3.2 Boiling Initiation and Evolution in Arrays of Parallel Mini- and Microchannels

Under low-flow conditions, the phenomenology of nucleate boiling initiation in parallel microchannels appears to be different than in large flow systems. Recall that in large systems the ONB phenomenon is associated with the appearance of bubbles on wall crevices, which must remain small and within a thin bubble layer next to the wall to avoid condensation. This leads to the concept of a wall voidage zone between the ONB and OSV points, as discussed in Section 12.3. Recent experimental data indicate that, in mini- and microchannels, the basic incipience process (i.e., the first appearance of microbubbles on crevices) is similar to that in large channels (Liu *et al.* 2005; Tian *et al.*, 2005; Zhang *et al.*, 2005). However, microbubbles do not remain small and wall-bound, and they do not condense as they grow large. Instead, they grow to become large in comparison with the channel cross section, and by the time they are detached from the heated wall they can be elongated vapor slugs (Qu and Mudawar, 2002; Hetsroni *et al.*, 2003).

In current established theories for ONB it is assumed that boiling incipience is caused by the growth of nucleation embryos to a size that is thermally and mechanically stable. These theories lead to the tangency models for ONB, described in Section 12.4. Liu *et al.* (2005) performed experiments using 25 parallel rectangular minichannels, each 275 µm wide and 363 µm deep, with deionized water. The minichannels were fabricated from a copper block, and had a fiberglass cover. The minichannels were connected to common inlet and outlet plena, and the experiments covered the $j_{L0} \approx 0.32$–0.92 m/s range. Their incipience data, defined as the first appearance of microbubbles detected by a microscope and high-speed camera, agreed with the predictions of a tangency model. Further evidence supporting the applicability of the heterogeneous nucleation theory to the ONB phenomenon in microchannels has been provided by Zhang *et al.* (2005), who performed experiments with deionized water in microchannels with hydraulic diameters in the 27–171 µm range. A surfactant was used for the reduction of surface tension. Tests were performed with rectangular, flat-wall silicon channels, where the wall features had a size range of 0.1–0.4 µm with a few defects in the 2–5 µm range. Experiments were also performed using microchannels with $D_H = 28$–72.5 µm, in which the walls were enhanced with 4–8 µm notches. The measured wall superheats at ONB for the tested channels agreed with the heterogeneous nucleation theory of Hsu (1962) (see Section 11.2.1).

Further evidence showing that conventional bubble nucleation theory is consistent with ONB phenomenology in small channels, at least under-low flow conditions, has been provided by Lee *et al.* (2011) and Mosyak *et al.* (2012). Lee *et al.* (2011) performed experiments with water using single square or rectangular cross-section microchannels with hydraulic diameters smaller than 100 µm. The mass flux was kept below 138.9 kg/m²·s. Bubble nucleation and growth on a single artificial cavity was investigated. The cavity radius was in the 12.5–16.4 µm range. Bubble inception on the cavity agreed with the conventional theory. The effect of confinement on bubble growth, expansion, and detachment could be clearly seen afterwards, however. Mosyak *et al.* (2012) investigated the conditions leading to ONB in the aforementioned (see previous section) 25 parallel rectangular channels, which had 200 µm × 580 µm cross sections, using low mass fluxes of subcooled water flow (at mass flux of 15.4–77.1 kg/m²). They measured the characteristic roughness of their channel walls. The surface RMS characteristic roughness (with respect to the mean line) was 8.7 µm. Analysis of their ONB data showed that Eq. (14.9) predicted $R_C^* \approx 12$–30 µm for their ONB experiments. The proximity of the critical bubble radius to their measured surface roughness indicated that the conventional bubble nucleation theory that is the basis of ONB models in large flow channels agrees well with their data.

For microchannels the experimental data of Lee *et al.* (2005) indicate that the critical nucleation size for ONB decreases with decreasing channel size when the channel spacing is small enough to be comparable with the critical bubble size. Lee *et al.* (2005) performed experiments in rectangular channels with channel heights in the 5–500 µm range, using methanol, ethanol, and water. Artificial nucleation sites with radii (defined by dividing the nucleation site base area by one-half of its perimeter) in the 2.5–370 µm range were installed. For methanol and ethanol the critical cavity size was a function of channel height for channel heights up to about 200 µm. For water the height effect was noticeable with channel widths up to 165 µm.

The thermofluid phenomenology following the initial incipience in parallel minichannels appears to be strongly influenced by flow fluctuations associated with parallel-channel instability. The appearance and growth of bubbles on one or more sites in a channel causes a partial flow blockage in that channel and leads to a reduction in the channel's flow rate. The reduction in the flow rate in turn leads to an increase in the liquid temperature in the vicinity of the bubble growth sites, causing a faster growth of the bubbles. The experimental observations of Hetsroni *et al.* (2003, 2005), conducted with deionized water in parallel triangular minichannels with $D_H = 103$–161 µm, show that, when a multitude of parallel channels is present in the system, boiling incipience first occurs only in some of them, with each of the active channels supporting only a few bubble-generation sites. Hetsroni *et al.* (2005) noted that under low flow conditions (95–340 kg/m²·s), with heat fluxes in the 80–330 kW/m² range, following ONB the generated bubble undergoes explosive growth and extends in both directions causing strong pressure fluctuations and even periodic dryout and rewetting. As a result of the explosive growth of bubbles flow reversal can take place in the microchannel(s) that undergo such bubble growth. The resulting flow regime has been termed explosive boiling by Hetsroni *et al.* (2005, 2006). According to Qu and Mudawar (2002), who performed experiments with deionized water in the range $j_{L0} \approx 0.13$–1.44 m/s using 21 parallel channels with 231 µm × 713 µm cross sections, on an active site a bubble remains attached to the wall until it grows to a size comparable with the channel hydraulic diameter. The bubble is then detached from the wall and is carried to the outlet plenum. The same few active sites for bubble growth continue releasing bubbles with a frequency that increases as the heat load to the system is increased. An essentially similar phenomenology was observed by Hetsroni *et al.* (2003), whose experiments were performed with water, with $j_{L0} = 0.046$ m/s. However, because of the extremely low liquid flow rate, the bubbles remained attached to the wall and grew into elongated bubbles with preferential growth in the axial direction, which eventually moved to the outlet plenum. The rapid growth of bubbles in the axial direction contributed to parallel-channel oscillations.

Qu and Mudawar (2002) have developed a bubble-detachment model based on their own experimental data. The model does not address the potential effects of parallel-channel fluctuations.

The foregoing discussion shows that for practical purposes there may not be a clear distinction between ONB and OSV in parallel mini- and microchannels with soft inlet conditions. Indeed, it can be argued that the ONB and OSV points coincide in these channels. Based on the limited available experimental data in parallel mini- and microchannel arrays, when severe pressure-drop oscillations are absent, one may conclude the following.

(1) ONB is caused by the growth and departure of bubbles generated on wall cavities.
(2) The thermofluid phenomenology associated with the evolution of boiling is different from that in large channels. The bubbles generated on wall cavities grow and become comparable to the channel hydraulic diameter before they are released.
(3) Parallel-channel oscillations play a crucial role and are likely to be at least partly responsible for the apparent differences with large channels with respect to the phenomenology of boiling initiation and evolution.

14.4 Boiling Heat Transfer

14.4.1 Background and Experimental Data

Table 14.1 is a summary of some of the widely acclaimed studies dealing with boiling in mini- and microchannels. More extensive tabulations of experimental studies can be found in Saha and Celata (2011), and Baldassari and Marengo (2013). Some of the more recent investigations were motivated by the need to substitute the ozone-depleting chlorofluorocarbon (CFC) (e.g., R-11) and hydrochlorofluorocarbon (HCFC) refrigerants (e.g., R-22) with hydrofluorocarbon (HFC) refrigerants. The latter group of refrigerants includes R-134a and R-407C and has essentially zero ozone-depletion potential (but are not completely benign, however, and have global warming potential).

The common practice in the literature is to include the experimental data representing channel hydraulic diameters in the $0.5 \lesssim D_H \lesssim 3\,mm$ range in the same (minichannel) category. Recall from Chapter 10 that, for circular and near-circular channel cross sections, $D_H \approx 0.3\lambda_L$ can be considered as an important threshold at least with respect to the effect of channel orientation on the two-phase flow hydrodynamics.

The available experimental data include single heated channels, as well as arrays of parallel minichannels connected to common inlet and outlet plena or headers. The significance of flow instability in the latter type of heated channel has already been discussed. It is likely that most of the multichannel experimental data include the effect of at least moderate parallel-channel oscillations, as mentioned by some authors (Qu and Mudawar, 2004a; Lie et al., 2006). The data of Yun et al. (2006) indicate that the flow oscillations associated with parallel-channel-type instability can increase the boiling heat transfer coefficients by about a factor of 2. It should also be emphasized that instability and flow oscillations may have been present in some single-channel data. Saitoh et al. (2005), for example, could detect flow oscillations with periods of 0.8–4 s during some experiments with their 1.12-mm-diameter test section. The oscillations appeared to be due to density waves. Yen et al. (2003) also detected pressure fluctuations that implied low-frequency flow oscillations with periods of 50–100 s.

The experimental data of Pettersen (2004) deal with two-phase flow and boiling of CO_2 (also referred to as R-744), which is attractive as a relatively benign refrigerant with respect to the environment. The critical pressure and temperature for CO_2 are $T_{cr} = 31\,°C$ and $P_{cr} = 7.39\,MPa$, respectively. Because of its low critical temperature and high critical pressure CO_2 must undergo phase change at high reduced pressures in common refrigeration systems. For example, for evaporation to take place at $0\,°C$, a reduced pressure of $P_r = P/P_{cr} = 0.47$ is needed. The operation at high P_r values leads to low surface tension, low liquid viscosity, and a large ρ_g/ρ_f ratio, and these imply that two-phase flow and boiling with CO_2 may be different than what is known about other commonly applied fluids.

Experimental data for boiling heat transfer in microchannels (i.e., channels with $D_H \lesssim 100\ \mu m$) are scarce. The experiments of Jiang et al. (1999, 2001), Zhang et al. (2005), and Lee et al. (2011) were discussed earlier with regard to bubble nucleation and boiling flow patterns. Jiang et al. (2001) have defined a boiling curve based on their device temperature and the q''_w/q''_{CHF} ratio. Zhang et al. (2005) have shown the crucial role of wall cavity size characteristics on the boiling mechanisms and the resulting two-phase flow regimes, and they have developed a boiling flow regime map using the cavity size and hydraulic diameter as coordinates. However, there is some uncertainty about these experiments, including the effect of flow oscillations, and even the effect of dissolved noncondensables (in the case of the investigation of Zhang et al.). The aforementioned experimental study of Lee et al. (2011) confirmed the applicability of the conventional bubble inception theory to ONB in microchannels.

14.4.2 Boiling Heat Transfer Mechanisms

The mechanism of flow boiling heat transfer in minichannels has been a subject of disagreement, owing to the sometimes contradictory trends in the various experimental data.

The parametric dependencies associated with flow boiling in large channels can be deduced from Figs. 12.10(a) and (b). It is noted from Fig. 12.10(a), for example, that in the partial boiling regime the heat transfer coefficient increases with mass flux G. In the fully developed nucleate boiling region, where nucleate boiling is the predominant heat transfer mechanism, the heat transfer coefficient H is insensitive to G and equilibrium quality x_{eq}, but it increases rather strongly with increasing wall heat flux q''_w. In the forced convective evaporation regime, as noted in Fig. 12.10(b), the heat transfer coefficient depends rather strongly on G and x_{eq}, in addition to its dependence on the heat flux.

The trends in the experimental data of some investigators are consistent with the predominance of nucleate boiling and indicate that H is a strong function of q''_w, being essentially independent of G and x_{eq}. The data of Bao et al. (2000), displayed in Fig. 14.11, are typical of these experiments. The experimental data of another group of investigators, in contrast, show that the heat transfer coefficient increases with increasing mass flux G and is sensitive to the equilibrium quality x_{eq}. These trends are consistent with the forced convective evaporation heat transfer mechanism in the annular-dispersed flow regime. The data of Lee and Lee (2001b), displayed in Fig. 14.12, are a good example.

These observations are both correct and simply refer to two different heat transfer regimes that can in fact simultaneously occur in different parts of the same heated channel. Thus, the nucleate-boiling-dominated mechanism represents the bubbly and confined bubble flow patterns in accordance with the flow regime definitions of Kew and Cornwell (1992), and more recently by Harirchian and Garimella (2010),

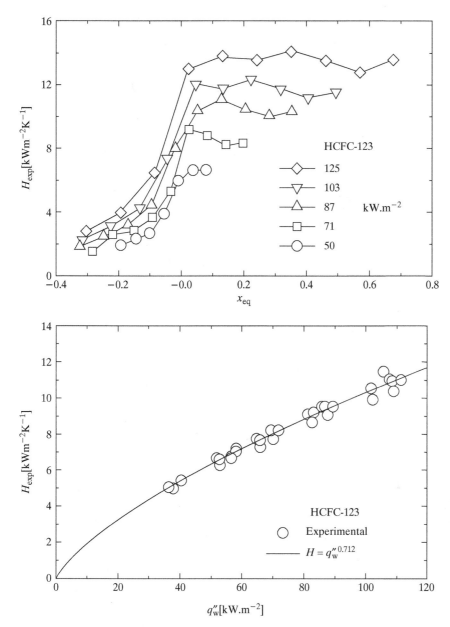

Figure 14.11. Typical parametric trends in the experiments of Bao *et al.* (2000). (HCFC-123, $G = 452$ kg/m^2·s, $P_{in} = 450$ kPa.)

whereas the convection-dominated heat transfer is associated with the slug-annular two-phase flow pattern. Experimental support for these observations can be found in Sumith *et al.* (2003), Lee and Mudawar (2005a, b), Agostini and Bontemps (2005), and Thome and co-workers (Szczukiewicz *et al.*, 2013a, b, 2014). The more recent investigations by Szczukiewicz *et al.* (2013a, b, 2014) have provided a clear picture of boiling flow regimes in minichannels. Figure 14.13 depicts the variation of the heat transfer coefficient as a function of quality in a 100 μm × 100 μm flow channel (Szczukiewicz *et al.*, 2014). The experiments were carried out using 67 parallel channels. Each channel was equipped with a flow restriction at its inlet, however,

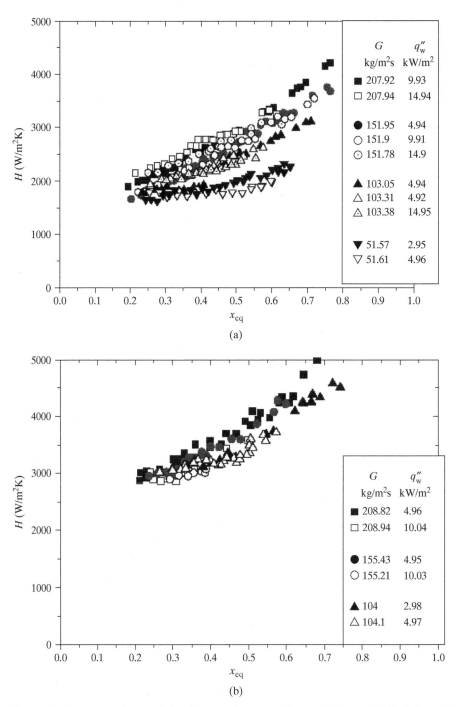

Figure 14.12. Parametric trends in the experiments of Lee and Lee (2001b): (a) $\delta = 1$ mm; (b) $\delta = 0.4$ mm.

which eliminated time-average temperature nonuniformities resulting from flow maldistribution among the channels. In the isolated and elongated bubble regime the heat transfer coefficient increases with quality. Once the flow regime is changed to coalesced bubbles (plug or slug flow regime) the trend reverses, and the heat

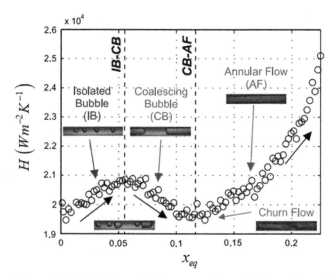

Figure 14.13. Variation of the heat transfer coefficient with quality in a 100 μm × 100 μm flow channel for flow boiling of R-236fa (mass flux = 2299 kg/m²·s, heat flux at channel base = 48:6 W/cm²) (Szczukiewicz *et al*., 2014).

transfer coefficient decreases with increasing quality. The phenomenological reason is that in this regime evaporation from the thin liquid film that separates the coalesced bubbles from the wall is the major contributor to heat transfer. Periodic local dryout resulting from the breakdown of the liquid film caused by evaporation, followed by rewetting, is common in this regime. The dryout process reduces the time-average heat transfer coefficient. With increasing quality, the frequency of dryout/rewetting process also increases, leading to the downward trend in the heat transfer coefficient-quality profile in the coalesced bubbly regime. The liquid film that separates a coalesced bubble from the channel walls is replenished by the liquid contained in the liquid slugs that separate a bubble from its adjacent bubbles. The coalesced bubbly regime eventually collapses once the liquid slugs are nearly completely depleted. A churn flow regime then develops, which coincides with a minimum that occurs in the heat transfer coefficient profile. The churn flow regime prevails over a small quality range, however, and is replaced with annular flow regime. The trend in the heat transfer coefficient–quality profile changes again. With increasing quality in the annular flow regime the liquid film thickness decreases and as a result the heat transfer coefficient also increases. Experimental data of Ong and Thome (2011b), obtained with R-134a in a copper test section, show that in the isolated or elongated bubble regime, as well as coalesced bubble regime, the heat transfer coefficient is strongly affected by heat flux, indicating that these regimes coincide with the aforementioned nucleate boiling-dominated boiling regime. The same data also show that in the annular flow regime the heat transfer coefficient is essentially independent of the heat flux.

The conditions leading to the heat transfer regime transition from nucleate-boiling dominated to convection dominated depends on a number of parameters, including the flow quality, channel size, and heat flux. This transition is important, since its correct prediction is crucial for the development of mechanistic or flow-regime-dependent boiling heat transfer models. The data of Lee and Mudawar

(2005b) and Sumith *et al.* (2003) show that the nucleate-boiling-dominated regime occurred for $x_{eq} \leq 0.05$. The data of Szczukiewicz *et al.* (2013a, b) also suggest that transition from the isolated or elongated bubble to the coalesced bubble regime occurs at $x_{eq} \approx 0.05$. The data of Agostini and Bontemps (2005) indicated that the transition occurs when $q''_w(1 - x_{eq})/Gh_{fg} = 2.2 \times 10^{-4}$. For the transition from coalesced bubbly regime to annular flow regime Costa-Patry and Thome (2013) have proposed

$$x_{eq} = 425 \left(\frac{\rho_g}{\rho_f} \right)^{0.1} \frac{Bo^{1.1}}{N_{con}{}^{0.5}} \tag{14.10}$$

where Bo and N_{con} represent the boiling number and confinement number, respectively.

14.4.3 Flow Boiling Correlations

In predictive methods for flow boiling, investigators generally seek agreement with data representing channel hydraulic diameters over the entire $0.1 \lesssim D_H \lesssim 3\,\text{mm}$ range. The data bases of widely used conventional flow boiling correlations often include experimental data that fall in this size range. The conventional flow boiling correlations described in Sections 12.11 and 12.13 can in fact predict the boiling heat transfer coefficients in minichannels within about an order of magnitude. In general, no single conventional correlation has been successful in predicting a large minichannel data set. A number of recent correlations that are based on wide ranges of experimental data will be discussed shortly. Good agreement between conventional correlations and specific minichannel data has been demonstrated in a few cases only, often after some minor modifications are introduced into the correlations. There are two major complications that cause disagreement between the conventional correlations and miniature flow passage experimental data:

(1) the occurrence of very low liquid Reynolds numbers in miniature flow passages, which often correspond to laminar flow when liquid-only convection heat transfer is considered, and
(2) the difference between boiling flow regimes in large channels and in miniature flow passages.

Some examples of agreement between conventional correlations and experimental data are as follows. The correlation of Chen (1966) (see Eqs. (12.92)–(12.100)) agreed with the experimental data of Bao *et al.* (2000) reasonably well, provided that H_{NB} was calculated by multiplying the prediction of the pool boiling correlation of Cooper (1984) (see Eq. (12.117)) by Chen's suppression factor S. Sumith *et al.* (2003) noted that the correlations of Chen (1966) and Klimenko (1988, 1990) agreed with their data reasonably well, at wall heat fluxes larger than about $100\,\text{kW/m}^2$. At low heat fluxes there was significant data scatter, however. The correlation of Kandlikar (1990, 1991), displayed in Eqs. (12.102)–(12.106), agreed with the experimental data of Yen *et al.* (2003) well, provided that only the nucleate boiling component of the correlation (the second term on the right side of Eq. (12.103) or (12.104)) was used and the term representing the evaporative forced convection (the first term on the right side of the latter equations) was completely left out.

Most of the investigators have noted that conventional correlations do not show satisfactory agreement with minichannel data, however, and a multitude of empirical correlations have been proposed for flow boiling in microchannels. The following general comments can be made about these correlations.

(1) Most of the correlations were originally developed based on a single data set only, often the data obtained by the authors themselves. When applied to the experimental data of other investigators, these correlations often perform poorly.
(2) Some of the correlations are based on experimental data representing either nucleate-boiling-dominated heat transfer or evaporative convection heat transfer only. This further limits their applicability.
(3) Some of the correlations are based on data obtained with arrays of parallel heated channels. These may thus suffer from the complications caused by parallel-channel flow oscillations.

Bertsch et al. (2008) have performed a critical review of the most widely referenced correlations dealing with boiling in small channels that had been published prior to 2008, which clarifies the surprising shortcomings of these correlations. Twelve correlations (items in Table 2 of Berstch et al., 2008) were examined against ten sets of experimental data. The correlations include three (Liu and Winterton, 1991, Eqs. (12.113)–(12.121); Cooper (pool boiling, 1984), Eq. (11.40), and Gorenflo (pool boiling, 1993), Eqs. (11.41)–(11.44)) that were originally not developed for small channels. The sources of the experimental data are specified in Table 14.1. The hydraulic diameters in the experimental data were in the 0.16 mm $< D_H < 2$ mm range, and the vapor quality range was from zero to one. The outcome of the review can be summarized as follows:

(a) Overall, the 12 correlations that were developed based on small-channel data did not perform any better than the aforementioned three correlations that were based on data with conventional channels.
(b) The correlation of Cooper (1984) for pool boiling (Eqs. (11.40)) provided the lowest overall deviation from the experimental data. The correlation of Gorenflo (1993) (Eqs. (11.41)–(11.44)), another pool boiling correlation, also performed better than most of the flow boiling correlations.
(c) Among the flow boiling correlations, the correlation of Haynes and Fletcher (2003) (see Eqs. (14.14)), Liu and Winterton (1991) (Eqs. (12.113)–(12.121), a correlation based on conventional flow boiling data), Thome et al. (the three-zone model, to be explained later), and Tran et al. (1993) (the forthcoming Eq. (14.13)) showed reasonable agreement with experimental data.

Single-channel experiments are often designed such that the flow channel is heated around its entire perimeter. That is rarely the case for multi-channel systems. Multi-channel applications, as well as experiments, in practice are often designed to be heated from one side. An example of such a system is depicted in Fig. 14.14 (Chen and Garimella, 2012). In their test section 60 parallel microchannels with rectangular cross sections were machined in a silicon die plate, and heating was applied to the bottom surface of the PCB plate to simulate cooling of a circuit board. The parallel channels are separated from each other by 100-μm-thick fins. For this type of flow passage the hydraulic diameter is evidently not the appropriate length scale for

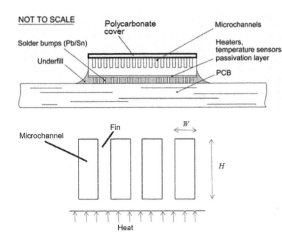

Figure 14.14. Cross-section and mounting details of the test section and the printed circuit board (PCB) of Chen and Garimella (2012). Each microchannel is 100 μm × 389 μm. The width of the fins between the microchannels is 100 μm.

heat transfer. The *heated equivalent diameter*, defined as $D_{\text{eff}} = 4A/p_{\text{heat}}$, where A is the flow area and p_{heat} is the heated perimeter, can be used in the definition of Nusselt number in these cases. For the geometry shown in Fig. 14.14, given that the fins that separate the microchannels are very thin and conductive, the heated equivalent diameter is often defined as (Qu and Mudawar, 2004b; Revellin *et al.*, 2009; Chen and Garimella, 2012)

$$A = WH$$

$$p_{\text{heat}} = W + 2H$$

$$D_{\text{eff}} = \frac{4WH}{W + 2H}. \tag{14.11}$$

The correlation of Lazarek and Black (1982) for nucleate-boiling-dominated heat transfer in saturated flow boiling is

$$H = 30\,\text{Re}_{\text{L0}}^{0.857}\,\text{Bo}^{0.714}(k_{\text{f}}/D_{\text{H}}), \tag{14.12}$$

where $\text{Bo} = q_{\text{w}}''/Gh_{\text{fg}}$ is the boiling number. The correlation is based on data obtained with R-113, covering the parameter range displayed in Table 14.1. This correlation predicts the R-113 data of Bao *et al.* (2000) reasonably well but deviates from their HCFC-123 data.

The correlation of Tran *et al.* (1993), whose data base is summarized in Table 14.1, also deals with nucleate-boiling-dominated heat transfer, in saturated flow boiling:

$$H = (8.4 \times 10^5)(\text{Bo}^2\,\text{We}_{\text{f0}})^{0.3}(\rho_{\text{f}}/\rho_{\text{g}})^{-0.4}, \tag{14.13}$$

where $\text{We}_{\text{f0}} = G^2 D_{\text{H}}/\rho_{\text{f}}\sigma$ and H is in watts-per-meter squared per kelvin. (Note that the correlation is not dimensionally balanced.)

The correlation of Haynes and Fletcher (2003) is a composite correlation and accounts for the contributions of both nucleate boiling and evaporative convection. It is based on about 2000 data points representing R-11 and HCFC-123 as working

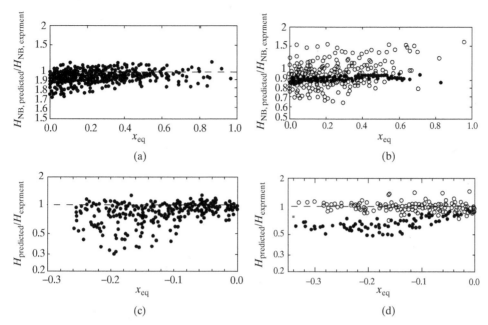

Figure 14.15. Comparison between the correlation of Haynes and Fletcher, Eq. (14.5), and experimental data (from Haynes and Fletcher, 2003). Open symbols represent $D = 0.92$ mm, and filled symbols represent $D = 1.95$ mm. (a) Saturated flow boiling of R-11; (b) saturated flow boiling of HCFC-123; (c) subcooled flow boiling of R-11; (d) subcooled flow boiling of HCFC-123.

fluids. The data base covers the following ranges:

$$D = 0.92, 1.95 \,\text{mm},$$
$$110 \leq G \leq 1,840 \,\text{kg/m}^2\text{·s},$$
$$11 \leq q_w'' \leq 170 \,\text{kW/m}^2.$$

The correlation of Haynes and Fletcher (2003) is

$$q_w'' = H_{\text{NB}}(T_w - T_{\text{sat}}) + H_{\text{L0}}(T_w - \overline{T}_\text{L}), \qquad (14.14)$$

where H_{NB} is to be calculated from the pool boiling correlation of Gorenflo (1993), Eqs. (11.41)–(11.44), and H_{L0} is to be found from appropriate single-phase forced-flow correlations. Haynes and Fletcher calculated H_{L0} from a fit to the analytical entrance-region solution for laminar flow and from the correlation of Gnielinski (1976), Eq. (12.86), for turbulent flow ($\text{Re}_{\text{L0}} > 2300$). For the laminar–turbulent transition range, they used the larger of the laminar and turbulent heat transfer coefficients. Figure 14.15 displays comparisons between the correlation and experimental data. Note that in Figs. 14.15(a) and (b) only the model-based and experimental nucleate boiling components of the heat transfer coefficients are compared. The experimental nucleate boiling heat transfer coefficient was found by Fletcher and Haynes by first calculating H_{L0} based on the method just described, and then calculating H_{NB} from Eq. (14.14).

The method of Kandlikar and co-workers (Kandlikar, 2004; Kandlikar and Steinke, 2003; Kandlikar and Balasubrahmanian, 2004) is an adaptation of the correlation of Kandlikar (1990, 1991) to make the correlation suitable for low-flow conditions that lead to laminar single-phase liquid flow. Accordingly, Eqs. (12.102)–(12.106) should be used. However, H_{L0} should be calculated from an appropriate fully developed laminar flow correlation when $Re_{L0} \leq 1600$ and a correlation suitable for convection in the laminar–turbulent transition regime for $1600 \leq Re_{L0} \leq 3000$.

However, given that flow stratification does not occur in minichannels with $D_H \leq 0.3\sqrt{\sigma/g\Delta\rho}$ (see the discussion in Section 3.7), $f_2(Fr_{L0}) = 1$ should be used. For $Re_{f0} < 450$, furthermore, Kandlikar and Balasubramanian (2004) have recommended that only the nucleate boiling part of the Kandlikar correlation (i.e., Eq. (12.103)) be used. The method of Kandlikar and co-workers can thus be summarized as follows.

$$H = \max(H_{NBD}, H_{CBD}) \tag{14.15}$$

$$H_{NBD} = \{0.6683\,Co^{-0.2}(1-x)^{0.8} + 1058.0\,Bo^{0.7}(1-x)^{0.8}F_{fl}\}H_{f0} \tag{14.16}$$

$$H_{CBD} = \{1.136\,Co^{-0.9}(1-x)^{0.8} + 667.2\,Bo^{0.7}(1-x)^{0.8}F_{fl}\}H_{f0} \tag{14.17}$$

The all-liquid heat transfer coefficient H_{f0} is to be found from the correlation of Gnielinski (1976) for $Re_{f0} > 3000$:

$$Nu_{f0} = \frac{H_{f0}D_H}{k_f} = \frac{(Re_{f0} - 1000)Pr_f\frac{f_{f0}}{2}}{1.0 + 12.7\sqrt{\frac{f_{f0}}{2}}(Pr_f^{2/3} - 1)} \tag{14.18}$$

where the all-liquid Fanning friction factor is to be found from an appropriate fully developed turbulent flow correlation. For $150 < Re_{f0} < 1600$, Eqs. (14.16) and (14.17) can be used, with Nu_{f0} calculated from an appropriate fully developed laminar flow correlation. For $1600 \lesssim Re_{f0} < 3000$, Eqs. (14.16) and (14.17) can be used with Nu_{f0} found by interpolation between Gnielinski's correlation for turbulent flow convection and an appropriate fully developed laminar flow convection heat transfer correlation. Finally, for $Re_{f0} \lesssim 100$, the heat transfer mechanism can be assumed to be nucleate boiling dominated and therefore $H = H_{NBD}$, with H_{NBD} found from Eq. (14.16) and Nu_{f0} found from a correlation representing fully developed laminar flow.

Kandlikar and Balasubramanian (2004) validated the above method against the data of Yen et al. (2003) for HCFC-123 for $Re_{f0} \approx 410$, and $Re_{f0} \approx 105$.

The flow boiling correlation of Chen (1966) has been presented and discussed in Section 12.13. Some investigators have modified the correlation of Chen to extend its applicability to small flow passage (Saitoh et al., 2007; Bertsch et al., 2009a, b). These correlations use Eqs. (12.92) through (12.95):

$$H = H_{NB}S + H_{FC}F, \tag{12.92}$$

where the liquid-flow forced convective heat transfer coefficient is found from

$$H_{FC}D_H/k_f = 0.023\,Re_f^{0.8}Pr_f^{0.4} \tag{12.93}$$

where:

$$Re_f = G(1-x)D_H/\mu_f \tag{12.94}$$

$$Pr_f = (\mu C_P/k)_f. \tag{12.95}$$

Based on data obtained with R-134a in circular tube diameters of 0.51 to 10.9 mm diameter, Saitoh *et al.* (2007) used the correlation of Stephan and Abdelsalam (1980) for refrigerants (Eq. (11.38)) for the nucleate boiling component:

$$H_{\mathrm{NB}} = 207 \frac{k_{\mathrm{f}}}{d_{\mathrm{Bd}}} \left(\frac{q_{\mathrm{w}}'' d_{\mathrm{Bd}}}{k_{\mathrm{f}} T_{\mathrm{sat}}} \right)^{0.745} \left(\frac{\rho_{\mathrm{g}}}{\rho_{\mathrm{f}}} \right)^{0.581} \mathrm{Pr}_{\mathrm{f}}^{0.533} \tag{14.19}$$

where the bubble departure diameter d_{Bd} is found from the correlation of Fritz (1935) (Eq. (11.22)). The enhancement factor (for forced convective evaporation heat transfer) and suppression factor (for nucleate boiling) were curve-fitted by Saitoh *et al.* according to

$$F = 1 + \frac{X^{-1.05}}{1 + \mathrm{We}_{\mathrm{g}}^{-0.4}} \tag{14.20}$$

$$S = \frac{1}{1 + 0.4(\mathrm{Re}_{\mathrm{tp}} \times 10^{-4})^{-1.4}}. \tag{14.21}$$

The two-phase Reynolds number and vapor-phase Weber number are found from

$$\mathrm{Re}_{\mathrm{tp}} = \mathrm{Re}_{\mathrm{f}} F^{1.25} \tag{14.22}$$

$$\mathrm{We}_{\mathrm{g}} = \frac{(Gx)^2 D}{\sigma \rho_{\mathrm{g}}}. \tag{14.23}$$

The Lockhart–Martinelli parameter, X, depends on the liquid and vapor flow regimes. For $\mathrm{Re}_{\mathrm{f}} > 1000$ and $\mathrm{Re}_{\mathrm{g}} > 1000$, liquid and vapor are both assumed to be turbulent and therefore $X = X_{\mathrm{tt}}$, with X_{tt} calculated from Eq. (12.91). In microchannels, however, we often deal with turbulent vapor and laminar liquid flows. For $\mathrm{Re}_{\mathrm{f}} < 1000$ and $\mathrm{Re}_{\mathrm{g}} > 1000$, Saitoh *et al.* (2007) proposed

$$X = \left(\frac{C_{\mathrm{f}}}{C_{\mathrm{g}}} \right)^{0.5} \mathrm{Re}_{\mathrm{g}}^{-0.4} \left(\frac{\rho_{\mathrm{g}}}{\rho_{\mathrm{f}}} \right)^{0.5} \left(\frac{\mu_{\mathrm{f}}}{\mu_{\mathrm{g}}} \right)^{0.5} \left(\frac{1-x}{x} \right)^{0.5}, \tag{14.24}$$

where $C_{\mathrm{f}} = 16$ and $C_{\mathrm{g}} = 0.046$. The correlation of Bertsch *et al.* (2009a, b) is a modification of Chen's correlation and is based on a vast data base representing water, R-134a, R-236fa, R-245fa, R-113, R-11, R-12, R-123, R-141b, R-410a, FC-77, and liquid nitrogen. The channels in the data base had circular and rectangular cross sections. The ranges of parameters that are covered by the data base are:

$$20 < G < 3000 \ \mathrm{kg/m^2 \cdot s}$$
$$4 \times 10^3 < q_{\mathrm{w}}'' < 1.15 \times 10^6 \ \mathrm{W/m^2}$$
$$194 < T_{\mathrm{sat}} < 97°\mathrm{C}$$
$$0.0 < x_{\mathrm{eq}} < 1.0.$$

In the correlation of Bertsch *et al.* (2009a, b), Eq. (12.92) is recast as

$$H = H_{\mathrm{NB}}(1 - x_{\mathrm{eq}}) + H_{\mathrm{FC,TP}} \left[1 + 80 \left(x_{\mathrm{eq}}^2 - x_{\mathrm{eq}}^6 \right) e^{-0.6 N_{\mathrm{con}}} \right] \tag{14.25}$$

where $N_{\mathrm{con}} = \frac{\sqrt{\frac{\sigma}{g(\rho_{\mathrm{f}} - \rho_{\mathrm{g}})}}}{D_{\mathrm{H}}}$ is the confinement number. The nucleate boiling heat transfer coefficient, H_{NB}, is found using the correlation of Cooper (1984) (Eq. (11.40)) for pool

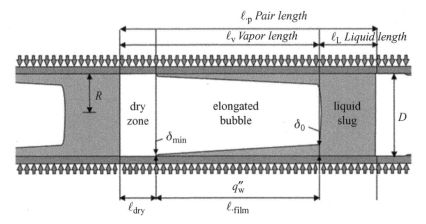

Figure 14.16. Schematic of a triplet comprised of a liquid slug, an elongated bubble, and a vapor slug in the three-zone heat transfer model for elongated bubble flow regime in microchannels (Thome *et al.*, 2004).

nucleate boiling. The two-phase mixture forced convection heat transfer coefficient, $H_{\text{FC,TP}}$, is found from:

$$H_{\text{FC,TP}} = H_{\text{FC,f0}}(1 - x_{\text{eq}}) + H_{\text{FC,g0}}x_{\text{eq}}. \tag{14.26}$$

The all-liquid and all-vapor forced convective heat transfer coefficients $H_{\text{FC,f0}}$ and $H_{\text{FC,g0}}$ can be found from appropriate laminar forced convection correlations. For $H_{\text{FC,f0}}$, for example, Bertsch *et al.* (2009a, b) use the correlation of Hausen (1983) for developing laminar flow

$$\text{Nu}_{\text{f0}} = \frac{H_{\text{FC,f0}}D_{\text{H}}}{k_{\text{f}}} = 3.66 + \frac{0.0668\text{Re}_{\text{f0}}\text{Pr}_{\text{f}}\frac{D}{l}}{1 + 0.045\left[\text{Re}_{\text{f0}}\text{Pr}_{\text{f}}\frac{D}{l}\right]^{0.66}}, \tag{14.27}$$

where $\text{Re}_{\text{f0}} = GD_{\text{H}}/\mu_{\text{f}}$ and l is the channel length. The all-vapor heat transfer coefficient, $H_{\text{FC,g0}}$, can be calculated similarly by using vapor properties in the above expressions. (The correlation of Hausen is for the average heat transfer coefficient for an entrance-dominated short channel.)

Lee and Mudawar (2005a, b) and Kosar *et al.* (2005) have emphasized the importance of the aforementioned boiling heat transfer mechanisms (i.e., nucleate-boiling dominated and convection-dominated). Lee and Mudawar have accordingly developed a set of correlations by defining three different regimes: nucleate-boiling dominated ($x_{\text{eq}} \leq 0.05$), transition ($0.05 < x_{\text{eq}} < 0.55$), and convection-dominated ($x_{\text{eq}} \geq 0.55$). Their correlations, however, are based on their own data, which were obtained with an array of parallel channels. A more elaborate and successful mechanistic model (the three-zone model) has also been developed by Thome *et al.* (2004) and Dupont *et al.* (2004) for saturated boiling in the elongated bubbly (slug) flow regime. The model considers the transient and periodic passage of elongate bubbles and liquid slugs over the heated channel surface. Any point on the channel surface is periodically subject to the passage of the three-zone system that is shown in Figure 14.16. The zones include:

(a) the liquid slug zone, in which liquid forced convection is the heat transfer mechanism;

(b) the elongated bubble zone, where liquid film evaporation occurs;

(c) the dry zone (when present), where vapor forced convection occurs.

Heat transfer in each zone is then found by applying channel flow correlations (for the slug and dryout zones) and simple mechanistic models (for the elongated bubble zone). Simple mechanistic models are also used for predicting the elongated bubble and total three-zone unit lengths, liquid film thickness, and conditions leading to transition from one zone to another. (The liquid and vapor are assumed to move at the same velocity.) The three-zone model has subsequently been improved and enhanced by Thome and co-workers (Agostini *et al.*, 2008a; Thome and Consolini, 2010; Costa-Patry and Thome, 2013).

Using a data base of 4228 data points representing 12 different working fluids that included water, CO_2, and several refrigerants, Li and Wu (2010b) proposed the following remarkably simple correlation for saturated boiling in confined channels

$$\frac{HD_H}{k_f} = 22.9\left[\text{Bd }\text{Re}_f^{0.5}\right]^{0.355}. \tag{14.27a}$$

The correlation is applicable when the flow confinement condition of Eqs. (3.81a) and (3.81b) are satisfied. The data base for the correlation covered the range:

$$23.4 < G < 1500 \text{ kg/m}^2\text{·s}$$
$$3 \times 10^3 < q_w'' < 3.75 \times 10^6 \text{ W/m}^2$$
$$0.16 < D_h < 3.1 \text{ mm}$$
$$0.0 < x_{eq} < x_{eq,\text{CHF}}$$
$$0.023 < P_r < 0.61.$$

EXAMPLE 14.2. A horizontal circular heated tube with 1.95 mm inner diameter is cooled with refrigerant R-11. The following local parameters were deduced from some tests:

(a) $G = 1841.6 \text{ kg/m}^2\text{·s}$, $P = 0.3473 \text{ MPa}$, $q_w'' = 86.16 \text{ kW/m}^2$, $x_{eq} = -0.058$, $T_w = 69.82°C$,

(b) $G = 390.7 \text{ kg/m}^2\text{·s}$, $P = 0.4215 \text{ MPa}$, $q_w'' = 87.55 \text{ kW/m}^2$, $x_{eq} = 0.038$, $T_w = 79.24°C$.

Compare these data with the predictions of the method of Haynes and Fletcher and that of Kandlikar and Steinke, where appropriate.

SOLUTION.

(a) The relevant thermophysical properties are

$$P_{cr} = 4.408 \times 10^6 \text{ Pa}, \sigma = 0.0129 \text{ N/m}, C_{Pf} = 889.2 \text{ J/kg} \geq \text{K}, h_f = 0.900$$
$$\times 10^5 \text{ J/kg}, h_{fg} = 1.643 \times 10^5 \text{ J/kg, and } T_{sat} = 337.3 \text{ K}.$$

The local liquid bulk temperature is found from

$$\overline{T}_L \approx T_{sat} - x_{eq}h_{fg}/C_{Pf} = 321.5 \text{ K}.$$

Let us use the method of Haynes and Fletcher. At $\overline{T}_L = 321.5$ K, we have $\mu_L = 3.43 \times 10^{-4}$ kg/m·s, $Pr_L = 3.72$, and $k_L = 0.081$ W/m·K. The all-liquid Reynolds number will be

$$Re_{L0} = GD/\mu_L = 10,472.$$

Equations (12.89) and (12.86), respectively, then give $f_{L0} = 0.0078$ and $Nu_{L0} = 69.48$, and from there one gets $H_{L0} = 2,890$ W/m²·K. The reduced pressure is $P_r = P/P_{cr} = 0.0788$. Equations (11.43) and (11.44) then give $F_{PR} = 0.887$ and $n = 0.76$. From Table 11.2 we have $H_0 = 2,800$ W/m²·K. With $q_0'' = 20,000$ W/m² and $R_P/R_{P0} = 1$, Eq. (11.41) gives $H_{NB} = 736.1$ W/m²·K.

The heat flux can now be calculated by using Eq. (14.6), and that leads to

$$q_w'' = 65,670 \text{ W/m}^2.$$

This result is about 25% lower than experimental data, and that may at least partially be due to the underestimation of surface roughness in the test section.

(b) In this case we are dealing with saturated flow boiling. Let us use the method of Kandlikar and Steinke (2003). The properties that are needed are

$$\rho_f = 1,358 \text{ kg/m}^3, \ \rho_g = 22.48 \text{ kg/m}^3, \ k_f = 0.0554 \text{ W/m·K},$$
$$\mu_f = 0.757 \times 10^{-4} \text{ kg/m·s}, \ \sigma = 0.0120 \text{ N/m, and}$$
$$h_{fg} = 1.610 \times 10^5 \text{ J/kg.}$$

We next find $Re_{f0} = GD/\mu_f = 10,060$. Equations (12.89) and (12.87) then give

$$f = 0.0078,$$
$$H_{f0} = 1,160 \text{ W/m}^2 \cdot \text{K.}$$

From Eqs. (12.105) and (12.107), respectively, we get

$$Co = 1.707,$$
$$Fr_{f0} = 4.329.$$

Since $Fr_{f0} > 0.4$, $f_2(Fr_{f0}) = 1$. From Table 12.1, we have $F_{fi} = 1.30$. Equations (12.102)–(12.104), along with (12.106), are then applied, bearing in mind that

$$q_w'' = H(T_w - T_{sat}).$$

The iterative solution of the aforementioned equations leads to

$$H_{NBD} = 32\,300 \text{ W/m}^2 \cdot \text{K,}$$
$$H_{CBD} = 20\,733 \text{ W/m}^2 \cdot \text{K,}$$
$$q_w'' = \max(H_{NBD}, H_{CBD})(T_w - T_{sat}) = 2.43 \times 10^5 \text{ W/m}^2.$$

This result does not agree well with the experimental data. However, it must be noted that the performance of a correlation should never be assessed based on one data point only. A statistical analysis of the results from many data points is needed instead.

14.5 Critical Heat Flux in Small Channels

14.5.1 General Remarks and Parametric Trends in the Available Data

Table 14.2 present a summary of some experimental investigations of CHF in mini- and microchannels. Interest in the safety and operational thresholds for tight-lattice rod bundles or plate-type rector cores and the cooling systems of the first walls in fusion reactors provided the impetus for the early CHF experiments with small channels. Some CHF experimental investigations in the past have thus included channels with $D_H \lesssim 1$ mm (Ornatskiy, 1960; Ornatskiy and Kichigan, 1962; Ornatskiy and Vinyarsky, 1964; Loomsmore and Skinner, 1965; Daleas and Bergles, 1965). More recent experimental studies in this group include those by Hosaka et al. (1990), Katto and Yokoya (1984), Nariai et al. (1987, 1989), Inasaka and Nariai (1993), Celata et al. (1993a), Oh and Englert (1993), and Vandervort et al. (1994). These investigations generally used single heated channels with $D \approx 1 - 3$ mm with large L/D ratios and focused on flows of highly subcooled liquid with high mass fluxes in channels subjected to large heat fluxes. Oh and Englert (1993) were concerned with natural convection boiling in reactors with plate-type fuel elements. Minichannel CHF data have in fact been included in the data bases of some of the widely used CHF correlations that were discussed in Chapter 13, including the correlations of Caira et al. (1995) (Eqs. (13.22)–(13.25)), Shah (1987) (Eqs. (13.26)–(13.45)), Katto (1992) (Eqs. (13.60)–(13.70)), and Celata et al. (1994b) (Eqs. (13.71)–(13.75)). The most recent investigations, however, have focused on minichannel- and microchannel-based heat sinks and therefore address low-flow conditions for the reasons explained earlier in Section 14.1 (Bowers and Mudawar, 1994; Roach et al., 1999a, b; Jiang et al., 1999; Qu and Mudawar, 2004b; Wojtan et al., 2006). The vast majority of these experiments are based on multiple parallel channels, with non-aqueous fluids, including various refrigerants, CO_2, and liquid nitrogen. The experiments by Yen et al. (2003) and Chen and Garimella (2012) were performed using the perfluorinated dielectric fluid FC-77.

The importance of the flow channel boundary conditions ("hard" boundary conditions for single-channel experiments versus "soft" boundary conditions for multiple parallel channels connected to common inlet and outlet volumes) must again be emphasized. Multichannel test rigs are prone to flow oscillations. Furthermore, little is known about CHF in microchannels. *The parametric trends in the CHF data to be discussed next are thus based on single minichannel experiments.*

CHF is sensitive to channel diameter. The experimental data of Nariai et al. (1987, 1989) include subcooled as well as saturated (two-phase) CHF data. The dependence of CHF on channel diameter was found to vary with the local quality. When CHF occurred in the subcooled bulk liquid, CHF monotonically increased as D decreased. When CHF occurred under $x_{eq} > 0$ conditions, however, the trend was reversed and CHF decreased with decreasing D. The aforementioned trend (i.e., increasing CHF in subcooled forced flow as D is decreased) had been noted earlier by Bergles (1962), who suggested that the increase in CHF as D becomes smaller can be attributed to three mechanisms, all of which involve vapor bubbles as they grow and are released from wall crevices. As D is decreased, (a) the vapor bubble terminal diameter (the

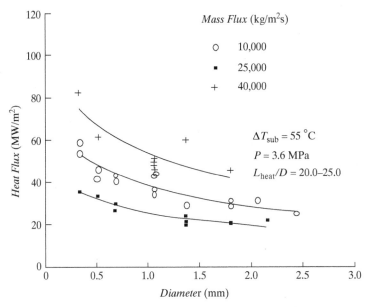

Figure 14.17. Parametric dependencies in the CHF data of Vandervort *et al.* (1994).

diameter of bubbles detaching from the wall) decreases as a result of larger liquid velocity gradient, (b) the bubble velocity relative to the liquid is increased, and (c) condensation at the tip of bubbles is stronger because of the large temperature gradient in the liquid. Nariai *et al.* (1987, 1989) thus explained the aforementioned trend of increasing CHF with decreasing D in subcooled liquids by arguing that smaller bubbles imply a thinner bubble layer and a smaller void fraction and thus lead to a higher CHF. Celata *et al.* (1993a) systematically assessed the effect of channel diameter on CHF in subcooled flow, based on experimental data from several sources, and confirmed the aforementioned mechanism. Hosaka *et al.* (1990), in their experiments with R-113, observed a similar trend and attributed the increase in CHF associated with decreasing D to the decreasing bubble terminal size.

The trends of the available data, however, indicate the existence of a threshold diameter, beyond which the effect of channel diameter on subcooled flow CHF becomes negligible. The data of Vandervort *et al.* (1994) are depicted in Fig. 14.17. Below a threshold diameter (about 2 mm for the depicted data), CHF in subcooled flow increases with decreasing D, whereas for larger diameters the influence of variations in D on the CHF is small. CHF is more sensitive to D at higher values of G. Similar trends have been noted by some other investigators (Celata *et al.* 1993).

The dependence of CHF on pressure is in general non-monotonic. At pressures well below the critical pressure P_{cr}, CHF is expected to increase with increasing pressure. The available data relevant to subcooled CHF in minichannels (which virtually all represent $P < P_{cr}$) indicate that CHF is insensitive to pressure (Vandervort *et al.*, 1994; Celata, 1993). However, a slight decreasing trend in CHF with respect to increasing pressure has also been noted by Vandervort *et al.* (1994) and Hosaka *et al.* (1990).

The CHF monotonically increases with increasing mass flux; it increases monotonically, and approximately linearly, with increasing local subcooling.

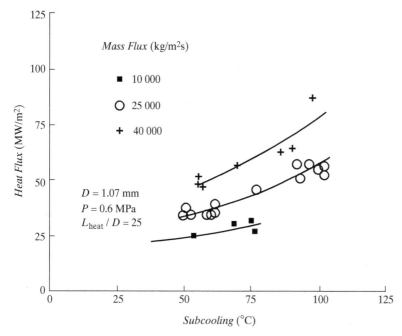

Figure 14.18. The effect of subcooling on CHF in the experiments of Vandervort *et al.* (1994).

Figure 14.18 depicts the effect of subcooling on CHF in the experiments of Vandervort *et al.* (1994).

Vandervort *et al.* (1994) and Roach *et al.* (1999a, b) attempted to measure the effect of dissolved air in subcooled water on CHF. The experimental results of both investigations indicated a negligibly small effect of dissolved air on CHF. Since the solubility of air in water is very low, and in view of the fact that considerable evaporation from boiling occurs in CHF, the insignificant contribution of dissolved air to CHF is not surprising. It should be noted, however, that for other fluid–noncondensable pairs for which the solubility of the noncondensable in the liquid is high, the impact of the dissolved noncondensable may not be negligible.

The parametric trends cited here were for high-flow conditions. The experiments of Roach *et al.* (1999a, b) involved CHF in subcooled water at low mass fluxes in heated single minichannels with hard boundary conditions. CHF occurred when $x_{eq} \gtrsim 0.36$ at the exit of their test sections, suggesting the occurrence of dryout; and $x_{eq} \lesssim 1$ was noted in many of their tests.

The parametric trends reviewed thus far dealt with very high coolant mass fluxes, which are representative of cooling of objects that are under extremely high thermal loads. More recent studies are concerned with cooling of heat sinks used in microelectronic devices. These experiments use multiple parallel channels connected to common plena or manifolds at their inlets and outlets, utilize non-aqueous fluids, and deal with relatively low coolant mass fluxes (Lee *et al.*, 2005; Hetsroni *et al.*, 2006; Wojtan *et al.*, 2006; Agostini *et al.*, 2008b; Park and Thome, 2010; Chen and Garimella, 2012). The mass fluxes are usually in the 10–10^3 kg/m^2·s range. Flow instabilities described in the previous section may have occurred in these experiments. These instabilities and flow oscillations generally lower the critical heat flux. In terms of parametric trends, the following general observations can be mentioned (Revellin *et al.*, 2009; Roday and Jensen, 2009). Given the low mass fluxes the CHF is predominantly due to dryout. With all other parameters maintained the same:

- critical heat flux increases with increasing mass flux
- critical heat flux increases with increasing subcooling
- critical heat flux increases with increasing saturation pressure
- critical heat flux increases with decreasing quality at the channel exit.

The dependence of critical heat flux on channel heated length and hydraulic diameter is complicated. Most studies indicate that critical heat flux increases as the hydraulic diameter is reduced (Revellin *et al.*, 2009), including the experiments of Roday and Jensen (2009). By analyzing their own data obtained with water in single stainless tubes with 0.286 mm and 0.427 mm diameters, and by comparing their data with the data of Wojtan *et al.* (2006), Roday and Jensen (2009) have speculated that a transition tube diameter, approximately equal to 0.5 mm for water, may exist. For channel diameters smaller than the aforementioned transition diameter the critical heat flux will increase with decreasing size, while for larger channels the critical heat flux is insensitive to the channel size. With respect to the dependence on heated channel length, most observations indicate that critical heat flux decreases with increasing channel length. However, the trend is not monotonic. By examining a large data base Wu *et al.* (2011) have noted the following trends.

- For $L_{\text{heat}}/D_{\text{eff}} \lesssim 50$, CHF (or, equivalently, the boiling number at CHF point, $\frac{q''_{\text{CHF}}}{G h_{\text{fg}}}$) is quite sensitive to the heated length and decreases with increasing $L_{\text{heat}}/D_{\text{eff}}$.
- For $L_{\text{heat}}/D_{\text{eff}} > 150$, CHF is insensitive to $L_{\text{heat}}/D_{\text{eff}}$.
- The range $50 \leq L_{\text{heat}}/D_{\text{eff}} < 150$ represents a transition where CHF decreases with increasing $L_{\text{heat}}/D_{\text{eff}}$, but with a lower sensitivity as $L_{\text{heat}}/D_{\text{eff}}$ increases.

The complicated effect of flow instability in multi-channel heat sinks should be emphasized. The instability leads to significant differences between multi-channel heat sinks and single-channel heated channels.

Flow oscillations, discussed earlier in Section 14.2.2, and the fact that parallel-channel systems are generally designed to operate under low-flow conditions are the main reasons for these differences. Qu and Mudawar (2004b) used a heat-sink module consisting of 21 parallel channels connected to inlet and outlet plena (see Table 14.2). A throttle valve was installed upstream of the inlet plenum to eliminate the severe-pressure instability type (see Section 14.2.2). Parallel-channel flow oscillations occurred, however. The oscillations became severe as CHF was approached, leading to significant backflow of vapor into the inlet plenum. Figure 14.19 depicts a schematic of the vapor backflow in their experiments. The resulting CHF data indicated that CHF increased with increasing mass flux but was insensitive to the subcooling of the coolant that entered the inlet plenum. The latter trend is contrary to the trends in single-channel data and arises because the vapor backflow into the inlet plenum ensures that the coolant that flows into the channels is saturated. The experimental results of Jiang *et al.* (1999) indicated that no stable saturated boiling took place in their parallel microchannels, and CHF appeared to occur when subcooled boiling terminated.

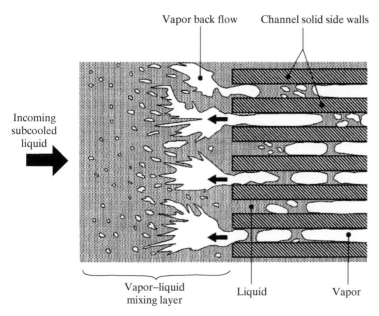

Figure 14.19. Schematic of the vapor backflow in parallel minichannels at near-CHF conditions. (From Qu and Mudawar, 2004b.)

14.5.2 Models and Correlations

As mentioned earlier, the data bases of some of the conventional CHF correlations include minichannel CHF data, implying that these correlations may apply to minichannels. Some of the empirical correlations that have recently been applied to these data and have been successful in predicting at least some minichannel CHF data will be discussed in this section.

Nariai *et al.* (1987, 1989) applied the correlation of Tong (1969), Eqs. (13.50)–(13.52), to their experimental data and noted that to achieve agreement they needed to correlate the parameter C separately for low- and high-heat-flux conditions. Celata *et al.* (1994a) noted that the data used by Nariai *et al.* were limited to relatively low heat flux and low pressure, and their modification of the correlation of Tong was inadequate. Celata *et al.* (1994a) correlated the parameter C according to Eqs. (13.53)–(13.57). Celata *et al.* indicated that the modified correlation could predict 98.1% of their compiled data points (which covered $0.1 < P < 9.4$ MPa, $0.3 < D < 25.4$ mm, $0.1 < L < 0.61$ m, $2 < G < 90$ Mg/m^2·s, and $90 < \Delta T_{sub} < 230$ K) within ±50%. The agreement of the correlation with the minichannel data included in the data base of Celata *et al.*, furthermore, appeared to be satisfactory. Celata *et al.* (1993a, 1994a) also compared their compiled data base with the predictions of several other empirical correlations, generally with poor agreement in comparison with the aforementioned modified-Tong correlation.

Hall and Mudawar (1997) assessed the validity of previously published CHF experimental data and have compiled a qualified CHF data base (referred to as the PU-BTPFL CHF Database). This data base includes experiments with water representing $0.3 \leq D \leq 45$ mm, $10 \leq G \leq 2484$ kg/m^2·s, and $-2.25 \leq x_{eq} \leq 1.0$, in vertical, upward flow tubes, with x_{eq} representing the local equilibrium quality at the

CHF point (i.e., the end of the heated segment of the test section). They compared 25 widely referenced correlations dealing with CHF in vertical, upward flow channels with their data base and showed that the empirical correlations of Caira et al. (1995) (Eqs. (13.22)–(13.25)) provided the most accurate predictions. The correlation of Bowring (1972), presented in Eqs. (13.10)–(13.18b), also showed relatively good agreement with the data. Although the PU-BTPFL Database and the aforementioned correlations generally address vertical channels with upward flow, the data included in the data base that represent small channels ($D \lesssim 1$ mm) may represent horizontal minichannels as well because of the small influence of channel orientation with respect to gravity on two-phase flow in such minichannels. Roach et al. (1999a, b) also noted that these two correlations provided reasonably close agreement with their data. The correlation of Bowring systematically underpredicted their data, however, on average by 36%. Caira's correlation agreed with their data better, with an average overprediction of the data by only 18%.

The data base for the correlation of Shah (1987), Eqs. (13.26)–(13.45), also includes minichannels. Hosaka et al. (1990) compared the predictions of Shah's correlation with their data. On average, the correlation overpredicted the data only slightly. Zhang et al. (2006) compared several empirical correlations with a vast data base dealing with CHF of water in small channels. They noted that for CHF when the working fluid is saturated the correlation of Shah (1987) predicted the experimental data well.

The DNB models of Katto (1992) and Celata et al. (1994b) were described in Section 13.5. Celata et al. (1994b) compared their model with experimental data covering the $0.2 < D < 25.4$ mm range with good agreement between model and data. The force balance on vapor blankets in the model of Celata et al., however, is based on bubble behavior in large systems and, furthermore, implies dependence on channel orientation.

Xie (2004) has examined the applicability of the correlations of Shah (1987) and Caira et al. (1995) to CHF in minichannels cooled with water. The experimental data of Lowdermilk et al. (1958), Weatherhead (1963), Lezzi et al. (1994), and Roach et al. (1999a, b) were used, and since dryout was of primary interest, only data with relatively high x_{eq} were selected. The selected data covered the $x_{eq} = 0.11$–0.99 range. Figures 14.20 and 14.21 displays his results. As noted, both correlations performed rather poorly.

As noted earlier, cooling of heat sinks by multiple parallel microchannels has been the primary subject of interest in the recent past. Most studies dealing with boiling heat transfer in mini- and microchannels have dealt with multiple parallel channels, subject to relatively low mass flux cooling by non-aqueous fluids, as reviewed in the previous section. A large number of empirical correlations, as well as a few mechanistic models, have been proposed. Most of the correlations are limited in their scope of applicability, however, and fail to predict data from a reasonably wide number of sources.

Revellin et al. (2009) have performed a careful review of several predictive methods, by comparing their predictions with 2996 experimental data points from 19 different research groups. Most of the data were based on single channels with circular cross sections, but the data base include four multi-channel data sets, three with parallel rectangular cross-section channels, and one with parallel circular cross-section channels. The cooling fluids in the data base included water, several refrigerants, CO_2,

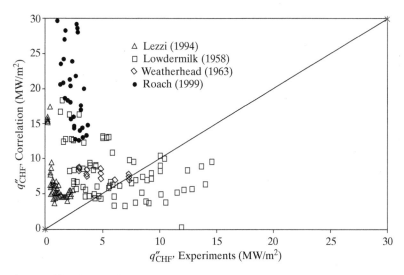

Figure 14.20. Comparison between some minichannel CHF data and the correlation of Shah (1987). (From Xie, 2004.)

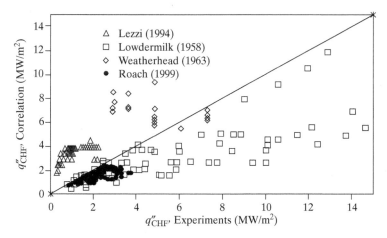

Figure 14.21. Comparison between some minichannel CHF data and the correlation of Caira *et al.* (1995). (From Xie, 2004.)

and liquid nitrogen. The range of channel hydraulic diameters was 0.29–6.1 mm, but values were mostly in the minichannel range. The correlations examined included those of Katto and Ohno (1984), Shah (1987), Wojtan *et al.* (2006), Qu and Mudawar (2004b), Zhang *et al.* (2006), and Qi *et al.* (2007), and the mechanistic model of Revellin and Thome (2008). For water, the correlation of Zhang *et al.* (2006) was found to be the most accurate. Among these predictive methods the model of Revellin and Thome (to be presented shortly) performed best for non-aqueous fluids. The experimental data of Bowers and Mudawar (1994) and Qu and Mudawar (2004b), as mentioned earlier, were obtained in heat-sink modules consisting of multiple parallel channels and operating under low-flow conditions. Qu and Mudawar (2004b) have developed the following correlation based on both data sets:

$$\frac{q''_{\text{CHF}}}{Gh_{\text{fg}}} = 33.43 \left(\frac{\rho_g}{\rho_f} \right)^{1.11} \text{We}_{\text{f0}}^{-0.21} (L_{\text{heat}}/D_{\text{eff}})^{-0.36}, \tag{14.28}$$

where $D_{\text{eff}} = 4A/p_{\text{heat}}$ is the heated equivalent diameter of the channel, all fluid properties are to be found at inlet pressure, and the all-liquid Weber number is defined as

$$\text{We}_{f0} = \frac{G^2 D}{\rho_f \sigma}. \tag{14.29}$$

Qu and Mudawar did not notice any effect of inlet subcooling on their CHF data; therefore the correlation does not include the effect of inlet quality. Their correlation is remarkable for the implied recognition of the importance of surface tension. Kosar *et al.* (2005), whose experiments were performed with water in 200 μm × 246 μm channels, found good agreement between their data and Qu and Mudawar's correlation. This correlation also agreed well with the experimental data of Chen and Garimella (2012). The latter data represent CHF in 60 parallel rectangular channels, each 100 μm × 389 μm in cross section, with FC-77 as coolant. The correlation is unphysical in its monotonic dependence on D_{eff}, and therefore it may not be appropriate for extrapolation to diameters significantly different than its data base.

Perfluorinated dielectric fluids, including FC-77 and FC-87, are suitable for microelectronic cooling and other applications because they are dielectric, nontoxic, stable, compatible with sensitive materials, nonflammable, and leave little residue upon evaporation. Experimental data with these fluids are absent from the data bases used for the derivation of most empirical CHF correlations. Chen and Garimella (2012) performed experiments with FC-77, using 60 parallel 100 μm × 389 μm rectangular channels, and developed the following correlation:

$$\frac{q''_{\text{CHF}}}{G h_{\text{fg}}} = 40.0 \left(\frac{\rho_g}{\rho_f} \right)^{1.12} \text{We}_{f0}^{-0.24} (L_{\text{heat}}/D_{\text{eff}})^{-0.34}, \tag{14.30}$$

where the all-liquid Weber number We_{f0} is defined according to Eq. (14.5.29). This correlation evidently represents only a slight modification of the aforementioned correlation of Qu and Mudawar.

Based on an extensive data base including about 2500 data points dealing with CHF for water-cooled small channels, Zhang *et al.* (2006) developed the following correlation:

$$\frac{q''_{\text{CHF}}}{G h_{\text{fg}}} = 0.0352 \left[\text{We}_{f0} + 0.0119(L_{\text{heat}}/D_{\text{H}})^{2.31} (\rho_g/\rho_f)^{0.361} \right]^{-0.295} (L_{\text{heat}}/D_{\text{H}})^{-0.311}$$

$$\times \left[2.05(\rho_g/\rho_f)^{0.17} - x_{\text{eq,in}} \right]. \tag{14.31}$$

The data base of the above correlation is limited to water, but includes several channels cross-section geometries. The range of parameters covered by the experimental data base included

$$D_{\text{H}} = 0.33\text{–}6.2\,\text{mm}$$
$$G = 5.33\text{–}1.34 \times 10^5 \,\text{kg/m}^2\text{·s}$$
$$L_{\text{heat}}/D_{\text{H}} = 1.0\text{–}975$$
$$x_{\text{eq,in}} = -2.35\text{–}0.0$$
$$x_{\text{eq,exit}} = -1.75\text{–}0.999.$$

Based on an analysis of 692 experimental data points from 17 sources that included water, several refrigerants and liquid nitrogen, and based on the argument that the local vapor quality is crucially important for the occurrence of CHF, Wu and Li (2011) have proposed the following correlations for saturated CHF. These correlations are only applicable when $Bd^{0.5}Re_f < 200$, which is the micro–macro channel transition criterion according to Li and Wu (2010b) and Wu and Sundén (2011). For $(L_{heat}/D_{eff}) < 150$,

$$\frac{q''_{CHF}}{Gh_{fg}} = 0.62(L_{heat}/D_{eff})^{-1.19}x_{eq,CHF}^{0.817}. \qquad (14.32a)$$

For $(L_{heat}/D_{eff}) > 150$,

$$\frac{q''_{CHF}}{Gh_{fg}} = 1.16 \times 10^{-3}(We_mCa^{0.8})^{-0.16}, \qquad (14.32b)$$

where Ca is the capillary number and We_m is the Weber number defined based on homogeneous flow assumption

$$We_m = \frac{G^2D_{eff}}{\rho_h\sigma}$$

$$\rho_h = \left[\frac{x_{eq,CHF}}{\rho_g} + \frac{1 - x_{eq,CHF}}{\rho_f}\right]^{-1}.$$

The parameter $x_{eq,CHF}$ is the local equilibrium vapor quality (typically the exit of a heated channel) and can be found from an energy balance on the heated channel. The range of applicability of the correlation is

$$23.4 < G < 5200 \text{ kg/m}^2\cdot\text{s}$$
$$0.223 < D_{eff} < 6.92 \text{ mm}$$
$$0 < x_{eq,CHF} < 1$$
$$0.005 < P_r < 0.627.$$

According to Wu and Li (2011), this correlation predicted 94% of the mini- and microchannel data representing water, and 97% of the data representing non-aqueous fluids in their data base within ±30% error.

As mentioned earlier, the predominant mechanism of CHF in mini- and microchannels that are commonly considered for microelectronic cooling is dryout, given that these systems often operate under relatively low heat and mass fluxes. The dryout process has been successfully modeled using a mechanistic approach (see Section 13.6). Revellin and Thome (2008) have applied a simplified mechanistic model for dryout in microchannels, which appears to be successful in predicting saturated CHF data (Revellin et al., 2009). This model is now described.

Consider steady, one-dimensional condensing flow of a pure fluid. Assume that the flow regime is ideal annular, and neglect the droplet entrainment process. Using the separated flow model, the conservation equations would be Eqs. (5.61) and (5.62) for liquid and vapor mass, respectively; (5.69) and (5.70) for liquid and vapor momentum, respectively; and (5.86) for the mixture energy. (The time-dependent terms in these equations should of course be neglected). Now, (a) assume that the mixture is always saturated, (b) neglect momentum exchange between the liquid and vapor phases due to phase change, (c) neglect the effect of gravity, (d) neglect the contribution of all mechanical energy terms to the energy conservation, and (e) neglect the virtual mass effect in the momentum conservation equations (the virtual mass effect is negligible in the annular flow regime), (f) assume that all the heat transferred to the flow field is used up for evaporation (consistent with neglecting the changes in kinetic and potential energy in the flow field), and (g) include the effect of interfacial tension on pressure difference between the two phases. What remains from the aforementioned conservation equations can then be cast as (note that q''_w is defined to be positive for a heated pipe):

$$\frac{d}{dz}[A\rho_f U_f(1-\alpha)] = -\frac{p_{heat}q''_w}{h_{fg}} \tag{14.33}$$

$$\frac{d}{dz}[A\rho_g U_g \alpha] = +\frac{p_{heat}q''_w}{h_{fg}} \tag{14.34}$$

$$\frac{\partial}{\partial z}[A\rho_L(1-\alpha)U_L^2] = -A(1-\alpha)\frac{\partial P_f}{\partial z} + AF_I - \dot{A}F_{WL} \tag{14.35}$$

$$\frac{\partial}{\partial z}(A\rho_G \alpha U_G^2) = -A\alpha\frac{\partial P_g}{\partial z} - AF_{wG} - AF_I. \tag{14.36}$$

An additional equation is obtained by differentiating the Laplace–Young equation (see Eq. (2.124))

$$\frac{dP_g}{dz} - \frac{dP_f}{dz} = \frac{d}{dz}\left(\frac{\sigma}{R_I}\right) \tag{14.37}$$

where $R_I = R - \delta_{film}$ is the radius of the vapor core. There is no need for an additional equation for conservation of energy, because Eqs. (14.33) and (14.34) already account for energy conservation.

The constitutive and closure equations are:

$$\alpha = (R_I/R)^2$$
$$AF_I = \tau_I(2\pi R_I)$$
$$AF_{WL} = \tau_w(2\pi R)$$
$$\tau_w = f_w\frac{1}{2}\rho_f U_f^2$$
$$\tau_I \approx f_I\frac{1}{2}\rho_f U_g^2.$$

The skin friction coefficients (Fanning friction factors) are found from appropriate channel flow correlations for laminar and turbulent flow (whichever is applicable) by defining the liquid and vapor Reynolds numbers, respectively, as $Re_f = \frac{2\rho_f U_f \delta_{film}}{\mu_f}$ and $Re_g = \frac{2\rho_g U_g R_I}{\mu_g}$.

Equations (14.35) and (14.36) are further simplified by assuming that the liquid and vapor are (locally) incompressible. Equations (14.33) through (14.37) then represent a set of five ordinary differential equations with P_f, P_g, U_f, U_g, and R_I as state variables. These equations can be numerically integrated starting from the inlet ($z = 0$) when the fluid is saturated at the inlet. Dryout occurs when the liquid

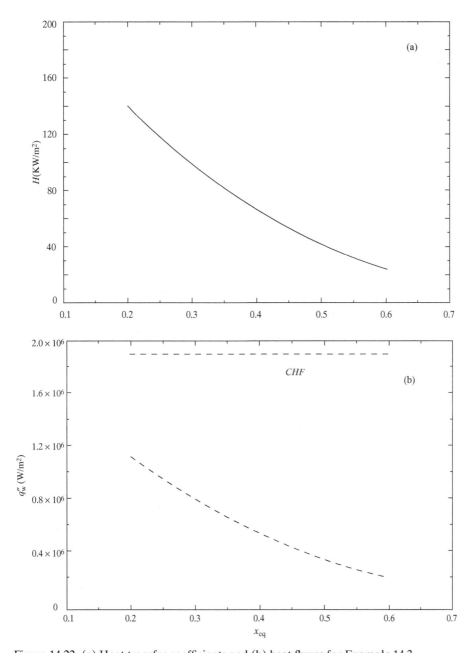

Figure 14.22. (a) Heat transfer coefficients and (b) heat fluxes for Example 14.3.

film breaks down. Revellin and Thome (2008) proposed that film breakdown occurs when

$$\delta_{\text{film}} \leq 0.15 R \left(\frac{\rho_g}{\rho_f} \right)^{-3/7} \left\{ \frac{R^2}{\sigma / [g(\rho_f - \rho_g)]} \right\}^{-1/7}. \tag{14.38}$$

The derivation of this empirical expression is based on the assumption that the liquid film oscillates with an amplitude that is proportional to the neutral wavelength in Rayleigh–Taylor instability (see Eq. (2.175)).

EXAMPLE 14.3. A horizontal, heated circular channel is cooled by refrigerant R-134a. The channel inner diameter is 200 µm, and the heated length is 1.2 cm. The refrigerant mass flux is $G = 340$ kg/m^2·s. Near the exit the local pressure is 4.15 bars and the wall temperature is 8 K above saturation temperature. For the exit equilibrium qualities in the range $x_{eq} = 0.20$–0.6 estimate and plot as a function of x_{eq} the heat transfer coefficient, and assess whether there is the risk of CHF.

SOLUTION. The relevant properties at 4.15-bar pressure are $\rho_f = 1{,}261$ kg/m^3, $\rho_g = 20.22$ kg/m^3, $k_f = 0.0903$ W/m·K, $h_{fg} = 1.908 \times 10^5$ J/kg, $\mu_f = 2.34 \times 10^{-4}$ kg/m·s, and $\sigma = 0.0101$ N/m. We will apply the method of Kandlikar and Steinke. Table 12.2 indicates $F_{rl} = 1.63$. We next calculate

$$\mathrm{Re}_{f0} = GD/\mu_f = 290.3.$$

We thus deal with laminar flow in the reference single-phase flow conditions. We will therefore write

$$\mathrm{Nu}_{f0} = H_{f0}D/k = 3.66 \Rightarrow H_{f0} = 1{,}652 \text{ W/m}^2\text{·K}.$$

From Eq. (12.107) we get

$$\mathrm{Fr}_{f0} = 37.04.$$

We now need to iteratively solve Eqs. (12.101)–(12.106), along with the following equation, for every selected value of x_{eq}:

$$q_w'' = H(T_w - T_{sat}).$$

The calculated heat transfer coefficients and the heat fluxes are plotted in Figs. 14.22(a) and (b), respectively.

To assess the risk of CHF, we can use the correlation of Qu and Mudawar, Eq. (14.7), for the estimation of the local critical heat flux. This equation gives $q_{CHF}'' = 1.89 \times 10^6$ W/m^2. Clearly, as noted in Fig. 14.3(b), the local heat flux is significantly lower than the estimated q_{CHF}'', indicating that there is little risk of critical heat flux conditions occurring in the heated tube.

PROBLEMS

14.1 Refrigerant R-123 at a pressure of 3.77 bars flows into a tube with 1.0-mm inner diameter. The pipe is uniformly heated, and the water at the inlet to the heated tube is subcooled by 24 °C. The mean liquid velocity in the tube is 5.25 m/s. Calculate the heat flux that would lead to ONB at a distance of 3.5 cm from the inlet using the model of Davis and Anderson for hemispherical bubbles.

14.2 In Problem 14.1, calculate the local equilibrium quality, the heat transfer regime, and the heat transfer coefficient at a location 1.5 cm downstream from the ONB point.

14.3 A 1.25-mm-diameter heated tube made from copper is cooled with R-134a at a pressure corresponding to $T_{sat} = 31$ °C. The heat flux is 20 kW/m^2 and the mass flux is 100 kg/m^2·s.

(a) Determine the local heat transfer regime for the following local equilibrium qualities: $x_{eq} = 0.08, 0.19$, and 0.32.

(b) For the conditions of part (a) calculate the local heat transfer coefficients. If appropriate, do these calculations using the correlations of Haynes and Fletcher (2003) and Kandlikar and Steinke (2003).

14.4 For the conditions of Problem 14.3, repeat the calculations for $x_{eq} = 0.08$, this time for channel diameters $D = 0.5, 1.0$, and 1.5 mm. Compare and discuss the dependence of the local heat transfer coefficients on tube diameter.

14.5 The data points in Table P14.5, which are from Lezzi *et al.* (1994), are included in the PU-BTPFL CHF Database (Hall and Mudawar, 1997). The test section is a uniformly heated horizontal tube. For these points, using the most appropriate methods:

(a) determine the location where ONB and OSV occur;

(b) determine the heat transfer regime, and the heat transfer coefficient, at $z = L/3$ and $L/2$, where z is the axial distance from the inlet.

Table P14.5. *Selected data from Lezzi* et al. *(1994) for water and $D = 1$ mm.*

L/D	$G(\text{kg/m}^2\cdot\text{s})$	$P(\text{bar})$	$x_{eq,in}$	$x_{eq,exit}$	$q''_w(\text{MW/m}^2)$
239	1.48×10^3	70.1	−0.1	0.82	2.052
241	1.496×10^3	69.6	−0.19	0.80	2.206
502	1.464×10^3	70.3	−0.25	0.71	1.021

14.6 For the data points in Table P14.5, using the local conditions at the channel middle and exit, calculate the local heat fluxes that would cause CHF to occur at those points.

14.7 For the data points in Table P14.5, assuming that the heat flux values provided in the table represent experimental CHF at exit, what type of CHF (DNB or dryout) is likely to have occurred? Using appropriate CHF correlations that are based on global as well as local conditions, calculate the heat fluxes that would cause CHF conditions to occur at the test channel exit. Also, assuming that the heat flux values provided in the table represent experimental CHF at exit, compare your predictions with these experimental data.

14.8 The data points in Table P14.8, which are from Nariai *et al.* (1987), are included in the PU-BTPFL CHF Database (Hall and Mudawar, 1997). The test section is a uniformly heated vertical tube. For these points, using the most appropriate methods:

(a) determine the location where ONB and OSV occur;

(b) determine the heat transfer regime, and the heat transfer coefficient, at $z = L/3$, where z is the axial distance from the inlet.

Table P14.8. *Selected data from Nariai* et al. *(1987) for water and D = 1.0 mm.*

L/D	$G(\text{kg/m}^2 \cdot \text{s})$	$P_{\text{exit}}(\text{bar})$	$T_{\text{in}}(^\circ\text{C})$	$x_{\text{eq,exit}}$	$q_w''(\text{MW/m}^2)$
10	6.71×10^3	1.01	37.7	−0.038	26.67
29.8	7.02×10^3	1.01	61.2	−0.016	6.68
50.40	13.48×10^3	1.01	64.0	0.002	11.71
50.20	20.16×10^3	1.01	22.5	0.003	30.86

14.9 For the data points in Table P14.8, assuming that the heat flux values provided in the table represent experimental CHF at exit, what type of CHF (DNB or dryout) is likely to have occurred? Using appropriate CHF correlations that are based on global as well as local conditions, calculate the heat fluxes that would cause CHF conditions to occur at the test channel exit. Also, assuming that the heat flux values provided in the table represent experimental CHF at exit, compare your predictions with these experimental data.

14.10 The circular ring shown in the figure is made of aluminum. It receives a total thermal load of 35 W through its back and front surfaces. Its surfaces must remain at a temperature not higher than 85 K. A cooling system involving a rectangular miniature channel similar to the one shown in the figure has been suggested, where cooling is provided by boiling of liquid nitrogen at atmospheric pressure.

Prepare a computer program for the purpose of designing this cooling system, which includes the calculation of local microchannel surface temperatures and total pressure drop, and perform parametric calculations to show the effects of various parameters including channel size, number of bends, liquid nitrogen flow rate, etc., on total pressure.

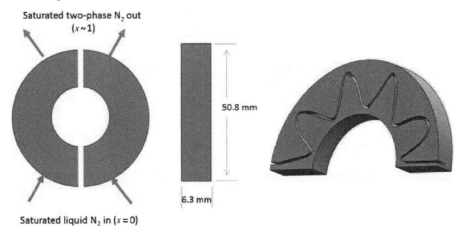

Saturated two-phase N_2 out
$(x \sim 1)$

50.8 mm

6.3 mm

Saturated liquid N_2 in $(x = 0)$

Figure P14.10.

14.11 Saturated R-134a at a temperature of 30 °C flows in a heated stainless steel microtube with 0.2 mm inner diameter. The wall heat flux is 10 kW/m². The mass flux is 100 kg/m²·s.

(a) Using the criteria of Cornwell and Kew (1992) and Garimella and Harirchian (2010) determine whether boiling occurs under confined space conditions. What channel diameter would define the threshold for spatial confinement?

(b) Find the flow regime using the flow regime model of Garimella and Harirchian (2010).

(c) Repeat parts (a) and (b), this time assuming that mass flux is 250 kg/m²·s, and tube diameter is 0.8 mm.

14.12 Consider the flow boiling of R-245fa in a 0.8-mm-inner-diameter microtube. The mass flux is 150 kg/m²·s, and a uniform wall heat flux equal to 10 kW/m² is imposed on the tube. At the inlet to the microtube the refrigerant is saturated liquid, and is at 303 K temperature. Using the flow regime transition models of Ong *et al.*, find the locations along the microtube where all the major and relevant flow regime transitions take place.

For simplicity neglect the effect of pressure drop on fluid properties.

14.13 Consider the flow boiling of R-245fa in a 0.8-mm-inner-diameter microtube. The mass flux is 150 kg/m²·s, and a uniform wall heat flux equal to 10 kW/m² is imposed on the tube. At inlet to the microtube the refrigerant is saturated liquid, and is at 303 K temperature. For qualities 0.01 and 0.1, find the wall surface temperature by calculating the heat transfer coefficient using the correlations of Haynes and Fletcher (2003) and the correlation of Bertsch *et al.* (2009b). In your opinion, the predictions of which correlation are more reliable?

Note: For the nucleate boiling component in the correlation of Haynes and Fletcher (2003), you may use the correlation of Cooper (1984).

14.14 Saturated R-134a at a temperature of 30 °C flows in a uniformly heated stainless steel microtube that has an inner diameter of 0.8 mm, and is 95 mm long. For mass flux of 100 and 250 kg/m²·s do the following.

(a) Find the maximum wall heat flux that can be imposed without CHF happening anywhere in the tube, using the correlation of Wu *et al.* (2011).

(b) Suppose we would like to replace R-134a with FC-87. Assuming that the inlet conditions (saturated liquid, 30 °C) are to be maintained, what mass flux would be needed to dispose of the same thermal load as in part (a) without reaching CHF anywhere in the tube? Use the correlation of Wu *et al.* (2011) for convenience.

(c) Based on experimental data with FC-77, obtained using 60 parallel 100 μm × 389 μm rectangular channels, Chen and Garimella (2012) have proposed

$$\frac{q''_{CHF}}{G h_{fg}} = 40.0 \left(\frac{\rho_g}{\rho_f} \right)^{1.12} We_{f0}^{-0.24} (L_{heat}/D_{eff})^{-0.34},$$

where $D_{eff} = 4A/p_{heat}$, and $We_{f0} = \frac{G^2 D}{\rho_f \sigma}$. Repeat part (b), using this correlation.

14.15

(a) Apply the CHF correlation of Zhang *et al.* (2006) to the data points tabulated in Table P14.5 (Problem 14.5).

(b) Repeat part (a), this time using the correlation of Wu *et al.* (2011).

(c) Using parts (a) and (b), comment on the applicability of the two correlations.

(d) The arrangements in parts (a) and (b) needed pumping power. Calculate and compare the frictional pressure losses in the tube for parts (a) and (b).

15 Fundamentals of Condensation

15.1 Basic Processes in Condensation

Condensation is a process in which the removal of heat from a system causes a vapor to convert into liquid. Condensation plays an important role in nature, where it is a crucial component of the water cycle, and in industry. Condensation processes are numerous, taking place in a multitude of situations. In view of their diversity, a classification of condensation processes is helpful. Classification can be based on various factors, including the following.

- Mode of condensation: homogeneous, dropwise, film, or direct contact.
- Conditions of the vapor: single-component, multicomponent with all components condensable, multicomponent including noncondensable component(s), etc.
- System geometry: plane surface, external, internal, etc.

There are of course overlaps among the categories from different classification methods. Classification based on mode of condensation is probably the most useful, and modes of condensation are now described.

Homogeneous Condensation

Homogeneous condensation can happen when vapor is sufficiently cooled below its saturation temperature to induce droplet nucleation; it may be caused by mixing of two vapor streams at different temperatures, radiative cooling of vapor–noncondensable mixtures (fog formation in the atmosphere), or sudden depressurization of a vapor. In fact, cloud formation in the atmosphere is a result of adiabatic expansion of warm and humid air masses that rise and cool. According to classical nucleation theory, in a pure, supersaturated vapor homogeneous condensation occurs when droplets of critical radius r^* (i.e., droplets just large enough so that the pressure difference between their interior and exterior can balance the surface tension force) are produced in significant number. The critical radius can be found from (see Section 2.5)

$$r^* = \frac{2\sigma v_{\mathrm{f}}}{\frac{R_{\mathrm{u}}}{M} T_{\mathrm{g}} \ln(P_{\mathrm{g}}/P_{\mathrm{sat}})}. \tag{15.1}$$

The rate of generation of droplets with radius r^* in a unit volume is

$$\frac{dn}{dt} = N \frac{v_{\mathrm{f}}}{v_{\mathrm{g}}} \left(\frac{2\sigma}{\pi m}\right)^{1/2} \exp\left[-4\pi\sigma \frac{r^{*2}}{3k_{\mathrm{B}} T_{\mathrm{g}}}\right], \tag{15.2}$$

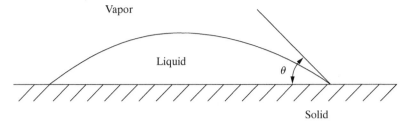

Figure 15.1. Condensation on a clean and dry surface.

where T_g and P_g are the temperature and pressure of vapor, respectively, and N is the number density of vapor molecules. Significant nucleation requires that $dn/dt \geq 10^{17}$–10^{22} m^{-3}s^{-1}.

Although homogeneous nucleation in pure vapors is possible, in practice dust and other particles act as droplet nucleation embryos. In the atmosphere, fog formation usually relieves the supersaturation [defined as $\varphi - 1$, where $\varphi = P_v/P_{sat}(T)$ is the relative humidity], when the supersaturation approaches a maximum value of about 1% (Friedlander, 2000). The droplets in fog have diameters in the 1–10 μm range.

Heterogeneous Condensation

The overwhelming majority of condensation processes are heterogeneous, where droplets form and grow on solid surfaces.

Significant subcooling of vapor is required for condensation to start when the surface is smooth and dry. The rate of generation of embryo droplets in heterogeneous condensation can be modeled by using kinetic theory, according to which dn''/dt, the rate of generation of droplets with the critical size on a unit surface area of clean and dry surface, is (see Fig. 15.1 (Carey, 2008)

$$\frac{dn''}{dt} = \left(\frac{2\sigma C}{\pi m}\right)^{1/2} \left(\frac{P_g}{\frac{R_u}{M} T_g}\right)^{5/3} (m^{-2/3}).v_f C \frac{1 - \cos\theta}{2} \exp\left[\frac{-16\pi\left\{\frac{\sigma C}{(R_u/M)T}\right\}^3 v_f^2}{3m\{\ln[P_g/P_{sat}]\}^2}\right]$$

(15.3)

where

$$C = \frac{2 - 3\cos\theta + \cos^3\theta}{4}.$$

(15.4)

This model predicts that, for nucleation to occur at a sufficiently large number of sites, considerable surface subcooling is needed. In practice, however, preexisting nucleation embryos in the form of surface contaminants make it possible for condensation to initiate at low wall subcooling temperatures. When liquid is present on surface cracks, rapid condensation occurs at much lower subcooling. Many oxides and corrosion products are hydrophilic. Absorbed water vapor molecules on these contaminants can serve as nuclei for condensation on metallic surfaces.

Heterogeneous condensation can lead to *dropwise* or *film condensation* modes (see Fig. 15.2).

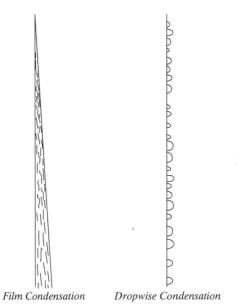

Film Condensation   *Dropwise Condensation*

Figure 15.2. Film and dropwise condensation.

Dropwise Condensation

Stable dropwise condensation takes place when the condensate liquid fails to wet the surface and form a film. The condensate forms droplets that stick to the surface.

Dropwise condensation is an extremely efficient heat transfer mode, providing condensation rates that are typically an order of magnitude larger than filmwise condensation. To facilitate dropwise condensation, industrial surfaces are made nonwetting ($\theta > 90°$) by promoters (e.g., long-chain fatty acids). Droplets then form and grow rapidly, larger droplets are removed by gravity or vapor shear, and the process resumes. Providing and maintaining the nonwetting surface characteristics can be difficult, however. The condensate liquid often gradually removes the promoters. Furthermore, the accumulation of droplets on a surface can eventually lead to the formation of a liquid film.

Film Condensation

Film condensation is the prevailing mode of condensation in most systems. The condensate, originally in the form of droplets, wets the surface, and drops coalesce and form a contiguous liquid film. The liquid film flows as a result of gravity, vapor shear, etc. Film condensation occurs in the majority of engineering applications. The flow of liquid condensate is governed by the same laws as other flow fields and involves phenomena such as laminar flow, wavy flow, laminar–turbulent transition, and droplet entrainment at the film surface.

Direct-Contact Condensation

Dropwise and film condensation processes both involve condensation on a cold solid surface. In condensers, cooling of the surface is provided by a secondary flow. The condenser is thus a heat exchanger, in which the condensing fluid stream is separated from a secondary coolant flow by a solid wall. The solid wall imposes a thermal resistance on the heat transfer between the two fluids. In the majority of applications the thermal resistance is relatively small, and it is tolerated because it is necessary to keep the condensing fluid separate from the secondary coolant. In some applications, however, the two fluid streams come into direct contact. A good example is the condensation on subcooled liquid sprays. Another example is the condensers of some open Rankine cycles, such as the direct-contact condensers in the ocean thermal energy conversion concept (Ghiaasiaan *et al.*, 1991). Direct-contact condensation is very efficient. The efficiency is not only due to the elimination of the wall resistance but more importantly due to the fact that the two streams can be mixed, resulting in large interfacial surface areas. However, direct-contact condensers are only used in special applications, because the condensate and the coolant end up mixed. In most applications it is necessary to keep the condensate and the secondary coolant separate.

15.2 Thermal Resistances in Condensation

The interfacial thermal resistance during phase change was discussed in Section 2.6, where it was explained that the interfacial thermal resistance is negligibly small in virtually all applications, except where microscale phenomena or condensation involving a liquid metal is concerned.

Let us focus on film condensation of a nonmetallic vapor in an arbitrary condensation system. Except for the very early stage of condensation, the cooled surface is typically completely covered by the condensate. The condensing vapor therefore does not directly interact with the cooled surface; instead, it faces a liquid–vapor interphase. Furthermore, the gas/vapor–liquid interfacial thermal resistance is negligible in virtually all problems of practical importance (see Section 2.6.1).

Figure 15.3 displays a condensate liquid film. For the liquid film, three convective heat transfer coefficients can in general be defined. Equivalently, we can envision three thermal resistances in series. These include heat transfer between the liquid–gas interface and film bulk, represented by H_{Fi}; heat transfer through the liquid film bulk, represented by H_{Fb}; and heat transfer between film bulk and solid surface, represented by H_{Fw}. In the condensation literature, the film heat transfer coefficient usually represents the overall liquid-side heat transfer coefficient, which can precisely be represented as

$$\frac{1}{H_F} = \frac{1}{H_{FI}} + \frac{1}{H_{Fb}} + \frac{1}{H_{Fw}}, \tag{15.5}$$

The temperature profile in a condensate film is also qualitatively shown in Fig. 15.3. In some configurations, such as condensation on a vertical surface when the condensate is a turbulent film, the thermal resistance associated with convection through the film bulk, $1/H_{Fb}$, can be much smaller than the other two terms in Eq. (15.5) and can be neglected. In that case, a mean liquid bulk temperature $\overline{T}_F$ can be defined, and there will be two main thermal resistances in the liquid.

The gas-side temperature profile near the liquid–vapor interphase depends on the conditions of the vapor phase. The case of condensation of a pure, saturated vapor is shown in Fig. 15.4. In this case the temperature in the vapor phase remains

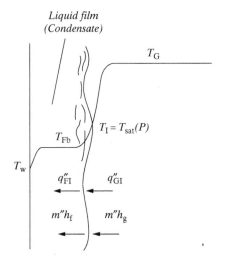

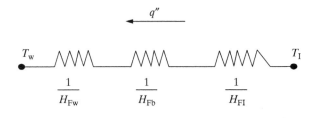

Figure 15.3. The temperature profile and thermal resistances in a condensate film, and condensation or evaporation of a pure superheated vapor.

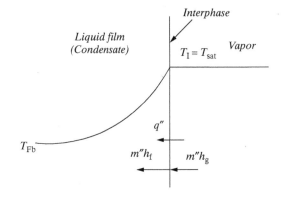

Figure 15.4. Condensation of a pure saturated vapor.

uniform at $T_{\text{sat}}(P)$, with P representing the local pressure, and the heat transfer process is *liquid-side controlled*, meaning that the bulk of the thermal resistances reside on the liquid side, and the thermal resistance on the gas side is insignificant. The temperature variations in the vicinity of the liquid–vapor interphase are similar to what is shown in Fig. 15.4, and the heat and condensation mass fluxes are related according to

$$q'' = m'' h_{\text{fg}}. \tag{15.6}$$

When a pure superheated vapor is in contact with its own liquid, the gas-side and liquid-side thermal resistances can both be important. The temperature profile at the vicinity of the interphase will be similar to that shown in Fig. 15.3. The energy

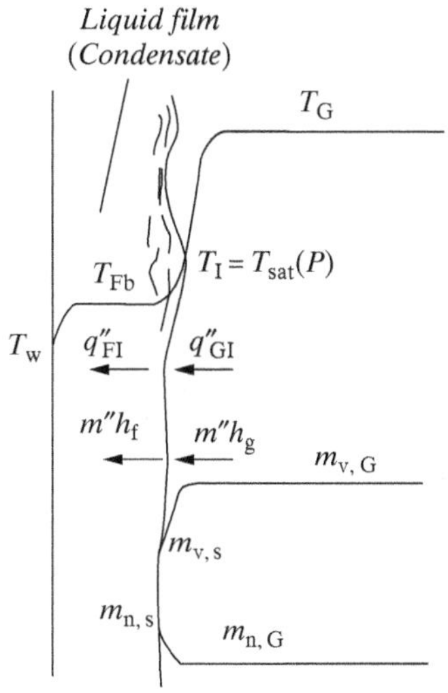

Figure 15.5. Temperature and concentration profiles in the vicinity of the interphase during condensation in the presence of a noncondensable.

fluxes associated with the interphase are also shown in Fig. 15.3. Either condensation or evaporation can take place in this case, depending on the magnitude of the energy flux terms. An energy balance on the interphase gives

$$m'' h_{fg} = q''_{FI} - q''_{GI}, \tag{15.7}$$

where

$$q''_{FI} = \dot{H}_{FI}(T_I - T_{Fb}), \tag{15.8}$$

$$q''_{GI} = \dot{H}_{GI}(T_G - T_I). \tag{15.9}$$

Evidently, condensation occurs when $m'' > 0$, and evaporation takes place when $m'' < 0$.

The thermal resistances for vapor–noncondensable mixtures are now discussed. A small quantity of a noncondensable reduces the condensation rate significantly, because a noncondensable-rich film forms near the condensate–vapor interphase. Vapor molecules have to diffuse through this film before they can reach the liquid phase, as shown qualitatively in Fig. 15.5. The vapor partial pressure at the condensate film surface is reduced by the accumulation of the noncondensable. The interphase temperature remains at saturation temperature, corresponding to the local vapor partial pressure. The vapor partial pressure near the interphase can be significantly lower than the vapor pressure away from the interphase. In this case the gas-side thermal and mass transfer resistances are both important, and gas- and liquid-side resistances should both be considered. In fact, often the gas-side resistances are predominant.

15.3 Laminar Condensation on Isothermal, Vertical, and Inclined Flat Surfaces

In this section, and the forthcoming Sections 15.4 through 15.6, analytical solutions and correlations dealing with condensation of a pure saturated vapor are discussed, starting with the classical solution of Nusselt (1916). The analytical solutions all lead to correlations for the film heat transfer coefficient, H_F. When condensation of pure

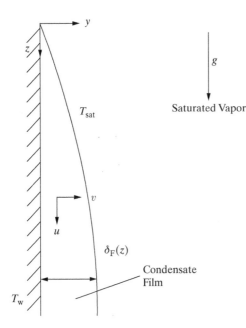

Figure 15.6. Laminar condensation on an isothermal flat vertical surface (Nusselt's analysis).

saturated vapor occurs, $1/H_F$ in fact represents the total thermal resistance between the solid surface and the vapor bulk. It should be emphasized, however, that the same correlations can be used when the vapor is not saturated or when noncondensables are present. However, in these cases $1/H_F$ will represent only the total liquid-side thermal resistance, and the calculation of the condensation rate also needs to account for the gas-side thermal and mass transfer resistance (and the gas-side mass transfer resistance as well when noncondensable are present).

Nusselt's (1916) Integral Analysis for Free Convection Condensation on a Vertical Surface

Consider laminar condensation on an isothermal, vertical flat surface (see Fig. 15.6), and assume the following:

(1) steady-state, laminar film flow;
(2) constant wall temperature;
(3) stagnant, pure, saturated vapor;
(4) zero shear stress at the film surface;
(5) a constant-property condensate liquid;
(6) negligible liquid inertia; and
(7) a linear temperature profile across the condensate.

The momentum conservation equation in the z-direction for the liquid film then becomes

$$\rho_L \underbrace{\left(u\frac{\partial u}{\partial z} + v\frac{\partial u}{\partial y} \right)}_{=0} = -\frac{dP}{dz} + \mu_L \frac{\partial^2 u}{\partial y^2} + \rho_L g. \tag{15.10}$$

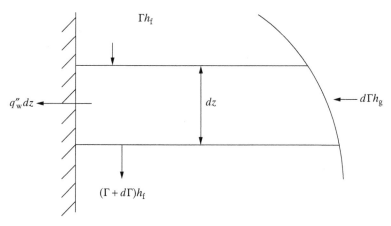

Figure 15.7. Energy flow terms for a slice of the condensate film.

Away from the surface, we must have $0 = -dP/dz + \rho_g g$, which can be used for the elimination of the pressure gradient from Eq. (15.10), giving

$$\frac{\partial^2 u}{\partial y^2} = -\frac{(\rho_L - \rho_g)g}{\mu_L}. \tag{15.11}$$

The boundary conditions for Eq. (15.11) are as follows: $u = 0$ at $y = 0$ and $\partial u/\partial y = 0$ at $y = \delta_F$.

The solution of Eq. (15.10) with these boundary conditions gives,

$$u(y) = \frac{(\rho_L - \rho_g)}{\mu_L}\left[y\delta_F - \frac{y^2}{2}\right]g. \tag{15.12}$$

The condensate film mass flux (in kilograms per meter per seconds in SI unit) will be

$$\Gamma_F = \rho_L \int_0^{\delta_F} u(y)dy = \frac{\rho_L(\rho_L - \rho_g)g}{3\mu_L}\delta_F^3. \tag{15.13}$$

Equation (15.13) contains two unknowns, Γ_F and δ_F, and therefore it cannot be solved. Closure of the equations can be achieved by performing an energy balance on an infinitesimally thin slice of the condensate film (see Fig. 15.7), the result of which will be

$$h_{fg}\frac{d\Gamma_F}{dz} = q_w''. \tag{15.14}$$

With the addition of Eq. (15.14), one new unknown q_w'' has been introduced, and closure has not been achieved yet. However, the assumption of a linear temperature profile across the film gives

$$q_w'' = k_L\frac{T_{sat} - T_w}{\delta_F}. \tag{15.15}$$

Equations (15.13)–(15.15) are now closed, since they contain three unknowns. A closed-form solution can be obtained by combining Eqs. (15.14) and (15.15) to get

$$\frac{d\Gamma_F}{dz} = \frac{k_L}{h_{fg}\delta_F}(T_{sat} - T_w).$$ (15.16)

Now, eliminating Γ_F between Eqs. (15.13) and (15.16), and solving the resulting equation for δ_F, gives

$$\delta_F = \left[\frac{4\mu_L k_L (T_{sat} - T_w)z}{gh_{fg}\rho_L(\rho_L - \rho_g)}\right]^{1/4}.$$ (15.17)

The local heat transfer coefficient at location z can now be found from

$$H_{F,z} = \frac{q_w''}{T_{sat} - T_w} = \frac{k_L}{\delta_F},$$

and it can be used in the definition of a Nusselt number $Nu_{F,z} = H_{F,z}z/k_L$ to get

$$Nu_{F,z} = \left[\frac{gh_{fg}\rho_L(\rho_L - \rho_g)z^3}{4\mu_L k_L (T_{sat} - T_w)}\right]^{1/4}.$$ (15.18)

An expression for the Nusselt number based on the average heat transfer coefficient, defined according to

$$\overline{H}_F = \frac{1}{T_{sat} - T_w}\frac{1}{l}\int_0^l H_{F,z}(T_{sat} - T_w)dz,$$

can now be easily derived:

$$\overline{Nu}_F = \frac{\overline{H}_F l}{k_L} = 0.943\left[\frac{gh_{fg}\rho_L(\rho_L - \rho_g)l^3}{\mu_L k_L (T_{sat} - T_w)}\right]^{1/4}.$$ (15.19)

Nusselt's analysis is for constant wall temperature. Equation (15.19) can be applied for the case of uniform wall heat flux, provided that $(T_{sat} - T_w)$ is replaced with (Fujii et al., 1972; Rose et al., 1999)

$$(\overline{T_{sat} - T_w}) = \frac{1}{l}\int_0^l (T_{sat} - T_w)dz.$$

Equation (15.19) can be cast in other forms, including the following:

$$\frac{\overline{H}_F}{k_L}\left[\frac{\mu_L^2}{\rho_L(\rho_L - \rho_g)g}\right]^{1/3} = 1.47Re_F^{-1/3},$$ (15.20)

$$\overline{Nu} = 0.943\left[\frac{Gr_1 Pr_L}{Ja}\right]^{1/4},$$ (15.21)

where $Re_F = 4\Gamma_F/\mu_L$ is the film Reynolds number, $Pr_L = (\mu C_P/k)_L$ is the film Prandtl number, $Ja = C_{PL}(T_{sat} - T_w)/h_{fg}$ is the Jakob number, and Gr_1 is a form of Grashof number defined as

$$Gr_1 = \rho_L(\rho_L - \rho_g)\frac{gl^3}{\mu_L^2}.$$ (15.22)

For condensation on an inclined, upward-facing flat surface the assumptions underlying Nusselt's analysis are acceptable and one can use Nusselt's analysis and everywhere replace g with $g\sin\theta$, where θ is the angle of inclination with respect to the horizontal plane.

Improvements to Nusselt's Analysis

Nusselt's analysis has three major shortcomings. First, because a saturated condensate film is assumed, the model does not account for the effect of the subcooling of the condensate film. The temperature distribution in the film varies between $T_I = T_{sat}$ and T_w, and so the condensate film is always subcooled. Second, condensation under natural convection conditions has been assumed, which is not appropriate for forced convection condensation where the shear stress at the film–vapor interphase is significant. Third, the model uses a laminar film without ripples and waves at the interphase, an assumption that is valid only for $Re_F \leq 33$ (Kapitza, 1948) (see Section 3.8). The interfacial waves cause mixing and significantly enhance the heat transfer in the film. Nusselt's analysis typically has $\pm 50\%$ inaccuracy.

Among these shortcomings, the first is very easy to fix. To correct for the effect of condensate subcooling, one can replace Eq. (15.14) with

$$q''_w = h_{fg}\frac{d\Gamma_F}{dz} + \frac{d}{dz}\int_0^{\delta_F} \rho_L C_{PL} u(T - T_{sat})dy. \tag{15.23}$$

A film temperature profile is needed. The assumption of a linear profile,

$$\frac{T_{sat} - T}{T_{sat} - T_w} = 1 - \frac{y}{\delta_F},$$

consistent with Nusselt's analysis, results in Eqs. (15.18) and (15.19), provided that h_{fg} is replaced with

$$h_{fg}^* = h_{fg}\left[1 + \frac{3}{8}\frac{C_{PL}(T_{sat} - T_w)}{h_{fg}}\right].$$

A more detailed integral analysis gives (Rohsenow, 1952)

$$h_{fg}^* = h_{fg}\left[1 + 0.68\frac{C_{PL}(T_{sat} - T_w)}{h_{fg}}\right]. \tag{15.24}$$

To account for the effect of interfacial shear and vapor motion on laminar film condensation, let us perform a momentum balance on a slice of the film shown in Fig. 15.8, where for simplicity the gravitational and pressure forces have not been shown:

$$(\delta_F - y)dz\left(\rho_L g - \frac{dP}{dz}\right) + \tau_I dz = \left(\mu_L\frac{du}{dy}\right)dz. \tag{15.25}$$

The ambient pressure gradient can in general be represented as

$$\frac{dP}{dz} = \rho_g g + \left(\frac{dP}{dz}\right)_m, \tag{15.26}$$

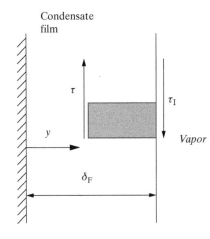

Condensate
film

Figure 15.8. Falling condensate film with interfacial
shear.

where $(dP/dz)_m$ is the pressure gradient in the vapor phase that is responsible for the vapor motion. This pressure gradient should be found from the solution of conservation equations governing the vapor phase flow field. For convenience, define a modified vapor density as

$$\rho_g^* g = \rho_g g + \left(\frac{dP}{dz}\right)_m . \tag{15.27}$$

We can now combine Eqs. (15.26) and (15.27), eliminate dP/dz in Eq. (15.25), and integrate the resulting equation, noting that $u = 0$ at $y = 0$, to get

$$u(y) = \frac{(\rho_L - \rho_g^*)g}{\mu_L}\left(y\delta_F - \frac{y^2}{2}\right) + \frac{\tau_I y}{\mu_L} . \tag{15.28}$$

The steps of Nusselt's analysis can now be repeated to get

$$\Gamma_F = \rho_L \int_0^{\delta_F} u(y)dy = \frac{\rho_L(\rho_L - \rho_g^*)g}{3\mu_L}\delta_F^3 + \frac{\tau_I \rho_L}{2\mu_L}\delta_F^2 , \tag{15.29}$$

$$h_{fg}^* \frac{d\Gamma_F}{dz} = \frac{k_L(T_{sat} - T_w)}{\delta_F} . \tag{15.30}$$

The condensation flux Γ_F can now be eliminated between Eqs. (15.29) and (15.30) to get

$$\frac{d\delta_F}{dz} = \frac{k_L\mu_L(T_{sat} - T_w)}{\rho_L(\rho_L - \rho_g^*)gh_{fg}^*\delta_F^* + \tau_I\rho_L h_{fg}^*\delta_F^2} . \tag{15.31}$$

Integration of Eq. (15.31), with $\delta_F = 0$ at $z = 0$ and with constant pressure gradient and interfacial shear stress assumed, gives

$$\frac{4zk_L\mu_L(T_{sat} - T_w)}{\rho_L(\rho_L - \rho_g^*)gh_{fg}^*} = \delta_F^4 + \frac{4\tau_I\delta_F^3}{3(\rho_L - \rho_g^*)g} . \tag{15.32}$$

Dimensionless representation of the solution gives (Rohsenow *et al.*, 1956)

$$z^* = (\delta_F^*)^4 + \frac{4}{3}(\delta_F^*)^3 \tau_I^*, \tag{15.33}$$

$$\delta_F^* = \delta_F \left[\frac{\rho_L(\rho_L - \rho_g^*)g}{\mu_L^2} \right]^{1/3}, \tag{15.34}$$

$$\tau_I^* = \left[\frac{\tau_I}{(\rho_L - \rho_g^*)g} \right] \left(\frac{\delta_F^*}{\delta_F} \right), \tag{15.35}$$

$$z^* = \left[\frac{4k_L z(T_{sat} - T_w)}{h_{fg}^* \mu_L} \right] \left(\frac{\delta_F^*}{\delta_F} \right). \tag{15.36}$$

The final correlations for a surface with length l are

$$Re_{F,l} = \frac{4\Gamma_{F,l}}{\mu_L} = \frac{4}{3}(\delta_{F,l}^*)^3 + 2\tau_I^*(\delta_{F,l}^*)^2, \tag{15.37}$$

$$\frac{\overline{H}_F}{K_L} \left[\frac{\mu_L^2}{\rho_L(\rho_L - \rho_g^*)g} \right]^{1/3} = \frac{4}{3} \frac{\delta_{F,l}^*}{l^*} + 2\frac{\tau_I^*(\delta_{F,l}^*)^2}{l^*}, \tag{15.38}$$

where parameters with subscript l correspond to $z = l$. Equations (15.33), (15.37), and (15.38) can be solved for three of the following five parameters: l^*, τ_I^*, $\delta_{F,l}^*$, $\overline{H}_{F,l}$, and $Re_{F,l}$. For example, when l, T_w, fluid properties, and interfacial shear stress are known, one can employ the following procedure to calculate the total condensation rate.

- Find l^* and τ_I^* from Eqs. (15.36) and (15.35), respectively.
- Solve Eqs. (15.33) and (15.38) for $\delta_{F,1}^*$ and $\overline{H}_F$, respectively.
- Find $Re_{F,l}$ from Eq. (15.37).

When T_w, fluid properties, τ_I, and $Re_{F,l}$ are specified, then the total length l can be found by the following procedure.

- Find τ_I^* from Eq. (15.35).
- Find $\delta_{F,l}^*$ from Eq. (15.37).
- Use $\delta_{F,l}^*$ in Eq. (5.33) and solve the equation for z^*, noting that $l^* = z^*$.
- Knowing l^*, find l from Eq. (15.36).

15.4 Empirical Correlations for Wavy-Laminar and Turbulent Film Condensation on Vertical Flat Surfaces

As mentioned earlier, falling laminar films become wavy at $Re_F \gtrsim 30$ (see Section 3.8). The interfacial waves enhance heat transfer in the film. Transition from the wavy-laminar to the turbulent flow regime occurs in falling films over the range $1000 \lesssim Re_F \lesssim 1800$ (Edwards *et al.*, 1979). Turbulent falling films are wavy, and the amplitude of their waves is typically 2–5 times the average thickness of the film. The hydrodynamics of turbulent falling film are complicated and preclude a simple, closed-form analytical solution.

The subject of mass transfer in falling films is of great interest in the chemical and process industries, because falling films are relatively easy to generate and control and have extremely large surface-to-volume ratios. Consequently, mass transfer between falling liquid films and their surrounding gas has been extensively studied in the past. In some studies, it has been shown that the behavior of the film can be predicted relatively well if it is idealized as a film with a uniform thickness (i.e., no waves) and an appropriate eddy diffusivity profile is assumed for the liquid in the film. In this case an integral analysis similar to Nusselt's analysis can be performed. The analysis of course needs a numerical solution. For example, Seban (1954) performed such an analysis using the universal law-of-the-wall profile in the liquid film. Turbulent eddies are damped by the wall as well as the film–gas interphase, however. More recent studies of mass or heat transfer in turbulent films include those by Chun and Seban (1971), Mills and Chung (1973), Sandal (1974), Habib and Na (1974), Subramanian (1975), Hubbard *et al.* (1976), Seban and Faghri (1976), and Shmerler and Mudawar (1988). Nevertheless, for turbulent condensate films, this type of analysis is cumbersome and time consuming. Empirical correlations are often used instead. Some of the widely applied correlations are now reviewed.

The correlation of Chen *et al.* (1987) is valid for $Re_F > 30$:

$$\frac{\overline{H}_F}{k_L}\left(\frac{v_L^2}{g}\right)^{1/3} = \left[Re_{F,l}^{-0.44} + 5.82 \times 10^{-6}Re_{F,l}^{0.8}\,Pr_L^{1/3}\right]^{1/2}. \tag{15.39}$$

This correlation applies to wavy–laminar and turbulent condensate conditions.

A correlation by Grober *et al.* (1961) that is recommended for $Re_F \geq 1{,}400$ and applies for $1 < Pr_L < 5$, is

$$\frac{\overline{H}_F}{k_L}\left(\frac{v_L^2}{g}\right)^{1/3} = 0.0131Re_{F,l}^{1/3}. \tag{15.40}$$

Note that in correlations that provide $\overline{H}_F$ in turbulent film, such as the two correlations just quoted, $\overline{H}_F = \frac{1}{l}\int_{z=0}^{l} H_{F,z}dz$, and the effects of flow regime transitions have been accounted for.

Uehara and Kinoshita (1994, 1997) have performed experiments with CFC11, CFC123, and HCFC123, where condensation was body-force dominated (i.e., there was little effect of interfacial shear). They classified the condensate film flow regimes into laminar, sine wave flow, harmonic wave flow, and turbulent flow. They developed flow-regime-dependent correlations for the local condensation heat transfer coefficients that agreed well with experimental data with water and several refrigerants (Rose *et al.*, 1999). The flow regimes and their corresponding correlations for local film condensation heat transfer coefficients are as follows. Define

$$Nu_z = H_{F,z}z/k_L, \tag{15.41}$$

$$Gr_z = \left(\frac{gz^3}{v_L^2}\right)\left(\frac{\rho_L - \rho_g}{\rho_L}\right), \tag{15.42}$$

$$So = \left[\frac{3\sigma^3}{\rho_L^3 g v_L^4}\right]^{1/5} \quad \text{(Sofiata number)}, \tag{15.43}$$

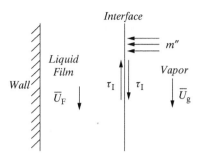

Figure 15.9. Interfacial shear.

where subscript z represents a local parameter. In the laminar film regime,

$$\text{Nu}_z = 0.707(\text{Gr}_z\text{Pr}_L/\text{Ja})^{1/4}, \tag{15.44}$$

where $\text{Ja} = C_{PL}(T_{sat} - T_w)/h_{fg}$ is the Jakob number.

Transition from the laminar flow to the sine wave flow occurs when

$$\text{Gr}_z\text{Pr}_L/\text{Ja} \geq 3.83 \times 10^{-5}\text{So}^4(\text{Ja}/\text{Pr}_L)^{-4}. \tag{15.45}$$

For the sine wave flow regime,

$$\text{Nu}_z = 1.65\text{So}^{-1/3}(\text{Ja}/\text{Pr}_L)^{1/3}(\text{Gr}_z\text{Pr}_L/\text{Ja})^{1/3}. \tag{15.46}$$

The transition from the sine wave regime to the harmonic wave regime occurs when

$$\text{Gr}_z\text{Pr}_L/\text{Ja} \geq 4.39 \times 10^{-6}\text{So}^5(\text{Ja}/\text{Pr}_L)^{-4}. \tag{15.47}$$

For the harmonic wave flow regime,

$$\text{Nu}_z = 0.725(\text{Ja}/\text{Pr}_L)^{1/15}(\text{Gr}_z\text{Pr}_L/\text{Ja})^{4/15}. \tag{15.48}$$

The transition from the harmonic wave regime to the turbulent film regime occurs when

$$\text{Gr}_z\text{Pr}_L/\text{Ja} = 1.59 \times 10^9\text{Ja}^{-4}\,\text{Pr}_L^{2.5}. \tag{15.49}$$

For the turbulent film regime,

$$\text{Nu}_z = 0.043\text{Ja}^{0.2}\text{Gr}_z^{0.4}. \tag{15.50}$$

15.5 Interfacial Shear

The sensible heat transfer and friction at the condensate–gas interface are all influenced by the waves as well as strong mass transfer resulting from condensation (see the discussions in Section 1.8 for the effect of mass transfer). The issue of interfacial heat transfer will be discussed later in Section 15.8. The interfacial shear will be discussed here.

Consider Fig. 15.9. The dependence of the interfacial shear on various parameters can formally be represented as $\tau_I = f(\overline{U}_F, \overline{U}_g, \text{Re}_F, \rho_g, \mu_g, m'')$, where $\overline{U}_F$ and $\overline{U}_g$ are the reference (usually the bulk) liquid film and gas velocities, respectively. In the absence of phase change (therefore, under conditions that the interfacial mass flux is very small or nonexistent), one can estimate τ_I from

$$\tau_I = f\frac{1}{2}\rho_g\left|\overline{U}_g - U_I\right|(\overline{U}_g - U_I), \tag{15.51}$$

where f, the Fanning friction factor, can be estimated from relevant correlations for single-phase adiabatic flow with consideration of the effect of interfacial ripples and waves. When a change of phase takes place, the effect of mass flux on shear stress must be considered.

The Couette flow film model (stagnant film model) provides a good engineering method in quasi-steady-state situations, whereby

$$\tau_I = \dot{f}\frac{1}{2}\rho_g\left|\overline{U}_g - U_I\right|(\overline{U}_g - U_I), \tag{15.52}$$

$$\frac{\dot{f}}{f} = \frac{\beta}{e^\beta - 1}, \tag{15.53}$$

$$\beta = \frac{-2m''}{\rho_g\left|\overline{U}_g - U_I\right|f}, \tag{15.54}$$

where m'' is the mass flux at the interface ($m'' > 0$ for condensation and $m'' < 0$ for evaporation).

15.6 Laminar Film Condensation on Horizontal Tubes

Condensation on the outside of horizontal tubes that carry a secondary coolant is important and occurs in most power plant condensers.

Condensation on a Single Horizontal Tube

For condensation under natural convection conditions, and when the condensate film thickness is small compared with the tube outer radius, Nusselt's analysis for vertical surfaces can be easily modified and applied.

Consider the schematic in Fig. 15.10. To modify Nusselt's analysis for vertical surfaces, assume that at any location the condensate film is similar to a film on an inclined flat surface with the local angle of inclination Ω, therefore,

$$h_{fg}\frac{d\Gamma_F}{dx} = q_w'', \tag{15.55}$$

$$q_w'' = k_L\frac{T_{sat} - T_w}{\delta_F}, \tag{15.56}$$

$$\Gamma_F = \int_0^{\delta_F} \rho u(y)dy, \tag{15.57}$$

$$u(y) = \frac{(\rho_L - \rho_g)}{\mu_L}g\sin\Omega\left[y\delta_F - \frac{y^2}{2}\right]. \tag{15.58}$$

Combining Eqs. (15.57) and (15.58), one gets

$$\Gamma_F = \rho_L(\rho_L - \rho_g)\frac{g\sin\Omega}{3\mu_L}\delta_F^3. \tag{15.59}$$

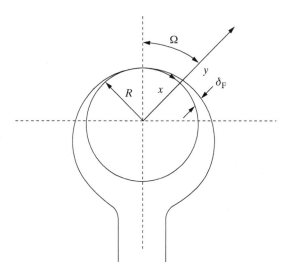

Figure 15.10. Condensation on the outside surface of a horizontal pipe.

Now, noting that $x = R\Omega$, combining Eqs. (15.55) and (15.56) results in

$$\frac{d\Gamma_F}{dx} = \frac{1}{R}\frac{d\Gamma_F}{d\Omega} = \frac{k_L(T_{sat} - T_w)}{\delta_F h_{fg}}. \tag{15.60}$$

From Eqs. (15.59) and (15.60) we can easily get

$$\Gamma_F^{1/3}d\Gamma_F = \frac{Rk_L(T_{sat} - T_w)}{h_{fg}}\left[\frac{(\rho_L - \rho_g)g}{3\nu_L}\right]^{1/3}(\sin\Omega)^{1/3}d\Omega. \tag{15.61}$$

To get the condensation rate per unit length over half of the tube surfaces we can apply $\int_0^{\Gamma_F}$ to the left side and $\int_0^{\pi}$ to the right side. Noting that $\int_0^{\pi/2}(\sin\Omega)^{1/3}d\Omega = 1.2936$ and $\int_0^{\pi}(\sin\Omega)^{1/3}d\Omega = 2\int_0^{\pi/2}(\sin\Omega)^{1/3}d\Omega$, we get

$$\Gamma_{F,1/2} = 1.923\left[\frac{R^3 k_L^3(T_{sat} - T_w)^3(\rho_L - \rho_g)g}{h_{fg}^3\nu_L}\right]^{1/4}, \tag{15.62}$$

where $\Gamma_{F,1/2}$ is one-half of the condensate mass flow rate, per unit length of the tube. An average heat transfer coefficient for the tube can now be defined, by neglecting film subcooling, from $2\pi R\overline{H}_F(T_{sat} - T_w) = 2\Gamma_{F,1/2}h_{fg}$, leading to

$$\overline{\mathrm{Nu}}_F = \frac{\overline{H}_F D}{k_L} = 0.728\left[\frac{gh_{fg}\rho_L(\rho_L - \rho_g)D^3}{\mu_L k_L(T_{sat} - T_w)}\right]^{1/4}. \tag{15.63}$$

An equivalent form of this expression is

$$\frac{\overline{H}_F}{k_L}\left(\frac{\mu_L^2}{g\rho_L(\rho_L - \rho_g)}\right)^{1/3} = 1.52\mathrm{Re}_F^{-1/3}, \tag{15.64}$$

where $\mathrm{Re}_F = 4\Gamma_F/\mu_L = 8\Gamma_{F,1/2}/\mu_L$.

Equations (15.63) and (15.64) are of course valid for laminar film. To estimate the range of validity of these derivations, one can use the regime transition for falling films discussed in Sections 3.8 and 3.9. Accordingly, the condensate film should remain laminar and smooth for $Re_F \lesssim 60$, be laminar and wavy for $60 \lesssim Re_F \lesssim 2000$ undergo laminar–turbulent transition for $2000 \lesssim Re_F \lesssim 3600$, and become turbulent for $Re_F \gtrsim 3600$. (Note that the regime transition values for Re_F are twice the corresponding values for flat surfaces because two identical films form on the two sides of the cylinder.) Condensers typically operate with film Reynolds numbers that rarely exceed the laminar–turbulent transition limit, however, and Eq. (15.63) predicts most of the relevant experimental data within 15%. However, the correlation overpredicts the condensation rate of liquid metals on horizontal tubes.

The foregoing expressions apply to natural convection conditions or situations involving low-velocity vapor flow. For forced-convection conditions where vapor flows vertically downward across a horizontal pipe, the following expression has been derived by Rose (1984) as an approximation to the solution of Shekriladze and Gomelauri (1966):

$$\frac{\overline{H}_F D}{k_L} = Re^{1/2} \frac{0.9 + 0.728 F_D^{1/2}}{\left(1 + 3.44 F_D^{1/2} + F_D\right)^{1/4}}, \tag{15.65}$$

where

$$F_D = \frac{(\rho_L - \rho_g)\mu_L h_{fg} g D}{\rho_L k_L (T_{sat} - T_w) U_\infty^2}, \tag{15.66}$$

$Re = \rho_L U_\infty D / \mu_L$, and U_∞ is the average vapor velocity. It agrees fairly well with experimental data (Rose et al., 1999).

Condensation of Pure Vapor on an In-line Bank of Horizontal Tubes

Let us consider the condensation of downward-flowing, low-velocity vapor on an in-line bank of n tubes. The analysis leading to Eqs. (15.62) through (15.64) can be extended to this problem, as displayed in Fig. 15.11, by noting that the condensate from tube 1 drains onto tube 2, and so on, such that the condensate from tubes $1, 2, \ldots,$ $n - 1$ drains onto tube n. At low inundation rates the condensate that drains from a tube falls as droplets onto the tube below. At medium inundation rates the draining condensate forms columns. At high inundation rates the draining condensate forms a falling sheet. Thus, except for the first tube, for every tube Γ is finite at $\Omega = 0$.

The derivations leading to Eqs. (15.62) through (15.64) evidently apply to the first tube. For tube 2, we must apply $f_{\Gamma_{F,1/2,1}}^{\Gamma_{F,1/2,2}}$ to the left side of Eq. (15.61) and f_0^π to the right side of that equation, leading to

$$\Gamma_{F,1/2,2}^{4/3} = \Gamma_{F,1/2,1}^{4/3} + B = 2B, \tag{15.67}$$

where $\Gamma_{F,1/2,2}$ represents half the condensate flux, per unit length, leaving tube 2, and

$$B = 2.393 \left[\frac{R^3 k_L^3 (T_{sat} - T_w)^3 \rho_L (\rho_L - \rho_g) g}{h_{fg}^3 \mu_L} \right]^{1/3}. \tag{15.68}$$

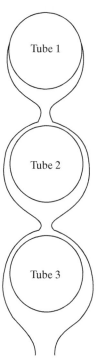

Figure 15.11. Condensation on the outside surface of a horizontal pipe.

One can proceed for tubes 3, 4, and so on. For the nth tube,

$$\Gamma_{F,1/2,n}^{4/3} = nB. \tag{15.69}$$

An average heat transfer coefficient for the entire tube bank can be defined according to

$$\overline{H}_F = \frac{2\Gamma_{F,1/2,n}h_{fg}}{n(2\pi R)(T_{sat} - T_w)},$$

and that leads to

$$\overline{Nu} = \frac{\overline{H}_F(nD)}{k_L} = 0.728\left[\frac{g\rho_L(\rho_L - \rho_g)(nD)^3 h_{fg}}{k_L\mu_L(T_{sat} - T_w)}\right]^{1/4}. \tag{15.70}$$

Alternatively, one can write

$$\frac{\overline{H}_F}{\overline{H}_{F,1}} = n^{-1/4} \tag{15.71}$$

and

$$\frac{\overline{H}_{F,n}}{\overline{H}_{F,1}} = n^{1/4} - (n-1)^{3/4}, \tag{15.72}$$

where

$\overline{H}_F$ is the average heat transfer coefficient for the entire row,
$\overline{H}_{F,1}$ is the average heat transfer coefficient for the first (top) tube (to be found from Eq. (15.63)), and
$\overline{H}_{F,n}$ is the average heat transfer coefficient for the nth tube.

Effect of Condensate Subcooling

All the previous derivations have neglected the effect of condensate subcooling. However, the effect of condensate film subcooling can be easily accounted for by replacing h_{fg} with

$$h_{fg}^* = h_{fg}\left[1 + 0.68\frac{C_{PL}(T_{sat} - T_w)}{h_{fg}}\right]. \tag{15.73}$$

EXAMPLE 15.1. Saturated, pure water vapor flows across and condenses on the outside surface of a vertical pipe that has a diameter of 6.03 cm. The pipe surface temperature is 50 °C, and the vapor phase is at 1 bar pressure. At a particular location along the pipe, the condensate flows at $\text{Re}_F = 2450$. Calculate the local condensation rate for $\mu_g = 1.23 \times 10^{-5}$ kg/m·s at that location.

SOLUTION. Let us assume the following. (1) The falling condensate film is in quasi-steady state. (2) There is symmetry around the centerline of the cylinder. The vapor properties will correspond to saturation at 100 °C. For liquid properties, let us use $\frac{1}{2}(T_w + T_{sat}) = 75$ °C as the reference temperature. Thus, $\rho_g = 0.6$ kg/m^3, $k_g = 0.025$ W/m·K, $h_{fg} = 2.257 \times 10^6$ J/kg, and $\text{Pr}_g = 1.0$. The liquid properties are $\rho_L = 975$ kg/m^3, $C_{PL} = 4{,}190$ J/kg·K, $k_L = 0.653$ W/m·K, $\mu_L = 3.79 \times 10^{-4}$ kg/m·s, and $\text{Pr}_L = 2.43$.

For the thermal resistance between the vapor–condensate interface and the cooled wall surface, we can use the falling film correlation in Eq. (3.126), whereby

$$\text{Nu}_{F,z} = 3.8 \times 10^{-3}\text{Re}_F^{0.4}\,\text{Pr}_L^{0.65} = 0.1535,$$

$$H_{F,z}\left(v_L^2/g\right)^{1/3}/k_L = 0.1535 \Rightarrow H_{F,z} = 3990 \text{ W/m}^2\text{·K}.$$

The condensation mass flux, when the vapor is saturated and pure, is then

$$m'' = H_{F,z}(T_{sat} - T_w)/h_{fg}^* = 0.0833 \text{ kg/m}^2\text{·s}.$$

Note that $h_{fg}^* = 2.40 \times 10^6$ J/kg from Eq. (15.24).

15.7 Condensation in the Presence of a Noncondensable

As mentioned, noncondensables reduce the heat transfer and condensation rates drastically, even when they are present in the vapor bulk in a small amount. Because of the noncondensables, the condensation process often becomes gas-side controlled, meaning that the thermal and mass transfer resistances on the gas side will be much larger than the liquid-side thermal resistance.

Noncondensables in fact reduce the condensation rate in quiescent as well as forced-flow situations. Their effects have been demonstrated experimentally in numerous studies (see, e.g., Al-Diwani and Rose, 1973; Lee and Rose, 1984). Old condenser design methods account for the effect of noncondensables by using purely empirical correlations. Some examples will follow. Note that in these equations the heat transfer coefficient when noncondensables are present is shown as H, and not H_F. This is because with noncondensables present H actually is an effective heat transfer coefficient accounting for all the liquid and gas-side thermal and mass transfer resistances.

The correlation of Meisenburg $et\ al.$ (1935) for condensation in vertical tubes is

$$\frac{H}{H_{F,Nu}} = \frac{1.17}{C^{0.11}}, \tag{15.74}$$

where $H_{F,Nu}$ is the local heat transfer coefficient according to Nusselt's method and $C\ (0 < C < 0.4)$ is the noncondensable (air) weight per cent in the bulk vapor–noncondensable mixture.

The correlation of Hampton (1951) for condensation on a flat plate is

$$\frac{H}{H_{F,Nu}} = 1.2 - 20\overline{m}_{n,G}, \tag{15.75}$$

where the air mass fraction in the bulk vapor–noncondensable mixture, $\overline{m}_{n,G}$, lies in the range $0 < \overline{m}_{n,G} < 0.02$.

The following correlation has been derived based on the work of Berman and Fuks (1958), for the mean gas-side heat transfer coefficient for condensation on tube rows (Chisholm, 1981):

$$\overline{H} = \frac{c_1 D_{1,2}}{D} \text{Re}_G^{1/2} \left(\frac{P}{P - P_g}\right)^{c_2} P^{1/3} \left(\frac{\rho_g h_{fg}}{T_G}\right)^{2/3} \left(\frac{1}{T_G - T_I}\right)^{1/3}, \tag{15.76}$$

where P_g is the vapor partial pressure, $T_G = T_{sat}(P_g)$ (a result of the assumption that the vapor–noncondensable mixture is saturated), $D_{1,2}$ is the vapor–noncondensable binary mass diffusivity, D is the tube outer diameter, and Re_G is the gas Reynolds number defined based on D. The coefficients c_1 and c_2 depend on Re_G and the position of the tube. The correlation is valid for downward flow, as well as horizontal flow, across a tube bundle (Marto, 1984).

For $\text{Re}_G > 350$, these coefficients are

$$c_2 = 0.6,$$
$$c_1 = \begin{cases} 0.52 & \text{(first tube row)}, \\ 0.67 & \text{(second tube row)}, \\ 0.82 & \text{(third tube row)}. \end{cases}$$

For $\text{Re}_G < 350$,

$$c_2 = 0.7,$$
$$c_1 = 0.52.$$

The Couette flow film model described in Section 1.6 is a useful tool for modeling condensation in the presence of noncondensables. Since the noncondensable gases are typically soluble in the liquid only at a trace level, we can assume that the liquid phase is compeletely impermeable to the noncondensable. Referring to Fig. 15.5, we see that the interphase temperature represents saturation with respect to the local vapor partial pressure, and we can write

$$T_I = T_{sat}(X_{v,s}P). \tag{15.77}$$

Energy balance on the interphase gives

$$\dot{H}_{GI}(\overline{T}_G - T_I) - \dot{H}_{FI}(T_I - \overline{T}_F) + m''h_{fg} = 0, \tag{15.78}$$

where $\dot{H}_{GI}$ and $\dot{H}_{FI}$ are to be found from Eqs. (2.77) and (2.78), or Eqs. (2.81) and (2.82). The condensation mass flux, m'', has to be convected through the noncondensable-rich film adjacent to the interphase, and according to Eq. (1.139) we can write

$$m'' = -K_{GI} \ln \frac{1 - m_{v,G}}{1 - m_{v,s}}, \qquad (15.79)$$

where K_{GI} is the mass transfer coefficient for the $m'' \to 0$ limit between the gas bulk and the interphase. From Eq. (1.44), the mass and mole fractions at the s surface are related according to

$$m_{v,s} = \frac{X_{v,s}M_v}{X_{v,s}M_v + (1 - X_{v,s})M_n}. \qquad (15.80)$$

Equations (15.77)–(15.80) can now be solved iteratively for the four unknowns T_I, $X_{v,s}$, $m_{v,s}$, and m''.

The preceding formulation is made on a mass flux basis. It is sometimes more convenient to use a molar-flux-based formulation. In that case, Eqs. (15.78) and (15.79) are replaced with

$$\dot{H}_{GI}(\overline{T}_G - T_I) - \dot{H}_{FI}(T_I - \overline{T}_F) + N''\tilde{h}_{fg} = 0, \qquad (15.81)$$

$$N'' = -\tilde{K}_{GI} \ln \frac{1 - X_{v,G}}{1 - X_{v,s}}, \qquad (15.82)$$

where N'' is the molar condensation flux and $\tilde{K}_{GI}$ is the molar-based mass transfer coefficient.

EXAMPLE 15.2. In Example 15.1, assume that the vapor contains air at a mass fraction of 1.0% and has a velocity of 0.85 m/s. Calculate the local condensation rate using the Couette flow film model. Compare the condensation rate with the condensation rate when the water vapor is pure.

SOLUTION. The bulk vapor temperature is assumed to be 100 °C, in view of the small noncondensable mass fraction. The thermophysical properties calculated in Example 15.1 are then applicable. The binary mass diffusivity of air and water vapor, in accordance with Appendix E, is

$$D_{12} = 0.26 \times 10^{-4}(373/298)^{3/2} = 3.64 \times 10^{-5} \text{ m}^2/\text{s}.$$

We need to deal with the vapor-side heat and mass transfer. The gas-side (vapor-side) Reynolds number is

$$\text{Re}_g \approx \rho_g \overline{U}_G D / \mu_g = 2{,}494.$$

Note that subscript G now represents the vapor and noncondensable mixture. We can use the correlation of Churchill and Bernstein (1977), which is recommended for $\text{Re}_G \text{Pr}_G \geq 0.2$ and $\text{Re}_G \text{Sc}_G \geq 0.2$ (Incropera et al., 2007):

$$\text{Nu}_{GI} = H_{GI}D/K_G = 0.3 + \frac{0.62\text{Re}_G^{1/2}\,\text{Pr}_G^{1/3}}{[1 + (0.4/\text{Pr}_G)^{2/3}]^{1/4}}\left[1 + \left(\frac{\text{Re}_G}{282{,}000}\right)^{5/8}\right]^{4/5}.$$

As an approximation, we will use Pr_g and Re_g for Pr_G and Re_G, respectively, and find

$$Nu_{GI} = 29.2,$$
$$H_{GI} = 12.2 \text{ W/m}^2 \cdot \text{s},$$

where H_{GI} is the average heat transfer coefficient over the perimeter of the circular tube.

The mass transfer coefficient can be obtained from the expression of Churchill and Bernstein by replacing Pr_G with Sc_G, where $Sc_G = \nu_G/D_{12} = 0.564$, and $\overline{Nu}_G$ with $\overline{Sh}_G$, where $Sh_{GI} = K_{GI}D/\rho_G D_{12}$. We thus get

$$Sh_{GI} = 23.3,$$
$$K_{GI} = 8.41 \times 10^{-3} \text{ kg/m}^2 \cdot \text{s}.$$

The condensation rate can now be found by the iterative solution of the following equations:

$$\dot{H}_{GI}(\overline{T}_g - T_I) - H_F(T_I - T_w) + m'' h_{fg} = 0,$$
$$T_I = T_{sat}(P_{v,s}),$$
$$P_{v,s} = X_{v,s}P,$$
$$m'' = -K_{GI} \ln \frac{1-m_{v,G}}{1-m_{v,s}},$$
$$m_{v,s} = \frac{X_{v,s}M_v}{X_{v,s}M_v + (1-X_{v,s})M_n},$$
$$\frac{\dot{H}_{GI}}{H_{GI}} = \frac{-m'' C_{Pg}}{\exp(-m'' C_{Pg}/H_{GI}) - 1},$$

where we have assumed that $\dot{H}_F \approx H_F$. Note that M_v and M_n are 18 and 29 kg/kmol, respectively, and $m_{v,G} = 0.99$. Also note that h_{fg} should now correspond to saturation at T_I. The iterative solution leads to

$$T_I = 349.2 \text{ K},$$
$$m_{v,s} = 0.291,$$
$$m'' = 0.0358 \text{ kg/m}^2 \cdot \text{s}.$$

As noted, the presence of only 1% mass of noncondensable in the bulk vapor resulted in about a 55% reduction in the condensation mass flux. The mass fraction of air at the interphase is now about 70%.

The film model described here evidently requires an iterative solution. Peterson *et al.* (1993) have developed a model (the diffusion layer model) that, unlike the stagnant film model or Couette flow film model, does not require an iterative solution for the interphase temperature. The diffusion layer model of Peterson has subsequently been used by Kagayama *et al.* (1993), Peterson (1996), and Munoz-Cobo *et al.* (1996). However, it has been shown that the diffusion layer model is in fact a weaker version of the molar-flux-based version of the stagnant film model (Ghiaasiaan and Eghbali, 1997).

15.8 Fog Formation

Fog formation occurs when a vapor–noncondensable mixture is cooled below the dew point of the vapor component. Fog formation is common in nature, resulting for example from mixing of warm and moist air with cool and relatively dry air. It can also occur in a noncondensable-rich vapor–gas boundary layer adjacent to a liquid–vapor interphase during condensation. The latter situation is the subject of discussion in this section. The liquid droplets that are generated as a result of fogging are typically very small (0.1–10 μm) and tend to remain suspended in the gas phase. Fog formation reduces the condensation rate.

Based on the stagnant film model, it is possible to predict the conditions when fog formation takes place (Brouwers 1991, 1992, 1996; Karl, 2000). According to the stagnant film model, the gas-side temperature and vapor concentration distributions in the vicinity of the liquid–gas interphase are as follows:

$$X_v(y) = X_{v,G} - (X_{v,G} - X_{v,s})e^{-\frac{N''}{CD_{v,n}}y},$$ (15.83)

$$T_G(y) = \overline{T}_G - (\overline{T}_G - T_I)e^{-\frac{N''C_{P,v}}{k_v}y},$$ (15.84)

where y is the distance from the interphase, N'' is the condensation molar flux, $\mathbf{C}$ is the total molar concentration in the gas–vapor mixture, $\mathbf{D}_{v,n}$ is the vapor–noncondensable binary mass diffusivity, and $\overline{T}_G$ is the gas bulk temperature. Fog formation takes place when the following condition is met over some portion of the gas-side boundary layer:

$$T_G(y) < T_{sat}[P_v(y)],$$ (15.85)

where $P_v(y) = PX_v(y)$. This condition is only possible when, on the gas side of the interphase,

$$\left.\frac{\partial T_G}{\partial y}\right|_{y=0} < \left.\frac{\partial T_{sat}}{\partial y}\right|_{y=0}.$$ (15.86)

Using Eq. (15.82), and the property function $T_{sat}(P_v) = f(P_v)$ for the vapor, we can examine Eq. (15.86) during the solution of the boundary layer equations to determine whether or not fog formation takes place.

Once fog formation occurs, as mentioned, the condensation rate is reduced. The film model described in Section 15.7 will then overpredict the condensation rate. Brouwer (1992, 1996) has derived a simple method for correcting the predictions of the film model for the effect of fog formation. Accordingly, in the molar-based formulation, $\tilde{K}_{GI}$ is replaced with $\tilde{K}_{GI}\theta_{cf}$, and $\dot{H}_{GI}$ is replaced with $\dot{H}_{GI}\theta_{tf}$, where θ_{cf} and θ_{tf} are correction factors to account for the effect of fog on concentration and temperature profiles, respectively, and are found from

$$\theta_{cf} = \frac{1 + \left[\frac{\tilde{h}_{fg}}{\tilde{C}_{Pv}}\frac{1}{Le}\frac{X_{v,G}-X_{v,s}}{\overline{T}_G-T_I}\frac{Sh_{GI}}{Nu_{GI}}\right]^{-1}}{1 + \left[\frac{\tilde{h}_{fg}}{\tilde{C}_{Pv}}\frac{1}{Le}\frac{dF}{dT}\Big|_{T_I}\right]^{-1}},$$ (15.87)

$$\theta_{tf} = \frac{1 + \left[\frac{\tilde{h}_{fg}}{\tilde{C}_{Pv}}\frac{1}{Le}\frac{X_{v,G}-X_{v,s}}{\overline{T}_G-T_I}\frac{Sh_{GI}}{Nu_{GI}}\right]}{1 + \frac{\tilde{h}_{fg}}{\tilde{C}_{Pv}}\frac{1}{Le}\frac{dF}{dT}\Big|_{T_I}},$$ (15.88)

where $F(T) = P_v(T)/P$ and $Le = \alpha_v/\mathbf{D}_{v,n}$.

The fog formation in a gas–vapor mixture that undergoes cooling can also be modeled by solving the energy and mass species conservation equations and assuming that no supersaturation occurs, that is, assuming immediate and complete condensation of the extra vapor (Epstein and Hauser, 1991).

15.9 Condensation of Binary Fluids

Condensation of multicomponent vapors is common in the petrochemical and cryogenic industries. The recent application of binary and ternary refrigerant mixtures, in particular, has led to intensive studies about the behavior of such mixtures. In this section we will limit the scope of the discussion to condensation of miscible binary mixtures. However, the Coulburn–Drew method, which will be the primary modeling technique discussed, can be applied to the condensation of multicomponent mixtures as well (Krishna *et al.*, 1976a, b; Webb and Sardesai, 1981; Butterworth, 1983; Taylor *et al.*, 1986; Furno *et al.*, 1986).

The liquid–vapor equilibrium for miscible binary mixtures was discussed earlier in Section 1.4, and in Section 11.8 in relation to boiling. Boiling of miscible binary liquids was also discussed in Sections 11.8 and 13.9, where phenomena that make boiling of binary mixtures more complicated than boiling in single-component liquids were discussed. Essentially the same phenomena cause the condensation of binary vapors (and of course multicomponent vapors) to be considerably more complicated than the condensation of single-component vapors. The main difference between boiling and condensation of binary mixtures is that during condensation the vapor-side resistances usually control the heat and mass transfer rate processes. In boiling the opposite is often true, and liquid-side resistances dominate the heat and mass transfer rate processes.

A distinction must be made between azeotropic and zeotropic (non-azeotropic) mixtures. In near-azeotropic mixtures the bubble and dew point temperatures are equal and the vapor and liquid phases have the same composition. These mixtures act like a single-component fluid. All theory and practice associated with single-component vapor condensation applies to these mixtures provided that all relevant mixture thermophysical properties are correctly defined and used. The single-component design and analysis techniques can also be applied for nearly azeotropic mixtures, for which the bubble and dew point temperatures are approximately equal. For zeotropic mixtures the temperature glide (i.e., the difference between bubble and dew point temperatures) and the occurrence of mass transfer resistances complicate the condensation process.

Consider the filmwise condensation of a binary vapor mixture at constant pressure, where the mole fraction of the more volatile species (species 1) in the vapor bulk is $Y_{1,v}$. Figure 15.12 shows the equilibrium phase diagram of the binary mixture at the given pressure. The bubble and dew point lines in fact specify the conditions at the liquid–vapor interphase, which is commonly assumed to be always at equilibrium. The situation with respect to equilibrium at the liquid–vapor interphase is thus similar to what was described in Section 2.6.2 with respect to the transfer of sparingly soluble species in a liquid (see Fig. 15.1).

If the condensation process rate is vanishingly slow, so that quasi-equilibrium can be assumed in the entire system, then the mole fraction of species 1 will be $Y_{1,v}$ in the entire vapor phase, and the mole fraction of species 1 in the

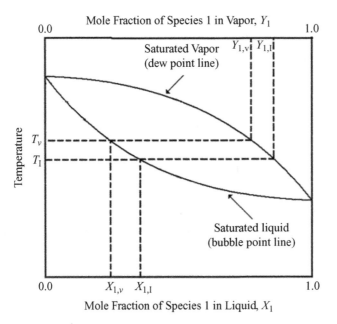

Figure 15.12. Phase diagram of a miscible binary mixture.

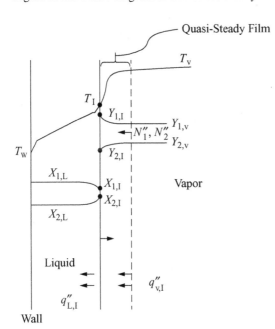

Figure 15.13. Profiles of temperature and mole fractions during filmwise condensation of a binary mixture.

condensing mass flux will be equal to $X_{1,v}$. However, the obvious fact that the condensing mass flux has a lower mole fraction of species 1 in comparison with the vapor bulk means that the vapor phase near the interphase will be partially depleted in species 2. The gas-side mole fraction of species 1 at the vicinity of the interphase will thus move to $Y_{1,I}$ in Fig. 15.12, and as a result T_I, the interphase temperature, will be lower than the vapor bulk temperature T_v. We will thus encounter a situation similar to Fig. 15.13. The condensation becomes a combined heat and mass transfer process and the condensation rate is reduced. Mass transfer resistance will play an important role because species 2, which condenses faster, needs to overcome a mass transfer resistance and diffuse towards the interphase through a species 1-rich layer. The transport of species 1 towards the interphase in the vapor phase in fact occurs in the direction of increasing concentration of that species (up the concentration gradient of species 1). Advection in this case overcomes the backward diffusion (diffusion away from the interphase), and leads to a positive condensation rate for species 1 as well.

As noted in Fig. 15.13, combined heat and mass transfer processes occur in the liquid phase as well. However, it turns out that the liquid-side mass transfer resistance is often small in comparison with the vapor-side resistance (Webb and Sardesai, 1981). In fact the entire condensation process is often gas-side controlled. A commonly applied method of analysis is based on the film theory (Colburn and Drew, 1937), which was described in Sections 1.8 and 2.7.

Consider the film condensation process depicted in Fig. 15.13. Assume that a stagnant and quasi-steady film of vapor separates the vapor bulk from the liquid–vapor interphase. The energy conservation for the film then gives:

$$q''_{L,I} = N''_1 \tilde{h}_{Lv,1} + N''_2 \tilde{h}_{Lv,2} + q''_{v,I} \tag{15.89}$$

where $\tilde{h}_{Lv,1}$ and $\tilde{h}_{Lv,2}$ are the molar-based latent heats of vaporization of the two components in the mixture. We have chosen to write the equations in terms of molar fluxes (rather than mass fluxes) for convenience here. All these equations can easily be cast in terms of mass fluxes. The sensible interfacial heat fluxes can be calculated from

$$q''_{L,I} = \dot{H}_{LI} (T_I - T_L) \tag{15.90}$$

$$q''_{v,I} = \dot{H}_{vI} (T_v - T_I). \tag{15.91}$$

The sensible heat flux between the vapor bulk and the interphase must account for the mass transfer effect. Using the film model, which is described in many textbooks (see Bird et al. (2002), Edwards et al. (1979), Ghiaasiaan (2011)), one can write

$$\dot{H}_{vI} = H_{vI} \frac{-\beta_{th,v}}{\exp(-\beta_{th,v}) - 1} \tag{15.92}$$

$$\beta_{th,v} = \left(N''_1 \tilde{C}_{Pv,1} + N''_2 \tilde{C}_{Pv,2} \right) / H_{vI}. \tag{15.93}$$

The parameter H_{vI} is the vapor–interphase heat transfer coefficient when the mass transfer rate vanishes, and can be found from models/correlations representing single-phase flow heat transfer for the prevailing geometry and flow conditions. The correction factor that is multiplied by H_{vI} on the right side of Eq. (15.92), also referred to as Ackermann's correction factor (Ackermann, 1937), is in fact identical to the correction factor on the right side of Eq. (2.77). The sensible heat transfer between the interphase and the liquid bulk should also account for the effect of mass transfer. The effect of mass transfer on the liquid-side heat transfer is usually small in filmwise condensation on vertical or inclined surfaces when the film thickness is small, however, in which case one can approximately write

$$q''_{L,I} = q''_w = H_F(T_I - T_w) \tag{15.94}$$

The heat transfer coefficient H_F represents the liquid-side heat transfer coefficient and accounts for the entire thermal resistance between the liquid–vapor interphase and the wall surface (see the discussion in Section 15.2). This heat transfer coefficient can be found by applying models and correlations representing condensation of single-component fluids, provided that correct liquid mixture thermophysical properties are applied.

The molar fluxes N_1'' and N_2'' can also be found using the film model, as described in Section 2.7, by writing

$$N_i'' = \dot{\tilde{K}}_{v,i}\left(Y_{i,v} - Y_{i,I}\right) + N_{tot}'' Y_{i,I}, \, i = 1, 2, \tag{15.95}$$

where

$$N_{tot}'' = N_1'' + N_2'' \tag{15.96}$$

$$\dot{\tilde{K}}_{v,i} = \tilde{K}_{v,i}\frac{-\beta_{ma,v}}{\exp(-\beta_{ma,v}) - 1} \tag{15.97}$$

$$\beta_{ma,v} = N_{tot}''/\tilde{K}_{v,i} \tag{15.98}$$

where $\tilde{K}_{v,i}$ is the molar-based mass transfer coefficient for component i between the vapor bulk and the vapor–liquid interphase in the limit of vanishingly small mass transfer rate. According to the film model the total molar flux follows

$$N_{tot}'' = C_{tot}\tilde{K}_{v,i}\ln\left(1 + B_m\right) \tag{15.99}$$

$$B_m = \frac{Y_{1,v} - Y_{1,I}}{Y_{1,I} - (N_1''/N_{tot}'')}. \tag{15.100}$$

In filmwise condensation the mass transfer resistance on the liquid side is often small and is neglected, leading to the approximation $X_{1,I} \approx X_{1,L}$, with $X_{1,L}$ representing the liquid bulk conditions. When the liquid side resistance is not negligible, one can write

$$N_i'' = \dot{\tilde{K}}_{L,i}\left(X_{i,I} - X_{i,L}\right) + N_{tot}''X_{i,I}, \, i = 1, 2 \tag{15.101}$$

$$\dot{\tilde{K}}_{L,i} = \tilde{K}_{L,i}\frac{\beta_{ma,L}}{\exp(\beta_{ma,L}) - 1} \tag{15.102}$$

$$\beta_{ma,L} = N_{tot}''/\tilde{K}_{L,i}, \tag{15.103}$$

where $\tilde{K}_{L,i}$ is the mass transfer coefficient between the vapor–liquid interphase and the liquid bulk at the limit of vanishing mass transfer rate.

Using Eqs. (15.89), (15.90), (15.92), and (15.94) one can derive for a filmwise condensation where the liquid-side mass transfer resistance is negligible

$$H_F\left(T_I - T_w\right) = N_1''\tilde{h}_{Lv,1} + N_2''\tilde{h}_{Lv,2} + H_{vI}\frac{\beta_{th,v}}{\exp(\beta_{th,v}) - 1}(T_v - T_I). \tag{15.104}$$

We can then write the following relation between the total condensation molar flux and the temperature difference across the liquid film

$$N_{tot}'' = H_F(T_I - T_w) \left/ \left\{\tilde{h}_{Lv,m} + \frac{\tilde{C}_{Pv,m}\left(T_v - T_I\right)}{1 - \exp(-\beta_{th,v})}\right\}\right. \tag{15.105}$$

$$\tilde{h}_{Lv,m} = \sum_{i=1}^{2}\frac{N_i''}{N_{tot}''}\tilde{h}_{Lv,i} \tag{15.106}$$

$$\tilde{C}_{Pv,m} = \sum_{i=1}^{2}\frac{N_i''}{N_{tot}''}\tilde{C}_{Pv,i}. \tag{15.107}$$

PROBLEMS

15.1 Using the method of Section 15.3, perform an analysis for laminar film condensation over the surface of the conical object shown in the figure. Note the similarity with Problem 11.7.

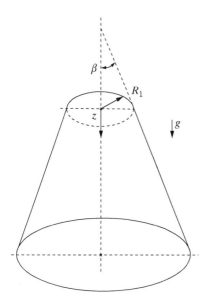

15.2 Modify the solution of Problem 15.1 for the case where a constant interfacial stress is imposed on the condensate.

15.3 On a flat, vertical surface at a distance of $x = 1.1$ m from the leading edge of the surface, a falling film of condensate water is flowing. The surface is at a uniform temperature of 50 °C. Consider film Reynolds numbers of $Re_F = 50, 250,$ and 1750.

(a) Suppose the condensation occurs in the presence of pure, saturated and quiescent water vapor at atmospheric pressure. Calculate the mean film thickness and the condensation mass flux.

(b) Suppose the vapor is pure and superheated at 120 °C and flows parallel to the condensate film with a mean velocity of 8 m/s. Determine the magnitude and nature of the phase-change mass flux (condensation or evaporation).

(c) Repeat part (b), assuming that the vapor is saturated and contains a bulk mass fraction of 4.5% of air.

15.4 The removal of condensate from a cooled surface can be crucial. Gravity is relied on in the great majority of condensation systems, but nongravitational removal is needed in some circumstances. Rotation of the condensation surface is a possible method for enhancement of condensation, where the centrifugal force facilitates the removal of condensate from a cooled surface.

Consider the rotating horizontal disk shown in the figure, where condensation of pure saturated vapor takes place on the upper surface of the disk. Assume for simplicity that the condensate moves only in the r direction and that inertial effects are negligible. Furthermore, assume that the vapor condensate film–vapor interfacial shear stress is negligible.

(a) By performing a force balance on the control volume shown in the figure, show that

$$\mu_L \frac{\partial U_L}{\partial y} = \rho_L(\delta_F - y)r\omega^2.$$

(b) Perform an analysis similar to Nusselt's analysis, and show that

$$\delta_F\sqrt{\omega/\nu_L} = 1.107\left[\frac{C_{PL}(T_{sat} - T_w)}{h_{fg}Pr_L}\right]^{1/2},$$

$$\mathrm{Nu}(r) = \frac{Hr}{k_L} = r\left[\frac{2h_{fg}\rho_L^2\omega^2}{3\mu_L k_L(T_{sat} - T_w)}\right]^{1/4},$$

where H and $\mathrm{Nu}(r)$ are the local heat coefficient and Nusselt number, respectively.

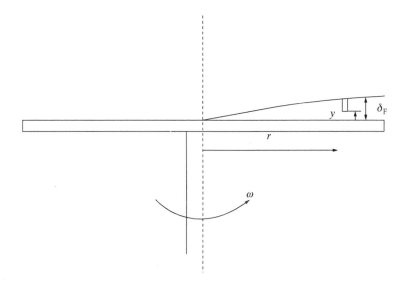

15.5 Saturated R-22 at 15.34 bars condenses on the outside of a horizontal tube that is 1.35 cm in outside diameter. The outer surface temperature of the tube is at 20 °C. Estimate the average condensation heat transfer for the following cases.

(a) A single horizontal tube in quiescent saturated vapor.

(b) A stack of seven parallel, in-line horizontal tubes in quiescent saturated vapor.

(c) A single horizontal tube, subject to a downward cross-flow of pure saturated vapor at a velocity of 2 m/s.

15.6 Saturated water vapor, at a pressure of 10 kPa, condenses on the outside of a horizontal copper tube that is 4.83 cm in outer diameter. The surface of the tube is maintained at a temperature of 27 °C.

(a) Estimate the condensation rate per unit length of tube, in kilograms per meter per second, assuming that the vapor is quiescent, pure, and saturated.

(b) Repeat part (a), this time assuming that the vapor contains a bulk concentration of 5% air by volume and flows across the tube with $Re_G = 550$, using the correlation of Berman and Fuks (1958).

15.7 Water vapor condenses on the outside of a vertical tube that is 2.13 cm in outer diameter and 1 m high. The pressure is 35 kPa, and the vapor is quiescent. The outer surface of the tube is at a uniform temperature of 275 K.

Calculate the total condensation rate, when the vapor is pure and saturated. Determine the condensate film flow regime at the vicinity of the bottom of the tube, and discuss the adequacy of ignoring the tube surface curvature in the treatment of the condensate film.

15.8 Consider the system described in Problem 15.7. Using the local conditions at the end of the tube provided by the solution of that problem, calculate the local condensation mass flux based on the Couette flow film model for the following conditions.

(a) The vapor is quiescent, saturated, and pure.

(b) The vapor bulk is quiescent and contains 5% air by volume.

(c) The vapor bulk contains 5% air by volume and moves downward, and parallel to the tube surface at a bulk velocity of 2 m/s.

For parts (b) and (c), address the likelihood of fog formation.

15.9 Condensation of atmospheric pure saturated water vapor is to take place on the outside of a vertical bundle of tubes. The tubes are configured in a square lattice, with a bundle unit cell hydraulic diameter of 8 cm. Each tube has an outer diameter of 1.37 cm. The outer surface temperature of the tubes is maintained at 65 °C. The water vapor flows upward and parallel to the tubes at a bulk velocity of 7.5 m/s. For simplicity, it can be assumed that the pressure and vapor bulk velocity remain constant in the tube bundle. Using the correlation of Eichhorn (1980), described in Problem 9.5, estimate the maximum tube height of the tube bundle to avoid flooding in the tube bundle. Discuss the adequacy of your solution with respect to the applicability of the assumptions leading to the methods you have used.

16 Internal-Flow Condensation and Condensation on Liquid Jets and Droplets

16.1 Introduction

Internal-flow condensation is encountered in refrigeration and air-conditioning systems and during some accident scenarios in nuclear reactor coolant systems. Internal-flow condensation leads to a two-phase flow with some complex flow patterns. The condensing two-phase flows have some characteristics that are different from other commonly encountered two-phase flows. Empirical correlations are available for pure vapors condensing in some simple basic geometries (e.g., horizontal circular channels). Heat transfer (condensation rate) and hydrodynamics are strongly coupled and are sensitive to the two-phase flow regime. The two-phase flow regimes themselves depend on the orientation of the flow passage with respect to gravity.

Internal-flow condenser passages are usually designed to support vertical downward flow, inclined downward flow, or horizontal flow. Configurations that can lead to unfavorable hydrodynamics (e.g., countercurrent flow limitation and loop seal effect) are avoided in these systems. As a result, most of the published experimental studies and analytical models cover vertical downflow, and horizontal flow. Condensation in unfavorable configurations can be encountered during off-normal and accident conditions of many systems, however.

Shell-side phenomena in shell-and-tube-type condensers will not be discussed in this chapter. Complex three-dimensional flow is encountered in large power plant condensers. In these condensers the condensing fluid (steam) typically flows in the shell side of the shell-and-tube-type heat exchangers, with the secondary coolant flowing inside the tubes. The shell-side flow and condensation processes have certain common features with both internal and external condensing flows. Marto (1984, 1988) has written some useful reviews of these condensers. Other studies include those by Huber *et al.* (1994a, b, c) and Huebsch and Pale (2004). Computational techniques based on the treatment of the shell-side flow field as an anisotropic porous medium have also been developed (Zhang, 1994; Ormiston *et al.*, 1995a, b).

In the forthcoming discussion and equations, the properties of the liquid phase will be designated with subscript L since slight subcooling of the liquid phase is common during condensation. It should be emphasized, however, that in most applications the properties of saturated liquid can be used as an approximation, because liquid subcooling levels that are encountered are typically small. For the same reason, the flow and equilibrium qualities (x and x_{eq}, respectively) can be used interchangeably in the flow condensation correlations.

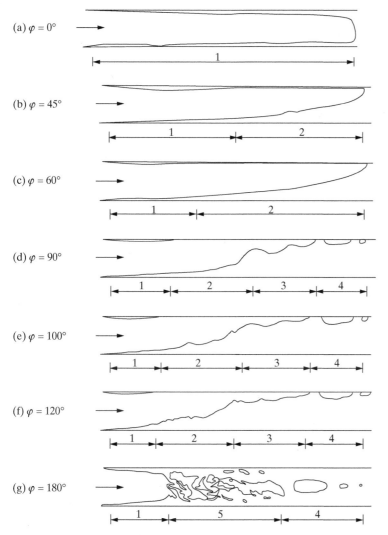

(a) $\varphi = 0°$

1

(b) $\varphi = 45°$

1 2

(c) $\varphi = 60°$

1 2

(d) $\varphi = 90°$

1 2 3 4

(e) $\varphi = 100°$

1 2 3 4

(f) $\varphi = 120°$

1 2 3 4

(g) $\varphi = 180°$

1 5 4

1 – Annular Flow; 2 – Stratified Flow; 3 – Half-Slug Flow; 4 – Slug Flow; 5 – Churn Flow

Figure 16.1. Condensation flow patterns for various angles of inclination.

16.2 Two-Phase Flow Regimes

The two-phase flow regimes depend strongly on the system orientation with respect to gravity. In downflow configuration the flow regime is predominantly annular and other flow regimes occur only near the lower end of the flow channel as a result of end effects.

In upflow configuration a variety of flow patterns occur, some resembling the flow patterns in a boiling channel. The flow regime schematics reported by Wang *et al.* (1998), shown in Fig. 16.1, are good examples. The experiments of Wang *et al.* were performed in a 1.2-m-long tube with 16-mm inner diameter, with refrigerant R-11 as the working fluid. In Fig. 16.1, φ is the angle of inclination of the channel with respect to the gravitational acceleration $\vec{g}$ ($\varphi = 0$ for vertical downward flow). As noted, among the displayed configurations, a complete sequence of the major

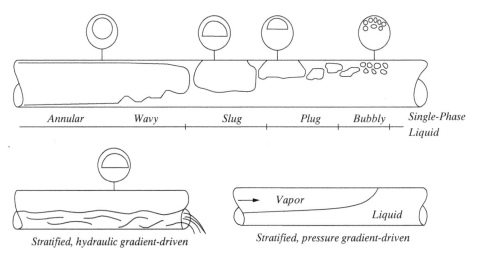

Figure 16.2. Condensation flow regimes in a horizontal tube.

flow regimes occurs only in the horizontal channel. These flow regimes are displayed schematically in more detail in Fig. 16.2.

For long tubes, commonly applied empirical flow regime maps such as those of Baker (1954) and Mandhane *et al.* (1974) (described in Chapter 4) may be used. These flow regime maps are not very accurate for condensing channels, however, in particular when they are applied to refrigerants that operate at high reduced pressure (P_r) conditions. These refrigerants have large ρ_g/ρ_f ratios, much larger than in typical air–water and steam–water applications, and therefore they should be expected to behave rather differently from air–water or steam–water mixtures. More accurate flow regime transition models that have been specifically developed for condensing pipe flows are available. Several flow regime maps have been proposed in the past and are used as the bases for regime-dependent correlations (Jaster and Kosky, 1976; Palen *et al.*, 1979; Breber *et al.*, 1980; Tandon *et al.*, 1982; Soliman, 1982, 1986; Coleman and Garimella, 2003; El Hajal *et al.*, 2003; Thome *et al.*, 2003).

Breber *et al.* (1980) argued that an important issue with respect to heat transfer (and therefore the condensation rate) during condensation in horizontal flow passages is whether the condensate flow process is gravity-controlled, or shear-controlled. In the gravity-dominated conditions one deals with wavy or stratified flow, whereas in shear-dominated flow conditions the annular flow regime is predominant, and the heat transfer is convection-like. They compared the flow regime models of Taitel and Dukler (1976a, b) for adiabatic horizontal flow with flow condensation data representing water, R-12, and R-113, and proposed the simple flow regime map depicted in Fig. 16.3. Accordingly, the flow regimes and their parameter ranges are as follows.

$$\text{Zone 1 (annular and annular-mist):} \quad j_{g,B}^* > 1.5, X_{tt} < 1.0 \tag{16.1}$$

$$\text{Zone 2 (stratified-smooth and stratified-wavy):} \quad j_{g,B}^* < 0.5, X_{tt} < 1.0 \tag{16.2}$$

$$\text{Zone 3 (bubbly):} \quad j_{g,B}^* < 1.5, X_{tt} > 1.5, \tag{16.3}$$

$$\text{Zone 4 (slug and plug):} \quad j_{g,B}^* > 1.5, X_{tt} > 1.5, \tag{16.4}$$

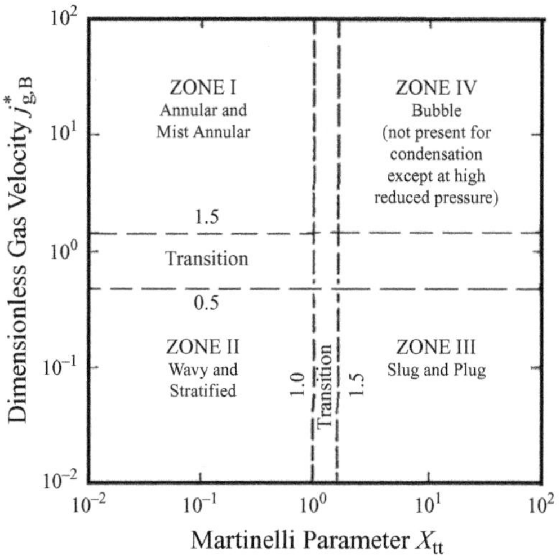

Figure 16.3. The flow regime map of Breber *et al.* (1980) for condensation in a horizontal tube.

where $j_{g,B}^*$ represents the vapor superficial velocity non-dimensionalized as in Wallis's correlation for flooding (see Eq. (9.2)), after it is combined with Martinelli's parameter defined for turbulent–turbulent flow conditions (see Eq. (8.27)):

$$j_{g,B}^* = \frac{G}{\sqrt{\rho_g(\rho_f - \rho_g)gD}}\left[\frac{K_P}{X_{tt}^{1.111} + K_P}\right] \qquad (16.5)$$

where K_P is a constant that depends on fluid properties only:

$$K_P = \left(\frac{\rho_g}{\rho_f}\right)^{0.555}\left(\frac{\mu_f}{\mu_g}\right)^{0.111}. \qquad (16.6)$$

The flow regime map evidently has relatively extensive transition zones. For calculating flow regime-dependent properties in these transition zones, including most importantly the heat transfer coefficient, Breber *et al.* (1980) recommend linear interpolation based on $j_{g,B}^*$ and X_{tt}.

The flow regime map of Tandon *et al.* (1982), displayed in Fig. 16.4, is simple and widely used. The map uses the dimensionless vapor superficial velocity, j_g^*, and the parameter $(1-\alpha)/\alpha$ as coordinates, where j_g^* represents the vapor superficial velocity non-dimensionalized as in Wallis's correlation for flooding (see Eq. (9.2)),

$$j_g^* = Gx/[gD\rho_g\Delta\rho]^{1/2}. \qquad (16.7)$$

In developing their map, Tandon *et al.* used the expression (Smith, 1969–1970):

$$\alpha = \left\{1 + (\rho_g/\rho_L)\left(\frac{1-x}{x}\right)\left[0.4 + 0.6\sqrt{\frac{\frac{\rho_L}{\rho_g} + 0.4\left(\frac{1-x}{x}\right)}{1 + 0.4\left(\frac{1-x}{x}\right)}}\right]\right\}^{-1} \qquad (16.8)$$

for predicting the void fraction.

The flow regimes and their parameter ranges are as follows.

Spray:

$$j_g^* \geq 6 \quad \text{and} \quad \frac{1-\alpha}{\alpha} \leq 0.5. \qquad (16.9)$$

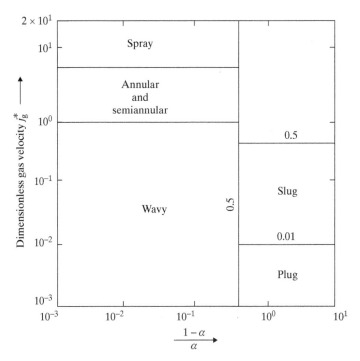

Figure 16.4. The flow regime map of Tandon *et al.* (1982) for condensing flow in a horizontal channel.

Annular semi-annular:

$$1.0 < j_g^* \leq 6 \quad \text{and} \quad \frac{1-\alpha}{\alpha} \leq 0.5. \tag{16.10}$$

Wavy:

$$j_g^* \leq 1 \quad \text{and} \quad \frac{1-\alpha}{\alpha} \leq 0.5. \tag{16.11}$$

Slug:

$$0.01 \leq j_g^* \leq 0.5 \quad \text{and} \quad \frac{1-\alpha}{\alpha} \geq 0.5. \tag{16.12}$$

Plug:

$$j_g^* \leq 0.01 \quad \text{and} \quad \frac{1-\alpha}{\alpha} \geq 0.5. \tag{16.13}$$

The original experimental data for the flow regime map of Tandon *et al.* were obtained with R-12 and R-113 refrigerants in tubes with internal diameters in the range 4.81–15.9 mm.

Doretti *et al.* (2013) have compared the predictions of several flow regime maps and flow regime transition models with extensive data representing condensing flows of refrigerants in plain and microfin tubes. The flow regime transition models that were examined included El Hajal *et al.* (2003), Breber *et al.* (1980), Taitel and Dukler (1976a, b), Tandon *et al.* (1982), and Cavallini *et al.* (2006). Among the flow regime transition models that were tested, the flow regime transition models of Breber *et al.*

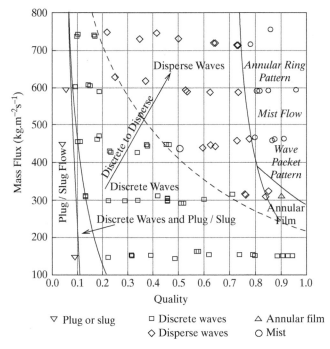

Figure 16.5. The flow regime map of Coleman and Garimella (2003) for condensation in a 4.91-mm-diameter horizontal round tube.

(1980) and Tandon *et al.* (1982) performed best. The flow regime maps, however, could not correctly predict the intermittent-annular flow regime transition for R-410A at high pressures.

Coleman and Garimella (2003) have developed the flow regime map depicted in Fig. 16.5, based on their flow condensation data obtained with R-134a refrigerant in horizontal, circular ($D = 4.91$ mm), and rectangular ($D_H = 4.8$ mm) channels. They defined the following four major flow regimes, each consisting of a number of finer flow patterns: dispersed (bubbly), intermittent (slug/plug), wavy (discrete and dispersed wavy), and annular (annular film, wave packet, wave ring, annular ring, and mist). Coleman and Garimella noted that the flow regime transition boundaries depended on channel hydraulic diameter, whereas the shape of the flow channel cross section had little effect.

The most important flow regime transition in near-horizontal channels is the boundary between the annular flow regime on the one hand and stratified/wavy/slug on the other. This is because, in the annular flow regime, which is shear-stress-dominated, the entire perimeter of the flow channel participates approximately uniformly in the heat transfer process. In the stratified/wavy or slug regimes, in contrast, the condensate flow field is dominated by gravity, and a relatively deep pool of condensate fills the bottom portion of the channel while often a thin condensate film covers the top segments of the channel. Heat transfer in this case occurs mostly through the top part of the channel, where the thermal resistance of the liquid phase is smaller. Accurate prediction of this flow regime transition is important, because it allows one to use flow-regime-dependent heat transfer correlations.

For high-pressure refrigerants, however, the convective heat transfer at the bottom becomes significant (Andresen, 2006).

A criterion for flow regime transition from stratified wavy (often simply called wavy) to annular flow during condensation in horizontal channels was derived by Soliman (1982, 1986), based on data with water, acetone, and some refrigerants in 4.8- and 25-mm-diameter tubes. The wavy flow regime defined by Soliman *et al.* included both the slug and wavy flow regimes. Accordingly, annular flow sets in when $Fr^* > 7$, and transition out of annular flow and into the aforementioned regimes takes place when $Fr^* < 7$, where Fr^* is a modified Froude number defined as

$$
Fr^* =
\begin{cases}
0.025\, Re_L^{1.59} \left[\dfrac{1 + 1.09 X_{tt}^{0.039}}{X_{tt}} \right]^{1.5} \dfrac{1}{\sqrt{Ga}} & \text{for } Re_L \le 1{,}250, \qquad (16.14) \\[4mm]
1.26\, Re_L^{1.04} \left[\dfrac{1 + 1.09 X_{tt}^{0.039}}{X_{tt}} \right]^{1.5} \dfrac{1}{\sqrt{Ga}} & \text{for } Re_L > 1{,}250, \qquad (16.15)
\end{cases}
$$

where $Re_L = G(1-x)D/\mu_L$, and the Galileo number is defined as $Ga = [\rho_L \Delta \rho g D^3 / \mu_L^2]$. Dobson *et al.* (1994) and Dobson and Chato (1998), however, have noted that the $Fr^* = 7$ threshold actually represents the change from the wavy to wavy-annular regimes. (Wavy-annular is a transitional zone on a flow regime map where both wavy and annular flow regime characteristics can be observed.) They noted that an annular flow regime represented by an approximately symmetric distribution of the condensate film over the channel periphery occurs when $Fr^* \ge 18$.

Soliman (1986) also proposed criteria for the annular to pure mist flow transition. Accordingly, the annular flow regime occurs when $We^* < 20$, purely mist flow with no liquid film can be assumed when $We^* > 30$, and annular-mist (a transitional flow regime that has the characteristics of both regimes) occurs when $20 < We^* < 30$, where We^* is a modified Weber number defined as

$$
We^* =
\begin{cases}
2.45 \dfrac{Re_g^{0.64}}{Su_g^{0.3} \left(1 + 1.09 X_{tt}^{0.039}\right)^{0.4}} & \text{for } Re_L \le 1{,}250, \quad (16.16) \\[4mm]
0.85 \left[\left(\dfrac{\mu_g}{\mu_L}\right)^2 \left(\dfrac{\rho_L}{\rho_g}\right) \right]^{0.084} \dfrac{Re_g^{0.79} X_{tt}^{0.157}}{Su_g^{0.3} \left(1 + 1.09 X_{tt}^{0.039}\right)} & \text{for } Re_L > 1{,}250, \quad (16.17)
\end{cases}
$$

where $Re_g = GxD/\mu_g$ and $Su_g = \rho_g D\sigma / \mu_g^2$ is the Suratman number. Dobson and Chato (1998), however, have indicated that a purely dispersed flow regime is unlikely in condensing channel flows, and a stable liquid film should be expected even when substantial liquid droplet entrainment is under way.

16.3 Condensation Heat Transfer Correlations for a Pure Saturated Vapor

Condensation of a pure saturated vapor in uniformly cooled flow passages has been studied extensively, leading to several well-known empirical correlations, to be discussed in this section. It is important to remember that these correlations provide the overall liquid-side heat transfer coefficient H_F, discussed in detail in the previous chapter. For a pure, saturated vapor, however, the vapor-side thermal resistance is often negligible. In the forthcoming discussion thus H and H_F are interchangeable.

16.3.1 Correlations for Vertical, Downward Flow

The Correlation of Soliman *et al.* (1968). This based on annular flow, which is the dominant flow regime in vertical downward flow, is

$$\frac{H_F(z)\mu_L}{k_L \rho_L^{1/2}} = 0.036 \mathrm{Pr}_L^{0.65} \tau_w^{1/2}, \tag{16.18}$$

where the wall shear stress τ_w has three components, representing the effects of the condensate–vapor interfacial shear, τ_I, external (gravitational) acceleration, τ_z, and flow acceleration, τ_a:

$$\tau_w = \tau_I + \tau_z + \tau_a. \tag{16.19}$$

The components are given by

$$\tau_I = \frac{D}{4}\left(-\frac{dP}{dz}\right)_{\mathrm{fr,TP}}, \tag{16.20}$$

$$\tau_z = \frac{D}{4}(1-\alpha)(\rho_f - \rho_g)g\sin\theta, \tag{16.21}$$

$$\tau_a = \frac{D}{4}\frac{G^2}{\rho_g}\left[\sum_{i=1}^{5} a_i(\rho_g/\rho_L)^{i/3}\right]\frac{dx}{dz}, \tag{16.22}$$

where θ, the angle of inclination, is defined in Fig. 5.4. The constants are defined as follows:

$$a_1 = 2x - 1 - \beta x,$$
$$a_2 = 2(1-x),$$
$$a_3 = 2(1 - x - \beta + \beta x),$$
$$a_4 = (1/x) - 3 + 2x,$$
$$a_5 = \beta[2 - (1/x) - x].$$

The parameter β represents the ratio of the interfacial velocity to the mean film velocity and is found from

$$\beta = \begin{cases} 2 & \text{(for laminar film)}, \\ 1.25 & \text{(for turbulent film)}. \end{cases}$$

The frictional pressure gradient is to be found from

$$\left(-\frac{dP}{dz}\right)_{\mathrm{fr,TP}} = \Phi_g^2\left(-\frac{dP}{dz}\right)_g, \tag{16.23}$$

where Φ_g is the two-phase multiplier. Soliman *et al.* (1968) developed a simple correlation for Φ_g as a substitute for Martinelli and Nelson's method (see Eq. (8.37)). The void fraction for substitution in the correlation of Soliman *et al.* is to be found by using Zivi's expression (1964) for the slip ratio in ideal annular flow (Eq. (6.52)) in the fundamental void-quality relation (Eq. (3.39)), which leads to

$$\alpha = \frac{1}{1 + \left(\frac{1-x}{x}\right)(\rho_g/\rho_L)^{2/3}}. \tag{16.24}$$

The Correlation of Chen *et al*. (1987). This accounts for the effect of interfacial shear and waves in vertical, downward flow:

$$\text{Nu} = H_{\text{F}}\left(\nu_{\text{L}}^2/g\right)^{1/3}/k_{\text{L}}\left[\left(\left(0.31\,\text{Re}_{\text{L}}^{-1.32} + \frac{\text{Re}_{\text{L}}^{2.4}\text{Pr}_{\text{L}}^{3.9}}{2.37 \times 10^{14}}\right)^{1/3}\right.\right.$$
$$\left.\left. + \frac{A_D\text{Pr}_{\text{L}}^{1.3}}{771.6}(\text{Re}_{\text{L0}} - \text{Re}_{\text{L}})^{1.4}\,\text{Re}_{\text{L}}^{0.4}\right]^{1/2}, \quad (16.25)$$

where

$$\text{Re}_{\text{L0}} = GD/\mu_{\text{L}}, \quad (16.26)$$

$$A_D = \frac{0.252\mu_{\text{L}}^{1.177}\mu_{\text{g}}^{0.156}}{D^2 g^{2/3}\rho_{\text{L}}^{0.553}\rho_{\text{g}}^{0.78}}. \quad (16.27)$$

EXAMPLE 16.1. Pure and saturated steam with a mass flux of $G = 80$ kg/m·s enters downward into a cooled vertical tube. The tube has an inner diameter of 18.8 mm. The pressure at the inlet is 20 bars. Assuming that the surface heat flux is uniform and equal to 6×10^4 W/m^2, calculate the total condensation rate in the top 2.5-m-long segment of the tube and the inner wall surface temperature at 2.5 m downstream from the inlet.

SOLUTION. In view of the high inlet pressure, the effect of pressure drop in the 2.5-m-long top section of the system on thermophysical properties will be small and is therefore neglected. The saturation properties that are needed are $\rho_{\text{g}} = 10.04$ kg/m^3, $\mu_{\text{g}} = 1.61 \times 10^{-5}$ kg/m·s, $k_{\text{g}} = 0.042$ W/m·K, $h_{\text{fg}} = 1.89 \times 10^6$ J/kg, $T_{\text{sat}} = 485.6$ K, and $\sigma = 0.035$ N/m. Let us also use a guessed mean liquid temperature of 483 K. The mean liquid properties will then be $\rho_{\text{L}} = 852.9$ kg/m^3, $\mu_{\text{L}} = 1.277 \times 10^{-4}$ kg/m·s, and $\text{Pr}_{\text{L}} = 0.90$.

The equilibrium quality at $z = 2.51$ m can be estimated by performing an energy balance, and neglecting the changes in kinetic and potential energy:

$$(\pi/4)D^2 G(1 - x_{\text{eq}})h_{\text{fg}} = \pi D q_{\text{w}}'' z \Rightarrow x_{\text{eq}} = 0.789.$$

The rate of condensation is

$$\dot{m}_{\text{con}} = (\pi/4)D^2 G(1 - x_{\text{eq}}) = 4.69 \times 10^{-3} \text{ kg/s}.$$

Let us apply the correlation of Chen *et al.* (1987), Eq. (16.25). The calculations will then proceed as follows:

$$\text{Re}_{\text{L0}} = GD/\mu_{\text{L}} = 11{,}780,$$
$$\text{Re}_{\text{L}} = G(1 - x_{\text{eq}})D/\mu_{\text{L}} = 2486.$$

From Eq. (16.27), we get $A_D = 2.88 \times 10^{-6}$. Equation (16.25) then leads to Nu $=$ 0.2208. The condensation heat transfer coefficient is then found as

$$H_{\text{F}} = \frac{k_{\text{L}}}{\left(\nu_{\text{L}}^2/g\right)^{1/3}}\text{Nu} = 10\,686 \text{ W/m}^2\text{·K}.$$

The local wall temperature is found from

$$T_w = T_{sat} - q_w'' / H_F = 480.6 \, \text{K}.$$

Evidently, the guessed value for the mean liquid temperature was reasonable.

16.3.2 Correlations for Horizontal Flow

Horizontal channels are among the most common configurations for internal condensing flows, and a multitude of correlations have been proposed for them. In this configuration gravitational and forced convective effects should both be considered. At high mass fluxes (typically $G \gtrsim 400 \, \text{kg/m}^2 \cdot \text{s}$) the heat transfer characteristics are dominated by forced convection and the effect of gravity is small. At mass fluxes in the $100 \lesssim G \lesssim 400 \, \text{kg/m}^2 \cdot \text{s}$ range gravity and forced convective effects are both important. Gravitational effects dominate the flow field at very low mass fluxes.

The Correlation of Akers et al. (1959). This correlation is recommended for condensation in horizontal tubes for $0 \le x \le 1$ and $G < 200 \, \text{kg/m}^2 \cdot \text{s}$ (Collier and Thome, 1994):

$$\frac{H_F D}{k_L} = C \, \text{Re}_{TP}^n \text{Pr}_L^{1/3}, \tag{16.28}$$

$$\text{Re}_{TP} = G[(1 - x) + x(\rho_L/\rho_g)^{1/2}]D/\mu_L \tag{16.29}$$

where

$$C = \begin{cases} 0.0265, n = 0.8 & \text{for } \text{Re}_{TP} > 50\,000, \tag{16.30} \\ 5.03, n = 1/3 & \text{for } \text{Re}_{TP} < 50\,000. \tag{16.31} \end{cases}$$

The Correlation of Shah (1979). Shah's correlation is among the most widely used, and it is particularly popular for forced-convection-dominated condensation (Koyama and Yu, 1999). The correlation uses the two-phase multiplier concept, whereby $H_F = H_{L0} f(x, \ldots)$, with H_{L0} the heat transfer coefficient if all the fluid mixture were pure liquid and the function $f(x, \ldots)$ a two-phase multiplier (correction factor). Shah's correlation is

$$\frac{H_F}{H_{L0}} = (1 - x)^{0.8} + \frac{3.8 x^{0.76}(1 - x)^{0.04}}{P_r^{0.38}}, \tag{16.32}$$

where $P_r = P/P_{cr}$ is the reduced pressure and H_{L0} can be found using the Dittus and Boelter (1930) correlation:

$$H_{L0} D / k_L = 0.023 (GD/\mu_L)^{0.8} \text{Pr}_L^{0.4}. \tag{16.33}$$

The recommended range of Shah's correlation is $10.8 < G < 1599 \, \text{kg/m}^2 \cdot \text{s}$, $\text{Re}_{L0} > 350$ for tubes and $\text{Re}_{L0} > 3000$ for annuli, $0.002 < P/P_{cr} < 0.44$, $\text{Pr}_L > 0.5$, and $0 < x < 1$ (Koyama and Yu, 1999).

The Correlation of Dobson and Chato (1998). This correlation is based on experimental data with R-12, R-22, R-134a, and R-32/R-125 mixtures, covering the quality range $x = 0.1$–0.9, in tubes with 3.14- to 7.04-mm inner diameter. When $G \ge 500 \, \text{kg/m}^2 \cdot \text{s}$, the following is recommended for all quality values:

$$\text{Nu}_F = H_F D/k_L = 0.023 \, \text{Re}_L^{0.8} \text{Pr}_L^{0.4} \left[1 + \frac{2.22}{X_{tt}^{0.89}}\right]. \tag{16.34}$$

Equation (16.34) in fact represents the purely annular flow regime. When $G <$ 500 kg/m^2·s, Dobson and Chato recommend Eq. (16.34) as long as Fr* > 20, where Fr* is defined in Eqs. (16.14) and (16.15). (Note that the latter condition ensures annular flow according to the observation of Dobson and Chato (1998) about the flow regime map of Soliman (1982, 1986); see the discussion below Eqs. (16.14) and (16.15).) For Fr* < 20 the flow regime is intermittent, wavy-stratified, or stratified, and the heat transfer coefficient is correlated as

$$\mathrm{Nu_F} = \frac{0.23\,\mathrm{Re}_{g0}^{0.12}}{1 + 1.1 X_{tt}^{0.58}} \left[\frac{\mathrm{GaPr_L}}{\mathrm{Ja_L}}\right]^{0.25} + \left(1 - \frac{\theta_L}{\pi}\right) \left[0.0195\,\mathrm{Re}_L^{0.8}\mathrm{Pr}_L^{0.4}\sqrt{1.376 + \frac{c_1}{X_{tt}^{c_2}}}\right],$$
(16.35)

where the liquid Jakob number is defined as $\mathrm{Ja_L} = C_{PL}(T_{sat} - T_w)/h_{fg}$, and

$$\theta_L \approx \pi - \cos^{-1}(2\alpha - 1). \tag{16.36}$$

The values of the constants c_1 and c_2 depend on the magnitude of $\mathrm{Fr_{L0}} = G^2/\rho_L^2 gD$. For $\mathrm{Fr_{L0}} \leq 0.7$,

$$c_1 = 4.172 + 5.48\mathrm{Fr_{L0}} - 1.564\mathrm{Fr_{L0}^2}, \tag{16.37}$$

$$c_2 = 1.773 - 0.169\mathrm{Fr_{L0}}. \tag{16.38}$$

For $\mathrm{Fr_{L0}} > 0.7$,

$$c_1 = 7.242, \tag{16.39}$$

$$c_2 = 1.655. \tag{16.40}$$

It should be noted that the transition between the different flow regimes at $G = 500$ kg/m^2·s is not smooth, which can lead to unrealistic jumps and discontinuities when the mass flux is close to 500 kg/m^2·s.

The Correlation of Shah (2009). Shah (2009) has improved and extended his aforementioned correlation for condensation in plane tubes described earlier (see Eqs. (16.32) and (16.33)). The revised correlation, which applies to all flow channel orientations (with some exceptions for upward flow, to be described later) is as follows.

First, a distinction is made between vertical or inclined tubes on one hand, and horizontal tubes on the other hand. For vertical and inclined tubes three regimes are defined, as shown in Fig. 16.6(a). For horizontal tubes only two regimes are defined, as displayed in Fig 16.6(b). Let us define Shah's parameter as

$$Z = \frac{(1 - x_{eq})^{0.8}}{x_{eq}^{0.8}} P_r^{0.4} \tag{16.41}$$

The boundaries among the regimes in Figs. 16.6(a) and (b) can be represented as follows:

Regime I and II in Fig. 16.6(a): $j_g^* \geq (2.4Z + 0.73)^{-1}$ (16.42)

Regime II and III in Fig. 16.6(a): $j_g^* \leq 0.89 - 0.93\exp(-0.087Z^{-1.17})$ (16.43)

Regimes I and II in Fig. 16.6(b): $j_g^* \geq 0.98(Z + 0.263)^{-0.62}$ (16.44)

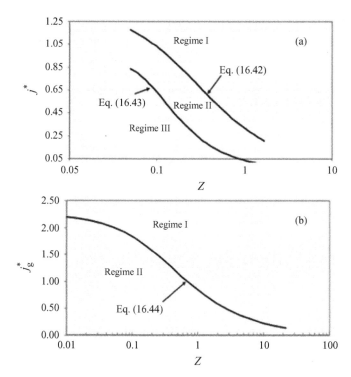

Fig. 16.6 Heat transfer regimes according to the correlation of Shah (2009): (a) vertical and inclined tube; (b) horizontal tube.

where j_g^* is defined similarly to Eq. (16.7). Now, define two single-phase heat transfer correlations:

$$H_{\mathrm{I}} = H_{f0}\left(\frac{\mu_f}{14\mu_g}\right)^{0.0058+0.557(P/P_{cr})}\left[(1-x)^{0.8} + \frac{3.8x^{0.76}(1-x)^{0.04}}{P_r^{0.38}}\right] \quad (16.45)$$

$$H_{\mathrm{Nusselt}} = 1.32\,\mathrm{Re}_{f0}^{-1/3}\left[\frac{\rho_f(\rho_f - \rho_g)gk_f^3}{\mu_f^2}\right]^{-1/3}. \quad (16.46)$$

Note that Eq. (16.46) is the same as laminar film condensation inside a vertical tube according to Nusselt's analysis.

The heat transfer coefficients can now be specified for each regime.

For regime I for all orientations,

$$H = H_{\mathrm{I}}. \quad (16.47)$$

For regime II for all orientations,

$$H = H_{\mathrm{I}} + H_{\mathrm{Nusselt}}. \quad (16.48)$$

This equation is recommended for horizontal tubes only when $\mathrm{Re}_{g0} \geq 3.5 \times 10^4$.

For regime III in vertical tubes

$$H = H_{\mathrm{Nusselt}}. \quad (16.49)$$

For vertical and inclined tubes the correlation applies to downward flow only, except for regime III in vertical tubes. In the latter case, because the shear stress at the liquid–vapor interphase is negligible, the condensing vapor flow can be downwards or upwards, and the liquid film will flow downward in both situations. Shah's correlation is based on a data base that includes 22 fluids (water, halocarbon and hydrocarbon refrigerants, and organics) covering the following range of parameters:

$$2 \, \text{mm} \le D \le 49 \, \text{mm}$$
$$8 \times 10^{-3} \le (P/P_{cr}) \le 0.9$$
$$4 \le G \le 820 \, \text{kg/m}^2 \cdot \text{s}$$
$$68 \le \text{Re}_{f0} \le 8.5 \times 10^4.$$

EXAMPLE 16.2. Pure steam with a mass flux of $G = 145$ kg/m^2·s condenses in a horizontal tube with an inner diameter of 4.75 cm. The steam is saturated at the inlet and is at a pressure of 3 bars. Assume for simplicity that the pressure remains uniform in the tube. The wall temperature is also uniform, at 120 °C. Specify the flow regime, and calculate the condensation rate in kilograms per meter at a location where $x_{eq} = 0.23$.

SOLUTION. The relevant properties are $T_{sat} = 133.7$ °C, $\rho_f = 931.9$ kg/m^3, $\rho_g = 1.65$ kg/m^3, $\mu_f = 2.07 \times 10^{-4}$ kg/m·s, $\mu_g = 1.34 \times 10^{-5}$ kg/m·s, $h_{fg} = 2.164 \times 10^6$ J/kg, and $Pr_f = 1.32$. For condensation of pure and saturated vapor $x_{eq} \approx x$. We therefore find

$$\text{Re}_{f0} = GD/\mu_f = 33\,288,$$
$$\text{Re}_f = G(1-x)D/\mu_f = 25,632.$$

Let us use the flow regime map of Tandon *et al.* (1982). From Eqs. (16.7) and (16.2), respectively, we will get $j_g^* = 1.248$ and $\alpha = 0.946$. From there, we will get $(1-\alpha)/\alpha = 0.058$. Thus, according to Eq. (16.10), the flow regime is annular/semi-annular.

We will now double check the correctness of the flow pattern by the method of Soliman (1982, 1986):

$$\text{Ga} = \rho_f \Delta \rho g D^3 / \mu_f^2 = 2.129 \times 10^{10},$$
$$X_{tt} = (\rho_f/\rho_g)^{0.5}(\mu_f/\mu_g)^{0.1}\left(\frac{1-x}{x}\right)^{0.9} = 0.164.$$

From Eq. (16.15), we get Fr$^* = 14.31$. Since Fr$^* > 7$, the flow regime should be annular.

To find the condensation rate, let us use the correlation of Shah (1979). Using Eq. (16.33), we get $H_{f0} = 1503$ W/m^2·K. Also,

$$P_r = P/P_{cr} = (3/220.6) = 0.0136.$$

Equation (16.32) then gives $H_F = 10\,693$ W/m^2·K. The condensation rate will then be found by writing

$$\pi D H_F (T_{sat} - T_w)/h_{fg} = 0.0101 \, \text{kg/m·s}.$$

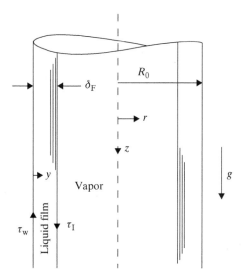

Figure 16.7. Schematic of annular flow.

16.3.3 Semi-Analytical Models for Horizontal Flow

The annular flow regime is predominant during flow condensation, as noted earlier. For condensation of a pure vapor, the bulk of the thermal resistance is in the liquid film covering the cooled surface. The relatively simple configuration of annular flow makes the semi-analytical modeling of the condensation process possible. Semi-analytical models have been relatively successful for this regime.

The Method of Kosky and Staub (1971)
Consider the annular flow depicted in Fig. 16.7. The liquid film thickness can be estimated from the following two correlations, which are for laminar and turbulent liquid films, respectively (Kosky and Staub, 1971):

$$\delta_F^+ = \begin{cases} \delta_F U_\tau/\nu_L = \sqrt{Re_L/2} & \text{for } Re_L = G(1-x)D/\mu_L \lesssim 1000, \qquad (16.50) \\ \\ 0.0504\, Re_L^{7/8} & \text{for } Re_L \gtrsim 1000, \qquad\qquad (16.51) \end{cases}$$

where $U_\tau = \sqrt{\tau_w/\rho_L}$. One can idealize the liquid film by assuming that its turbulent aspects are similar to the turbulence in a boundary layer. One can also assume that the liquid film thickness is much smaller than the radius of curvature of the cooled surface. The dimensionless velocity and temperature distributions in the liquid film will then follow the law-of-the-wall profiles. The dimensionless liquid temperature at the surface of the film is defined as

$$T_\delta^+ = \rho_L C_{PL} U_\tau (T_w - T_{\delta_F})/q_w'',$$

where T_{δ_F} is the temperature at the surface of the liquid film. Parameter $T_{\delta_F}^+$ can then be found from the turbulent boundary layer temperature law of the wall described

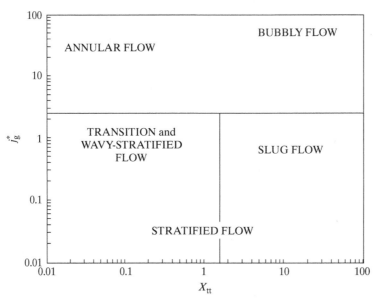

Figure 16.8. Condensation heat transfer regimes used in the method of Cavallini *et al.*

in Section 1.7 (Martinelli, 1947):

$$T_\delta^+ = \begin{cases} \delta_F^+ \mathrm{Pr_L} & \text{for } \delta_F^+ \leq 5, & (16.52) \\ 5\{\mathrm{Pr_L} + \ln[1 + \mathrm{Pr_L}(\delta_F^+/5 - 1)]\} & \text{for } 5 < \delta_F^+ < 30, & (16.53) \\ 5[\mathrm{Pr_L} + \ln(1 + 5\mathrm{Pr_L}) + 0.495\ln(\delta_F^+/30)] & \text{for } \delta_F^+ \geq 30. & (16.54) \end{cases}$$

The wall shear stress is needed for the closure of these equations. It can be found from $\tau_w = \frac{D}{4}(-dP/dz)_f$, where the two-phase frictional pressure gradient can be found from an appropriate correlation. Thus, knowing G and x, one can find δ_F^+ from Eq. (16.50) or (16.51), whichever is appropriate, and then calculate T_δ^+ from Eqs. (16.52)–(16.54), whichever is appropriate. Then, if T_w is known, q_w'' can be found from the definition of T_δ^+. Likewise, if q_w'' is known, then T_w can be found from the definition of T_δ^+. The local heat transfer coefficient will then be found from $H_F = q_w''/(T_w - T_{L,\delta_F})$. Note that $T_{\delta_F} = T_{\text{sat}}$ (P) when condensation of a pure vapor is considered.

Kosky and Staub (1971) applied the foregoing formulation, using a semi-empirical correlation for the frictional pressure gradient that was developed by Wallis (1969) based on the method of Lockhart and Martinelli (1948).

The Method of Cavallini *et al.*
The method of Cavallini *et al.* (2001, 2002, 2003, 2006) is based on the flow regime map shown in Fig. 16.8. The data base utilized by Cavallini *et al.* is extensive and included data obtained with R-22, R-32, R-125, R410A, R-236ea, R-134a, and R-407c. It covers $P_r < 0.75$, $\rho_f/\rho_g > 4$, and a wide range of mass fluxes and qualities, but its data base is limited to $D \gtrsim 2$ mm. The main elements of the method are correlations for the heat transfer coefficient in the stratified and annular regimes. For other regimes, the correlations for these two regimes are used in interpolation schemes. In

the method of Cavallini *et al.*, the void fraction is found from Eq. (16.24). For frictional pressure drop, the correlation of Friedel (1979), Eq. (8.38), is used for all flow regimes except annular flow. In annular flow, the following modified version of the correlation of Friedel (1979) is used:

$$\Phi_{L0}^2 = A + 1.262 x^{0.6978} \left(\frac{\rho_L}{\rho_G}\right)^{0.3278} \left(\frac{\mu_G}{\mu_L}\right)^{-1.181} \left(1 - \frac{\mu_G}{\mu_L}\right)^{3.477} We^{-0.1458}, \quad (16.55)$$

where $A = (1-x)^2 + x^2 \rho_L f_{G0}(\rho_G f_{L0})^{-1}$. In comparison with the original correlation of Friedel (see Eq. (8.38)), the Froude number Fr has not been included in this correlation. The single-phase Fanning friction factors f_{L0} and f_{G0} are calculated by using the following expression with $j = L$ or G, noting that $\text{Re}_j = GD/\mu_j$. For laminar flow, which is assumed to occur when $\text{Re}_j \le 2000$, $f_{j0} = 16/\text{Re}_j$ is to be used. Where turbulent flow is assumed, Cavallini *et al.* recommend $f_{j0} = 0.046 \, \text{Re}_j^{-0.2}$.

Annular flow is assumed when

$$j_g^* = \sqrt{\frac{\rho_g}{\Delta \rho g D}} \frac{Gx}{\rho_g} > 2.5. \quad (16.56)$$

Note that j_g^* is in fact Wallis's flooding parameter (see Chapter 9). In this regime, the aforementioned method of Kosky and Staub is used, with the following modification. For choosing between Eqs. (16.50) or (16.51), the criterion $\text{Re}_L = G(1-x)D/\mu_L = 1{,}145$ is used as the film Reynolds number at which laminar flow is replaced by turbulent flow.

In the stratified flow regime, the heat transfer coefficient is found from

$$H_{\text{strat}} = 0.725 \left[\frac{k_L^3 \rho_L \Delta \rho g h_{\text{fg}}}{\mu_L D \Delta T}\right]^{0.25} \left\{1 + 0.82 \left[\frac{1-x}{x}\right]^{0.268}\right\}^{-1} + H_{L0}(1-x)^{0.8}(1 - \theta_L/\pi), \quad (16.57)$$

where $\Delta T = T_{\text{sat}} - T_w$, and θ_L is to be found from Eq. (16.36). Equation (16.57) accounts for the condensation at the upper part of the tube, as well as the forced convection in the liquid pool at the bottom.

For the transition and wavy-stratified flow region, where $j_g^* < 2.5$ and $X_{\text{tt}} < 1.6$, the following interpolation scheme is to be used:

$$H = H_{\text{strat}} + (H_{\text{ann}}|_{j_g^*=2.5} - H_{\text{strat}})(j_g^*/2.5), \quad (16.58)$$

where H_{strat} represents the prediction of Eq. (16.57), and $H_{\text{ann}}|_{j_g^*=2.5}$ represents the heat transfer coefficient calculated from the annular flow regime method described here when $j_g^* = 2.5$.

For the region where $j_g^* < 2.5$ and $X_{\text{tt}} > 1.6$, representing the stratified–slug transition as well as the slug regimes,

$$H = H_{L0} + \frac{x}{x|X_{\text{tt}}=1.6}(H|_{X_{\text{tt}}=1.6} - H_{L0}), \quad (16.59)$$

where

$$x|X_{\text{tt}}=1.6 = \frac{(\mu_L/\mu_g)^{1/9}(\rho_g/\rho_L)^{5/9}}{1.686 + (\mu_L/\mu_g)^{1/9}(\rho_g/\rho_L)^{5/9}}. \quad (16.60)$$

The heat transfer coefficient $H|_{X_{tt}=1.6}$ should be found from Eq. (16.58) at quality $x|_{X_{tt}=1.6}$.

The Method of Moser et al. (1998)

Moser et al. (1998) developed a semi-analytical model for condensation in the annular flow regime, accounting for the effect of tube wall curvature. An equivalent Reynolds number is defined as

$$\mathrm{Re}_e = \mathrm{Re}_{L0}\Phi_{L0}^{8/7}, \tag{16.61}$$

where the frictional pressure-drop multiplier Φ_{L0} can be found from the correlation of Friedel (1979), Eq. (8.38). The condensation heat transfer coefficient is then found from

$$H_F/H_{L0} = F, \tag{16.62}$$

where H_{L0} is found from an appropriate single-phase flow correlation using Re_e, namely,

$$H_{L0} = \frac{k_L}{D_H}\mathrm{Nu}(\mathrm{Re}_e, \mathrm{Pr}_L). \tag{16.63}$$

The correction factor $F = (\overline{T}_L - T_w)/(T_{\delta_F} - T_w)$ is meant to account for the fact that, although the driving temperature difference for the *annular flow regime* is $T_{\delta_F} - T_w$ (with T_{δ_F} representing the temperature at the liquid film–vapor interphase, where $T_{\delta_F} = T_{sat}(P)$ for a pure vapor), the driving temperature difference in a *single-phase channel flow* is actually $\overline{T}_L - T_w$ with $\overline{T}_L$ representing the bulk fluid temperature. To find F, the turbulent velocity and temperature laws of the wall in a hydrodynamically fully developed and thermally developed turbulent pipe flow are utilized. For pipe flow with constant wall heat flux boundary conditions, the dimensionless temperature profile (the temperature law of the wall) is represented in terms of T^+, defined as (Kays et al., 2005)

$$T^+(y) = \frac{T_w - T(y)}{q_w''(\rho_L C_{PL}U_\tau)}, \tag{16.64}$$

where $U_\tau = \sqrt{\tau_w/\rho_L}$ and y is the distance from the wall. One can evidently write

$$F = \frac{\overline{T}_L^+}{T_\delta^+}. \tag{16.65}$$

Using Eq. (16.64), furthermore, one can easily show that

$$\overline{T}_L^+ = \frac{\sqrt{f/2}}{\mathrm{St}}, \tag{16.66}$$

where St is the Stanton number. Note that the Fanning friction factor f is to be calculated based on properties of the liquid and the effective Reynolds number Re_e. The Stanton number is found from the correlation of Petukhov (1970), Eq. (12.86). In terms of Stanton number, the correlation of Petukhov can be represented as

$$\mathrm{St} = \frac{f/2}{1.07 + 12.7\sqrt{f/2}(\mathrm{Pr}^{2/3} - 1)}. \tag{16.67}$$

Equations (16.64)–(16.67) lead to

$$F = \frac{1.07\sqrt{2/f} + 12.7\left(\mathrm{Pr}_L^{2/3} - 1\right)}{T_\delta^+}.$$

(16.68)

To close the equations, T_δ^+ is needed. This is found by using the velocity and temperature laws of the wall. First, the liquid film thickness in wall units, δ_F^+ is found from the correlations of Traviss *et al.* (1973):

$$\delta_F^+ = \begin{cases} 0.7071\,\mathrm{Re}_L^{0.5} & \text{for } \mathrm{Re}_L < 50, & (16.69) \\ 0.4818\,\mathrm{Re}_L^{0.585} & \text{for } 50 \leq \mathrm{Re}_L \leq 1125, & (16.70) \\ 0.095\,\mathrm{Re}_L^{0.812} & \text{for } \mathrm{Re}_L > 1125. & (16.71) \end{cases}$$

Then

$$T_\delta^+ = \begin{cases} \delta_F^+ \mathrm{Pr}_L & \text{for } \delta_F^+ < 5, & (16.72) \\ 5\mathrm{Pr}_L + \displaystyle\int_5^{\delta_F^+} \frac{\mathrm{Pr}_L}{1 + \mathrm{Pr}_L\left(\frac{y^+}{5} - 1\right)}\,dy^+ & \text{for } 5 < \delta_F^+ < 30, & (16.73) \\ (T^+)_{\delta_F^+=30} + \displaystyle\int_{30}^{\delta_F^+} \frac{1 - \frac{y^+}{R^+}}{\frac{1}{\mathrm{Pr}_L} - 1 + \frac{y^+}{2.5}\left(1 - \frac{y^+}{R^+}\right)}\,dy^+ & \text{for } \delta_F^+ > 30, & (16.74) \end{cases}$$

where $R^+ = \mathrm{Re}_e\sqrt{f/8}$. Using Blasius's equation, $f = 0.079\,\mathrm{Re}_e^{-0.25}$, and Eq. (16.61), one gets

$$R^+ = 0.0994\,\mathrm{Re}_e^{7/8}.$$

(16.75)

This procedure for calculating T_δ^+ is rather tedious and often needs numerical integration. For $\delta_F^+ > 30$, Moser *et al.* correlated the predictions of Eq. (16.74) as

$$F = 1.31(R^+)^{C_1}\,\mathrm{Re}_L^{C_2}\mathrm{Pr}_L^{-0.185},$$

(16.76)

with

$$C_1 = 0.126\mathrm{Pr}_L^{-0.448},$$

(16.77)

$$C_2 = -0.113\mathrm{Pr}_L^{-0.563}.$$

(16.78)

Cavallini *et al.* (2006) compared experimental data from a number of sources, covering many refrigerants including R-134, R22, R404A, and R-407C, with several correlations. Overall, the method of Moser *et al.* (1998) performed well. It could predict the data with an average error of 3.2% and a standard deviation of 18.2%.

EXAMPLE 16.3. Repeat the solution of Example 16.2, this time using the method of Moser *et al.* (1998).

SOLUTION. We need to find Φ_{f0}^2 based on the correlation of Friedel (1979), as described in Section 8.3. Applying Eq. (8.40) we get

$$f_{f0} = 0.005\,73,$$

$$f_{g0} = 0.003\,28.$$

Note that subscript f is used instead of L, because we are dealing with saturated liquid properties. Also,

$$\rho_h = 1 \Big/ \left(\frac{x}{\rho_g} + \frac{(1-x)}{\rho_f}\right) = 7.13 \, \text{kg/m}^3,$$

$$\text{We} = \frac{G^2 D}{\rho_h \sigma} = 2{,}661,$$

$$\text{Fr} = \frac{G^2}{gD\rho_h^2} = 888.6,$$

$$\text{A} = (1-x)^2 + x^2 \rho_f f_{g0}/(\rho_g f_{f0}) = 17.67.$$

Using these in Eq. (8.38) we find $\Phi_{f0} = 10.74$.

From Eq. (16.61) we get $\text{Re}_e = \text{Re}_{f0}\Phi_{f0}^{8/7} = 5.019 \times 10^5$. ($\text{Re}_{f0} = 33{,}288$, from Example 16.2.) Given that $\text{Re}_f > 1125$, we next use Eq. (16.71), and find $\delta_F^+ = 36.11$. Equation (16.75) gives $R^+ = 9670$. We then apply Eqs. (16.76), (16.77), and (16.78) to get

$$C_1 = 0.111,$$

$$C_2 = -0.0966,$$

$$F = 1.295.$$

We also need f, the Fanning friction factor based on Re_e. Using Blasius's correlation, we will have

$$f = 0.079 \, \text{Re}_e^{-0.25} = 0.00297.$$

We next need to apply Petukhov's correlation, Eq. (16.67), using Re_e, and that leads to $\text{St} = 0.00127$. Given that $\text{St} = \text{Nu}/\text{Re}_e\text{Pr}_f$, we find $\text{Nu} = 841$. Since $\text{Nu} = H_{L0}D/k_f = 841$ and $H_F = H_{L0}F$, we get

$$H_F = 1.54 \times 10^4 \, \text{W/m}^2\text{·K}.$$

The condensation rate, per unit length, will then be found by writing

$$\pi D H_F (T_{sat} - T_w)/h_{fg} = 0.0145 \, \text{kg/m·s}.$$

16.4 Effect of Noncondensables on Condensation Heat Transfer

As discussed in the previous chapter, for condensation of a pure vapor, the heat transfer coefficient H_F provided by various correlations represents the overall thermal resistance between the cooled wall and the liquid–vapor interphase (i.e., the total liquid-side thermal resistance). With a noncondensable present, however, the vapor-side thermal and mass transfer resistances become important. This is true for internal-flow condensation as well. Othmer (1929) performed one of the earliest experimental investigations into the subject, using steam–air mixtures in a 7.62-cm-diameter copper, vertical tube. He noted a 50% decrease in heat transfer rate resulting from only 0.5% air in the inlet steam. Experimental investigations dealing with condensation in tubes have also been reported by Borishanskiy *et al.* (1977,

1978), Morgan and Rush (1983), and more recently by Vierow (1990), Ogg (1991), Kagayama *et al.* (1993), and Siddique (1992), all confirming that small concentrations of a noncondensable in the vapor can significantly worsen the condensation rate. Borishanskiy *et al.* (1977) have proposed the following correlation, based on experimental data from condensation of steam in a 3-m-long, 10-cm-diameter vertical tube:

$$\overline{H}/\overline{H}_0 = 1 - 0.25\xi^{0.7}, \tag{16.79}$$

where ξ is the volume fraction of the noncondensable in the steam–noncondensable mixture at the inlet, and $\overline{H}$ and $\overline{H}_0$ represent the average overall heat transfer coefficients with and without noncondensable, respectively. (With condensation of a pure and saturated vapor being liquid-side-controlled, evidently $\overline{H}_0 = H_F$.) These overall average heat transfer coefficients are defined according to

$$\overline{H} = \frac{\overline{q}_w''}{T_{sat}(P_{v,in}) - \overline{T}_w}. \tag{16.80}$$

The investigations by Vierow (1990), Ogg (1991), and Siddique (1992), motivated by interest in the condensation phenomena in the containment of nuclear reactors have all addressed the condensation of downward-flowing steam in vertical tubes. Ogg (1991) and Siddique (1992) developed empirical correlations for the degradation of heat transfer caused by the noncondensables, based on their own data.

Unlike external-flow condensation, where far-field properties are often known or uniform, the flow field and its bulk properties change from location to location in internal-flow condensation. These changes are particularly significant when noncondensables are present, because the bulk mass fraction of the noncondensable increases as more condensation takes place. As a result, average heat transfer coefficients are of limited use, and local heat transfer needs to be considered. Models that are based on the solution of conservation equations along the channel are often used (Siddique *et al.*, 1994; Ghiaasiaan *et al.*, 1995; Yao *et al.*, 1996; Yao and Ghiaasiaan, 1996a, b; Munoz-Cobo *et al.*, 1996). In these and other similar models, the stagnant (Couette flow) film model described in the previous chapter has been successfully applied for condensation in the presence of a noncondensable.

16.5 Direct-Contact Condensation

Direct-contact (DC) condensation occurs when no solid wall separates the condensing vapor from the coolant fluid and is utilized when highly efficient condensation is required. Typically, DC condensers rely on the injection of subcooled liquid (as spray or jets) into its own vapor. DC condensation also occurs in emergency coolant injection systems of nuclear reactors.

As far as the processes at the vicinity of the vapor–condensate interphase are concerned, the principles for DC condensation are the same as those for condensation onto a condensate film or pool. The main difference is that in DC condensation the liquid bulk acts as the heat sink.

Condensation on Subcooled Liquid Droplets

Spray condensers are among the most efficient and widely used DC condensation systems. Modeling on the macroscopic scale should address the overall vapor, non-condensable, and liquid conservation equations. Droplet-level processes represent some of the critical and difficult issues and will be briefly reviewed.

Consider a subcooled spherical droplet with d_0 representing the droplet initial diameter and T_0 representing its initial (uniform) temperature. The droplet enters a pure and saturated vapor environment at uniform pressure P, where $T = T_{sat}(P) > T_0$. Assume spherical symmetry and no droplet internal circulation. The droplet grows in size as a result of condensation, and the maximum droplet diameter can be found from the following simple energy balance

$$\frac{\pi}{6}d_0^3 C_{PL}(T_{sat} - T_0) = \frac{\pi}{6}(d_{max}^3 - d_0^3)\rho_L h_{fg} \Rightarrow d_{max} = d_0[1 + C_{PL}(T_{sat} - T_0)/h_{fg}]^{1/3}. \tag{16.81}$$

For common fluids the change in diameter is typically very small. An analytical solution can therefore be easily derived by assuming that during the condensation process $d \approx d_0 = $ const. The problem then reduces to conduction of heat into a rigid sphere. Accordingly,

$$\frac{\partial T_L}{\partial t} = \alpha_L \frac{1}{r^2} \frac{\partial}{\partial r}\left(r^2 \frac{\partial T_L}{\partial r}\right), \tag{16.82}$$

where α_L is the thermal diffusivity of the liquid. The initial and boundary conditions are

$$T_L = \begin{cases} T_0 & \text{at } t = 0, \\ T_{sat} & \text{at } r = d_0/2, \end{cases}$$

$$\frac{\partial T_L}{\partial r} = 0 \quad \text{at } r = 0.$$

The average temperature of the droplet can be found from

$$\overline{T}_L = \frac{1}{\frac{\pi}{6}d_0^3} \int\limits_0^{d_0/2} 4\pi r^2 T \, dr. \tag{16.83}$$

The solution of this equations is

$$\frac{T_L - T_0}{T_{sat} - T_0} = 1 - \frac{d_0}{\pi r} \sum_{n=1}^{\infty} \frac{(-1)^n}{n} \sin\left(n\pi \frac{2r}{d_0}\right) \exp\left(-\alpha_L \eta_c \frac{4n^2\pi^2 t}{d_0^2}\right), \tag{16.84}$$

$$\frac{\overline{T}_L - T_0}{T_{sat} - T_0} = 1 - \frac{6}{\pi^2} \sum_{n=1}^{\infty} \frac{1}{n^2} \exp\left(-n^2 \eta_c \frac{4\pi^2 \alpha_L t}{d_0^2}\right), \tag{16.85}$$

where η_c is a convective enhancement factor. For a rigid droplet $\eta_c = 1$.

This solution neglects droplet internal circulation and oscillations, both of which enhance the heat transfer rate. Kronig and Brink (1950) solved the problem of transient diffusion inside a spherical fluid particle with Hills-vortex-type internal circulation. According to their solution, at the limit of $t \to \infty$, which represents a near-equilibrium solution, we will have

$$\mathrm{Nu}_{d0} = H_{\mathrm{LI}}d_0/k_{\mathrm{L}} = 17.9. \tag{16.86}$$

Calderbank and Korchinski (1956) showed that this near-equilibrium solution can be approximated with

$$\frac{\overline{T}_{\mathrm{L}} - T_0}{T_{\mathrm{sat}} - T_0} \approx \left[1 - \exp\left(-2.25\frac{4\pi^2\alpha_{\mathrm{L}}t}{d_0^2} \right) \right]^{1/2}. \tag{16.87}$$

In these solutions, as noted earlier, a constant droplet diameter has been assumed. Ford and Lekic (1973) used Eq. (16.84) for calculating the heat transfer rate to the droplet, but they used the resulting heat transfer rate to calculate the droplet growth rate. They could correlate the results of their parametric calculations with the following expression, which agreed well with their experimental data dealing with condensation of pure steam on subcooled water droplets:

$$d/d_0 = 1 + \zeta\left[1 - \exp\left(-4\pi^2\alpha_{\mathrm{L}}t/d_0^2 \right) \right]^{1/2}, \tag{16.88}$$

$$\zeta = \left[1 + \frac{C_{\mathrm{PL}}(T_{\mathrm{sat}} - T_0)}{h_{\mathrm{fg}}} \right]^{1/3} - 1. \tag{16.89}$$

A method for accounting for the effect of internal recirculation on transient condensation in a droplet is to introduce an empirical value for the convective enhancement factor η_c in Eqs. (16.84) and (16.85) (Pasamehmetoglu and Nelson, 1987). Celata *et al.* (1991) have reviewed an extensive data base and have suggested the following empirical correlation. Equations (16.84) and (16.85) are to be used, where

$$\eta_c = c_1(\mathrm{Pe}')^{c_2} \tag{16.90}$$

with $c_1 = 0.153$ and $c_2 = 0.454$, and the modified Péclet number is defined as

$$\mathrm{Pe}' = \frac{d_0 U}{\alpha_{\mathrm{L}}} \frac{\mu_{\mathrm{g}}}{\mu_{\mathrm{g}} + \mu_{\mathrm{L}}}, \tag{16.91}$$

where U is the droplet velocity relative to the surrounding vapor.

Condensation on Subcooled Liquid Jets

Condensation on slabs or cylindrical subcooled liquid jets is of interest in DC heat exchangers and during the emergency core coolant injection in light water nuclear reactors. It should be noted that liquid jets issuing from nozzles remain coherent only over a relatively short distance, shattering into droplets afterward. The following discussion addresses coherent jets.

The liquid film prior to leaving its nozzle can be in the laminar or the turbulent regime. The initial velocity profile in the liquid jet depends on the length of the nozzle and the flow regime of liquid in the nozzle. For jets issuing from short nozzles, the velocity profile at the exit from the nozzle can be relatively uniform. This is particularly true for turbulent jets. Furthermore, because of the typically very small gas-phase viscosity, the boundary condition for the jet is approximately a zero-velocity gradient at the liquid–gas interphase. As a result, the velocity profile in the jet tends to flatten as the distance from the nozzle exit increases.

For condensation on a cylindrical jet, assuming a constant jet radius ($R = R_0$) and uniform and constant velocity in the jet, and assuming that the effective thermal conductivity is constant, one can derive an analytical solution as follows. The energy equation in the jet can be written as

$$U_L(z)\frac{\partial T_L}{\partial z} = \alpha_{L,eff}\frac{1}{r}\frac{\partial}{\partial r}\left(r\frac{\partial T_L}{\partial r}\right), \tag{16.92}$$

where z is the longitudinal coordinate starting from the exit of the nozzle and $\alpha_{L,eff}$ represents the effective liquid thermal diffusivity. The boundary conditions are $T_L = T_I = T_{sat}$ at R_0 and $\partial T_L/\partial r = 0$, at $r = 0$, where subscript 0 represents the nozzle exit. Equation (16.92) can be solved by the separation of variables technique, leading to (Hasson *et al.*, 1964)

$$\frac{T_{sat} - T_L}{T_{sat} - T_{L0}} = \sum_{i=1}^{\infty} A_i j_0\left(\lambda_i\frac{r}{R_0}\right)\exp\left(-4\frac{\lambda_i^2}{Gz}\right), \tag{16.93}$$

where λ_i are the roots of the Bessel function of the first kind and zeroth order (i.e., $J_0(\lambda) = 0$), and $A_i = 2/[\lambda_i j_1(\lambda_i)]$. The Graetz number is defined as

$$Gz = \frac{4U_{L0}R_0^2}{\alpha_{L,eff}z} = 1/Fo, \tag{16.94}$$

where Fo is the Fourier number. The average liquid temperature can then be found from

$$\frac{T_{sat} - \overline{T}_L}{T_{sat} - T_{L0}} = \sum_{i=1}^{\infty}\frac{4}{\lambda_i^2}\exp\left(-4\frac{\lambda_i^2}{Gz}\right). \tag{16.95}$$

The liquid-side interfacial heat transfer coefficient can now be obtained from

$$H_{LI}(z) = \frac{-k_{L,eff}\frac{\partial T_L}{\partial r}\big|_{r=R_0}}{(T_{sat} - \overline{T}_L)}, \tag{16.96}$$

where $k_{L,eff}$ is the effective thermal conductivity of liquid. Equation (16.96) leads to

$$Nu(z) = \frac{H_{LI}(z)D_0}{k_{L,eff}} = \frac{\sum_{i=1}^{\infty}\exp\left(-4\frac{\lambda_i^2}{Gz}\right)}{\sum_{i=1}^{\infty}\frac{1}{\lambda_i^2}\exp\left(-4\frac{\lambda_i^2}{Gz}\right)}, \tag{16.97}$$

where D_0 is the initial diameter of the jet. This solution leads to the following limiting cases:

$$Nu(z) = 5.784 \qquad \text{for small Gz(or large} z/R_0), \tag{16.98}$$

$$Nu(z) \approx \sqrt{Gz/\pi} \quad \text{for large Gz (small } z/R_0). \tag{16.99}$$

With $k_{L,eff} = k_L$ (or equivalently $\alpha_{L,eff} = \alpha_L$) the jet is assumed to remain laminar and the solution here evidently can be considered as a conservative, lower bound on the jet heat transfer. For turbulent jets, Kutateladze (1952) proposed

$$\alpha_{L,\text{eff}} = \alpha_L + \varepsilon^* R U_L, \tag{16.100}$$

where R and U_L are the radius and velocity, respectively, of the jet at location z, and ε^* is an empirical parameter.

In a downward-flowing jet, the radius and velocity vary with z according to

$$U_L(z) = U_{L0}\left[1 + \frac{2gz}{U_{L0}^2}\right]^{1/2}, \tag{16.101}$$

$$R/R_0 = \sqrt{U_{L0}/U_L}. \tag{16.102}$$

The solution to the liquid-side energy conservation equation, accounting for the variation of R and U_L with z, is complicated. In a turbulent jet, however, U_{L0} is typically large and the reduction in the jet radius over the distance where the jet remains coherent is small. With the $R = R_0$ (and therefore $U_L = U_{L0}$) assumption, the solution leads to the following result for the local heat transfer coefficient (Kutateladze, 1952):

$$\text{Nu}_L(z) = (1 + \varepsilon^* R_0 U_{L0}/\alpha_L) \cdot \frac{\sum_{i=1}^{\infty} \exp\left(-4\lambda_i^2/\text{Gz}_{\text{turb}}\right)}{\sum_{i=1}^{\infty} \frac{1}{\lambda_i^2} \exp\left(-4\lambda_i^2/\text{Gz}_{\text{turb}}\right)}. \tag{16.103}$$

The turbulent Graetz number is defined as

$$\text{Gz}_{\text{turb}} = \frac{4}{z}\left(\frac{\alpha_L}{U_{L0}R_0^2} + \frac{\varepsilon^*}{R_0}\right)^{-1}. \tag{16.104}$$

Kutateladze proposed $\varepsilon^* = 5 \times 10^{-4}$. This solution leads to the following limiting cases:

$$\text{Nu}(z) = 5.784\left(1 + \frac{\varepsilon^* R_0 U_{L0}}{\alpha_L}\right) \quad \text{for small Gz}_{\text{turb}}, \tag{16.105}$$

$$\text{Nu}(z) = \sqrt{\text{Gz}_{\text{turb}}/\pi}\left(1 + \frac{\varepsilon^* R_0 U_{L0}}{\alpha_L}\right) \quad \text{for large Gz}_{\text{turb}}. \tag{16.106}$$

Kutateladze's model for the turbulence effect is evidently an oversimplification. Several empirical correlations have also been proposed in the past. Table 16.1 gives a summary of some of these correlations. The following definitions apply to these correlations:

$$\overline{\text{St}} = \frac{\overline{H}_{\text{FI}}}{\rho_L C_{\text{PL}} U_{L0}}, \tag{16.107}$$

$$\text{We}_g = \frac{\rho_g U_g^2 D_0}{\sigma}, \tag{16.108}$$

$$\text{We}_{L0} = \frac{\rho_L U_{L0}^2 D_0}{\sigma}, \tag{16.109}$$

$$\text{Ja}_L = \frac{C_{\text{PL}}(T_{\text{sat}} - T_{L0})}{h_{\text{fg}}}, \tag{16.110}$$

$$\text{Su} = \frac{\rho_L \sigma D_0}{\mu_L^2}. \tag{16.111}$$

Table 16.1. *Empirical correlations for condensation of a pure saturated vapor on a subcooled liquid jet.*

Author	Correlation
Isachenko *et al.* (1971)	$\overline{St} = 0.0335(L/D_0)^{-0.42} \, Re_{L0}^{-0.17} Pr_L^{-0.09} Ja_L^{-0.13} We_g^{0.35}$
Benedek (1976)	$\overline{St} = 0.00286(A_0/A_S)^{0.06} Ja_L^{-0.084}$ $A_0 = $ jet cross-sectional area at the inlet $A_s = $ jet total surface area
Sklover and Rodivilin (1975)	$\overline{St} = 2.7(L/D_0)^{-0.6} \, Re_{L0}^{-0.4} Pr_L^{-0.55} Ja_L^{-0.11} \, We_g^{0.4}$
De Salve *et al.* (1986)	$\overline{St} = 3.25(L/D_0)^{-0.52} \, Re_{L0}^{-0.38} Pr_L^{-0.52} Ja_L^{0.19}$
Kim and Mills (1989)	$\overline{St} = 3.2 \, Re_{L0}^{-0.2} Pr_L^{-0.7} Su^{-0.19} (L/D_0)^{-0.57} F^{0.18}$ $F = $ ratio of rough nozzle to smooth nozzle friction factors
Sam and Patel (1984)	$\overline{St} = 0.075 \, Re_{L0}^{-0.35}$ for $1 < We_g < 13$ $\overline{St} = 0.075 \, Re_{L0}^{-0.35} We_{L0}^{0.3}$ for $We_g < 1$

For planar (slab) jets, the Weber numbers in Eqs. (16.108) and (16.109) are based on the jet thickness.

The correlation of Benedek (1976) is based on steam–water data for freely falling jets, with $0.2 < P < 1$ bar and $3000 < Re_{L0} < 1.8 \times 10^5$. The correlation of Kim and Mills (1989) is based on experiments with coherent ethanol and water jets, with $3 \leq D \leq 7$ mm and $6{,}000 \leq Re_{L0} \leq 40{,}000$, using smooth and rough-surface glass tubes. The parameter F in the correlation of Kim and Mills (1989) stands for the ratio between the nozzle friction factor and the friction factor of a similar but smooth nozzle. Two different correlations, one in terms of dimensionless parameters and the other in terms of the primitive dimensional parameters (not included in the table), were developed by Kim and Mills. The correlation of Sam and Patel (1984) is based on water–steam data at low pressure, representing the operating range of the open-cycle ocean thermal energy conversion (OC-OTEC) systems.

Celata *et al.* (1989) have compared the predictions of several empirical correlations with experimental data with water jets issuing from nozzles with diameters in the range $D = 1$–5 mm and have proposed an empirical correlation based on an effective jet thermal conductivity.

16.6 Mechanistic Models for Condensing Annular Flow

Annular flow comprises the predominant regime in the majority of near-vertical internal-flow condensation processes. It is also predominant in horizontal channels as long as the vapor velocity is high. Mechanistic models do reasonably well for describing the annular two-phase flow regime in general, since the hydrodynamic and transport phenomena in the annular flow regime are understood relatively well. Nevertheless, mechanistic modeling involves uncertainties about a multitude of constitutive relations. Some examples of mechanistic models for dryout, which involves the annular flow regime, were discussed in Section 13.6.

In condensing internal flows, the flow parameters and the properties of the gas-phase vary with longitudinal position. The most important among these parameters are pressure, gas-phase mean velocity, interfacial shear stress, and partial pressure of the noncondensables (when they are present). Mechanistic models that account for all these variables lead to differential equations that need to be solved numerically.

EXAMPLE 16.4. Develop a mechanistic model based on the two-fluid modeling technique for steady-state condensation of a downward-flowing pure and saturated vapor in a cooled vertical pipe. Assume that the flow regime is annular and that droplet entrainment and deposition are negligible.

SOLUTION. It is reasonable to assume that the vapor remains saturated everywhere, while the liquid film will be slightly subcooled. The steady-state and one-dimensional two-fluid model equations for annular flow can be written as (see the schematic in Fig. 16.7)

$$\frac{d}{dz}[\rho_g U_g \alpha] = -a_I'' m'', \tag{16.112}$$

$$\frac{d}{dz}[\rho_L U_L (1-\alpha)] = a_I'' m'', \tag{16.113}$$

$$\rho_g \alpha U_g \frac{dU_g}{dz} = -\alpha \frac{dP}{dz} + \rho_g \alpha g - F_I - F_{wg} - a_I'' m'' (U_I - U_g)$$
$$- C_{vM}\left(U_g \frac{dU_g}{dz} - U_L \frac{dU_L}{dz}\right), \tag{16.114}$$

$$\rho_L(1-\alpha)U_L \frac{dU_L}{dz} = -(1-\alpha)\frac{dP}{dz} + \rho_L(1-\alpha)g + F_I - F_{wL}$$
$$+ a_I'' m'' (U_I - U_L) + C_{vM}\left(U_g \frac{dU_g}{dz} - U_L \frac{dU_L}{dz}\right), \tag{16.115}$$

$$\frac{d}{dz}\left[\rho_L(1-\alpha)U_L\left(h_L + \frac{U_L^2}{2} + gz\right) + \rho_g \alpha U_g\left(h_g + \frac{U_g^2}{2} + gz\right)\right]$$
$$= [(1-\alpha)U_L + \rho_g \alpha U_g]\frac{dP}{dz} - 2q_w''/R_0. \tag{16.116}$$

These equations include the virtual mass force (the last terms on the right side of Eqs. (16.114) and (16.115)), which is negligibly small in the annular flow regime but may be kept to help the numerical computations (see the discussion in Chapter 5). The term can be modeled following Watanabe *et al.* (1990). (See Eqs. (5.81) and (5.82).) Other useful and rather obvious relations are

$$\delta_F = R_0(1 - \sqrt{\alpha}), \tag{16.117}$$

$$a_I'' = \frac{2}{R_0}\sqrt{\alpha}, \tag{16.118}$$

$$F_I = a_I'' \tau_I, \tag{16.119}$$

$$F_{wL} = \frac{2}{R_0}\tau_w, \tag{16.120}$$

$$F_{wg} \approx 0, \tag{16.121}$$

$$T_I = T_{sat}(P), \tag{16.122}$$

$$q_w'' = \dot{H}_{FI}(T_{sat}(P) - T_L), \tag{16.123}$$

$$m'' = q_w''/h_{fg}. \tag{16.124}$$

Note that T_L represents the liquid bulk temperature here. Equation (16.123) considers the thermal resistance between the surface and bulk of the liquid film and accounts for the effect of mass transfer on the heat transfer process. The heat transfer coefficient for vanishingly small mass flux, H_{FI}, can be found from an appropriate correlation, such as the correlations discussed in Section 3.10. The relation between H_{FI} and $\dot{H}_{FI}$ can be found from the stagnant film model described in Section 1.8.

To close the equations, we still need to specify τ_w and τ_I. The wall friction can be found from

$$\tau_w = \frac{R_0}{2}\left(-\frac{dP}{dz}\right)_{fr},$$

where the frictional pressure gradient can be found from an appropriate correlation. The interfacial shear stress can be modeled by thinking of the vapor flow as flow through a rough-walled pipe that has an average radius of $R_0\sqrt{\alpha}$:

$$\tau_I = \dot{f}_I \frac{1}{2}|U_g - U_L|(U_g - U_L). \tag{16.125}$$

The friction factor $\dot{f}_I$ also accounts for the effect of mass transfer and is related to f_I, the friction factor when there is no mass transfer, as described in Section 1.8. The latter friction factor can be found by using appropriate smooth-pipe correlations and then correcting the result for the effect of the interfacial waves. For example, when $\mathrm{Re}_g = 2|U_g - U_L|R_0\sqrt{\alpha}/\nu_g > 2{,}300$, one can use Blasius's correlation to get the smooth-pipe friction factor, $f_{Is} = 0.079\,\mathrm{Re}_g^{-0.25}$. The effect of interfacial waves can then be accounted for by the following widely used method (Wallis, 1969):

$$f_{Is}/f_{Is} = (1 + 150\delta_F/R_0). \tag{16.126}$$

The following thermodynamic relations are also needed: $h_g(P)$, $T_L(P, h_L)$, and $T_{sat}(P)$. These are available from property routines. The equation set is now closed. Equations (16.112) through (16.116) now form five coupled ordinary differential equations with five unknowns (state variables): U_g, U_L, P, h_L, and α.

In the foregoing formulation, we treated α as a state variable. By using α, the flow quality x can then be found from the fundamental void-quality relation, Eq. (3.39).

Alternatively, one can use the flow quality as a state variable, in which case α can be found from the fundamental void–quality relation.

Wall Friction

The wall friction in annular two-phase flow can be modeled by using an empirical two-phase frictional pressure-drop correlation, as was done in the previous example, or alternatively based on the film velocity profile. In the model derived by Wallis (1969) the liquid film is assumed to be part of a single-phase liquid flow occupying the entire channel cross section. Accordingly, $\tau_w = f_w \frac{1}{2}\rho_L \overline{U}_L^2$, where $\overline{U}_L$ is a fictitious average liquid velocity that would have occurred had the liquid film been part of an entirely liquid flow field. For a laminar film, where $\overline{\mathrm{Re}}_L = \rho_L\overline{U}_L D/\mu_L < 2{,}300$,

$$\overline{U}_L = \frac{j_L}{(1-\alpha)^2}. \tag{16.127}$$

Using this velocity, we can write

$$f_w = 16/\overline{Re}_L. \tag{16.128}$$

For turbulent film ($\overline{Re}_L > 2{,}300$), the velocity profile is assumed to follow the 1/7 power dependence on distance from the wall, and that leads to

$$\overline{U}_L = \frac{j_L}{(1 - \sqrt{\alpha})^{8/7}(1 + \frac{8}{7}\sqrt{\alpha})}. \tag{16.129}$$

Blasius's correlation can now be used:

$$f_w = 0.079\overline{Re}_L^{-0.25}. \tag{16.130}$$

By comparing the predictions of a two-fluid model with experimental data representing the liquid film thickness in annular flow, Yao and Ghiaasiaan (1996a) showed that calculating the wall friction based on a semi-empirical two-phase pressure-drop correlation could lead to the underprediction of the liquid film thickness. With the film-velocity-profile-based method given here, however, the two-fluid model could predict the experimental data well.

Simplification to the Model

The liquid film differential conservation equations described in Example 16.4 can be avoided by assuming that the film is in quasi-equilibrium conditions, so that a well-defined velocity profile can be assigned to it. In the simplest approach, for example, for a laminar film the velocity profile can be assumed to follow Nusselt's analysis (Siddique *et al.*, 1994), whereby

$$\dot{m}_L(z) = \frac{2\pi g\rho_L \Delta\rho}{\mu_L}\left[\frac{R_0\delta_F^3}{3} - \frac{5\delta_F^4}{24}\right], \tag{16.131}$$

$$h_{fg}\frac{d}{dz}\dot{m}_L = 2\pi R_0 \dot{H}_F(T_I - T_w) - 2\pi(R_0 - \delta_F)\dot{H}_{GI}(\overline{T}_G - T_I), \tag{16.132}$$

where $\dot{m}_L$ is the total liquid mass flow rate. The latter equation includes the contribution of sensible heat transfer from the gas bulk to the liquid film. The sensible heat transfer rate vanishes when pure, saturated vapor undergoes condensation. The laminar film velocity profile can be easily modified to account for the effect of interfacial friction (Seban and Hodgson, 1982), as well as the longitudinal pressure gradient (Munoz-Cobo *et al.*, 1996). A similar approach can be followed for turbulent films. In this case, one can assume that the liquid film velocity profile follows von Kármán's universal velocity profile (Azer and Said, 1986). An alternative approach, however, is to use a turbulent viscosity model for the liquid film.

EXAMPLE 16.5. Develop a mechanistic model for steady-state condensation of a downward-flowing pure vapor in a cooled vertical pipe, by using an eddy diffusivity model for the liquid film. Assume that the flow regime is annular, the droplet entrainment and deposition are negligible, and the terms representing changes in kinetic and potential energy are negligible.

SOLUTION. Neglecting the effect of condensate film subcooling on the overall energy balance, we get for mixture energy conservation

$$\frac{dx}{dz} = -\frac{2q_w''}{GR_0 h_{fg}}, \tag{16.133}$$

At any location along the tube we have

$$\frac{R_0}{2} G(1-x) = \rho_L \int_0^{\delta_F} u_L(y)(1-y/R_0) dy. \tag{16.134}$$

where y is the distance from the wall. Equations (16.112)–(16.114) apply, and the void fraction is related to the film thickness from $\delta_F = R_0(1 - \sqrt{\alpha})$. Moreover, the virtual mass force term in Eq. (16.114) can be neglected.

A force balance on a cylindrical element of the liquid film gives

$$\tau = \tau_I \left(\frac{R_0 - \delta_F}{r} \right) + \frac{1}{2} \left[\frac{r^2 - (R_0 - \delta_F)^2}{r} \right] \left(\rho_L g - \frac{dP}{dz} \right). \tag{16.135}$$

By substituting for τ from $\tau = (\mu_L + \rho_L E)\partial u_L/\partial y$, where E is the eddy diffusivity, Eq. (16.135) gives

$$\frac{\partial u_L}{\partial y} = \frac{\tau_I}{\mu_L + \rho_L E} \left(\frac{R_0 - \delta_F}{R_0 - y} \right) + \frac{1}{2} \left(\rho_L g - \frac{dP}{dz} \right) \cdot \left(\frac{R_0 - y}{\mu_L + \rho_L E} \right) \left[1 - \left(\frac{R_0 - \delta_F}{R_0 - y} \right)^2 \right]. \tag{16.136}$$

For calculating the heat transfer through the liquid film, one should consider the subcooling in the liquid film. From the conservation of energy in the liquid the temperature profile in the film follows

$$\frac{\partial}{\partial y} \left[\left(k_L + \rho_L C_{PL} \frac{E}{Pr_{turb}} \right) \frac{\partial T}{\partial y} \right] = 0. \tag{16.137}$$

At $y = 0$, representing the channel wall surface, the typical boundary conditions are a known heat flux, q_w'', or a constant temperature, T_w. In the former case,

$$q_w'' = -k_L \frac{\partial T}{\partial y} \bigg|_{y=0}. \tag{16.138}$$

The conditions at the condensate–gas interface ($y = \delta_F$) are

$$\frac{\partial u}{\partial y} = \frac{\tau_I}{\mu_L + \rho_L E}, \tag{16.139}$$

$$T = T_{sat}(P). \tag{16.140}$$

These equations can be solved numerically, provided that appropriate closure relations for τ_i and E are used. For E, the eddy diffusivity, see Section 3.11. For a laminar film $E = 0$.

The discussion thus far has been limited to the condensation of a pure, saturated vapor. Noncondensables result in gas-side thermal and mass transfer resistances, reducing the condensation rate, as described in Section 15.7. As condensation proceeds along the channel, the concentration of noncondensable in the gas mixture increases, leading to further reduction in the condensation rate. The vapor–noncondensable mixture remains saturated with vapor everywhere, and its properties follow the discussion in Section 1.5.

Past studies have shown that the film model described in Section 15.7 agrees with the experimental data well (Siddique *et al.*, 1994; Ghiaasiaan *et al.*, 1995). The following example shows how the film model can be applied.

EXAMPLE 16.6. Modify the model in Example 16.4 for the case where the vapor is mixed with a noncondensable gas.

SOLUTION. Equations (16.112) through (16.116) all apply, provided that subscript g (which by our convention represents saturated pure vapor at pressure P) is replaced with G (which represents the gas phase in general). Subscript G thus represents the vapor–noncondensable mixture. The gas mixture thermodynamic properties are

$$\rho_G = \rho_n + \rho_g(P_v) = \frac{P - P_v}{\frac{R_u}{M_v} T_G} + \rho_g(P_v), \qquad (16.141)$$

$$h_G = m_n h_n + (1 - m_n) h_g(P_v), \qquad (16.142)$$

$$h_n = h_n(T_{ref}) + \int_{T_{ref}}^{T_G} C_{Pn} dT, \qquad (16.143)$$

where the noncondensable has been treated as an ideal gas. In addition to Eqs. (16.112) through (16.116), an equation representing the conservation of mass for the noncondensable species is needed. Assuming that the noncondensable is completely insoluble in the liquid, we have

$$\frac{d}{dz}[\alpha \rho_G m_n U_G] = 0. \qquad (16.144)$$

In comparison with the case where condensation of pure vapor was considered, we have added one new equation (Eq. (16.144)) and one new unknown (m_n). Among the closure relations, only Eqs. (16.122), (16.123), and (16.124) need modification. Equation (16.122) is replaced with Eq. (15.77), and Eq. (16.124) is replaced with Eq. (15.78). In Eq. (16.123), furthermore, $T_{sat}(P)$ must be replaced with T_I.

16.7 Flow Condensation in Small Channels

Minichannel-based heat exchangers can be used in residential and automotive air conditioning. In comparison with conventional heat exchangers, minichannel-based heat exchangers are compact, are more efficient, and use less construction material. Flat, multiport minichannel arrays can be manufactured by extruding aluminum, resulting in compact and light-weight heat exchangers. Current residential air-conditioning units typically use copper tubing a few millimeters in diameter, whereas brazed aluminum automotive condensers use extruded multiport aluminum minichannels with hydraulic diameters in the range 1–3 mm. The test section of Wang *et al.* (2002), displayed in Fig. 16.9, is a good example of a flat multiport array of

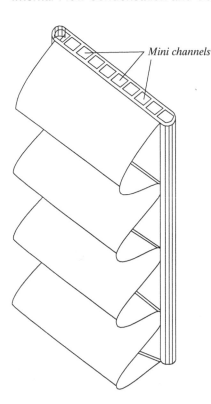

Figure 16.9. The test section of Wang *et al.* (2002): $\mathrm{Re}_j > 2000$.

condenser minichannels. The secondary coolant in the displayed design flows through a jacket against the condenser.

Table 16.2 summarizes some recent experimental investigations. Although there have been many minichannel-based experimental investigations, microchannel-based condensers have been studied by only a few investigators (Wu and Cheng, 2005; Chen and Cheng, 2005; Agarwal, 2006).

The flow boiling phenomena in arrays of parallel mini- and microchannels were discussed earlier in Section 14.2. As noted there, at least for boiling, flow maldistribution among the channels as well as flow instability and oscillations should be expected, unless the backflow of fluid from channels into their inlet plenum is prevented by means of flow resistances installed at channel inlets. Some investigators have expressed concern about flow oscillations and their impact on multichannel condensers (Baird *et al.*, 2003). Relatively severe flow instability in experiments with parallel microchannel condensers was observed by Cheng (2005) and Wu and Cheng (2005). However, little has been reported about flow oscillations in arrays of condenser minichannels. According to Webb and Zhang (1998), the experimental data of Zhang and Webb (1998) indicate that there was essentially no difference between the condensation heat transfer in a single copper tube and the condensation heat transfer in a multiport extruded aluminum tube of the same hydraulic diameter. Most of the recently developed correlations for condensation pressure drop and heat transfer in minichannels have indeed used data bases that included experiments with single channels as well as arrays of parallel channels.

To better understand the experimental trends and the limitations of empirical correlations, it is important to know how minichannel condensation data are

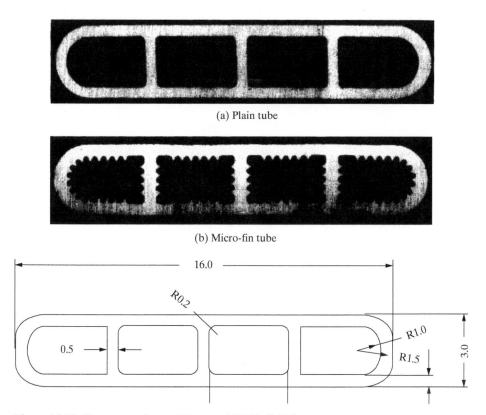

(a) Plain tube

(b) Micro-fin tube

Figure 16.10. The test sections of Yang and Webb (1996).

measured and interpreted. (For a detailed discussion, see Fronk and Garimella, 2013.) In this respect, the investigation by Yang and Webb (1996) can serve as a good example. The test section of Yang and Webb, displayed in Fig. 16.10, included plain (smooth) as well as microfin-enhanced minichannels. The working fluid was R-12. A secondary coolant (water) flowed through the annular water channel, acting as the heat sink for the condenser. In each test, the test section average refrigerant quality, x_{avg}, the heat exchanger overall heat transfer coefficient, U_0, and the average condensation heat transfer coefficient for the minichannels, $\overline{H}$, were found, respectively, from

$$x_{avg} = x_{in} - \frac{\dot{Q}}{2\dot{m}_R h_{fg}}, \qquad (16.145)$$

$$U_0 = \frac{\dot{Q}}{A_0 \Delta T_{LM}}, \qquad (16.146)$$

$$\overline{H} = \frac{1}{\left(\frac{1}{U_0} - \frac{1}{H_0} - \frac{t_w}{k_w}\right)\frac{A_i}{A_0}}, \qquad (16.147)$$

where $\dot{m}_R$ is the total refrigerant mass flow rate, x_{in} is the average refrigerant quality at the inlet to the channels, A_0 is the total outside surface area of the minichannel assembly (i.e., the total inner surface area of the annular water channel), H_0 is the heat transfer coefficient between the water flowing in the annular flow path and the

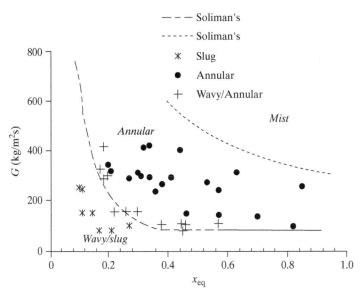

Figure 16.11. The flow regime data of Wang *et al.* (2002) for condensation in a horizontal tube compared with the flow regime transition correlations of Soliman (1982, 1986).

annular channel's inner surface, t_w is the tube assembly wall thickness, k_w is the thermal conductivity of the tube wall, ΔT_{LM} is the logarithmic mean temperature difference in the heat exchanger, and A_i is the total inner surface area of all the channels. Yang and Webb used x_{avg} and $\overline{H}$ for the parametric analysis of their data and correlation development.

This approach – namely, defining an average heat transfer coefficient based on the total inner surface area of all the condenser channels in a multiport assembly, and correlating it in terms of the average minichannel assembly properties including x_{avg} – has been applied by other investigators as well (Zhang and Webb, 1998; Yan and Lin, 1999; Wang *et al.*, 2002; Agarwal *et al.*, 2010; Zhang *et al.*, 2012; Fronk, 2014). Fronk (2014) used the latter technique for deriving semi-analytical methods for condensation of pure ammonia and zeotropic mixtures of water and ammonia in small channels. He then applied the derived methodology as part of a differential model, and showed that the model predicted his experimental data. The use of average test section properties for the interpretation and correlation of experimental data is at least partly due to the difficulty in measuring local parameters. Furthermore, correlations that are in terms of test section average properties are more convenient for design calculations.

Del Col *et al.* (2010) have measured the local condensation heat transfer coefficient in an experiment in which R-1234yf was the condensing fluid that flowed in a 0.96 mm tube. Their test rig was a counter flow heat exchanger in which the refrigerant flowed inside the aforementioned 0.96 mm diameter tube (the inner tube), while the coolant (water) flowed in a complex flow passage that provided for heat transfer between the coolant and the refrigerant. The calculation of the local heat transfer coefficient was made possible by measuring the profiles of coolant (water) and inner channel wall temperatures by using multiple thermocouple sensors along the test section.

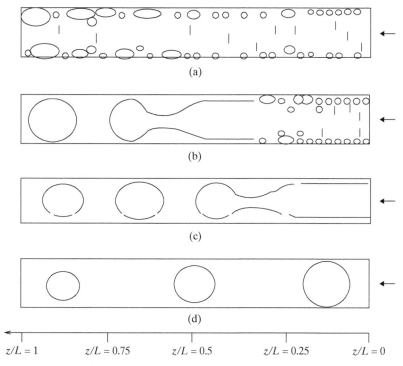

(a)

(b)

(c)

(d)

$z/L = 1$ $z/L = 0.75$ $z/L = 0.5$ $z/L = 0.25$ $z/L = 0$

Figure 16.12. The flow patterns in the middle channel of Wu and Cheng (2005): (a) fully droplet ($G = 47.5\,\mathrm{g/cm^2 \cdot s}$); (b) droplet/annular/injection/slug-bubbly ($G = 30.4\,\mathrm{g/cm^2 \cdot s}$); (c) annular/injection/slug-bubbly ($G = 23.6\,\mathrm{g/cm^2 \cdot s}$); (d) slug-bubbly ($G = 19.3\,\mathrm{g/cm^2 \cdot s}$).

16.8 Condensation Flow Regimes and Pressure Drop in Small Channels

16.8.1 Flow Regimes in Minichannels

Two-phase flow regimes in conventional condensing flows were discussed in Section 16.2. As noted there, at high quality and mass flux the annular flow is the dominant flow regime. Other flow patterns, including slug, plug, bubbly, stratified, and dispersed, also occur when favorable conditions are encountered. Annular flow is the dominant flow regime in minichannels as well and leads to slug flow only when the quality becomes very small, or under low mass flux and low and moderate qualities.

The flow regime transition criteria of Soliman (1982, 1986) were discussed earlier in Section 16.2. Soliman's transition lines are for wavy-slug to annular and for annular to mist flow. Wang *et al.* (2002) could distinguish stratified wavy, slug, slug-wavy (i.e., transition between stratified wavy and slug), annular, and wavy-annular in their experiments. Mist flow did not occur in their experiments, but that might have been because of their relatively low mass fluxes. Their data are compared with the flow regime transition correlations of Soliman in Fig. 16.11. Soliman's correlation for the wavy-slug to annular transition agreed well with the wavy-slug to annular regime transition data of Wang *et al.*, with a slight overprediction of the mass flux at low qualities. (Note that Soliman *et al.*'s definition of the wavy flow regime in fact includes the slug flow regime as well.)

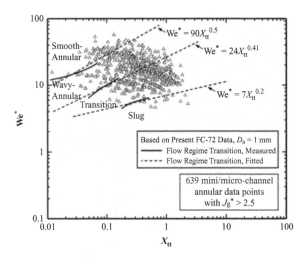

Figure 16.13. The flow regime map proposed by Kim and Mudawar (2012b).

The parametric trends in experimental condensation heat transfer coefficients provide indirect information about the two-phase flow regimes in channels. These trends, which will be discussed in the next section, confirm the predominance of the annular flow regime.

Based on experimental data with FC-72 fluid in 10 parallel rectangular channels with 1 mm hydraulic diameter, as well as minichannel data from various sources, Kim and Mudawar (2012b) proposed the flow map depicted in Figure 16.13, in which the modified Weber number We* is defined according to Eqs. (16.16) and (16.17) (Soliman, 1986). The vast majority of the experimental data that were examined showed that for $j_g^* > 2.5$ (where j_g^* is defined according to Eq. (16.7)) the boundaries among the major flow regimes were as follows

Smooth-annular to wavy – annular: $We^* = 90X_{tt}^{0.5}$
Wavy-annular to transition: $We^* = 24X_{tt}^{0.41}$
Transition to slug and bubbly: $We^* = 7X_{tt}^{0.2}$

Kim and Mudawar then used this flow regime map as the basis for the development of a correlation for predicting condensation heat transfer in minichannels, to be described in the next section.

Based on extensive data with R-134a in circular and rectangular channels with $1 < D_H < 4.91$ mm, Nema et al. (2014) have recently developed a semi-mechanistic flow regime transition model for condensing flows in horizontal channels. They make a distinction between small channels, for which the Bond number is smaller than a critical Bond number, and large channels where the Bond number is larger than a critical Bond number value. The critical Bond number, presented in the forthcoming equation, was derived by extending an analysis by Barnea et al. (1983) in which the condition where surface tension overcomes buoyancy and prevents the development of stratified flow regime is specified

$$Bo_{cr} = \left(\frac{\rho_f}{\rho_f - \rho_g} - \frac{\pi}{4} \right)^{-1}.$$

Nema *et al.* (2014) compared their flow regime transition models with the flow regime maps of Coleman and Garimella (2003), Triplett *et al.* (1999a, b) and Cavallini *et al.* (2002), with fair agreement for most of flow regime transition lines.

16.8.2 Flow Regimes in Microchannels

Experimental data dealing with condensation in parallel microchannels are scarce. Chen and Cheng (2005) and Wu and Cheng (2005) both studied the condensation of steam in 10 parallel silicon microchannels with trapezoidal cross sections. The parametric ranges of the experiments of Chen and Cheng are displayed in Table 16.2. The investigation by Chen and Cheng (2005) was conducted with microchannels with $D_H = 75\ \mu m$, which were covered by a Pyrex plate that allowed microscope-assisted viewing of the flow field. Chen and Cheng have reported that stable droplet condensation occurred near the entrance of their microchannels. The droplets coalesced further downstream, leading to intermittent flow. Annular, wavy, and dispersed flow did not occur in their tests.

The investigation of Chen and Cheng (2005) involved a systematic study of the flow patterns in microchannels with $D_H = 82.8\ \mu m$, using degassed and deionized water. The flow patterns in their microchannels depended on the mass flow rate. The major flow patterns recorded for their middle microchannel are schematically displayed in Fig. 16.12. Fully droplet flow occurred at high-mass-flux conditions when much of the steam would flow uncondensed through the channel. As the mass flux was lowered, other flow regimes appeared. In Fig. 16.12(b), for example, droplet flow near the inlet leads to annular flow, then to a flow regime called injection flow, and finally to slug-bubbly flow. With further reduction in mass flux, first droplet flow (Fig. 16.12(c)), and eventually the annular and injection regimes, completely disappeared (Fig. 16.12(d)). The regime shown in Fig. 16.12(d) was accompanied by flow oscillations that caused reverse condensate flow (i.e., periodic injection of condensate liquid into the inlet volume). Local fluctuations in flow patterns also occurred.

16.8.3 Pressure Drop in Condensing Two-Phase Flows

The two-phase pressure drop in minichannel flow condensation has been investigated by several research groups. Some have concluded that conventional two-phase pressure-drop methods are adequate for condensing minichannels, at least for refrigerants working under low relative pressure. (Recall that at high P_r, σ, ρ_f/ρ_g, and μ_f are small. Conventional models and correlations, however, are primarily based on data with fluids operating at low P_r.) The experimental data of Baird *et al.* (2003), which represented the flow of R-11 and R-123 in single minitubes (see Table 16.2), agreed well with the method of Lockhart and Martinelli (1949). Cavallini *et al.* (2006) also noted that the correlation of Friedel (1979) predicted their data representing the condensation of R-134a and R-236ea in 1.4-mm-diameter minitubes well. The method of Chisholm, as modified by Mishima and Hibiki (1996) (see Eq. (10.1)), also predicted the latter data of Cavallini well. However, Mishima and Hibiki's correlation, Friedel's correlation, and several other conventional correlations failed to

Table 16.2. *Summary of some experimental investigations dealing with condensation in mini- and microchannels.*

Author	Channel characteristics	D_H (mm)	Fluid	G (kg/m²·s)
Yang and Webb (1996)	Smooth, microfin, horizontal, M	2.64, 1.56	R-12	400–1400
Zhang (1998)	Smooth, horizontal, SP-C, M-C, M-R	6.25, 3.25, 2.13, 1.45, 0.96, 1.33	R-134a, R-22, R-404A	200–1000
Webb and Ermis (2001)	Smooth, microfin, horizontal, M, R	0.611, 1.564, 0.44, 1.33	R-134a	300–1000
Garimella and Bandhauer (2001)	Smooth, horizontal, M-S	0.76	R-134a	150–750
Wilson et al. (2003)	Microfin, horizontal, flattened	1.84, 4.4, 6.37, 7.79	R-134a, R-410A	75–400
Koyama et al. (2001)	Smooth horizontal, M, R	1.11, 0.807	R-134a	100–700
Koyama et al. (2003)	Smooth, microfin, horizontal, M, R	0.807, 0.889, 0.937, 1.062	R-134a	100–700
Yan and Lin (1999)	Smooth, horizontal, M	2.0	R-134a	100–200
Vardhan and Dunn (1997)	Smooth, horizontal, M-C	1.49	R-134a, R-22, R-407C	400–1100
Kim et al. (2003a)	Smooth, microfin horizontal, M-R	1.41, 1.56	R-22, R-410A	200–600
Wang (1999)	Smooth, horizontal, M-R	1.46	R 134a	79–760
Baird et al. (2003)	Smooth, horizontal, SP-C	0.92, 1.95	R-123, R-11	70–600
Cavallini et al. (2005, 2006)	Smooth, horizontal, M-R	1.4	R-134a, R-236ea, R-410A	200–1400
Kim et al. (2003b)	Smooth, horizontal, SP-C	0.691	R-134a	100–600
Wu and Cheng (2005)	Smooth, horizontal, M, trapezoidal	0.083	Water	193–475
Agarwal (2006)	Smooth, horizontal, M, rectangular	0.10–0.160	R-134a	200–800
Shin and Kim (2005)	Horizontal, copper, SP, C/R	0.493 – 1.076	R-134a	100–600
Andresen (2006)	Horizontal, aluminum/copper, MP, C/R	0.76, 1.52, 3.05	R410A	200–800
Bandhauer et al. (2006)	Horizontal, aluminum, MP, C	0.506, 0.761	R-134a	150–750
Marak (2009)	Stainless steel, vertical upward and downward flow, SP, C	1.0	Methane	162–701
Matkovic et al. (2009)	Horizontal, copper, SP, C	0.96	R-134a, R-32	100–1200

Table 16.2. *(cont.)*

Author	Channel characteristics	D_H (mm)	Fluid	G (kg/m$^2\cdot$s)
Park and Hrnjak (2009)	Horizontal, aluminum, MP, C	0.89	CO$_2$	200–800
Agarwal *et al.* (2010)	Horizontal, aluminum, MP, R	0.424, 0.762	R-134a	150–750
Bortolin (2010)	Horizontal, copper, SP, C/R	0.96, 1.23	R-245fa, R-134a	67–789
Del Col *et al.* (2010)	Horizontal, copper, SP, C	0.96	R-1234yf	200–1000
Huang *et al.* (2010)	Horizontal, copper, SP, C	1.6, 4.18	R-410A	200–600
Oh and Son (2011)	Horizontal, copper, SP, C	1.77	R-22, R-134a, R-410A	450–1050
Park *et al.* (2011)	Vertical downward flow, aluminum, MP, R	1.45	R-134a, R-236fa, R1234ze(E)	100–260
Derby *et al.* (2012)	Horizontal, copper, MP, R	1.0	R-134a	75–450
Kim and Mudawar (2012a, b)	Horizontal, copper, MP, R	1.0	FC-72	118–367
Heo *et al.* (2013)	Horizontal, aluminum, MP, R	1.5	CO$_2$	400–1000
Zhang *et al.* (2012)	Horizontal, stainless steel, SP, C	1.088, 1.289	R-22, R-410A, R-407C	300–600

M=multi port; SP=single port; S=square, R=rectangular, C=circular.
Note: Based in part on Cavallini *et al.* (2006) and Kim and Mudawar (2013).

predict the pressure-drop data representing the high relative-pressure refrigerant R-410A.

Based on experimental data representing R-22, R-134a, and R-404A in test sections that included single copper tubes ($D = 3.25$ and 6.25 mm) and an extruded aluminum array of six tubes ($D = 3.13$ mm), Zhang and Webb (2001) proposed the following correlation:

$$\Phi_{L0}^2 = (1 - x)^2 + 2.87x^2 P_r^{-1} + 1.68x^{0.8}(1 - x)^{0.25} P_r^{-1.64}. \tag{16.148}$$

This correlation agreed well with the aforementioned experimental data of Cavallini *et al.* (2006) for R-134a and R-410A, but it overpredicted the data with R-236ea representing conditions when high frictional pressure gradients occurred.

16.9 Flow Condensation Heat Transfer in Small Channels

The experimental heat transfer data discussed in the following, unless otherwise stated, are all in terms of average test section heat transfer coefficient (see

Eqs. (16.145)–(16.147)), and all parametric discussions are based on test section average properties, including the test section average quality. For convenience, however, the subscript denoting average will not be used.

Yang and Webb (1996), whose test section is shown in Fig. 16.10, noted that the condensation heat transfer coefficient increased monotonically with increasing mass flux G and quality x. Similar trends have been observed by many other investigators over a wide range of quality (see, e.g., Yan and Lin, 1999, and Baird et al., 2003).

These trends confirm the predominance of annular flow. The correlation of Shah (1979), described earlier in Eqs. (16.32) and (16.33), overpredicted the plain tube data of Yang and Webb (1996) systematically and significantly. The correlation of Akers et al. (1959), displayed in Eqs. (16.28) and (16.29), agreed with their plain channel data at low mass fluxes but overpredicted the measured heat transfer coefficients by about 10%–20% at high mass fluxes. The condensation heat transfer rate in the microfin-enhanced test section of Yang and Webb was considerably higher than the heat transfer rate in their plain test section, even when the increased surface area of the microfins was accounted for. Yang and Webb attributed the enhancement in the heat transfer coefficient to the effect of surface tension-induced pressure gradients that maintain a thin liquid film on the fin tips and prevent complete flooding of the fins by the condensate. Wang et al. (2002) also compared their condensation heat transfer data (see Table 16.2) with several correlations, noting that the correlation of Akers et al. (1959) performed reasonably well. The correlation slightly overestimated their data at low mass fluxes and underestimated the data at high mass fluxes, with a mean deviation of about 15%.

All these investigations were based on arrays of parallel minichannels. The potential effects of flow maldistribution and oscillation on these data are not clear. Baird et al. (2003) expressed concern about these effects and conducted experiments aimed at the measurement of local condensation heat transfer coefficients in single mini-tubes. Using an innovative experimental method, they measured the average heat transfer coefficients in 10 equal segments of two test sections that were each 300 mm long. Their flow control system, furthermore, ensured a stable mass flux in each test. The average flow quality in each of the 10 segments of the test sections was calculated based on energy balance considerations. The data of Baird et al. are evidently good approximations to local property measurements. Unambiguous parametric trends could therefore be deduced in these experiments. The data showed that, except at very low qualities, the condensation heat transfer coefficient increased with increasing mass flux and quality. The trends in their data thus confirmed the prevalence of annular flow, except at very low flow qualities where slug flow is likely to occur.

Baird et al. compared their heat transfer data with the correlations of Akers et al. (1959), Shah (1979), Dobson and Chato (1998), Moser et al. (1998), and several others. The correlations all showed poor agreement with the data. The correlation of Shah (1979), for example, overpredicted the data systematically. Baird et al. developed a semi-analytical, shear-driven annular flow model, similar in some important aspects to the model of Kosky and Staub (1971) (described after Eqs. (16.50) and (16.51)). The model could predict the dependence of the condensation heat transfer coefficient on mass flux, quality, channel size, and pressure but did not correctly predict the effect of heat flux.

With rising interest in the application of refrigerants and dielectric fluids in micro and minichannels, a significant number of studies have been reported in the past decade where flow condensation in minichannels were studied and correlated. Detailed reviews can be found in Kim and Mudawar (2012b, 2013), Awad et al. (2014), and Fronk and Garimella (2013), among others. Several correlations have also been developed that specifically address minichannels. Table 16.3 is a summary of a selected subset of the recently published correlations. Kim and Mudawar (2012b) critically reviewed and compared the predictions of a large number of existing correlations with a vast data base representing the minichannel data from various sources. They noted that, overall, correlations intended for macrochannels generally provide better predictions of minichannel data than correlations intended specifically for mini/microchannels. Kim and Mudawar (2013) subsequently developed the flow regime-dependent correlations presented below for flow condensation in minichannels based on extensive experimental data from various sources. The flow regimes are to be predicted using the aforementioned flow regime map in Fig. 16.13.

For annular flow (smooth-annular, wavy-annular, transition) where $We^* > 7X_{tt}^{0.2}$

$$\frac{HD_H}{k_f} = 0.048Re_f^{0.69}Pr_f^{0.34}\frac{\Phi_g}{X_{tt}} \qquad (16.149)$$

For slug and bubbly flows where $We^* < 7X_{tt}^{0.2}$

$$\frac{HD_H}{k_f} = \left[\left(0.048Re_f^{0.69}Pr_f^{0.34}\frac{\Phi_g}{X_{tt}}\right)^2 + \left(3.2\times10^{-7}Re_f^{-0.38}Su_{go}^{1.39}\right)^2\right]^{0.5} \qquad (16.150)$$

where $Su_{G0} = \frac{\rho_G\sigma D_H}{\mu_G^2}$ is the Suratman number; Φ_g is the square root of the two-phase multiplier defined in Eq. (8.12) and is to be found from (see Eq. (8.29))

$$\Phi_g^2 = 1 + CX_{tt} + X_{tt}^2. \qquad (16.151)$$

The turbulent–turbulent Martinelli parameter X_{tt} is found from Eq. (8.25), and the constant C is to be found from Table 8.3. The range of parameters for the correlation of Kim and Mudawar (2013) is as follows.

Working fluid: R-12, R-123, R-1234yf, R-1234ze(E), R-134a, R-22, R-236fa, R-245fa, R-32, R-404A, R-410A, R-600a, FC-72, methane, and CO_2:

$$0.424 \le D_H \le 6.22\,\text{mm}$$
$$53 \le G \le 1403\,\text{kg/m}^2\text{·s}$$
$$276 \le Re_{f0} \le 8.98\times10^4$$
$$0 \le Re_f \le 7.9\times10^4$$
$$0 \le Re_g \le 2.48\times10^5$$
$$0 < x_{eq} < 1$$
$$0.04 < (P/P_{cr}) < 0.91.$$

EXAMPLE 16.7. Modify the formulation of the semi-analytical model of Kosky and Staub (1971) to make it applicable to the situation where the vapor phase is superheated. Also, instead of using a two-phase pressure-drop correlation for calculating τ_w, use a model based on the shear stress at condensate film–vapor interphase.

Table 16.3. *Summary of some condensation heat transfer correlations for small flow passages.*

Author(s)	Correlation	Comments
Koyama *et al.* (2003)	$$\frac{HD_H}{k_f} = 0.0152\left(1 + 0.6\mathrm{Pr}_f^{0.8}\right)\frac{\Phi_g}{X_{tt}};$$ $$\Phi_g^2 = 1 + 21[1 - \exp(-0.319D_H)]X_{tt} + X_{tt}^2$$ D_H in mm	Based on data with R-134a, in multichannel test section with $D_H = 0.8$ and 1.11 mm
Huang *et al.* (2010)	$$\frac{HD_H}{k_f} = 0.0152\left(1 + 0.6\mathrm{Pr}_f^{0.8}\right)\frac{\Phi_g}{X_{tt}};$$ $$\Phi_g^2 = 1 + 0.05\left[\frac{G}{\sqrt{g\rho_g\left(\rho_f - \rho_g\right)D_H}}\right]^{0.35}X_{tt}^{0.35}$$	Based on data with R-410a and R-410a/oil, in horizontal smooth tubes with $D_H = 1.6$ and 4.8 mm
Bohdal *et al.* (2011)	$$\frac{HD_H}{k_f} = 25.084\,\mathrm{Re}_f^{0.258}\mathrm{Pr}_f^{-0.495}\,(P/P_{cr})^{-0.288}\left(\frac{x_{eq}}{1 - x_{eq}}\right)^{0.266}$$	Based on data with R-134a, R-404A, in horizontal smooth tubes with $D_H = 0.31$ to 3.3 mm
Park *et al.* (2011)	$$\frac{HD_H}{k_f} = 0.0055\mathrm{Pr}_f^{1.37}\frac{\Phi_g}{X_{tt}}\,\mathrm{Re}_f^{0.7};$$ $$\Phi_g^2 = 1 + 13.17(\rho_g/\rho_f)^{0.17}\left\{1 - \exp\left[-0.6\sqrt{\frac{g\left(\rho_f - \rho_g\right)D_H^2}{\sigma}}\right]\right\}X_{tt} + X_{tt}^2$$	Based on data with R-134a, R-236fa, R-1234ze(E), in multichannel test section with $D_H = 1.45$ mm, expression for Φ_g from Haraguchi *et al.* (1994)
Kim and Mudawar (2013)	See Eqs. (16.149) and (16.150)	Based on an extensive data base from multiple sources

SOLUTION. Equations (16.50)–(16.54), which assume that the condensate film temperature distribution follows the temperature law of the wall for turbulent boundary layers, apply. Alternately, one can use the correlation of Henstock and Hanratty (1976) for the liquid film thickness in annular flow:

$$\delta_F^+ = \left[(0.7071\,\mathrm{Re}_L^{0.5})^{2.5} + (0.0379\,\mathrm{Re}_L^{0.9})^{2.5}\right]^{0.4}. \tag{16.152}$$

The wall shear stress can be estimated by writing the following equation, which results from a force balance on the liquid film in the axial direction if one neglects the axial momentum transfer from condensation:

$$\tau_w D = \tau_I (D - 2\delta_F). \tag{16.153}$$

To formulate the condensation rate, we can follow the discussion leading to Eq. (5.92) in Chapter 5, noting that in steady state $\dot{H}_{LI}(T_I - \overline{T}_L) = q_w''$, where y is the distance from the wall. The result will be

$$m'' = -m_I'' = \frac{q_w'' - \dot{H}_{GI}(T_v - T_I)}{h_{fg}}. \tag{16.154}$$

Note that, for pure vapor, $T_I = T_{sat}(P)$. We now need to formulate τ_I and $\dot{H}_{GI}$. The former can be found by idealizing the vapor core as flow in a channel and writing

$$\tau_I = f_I \frac{1}{2}\rho_v |\overline{U}_v - \overline{U}_L|(\overline{U}_v - \overline{U}_L) \approx f_I \frac{1}{2}\rho_v \overline{U}_v^2, \tag{16.155}$$

where $\overline{U}_v = Gx/\rho_v$. The vapor–liquid heat transfer can be estimated similarly, by first finding H_{GI}, the heat transfer coefficient for vanishing mass transfer rate, from an appropriate single-phase tube flow correlation, and then correcting it for the effect of interfacial mass transfer caused by condensation. One can thus write

$$\mathrm{Nu}_{GI} = H_{GI}(D - 2\delta_F)/k_v = f(\mathrm{Re}_v, \mathrm{Pr}_v), \tag{16.156}$$

$$\mathrm{Re}_v = \rho_v \overline{U}_v (D - 2\delta_F)/\mu_v. \tag{16.157}$$

The skin friction coefficient in the absence of mass transfer, f_I, can be estimated similarly. For example, if the interphase remains smooth, $f_I = 16/\mathrm{Re}_v$ for laminar vapor flow, and for turbulent vapor flow one can use Blasius's correlation $f_I = 0.079\,\mathrm{Re}_v^{-1/4}$. If the interfacial waves are considered, the correlation of Wallis (1969), Eq. (5.76), might be a better choice. The effect of interfacial mass transfer on friction and heat transfer can be treated by using Ackerman's formulation described in Section 1.8.

This example in fact represents the essential elements of the model developed by Baird *et al.* (2003) for shear-driven condensing flow in a minitube.

16.10 Condensation in Helical Flow Passages

Condensation in the interior of helically coiled tubes is of interest in refrigeration systems for reasons that are similar to what makes helically coiled boiler channels

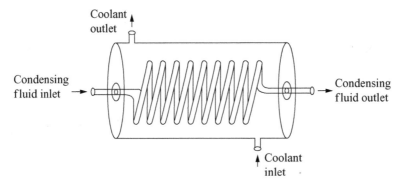

Figure 16.14. Helically coiled condenser in shell-and-tube configuration.

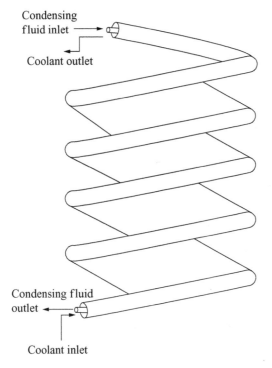

Figure 16.15. Helically coiled condenser in tube-in-tube configuration.

attractive. Helically coiled tubes are more compact, support higher heat transfer coefficients, and in comparison with straight channels can better withstand forces arising from axial thermal expansion or contraction. Published investigations on this topic include Kang *et al.* (2000), Yu *et al.* (2003), Wongwises and Polsongkram (2006a, b), Al Hajeri *et al.* (2007), Lin and Ebadian (2007), Li *et al.* (2007), El-Sayed Mosaad *et al.* (2009), Abd Raboh *et al.* (2011), and Gupta *et al.* (2014).

Condensers that use helical coils can have two different configurations. In one configuration shown schematically in Fig. 16.14, the coolant (secondary fluid) flows through a shell while the condensing fluid flows inside the helically coiled tube (Gupta *et al.*, 2014). However, the more popular configuration is the counter-current tube-in-tube, shown schematically in Fig. 16.15, where the condensing fluid flows through the interior of a helically coiled tube while the coolant flows through an annular space surrounding the helically coiled tube. The majority of the published studies deal with the tube-in-tube configuration.

Helically coiled condensers have generally higher average heat transfer coefficients than their equivalent straight-channel condensers. For R-134a condensing in a tube-in-tube with 9.4 mm and 12.7 mm inner and outer diameters of the inner tube, and 21.2 mm inner diameter of the outer tube, with a coil diameter of 177.8 mm and a coil pitch of 34.9 mm, Li *et al.* (2007) measured enhancements up to about 15% in the average heat transfer coefficient. They also noted that the enhancement decreased monotonically with average condensing fluid quality in the test section. Using the same refrigerant in a tube-in-tube with 8.3 mm and 9.5 mm inner and outer diameters of the inner tube, and 21.2 mm inner diameter of the outer tube, with a coil diameter of 305 mm and a coil pitch of 35 mm, Wongwises and Polsongkram (2006a, b) reported average heat transfer coefficient enhancements in the 33% to 53% range.

The following parametric trends have also been reported by various researchers with respect to the average heat transfer coefficient in a helicoidally coiled condenser tube, when the condensing fluid is a refrigerant (note that for parametric dependence on any single parameter, all other flow parameters and boundary conditions are assumed to be unchanged).

(1) The heat transfer coefficients monotonically increase with condensing fluid mass flux.
(2) The heat transfer coefficients monotonically increase with decreasing saturation temperature of the condensing fluid. This is because as the saturation temperature (and hence the saturation pressure) is decreased, the vapor specific volume increases, and as a result the mean velocity of the two-phase mixture tends to increase. Lower temperature may also lead to higher liquid thermal conductivity for the refrigerant, which evidently enhances heat transfer (Wongwises and Polsongkram, 2006a, b).
(3) The heat transfer coefficients monotonically increase with increasing the average vapor quality. The enhancement factor, defined as the ratio between the average heat transfer coefficient of the helical coil and the average heat transfer coefficient of the same heat exchanger when it is a straight tube, however, monotonically decreases with increasing tube average quality (Li *et al.*, 2007). Higher vapor quality implies higher vapor velocity and higher interfacial shear stress in annular flow regime. The higher interfacial shear stress enhances the interfacial waves and droplet entrainment, both of which enhance the heat transfer (Wongwises and Polsongkram, 2006a, b). This trend is not always monotonic, however. When the mass flux is low, at very high qualities (typically larger than about 60%) the average heat transfer coefficient may become insensitive to quality, or even decrease slightly with further increase in flow quality.

Empirical correlations have been proposed by several investigators (Wongwises and Polsongkram, 2006a, b; Lin and Ebadian, 2007; El-Sayed Mosaad *et al.*, 2009; Abd Rahob *et al.*, 2011; Colorado *et al.*, 2011). These correlations are based on very limited data, however. The correlation of Wongwises and Polsongkram (2006a, b) is based on experiments with R-134a in a tube-in-tube helicoidally coiled condenser, as described earlier (tube-in-tube, 8.3 mm and 9.5 mm inner and outer diameters of the inner tube, 21.2 mm inner diameter of the outer tube, 305 mm coil diameter and 35 mm coil pitch), with test section average saturation temperatures in the 40–50 °C range, and mass fluxes in the 400–800 kg/m^2·s range:

$$\overline{\mathrm{Nu}} = \overline{H}D_H/k_f = 0.1352\mathrm{Dn}_{\mathrm{Eq}}^{0.7654}\mathrm{Pr}_f^{0.8144}X_{tt}^{0.0432}(P/P_{cr})^{-0.3356}(\mathrm{Bo}\times10^4)^{0.112}$$

$$(16.158)$$

where X_{tt} is the turbulent–turbulent Martinelli factor and is defined as in Eq. (8.29), and the equivalent Dean number Dn is defined as

$$\mathrm{Dn}_{\mathrm{Eq}} = \left[\mathrm{Re}_f + \mathrm{Re}_g\left(\frac{\rho_f}{\rho_g}\right)^{0.5}\left(\frac{\mu_g}{\mu_f}\right)\right]\left(\frac{D_i}{D_{cl}}\right)^{0.5}. \qquad (16.159)$$

Gupta *et al.* (2014) have compared this correlation with their helically coiled tube-in-shell condensation data with reasonable agreement. Their data were obtained using R-134a (8.33 mm and 9.52 mm inner and outer tube diameters, respectively, 90.48 mm coil diameter, 22.5 mm coil pitch, test section average saturation temperatures in the 35–40 °C range).

16.11 Internal Flow Condensation of Binary Vapor Mixtures

Flow condensation of multicomponent vapor mixtures is encountered in petrochemical, power, and cryogenics industries, and has been of interest for decades. More recently, however, the application of refrigerant mixtures in vapor compression and absorption refrigeration cycles, as well as certain types of power cycles, has led to intense interest in multicomponent vapor condensation.

The condensation of multicomponent vapor mixtures involves complicated mass transfer and is beyond the scope of this book. Useful discussions about the subject can be found elsewhere (Modine, 1963; Krishna, 1981; Furno *et al.*, 1986; Taylor *et al.*, 1986). Instead, we will focus on condensation of binary vapor mixtures.

Flow condensation inside and outside vertical tubes has been studied for decades because such vertical passages are used in process equipment such as wetted towers and distillation towers. The published studies for these flow configurations cover a wide range of flow rates and tube diameters. A detailed review of the literature can be found in Fronk and Garimella (2013). Recent applications in HVAC&R industry, however, mostly involve flows in horizontal tubes under relatively high flow rates, and in relatively small inner diameter (≲25 mm) tubes.

We will discuss the internal flow condensation of binary vapors, and describe the Colburn–Drew (Colburn and Drew, 1937) film model for the analysis of the condensation process in such systems. Binary and miscible fluid mixtures have extensive applications. Although the technique will be discussed for binary mixtures, it is widely used for modeling the condensation of multicomponent mixtures as well (Butterworth, 1983; Krishna *et al.*, 1976; Krishna and Standard, 1976).

The hydrodynamic aspects of internal condensing flows of binary mixtures are similar to those for single-component fluids. The discussions and predictive methods presented earlier in this chapter, with respect to such aspects as wall friction and two-phase flow regimes, are all applicable to binary mixtures as well. Thus, in vertical flow

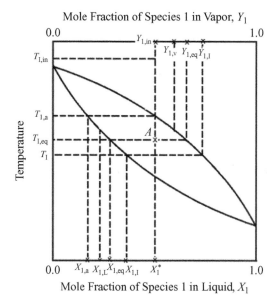

Figure 16.16. Phase diagram of a miscible binary mixture during internal flow condensation.

passages the dominant flow regime is annular, while in horizontal flow passages various flow regimes that were depicted in Figs. 16.1 and 16.2 can occur. The flow regime maps and the prediction methods for flow regime transitions described earlier in this chapter can all be applied, provided that the mixture thermophysical properties are correctly applied.

With respect to condensation heat transfer, the complications that result from the coupling between heat and mass transfer processes, as well as the preferential condensation of the less volatile component in the mixture (i.e., the component with a higher dew point temperature) are all similar to those discussed in Section 15.9 in relation to external flow condensation of binary mixtures. These complications are enhanced due to the stronger effect of flow in a confined passage. A major difference between external and internal flow condensation processes is that in internal flow the conditions of both phases change along the flow passage, and these changes must be quantified in order to calculate the condensation heat transfer in any location along the flow passage.

Consider condensation in a flow passage, where the vapor at inlet is a binary vapor mixture. Referring to Fig. 16.16, the incoming vapor can be assumed to have a mole fraction of species 1 (the more volatile species) equal to $Y_{1,\text{in}}$, and a uniform temperature equal to T_{in}. The assumed inlet vapor condition is obviously superheated, and the vapor undergoes cooling by convection heat transfer with the flow passage wall until its temperature in the vicinity of the cooled wall reaches $T_{1,\text{a}}$. (Note that this can happen while the bulk vapor temperature is still higher than $T_{1,\text{a}}$, due to the potential temperature nonuniformity in the flow cross section.) Condensation will then start, and the liquid that condenses at this point (i.e., the condensing mass flux) will contain species 1 at a mole fraction of $X_{1,\text{a}}$. (Subscript a represents the location where condensation starts.) The condensing mass flux is thus richer than the vapor bulk with respect to the mole fraction of species 2. This preferential condensation of species 2 implies that as we move along the flow passage the vapor in the vicinity of the liquid–vapor interphase will be partially depleted from species 2, and it will contain a higher mole fraction of species 1 than $Y_{1,in}$. The vapor–condensate interphase temperature will be lowered and as a result the condensation heat transfer rate diminishes. Because of the nonuniformity of the concentration of either component in the vapor phase that results from the aforementioned preferential condensation of species 2, a mass transfer resistance is generated which tends to slow the condensation rate, further deteriorating the heat transfer and condensation rates.

The liquid–condensate interphase will qualitatively look like Fig. 15.13. The vapor–condensate interphase is always assumed to be at equilibrium with respect to the interphase temperature as well as the concentrations of the two species. In other words, both phases are at the same temperature where they meet, and for any given interphase temperature the vapor-side interphase concentration of species 1 is located on the dew point line while the liquid-side interphase concentration of species 1 is on the bubble point line. As we march along the flow passage the average mole fraction of the entire flow mixture will of course remain equal to $Y_{1,in}$. At an arbitrary point represented by point A in Fig. 16.16, had the two phases been at equilibrium, we would have $T_{I,eq}$ as the vapor–condensate interphase temperature, and the mole fractions of species 1 in the liquid and vapor phases would be equal to $X_{1,eq}$ and $Y_{1,eq}$, respectively. However, primarily because of the aforementioned preferential condensation of species 2, the interphase will be at T_I. The mole fractions of species 1 on the liquid and vapor sides of the interphase will then be equal to $X_{1,I}$ and $Y_{1,I}$, respectively. The bulk concentration of species 1 in the vapor will be equal to $Y_{1,v}$, where $Y_{1,v} < Y_{1,I}$; and the bulk concentration of species 1 in the liquid will be equal to $X_{1,L}$, where $X_{1,L} > X_{1,I}$. The condition $Y_{1,v} < Y_{1,I}$ is evidently equivalent to $Y_{2,v} > Y_{2,I}$, which is necessary for the diffusion of species 2 from the vapor bulk towards the interphase in order to make up for the depletion of the near-interphase vapor from species 2 caused by its preferential condensation. Likewise, $X_{1,L} > X_{1,I}$, which implies $X_{2,I} > X_{2,I}$, that is necessary for the diffusion of species 2 from the interphase towards the liquid bulk in order to make equilibrium at the interphase possible.

One critical issue in modeling internal flow condensation is how to calculate the variation of liquid- and vapor-side concentrations along the flow passage, which requires that, along with other flow conservation equations, the mass species conservation equations for the two phases be solved. The Colburn–Drew film model described in Section 15.9, along with a one-dimensional two-phase flow model, has been found to be quite good for the analysis of binary flow condensation (Butterworth, 1983; Taylor *et al.*, 1986; Jin *et al.*, 2003). At any location along the flow passage Eqs. (15.89) through (15.93), as well as (15.95) through (15.103), all apply.

EXAMPLE 16.8. Consider the flow condensation of a miscible binary vapor mixture inside a long, vertical tube with downward vapor flow. Assume steady state and an annular flow regime. Using the separated flow modeling method, write the complete set of one-dimensional conservation equations that are required, as well as the constitutive and closure relations that are needed. For simplicity use an appropriate void-quality relation, and assume that the effect of subcooling of the condensate on energy conservation is negligible.

SOLUTION. Let us assume annular flow regime remains annular without entrained droplets, and assume that the liquid film thickness is uniform around the tube perimeter. From Eqs. (5.61) and (5.63), respectively

$$\frac{\partial}{\partial z}[\rho_L U_L (1 - \alpha)] = -\Gamma_1 - \Gamma_2 \tag{16.160}$$

$$\frac{\partial}{\partial z}[\rho_L (1 - \alpha) U_L + \rho_v \alpha U_v] = 0, \tag{16.161}$$

where Γ_1 and Γ_2 are the volumetric rates of phase change of species 1 and 2, respectively, and consistent with the convention is Chapter 5 they are defined positive for evaporation. The mass species conservation equations for component 1 can be written as:

$$\frac{\partial}{\partial z}[\rho_L U_L (1 - \alpha) m_{1,L}] = -\Gamma_1 \tag{16.162}$$

$$\frac{\partial}{\partial z}[\rho_L (1 - \alpha) U_L m_{1,L} + \rho_v \alpha U_v m_{1,v}] = 0, \tag{16.163}$$

where $m_{1,L}$ and $m_{1,v}$ are the mean mass fractions of species 1 in the liquid and vapor phases, respectively. The derivation of these equations is simple and can be done following the technique for the derivation of mass conservation equations in Section 5.7. Equation (16.162) represents the conservation of species 1 in the liquid phase and neglects the diffusion of species 1 in the axial direction, an assumption that is perfectly fine except at creep flow conditions. Equation (16.163) represents the conservation of species 1 in the entire flow field.

The momentum conservation equation can be derived by simplifying Eq. (5.72) and replacing F_w with $\left(-\frac{\partial P}{\partial z}\right)_{fr}$

$$G^2 \frac{\partial}{\partial z}\left[\frac{(1-x)^2}{\rho_L(1-\alpha)} + \frac{x^2}{\rho_v \alpha}\right] = -\frac{\partial P}{\partial z} + [\rho_v \alpha + \rho_L(1-\alpha)]g\sin\theta - \left(-\frac{\partial P}{\partial z}\right)_{fr}. \tag{16.164}$$

The assumption that the subcooling of the condensate is negligibly small means that the specific enthalpy of the liquid phase will be equal to the specific enthalpy of liquid at the liquid–condensate interphase temperature and local pressure. This assumption makes it possible to use only one energy equation. We choose the mixture energy conservation equation. The energy conservation equation can thus be derived by simplifying Eq. (5.86)

$$G\frac{\partial}{\partial z}[xh_v(m_{1,v}, T_v, P) + (1-x)h_L(m_{1,L}, T_I, P)] + \frac{G^3}{2}\frac{\partial}{\partial z}\left[\frac{(1-x)^3}{\rho_L^2(1-\alpha)^2} + \frac{x^3}{\rho_v^2\alpha^2}\right]$$

$$- g\sin\theta G - \frac{\partial P}{\partial t} = \frac{4}{\pi D}q_w'', \tag{16.165}$$

where, consistent with the convention followed in Chapter 5, the wall heat flux is defined positive when it flows into the fluid from the wall. Note that the assumption of $h_L \approx h_L(m_{1,L}, T_I, P)$ is an approximation and does not mean that the liquid temperature is assumed to be uniform.

We can now focus on constitutive and closure relations. For the void-quality relation we can choose the correlation of Smith (1969–1970), Eq. (16.8), which is based on annular regime in condensing pipe flows. We can also write:

$$m_{1,v} + m_{2,v} = 1 \tag{16.166}$$

$$m_{1,L} + m_{2,L} = 1 \tag{16.167}$$

$$X_I = X_I(T_I) \tag{16.168}$$

$$Y_I = Y_I(T_I) \tag{16.169}$$

$$\Gamma_1 = -N_1'' M_1 a_I'' \tag{16.170}$$

$$\Gamma_2 = -N_2'' M_2 a_I''. \tag{16.171}$$

Note that the mass fractions in Eqs. (16.166) and (16.167) are related to their respective mole fractions according to Eqs. (1.44) or (1.45). The same can be said about the mole fractions that appear in Eqs. (16.168) and (16.169). For an ideal annular flow regime, furthermore,

$$a_I'' = \frac{4}{D}\alpha. \tag{16.172}$$

Equation (15.89), which represents the conservation of energy principle for the condensate–vapor interphase, applies. In addition, Eqs. (15.95) for $i = 1$ and 2, along with Eqs. (15.96) and (15.97), all apply. Equations (15.101) through (15.105) apply as well.

We need a relation between the wall heat flux and wall temperature, so that knowing one would allow us to specify the other. We can simplify the analysis by using Eq. (15.94), which is consistent with the assumption that the effect of subcooling of the condensate on the overall energy balance is negligible.

It remains to specify $(-\frac{\partial P}{\partial z})_{fr}$, and specify closure relations for all the transfer coefficients, H_{vI}, H_{LI}, $\tilde{K}_{v,i}$, and $\tilde{K}_{L,i}$. The frictional pressure gradient $(-\frac{\partial P}{\partial z})_{fr}$ can be modeled using one of the methods reviewed in Chapter 8, for example the correlation of Friedel (1979), Eq. (8.39). The liquid-side heat transfer coefficient H_F can be calculated from relevant correlations representing condensation of a pure single-component liquid. An example would be the correlation of Chen *et al.* (1987), presented in Eqs. (15.19)–(15.21). Correlations for the calculation of other transfer coefficients can be borrowed from the single-phase internal flow literature.

PROBLEMS

16.1 Saturated water vapor with a mass flow rate of 0.0202 kg/s flows downward into a cooled vertical tube. The pressure in the tube is 2.7 bars and for simplicity can be assumed to be uniform. The tube has an inner diameter of 4.9 cm and a total cooled length of 2.44 m. At points that are located 0.5 and 1.0 m downstream from the inlet, the vapor mass flow rates are estimated to be 13.5×10^{-3} and 5.0×10^{-3} kg/s, respectively.

(a) Find the condensation heat transfer coefficient and heat flux at these two locations, assuming that the wall surface temperature is lower than the saturation temperature by 10 °C.

(b) Assuming that the flow regime is annular everywhere, calculate the liquid film thickness at 0.5 m and 1.0 m downstream from the inlet of the tube.

16.2 Pure refrigerant R-134a undergoes condensation in a vertical tube that is 8 mm in diameter. The flow configuration is downward, and the pressure at the inlet is 4.9 bars, which for simplicity can be assumed to apply to the entire channel. The total mass flux is 400 kg/m²·s.

(a) Using an appropriate correlation, calculate the local heat transfer coefficient at a location along the channel where $x_{eq} = 0.7$.

(b) Assuming that the wall surface is 11 °C below saturation temperature, calculate the local heat flux.

(c) Assuming that the heat flux is uniform, calculate the total length needed for an exit quality of 0.95. Calculate the wall temperature at several locations along the tube, and plot the variation of the wall temperature along the tube.

16.3 In Problem 16.1 assume that at the inlet the steam contains 1% by weight helium and the measured mass flow rate of water vapor 0.5 m downstream from the inlet is 1.52×10^{-2} kg/s.

(a) Calculate the local bulk mass fraction of helium and the liquid film thickness 0.5 m downstream from the inlet.

(b) Assuming that the wall inner surface temperature is 10 °C below the saturation temperature associated with the system pressure, using an appropriate correlation calculate the condensation mass flux, 0.5 m downstream from the inlet, if the gas–vapor mixture were pure saturated vapor.

(c) Repeat part (b), this time using the Couette flow film model.

(d) Repeat part (c), this time including the effect of the helium gas. Compare the results of parts (c) and (d), and comment on the impact of the noncondensable on the local condensation rate.

16.4 In an experiment, saturated water vapor with a flow rate of 5.67×10^{-3} kg/s flows downward in a vertical tube that is 4.6 cm in diameter and is cooled over a length of 2.44 m. The pressure is 2.14 bars, and is assumed to be uniform over the entire length of the channel. At distances of 0.5 and 1 m from the inlet, the vapor mass flow rates are reduced to 3.75×10^{-3} and 1.9×10^{-3} kg/s, respectively. it is assumed that the tube inner wall temperature is 10 °C below the saturation temperature associated with the system pressure.

(a) Find the heat fluxes and the local heat transfer coefficients at 0.5 m and 1.0 m from the inlet, assuming that the condensing vapor is pure.

(b) Repeat part (a), this time assuming that at the inlet the vapor contains air at a concentration of 0.1% by weight.

16.5 Refrigerant R-123 flows through a horizontal cooled tube, of inside diameter 3.25 cm, with a mass flux of 400 kg/m^2·s. The pressure at the inlet to the tube is 6.24 bars and for simplicity is assumed to remain uniform in the tube.

(a) Using the method of Tandon *et al.* (1982), determine the qualities at which major flow regime transitions occur in the tube.

(b) Repeat part (a), this time using the flow regime transition criteria of Soliman (1982, 1986).

16.6 Pure steam with a mass flux of $G = 200$ kg/m^2·s condenses in a horizontal tube with an inner diameter of 1.71 cm. The steam is saturated at the inlet and is at a pressure of one bar. Assume for simplicity that the pressure remains uniform in the tube.

(a) Specify the flow regimes at locations where $x_{eq} = 0.9, 0.7,$ and 0.5.

(b) Calculate the heat transfer coefficients at the locations mentioned in part (a), using the following correlations: Akers *et al.* (1959), Shah (1979), and Dobson and Chato (1998). Compare the predictions of the three correlations, and comment on the results.

16.7 Pure, saturated ammonia vapor (refrigerant R-717) at a pressure of 33.1 bars flows into a cooled horizontal tube that has an inner diameter of 2.79 cm. The mass flux is 50 kg/m^2·s. Assume for simplicity that the pressure remains constant in the tube.

(a) Determine the flow regimes at locations where $x_{eq} = 0.95, 0.8,$ and 0.45.

(b) Over what quality range do stratified or wavy flow regimes occur?

16.8 In Problem 16.6, assuming that the wall surface temperature is fixed at 72 °C, calculate the heat transfer coefficients at the locations where $x_{eq} = 0.8, 0.55,$ and 0.1.

16.9 Pure, saturated R-134 vapor is introduced into a horizontal tube with an inner diameter of 18 mm, at a pressure of 16.8 bars and with a mass flux of 300 kg/m^2·s. The tube inner wall surface temperature is at 51 °C. Assuming for simplicity that the pressure remains uniform in the tube, determine the flow regimes, the heat transfer coefficients, and the wall heat fluxes at locations where $x_{eq} = 0.9, 0.2,$ and 0.05.

16.10 Pure saturated steam at 160 °C with a mass flux of $G = 400$ kg/m^2·s condenses in a horizontal tube with 8.5-mm inner diameter. At a location where $x_{eq} = 0.92$, the inner surface temperature is 148 °C.

(a) Assuming that the pressure remains uniform in the tube, find the flow regime, the frictional pressure gradient, the heat transfer coefficient, and the condensation rate (in kilograms per meter per second) at this location.

(b) Assuming that the wall heat flux is constant and equal to the heat flux at this location, estimate the total length of the condenser tube needed for achieving $x_{eq} = 0.5$ at the exit.

(c) Assess the error associated with the assumption of constant pressure in the tube, by estimating the total pressure drop in the test section. Do this by dividing the test section into several segments, and calculating the pressure drop in each segment based on its inlet conditions.

16.11 Refrigerant R-22 with a mass flux of 495 kg/m²·s undergoes condensation in a horizontal tube with 4.57-mm inner diameter. Using the method of Moser *et al.* (1998), calculate the heat transfer coefficient at locations where $T_{sat} = 35\,°C$ and $x_{eq} = 0.25, 0.6,$ and 0.85. Repeat the calculations using the correlation of Shah (1979). Compare and comment on the predictions of the two methods.

16.12 Uniform-size water droplets with $d_0 = 0.72$ mm initial diameter are sprayed into a steam-conditioning device containing saturated steam that is at 0.41-MPa pressure. The droplets are initially at 80 °C. When the droplets leave the device, they have an average temperature of 135 °C.

(a) Estimate the growth in droplet diameter and the residence time of droplets in the steam-conditioning device using the method of Ford and Lekic (1973).

(b) Repeat part (a) assuming that the spray droplets move in the steam-conditioning device with a velocity of 2 m/s, and using the method of Celata *et al.* (1991).

16.13 A direct-contact condenser is to be designed for an OC-OTEC system in which saturated steam at 20 °C is to condense on subcooled water that is initially at 5 °C. As a figure of merit for the designed condenser, a condensation efficiency can be defined according to

$$\eta = 1 - \frac{T_{sat} - \overline{T}_{L,out}}{T_{sat} - T_{L,in}},$$

where $T_{L,in}$ and $\overline{T}_{L,out}$ represent the liquid temperatures at the inlet and outlet of the condenser. The water and vapor are both assumed to be pure, and they contain no noncondensables. Calculate η as a function of the condenser height, assuming that water is sprayed downward into the condenser, for a droplet diameter of $d_0 = 0.65$ mm and zero inlet downward droplet velocity. Determine the height necessary to obtain $\eta = 0.95$.

16.14 Pure, saturated ammonia (R-717) vapor at a pressure of 33.1 bars flows into a cooled horizontal tube that has an inner diameter of 1.2 mm. The mass flux is 100 kg/m²·s. Assume for simplicity that the pressure remains constant in the tube. Using an appropriate method, determine the flow regimes and the heat transfer coefficients at locations where $x_{eq} = 0.95, 0.8,$ and 0.45.

16.15 Pure, saturated R-134a vapor is introduced into a horizontal minitube with an inner diameter of 0.65 mm. The pressure is 21.17 bars and the mass flux is 300 kg/m²·s. The tube inner wall surface temperature is at 60 °C. Assuming for simplicity that the pressure remains uniform in the tube, determine the flow regimes, the heat transfer coefficients, and the wall heat fluxes at locations where $x_{eq} = 0.9, 0.2,$ and 0.05.

16.16 In Problem 16.13, the vapor and liquid droplets were assumed to be pure water. The assumption is not realistic, however, because seawater contains dissolved noncondensables, which are partially released into the system and accumulate in the condenser. Assume that the vapor in the condenser contains air at a known concentration (say, 5% by volume). Also, assume that the subcooled droplets are originally under atmospheric pressure and are saturated with dissolved air before they enter the condenser. Formulate thoroughly the solution of Problem 16.13, by writing all the necessary conservation equations (in the Lagrangian frame) and closure relations for a droplet, so that the droplet conditions at the exit from the condenser (including the bulk concentration of dissolved air in the droplet) can be found. (You do not need to numerically solve the equations.)

16.17 Saturated steam at 30 °C, flowing downward, condenses inside a vertical tube that is 19.05 mm in diameter. The tube inner surface is at 24 °C temperature. The total mass flux is 1100 kg/m²·s, and the local quality is 7.5%. Determine the local condensation heat transfer coefficient, and condensation rate (in kg/s), and estimate dx/dz.

16.18 In Problem 16.7, assume that the tube inner surface temperature is 10 °C below the saturation temperature of ammonia at the given pressure. For mass fluxes equal to 50 kg/m²·s and 150 kg/m²·s, determine the wall heat transfer coefficients at locations where $x_{eq} = 0.95, 0.8$, and 0.45, using the correlation of Shah (2009).

16.19 Repeat the solution of Problem 16.15, this time using the method of Kim and Mudawar (2013).

16.20 Saturated R-134a at a temperature of 30 °C flows in a uniformly heated stainless steel microtube that has an inner diameter of 0.8 mm, and is 20 mm long. For mass fluxes of 100 and 250 kg/m²·s, at locations where the equilibrium quality is 0.05 and 0.15, find the condensation heat transfer coefficient using the correlations of Kim and Mudawar (2013).

17 Choking in Two-Phase Flow

17.1 Physics of Choking

Choking can happen when a fluid is discharged through a passage from a pressurized chamber into a chamber that is at a significantly lower pressure. When a flow passage is choked, it supports the maximum possible fluid discharge rate for the given system conditions.

Choking can be better understood by the simple experiment shown in Fig. 17.1, where a chamber containing a fluid at an elevated pressure P_0 is connected to another chamber that is at a lower pressure P_{out} by a flow passage. Suppose that the upstream conditions are maintained unchanged in the experiment, while the pressure in the downstream chamber, P_{out}, is gradually reduced, and the mass flow rate is continuously measured. It will be observed that the mass flux increases as P_{out} is reduced, until P_{out} reaches a critical value P_{ch}. Further reduction of P_{out} will have no impact on mass flux or anything else associated with the channel interior.

The physical explanation of critical flow is as follows. A flow is critical (choked) when disturbances (or hydrodynamic signals) initiated downstream of some critical cross section cannot propagate upstream of the critical cross section. In single-phase flow, infinitesimally small disturbances (hydrodynamic signals) travel with the speed of sound. In a straight channel often the critical cross section occurs at the exit. In nozzles and other converging–diverging channels, the throat acts as the critical cross section.

Critical flow is an important process. Predictive methods for critical flow are needed since drainage of high-pressure fluids through passages, breaks, cracks, etc., is important in many applications. Two-phase critical flow is particularly important for the modeling of a number of nuclear reactor accident scenarios.

In choking problems we often know the conditions upstream of the entrance and downstream from the outlet of the flow passage. Using the fluid properties in the chamber with higher pressure as well as the characteristics of the channel itself, we must be able to calculate P_{ch}. For $P_{out} > P_{ch}$, the fluid momentum conservation equation must be solved for the channel using P_0 and P_{out} as the boundary conditions, to determine the flow rate. However, for $P_{out} < P_{ch}$, the flow rate through the channel will no longer depend on the conditions downstream from the channel exit.

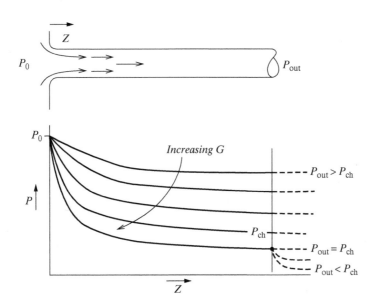

Figure 17.1. Critical flow in a channel.

17.2 Velocity of Sound in Single-Phase Fluids

Consider a steady-state, one-dimensional flow of an inviscid fluid. Let us use a Lagrangian coordinate system moving with the fluid's steady-state velocity. The fluid is then quiescent in the coordinate system.

The flow resulting from a small perturbation in density is to be analyzed; therefore,

$$\rho = \rho_0 \left[1 + \varepsilon(t, z) \right].\tag{17.1}$$

The fluid continuity and momentum equations are

$$\frac{\partial \rho}{\partial t} + \frac{\partial}{\partial z}(\rho U) = 0,\tag{17.2}$$

$$\frac{\partial U}{\partial t} + U \frac{\partial U}{\partial z} = -\frac{1}{\rho} \frac{\partial P}{\partial z} = -\frac{1}{\rho} \left(\frac{\partial P}{\partial \rho} \right) \frac{\partial \rho}{\partial z}.\tag{17.3}$$

Substituting from (17.1) into (17.2) and (17.3), and neglecting second-order terms, we will get

$$\frac{\partial \varepsilon}{\partial t} = -\frac{\partial U}{\partial z},\tag{17.4}$$

$$\frac{\partial U}{\partial t} = -\left(\frac{\partial P}{\partial \rho} \right)_s \frac{\partial \varepsilon}{\partial z}.\tag{17.5}$$

The subscript s implies isentropic and results from the reversible nature of the flow process. Now, apply $\frac{\partial}{\partial z}$ to Eq. (17.4), apply $\frac{\partial}{\partial t}$ to Eq. (17.5), and then eliminate ε between the two resulting equations to derive the "wave equation":

$$\frac{\partial^2 U}{\partial t^2} = \left(\frac{\partial P}{\partial \rho} \right)_s \frac{\partial^2 U}{\partial z^2}.\tag{17.6}$$

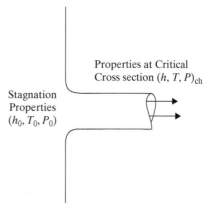

Figure 17.2. Critical discharge for an ideal gas.

Stagnation Properties (h_0, T_0, P_0)

Properties at Critical Cross section $(h, T, P)_{\text{ch}}$

Evidently, therefore, the velocity of sound is

$$a = \sqrt{(\partial P/\partial \rho)_s} = -\frac{1}{\rho}\sqrt{(-\partial P/\partial v)_s}. \tag{17.7}$$

For an isentropic process involving an ideal gas,

$$P\rho^{-\gamma} = \text{const.}, \tag{17.8}$$

$$P = \rho \left(\frac{R_u}{M}\right) T, \tag{17.9}$$

where $\gamma = C_P/C_V$, R_u is the universal gas constant, M is the gas molecular mass, and C_P and C_V are the constant-pressure and constant-volume specific heats, respectively. As a result,

$$a = \sqrt{\gamma (R_u/M) T}. \tag{17.10}$$

The speed of sound in solids and liquids can also be calculated from

$$a = \sqrt{\frac{E_B}{\rho}}, \tag{17.11}$$

where E_B is the bulk modulus of elasticity.

17.3 Critical Discharge Rate in Single-Phase Flow

For a single-phase fluid, the prediction of critical discharge rate is relatively straightforward. First let us consider the steady-state critical discharge of an ideal gas through the adiabatic and frictionless short break displayed in Fig. 17.2. Energy conservation in the flow channel requires

$$h_0 = \left(h + \frac{1}{2}U^2\right). \tag{17.12}$$

Mass continuity requires

$$G = (\rho U). \tag{17.13}$$

Now solving for U from Eq. (17.12), substituting the result for U in the continuity equation, and assuming that $C_P =$ const. over the temperature range of interest lead to

$$G = \rho\sqrt{2C_P(T_0 - T)} = \rho\sqrt{2C_P T_0\left(1 - \frac{T}{T_0}\right)}. \tag{17.14}$$

For an isentropic process $P\rho^{-\gamma} =$ const. and $TP^{\frac{1-\gamma}{\gamma}} =$ const. Using these relations, we can recast Eq. (17.14) as

$$G = \rho_0\left\{2C_P T_0\left[(P/P_0)^{2/\gamma} - (P/P_0)^{\frac{\gamma+1}{\gamma}}\right]\right\}^{1/2}. \tag{17.15}$$

If the pressure P at the exit of the channel is known, then Eq. (17.15) would determine the flow rate. This equation cannot yet be solved for choked flow, because in choked flow $P = P_{ch}$ and the latter is not known. To find P_{ch}, we can use one of three equivalent methods.

(1) For critical flow, as shown in Fig. 17.1, at the channel exit we must have

$$\left.\frac{dG}{dP}\right|_{exit} = 0. \tag{17.16}$$

Using Eqs. (17.15) and (17.16), one gets

$$(P/P_0)_{ch} = \left(\frac{2}{\gamma+1}\right)^{\frac{\gamma}{\gamma-1}}. \tag{17.17}$$

Substituting for $(P/P_0)_{ch}$ from Eq. (17.17) in Eq. (7.15) gives

$$G_{ch} = \rho_0\left\{2C_P T_0\left[\left(\frac{2}{\gamma+1}\right)^{\frac{2}{\gamma-1}} - \left(\frac{2}{\gamma+1}\right)^{\frac{\gamma+1}{\gamma-1}}\right]\right\}^{1/2}. \tag{17.18}$$

(2) In Eq. (17.12), replace U with the speed of sound, a, noting that for an ideal gas

$$C_P = \frac{\gamma}{\gamma-1}\frac{R_u}{M},$$

to get

$$(T/T_0)_{ch} = \frac{2}{\gamma+1}. \tag{17.19}$$

Combining Eq. (17.19) with $TP^{\frac{1-\gamma}{\gamma}} =$ const. (which applies to any isentropic process for an ideal gas) will reproduce Eq. (17.17). Substitution from Eq. (17.17) into Eq. (17.15) will reproduce Eq. (17.18).

(3) Solve the compressible, steady-state, one-dimensional mass, momentum, and energy conservation equations, and by iteratively varying G, obtain $G = G_{ch}$, which would lead to $|dP/dz| \to \infty$ at the exit. For an ideal gas the result will be identical to Eq. (17.18). It is important, however, to note that this method is quite general and is not limited to isentropic and inviscid flow.

17.4 Choking in Homogeneous Two-Phase Flow

Let us now consider a one-dimensional, homogeneous gas–liquid flow in which phasic densities are functions of pressure only (as, for example, in a saturated vapor–liquid mixture); thus $\rho_f = \rho_f(P)$ and $\rho_g = \rho_g(P)$. One can then write

$$\frac{d\rho_g}{dz} = \left(\frac{d\rho_g}{dP}\right)\frac{dP}{dz}, \quad \frac{d\rho_f}{dz} = \left(\frac{d\rho_f}{dP}\right)\frac{dP}{dz}. \tag{17.20}$$

Equivalently, one can write

$$\frac{dv_g}{dz} = \left(\frac{dv_g}{dP}\right)\frac{dP}{dz}, \quad \frac{dv_f}{dz} = \left(\frac{dv_f}{dP}\right)\frac{dP}{dz}. \tag{17.21}$$

The HEM momentum equation (Eq. (5.36)) can then be easily recast as

$$-\frac{dP}{dz} = \frac{\frac{2f}{D}v_h G^2 + G^2 v_{fg}\frac{dx}{dz} - G^2 v_h(dA/dz)/A + g\cos\theta/v_h}{1 + G^2\left[x\frac{dv_g}{dP} + (1-x)\frac{dv_f}{dP}\right]}, \tag{17.22}$$

where f is the Fanning friction factor. Note that for homogeneous flow $\rho_h = \alpha\rho_g + (1-\alpha)\rho_f$. Equation (17.22) is similar to Eq. (5.44), with the difference that in Eq. (5.44) the liquid has been assumed to be incompressible and $A = \text{const.}$ has been assumed. At the critical (choking) point we must have $|dP/dz| \to \infty$, and this condition can be met by equating the denominator of Eq. (17.22) with zero. The result will be

$$G_{ch} = \left\{-\left[x\left(\frac{dv_g}{dP}\right) + (1-x)\left(\frac{dv_f}{dP}\right)\right]\right\}^{-1/2}. \tag{17.23}$$

Bearing in mind that G_{ch} is related to the speed of sound at the critical cross section (throat) according to $a_h = (G/\rho_h)_{ch}$, one gets

$$a_h = \left[-\frac{x(dv_g/dP) + (1-x)(dv_f/dP)}{(v_f + xv_{fg})^2}\right]_{ch}^{-1/2}. \tag{17.24}$$

Equation (17.24) can be recast in another interesting form by noting that

$$dP/dv_g = -\rho_g^2(dP/d\rho_g) = -\rho_g^2 a_g^2, \tag{17.25}$$

$$dP/dv_f = -\rho_f^2(dP/d\rho_f) = -\rho_f^2 a_f^2, \tag{17.26}$$

where a_g and a_f are the pseudo-speeds of sound in the gas (vapor) and liquid, respectively, following the thermodynamic path consistent with the two-phase flow. (Remember that a_g and a_f would be the conventional speeds of sound if the processes were isentropic.) Substitution of these relations in Eq. (17.24) then gives

$$a_h = \left[\rho_h^2\left(\frac{x}{\rho_g^2 a_g^2} + \frac{1-x}{\rho_f^2 a_f^2}\right)\right]^{-1/2}. \tag{17.27}$$

Equivalently, using the void–quality relation for homogeneous flow, one has

$$a_h = \left[\rho_h\left(\frac{\alpha}{\rho_g a_g^2} + \frac{1-\alpha}{\rho_f a_f^2}\right)\right]^{-1/2}. \tag{17.28}$$

Interestingly, often $a_h < a_g$ and $a_h < a_f$.

EXAMPLE 17.1. Estimate the velocity of sound in a homogeneous mixture of air and water at atmospheric pressure and 20 °C temperature for void fractions in the 0.1–0.7 range.

SOLUTION. The speed of sound for air can be found from Eq. (17.10) by noting that for air $\gamma \approx 1.4$ and $M = 22$ kg/kmol, and assuming isentropic expansion for each phase. Furthermore, $R_u = 8,314$ N·m/kmol·K. With $T = 293$ K, we will get

$$a_G = 393.7 \text{ m/s}$$

The bulk module of water is $\mathbf{E}_B = 2.15 \times 10^9$ N/m². From Eq. (17.11) we then get for the speed of sound in water

$$a_L = 1468 \text{ m/s}.$$

We can now find the speed of sound in the mixture from Eq. (17.28), noting that for the homogeneous mixture $\rho_h = (1 - \alpha)\rho_L + \alpha\rho_G$, where

$$\rho_G = P/[(R_u/M)T] = 1.20 \text{ kg/m}^3,$$
$$\rho_L = 998.3 \text{ kg/m}^3.$$

The calculations, when plotted, lead to the figure below.

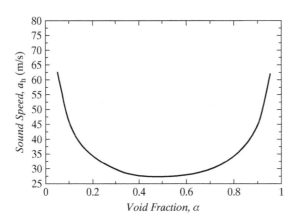

17.5 Choking in Two-Phase Flow with Interphase Slip

A simple analysis similar to what was discussed for single-phase flow can be performed to better understand the relationship between momentum conservation and choking. Recall from Chapter 8 that the one-dimensional two-phase mixture momentum equation can be cast as the summation of various pressure gradient terms (see Eq. (8.1)). For steady-state conditions

$$(-dP/dz) = (-dP/dz)_{sa} + (-dP/dz)_{fr} + (-dP/dz)_g, \quad (17.29)$$

where the terms on the right side represent the spatial acceleration, frictional, and gravitational pressure gradients, respectively. The spatial acceleration component of the pressure drop can be shown as

$$(-dP/dz)_{sa} = \frac{d}{dz}[G^2/\rho'], \tag{17.30}$$

$$\rho' = \left[\frac{(1-x)^2}{\rho_f(1-\alpha)} + \frac{x^2}{\rho_g\alpha}\right]^{-1}. \tag{17.31}$$

If we assume that $\rho_f = \rho_f(P)$ and $\rho_g = \rho_g(P)$, Eq. (17.30) will give

$$(-dP/dz)_{sa} = \left[\frac{d}{dP}(G^2/\rho')\right](dP/dz). \tag{17.32}$$

We can now substitute from Eq. (17.32) into Eq. (17.29) and factor out $(-dP/dz)$. The mixture momentum equation for a channel with uniform cross section then becomes

$$\left(-\frac{dP}{dz}\right) = \frac{(-dP/dz)_{fr} + (-dP/dz)_g}{1 + G^2\frac{d}{dP}(1/\rho')}. \tag{17.33}$$

At the critical (choked) point the denominator will be equal to zero, leading to

$$G_{ch} = \left\{-\frac{d}{dP}\left[\frac{(1-x)^2}{\rho_f(1-\alpha)} + \frac{x^2}{\rho_g\alpha}\right]\right\}_{ch}^{-1/2}. \tag{17.34}$$

Equation (17.34) of course cannot be applied unless the local slip, the extent of local thermodynamic nonequilibrium, etc., are all known.

17.6 Critical Two-Phase Flow Models

Experimental results and physical insight show that velocity slip and thermodynamic nonequilibrium are likely in critical flow, in view of the significant spatial accelerations that take place. The magnitudes of velocity slip and thermodynamic nonequilibrium are difficult to predict, and they vary from case to case. Uncertainty with respect to the extent of thermal and mechanical nonequilibrium, in general, makes it impossible to calculate the two-phase critical discharge rate in terms of stagnation properties only. Modeling the critical discharge in two-phase flow is thus considerably more complicated than in single-phase flow. Various assumptions have been made by investigators for the estimation of thermodynamic and velocity nonequilibrium at the throat, leading to a number of relatively accurate two-phase critical flow models. Some of the most widely used models are described in the following.

17.6.1 The Homogeneous-Equilibrium Isentropic Model

This is among the simplest models for saturated liquid–vapor mixture flow. It is assumed that no thermal or mechanical nonequilibrium occurs anywhere in the flow and that the expansion of the mixture is isentropic. The model is robust and handy for two reasons. First, besides the aforementioned assumptions, no other arbitrary assumption is needed for the model. Second, the model leads to a closed-form solution.

Consider the system shown in Fig. 17.3, where s_0, h_0, and P_0 are the stagnation-state properties of the fluid. The channel obviously needs to be adiabatic and

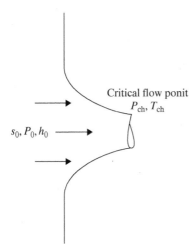

Figure 17.3. Homogeneous-equilibrium isentropic critical two-phase flow.

frictionless (otherwise isentropic flow would not be valid), and therefore energy conservation would require

$$h + \frac{1}{2}\frac{G^2}{\rho_h} = h_0 \tag{17.35}$$

$$\Rightarrow G = \frac{1}{v}\sqrt{2(h_0 - h)}. \tag{17.36}$$

Here $v = 1/\rho_h$. Critical flow requires that at the exit $\partial G/\partial P|_{ch} = 0$. Applying this condition to Eq. (17.36) gives

$$\frac{2(h_0 - h)}{v^2} = -\frac{1}{v}\frac{(\partial h/\partial P)_s}{(\partial v/\partial P)_s}. \tag{17.37}$$

From the second $T\,ds$ relation of thermodynamics, $T\,ds = dh - v\,dP$, one can get

$$\left(\frac{\partial h}{\partial P}\right)_s = v. \tag{17.38}$$

Combining Eqs. (17.37) and (17.38) yields the final result

$$2\left[h_0 - h(s_0, P_{ch})\right]\frac{\partial v}{\partial P}\bigg|_{s_0, P_{ch}} + \left[v(s_0, P_{ch})\right]^2 = 0, \tag{17.39}$$

where

$$v = xv_g(P_{ch}) + (1 - x)v_f(P_{ch}), \tag{17.40a}$$

$$h = xh_g(P_{ch}) + (1 - x)h_f(P_{ch}), \tag{17.41a}$$

$$s = xs_g(P_{ch}) + (1 - x)s_f(P_{ch}). \tag{17.42a}$$

Expressions similar to Eqs. (17.40a)–(17.42a) can of course be written with P_{ch} replaced with P_0, and with $v, h,$ and s replaced with $v_0, h_0,$ and s_0, respectively, namely,

$$v_0 = x_0 v_g(P_0) + (1 - x_0)v_f(P_0), \tag{17.40b}$$

$$h_0 = x_0 h_g(P_0) + (1 - x_0)h_f(P_0), \tag{17.41b}$$

$$s_0 = x_0 s_g(P_0) + (1 - x_0)s_f(P_0). \tag{17.42b}$$

Note that $s = s_0$ in this model. Equations (17.39)–(17.42b), will be closed, provided that the necessary fluid property routines are used. The iterative solution of these equations (most conveniently performed by iteratively varying P_{ch}) will give all properties at the critical cross section, including h. Equation (17.36) then provides G_{ch}.

Moody (1975, 1979) compared the predictions of the homogeneous-equilibrium isentropic model with data representing the blowdown of pressurized tanks by means of critical flow in tubes. He showed that the model well predicted the data, indicating that homogeneous-flow choking near the entrance of long tubes controls the mass flux in them, provided that the tubes are long enough so that the residence time of liquid in the flow passage is sufficient for the nucleation of bubbles. For water, the required minimum length was about 13 cm, which corresponds to about one millisecond residence time for the nucleation of bubbles and the occurrence of critical flow in liquid water. He also argued that when blowdown of a pressurized tank takes place as a result of critical flow in a long tube, critical flow conditions will develop near both ends of the tube. Near the inlet, nucleation of bubbles resulting from flashing causes choking, and the conditions agree with the homogeneous-equilibrium isentropic model. Phase separation develops further downstream as the void fraction increases, however, and near the exit of the tube another choking takes place. The choking conditions in the latter point are consistent with the slip-flow models discussed below. He showed that these slip-flow models are appropriate for critical flow calculations when the local conditions and properties near the exit are used. Moody also performed extensive numerical calculations for water. The model did particularly well in predicting the blowdown of saturated liquid–vapor mixtures. Moody's parametric calculation results for water are depicted in Figs. 17.4 and 17.5 (Moody, 1979, 1990).

17.6.2 Critical Flow Model of Moody

The critical flow model of Moody (1965) is among the most popular liquid–vapor two-phase critical flow models. The model assumes thermodynamic equilibrium between the two phases everywhere, but it allows for velocity difference between the two phases at the throat. It is based on the following assumptions.

(a) The flow is a saturated vapor–liquid (one-component) mixture everywhere.
(b) The flow is one-dimensional, with a uniform velocity distribution for each phase.
(c) The mixture expansion is isentropic.
(d) At the critical cross section, we must have

$$\left(\frac{\partial G}{\partial P}\right)_{S_r} = 0, \tag{17.43}$$

$$\left(\frac{\partial G}{\partial S_r}\right)_{P} = 0, \tag{17.44}$$

where $S_r = U_g/U_f$ is the slip ratio. Note that the model allows for unequal phasic velocities, and because S_r is an added unknown, two conditions for choking at the throat are imposed.

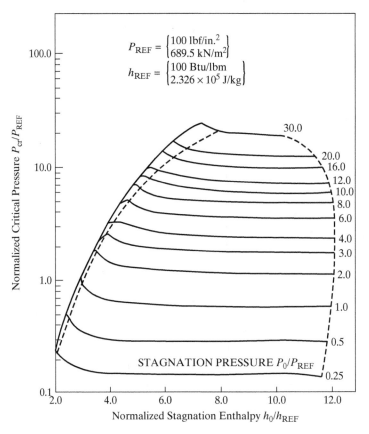

Figure 17.4. Stagnation and critical discharge properties for steam–water according to the homogeneous-equilibrium isentropic model. (After Moody, 1979.)

The derivation of the model proceeds as follows. At any point, including the throat,

$$U_g = xG/(\alpha \rho_g), \qquad (17.45)$$

$$U_f = (1 - x)G/\left[(1 - \alpha)\rho_f\right], \qquad (17.46)$$

$$h_0 = x\left(h_g + \frac{1}{2}U_g^2\right) + (1 - x)\left(h_f + \frac{1}{2}U_f^2\right), \qquad (17.47)$$

$$s_0 = s_f + xs_{fg}, \qquad (17.48)$$

$$\alpha = \frac{1}{1 + S_r \frac{1-x}{x}\frac{v_f}{v_g}}. \qquad (17.49)$$

Elimination of U_g, U_f, x, and α among the above five equations leads to

$$G = \left\{ \frac{2\left[h_0 - h_f - \frac{h_f}{s_{fg}}(s_0 - s_f)\right]}{\left[\frac{S_r(s_g - s_0)}{s_{fg}}v_f + \frac{(s_0 - s_f)v_g}{s_{fg}}\right]^2 \left[\frac{s_0 - s_f}{s_{fg}} + \frac{s_g - s_0}{S_r^2 s_{fg}}\right]} \right\}^{1/2}. \qquad (17.50)$$

Conditions represented by Eqs. (17.43) and (17.44) must now be applied. The application of the condition of Eq. (17.44) to Eq. (17.50) gives

$$S_r = (v_g/v_f)^{1/3}. \qquad (17.51)$$

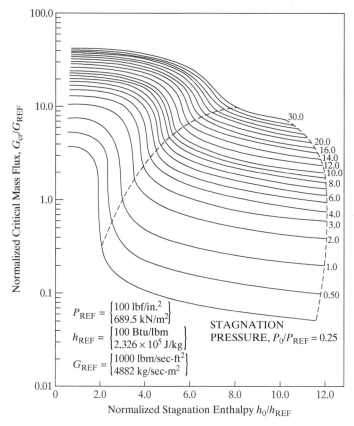

Figure 17.5. Critical mass flux for steam–water homogeneous-equilibrium isentropic flow according to Moody. (After Moody, 1979.)

This expression for S_r agrees with an analysis by Zivi (1964), who derived the same result for equilibrium annular flow, based on the minimum entropy production postulate.

Using Eq. (17.51), we can now eliminate S_r from Eq. (17.50). The resulting equation, along with Eq. (17.43), can then be iteratively solved.

The range of validity of Moody's critical flow model for water is $x_0 \approx 0.01$–1.0 and $P_0 \approx 14.7 - 400$ psia (1.0–27.2 bars).

17.6.3 Critical Flow Model of Henry and Fauske

A schematic of the flow field, which also displays the basic assumptions regarding nonequilibrium between the two phases in this model of Henry and Fauske (1971), is shown in Fig. 17.6. The two phases are assumed to be at thermodynamic and velocity equilibrium, except at the throat. The basic assumptions in this model are as follows.

(a) The mixture flow is adiabatic and frictionless.
(b) There is no energy transfer between the two phases, up to the throat.
(c) There is no velocity slip between the two phases, up to the throat.
(d) Each phase expands isentropically, up to the throat.
(e) The liquid is incompressible.

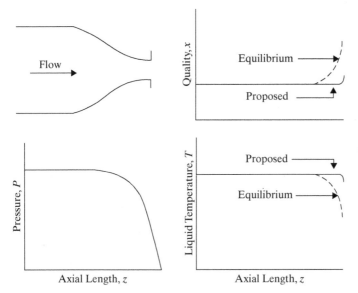

Figure 17.6. Temperature and quality profiles in the critical flow model of Henry and Fauske (1971).

The derivation starts by writing, for the mixture momentum,

$$-dP = Gd\left[xU_v + (1-x)U_L\right].$$ (17.52)

This expression can be solved for G, and for the critical conditions,

$$G_{ch}^{-1} = -\frac{d}{dP}[xU_v + (1-x)U_L]_{ch}.$$ (17.53)

This equation can be expanded by taking the term-by-term derivatives on the right side. The result will be several unknown derivative terms, which will have to be quantified. The following two conditions are assumed to apply:

$$(\partial G/\partial P)_{ch} = 0,$$ (17.54)

$$(\partial S_r/\partial P)_{ch} = 0.$$ (17.55)

The first of these relations of course ensures that critical flow occurs at the throat. The second relation is somewhat arbitrary. The deviation from thermodynamic and mechanical equilibrium starts at the throat, where the vapor expansion is assumed to be polytropic according to

$$\left(\frac{dv_v}{dP}\right)_{ch} = \left(\frac{v_v}{nP}\right)_{ch},$$ (17.56)

where n is the thermal equilibrium polytropic exponent for the two-phase mixture. If the vapor expands as an ideal gas during the polytropic process, then

$$\left(\frac{ds_v}{dP}\right)_{ch} = -\frac{C_{Pg}}{P_{ch}}\left(\frac{1}{n} - \frac{1}{\gamma}\right).$$ (17.57)

Following Tangren *et al.* (1949), we have for the thermal equilibrium polytropic exponent

$$n = \frac{(1-x)C_{Pf}/C_{Pg} + 1}{(1-x)C_{Pf}/C_{Pg} + 1/\gamma}, \tag{17.58}$$

where γ is the specific heat ratio for vapor. Also, quality is assumed to vary at the throat according to

$$\left(\frac{dx}{dP}\right)_{ch} = N\left(\frac{dx_{eq}}{dP}\right)_{ch}, \tag{17.59}$$

where

$$N = x_{eq,ch}/0.14. \tag{17.60}$$

Manipulation of Eq. (17.53), and substitution of all these conditions, leads to

$$G_{ch}^2 = \left\{\frac{x_0 v_v}{nP_{ch}} + (v_v - v_{L0})\left[\frac{(1-x_0)N}{S_{v,eq} - S_{L,eq}}\frac{ds_{L,eq}}{dP} - \frac{x_0 C_{Pg}\left(\frac{1}{n} - \frac{1}{\gamma}\right)}{P_{ch}(s_{g0} - s_{L0})}\right]\right\}_{ch}^{-1}, \tag{17.61}$$

where subscript g represents saturated vapor. Equation (17.61) cannot be solved alone, because it has two unknowns, P_{ch} and G_{ch}. An additional equation can be obtained by integrating the one-dimensional mechanical energy conservation equation between the inlet and the throat to get

$$(1-x_0)v_{L0}(P_0 - P_{ch}) + \frac{x_0\gamma}{\gamma - 1}\left[P_0 v_{g0} - (Pv_v)_{ch}\right] = \frac{1}{2}[(1-x_0)v_{L0} + x_0 v_{v,ch}]^2 G_{ch}^2. \tag{17.62}$$

Equations (17.61) and (17.62) can now be iteratively solved for the two unknowns. These equations can also be combined and expressed in the following form:

$$\eta_{ch} = \left\{\frac{\frac{1-\alpha_0}{\alpha_0}(1 - \eta_{ch}) + \frac{\gamma}{\gamma - 1}}{\frac{1}{2\beta\alpha_{ch}^2} + \frac{\gamma}{\gamma - 1}}\right\}^{\frac{\gamma}{\gamma - 1}}, \tag{17.63}$$

$$\eta_{ch} = P_{ch}/P_0, \tag{17.64}$$

$$\beta = \left\{\frac{1}{n} + \left(1 - \frac{v_{L0}}{v_{v,ch}}\right)\left[\frac{(1-x_0)NP_{ch}}{x_0 s_{fg,ch}}\frac{ds_f}{dP}\right]_{ch} - \frac{C_{Pv}\left(\frac{1}{n} - \frac{1}{\gamma}\right)}{s_{g0} - s_{L0}}\right\}, \tag{17.65}$$

$$v_{v,ch} = v_{g0}\eta_{ch}^{-\frac{1}{\gamma}}. \tag{17.66}$$

The void fractions are related to local qualities according to the requirements of homogeneous flow:

$$\alpha_0 = \frac{v_{g0}x_0}{v_{g0}x_0 + v_{L0}(1 - x_0)}, \tag{17.67}$$

$$\alpha_{ch} = \frac{x_0 v_{v,ch}}{x_0 v_{v,ch} + (1 - x_0)v_{L0}}. \tag{17.68}$$

Extended Henry and Fauske Model

This extension can be considered as a semi-empirical modification of the Henry and Fauske (1971) model to simplify the calculations and expand the applicability of the model to subcooled liquid inlet conditions (McFadden *et al.*, 1992). In Eqs. (17.61) and (17.62), by replacing $v_{v,ch}$ with $v_{g,ch}$, one gets

$$G_{ch}^2 = \left\{ \frac{x_0 v_g}{n P_{ch}} + (v_g - v_{L0}) \left[\frac{(1-x_0)N}{s_{fg,eq}} \frac{ds_f}{dP} - \frac{x_0 C_{Pg}\left(\frac{1}{n} - \frac{1}{\gamma}\right)}{P_{ch} s_{fg0}} \right]^{-1} \right\}_{ch}, \quad (17.69)$$

$$(1-x_0)v_{L0}(P_0 - P_{ch}) + \frac{x_0 \gamma}{\gamma - 1}\left[P_0 v_{g0} - P_{ch} v_{g,ch}\right] = \frac{1}{2}\left[(1-x_0)v_{L0} + x_0 v_{g,ch}\right]^2 G_{ch}^2, \quad (17.70)$$

where subscript eq implies equilibrium. These equations apply to subcooled liquid at the inlet provided that $x_0 = 0$ is used.

The range of validity of the model of Henry and Fauske is as follows. With a steam–water mixture at the inlet, $P_0 \approx 17.6$–882 psia and $x_0 \approx 10^{-3}$–1.0. For saturated water at the inlet, $P_0 \approx 50$–200 psia. With subcooled water at the inlet, where the extended Henry–Fauske model applies, $P_0 \approx 50 - 200$ psia and $T_0 \approx 250$–350 °F.

17.7 RETRAN Curve Fits for Critical Discharge of Water and Steam

The isentropic-HEM (isoenthalpic) model, Moody's model, and the extended Henry–Fauske method all require iterative solutions of nonlinear sets of equations along with water–steam property routines or functions. Extensive parametric numerical calculations with these models have been performed (e.g., De Young, 1975). The results have been fitted to simple polynomial-type correlations that are explicit with respect to G_{ch}. Thus, generically,

$$G_{ch} = f(P_0, h_0). \quad (17.71)$$

These correlations are used in RETRAN-03 and RETRAN-3D computer codes (McFadden *et al.*, 1992). In the correlations everywhere the English unit system is to be used; therefore h_0 should be in British thermal units per pound-mass, P_0 should be in pounds per square inch absolute, and will be in pound-mass per foot squared per second. The functions $f(P_0, h_0)$ for the three critical discharge models, and their recommended ranges of application, are as follows.

For the critical mass flux according to the model of Moody (1965),

$$G_{ch}^{Moody}(P_0, h_0) = \begin{cases} \exp\left[\sum_{j=0}^{5}\sum_{i=0}^{5} M1_{i,j} P_0^j h_0^i\right] & \text{for } 15 \le P_0 \le 200 \text{ psia}, \quad (17.72) \\ \sum_{j=0}^{5}\sum_{i=0}^{5} M2_{i,j} P_0^j h_0^i & \text{for } 200 < P_0 \le 3{,}000 \text{ psia}. \quad (17.73) \end{cases}$$

The recommended range for the application of Moody's curve fits is

$$0.01 \le x_{eq} \le 1,$$
$$15 \le P_0 \le 3{,}000 \text{ psia}.$$

(Note that a saturated mixture is required.) For the extended Henry–Fauske model,

$$G_{ch}^{EHF}(P_0, h_0) = \begin{cases} \sum_{j=0}^{5} \sum_{i=0}^{5} H1_{i,j} P_0^j h_0^i & \text{for } 15 \le P_0 \le 300\,\text{psia}, & (17.74) \\ \sum_{j=0}^{5} \sum_{i=0}^{5} H2_{i,j} P_0^j h_0^i & \text{for } 300 < P_0 \le 3{,}000\,\text{psia}. & (17.75) \end{cases}$$

The extended Henry–Fauske method is recommended by the developers of the RETRAN code for the following range, as long as the fluid at stagnation is subcooled liquid:

$$h_0 \ge 170\,\text{Btu/lb}_m,$$
$$15 \le P_0 \le 3{,}000\,\text{psia}.$$

For the isentropic mixture model (also referred to as the isoenthalpic model), the entire parameter range is divided into three pressure zones. For the low-pressure range, $14.7 \le P_0 \le 100$ psia, for the subcooled liquid,

$$8\,\text{Btu/lb}_m \le h_0 \le h_f,$$
$$G_{ch}^{ISO}(P_0, h_0) = \sum_{j=0}^{5} \sum_{i=0}^{5} I1_{i,j} P_{10}^j h_{10}^i, \tag{17.76}$$

where

$$P_{10} = 750 P_0^{0.5}, \tag{17.77}$$
$$h_{10} = 5500 - 2.3 \times 10^{-11} h_0^6, \tag{17.78}$$

and for the saturated mixture or superheated vapor,

$$h_f \le h_0 \le 1{,}750\,\text{Btu/lb}_m,$$
$$G_{ch}^{ISO}(P_0, h_0) = \sum_{j=0}^{4} \sum_{i=0}^{5} I2_{i,j} P_0^j h_{20}^i, \tag{17.79}$$

where

$$h_{20} = 5900 \exp(-0.001\,3733 h_0). \tag{17.80}$$

For the medium pressure range, $100 < P_0 \le 2800$ psia, for the subcooled liquid,

$$8\,\text{Btu/lb}_m \le h_0 \le h_f,$$
$$G_{ch}^{ISO}(P_0, h_0) = \sum_{j=0}^{4} \sum_{i=0}^{5} I3_{i,j} P_{30}^j h_{30}^i, \tag{17.81}$$

where

$$P_{30} = 58.863 P_0^{0.75}, \tag{17.82}$$
$$h_{30} = 20\,000 - 0.053\,98 h_0^2, \tag{17.83}$$

and for the saturated mixture or superheated vapor,

$$h_f \le h_0 \le 1750\,\text{Btu/lb}_m,$$
$$G_{ch}^{ISO}(P_0, h_0) = \sum_{j=0}^{3} \sum_{i=0}^{4} I4_{i,j} P_0^j h_{40}^i, \tag{17.84}$$

where

$$h_{40} = 50\,000 \exp(-0.002 h_0). \tag{17.85}$$

For the high pressure range, $2800 < P_0 \leq 6000$ psia, for the subcooled liquid,

$$250 \leq h_0 \leq 800 \text{ Btu/lb}_m,$$

$$G_{ch}^{ISO}(P_0, h_0) = \sum_{j=0}^{3} \sum_{i=0}^{4} I6_{i,j} P_0^j h_{60}^i, \tag{17.86}$$

where

$$h_{60} = 59\,000 \exp(-0.001\,3733 h_0), \tag{17.87}$$

and for the saturated mixture or superheated vapor,

$$800 \leq h_0 \leq 1750 \text{ Btu/lb}_m,$$

$$G_{ch}^{ISO}(P_0, h_0) = \sum_{j=0}^{3} \sum_{i=0}^{4} I5_{i,j} P_{50}^j h_{50}^i, \tag{17.88}$$

where

$$P_{50} = 22\,000 + 7.36842 P_0, \tag{17.89}$$

$$h_{50} = 50\,000 - 0.06173 h_0^2. \tag{17.90}$$

The coefficients in these equations are summarized in Tables 17.1–17.3 (Elias and Lelluche, 1994; McFadden *et al.*, 1992).

17.8 The Omega Parameter Methods

These methods are based on the application of an approximate equation of state for homogeneous-equilibrium two phase mixtures, for modeling the critical discharge of two-phase mixtures. The approximate equation of state, originally suggested by Epstein *et al.* (1983), includes a parameter that is often shown as ω (hence the name of the method). The omega parameter methods often lead to explicit expressions for predicting the choked flow rates, and are thus very simple. The assumption of homogeneous and equilibrium flow of course limits the accuracy of these models for some cases; nevertheless, even for flow conditions where significant velocity slip and thermal non-equilibrium are likely the omega parameter method provides a good estimate of the lower bound of the critical discharge rate.

Critical Discharge of a Saturated Liquid

Consider isentropic choked flow of a single-component fluid, and assume that homogeneous-equilibrium conditions apply (see Fig. 17.3). Equations (17.35) and (17.36) will then apply. Starting from the thermodynamics second Tds relation, we can write

$$Tds = dh - vdP. \tag{17.91}$$

Table 17.1. *The constant coefficients in the RETRAN curve fits for the critical flow model of Moody.*

	$M1_{i,j}$					
i/j	0	1	2	3	4	5
0	0.7539883E1	0.48635447E0	−0.14054847E-1	0.18252651E-3	−0.10492510E-5	0.21537617E-8
1	−0.27349762E-1	−0.25172525E-2	−0.90215743E-4	−0.12541056E-5	0.74318133E-8	−0.15500316E-10
2	0.77033614E-4	0.66198222E-5	−0.25893147E-6	0.36860723E-8	−0.22032243E-10	0.46118179E-13
3	−0.11597165E-6	−0.84665377E-8	0.35752133E-9	−0.51915992E-11	0.31249946E-13	−0.65601247E-16
4	0.85314613E-10	0.52492322E-11	−0.23790918E-12	0.35151649E-14	−0.21287482E-16	0.44796907E-19
5	−0.24171538E-13	−0.12636194E-14	0.61256651E-16	−0.91925726E-18	0.55968113E-20	−0.11802990E-22

	$M2_{i,j}$					
i/j	0	1	2	3	4	5
0	0.64582892E4	0.15724915E3	−0.16101131E0	−0.14741137E-4	0.61947560E-7	−0.19206668E-10
1	−0.59379818E2	−0.76566784E0	0.11162190E-2	−0.18726780E-6	−0.22806979E-9	0.88455810E-13
2	0.19040194E0	0.14863253E-2	−0.27575375E-5	0.85169692E-9	0.30284416E-12	−0.16287505E-15
3	−0.27991119E-3	−0.13833255E-5	0.31653794E-8	−0.12626170E-11	−0.16803993E-15	0.15109240E-18
4	0.19327093E-6	0.60309542E-9	−0.17199274E-11	0.79444415E-15	0.29596473E-19	−0.71294533E-22
5	−0.50859277E-10	−0.95031339E-13	0.35699646E-15	−0.18158912E-18	0.17431814E-23	0.13825335E-25

Table 17.2. *The constant coefficients in the RETRAN curve fits for the extended Henry–Fauske critical flow model.*

$H1_{i,j}$

i/j	0	1	2	3	4	5
0	0.11971131E5	−0.25664444E3	−0.11154016E3	0.14181940E1	−0.57956498E-2	0.77436012E-5
1	−0.29275019E3	0.40495998E2	0.17440962E1	−0.25888201E-1	0.11065153E-3	−0.15104216E-6
2	0.13088631E1	−0.60219912E0	−0.93144283E-2	0.17982760E-3	−0.81656598E-6	0.11454533E-8
3	0.96555931E-2	0.34802345E-2	0.15567712E-4	−0.57489343E-6	0.28585300E-8	−0.41616729E-11
4	−0.85871644E-4	−0.85504973E-5	0.13551325E-7	0.81339877E-9	−0.46830222E-11	0.71890283E-14
5	0.16193695E-6	0.71823710E-8	−0.41240653E-10	−0.38055461E-12	0.28432048E-14	−0.46890691E-17

$H2_{i,j}$

i/j	0	1	2	3	4	5
0	0.11996419E5	0.33614071E2	−0.34555139E-1	0.27341308E-4	−0.10430915E-7	0.13430502E-11
1	−0.66516773E2	−0.21670146E0	0.48840252E-3	−0.43249560E-6	0.16802644E-9	−0.21923832E-13
2	0.25523516E0	0.18031256E-2	−0.35661004E-5	0.28854007E-8	−0.10625928E-11	0.13604749E-15
3	−0.44088343E-3	−0.57669789E-5	0.10431626E-7	−0.78849169E-11	0.27918999E-14	−0.35206568E-18
4	−0.53252751E-6	0.98378584E-8	−0.15532591E-10	0.10650136E-13	−0.35327078E-17	0.43073516E-21
5	0.76399011E-9	−0.64276860E-11	0.93866385E-14	−0.59412445E-17	0.18329776E-20	−0.21322858E-24

Table 173. *The constant coefficients in the RETRAN curve fits for the isenthalpic critical flow model.*

$I1_{i,j}$

i/j	0	1	2	3	4	5
0	−0.61079017E6	0.35835693E3	−0.82113001E-1	0.92074993E-5	−0.50256939E-9	0.10656235E-13
1	0.17723792E3	−0.85173920E-1	0.15133113E-4	−0.12069593E-8	0.40839937E-13	−0.39301326E-18
2	−0.15259033E-1	0.46248520E-5	−0.14319090E-9	−0.77321077E-13	0.87190503E-17	−0.26645196E-21
3	0.59775167E-6	−0.16855193E-9	0.16596792E-13	−0.94175361E-18	0.54924497E-22	−0.18513150E-26
4	−0.33084180E-11	0.49957406E-15	0.10894667E-19	−0.39065314E-23	0.75466939E-28	0.35921080E-32
5	−0.27603377E-16	0.33248852E-20	−0.18476858E-24	0.14786815E-28	0.20499051E-33	−0.53964250E-37

$I2_{i,j}$

i/j	0	1	2	3	4
0	−0.21851768E1	0.82610264E0	−0.39296656E-1	0.24894000E-3	−0.38481513E-6
1	0.77226019E-2	0.45429711E-3	0.15811982E-3	−0.98756816E-6	0.15181668E-8
2	−0.94283873E-5	0.14299562E-5	−0.22417512E-6	0.13771995E-8	−0.21029987E-11
3	0.51095086E-8	−0.10919498E-8	0.14040350E-9	−0.84745729E-12	0.12845429E-14
4	−0.12429312E-11	0.31314774E-12	−0.39756584E-13	0.23478555E-15	−0.35282311E-18
5	0.11033567E-15	−0.30232812E-19	0.41644534E-17	−0.23853748E-19	0.35476040E-22

$I3_{i,j}$

i/j	0	1	2	3	4
0	0.23396109E6	−0.52765454E2	0.43211218E-2	−0.14844802E-6	0.18711342E-11
1	−0.10755023E3	0.20440480E-1	−0.14090500E-5	0.42549018E-10	−0.47722081E-15
2	0.14546479E-1	−0.22721603E-5	0.12791578E-9	−0.30319026E-14	0.24694464E-19
3	−0.87942880E-6	0.11159414E-9	−0.48684729E-14	0.78514401E-19	−0.24454354E-24
4	0.25418710E-10	−0.26125639E-14	0.93851532E-19	−0.15480205E-23	0.14229046E-28
5	−0.29079157E-15	0.25300892E-19	−0.10895948E-23	0.41673812E-28	−0.80302624E-33

(cont.)

Table 173. (cont.)

$I4_{i,j}$

i/j	0	1	2	3
0	0.94440375E2	0.68202161E0	0.22277618E-3	-0.68481663E-7
1	-0.70532888E-1	0.47739930E-3	-0.16796720E-6	0.53600389E-10
2	0.14151552E-4	-0.49292500E-7	0.36261394E-10	-0.11953664E-13
3	-0.94543999E-9	0.22344108E-11	-0.30279729E-14	0.10518184E-17
4	0.19256386E-13	-0.26353127E-16	0.11323359E-18	-0.33126657E-22

$I5_{i,j}$

i/j	0	1	2	3
0	0.13922324E6	-0.85183342E1	0.17344230E-3	-0.10770670E-8
1	0.55483258E0	-0.67033465E-4	0.19159431E-8	-0.13962161E-13
2	-0.10206096E-3	0.50562629E-8	-0.79809531E-13	0.42688713E-18
3	-0.14516886E-8	0.76341357E-13	-0.12984123E-17	0.73587201E-23
4	-0.52974405E-14	0.28362334E-18	-0.49216747E-23	0.28301033E-28

$I6_{i,j}$

i/j	0	1	2	3
0	0.67095585E7	-0.46985672E4	0.10797070E1	-0.80901658E-4
1	-0.87672192E3	0.61458710E0	-0.14215439E-3	0.10722893E-7
2	0.41252890E-1	-0.29071947E-4	0.67976539E-8	-0.51734035E-12
3	-0.83260068E-6	0.59326327E-9	-0.14056658E-12	0.10810246E-16
4	0.61173830E-11	-0.44261205E-14	0.10648646E-17	-0.82851421E-22

Assuming isentropic expansion, then

$$h_0 - h = \int_{P_0}^{P} -v dP. \tag{17.92}$$

We can combine this equation with Eq. (17.36) to get

$$G = \frac{1}{v} \left[2 \int_{P_0}^{P} -v dP \right]^{1/2} \tag{17.93}$$

To find the critical discharge rate, G_{ch}, we must look for P_{ch}, where $P = P_{ch}$ causes G to become maximum. We evidently need an equation of state for the homogeneous fluid mixture in order to calculate v.

For isentropic flashing of an initially saturated liquid, Epstein *et al.* (1983) proposed the following simple equation of state that is conveniently explicit in terms of the specific volume

$$\frac{v}{v_{f0}} = \omega \left(\frac{P_0}{P} - 1 \right) + 1. \tag{17.94}$$

As noted, this equation shows the variation of the flashing liquid–vapor mixture specific volume as a function of pressure, using the stagnation conditions as the reference. For isentropic expansion, Epstein *et al.* (1983) derived

$$\omega = \frac{C_{f0} T_0 P_0}{v_{f0}} \left(\frac{v_{fg0}}{h_{fg0}} \right)^2 \tag{17.95}$$

where all properties with subscript 0 refer to the stagnation conditions, and C_{f0} represents the specific heat of the liquid phase.

We can now substitute from Eq. (17.94) in Eq. (17.93) and apply $\frac{dG}{dP} = 0$ in order to find P_{ch}. This will lead to the following transcendental equation for η_{ch}, where $\eta = P/P_0$

$$\eta_{ch}^2 + (\omega^2 - 2\omega)(1 - \eta_{ch})^2 + 2\omega^2 \ln \eta_{ch} + 2\omega^2 (1 - \eta_{ch}) = 0. \tag{17.96}$$

Knowing η_{ch}, G_{ch} can be found from Eq. (17.93), and that leads to

$$G_{ch} = \eta_{ch} \left(\frac{P_0}{\omega v_{f0}} \right)^{1/2}. \tag{17.97}$$

Thus, in summary, for choked flow of a flashing liquid that is a saturated liquid at the stagnation conditions, one first calculates ω from Eq. (17.95). Equation (17.96) can then be used for calculating η_{ch}. The critical discharge mass flux G_{ch} can then be found from Eq. (17.97).

Generalization of the Model

The above analysis was limited to saturated liquid stagnation conditions. The analysis is now modified to address a saturated two-phase mixture as the stagnation condition (Leung, 1986).

Let us replace the equation of state (Eq. (17.94)) with

$$\frac{v}{v_0} = \omega \left(\frac{P_0}{P} - 1 \right) + 1, \tag{17.98}$$

where

$$v_0 = v_{f0} + x_0 v_{fg0}. \tag{17.99}$$

For the derivation of an expression for ω in terms of stagnation properties, let us replace the assumption of isentropic expansion (which was the basis for the derivation of Eq. (17.95)) with the assumption of isoenthalpic mixture expansion. Therefore, further assuming that $h_{fg} \approx h_{fg0}$ we can write

$$x = \frac{h_0 - h_f}{h_{fg}} \approx x_0 + \frac{h_{f0} - h_f}{h_{fg}} = x_0 + \frac{\overline{C}_f (T_0 - T)}{h_{fg}} \tag{17.100}$$

where the overbar represents an averaged property. Assume, furthermore,

$$v_g \gg v_f; \ v_{g0} \gg v_{f0}; \ P v_{fg} = P_0 v_{fg0}.$$

Also, from Clapeyron's relation (see Eq. (1.9))

$$\frac{P_0 - P}{T_0 - T} \approx \frac{h_{fg}}{\bar{v}_{fg} T} \approx \frac{h_{fg0}}{v_{fg0} T_0}. \tag{17.101}$$

Using these approximations, Eq. (17.94) can be manipulated to derive (see Problem 17.12)

$$\omega = \frac{x_0 v_{fg0}}{v_0} + \frac{C_{f0} T_0 P_0}{v_{f0}} \left(\frac{v_{fg0}}{h_{fg0}} \right)^2. \tag{17.102}$$

Note that with $x_0 = 0$, this equation will reduce to Eq. (17.95).

The procedure for calculating the critical discharge is as follows. Knowing the stagnation conditions, find ω from Eq. (17.102), and use ω to calculate η_{ch} by solving Eq. (17.96). The critical discharge can then be found from Eq. (17.97). Leung (1986) compared the predictions of this method with data representing ten different fluids including water, ammonia, nitrogen, and propane, with very good agreement between model calculations and data. The method is applicable for reduced temperatures up to about 0.9 ($T_r < 0.9$) and reduced pressures up to about 0.5 ($P_r < 0.5$). Property variations become significant as critical pressure and temperature are approached, and assumptions and approximations that led to the aforementioned derivations become unrealistic.

The aforementioned procedure includes an iterative solution of the transcendental equation, Eq. (17.96). Leung (1986) derived the following curve fit which closely approximates the solution of Eq. (17.96)

$$\eta_{ch} = 0.55 + 0.217 \ln \omega - 0.046 (\ln \omega)^2 + 0.004 (\ln \omega)^3. \tag{17.103}$$

Extension of the Methos to Subcooled Liquid Stagnation Conditions

The omega parameter method has been extended to subcooled liquid stagnation conditions (Leung and Grolmes, 1988) by assuming the following.

(a) Isentropic flow.
(b) Equal liquid and vapor velocities.
(c) Thermal equilibrium between vapor and liquid.

Because thermal nonequilibrium as well as velocity slip are both neglected, the method should be expected to provide a lower bound for the critical discharge rate. (Velocity slip and thermal nonequilibrium both increase the critical discharge rate.) The outline of the method is now presented.

The fluid equation of state, once flashing starts (i.e., beyond the point where saturation is reached) is, approximately,

$$\frac{v}{v_{f0}} = \omega \left(\frac{P_{sat}}{P} - 1 \right) + 1 \tag{17.104}$$

where ω is found from Eq. (17.95), and v_{f0} and P_{sat} correspond to T_0. Equation (17.93) now gives

$$G = \frac{1}{v} \left[-2 \int_{P_0}^{P_{sat}} v_{f0} dP - 2 \int_{P_{sat}}^{P} v dP \right]^{1/2}. \tag{17.105}$$

Substitution from Eq. (17.104) in this equation then leads to

$$(Gv)^2 = 2v_{f0} \left\{ (P_0 - P_{sat}) + \omega P_{sat} \ln \frac{P_{sat}}{P} - (\omega - 1)(P_{sat} - P) \right\}. \tag{17.106}$$

Choked flow occurs when $\frac{dG}{dP} = 0$. This leads to the following transcendental equation for η_{ch}

$$\frac{\left(\omega + \frac{1}{\omega} - 2 \right)}{2\eta_{sat}} \eta_{ch}^2 - 2(\omega - 1)\eta_{ch} + \omega \eta_{sat} \ln \frac{\eta_{ch}}{\eta_{sat}} + \frac{3}{2}\omega \eta_{sat} - 1 = 0. \tag{17.107}$$

The following simple and accurate curvefit to the solution of this transcendental equation has been derived by Leung and Grolmes (1988)

$$\eta_{ch} = \eta_{sat} \left(\frac{2\omega}{2\omega - 1} \right) \left\{ 1 - \left[1 - \frac{1}{\eta_{sat}} \left(\frac{2\omega - 1}{2\omega} \right) \right]^{1/2} \right\}. \tag{17.108}$$

The critical discharge mass flux will then be

$$G_{ch} = \eta_{ch} \left(\frac{P_0}{\omega v_{f0} \eta_{sat}} \right)^{1/2}. \tag{17.109}$$

Equation (17.108) indicates that a solution for η_{ch} exists if $\eta_{sat} \geq \eta_{tr}$, where

$$\eta_{tr} = \frac{2\omega - 1}{2\omega}. \tag{17.110}$$

When $\eta_{sat} < \eta_{tr}$, flashing does not occur in the flow passage and choking is due to the initiation of flashing right at the exit (i.e., where the critical cross section is reached), therefore

$$G_{ch} = [2\rho_{f0}(P_0 - P_{sat})]^{1/2}. \tag{17.111}$$

On the other hand, when $\eta_{\text{sat}} > \eta_{\text{tr}}$, then flashing occurs in a part of the flow passage upstream from the exit. In this case η_{ch} is found from Eq. (17.108) and the critical discharge mass flux is found from Eq. (17.109).

The omega parameter method has also been extended to the critical discharge of flashing two-phase flows in long horizontal as well as inclined ducts (Leung and Grolmes, 1988; Leung and Epstein, 1990a, b), where the effect of frictional loss in the long duct is included in the analysis.

Critical Discharge of a non-Flashing Two-Phase Mixture

The omega parameter method has been extended to the choking of isentropic non-flashing homogeneous mixtures in frictionless flow passages. Examples of such flows are liquid–noncondensables (e.g., air–water) and gas–powder mixtures. The neglect of the slip between the two phases of course limits the accuracy of the method. Nevertheless, the method is simple and easy to apply, at least for scoping calculations. A brief summary of the method is presented below, and more details can be found in Leung and Epstein (1990a, b).

The critical discharge mass flux can be found from

$$\frac{G_{\text{ch}}}{\sqrt{P_0 \rho_0}} = \frac{\left\{\frac{2}{\alpha_0}\left[\left(\frac{1-\alpha_0}{\alpha_0}\right)(1-\eta_{\text{ch}}) + \frac{n}{n-1}\left(1 - \eta_{\text{ch}}^{\frac{n-1}{n}}\right)\right]\right\}^{1/2}}{\eta_{\text{ch}}^{-1/n} + \left(\frac{1-\alpha_0}{\alpha_0}\right)} \tag{17.112}$$

where P_0, ρ_0, and α_0 represent stagnation mixture conditions, n is Tangren's polytropic exponent (Tangren et al., 1949) (see Eq. (17.58)), and the critical pressure ratio η_{ch} is found from

$$\eta_{\text{ch}} = \left[\left(\frac{n+1}{2}\right)^{\frac{an}{n-1}} + \left(\frac{n}{2}\frac{1-\alpha_0}{\alpha_0}\right)^{\frac{an}{n+1}}\right]^{-1/a}, \ a = \frac{4}{3}. \tag{17.113}$$

This equation is in fact a curve fit to the solutions of a transcendental equation.

More recent applications and extensions of the omega parameter method can be found in Grolmes and Yue (1993), Lenzing et al. (1998), and Monvalco and Friedel (2007, 2010).

EXAMPLE 17.2. The primary coolant system of a pressurized water reactor contains subcooled water at 150 bar pressure, with 25 °C subcooling. The water leaks through a break that is 1 cm² in area into the containment. The pressure in the containment is 2 bars. Does critical flow occur? Calculate the leakage rate through the break. What would the solution be if the initial subcooling was 100 °C?

SOLUTION. The saturation temperature at the stagnation point is $T_{\text{sat}} = 615.3 \, \text{K}$. Therefore, $T_0 = 590.3 \, \text{K}$ and

$$C_{\text{Pf0}} = C_{\text{Pf}}\big|_{T_{\text{sat}}=590.3\,\text{K}} = 6{,}047 \, \text{J/kg·K} \ ; \ v_{\text{f0}} = v_{\text{f}}\big|_{T_{\text{sat}}=590.3\,\text{K}} = 0.001\,483 \, \text{m}^3/\text{kg}$$

$$v_{\text{g0}} = v_{\text{g}}\big|_{T_{\text{sat}}=590.3\,\text{K}} = 0.016\,24 \, \text{m}^3/\text{kg}; \ h_{\text{fg0}} = h_{\text{fg}}\big|_{T_{\text{sat}}=590.3\,\text{K}} = 1.264 \times 10^6 \, \text{J/kg}$$

$$v_{\text{fg0}} = v_{\text{g0}} - v_{\text{f0}} = 0.001\,483 \, \text{m}^3/\text{kg}$$

$$\omega = \left(\frac{C_{\text{Pf0}} T_0 P_0}{v_{\text{f0}}}\right) \cdot \left(v_{\text{fg0}}/h_{\text{fg0}}\right)^2 = 4.92.$$

Because we have subcooled liquid at stagnation, we will examine whether saturation is reached within the flow passage. From Eq. (17.110):

$$\eta_{tr} = \frac{2\omega - 1}{2\omega} = \frac{2 \times 4.92 - 1}{2 \times 4.92} = 0.8984$$

$$\eta_{sat} = P_{sat}|_{T_0}/P_0 = 0.7244.$$

Because $\eta_{sat} < \eta_{tr}$, saturation occurs right at the exit of the flow passage. The cricitcal mass flux is then found from Eq. (17.111):

$$G_{ch} = [(2/v_{f0})(P_0 - P_{sat})]^{1/2}$$

$$= [(2/0.001\,483\,\text{m}^3/\text{kg})(1.5 \times 10^7\,\text{Pa} - 1.087 \times 10^7\,\text{Pa})]^{1/2} = 74\,664\,\text{kg/m}^2\text{·s}.$$

$$\dot{m}_{ch} = G_{ch}A = (74\,664\,\text{kg/m}^2\text{·s}) \times 10^{-4}\,\text{m}^2 = 7.466\,\text{kg/s}.$$

Let us now visit the case with 4 K subcooling at stagnation. We then have $T_0 = T_{sat}|_{P_0} - 4K = 611.3\,\text{K}$, and from there:

$$C_{Pf0} = C_{Pf}|_{T_{sat}=611.3\,\text{K}} = 7776\,\text{J/kg·K}\,; v_{f0} = v_f|_{T_{sat}=611.3\,\text{K}} = 0.009\,545\,\text{m}^3/\text{kg}$$

$$v_{g0} = v_g|_{T_{sat}=611.3\,\text{K}} = 0.009\,545\,\text{m}^3/\text{kg}; h_{fg0} = h_{fg}|_{T_{sat}=611.3\,\text{K}} = 1.049 \times 10^6\,\text{J/kg}$$

$$v_{fg0} = v_{g0} - v_{f0} = 0.009\,545\,\text{m}^3/\text{kg}.$$

Equations (17.102) and (17.110) then give

$$\omega = 3.64$$

$$\eta_{tran} = 0.8626.$$

We then get

$$\eta_{sat} = P_{sat}|_{611.3K}/P_0 = 0.9511 > \eta_{tran}.$$

Thus evaporation occurs inside the flow passage (i.e., upstream the exit). The critical mass flux can then be found from Eq. (17.109):

$$G_{ch} = \eta_{ch}\left(\frac{P_0}{\omega v_{f0}\eta_{sat}}\right)^{1/2} = 0.7663\left(\frac{1.5 \times 10^7\,\text{Pa}}{3.64 \times 0.001622\,\text{m}^3/\text{kg} \times 0.9511}\right)^{1/2}$$

$$= 39\,614\,\text{kg/m}^2\text{·s}$$

$$\dot{m}_{ch} = G_{ch}A = 3.96\,\text{kg/s}.$$

17.9 Choked Two-Phase Flow in Small Passages

Critical flow of initially subcooled liquid through small holes, cracks, and slits is of interest in relation to the safety of nuclear and chemical reactors.

Cracks often have sub-millimeter hydraulic diameters, large cross-sectional aspect ratios, and large length-to-hydraulic-diameter ratios. Cracks also often have highly tortuous and rough flow passages. Irreversible form and frictional pressure losses are usually very significant in cracks. The widely used semi-analytical methods that are mostly based on isentropic flow are therefore not applicable to cracks, as they overpredict data by up to an order of magnitude. The "leak-before-break" issue inspired an intensive investigation of this subject during the 1980s and early 1990s

(Collier and Norris, 1983; Collier *et al.*, 1984; Amos and Schrock, 1984; Nabarayashi *et al.*, 1989; Matsumoto *et al.*, 1989; Lee and Schrock, 1988; Schrock *et al.*, 1988; Feburie *et al.*, 1993; Ghiaasiaan and Geng, 1997; Geng and Ghiaasiaan, 1998). The safety concerns over intergranular stress corrosion cracking in stainless steel piping of BWRs and steam generator tubes in PWRs was the impetus behind these studies. The purpose was to determine whether the leak-before-break phenomenon would indeed lead to failure in the piping. The aforementioned characteristics of cracks are to a great extent true for mini- and microchannels as well. The experience and the developed modeling methods that resulted from those studies can thus be very useful for critical flow in minichannels.

Some important observations regarding critical two-phase flow in cracks or minichannels are as follows.

(a) For small and very short passages, where frictional pressure losses are negligible, the homogeneous-equilibrium isentropic expansion model and the approximate model of Leung and Grolmes (1988) perform well, provided that the entrance pressure loss is correctly accounted for (Ghiaasiaan *et al.*, 1997b). As an example, to account for the entrance losses, in the homogeneous-equilibrium isentropic expansion model, replace $s_0(P_0, h_0)$ with $s(P_0^*, h_0^*)$, where

$$P_0^* = P_0 - K_{ent} \left(\frac{G^2}{2\rho_0} \right)_{inlet}, \tag{17.114}$$

$$h_0^* = h_{inlet} + K_{ent} \left(\frac{G^2}{2\rho_0} \right)_{inlet}. \tag{17.115}$$

(b) Flashing (bubble nucleation) plays a crucial role in causing choking. In some simple models it is assumed that the required condition for choking is the initiation of nucleation at the critical cross section.

(c) Delayed flashing and thermodynamic nonequilibrium (i.e., the occurrence of metastable superheated liquid and saturated vapor) should be expected in deep cracks and slits.

(d) Mechanistic models based on homogeneous-equilibrium flow can predict experimental data well, provided that pressure losses and friction are correctly modeled.

Figure 17.7 shows the pressure profiles along a slit during a critical flow experiment ($L = 6.35$ mm; $D_H = 0.75$ mm) (Amos and Schrock, 1984), where the horizontal dotted line represents the saturation pressure corresponding to the inlet conditions, the broken line depicts the measurements, and the straight segment of the solid curve represents the pressure profile associated with liquid single-phase flow. Flashing starts at the location where the solid and broken curves start to deviate from one another. These and other experiments show the occurrence of considerable temperature undershoot before flashing actually takes place; see also Fig. 17.8.

The following correlation for the pressure undershoot at which flashing occurs during the flow of superheated water has been proposed by Alamgir and Lienhard (1981):

$$P_{sat} - P = 0.252 \frac{\sigma^{3/2} (T_{in}/T_{cr})^{13.73} (1 + 14\Sigma'^{0.8})^{0.5}}{\sqrt{\kappa_B T_{cr}} \left(1 - \frac{v_f}{v_g} \right)}, \tag{17.116}$$

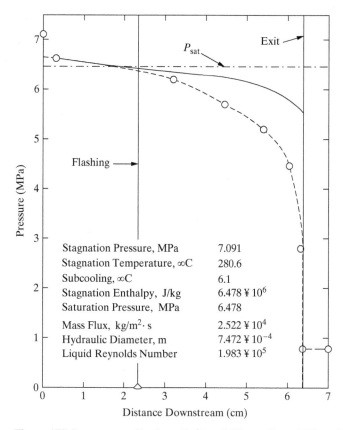

Figure 17.7. Pressure profiles in a slit ($L = 6.35$ mm; $D_h = 0.75$ mm) during a critical flow experiment. (From Amos and Schrock, 1984.)

The figure contains the following legend:

Stagnation Pressure, MPa	7.091
Stagnation Temperature, ∞C	280.6
Subcooling, ∞C	6.1
Stagnation Enthalpy, J/kg	$6.478 ¥ 10^6$
Saturation Pressure, MPa	6.478
Mass Flux, kg/m$^2 \cdot$ s	$2.522 ¥ 10^4$
Hydraulic Diameter, m	$7.472 ¥ 10^{-4}$
Liquid Reynolds Number	$1.983 ¥ 10^5$

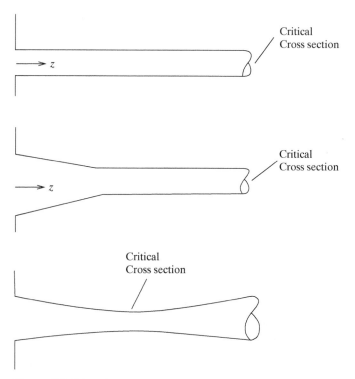

Figure 17.8. The critical cross section in some flow passage geometries.

where T_{cr} is the critical temperature of the fluid, κ_B is Boltzmann's constant, and Σ' is the depressurization rate, in million atmospheres per second (Matm/s). Except for Σ', the remainder of this equation is dimensionally consistent.

Because of the large length-to-diameter ratio, friction losses are important. A useful liquid single-phase flow friction factor correlation, developed based on experimental data in slits with 0.2 to 0.6 mm gaps, is (John et al., 1988)

$$f' = \left(3.39\log_{10}\frac{D_H}{\varepsilon_D} - 0.866\right)^{-2}, \tag{17.117}$$

where ε_D represents the crack surface roughness average height.

Abdollahian et al. (1983) proposed the following simple semi-empirical method for the critical discharge rate in long cracks:

$$G_{ch} = \left\{\frac{2[P_0 - P_{sat}(T_0)]}{v_m\left(1 + f'\frac{L}{D_H} + K_{ent}v_0\right)}\right\}^{1/2}, \tag{17.118}$$

$$v_m = \bar{v}_f + \bar{x}(\bar{v}_{fg}), \tag{17.119}$$

$$K_{ent} = 2.7, \tag{17.120}$$

where the averaged (barred) properties are to be calculated at $\bar{P} = [P_0 + P_{sat}(T_0)]/2K_{ent} = 2.7$, thus,

$$\bar{x} = \frac{h_0 - \bar{h}_f}{\bar{h}_{fg}}. \tag{17.121}$$

Equations (17.119) and (17.121) evidently make sense as long as $h_0 > \bar{h}_f$; otherwise one may need to use liquid specific volume for v_m (see Example 17.3). Abdollahaian et al. obtained the friction factor from the following modified von Kármán correlation:

$$f' = \left(2\log_{10}\frac{D_H}{2\varepsilon_D} + 1.74\right)^{-2}. \tag{17.122}$$

EXAMPLE 17.3. In an experiment, subcooled pressurized water flows through an artificial crack. The stagnation properties for the pressurized water are $P_0 = 8.80$ MPa and $T_0 = 205.8\,°C$. The crack is a slit with $\delta = 0.23$ mm, $W = 6.35$ cm, and $l/D_H = 128$. The mean surface roughness in the crack is $\varepsilon_D = 6.2\,\mu m$. Calculate the mass flow rate through the crack.

SOLUTION. The relevant properties are $T_{sat}(P_0) = 574.9$ K, $h_0 = 8.806 \times 10^5$ J/kg, $v_{f0} = 1.165 \times 10^{-3}$ m³/kg, $P_{sat}(T_0) = 1.746 \times 10^6$ Pa, and $\bar{P} = (1/2)[P_{sat}(T_0) + P_0] = 5.273 \times 10^6$ Pa.

This leads to $\bar{h}_f = 1.17 \times 10^6$ J/kg. We thus have $h_0 < \bar{h}_f$, and this would lead to $\bar{x} < 0$. We will therefore apply Eq. (17.118), using v_{f0} instead of v_m.

We need to find f', and for that

$$D_H = 2\delta = 0.46 \times 10^{-3} \text{ m}.$$

Equation (17.122) now gives $f' = 0.042$. (The correlation of John *et al.* (1988), Eq. (17.117), would give $f' = 0.033$.) Substitution in Eq. (17.118) results in

$$G_{ch} = 41\,317\,\text{kg/m}^2{\cdot}\text{s}.$$

If we use $f' = 0.033$ in Eq. (17.118), the result will be $G_{ch} = 45\,447\,\text{kg/m}^2{\cdot}\text{s}$.

This example actually represents an experimental data point of Collier and Norris (1983). The experimentally measured mass flux was $3.5 \times 10^4\,\text{kg/m}^2{\cdot}\text{s}$.

EXAMPLE 17.4. In an experiment, pressurized and highly subcooled water leaks through a short nozzle into an atmospheric-pressure chamber. The nozzle is 0.78 mm in diameter and has a total length of 0.78 mm. At the inlet, the configuration resembles a sudden contraction. Calculate the leak mass flux for the following test conditions: $P_0 = 3.453\,\text{MPa}$ and $T_0 = 321\,\text{K}$. Assume that $\varepsilon_D/D = 0.01$.

SOLUTION. This example represents a data point of Ghiaasiaan *et al.* (1997b). The measured mass flux was $64\,651\,\text{kg/m}^2{\cdot}\text{s}$. The properties that are needed are $T_{sat}(P_0) = 515\,\text{K}$ and $P_{sat}(T_0) = 11\,087\,\text{Pa}$.

Let us first obtain a solution based on the method of Abdollahian *et al.* (1983). Accordingly, we will have $h_0 = 2.003 \times 10^5\,\text{J/kg}$, $v_{f0} = 1.011 \times 10^{-3}\,\text{m}^3/\text{kg}$, $\bar{P} = (1/2)[P_{sat}(T_0) + P_0] = 1.732 \times 10^6\,\text{Pa}$, and $\bar{h}_f = 8.76 \times 10^5\,\text{J/kg}$.

Since $h_0 < \bar{h}_f$, we will apply Eq. (17.118) using v_{f0} in place of v_m With $D_H = D = 0.78 \times 10^{-3}\,\text{m}$, Eq. (7.122) gives $f' = 0.038$. Equation (17.118), with $K_{ent} = 2.7$, gives $G_{ch} = 75\,362\,\text{kg/m}^2{\cdot}\text{s}$.

We now repeat the solution, this time assuming homogeneous-equilibrium flow, and using the corrections presented in Eqs. (17.95) and (17.111). The solution follows the first part of Example 17.2. With $T_0 = 321\,\text{K}$ we have

$$C_{Pf0} = C_{Pf}\big|_{T_{sat}=321\,\text{K}} = 7,174\,\text{J/kg}{\cdot}\text{K}; \quad v_{f0} = v_f\big|_{T_{sat}=321\,\text{K}} = 0.001011\,\text{m}^3/\text{kg}$$

$$v_{g0} = v_g\big|_{T_{sat}=321\,\text{K}} = 13.2\,\text{m}^3/\text{kg}; \quad h_{fg0} = h_{fg}\big|_{T_{sat}=321\,\text{K}} = 2.387 \times 10^6\,\text{J/kg}$$

$$v_{fg0} = v_{g0} - v_{f0} \approx 0.001483\,\text{m}^3/\text{kg}$$

$$\omega = \left(\frac{C_{Pf0}\,T_0\,P_0}{v_{f0}}\right) \cdot \left(v_{fg0}/h_{fg0}\right)^2 \approx 1.42 \times 10^5.$$

Because we have subcooled liquid at stagnation, we will examine whether saturation is reached within the flow passage. From Eq. (17.110):

$$\eta_{tr} = \frac{2\omega - 1}{2\omega} \approx 1$$

$$\eta_{sat} = P_{sat}\big|_{T_0}/P_0 = 0.003\,211.$$

Because $\eta_{sat} < \eta_{tr}$, saturation occurs right at the exit of the flow passage. The critical mass flux is then found from Eq. (17.111):

$$G_{ch} = \left[(2/v_{f0})\,(P_0 - P_{sat})\right]^{1/2}$$
$$= \left[\left(2/0.001011\,\text{m}^3/\text{kg}\right)\left(3.453 \times 10^6\,\text{Pa} - 1.109 \times 10^4\,\text{Pa}\right)\right]^{1/2} = 82,509\,\text{kg/m}^2{\cdot}\text{s}$$

Ghiaasiaan *et al.* (1977b) found that the method of Abdollahian *et al.* consistently disagreed with their data, and they concluded that the method is inadequate for short passages. The homogeneous-equilibrium method, however, agreed with the data, provided that the corrections of Eqs. (17.114) and (17.115) were made.

17.10 Nonequilibrium Mechanistic Modeling of Choked Two-Phase Flow

When critical discharge occurs under circumstances where a one-dimensional two-phase flow can be assumed at the vicinity of the critical cross section, the critical flow can be mechanistically modeled. Examples include critical discharge through a long flow passage and deep cracks and slits. In this method, the steady-state, one-dimensional differential conservation equations are numerically solved in the break vicinity, using the upstream boundary conditions. The discharged mass flux is iteratively varied until critical conditions occur at the critical cross section. The method has been successfully demonstrated by several authors (Ardron, 1978; Elias and Chambre, 1984; Richter, 1983; Dobran, 1987; Schwellnus and Shoukri, 1991; Blinkov et al., 1993; Dagan et al., 1993; Ghiaasiaan and Geng, 1997).

Although published mechanistic models differ in some important details, they are generally based on the following procedure.

(a) Write the steady-state, one-dimensional two-phase conservation equations (homogeneous-nonequilibrium (Geng and Ghiaasiaan, 1998), drift flux (Elias and Chambre, 1984), two-fluid (Richter, 1983; Dobran, 1987; Schwellnus and Shoukri, 1991), or homogeneous or two-fluid with noncondensables (Ghiaasiaan and Geng, 1997; Geng and Ghiaasiaan, 1998).

(b) Use constitutive and closure relations for the conservation equations, borrowed from the literature dealing with common low-flow conditions.

(c) Use a criterion for the initiation of nucleation and flashing (pressure undershoot).

(d) Assume the a priori existence, or generation, of microbubbles in the liquid at and beyond the point where nucleation and flashing start.

(e) Determine the critical throat location, based on physical insight.

(f) Numerically, and iteratively, solve the one-dimensional differential conservation equations. Perform iterations on the mass flux, and seek the mass flux that causes critical flow conditions at the throat.

The pressure undershoot that causes flashing can be modeled by the aforementioned criterion of Alamgir and Lienhard (1981), Eq. (17.116) (Dobran, 1987). However, to obtain agreement with their experimental data, Amos and Schrock (1984) included a multiplier in the correlation. The multiplier was empirically correlated in terms of local velocity by Amos and Schrock (1984), and as a function of Reynolds and Jakob numbers by Lee and Schrock (1988). Some authors have used the mechanical equilibrium criterion $P - P_{sat} = -2/R_{B0}$, where R_{B0} is the radius of the naturally occurring microbubbles that are assumed to be present in the metastable superheated liquid (Richter, 1983; Schwellnus and Shoukri, 1991; Ghiaasiaan and Geng, 1997).

A major uncertainty in this method is associated with the treatment of the poorly understood bubble nucleation processes following the initiation of flashing. Most of the published models use uniform-size microbubbles with a fixed number density in the subcooled liquid, which accounts for the combined effects of the suspended nucleation sites and the nucleation sites on the wall crevices. The assumed initial bubble size and number density vary among models. Ardron (1978) assumed 10^6 m^{-3}, but several others used 10^{11} m^{-3} (Richter, 1983; Dobran, 1987; Schwellnus and Shoukri, 1991; Ghiaasiaan and Geng, 1997). Dagan et al. (1993) correlated the number density of microbubbles in terms of the channel length-to-diameter ratio. The typical assumed initial diameter for the microbubbles is 25 μm.

Following flashing, a bubbly flow regime is established, and rapid evaporation into the suspended bubbles occurs, leading to increasing voidage and the possibility of flow regime transitions. The calculation of the evaporation rate thus requires a flow regime transition method and models for interfacial heat and mass transfer processes.

The critical flow conditions are now described. Based on physical arguments, critical flow can be assumed when $|dP/dz| \to \infty$ (Richter, 1983; Schwellnus and Shoukri, 1991; Dagan et al., 1993), or when the flow rate becomes independent of the downstream pressure (Rivard and Travis, 1980; Blinkov et al., 1993). The critical flow conditions can also be modeled based on the mathematics of the conservation equations. The one-dimensional conservation equations in general can be cast in the form

$$\mathbf{A}_t \frac{\partial \mathbf{X}}{\partial t} + \mathbf{A}_z \frac{\partial \mathbf{X}}{\partial z} = \mathbf{B}, \tag{17.123}$$

where $\mathbf{A}_t$ and $\mathbf{A}_z$ are $n \times n$ coefficient matrices, $\mathbf{B}$ is a column vector with n elements, and $\mathbf{X}$ represents an n-element column vector containing the state variable. For the case of the full two-fluid model without a noncondensable, for example,

$$\mathbf{X} = (\alpha, x, U_G, U_L, P, h_L)^T. \tag{17.124}$$

The characteristics of this set of n quasi-linear, hyperbolic partial differential equations are represented by the n roots of the following polynomial equation:

$$\det(\mathbf{A}_t \lambda - \mathbf{A}_z) = 0, \tag{17.125}$$

where $\lambda_i = 1, 2, ..., n$ are the velocities of propagation of hydrodynamic signals. Critical flow occurs when $\lambda_i \geq 0$ for all i, implying that hydrodynamic signals at and beyond the throat are unable to move upstream. Let us rewrite Eq. (17.123) as

$$\mathbf{A}_z \frac{\partial \mathbf{X}}{\partial z} = \mathbf{B} - \mathbf{A}_t \frac{\partial \mathbf{X}}{\partial t} = \mathbf{B}^*. \tag{17.126}$$

The critical condition can then be assumed when

$$\det(\mathbf{A}_z) = 0, \tag{17.127}$$

$$\det(\mathbf{A}_z^*) = 0, \tag{17.128}$$

where $\mathbf{A}_z^*$ is the $(n \times n)$ matrix formed by replacing the ith column of $\mathbf{A}_z$ with $\mathbf{B}^*$.

Once Eq. (17.128) is satisfied for one i, it will be satisfied for all other i as well (Bouré, 1977). Condition (17.128) is difficult to implement in practice, however. Instead, when Eq. (17.127) is satisfied, it can be argued that the flow is pseudo-critical if the flow rate is unaffected in response to small changes downstream of the critical cross section (Bouré, 1977). Also, when condition (17.127) is met, the numerical solution can be repeated with a slightly higher mass flow rate to make sure that an impossible condition is then predicted at the critical cross section.

EXAMPLE 17.5. An alternative method to the calculation of the evaporation rate by Eq. (5.92) is the nonequilibrium relaxation method, whereby it is assumed that during flashing the volumetric evaporation rate follows (Downar-Zapolski *et al.*, 1996)

$$\frac{\Gamma}{\bar{\rho}} = \frac{h_L - h_f(P)}{h_{fg}\theta_{rel}},$$

where θ_{rel} is a relaxation time; for water it is of the order of 1 s when evaporation starts and reduces to about 0.01 s at high void fractions. Assuming homogeneous nonequilibrium flow (i.e., $S_r = 1$), formulate a mechanistic, nonequilibrium model for critical flow through a straight pipe, by applying the thermal relaxation method.

SOLUTION. We assume steady state. We also assume that the two phases move at the same velocity, the vapor and gas are at the same pressure at any cross section, and the vapor is always saturated. The conservation equations can be written as

$$\frac{d}{dz}[\rho_L(1 - \alpha)U] = -\Gamma,$$

$$\frac{d}{dz}[\rho_g \alpha U] = \Gamma,$$

$$\frac{d}{dz}\left\{[\rho_L(1 - \alpha) + \rho_g \alpha]U^2\right\} + \frac{dP}{dz} = -[\rho_L(1 - \alpha) + \rho_g \alpha]g\sin\theta - \tau_w,$$

$$\frac{d}{dz}\left\{[\rho_L(1 - \alpha)h_L + \rho_g \alpha h_g + \bar{\rho}gz\sin\theta]U\right\} = 4q''_w,$$

where θ is the angle of inclination with the horizontal plane. We have neglected the kinetic energy terms in the energy equation. Usually little evaporation is needed for critical flow to occur. The liquid can be in a metastable superheated state. The unknowns in these equations are P, U, α, and h_L. The liquid density can be represented by

$$\rho_L \approx \rho_f[1 - \beta(T_L - T_{sat})],$$

where β is the thermal expansion coefficient and T_{sat} corresponds to pressure P.

The aforementioned set of four differential equations should now be expanded similar to Example 5.3, and cast in the form of Eq. (17.126) or, equivalently, Eq. (5.117).

PROBLEMS

17.1 For a homogeneous mixture of water and ethylene glycol at room temperature, calculate the speed of sound in the mixture for $\alpha_{H_2O} = 0.1$–0.8, where α_{H_2O} is the volume fraction of water in the mixture.

17.2 A saturated water–steam mixture with $x_{eq} = 4\%$ and at 80 bars is discharged through a short break that is 3 cm² in area and opens to a containment that is at 1.5 bars.

(a) Estimate the discharge mass flow rate using the graphical solutions of the model of Moody.

(b) Repeat part (a), this time using the method of Leung and Grolmes.

17.3 High-pressure subcooled pure water with stagnation pressure and temperature (P_0, T_0) flows out of a container through a break that can be idealized as a circular channel with $D = 0.78$ mm diameter and $L = 0.78$ mm length. The break opens to the atmosphere. For the stagnation conditions summarized in Table P17.3, calculate the flow rate through the break using the method of Abdollahian *et al.* (1983) and that of Leung (1986) and Leung and Grolmes (1988). Compare the model predictions with the experimental data, and comment on the results.

Table P17.3.

Test number	P_0 (MPa)	T_0 (K)	$G_{ch} \times 10^{-3}$ (kg/m²·s)
5	0.731	338.5	30.00
10	1.849	334.2	45.35
21	2.084	326	48.24
41	5.103	333.4	76.71

17.4 The content of a pressurized tank flows through a long tube that opens to the atmosphere. The tank is at 4.136 MPa (600 psia) pressure, and it contains a saturated mixture of water and steam with a quality of $x_0 = 0.08$. Does critical flow occur in the tube? If so, calculate the mass flux and the pressure at the critical cross section. Perform the calculations using the graphs provided by Moody (Figs. 17.4 and 17.5) and the method of Leung (1986) and Leung and Grolmes (1988).

17.5 The critical flow of a liquid–noncondensable mixture flowing through a short passage can be predicted by the "frozen composition model," in which it is assumed that the gas and liquid phases have no time to exchange heat in the flow passage and the gas undergoes an isentropic expansion. If it is assumed that the two phases also move at the same velocity, the model results in (Moody, 1990)

$$b_0^2 + \frac{2}{\gamma}\left[(\gamma + 1)\left(\frac{P_{ch}}{P_0}\right)^{-\frac{1}{\gamma}} - \left(\frac{P_{ch}}{P_0}\right)^{-\frac{(\gamma+1)}{\gamma}} \right] b_0$$

$$+ \frac{1}{\gamma - 1}\left[(\gamma + 1)\left(\frac{P_{ch}}{P_0}\right)^{-\frac{2}{\gamma}} - 2\left(\frac{P_{ch}}{P_0}\right)^{-\frac{(\gamma+1)}{\gamma}} \right] = 0, \qquad \text{(a)}$$

$$\frac{G_{ch}^2 x(\gamma - 1)}{2\gamma P_0} = \frac{1 - (P_{ch}/P_0)^{\frac{(\gamma-1)}{\gamma}} + [(\gamma - 1)\gamma](1 - P_{ch}/P_0)b_0}{\left(\frac{P_{ch}}{P_0}\right)^{-\frac{1}{\gamma}} + b_0}, \qquad \text{(b)}$$

where

$$b_0 = \frac{1 - x}{x}\frac{v_L}{v_{G_0}}.$$

Equation (b) can be used for calculating the critical pressure. Equation (a) predicts the total mass flux, once P_{ch} is known. (Equation (a) in fact applies to a noncritical flow between the stagnation point and an arbitrary point where the pressure is P, when P_{ch} is replaced with P in that equation.)

In an experiment, a saturated steam–water mixture at a stagnation pressure of 2.76 MPa (400 psia) and stagnation quality of $x_0 = 20\%$ flows through a short passage into an atmospheric container. Calculate and compare the mass fluxes using the homogeneous-equilibrium isentropic method and the frozen composition model described here.

17.6 Based on experimental data representing critical flow of initially subcooled pure water in a pipe with 4.6-mm inner diameter and $L/D = 325$, Celata et al. (1983, 1988) proposed the following correlation:

$$G_{ch} = \left\{ \frac{2[P_0 - P_{sat}(T_0)]}{v_{f0}(K_{ent} + f'L/D)} \right\}^{1/2},$$

where K_{ent} is the entrance loss coefficient. In Example 17.2, assume that the break is a 20-cm-long straight tube, with a sharp entrance. Calculate the critical mass flux using Celata et al.'s correlation, and compare the result with those provided in the example.

17.7 Using the RETRAN curve fits for Moody's model, calculate the mass flow rate of a saturated water and steam mixture that is initially at a pressure of 100 bars and has a quality of 0.8% and is discharging into the atmosphere through a short break that is 5 cm^2 in cross section.

17.8 Using the RETRAN curve fits for Moody's model, calculate the mass fluxes of saturated water and steam that is initially at a pressure of 50 bars for upstream qualities in the 0.1%–11% range and is discharging into the atmosphere through a short break. Compare the results with the mass flux associated with the flow of superheated water vapor initially at 50 bars and 400 °C.

17.9 Repeat Example 17.3, this time for $P_0 = 8.36$ MPa and $T_0 = 265$ °C. The crack here is a slit with $\delta = 1.12$ mm, $W = 6.35$ cm, and $l/D_H = 30$. The mean surface roughness in the crack is $\mu_D = 0.3$ μm. Compare your calculation results with the experimentally measured mass flux of 2.86×10^4 kg/m^2·s.

17.10 For the nozzle in Example 17.4, calculate the critical mass flux for the following conditions.

(a) $P_0 = 0.793$ MPa and $T_0 = 334.2$ K. Assume that $\varepsilon_D/D = 0.01$.

(b) $P_0 = 5.006$ MPa and $T_0 = 317.4$ K. Assume that $\varepsilon_D/D = 0.01$.

Compare your calculation results with experimental results, which are $G_{ch} = 31\,759$ kg/m^2·s and $G_{ch} = 76\,711$ kg/m^2·s, respectively, and comment.

17.11 Guided by the discussion in Section 17.10, write a set of one-dimensional two-fluid conservation equations that can be numerically and iteratively solved for the determination of critical two-phase flow in a channel. Note that the equation set must include a mixture energy equation in which the vapor is saturated, but the liquid can be superheated. The evaporation mass flux will thus need to be modeled as

$$m''_{ev} = \dot{H}_{LI}(T_L - T_{sat}).$$

Choose simple but reasonable closure relations that are needed for the solution of the conservation equations.

17.12 Derive Eq. (17.82).

17.13 A large pressure vessel contains saturated liquid water at a pressure of 1.1 MPa. A crack develops in the wall of the pressure vessel that can be idealized as a hole with a flow area of 1 cm^2. The tank is surrounded by ambient air.

(a) Calculate the discharge flow rate of water, using the method of Henry and Fauske (1971), and the omega parameter technique.

(b) Repeat part (a), this time assuming that the water in the tank is subcooled by 10 K.

17.14 In an experiment dealing with small-break loss of coolant accidents in PWR reactors, the leakage rate from a pressurized tank is to be measured. The break is 0.5 cm^2 in flow area. The tank contains water at 110 bar pressure and a temperature of 300 °C. Calculate and compare the discharge rate using the extended Henry–Fauske method, and the omega parameter method.

17.15 Saturated liquid nitrogen is stored in a tank.

(a) For stagnation pressures of 1, 1.5, and 2 MPa, calculate the discharge flow rate through a valve that can be idealized as a short flow passage with a flow area of 0.5 cm^2.

(b) Repeat part (a), this time assuming that the liquid nitrogen in the tank is at a temperature of 100 K.

Thermodynamic Properties of Saturated Water and Steam

T (°C)	P (bar)	v_f (m³/kg)	v_g (m³/kg)	u_f (kJ/kg)	u_g (kJ/kg)	h_f (kJ/kg)	h_g (kJ/kg)	s_f (kJ/kg·K)	s_g (kJ/kg·K)
0.01	0.006117	0.00100	205.99	0.00	2374.5	0.00	2500.5	0.0000	9.1541
5	0.00873	0.00100	147.02	21.02	2381.4	21.02	2509.7	0.0763	9.0236
10	0.01228	0.00100	106.32	62.99	2388.3	41.99	2518.9	0.1510	8.8986
15	0.01706	0.00100	77.90	62.92	2395.2	62.92	2528.0	0.2242	8.7792
20	0.02339	0.00100	57.778	83.83	2402.0	83.84	2537.2	0.2962	8.6651
25	0.03169	0.00100	43.361	104.75	2408.9	104.75	2546.3	0.3670	8.556
30	0.04245	0.00100	32.90	125.67	2415.7	125.67	2555.3	0.4365	8.4513
35	0.05627	0.00100	25.222	146.58	2422.5	146.59	2564.4	0.5050	8.3511
40	0.07381	0.00100	19.529	167.50	2429.2	167.50	2573.4	0.5723	8.255
45	0.09590	0.00100	15.263	188.41	2435.9	188.42	2582.3	0.6385	8.1629
50	0.12344	0.00101	12.037	209.31	2442.6	209.33	2591.2	0.7037	8.0745
55	0.15752	0.00101	9.572	230.22	2449.2	230.24	2600.0	0.7679	7.9896
60	0.19932	0.00102	7.674	251.13	2455.8	251.15	2608.8	0.8312	7.9080
65	0.2502	0.00102	6.199	272.05	2462.4	272.08	2617.5	0.8935	7.8295
70	0.3118	0.00102	5.044	292.98	2468.8	293.01	2626.1	0.9549	7.7540
75	0.3856	0.00103	4.133	313.92	2475.2	313.96	2634.6	1.0155	7.6812
80	0.4737	0.00103	3.409	334.88	2481.6	334.93	2643.1	1.0753	7.6111
85	0.57815	0.00103	2.829	355.86	2487.9	355.92	2651.4	1.1343	7.5436
90	0.70117	0.00104	2.362	376.86	2494.0	376.93	2659.6	1.1925	7.4783
95	0.8453	0.00104	1.983	397.89	2500.1	397.98	2667.7	1.2501	7.4154
100	1.0132	0.00104	1.674	418.96	2506.1	419.06	2675.7	1.3069	7.3545
105	1.2079	0.00105	1.420	440.05	2512.1	440.18	2683.6	1.3630	7.2956
110	1.4324	0.00105	1.211	461.19	2517.9	461.34	2691.3	1.4186	7.2386
115	1.6902	0.00106	1.037	482.36	2523.5	482.54	2698.8	1.4735	7.1833
120	1.9848	0.00106	0.8922	503.57	2529.1	503.78	2706.2	1.5278	7.1297
130	2.7002	0.00107	0.6687	546.12	2539.8	546.41	2720.4	1.6346	7.0272
140	3.6119	0.00108	0.5090	588.85	2550.0	589.24	2733.8	1.7394	6.9302
150	4.7572	0.00109	0.3929	631.80	2559.5	632.32	2746.4	1.8421	6.8381
160	6.1766	0.00110	0.3071	674.97	2568.3	675.65	2758.0	1.9429	6.7503
170	7.9147	0.00111	0.2428	718.40	2576.3	719.28	2768.5	2.0421	6.6662
180	10.019	0.001127	0.1940	762.12	2583.4	763.25	2777.8	2.1397	6.5853
190	12.542	0.00114	0.1565	806.17	2589.6	807.60	2785.8	2.2358	6.5071
200	15.536	0.00116	0.1273	850.58	2594.7	852.38	2792.5	2.3308	6.4312
220	23.178	0.00119	0.08616	940.75	2601.6	943.51	2801.3	2.5175	6.2847
240	33.447	0.00123	0.05974	1033.12	2603.1	1037.24	2803.0	2.7013	6.1423
260	46.894	0.001276	0.04219	1128.40	2598.4	1134.38	2796.2	2.8838	6.0010
280	64.132	0.001332	0.03016	1227.53	2585.7	1236.08	2779.2	3.0669	5.8565
300	85.838	0.001404	0.02167	1332.01	2562.8	1344.05	2748.7	3.2534	5.7042
320	112.79	0.001498	0.01548	1444.36	2525.2	1461.26	2699.7	3.4476	5.5356
340	145.94	0.001637	0.01079	1569.9	2463.9	1593.8	2621.3	3.6587	5.3345
360	186.55	0.001894	0.00696	1725.6	2352.2	1761.0	2482.0	3.9153	5.0542
373.98	220.55	0.003106	0.003106	2017	2017	2086	2086	4.409	4.409

Transport Properties of Saturated Water and Steam

Transport Properties of Saturated Water and Steam[a]

Temperature (K)	Pressure (bars)	$v_f \times 10^3$ (m³/kg)	v_g (m³/kg)	c_{Pf} (kJ/kg·K)	c_{Pg} (kJ/kg·K)	$\mu_f \times 10^6$ (kg/m·s)	$\mu_g \times 10^6$ (kg/m·s)	$k_f \times 10^3$ (W/m·K)	$k_g \times 10^3$ (W/m·K)	Pr_f	Pr_g
273.15	0.00611	1.000	206.3	4.217	1.854	1750	8.02	569	18.2	12.99	0.815
275.	0.00697	1.000	181.7	4.211	1.855	1652	8.09	574	18.3	12.22	0.817
280	0.0099	1.000	130.4	4.198	1.858	1422	8.29	582	18.6	10.26	0.825
285	0.01387	1.000	99.4	4.189	1.861	1225	8.49	590	18.9	8.81	0.833
290	0.01917	1.001	69.7	4.184	1.864	1080	8.69	598	19.3	7.56	0.841
295	0.02616	1.002	51.94	4.181	1.868	959	8.89	606	19.5	6.62	0.849
300	0.03531	1.003	39.13	4.179	1.872	855	9.09	613	19.6	5.83	0.857
310	0.06221	1.007	13.98	4.178	1.882	695	9.49	628	20.4	4.62	0.873
320	0.1053	1.011	13.98	4.18	1.895	577	9.89	640	21	3.77	0.894
330	0.1719	1.016	8.82	4.184	1.895	489	10.29	650	21.7	3.15	0.908
340	0.2713	1.021	5.74	4.188	1.930	420	10.69	660	22.3	2.66	0.925
350	0.4163	1.027	3.846	4.195	1.954	365	11.09	668	23	2.29	0.942
360	0.6209	1.034	2.645	4.203	1.983	324	11.49	674	23.7	2.02	0.960
370	0.9040	1.041	1.861	4.214	2.017	289	11.89	679	24.5	1.80	0.978
373.15	1.0113	1.044	1.679	4.217	2.029	279	12.02	680	24.8	1.76	0.984
380	1.2869	1.049	1.337	4.226	2.057	260	12.29	683	25.4	1.61	0.999
390	1.794	1.058	0.98	4.239	2.104	237	12.69	686	26.3	1.47	1.013
400	2.455	1.067	0.731	4.256	2.158	217	13.05	688	27.2	1.34	1.033
420	4.37	1.088	0.425	4.302	2.291	185	13.79	688	29.8	1.16	1.075
440	7.333	1.11	0.261	4.36	2.46	162	15.4	682	31.7	1.04	1.12
460	11.71	1.137	0.167	4.44	2.68	143	15.19	673	24.6	0.95	1.17
480	17.19	1.167	0.111	4.53	2.94	129	15.88	660	38.1	0.89	1.23
500	26.40	1.203	0.0766	4.66	3.27	118	16.59	642	42.3	0.86	1.28
520	37.7	1.244	0.0525	4.84	3.70	108	7.33	621	47.5	0.84	1.35
540	52.38	1.294	0.0375	5.08	4.27	101	18.1	594	54.0	0.86	1.43
560	71.08	1.355	0.0269	5.43	5.09	94	19.1	563	63.7	0.90	1.52
580	94.51	1.433	0.0193	6.00	6.40	88	20.4	528	76.7	0.982	1.68
600	123.5	1.541	0.00137	7.00	8.75	81	22.7	497	92.9	1.14	2.15
620	159.1	1.705	0.0094	9.35	15.4	72	25.9	444	114	1.52	3.46
640	202.7	2.075	0.0057	26	42	59	32	367	155	4.2	9.6
647.3[b]	221.2	3.17	0.0032	∞	∞	45	45	238	238	∞	∞

[a] Based on Incropera et al. (2007).
[b] Critical temperature.

Thermodynamic Properties of Saturated Liquid and Vapor for Selected Refrigerants

Thermodynamic Properties of Saturated Liquid and Vapor for Selected Refrigerants

T (°C)	P (MPa)	ρ_f (kg/m³)	ρ_g (kg/m³)	h_f (kJ/kg)	h_g (kJ/kg)	C_{Pf} (kJ/kg·K)	C_{Pg} (kJ/kg·K)	μ_f (μPa/s)	μ_g (μPa/s)	k_f (W/m·K)	k_g (W/m·K)	σ (N/m)
				R-11 (CCl₃F) (P_{cr} = 44.08 bar, T_{cr} = 197.96 °C; h = 0 at 26.86 °C and 1 bar)								
−50.00	0.00265	1643	0.1966	157.9	364.6	0.8261	0.5070	1063	7.638	0.1099	0.00547	0.02779
−40.00	0.00509	1622	0.3624	166.2	369.5	0.8322	0.5209	897.7	7.982	0.1066	0.00584	0.02640
−30.00	0.00919	1600	0.6293	174.5	374.5	0.8383	0.5347	772.1	8.327	0.1034	0.00622	0.02502
−20.00	0.01573	1579	1.038	183.0	379.5	0.8447	0.5484	673.3	8.672	0.1002	0.00661	0.02366
−10.00	0.02567	1556	1.636	191.4	384.6	0.8516	0.5621	593.5	9.016	0.09713	0.00701	0.02231
0.0	0.04020	1534	2.480	200.0	389.8	0.8590	0.5756	527.6	9.359	0.09411	0.00741	0.02098
10.00	0.06068	1511	3.634	208.6	394.9	0.8671	0.5892	472.4	9.703	0.09115	0.00782	0.01966
20.00	0.08868	1488	5.170	217.4	400.1	0.8760	0.6028	425.3	10.05	0.08825	0.00824	0.01836
30.00	0.1260	1464	7.169	226.2	405.2	0.8858	0.6167	384.6	10.35	0.08539	0.00866	0.01708
40.00	0.1744	1440	9.718	235.1	410.3	0.8965	0.6310	349.0	10.68	0.08258	0.00910	0.01582
50.00	0.2361	1415	12.92	244.2	415.3	0.9082	0.6460	317.6	11.01	0.07980	0.00954	0.01458
60.00	0.3133	1389	16.88	253.3	420.3	0.9211	0.6619	289.6	11.35	0.07706	0.01000	0.01336
70.00	0.4082	1362	21.73	262.6	425.1	0.9354	0.6791	264.4	11.69	0.07434	0.01047	0.01216
80.00	0.5232	1335	27.63	272.1	429.8	0.9513	0.6981	241.6	12.04	0.07163	0.01096	0.01098
90.00	0.6609	1306	34.73	281.7	434.3	0.9691	0.7194	220.8	12.40	0.06894	0.01148	0.00983

T (°C)	P (MPa)	ρ_f (kg/m³)	ρ_g (kg/m³)	h_f (kJ/kg)	h_g (kJ/kg)	C_{Pf} (kJ/kg·K)	C_{Pg} (kJ/kg·K)	μ_f (μPa/s)	μ_g (μPa/s)	k_f (W/m·K)	k_g (W/m·K)	σ (N/m)
\multicolumn{13}{l}{R-22 (CHClF₂) (P_{cr} = 49.90 bar, T_{cr} = 96.15 °C; h = 0 at 0 °C and 1 bar)}
−60.00	0.3750	1464	1.863	133.3	378.6	1.071	0.5637	441.4	8.942	0.1226	0.00612	0.02124
−50.00	0.6453	1436	3.088	144.0	383.4	1.079	0.5847	387.5	9.364	0.1178	0.00659	0.01958
−40.00	1.052	1407	4.873	154.9	388.1	1.091	0.6083	342.6	9.786	0.1131	0.00709	0.01794
−30.00	1.639	1377	7.379	165.9	392.7	1.105	0.6349	304.6	10.21	0.1085	0.00761	0.01634
−20.00	2.453	1347	10.79	177.0	397.1	1.123	0.6650	271.9	10.63	0.1039	0.00817	0.01476
−10.00	3.548	1315	15.32	188.4	401.2	1.144	0.6994	243.4	11.06	0.09934	0.00877	0.01321
0.0	4.980	1282	21.23	200.0	405.0	1.169	0.7390	218.2	11.50	0.09484	0.00942	0.01170
10.00	6.809	1247	28.82	211.9	408.6	1.199	0.7852	195.7	11.96	0.09037	0.01014	0.01022
20.00	9.100	1210	38.48	224.1	411.7	1.236	0.8404	175.3	12.43	0.08589	0.01095	0.00878
30.00	11.92	1171	50.70	236.6	414.3	1.281	0.9081	156.7	12.95	0.08140	0.01189	0.00738
40.00	15.34	1129	66.19	249.6	416.2	1.339	0.9948	139.4	13.52	0.07688	0.01302	0.00604
50.00	19.43	1082	85.95	263.2	417.4	1.419	1.113	123.1	14.18	0.07229	0.01445	0.00474
60.00	24.27	1030	111.6	277.6	417.5	1.539	1.287	107.6	14.98	0.06762	0.01636	0.00351
70.00	29.97	969.7	146.0	293.1	416.1	1.743	1.584	92.36	16.02	0.06292	0.01916	0.00236
80.00	36.64	893.7	195.4	310.4	412.0	2.181	2.231	76.64	17.55	0.05860	0.02387	0.00130
90.00	44.42	780.1	280.6	332.1	401.9	3.981	4.975	58.33	20.48	0.05935	0.03455	0.00040

(*cont.*)

(cont.)

R-123 (CHCl$_2$CF$_3$) (P_{cr} = 36.62 bar, T_{cr} = 183.68 °C; h = 0 at 0 °C and 1 bar)

T (°C)	P (MPa)	ρ_f (kg/m³)	ρ_g (kg/m³)	h_f (kJ/kg)	h_g (kJ/kg)	C_{Pf} (kJ/kg·K)	C_{Pg} (kJ/kg·K)	μ_f (μPa/s)	μ_g (μPa/s)	k_f (W/m·K)	k_g (W/m·K)	σ (N/m)
-30.00	0.06748	1597	0.5136	170.8	363.7	0.9578	0.6011	848.0	8.699	0.09297	0.00605	0.02192
-20.00	0.1200	1574	0.8800	180.4	369.5	0.9682	0.6174	735.4	9.086	0.08985	0.00661	0.02066
-10.00	0.2025	1550	1.435	190.1	375.4	0.9790	0.6339	642.4	9.465	0.08674	0.00718	0.01941
0.0	0.3265	1526	2.242	200.0	381.4	0.9902	0.6508	564.6	9.838	0.08369	0.00774	0.01818
10.00	0.5057	1502	3.374	210.0	387.5	1.002	0.6682	498.8	10.20	0.08071	0.00831	0.01697
20.00	0.7561	1477	4.917	220.1	393.5	1.014	0.6861	442.6	10.56	0.07782	0.00889	0.01577
30.00	1.096	1451	6.966	230.3	399.5	1.026	0.7047	394.3	10.91	0.07504	0.00948	0.01459
40.00	1.545	1425	9.630	240.6	405.5	1.038	0.7243	352.4	11.26	0.07236	0.01008	0.01343
50.00	2.125	1398	13.03	251.1	411.5	1.052	0.7448	315.9	11.60	0.06979	0.01070	0.01228
60.00	2.859	1370	17.31	261.7	417.4	1.066	0.7667	283.9	11.94	0.06731	0.01134	0.01116
70.00	3.772	1341	22.63	272.4	423.2	1.082	0.7904	255.6	12.28	0.06493	0.01201	0.01005
80.00	4.891	1311	29.19	283.3	428.9	1.100	0.8162	230.5	12.63	0.06262	0.01273	0.00897
90.00	6.242	1280	37.21	294.5	434.4	1.120	0.8450	208.2	12.98	0.06038	0.01348	0.00792
100.0	7.855	1247	47.00	305.8	439.8	1.143	0.8780	188.1	13.37	0.05819	0.01429	0.00689
110.0	9.760	1212	58.91	317.3	444.9	1.172	0.9168	169.9	13.80	0.05603	0.01517	0.00588
120.0	11.99	1174	73.47	329.1	449.7	1.207	0.9643	153.4	14.29	0.05388	0.01614	0.00491
130.0	14.58	1134	91.38	341.3	454.1	1.254	1.026	138.1	14.89	0.05172	0.01722	0.00398
140.0	17.56	1088	113.7	353.9	457.9	1.318	1.111	123.8	15.65	0.04952	0.01844	0.00309

T (°C)	P (MPa)	ρ_f (kg/m³)	ρ_g (kg/m³)	h_f (kJ/kg)	h_g (kJ/kg)	C_{Pf} (kJ/kg·K)	C_{Pg} (kJ/kg·K)	μ_f (μPa/s)	μ_g (μPa/s)	k_f (W/m·K)	k_g (W/m·K)	σ (N/m)
					R-134a (CH_2FCF_3) (P_{cr} = 40.59 bar, T_{cr} = 101.06 °C; h = 0 at 0 °C and 1 bar)							
−60.00	0.1591	1474	0.9268	123.4	361.3	1.223	0.6924	663.1	8.304	0.1207	0.00656	0.02080
−50.00	0.2945	1446	1.650	135.7	367.7	1.238	0.7197	555.1	8.716	0.1156	0.00736	0.01918
−40.00	0.5121	1418	2.769	148.1	374.0	1.255	0.7490	472.2	9.122	0.1106	0.00817	0.01760
−30.00	0.8438	1388	4.426	160.8	380.3	1.273	0.7809	406.4	9.525	0.1058	0.00899	0.01604
−20.00	1.327	1358	6.784	173.6	386.6	1.293	0.8158	353.0	9.925	0.1011	0.00982	0.01451
−10.00	2.006	1327	10.04	186.7	392.7	1.316	0.8544	308.6	10.33	0.09649	0.01066	0.01302
0.0	2.928	1295	14.43	200.0	398.6	1.341	0.8972	271.1	10.73	0.09201	0.01151	0.01156
10.00	4.146	1261	20.23	213.6	404.3	1.370	0.9455	238.8	11.15	0.08761	0.01240	0.01014
20.00	5.717	1225	27.78	227.5	409.7	1.405	1.001	210.7	11.58	0.08328	0.01333	0.00876
30.00	7.702	1187	37.54	241.7	414.8	1.446	1.065	185.8	12.04	0.07899	0.01433	0.00742
40.00	10.17	1147	50.09	256.4	419.4	1.498	1.145	163.4	12.55	0.07471	0.01544	0.00613
50.00	13.18	1102	66.27	271.6	423.4	1.566	1.246	143.1	13.12	0.07042	0.01672	0.00489
60.00	16.82	1053	87.38	287.5	426.6	1.660	1.387	124.2	13.79	0.06609	0.01831	0.00372
70.00	21.17	996.2	115.6	304.3	428.6	1.804	1.605	106.4	14.65	0.06167	0.02045	0.00261
80.00	26.33	928.2	155.1	322.4	428.8	2.065	2.012	89.02	15.84	0.05717	0.02372	0.00160
90.00	32.44	837.8	216.8	342.9	425.4	2.756	3.121	70.86	17.81	0.05284	0.02991	0.00071
100.0	39.72	651.2	373.0	373.3	407.7	17.59	25.35	45.10	24.21	0.05995	0.06058	0.00004

(*cont.*)

(cont.)

T (°C)	P (MPa)	ρ_f (kg/m³)	ρ_g (kg/m³)	h_f (kJ/kg)	h_g (kJ/kg)	C_{Pf} (kJ/kg·K)	C_{Pg} (kJ/kg·K)	μ_f (μPa/s)	μ_g (μPa/s)	k_f (W/m·K)	k_g (W/m·K)	σ (N/m)
					R-141b (CH$_3$CCl$_2$F) (P_{cr} = 42.50 bar, T_{cr} = 204.20 °C; h = 0 at 0 °C and 1 bar)							
−40.00	0.03119	1357	0.1887	156.2	411.3	1.072	0.6538	1041	7.200	0.1168	0.00643	0.02660
−30.00	0.05857	1338	0.3405	167.0	417.7	1.083	0.6719	864.7	7.511	0.1130	0.00687	0.02526
−20.00	0.1038	1320	0.5808	177.9	424.3	1.095	0.6901	733.4	7.823	0.1094	0.00733	0.02394
−10.00	0.1747	1301	0.9437	188.9	431.0	1.107	0.7087	632.2	8.136	0.1058	0.00781	0.02263
0.0	0.2812	1282	1.470	200.0	437.7	1.120	0.7278	551.8	8.450	0.1024	0.00829	0.02133
10.00	0.4351	1263	2.207	211.3	444.5	1.133	0.7472	486.5	8.763	0.09906	0.00879	0.02005
20.00	0.6503	1244	3.208	222.7	451.4	1.147	0.7673	432.3	9.076	0.09582	0.00930	0.01878
30.00	0.9425	1224	4.533	234.2	458.3	1.162	0.7880	386.5	9.387	0.09266	0.00982	0.01753
40.00	1.329	1204	6.250	246.0	465.1	1.178	0.8094	347.4	9.699	0.08958	0.01036	0.01629
50.00	1.828	1184	8.432	257.8	472.0	1.195	0.8317	313.6	10.01	0.08657	0.01091	0.01508
60.00	2.461	1163	11.16	269.9	478.8	1.213	0.8551	283.9	10.32	0.08362	0.01148	0.01388
70.00	3.249	1141	14.53	282.1	485.6	1.232	0.8797	257.7	10.64	0.08073	0.01206	0.01270
80.00	4.214	1119	18.65	294.6	492.3	1.253	0.9059	234.2	10.96	0.07789	0.01266	0.01154
90.00	5.380	1095	23.63	307.3	498.8	1.276	0.9342	213.1	11.28	0.07509	0.01330	0.01041
100.0	6.771	1071	29.64	320.2	505.3	1.302	0.9651	193.9	11.61	0.07232	0.01396	0.00929
110.0	8.413	1046	36.83	333.3	511.5	1.331	0.9995	176.4	11.96	0.06957	0.01466	0.00820
120.0	10.33	1019	45.43	346.8	517.5	1.364	1.039	160.1	12.33	0.06685	0.01542	0.00714
130.0	12.55	990.9	55.74	360.6	523.2	1.403	1.085	145.1	12.72	0.06414	0.01624	0.00611
140.0	15.11	960.5	68.14	374.7	528.6	1.450	1.142	130.9	13.16	0.06144	0.01716	0.00511

T (°C)	P (MPa)	ρ_f (kg/m³)	ρ_g (kg/m³)	h_f (kJ/kg)	h_g (kJ/kg)	C_{Pf} (kJ/kg·K)	C_{Pg} (kJ/kg·K)	μ_f (μPa·s)	μ_g (μPa·s)	k_f (W/m·K)	k_g (W/m·K)	σ (N/m)
					R-717 (NH₃, ammonia) (P_{cr} = 113.3 bar, T_{cr} = 132.35 °C; h = 0 at 0 °C and 1 bar)							
−60.00	0.2189	713.6	0.2125	−68.06	1374	4.303	2.125	391.3	7.296	0.7570	0.01993	0.05505
−50.00	0.4084	702.1	0.3806	−24.73	1391	4.360	2.178	328.9	7.573	0.7223	0.02024	0.05111
−40.00	0.7169	690.2	0.6438	19.17	1408	4.414	2.244	281.2	7.859	0.6881	0.02064	0.04726
−30.00	1.194	677.8	1.037	63.60	1423	4.465	2.326	244.1	8.152	0.6546	0.02115	0.04352
−20.00	1.901	665.1	1.603	108.6	1438	4.514	2.425	214.4	8.449	0.6220	0.02177	0.03988
−10.00	2.907	652.1	2.391	154.0	1451	4.564	2.542	190.2	8.751	0.5901	0.02250	0.03634
0.0	4.294	638.6	3.457	200.0	1462	4.617	2.680	170.1	9.056	0.5592	0.02337	0.03291
10.00	6.150	624.6	4.868	246.6	1472	4.676	2.841	153.0	9.364	0.5291	0.02437	0.02959
20.00	8.575	610.2	6.703	293.8	1480	4.745	3.030	138.3	9.676	0.4999	0.02552	0.02638
30.00	11.67	595.2	9.053	341.8	1486	4.828	3.250	125.5	9.995	0.4714	0.02685	0.02328
40.00	15.55	579.4	12.03	390.6	1490	4.932	3.510	114.0	10.33	0.4436	0.02838	0.02029
50.00	20.34	562.9	15.79	440.6	1491	5.064	3.823	103.8	10.67	0.4164	0.03017	0.01743
60.00	26.16	545.2	20.49	492.0	1489	5.235	4.208	94.48	11.05	0.3898	0.03230	0.01469
70.00	33.13	526.3	26.41	545.0	1484	5.465	4.699	85.93	11.47	0.3635	0.03491	0.01208
80.00	41.42	505.7	33.89	600.3	1474	5.784	5.355	77.98	11.95	0.3376	0.03826	0.00961
90.00	51.17	482.8	43.48	658.6	1459	6.250	6.291	70.47	12.55	0.3119	0.04291	0.00730
100.0	62.55	456.6	56.12	721.0	1437	6.991	7.762	63.23	13.32	0.2862	0.05009	0.00515
110.0	75.78	425.6	73.55	789.7	1403	8.362	10.46	56.03	14.42	0.2608	0.06301	0.00320
120.0	91.12	385.5	100.1	869.9	1350	11.94	17.21	48.34	16.21	0.2370	0.09390	0.00150
130.0	109.0	312.3	156.8	992.0	1239	54.21	76.49	37.29	20.63	0.2559	0.1561	0.00018

(cont.)

(cont.)

R-744 (CO$_2$) ($P_{cr} = 73.84$ bar, $T_{cr} = 31.06\,°C$; $h = 0$ at $0\,°C$ and 1 bar)

T (°C)	P (MPa)	ρ_f (kg/m³)	ρ_g (kg/m³)	h_f (kJ/kg)	h_g (kJ/kg)	C_{Pf} (kJ/kg·K)	C_{Pg} (kJ/kg·K)	μ_f (µPa/s)	μ_g (µPa/s)	k_f (W/m·K)	k_g (W/m·K)	σ (N/m)
−40.00	10.04	1116	26.15	114.2	435.5	1.986	1.052	198.6	11.57	0.1481	0.00163	0.01314
−35.00	12.02	1096	31.23	124.2	436.4	2.005	1.101	184.3	11.84	0.1421	0.00248	0.01196
−30.00	14.26	1075	37.10	134.3	437.0	2.033	1.157	171.1	12.13	0.1358	0.00290	0.01082
−25.00	16.81	1054	43.86	144.5	437.3	2.071	1.222	158.6	12.43	0.1294	0.00325	0.00970
−20.00	19.67	1031	51.65	155.0	437.2	2.123	1.299	146.9	12.74	0.1230	0.00366	0.00860
−15.00	22.88	1008	60.64	165.7	436.6	2.190	1.390	135.9	13.08	0.1167	0.00417	0.00754
−10.00	26.45	983.2	71.04	176.7	435.6	2.277	1.501	125.3	13.45	0.1103	0.00482	0.00651
−5.000	30.42	956.7	83.14	188.1	434.0	2.389	1.642	115.2	13.85	0.1040	0.00567	0.00551
0.0	34.81	928.1	97.32	200.0	431.6	2.535	1.826	105.4	14.31	0.09755	0.00679	0.00455
5.000	39.65	896.7	114.1	212.5	428.4	2.731	2.077	95.84	14.83	0.09105	0.00833	0.00364
10.00	44.97	861.7	134.4	225.8	424.0	3.009	2.444	86.37	15.46	0.08445	0.01049	0.00277
15.00	50.81	821.8	159.6	240.1	418.0	3.442	3.035	76.84	16.24	0.07778	0.01371	0.00196
20.00	57.22	774.2	192.3	256.0	409.7	4.243	4.154	66.97	17.28	0.07124	0.01894	0.00121

T (°C)	P_L (bar)	P_v (bar)	ρ_f (kg/m³)	ρ_g (kg/m³)	h_f (kJ/kg)	h_g (kJ/kg)	C_{Pf} (kJ/kg·K)	C_{Pg} (kJ/kg·K)	μ_f (μPa/s)	μ_g (μPa/s)	k_f (W/m·K)	k_g (W/m·K)	σ (N/m)
							R-410A (50% R-32, CH_2F_2; 50% R-125, CHF_2CF_3; $h = 0$ at 0 °C and 1 bar)						
−60.00	0.6493	0.6477	1377	2.742	115.5	393.8	1.337	0.7623	353.0	9.427	0.1506	0.00767	0.01871
−50.00	1.098	1.095	1346	4.491	129.0	399.2	1.354	0.8052	307.7	9.859	0.1444	0.00819	0.01693
−40.00	1.763	1.759	1314	7.023	142.7	404.3	1.373	0.8530	270.1	10.29	0.1384	0.00873	0.01520
−30.00	2.709	2.702	1281	10.57	156.5	409.1	1.397	0.9057	238.3	10.74	0.1323	0.00933	0.01350
−20.00	4.009	3.998	1246	15.42	170.7	413.5	1.426	0.9646	210.9	11.24	0.1263	0.01003	0.01184
−10.00	5.741	5.724	1210	21.91	185.1	417.5	1.462	1.031	187.0	11.76	0.1202	0.01081	0.01023
0.0	7.991	7.967	1171	30.48	200.0	420.9	1.507	1.110	165.7	12.31	0.1140	0.01173	0.00867
10.00	10.85	10.82	1130	41.75	215.3	423.7	1.565	1.207	146.6	12.91	0.1077	0.01288	0.00716
20.00	14.43	14.38	1085	56.53	231.3	425.6	1.642	1.332	129.2	13.56	0.1013	0.01444	0.00571
30.00	18.83	18.77	1035	76.11	248.0	426.3	1.750	1.508	113.0	14.31	0.09461	0.01673	0.00433
40.00	24.17	24.10	979.0	102.6	265.7	425.5	1.919	1.783	97.69	15.21	0.08765	0.02039	0.00303
50.00	30.61	30.53	911.6	140.0	285.1	422.3	2.227	2.296	82.78	16.40	0.08025	0.02673	0.00183
60.00	38.33	38.25	823.0	198.6	307.3	414.5	3.030	3.671	67.30	18.28	0.07220	0.03952	0.00078

Properties of Selected Ideal Gases at 1 Atmosphere

Temperature, T (K)	Density, ρ (kg/m^3)	Specific heat, C_P (kJ/kg·K)	Viscosity, μ (kg/m·s) $\times 10^7$	Thermal conductivity, k (W/m·K) $\times 10^3$	Prandtl number, Pr
Air					
100	3.5562	1.032	71.1	9.34	0.786
150	2.3364	1.012	103.4	13.8	0.758
200	1.7458	1.007	132.5	18.1	0.737
250	1.3947	1.006	159.6	22.3	0.72
300	1.1614	1.007	184.6	26.3	0.707
350	0.995	1.009	208.2	30.0	0.700
400	0.8711	1.014	230.1	33.8	0.690
450	0.774	1.021	250.7	37.3	0.686
500	0.6964	1.03	270.1	40.7	0.684
550	0.6329	1.04	288.4	43.9	0.683
600	0.5804	1.051	305.8	46.9	0.685
650	0.5356	1.063	322.5	49.7	0.690
700	0.4975	1.075	338.8	52.4	0.695
800	0.4354	1.099	369.8	57.3	0.709
900	0.3868	1.121	398.1	62.0	0.720
1000	0.3482	1.141	424.4	66.7	0.726
1100	0.3166	1.159	449.0	71.5	0.728
1200	0.2902	1.175	473.0	76.3	0.728
1300	0.2679	1.189	496.0	82	0.719
1400	0.2488	1.207	530	91	0.703
1500	0.2322	1.230	557	100	0.685
1600	0.2177	1.248	584	106	0.688
1700	0.2049	1.267	611	113	0.685
1800	0.1935	1.286	637	120	0.683
1900	0.1833	1.307	663	128	0.677
2000	0.1741	1.337	689	137	0.672
2100	0.1658	1.372	715	147	0.667
2200	0.1582	1.417	740	160	0.655
2300	0.1513	1.478	766	175	0.647
2400	0.1448	1.558	792	196	0.630
2500	0.1389	1.665	818	222	0.613
3000	0.1135	2.726	955	486	0.536

(cont.)

Temperature, T (K)	Density, ρ (kg/m^3)	Specific heat, C_P (kJ/kg·K)	Viscosity, μ (kg/m·s) $\times 10^7$	Thermal conductivity, k (W/m·K) $\times 10^3$	Prandtl number, Pr
Nitrogen (N$_2$)					
100	3.4388	1.070	68.8	9.58	0.768
150	2.2594	1.050	100.6	13.9	0.759
200	1.6883	1.043	129.2	18.3	0.736
250	1.3488	1.042	154.9	22.2	0.727
300	1.1233	1.041	178.2	25.9	0.716
350	0.9625	1.042	200.0	29.3	0.711
400	0.8425	1.045	220.4	32.7	0.704
450	0.7485	1.050	239.6	35.8	0.703
500	0.6739	1.056	257.7	38.9	0.700
550	0.6124	1.065	274.7	41.7	0.702
600	0.5615	1.075	290.8	44.6	0.701
700	0.4812	1.098	321.0	49.9	0.706
800	0.4211	1.12	349.1	54.8	0.715
900	0.3743	1.146	375.3	59.7	0.721
1000	0.3368	1.167	399.9	64.7	0.721
1100	0.3062	1.187	423.2	70.0	0.718
1200	0.2807	1.204	445.3	75.8	0.707
1300	0.2591	1.219	466.2	81.0	0.701
1400	0.2438	1.229	486	87.5	0.709
1600	0.2133	1.250	510	97	0.71
Oxygen (O$_2$)					
100	3.945	0.962	76.4	9.25	0.796
150	2.585	0.921	114.8	13.8	0.766
200	1.930	0.915	147.5	18.3	0.737
250	1.542	0.915	178.6	22.6	0.723
300	1.284	0.920	207.2	26.8	0.711
350	1.100	0.929	233.5	29.6	0.733
400	0.9620	0.942	258.2	33.0	0.737
450	0.8554	0.956	281.4	36.3	0.741
500	0.7698	0.972	303.3	41.2	0.716
550	0.6998	0.988	324.0	44.1	0.726
600	0.6414	1.003	343.7	47.3	0.729
700	0.5498	1.031	380.8	52.8	0.744
800	0.4810	1.054	415.2	58.9	0.743
900	0.4275	1.074	447.2	64.9	0.740
1000	0.3848	1.090	477.0	71.0	0.733
1100	0.3498	1.103	505.5	75.8	0.736
1200	0.3206	1.115	532.5	81.9	0.725
1300	0.2960	1.125	588.4	87.1	0.721
Carbon dioxide (CO$_2$)					
220	2.4733	0.783	110.6	10.9	0.795
250	2.1657	0.804	125.7	12.95	0.780
280	1.9022	0.830	140	15.20	0.765
300	1.7730	0.851	149	16.55	0.766
320	1.6609	0.872	156	18.05	0.754
340	1.5618	0.891	165	19.70	0.746

(cont.)

(cont.)

Temperature, T (K)	Density, ρ (kg/m^3)	Specific heat, C_P (kJ/kg·K)	Viscosity, μ (kg/m·s) $\times 10^7$	Thermal conductivity, k (W/m·K) $\times 10^3$	Prandtl number, Pr
350	1.5362	0.900	174	20.92	0.744
360	1.4743	0.908	173	21.2	0.741
380	1.3961	0.926	181	22.75	0.737
400	1.3257	0.942	190	24.3	0.737
450	1.1782	0.981	210	28.3	0.728
500	1.0594	1.02	231	32.5	0.725
550	0.9625	1.05	251	36.6	0.721
600	0.8826	1.08	270	40.7	0.717
650	0.8143	1.10	288	44.5	0.712
700	0.7564	1.13	305	48.1	0.717
750	0.7057	1.15	321	51.7	0.714
800	0.6614	1.17	337	55.1	0.716
Carbon monoxide (CO)					
200	1.6888	1.045	127	17.0	0.781
220	1.5341	1.044	137	19.0	0.753
240	1.4055	1.043	147	20.6	0.744
260	1.2967	1.043	157	22.1	0.741
280	1.2038	1.042	166	23.6	0.733
300	1.1233	1.043	175	25.0	0.730
320	1.0529	1.043	184	26.3	0.730
340	0.9909	1.044	193	27.8	0.725
360	0.9357	1.045	202	29.1	0.725
380	0.8864	1.047	210	30.5	0.729
400	0.8421	1.049	218	31.8	0.719
450	0.7483	1.055	237	35.0	0.714
500	0.67352	1.065	254	38.1	0.710
550	0.61226	1.076	271	41.1	0.710
600	0.56126	1.088	286	44.0	0.707
650	0.51806	1.101	301	47.0	0.705
700	0.48102	1.114	315	50.0	0.702
750	0.44899	1.127	329	52.8	0.702
800	0.42095	1.140	343	55.5	0.705
900	0.3791	1.155	371	59.0	0.705
1000	0.3412	1.165	399	61.64	0.705
Hydrogen (H$_2$)					
100	0.24255	11.23	42.1	67.0	0.707
150	0.16156	12.60	56.0	101	0.699
200	0.12115	13.54	68.1	131	0.704
250	0.09693	14.06	78.9	157	0.707
300	0.08078	14.31	89.6	183	0.701
350	0.06924	14.43	98.8	204	0.700
400	0.06059	14.48	108.2	226	0.695
450	0.05386	14.50	117.2	247	0.689
500	0.04848	14.52	126.4	266	0.691
550	0.04407	14.53	134.3	285	0.685
600	0.04040	14.55	142.4	305	0.678
700	0.03463	14.61	157.8	342	0.675
800	0.03030	14.70	172.4	378	0.670

(cont.)

Temperature, T (K)	Density, ρ (kg/m^3)	Specific heat, C_P (kJ/kg·K)	Viscosity, μ (kg/m·s) $\times 10^7$	Thermal conductivity, k (W/m·K) $\times 10^3$	Prandtl number, Pr
900	0.02694	14.83	186.5	412	0.671
1000	0.02424	14.99	201.3	448	0.673
1100	0.02204	15.17	213.0	488	0.662
1200	0.02020	15.37	226.2	528	0.659
1300	0.01865	15.59	238.5	568	0.655
1400	0.01732	15.81	250.7	610	0.650
1500	0.01616	16.02	262.7	655	0.643
1600	0.0152	16.28	273.7	697	0.639
1700	0.0143	16.58	284.9	742	0.637
1800	0.0135	16.96	296.1	786	0.639
1900	0.0128	17.49	307.2	835	0.643
2000	0.0121	18.25	318.2	878	0.661

Helium (He)

50	0.9732	5.201	60.7	47.6	0.663
100	0.4871	5.193	96.3	73.0	0.686
120	0.4060	5.193	107	81.9	0.679
140	0.3481	5.193	118	90.7	0.676
160	0.309	5.193	129	99.2	0.674
180	0.2708	5.193	139	107.2	0.673
200	0.2437	5.193	150	115.1	0.667
220	0.2216	5.193	160	123.1	0.675
240	0.205	5.193	170	130	0.678
260	0.1875	5.193	180	137	0.682
280	0.175	5.193	190	145	0.681
300	0.1625	5.193	199	152	0.680
350	0.1393	5.193	221	170	0.663
400	0.1219	5.193	243	187	0.675
450	0.1084	5.193	263	204	0.663
500	0.09754	5.193	283	220	0.668
550	0.0894	5.193	300	234	0.665
600	0.08128	5.193	320	252	0.663
650	0.0755	5.193	332	264	0.658
700	0.06969	5.193	350	278	0.654
750	0.0653	5.193	364	291	0.659
800	0.06096	5.193	382	304	0.664
900	0.05419	5.193	414	330	0.664
1000	0.04879	5.193	454	354	0.654
1100	0.04434	5.193	495	387	0.664
1200	0.04065	5.193	527	412	0.664
1300	0.03752	5.193	559	437	0.664
1400	0.03484	5.193	590	461	0.665
1500	0.03252	5.193	621	485	0.665

Water vapor (H$_2$O)

373.15	0.5976	2.080	122.8	25.09	0.98
380	0.5863	2.060	127.1	24.6	0.98
400	0.5542	2.014	134.4	26.1	0.98
450	0.4902	1.980	152.5	29.9	0.97

(cont.)

(cont.)

Temperature, T (K)	Density, ρ (kg/m^3)	Specific heat, C_P (kJ/kg·K)	Viscosity, μ (kg/m·s) $\times 10^7$	Thermal conductivity, k (W/m·K) $\times 10^3$	Prandtl number, Pr
500	0.4405	1.985	173	33.9	0.96
550	0.4005	1.997	188.4	37.9	0.95
600	0.3652	2.026	215	42.2	0.94
650	0.3380	2.056	236	46.4	0.93
700	0.3140	2.085	257	50.5	0.92
750	0.2931	2.119	277.5	54.9	0.91
800	0.2739	2.152	298	59.2	0.9
850	0.2579	2.186	318	63.7	0.895
873.15	0.2516	2.203	326.2	79.90	0.897
900	0.241	2.219	339	84.3	0.89
973.15	0.2257	2.273	365.5	93.38	0.888
1000	0.217	2.286	378	98.1	0.883
1073.15	0.2046	2.343	403.8	107.3	0.881
1200	0.181	2.43	448	130	0.85
1400	0.155	2.58	506	160	0.82
1600	0.135	2.73	565	210	0.74
1800	0.12	3.02	619	330	0.57
2000	0.108	3.79	670	570	0.45

APPENDIX E

Binary Diffusion Coefficients of Selected Gases in Air at 1 Atmosphere[a]

Substance 1	T (K)	D_{12} (m^2/s)[b]
Ar	293	0.189×10^{-4}
CO_2	298	0.16×10^{-4}
H_2	298	0.41×10^{-4}
He	300	0.777×10^{-4}
O_2	298	0.21×10^{-4}
N_2	273	0.181×10^{-4}
H_2O	298	0.26×10^{-4}
CH_4	293	0.21×10^{-4}
NH_3	298	0.28×10^{-4}
CO	300	0.202×10^{-4}
NO	300	0.18×10^{-4}
SO_2	300	0.126×10^{-4}
SF_6	295	0.109×10^{-4}
Benzene	298	0.083×10^{-5}
Naphthalene	300	0.62×10^{-5}

[a] For ideal gases, $D_{12} \sim P^{-1} T^{3/2}$.
[b] Air is substance 2.

Henry's Constant of Dilute Aqueous Solutions of Selected Substances at 298.16 K Temperature and Moderate Pressures[a,b]

Solute	B (K)	C_{He} (bars)
O_2	4.41×10^6	4.31×10^6
H_2	7.16×10^6	7.09×10^6
D_2	7.20×10^6	7.09×10^6
N_2	8.87×10^6	8.61×10^6
NH_3	101.7	93.31
Br_2	7762	7.18×10^3
I_2	2167	2.00×10^3
H_2S	63 715	6.15×10^4
SO_2	4973	4.67×10^3
SF_6	2.48×10^7	2.33×10^7
He	1.44×10^7	1.44×10^7
Ar	4.12×10^6	4.00×10^6
Kr	2.32×10^6	2.24×10^6
Xe	1.40×10^6	1.33×10^6
Rn	6.42×10^5	6.09×10^5
CH_4	4.43×10^6	4.31×10^6
C_2H_6	3.09×10^6	2.95×10^6
C_3H_8	3.95×10^6	3.73×10^6
CO	5.95×10^6	5.77×10^6
CO_2	1.627×10^5	1.56×10^5

[a] From various sources (compiled by Sander, 2015).
[b] For other temperatures use:

$$\frac{1}{C_{He}(T)} = \frac{1}{C_{He}(T_0)} \cdot \exp\left[B \left(\frac{1}{T} - \frac{1}{T_0} \right) \right]$$

where $T_0 = 298.16$ K and $C_{He}(T_0)$ is Henry's constant at T_0.

Diffusion Coefficients of Selected Substances in Water at Infinite Dilution at 25 °C

Solute (Substance 1)	D_{12} (10^{-9} m^2/s)^a
Argon	2.00
Air	2.00
Carbon dioxide	1.92
Carbon monoxide	2.03
Chlorine	1.25
Ethane	1.20
Ethylene	1.87
Helium	6.28
Hydrogen	4.50
Methane	1.49
Nitric oxide	2.60
Nitrogen	1.88
Oxygen	2.10
Propane	0.97
Ammonia	1.64
Benzene	1.02
Hydrogen sulfide	1.41

^{a} Substance 2 is water.

Lennard-Jones (6–12) Potential Model Constants for Selected Molecules[a]

Molecule		$\tilde{\sigma}$ (Å)	$\frac{\tilde{\varepsilon}}{\kappa_B}$ (K)
Ar	Argon	3.542	93.3
He	Helium	2.551	10.22
Kr	Krypton	3.655	178.9
Ne	Neon	2.820	32.8
Xe	Xenon	4.047	231.0
Air	Air	3.711	78.6
CCl_4	Carbon tetrachloride	5.947	322.7
CF_4	Carbon tetrafluoride	4.662	134.0
CH_4	Methane	3.758	148.6
CO	Carbon monoxide	3.690	91.7
CO_2	Carbon dioxide	3.941	195.2
C_2H_2	Acetylene	4.033	231.8
C_2H_4	Ethylene	4.163	224.7
C_2H_6	Ethane	4.443	215.7
C_6H_6	Benzene	5.349	412.3
Cl_2	Chlorine	4.217	316.0
F_2	Fluorine	3.357	112.6
HCN	Hydrogen cyanide	3.630	569.1
HCl	Hydrogen chloride	3.339	344.7
HF	Hydrogen fluoride	3.148	330
HI	Hydrogen iodide	4.211	288.7
H_2	Hydrogen	2.827	59.7
H_2O	Water	2.641	809.1
H_2S	Hydrogen sulfide	3.623	301.1
Hg	Mercury	2.969	750
I_2	Iodine	5.160	474.2
NH_3	Ammonia	2.900	558.3
NO	Nitric oxide	3.492	116.7
N_2	Nitrogen	3.798	71.4
N_2O	Nitrous oxide	3.828	232.4
O_2	Oxygen	3.467	106.7
SO_2	Sulfur dioxide	4.112	335.4
UF_6	Uranium hexafluoride	5.967	236.8

[a] Based on Svehla (1962).

APPENDIX I

Collision Integrals for the Lennard-Jones (6–12) Potential Model[a]

$\frac{\kappa_B T}{\tilde{\varepsilon}}$	$\Omega_k = \Omega_\mu$	Ω_D	$\frac{\kappa_B T}{\tilde{\varepsilon}}$	$\Omega_k = \Omega_\mu$	Ω_D	$\frac{\kappa_B T}{\tilde{\varepsilon}}$	$\Omega_k = \Omega_\mu$	Ω_D
0.30	2.785	2.662	1.60	1.279	1.167	3.80	0.9811	0.8942
0.35	2.628	2.476	1.65	1.264	1.153	3.90	0.9755	0.8888
0.40	2.492	2.318	1.70	1.248	1.140	4.00	0.9700	0.8836
0.45	2.368	2.184						
0.50	2.257	2.066	1.80	1.221	1.116	4.20	0.9600	0.8740
0.55	2.156	1.966						
0.60	2.065	1.877	1.90	1.197	1.094	4.40	0.9507	0.8652
0.65	1.982	1.798	1.95	1.186	1.084	4.50	0.9464	0.8610
0.70	1.908	1.729	2.00	1.175	1.075	4.60	0.9422	0.8568
0.75	1.841	1.667	2.10	1.156	1.057	4.70	0.9382	0.8530
0.80	1.780	1.612	2.20	1.138	1.041	4.80	0.9343	0.8492
0.85	1.725	1.562	2.30	1.122	1.026	4.90	0.9305	0.8456
0.90	1.675	1.517	2.40	1.107	1.012	5.0	0.9269	0.8422
0.95	1.629	1.476	2.50	1.093	0.9996	6.0	0.8963	0.8124
1.00	1.587	1.439	2.60	1.081	0.9878	7.0	0.8727	0.7896
1.05	1.549	1.406	2.70	1.069	0.9770	8.0	0.8538	0.7712
1.10	1.514	1.375	2.80	1.058	0.9672	9.0	0.8379	0.7556
1.15	1.482	1.346	2.90	1.048	0.9576	10.0	0.8242	0.7424
						16.0	0.7683	0.6878
1.20	1.452	1.320	3.00	1.039	0.9490	20.0	0.7432	0.6640
						25	0.7198	0.6414
1.25	1.424	1.296	3.10	1.030	0.9406	30.0	0.7005	0.6232
						35	0.6854	0.6088
1.30	1.399	1.273	3.20	1.022	0.9328	40.0	0.6718	0.5960
1.35	1.375	1.253	3.30	1.014	0.9256	50.0	0.6504	0.5756
1.40	1.353	1.233	3.40	1.007	0.9186	60.0	0.6335	0.5596
1.45	1.333	1.215	3.50	0.9999	0.9120	70.0	0.6194	0.5464
						75.0	0.6140	0.5415
1.50	1.314	1.198	3.60	0.9932	0.9058	80.0	0.6076	0.5352
1.55	1.296	1.182	3.70	0.9870	0.8998	90.0	0.5973	0.5256
						100	0.5887	0.5180

[a] Based in part on Hirschfelder *et al.* (1954), and Bird *et al.* (2002).

APPENDIX J

Physical Constants

Universal gas constant:

$$R_u = 8314.3 \text{ J/kmol·K}$$
$$= 8.3143 \text{ kJ/kmol·K}$$
$$= 1545 \text{ lb}_f\text{ft/lbmol·°R}$$
$$= 8.205 \times 10^{-2} \text{ m}^3 \text{ atm/kmol·K}$$

Standard atmospheric pressure:

$$P = 101\,325 \text{ N/m}^2$$
$$= 101.325 \text{ kPa}$$
$$= 14.696 \text{ psi}$$

Standard gravitational acceleration:

$$g = 9.80665 \text{ m/s}^2$$
$$= 980.665 \text{ cm/s}^2$$
$$= 32.174 \text{ ft/s}^2$$

Atomic mass unit:

$$\text{amu} = 1.66043 \times 10^{-27} \text{ kg}$$

Avagadro's number:

$$N_{Av} = 6.022\,136 \times 10^{26} \text{ molecules/kmol}$$
$$= 6.022\,136 \times 10^{23} \text{ molecules/mol}$$

Boltzmann constant:

$$\kappa_B = 1.380\,658 \times 10^{-23} \text{ J/K}$$
$$= 1.380\,658 \times 10^{-16} \text{ erg/K}$$

Planck's constant:

$$h = 6.626\,08 \times 10^{-34} \text{ J}$$
$$= 6.626\,08 \times 10^{-27} \text{ erg}$$

Speed of light in vacuum:

$$C = 2.997\,92 \times 10^8 \text{ m/s}$$
$$= 2.997\,92 \times 10^{10} \text{ cm/s}$$

Stefan–Boltzmann constant:

$$\sigma = 5.670 \times 10^{-8} \text{ W/m}^2\text{·K}^4$$
$$= 1.712 \times 10^{-9} \text{ Btu/hr·ft}^2\text{·° R}^4$$

APPENDIX K

Unit Conversions

Density:

$$1 \, \text{kg/m}^3 = 10^{-3} \, \text{g/cm}^3$$
$$= 0.06243 \, \text{lb}_\text{m}/\text{ft}^3$$

Diffusivity:

$$1 \, \text{m}^2/\text{s} = 3.875 \times 104 \, \text{ft}^2/\text{hr}$$

Energy and work:

$$1 \, \text{J} = 107 \, \text{erg}$$
$$= 6.242 \times 10^{18} \, \text{eV}$$
$$= 0.2388 \, \text{cal}$$
$$= 9.4782 \times 10^{-4} \, \text{Btu}$$
$$= 0.7376 \, \text{lb}_\text{f} \, \text{ft}$$

Force:

$$1 \, \text{N} = 10^5 \, \text{dyn}$$
$$= 0.22481 \, \text{lb}_\text{f}$$

Heat flux:

$$1 \, \text{W/m}^2 = 0.3170 \, \text{Btu/hr·ft}^2$$
$$= 2.388 \times 10^{-5} \, \text{cal/cm}^2\text{·s}$$

Heat generation rate (volumetric):

$$1 \, \text{W/m}^3 = 0.096 \, 62 \, \text{Btu/hr·ft}^3$$

Heat transfer coefficient:

$$1 \, \text{W/m}^2\text{·K} = 0.17611 \, \text{Btu/hr· ft}^2\text{·°R}$$

Length:

$$1 \, \text{m} = 3.2808 \, \text{ft}$$
$$= 39.370 \, \text{in.}$$
$$= 10^6 \, \mu\text{m}$$
$$= 10^{10} \, \text{Å}$$
$$1 \, \text{mill} = 10^{-3} \, \text{in.}$$

Mass:

$$1\,kg = 10^3\,g$$
$$= 2.2046\,lb_m$$

Mass flow rate:

$$1\,kg/s = 7936.6\,lb_m/hr$$

Mass flux or mass transfer coefficient:

$$1\,kg/m^2{\cdot}s = 737.3\,lb_m/ft^2{\cdot}hr$$

Power:

$$1\,W = 10^{-3}\,kW$$
$$= 3.4121\,Btu/hr$$
$$= 1.341 \times 10^{-3}\,hp$$

Pressure or stress:

$$1\,N/m^2 = 1\,Pa = 10\,dyn/cm^2$$
$$= 10^{-5}\,bar$$
$$= 0.020\,885\,lb_f/ft^2$$
$$= 1.4504 \times 10^{-4}\,lb_f/in^2\ (psi)$$
$$= 7.501 \times 10^{-3}\,mm\,Hg$$
$$= 2.953 \times 10^{-4}\,in.\,Hg$$
$$1\,atm = 760\,torr$$

Specific enthalpy or internal energy:

$$1\,J/kg = 10^{-3}\,kJ/kg$$
$$= 4.299 \times 10^{-4}\,Btu/lb_m$$
$$= 2.393 \times 10^{-4}\,cal/g$$

Specific heat:

$$1\,J/kg{\cdot}K = 10^{-3}\,kJ/kg{\cdot}K$$
$$= 0.2388 \times 10^{-3}\,Btu/lb_m{\cdot}R$$
$$= 2.393 \times 10^{-4}\,cal/g{\cdot}K$$

Temperature:

$$T[K] = T[^\circ C] + 273.15[K]$$
$$T[^\circ R] = T[^\circ F] + 459.67[^\circ R]$$
$$1\,K = 1\,^\circ C = 1.8\,^\circ R = 1.8\,^\circ F$$

Thermal conductivity:

$$1\,W/m{\cdot}K = 0.57779\,Btu/hr{\cdot}ft{\cdot}^\circ{\cdot}R$$

Velocity:

$$1\,\text{m/s} = 3.28\,\text{ft/s}$$
$$= 3.600\,\text{km/hr}$$
$$1\,\text{km/hr} = 0.6214\,\text{mph}$$

Viscosity:

$$1\,\text{kg/m·s} = \text{N·s/m}^2$$
$$= 10\,\text{poise}$$
$$= 103\,\text{cp}$$
$$= 2419.1\,\text{lbm/ft·hr}$$
$$= 5.8015 \times 10^{-6}\,\text{lbf·hr/ft}^2$$
$$= 2.0886 \times 10^{-2}\,\text{lbf·s/ft}^2$$

Volume:

$$1\,\text{m}^3 = 103\,\text{L}$$
$$= 35.315\,\text{ft}^3$$
$$= 264.17\,\text{gal (US)}$$

References

Abd Raboh, M. A., Mostafa, H. M., Ali, M. A. A., and Hassan, A. M. (2011). Experimental study for condensation heat transfer inside helical coil. In *Evaporation, Condensation and Heat Transfer*, Ahsan, A., Ed., ISBN: 978-953-307-583-9, InTech, Available from: www.intechopen.com/books/evaporation-condensation-and-heat-transfer/experimental-study-forcondensation-heat-transfer-inside-helical-coil.

Abdelall, F., Hahn, G., Ghiaasiaan, S. M., Abdel-Khalik, S. I., Jeter, S. M., Yoda, M., and Sadowski, D. L. (2005). Pressure drop caused by abrupt flow area changes in microchannels, *Exp. Thermal Fluid Sci.*, **29**, 425–434.

Abdollahian, D., Chexal, B., and Norris, D. M. (1983). Prediction of leak rates through intergranular stress corrosion cracks. *Proc. CSNI Leak-Before-Break Conf.*, Nuclear Regulatory Commission Rep. NUREG/CP-00151, pp. 300–326.

Abdollahian, D., Healzer, J., Janssen, E., and Amos, C. (1982). Critical flow data review and analysis. Electric Power Research Institute Rep. EPRI NP-2192, Palo Alto, CA.

Abe, Y., Oka, T., Mori, Y., and Negashima, A. (1994). Pool boiling of a non-azeotropic mixture under microgravity. *Int. J. Heat Mass Transfer*, **37**, 2405–2413.

Ackerman, G. (1937). Heat transfer and molecular mass transfer in the same field at high temperatures and large partial pressure differences. *Forsch. Ing. Wes., VDI-Forschungesheft*, **8**, 232.

Acosta, R. E., Buller, R. H., and Tobias, C. W. (1985). Transport processes in narrow (capillary) channels. *AIChE J.*, **81**, 473–482.

Adams, T. M., Abdel-Khalik, S. I., Jeter, S. M., and Qureshi, Z. H. (1997). An experimental investigation of single-phase forced convection in microchannels. *Int. J. Heat Mass Transfer*, **41**, 851–857.

Adams, T. M., Ghiaasiaan, S. M., and Abdel-Khalik, S. I. (1999). Enhancement of liquid forced convection heat transfer in microchannels due to the release of dissolved noncondensables. *Int. J. Heat Mass Transfer*, **42**, 3563–3573.

Agarwal, A. (2006). Heat transfer and pressure drop during condensation of refrigerants in microchannels. PhD thesis, Georgia Institute of Technology, Atlanta.

Agarwal, A., Bandhauer, T. M., and Garimella, S. (2010). Measurement and modeling of condensation heat transfer in non-circular microchannels. *Int. J. Refrigeration*, **33**, 1169–1179.

Agostini, B., Bontemps, A., Watel, B., and Thonon, B. (2003). Boiling heat transfer in minichannels: influence of the hydraulic diameter. *Proc. Int. Congress of Refrigeration*, ICR0070, 2003.

Agostini, B., and Bontemps, A. (2005). Vertical flow boiling of refrigerant R134a in small channels. *Int. J. Heat Mass Transfer*, **26**, 296–306.

Agostini, B., Thome, J., Fabbri, M., Michel, B., Calmi, D., and Kloter, U. (2008a). High heat flux flow boiling in silicon multimicrochannels: part I. Heat transfer characteristics of refrigerant R236fa. *Int. J. Heat Mass Transfer*, **51**, 5400–5414.

Agostini, B., Thome, J. R., Fabbri, M., Michel, B., Calmi, D., and Kloter, U. (2008b). High heat flux flow boiling in silicon multi-microchannels: part II. Heat transfer characteristics of refrigerant R245fa. *Int. J. Heat Mass Transfer*, **51**, 5415–5425.

Agostini, B., Revellin, R., Thome, J. R., Fabbri, M., Michel, B., Calmi, D., and Kloter, U. (2008c). High heat flux flow boiling in silicon multi-microchannels: part III. Saturated critical heat flux of R-236fa and two-phase pressure drops. *Int. J. Heat Mass Transfer*, **51**, 5426–5442.

Ahmad, S. Y. (1970). Axial distribution of bulk temperature and void fraction in a heated channel with inlet subcooling. *Int. J. Heat Mass Transfer*, **92**, 595–609.

Ahmand, S. Y. (1973). Fluid to fluid modeling of critical heat flux: A compensated distortions model. *Int. J. Heat Mass Transfer*, **16**, 641–662.

Akbar, M. K., and Ghiaasiaan, S. M. (2006). Simulation of Taylor flow in capillaries based on the volume-of-fluid technique. *Ind. Eng. Chem. Res.*, **45**, 5396–5403.

Akbar, M. K., Plummer, D. A., and Ghiaasiaan, S. M. (2003). On gas–liquid two-phase flow regimes in microchannels. *Int. J. Multiphase Flow*, **29**, 855–865.

Akers, W. W., Deans, H. A., and Crosser, O. K. (1958). Condensing heat transfer within horizontal tubes. *Chem. Eng. Prog.*, **54**, 89–90.

Akers, W. W., Deans, H. A., and Crosser, O. K. (1959). Condensation heat transfer within horizontal tubes. *Chem. Eng. Prog. Symp. Ser.*, **55**, 171–176.

Alamgir, M. D., and Lienhard, J. H. (1981). Correlation of pressure undershoot during hotwater depressurization. *J. Heat Transfer*, **103**, 52–55.

Al-Diwani, H. K., and Rose, J. W. (1973). Free convection film condensation of steam in the presence of noncondensing gases. *Int. J. Heat Mass Transfer*, **16**, 1359–1369.

Alekseev, V. P., Poberezkin, A. E., and Gerasimov, P. V. (1972). Determination of flooding rates in regular packings. *Heat Transfer Sov. Res.*, **4**(6), 159–163.

Alekseev, V. P., Ilyukhin, Yu. N., Kukhtevich, V. O., Svetlov, S. V., and Sidoriv, V. G. (2000). Countercurrent flow of gas and water in perforated plates. *High Temperature*, **38**(5), 805–810.

Al-Hajeri, M. H., Koluib, A. M., Mosaad, M., and Al-Kulaib, S. (2007). Heat transfer performance during condensation of R-134a inside helicoidal tubes. *Energy Convers. Manage.*, **48**, 2309–2315.

Al-Hajeri, M. H., Koluib, A. M., Alajmi, R., and Kalim, S. P. (2009). Effect of coolant temperature on the condensation heat transfer in air-conditioning and refrigeration applications. *Exp. Heat Transfer*, **22**, 58–72.

Al-Hayes, R. A. M., and Winterton, R. H. S. (1981). Bubble diameter on detachment in flowing liquids. *Int. J. Heat Mass Transfer*, **24**, 223–230.

Ali, M. I., and Kawaji, M. (1991). The effect of flow channel orientation on two-phase flow in a narrow passage between flat plates. *Proc. ASME/JSME Thermal Eng. Conf.*, **2**, 183–190.

Ali, M. I., Sadatomi, M., and Kawaji, M. (1993). Adiabatic two-phase flow in narrow channels between two flat plates. *Can. J. Chem. Eng.*, **71**, 657–666.

Alimonti, C. (2014). Experimental characterization of globe and gate valves in vertical gas–liquid flows. *Exp. Therm. Fluid Sci.*, **54**, 259–266.

Alopaeus, V., Koshinen, J., and Keskinen, K. I. (1999). Simulation of the population balances for liquid–liquid systems in a nonideal stirred tank. Part 1: Description and qualitative validation of the model. *Chem. Eng. Sci.*, **54**, 5887–5899.

Aly, M. M. (1981). Flow regime boundaries for an interior subchannel of a horizontal 37 element bundle. *Can J. Chem. Eng.*, **59**, 158–163.

Ambrosini, W., Forgione, N., and Oriolo, F. (2002). Statistical characteristics of a water film falling down a flat plate at different inclinations and temperatures. *Int. J. Multiphase Flow*, **28**, 1521–1540.

Amos, C. N., and Schrock, V. E. (1984). Two-phase critical flow in slits. *Nucl. Sci. Eng.*, **88**, 261–274.

Andresen, U. (2006). Supercritical gas cooling and near-critical-pressure condensation of refrigerant blends in microchannels. PhD thesis, Georgia Institute of Technology, Atlanta, GA.

Ansari, M. R., and Nariai, H. (1989). Experimental investigation on wave initiation and slugging of air–water stratified flow in horizontal duct. *J. Nucl. Sci. Technol.*, **26**, 681–688.

Antal, S. P., Lahey, R. T., Jr., and Flaherty, J. E. (1991). Analysis of phase distribution in fully developed laminar bubbly two-phase flow. *Int. J. Multiphase Flow*, **17**, 635–652.

Ardron, K. H. (1978). A two-fluid model for critical vapor–liquid flow. *Int. J. Multiphase Flow*, **4**, 323–327.

Aria, H., Akhavan-Behabadi, M. A., and Shemirani, F. M. (2012). Experimental investigation on flow boiling heat transfer and pressure drop of HFC-134a inside a vertical helically coiled tube. *Heat Transfer Eng.*, **33**, 79–87.

Armand, A. A., and Treschev, G. G. (1946). The resistance during the movement of a two-phase system in horizontal pipes. *Izvestya Vsesoyuznogo Teplotekhnicheskogo Instituta*, **1**, 16–23.

Armand, A. A. (1959). The resistance during the movement of a two-phase system in horizontal pipes. *Atomic Energy Research Establishment (AERE) Libr. Trans.* **828**.

Asadolahi, A. N., Gupta, R., Fletcher, D. F., and Haynes, B. S. (2011). CFD approaches for the simulation of hydrodynamics and heat transfer in Taylor flow. *Chem. Eng. Sci.*, **66**, 5575–5584.

Asadolahi, A. N., Gupta, R., Leung, S. S. Y., Fletcher, D. F., and Haynes, B. S. (2012). Validation of a CFD model of Taylor flow hydrodynamics and heat transfer. *Chem. Eng. Sci.*, **69**, 541–552.

Asali, J. C., Hanratty, T. J., and Andreussi, P. (1985). Interfacial drag and film height for vertical annular flow. *AIChE J.*, **31**, 895–902.

Aussillous, P., and Quéré, D. (2000). Quick deposition of a fluid on the wall of a tube. *Phys. Fluids*, **12**, 2367–2371.

Awad, M. M., and Muzychka, Y. S. (2008). Effective property models for homogeneous twophase flows. *Exp. Therm. Fluid Sci.*, **33**, 106–113.

Awad, M. M., Dalkiliç, A. S., and Wongwises, S. (2014). A critical review on condensation heat transfer in microchannels and minichannels. *J. Nanotechnology in Engineering and Medicine*, **5**, paper # 010801, doi:10.1115/1.4028092.

Awwad, A., Xin, R. C., Dong, Z. F., Ebadian, M. A., and Soliman, H. M. (1995a). Flow patterns and pressure drop in air–water two-phase flow in horizontal helicoidal pipes. *J. Fluids Eng.*, **117**, 720–726.

Awwad, A., Xin, R. C., Dong, Z. F., Ebadian, M. A., and Soliman, H. M. (1995b). Measurement and correlation of the pressure drop in air–water two-phase flow in horizontal helicoidal pipes. *Int. J. Multiphase Flow*, **21**, 607–619.

Azer, N. H., and Said, S. A. (1986). Heat transfer analysis of annular flow condensation inside circular and semi-circular horizontal tubes. *ASHRAE Trans.*, **92**, 41–54.

Bailey, N. A. (1971). Film boiling on submerged vertical cylinders. UKAEA Rep. AEEW-M1051.

Baird, J. R., Fletcher, D. F., and Haynes, B. S. (2003). Local condensation heat transfer in fine passages. *Int. J. Heat Mass Transfer*, **46**, 4453–4466.

Baker, O. (1954). Simultaneous flow of oil and gas. *Oil Gas J.*, **53**, 185–195.

Balakrishnan, R., Santhappan, J. S., and Dhasan, M. L. (2009). Heat transfer correlation for a refrigerant mixture in a vertical helical coil evaporator. *Thermal Sci.*, **13**, 197–206.

Baldassari, C., and Marengo, M. (2013). Flow boiling in microchannels and microgravity. *Prog. Energy Combustion Sci.*, **39**, 1–36.

Bandhauer, T. M., Agarwal, A., and Garimella, S. (2006). Measurement and modeling of condensation heat transfer coefficient in circular microchannels. *J. Heat Transfer*, **128**, 1050–1059.

Banerjee, S., Rhodes, S., and Scott, D. S. (1967). Film inversion of co-current two-phase flow in helical coils. *AIChE J.*, **13**, 189–191.

Banerjee, S., Rhodes, E., and Scott, D. S. (1969). Studies on concurrent gas–liquid flow in helically coiled tubes – I. Flow pattern, pressure drop and holdup. *Can. J. Chem. Eng.*, **47**, 445–453.

Banerjee, S. (1980). Analysis of separated flow models. Electric Power Research Institute Rep. EPRI NP-1442, Palo Alto, CA.

Banerjee, S. B., and Chan, A. M. (1980). Separated flow models – I Analysis of the average and local instantaneous formulations. *Int. J. Multiphase Flow*, **6**, 1–24.

Bankoff, S. G. (1960). A variable density single-fluid model for two-phase flow with particular reference to steam–water flow. *Trans. ASME*, **82**, 265–272.

Bankoff, S. G. (1963). Asymptotic growth of a bubble in a liquid with uniform initial superheat. *Appl. Sci. Res.*, **12A**, 567.

Bankoff, S. G., and Lee, S. C. (1986). A critical review of the flooding literature. In *Multiphase Science and Technology*, Hewitt, G. F., Delhaye, J. M., and Zuber, N., Eds., Hemisphere, New York, Vol. 2, Chapter 2.

Bankoff, S. G., and Lee, S. C. (1987). Flooding and hysteresis effects in nearly-horizontal countercurrent stratified stead-water flow. *Int. J. Heat Mass Transfer*, **30**, 581–588.

Bankoff, S. G., Tankin, R. S., Yuen, M. C., and Hsieh, C. L. (1981). Countercurrent flow of air/water and steam/water through a horizontal perforated plate. *Int. J. Heat Mass Transfer*, **24**, 1381–1395.

Bao, Z.-Y., Bosnich, M. G., and Haynes, B. S. (1994). Estimation of void fraction and pressure drop for two-phase flow in fine passages. *Trans. Inst. Chem. Eng.*, **72A**, 625–532.

Bao, Z. Y., Fletcher, D. F., and Haynes, B. S. (2000). Flow boiling heat transfer of Freon R11 and HCFC123 in narrow passages. *Int. J. Heat Mass Transfer*, **43**, 3347–3358.

Bapat, P. M., and Tavlarides, L. L. (1985). Mass transfer in a liquid-liquid CFSTR. *AIChE J.*, **31**, 659–666.

Barajas, A. M., and Panton, R. L. (1993). The effect of contact angle on two-phase flow in capillary tubes. *Int. J. Multiphase Flow*, **19**, 337–346.

Barnea, D. (1986). Transition from annular and from dispersed bubble flow – Unified models for the whole range of pipe inclination. *Int. J. Multiphase Flow*, **12**, 733–744.

Barnea, D. (1987). A unified model for predicting flow-pattern transitions for the whole range of pipe inclinations. *Int. J. Multiphase Flow*, **13**, 1–12.

Barnea, D., Ben Yoseph, N., and Taitel, Y. (1986). Flooding in inclined pipes – Effects of entrance section. *Can. J. Chem. Eng.*, **64**, 177–184.

Barnea, D., Lulinkski, Y., and Taitel, Y. (1983). Flow in small diameter pipes. *Can. J. Chem. Eng.*, **61**, 617–620.

Barnea, D., Shoham, O., and Taitel, Y. (1982). Flow pattern transition for vertical downward two-phase flow. *Chem. Eng. Sci.*, **37**, 741–744.

Barnea, D., Shoham, O., and Taitel, Y. (1985). Gas–liquid flow in inclined tubes: Flow pattern transitions for upward flow. *Chem. Eng. Sci.*, **40**, 131–136.

Baroczy, C. J. (1963). Correlation of liquid fraction in two-phase flow with application to liquid metals. NAA-SR-8171.

Baroczy, C. J. (1965). Correlation of liquid fraction in two-phase flow with applications to liquid metals. *Chem. Eng. Prog. Symp. Ser.*, **61**, 179–191.

Basu, N., Warrier, G. R., and Dhir, V. K. (2002). Onset of nucleate boiling and active nucleation site density during subcooled flow boiling. *J. Heat Transfer*, **124**, 717–728.

Batchelor, G. K. (1970). *Theory of Homogeneous Turbulence*, Cambridge University Press, Cambridge, UK.

Baumeister, K. J., and Simon, F. F. (1973). Leidenfrost temperature – Its correlation for liquid metals, cryogens, hydrocarbons, and water. *J. Heat Transfer*, **95**, 166–173.

Beattie, D. R. H. (1973). A note on the calculation of two-phase pressure losses. *Nucl. Eng. Design*, **25**, 395–402.

Beattie, D. R. H., and Whalley, P. B. (1982). A simple two-phase frictional pressure drop calculation method. *Int. J. Multiphase Flow*, **8**, 83–87.

Beattie, D. R. H., and Sugawara, S. (1986). Steam–water void fraction for vertical upflow in a 73.9 mm pipe. *Int. J. Multiphase Flow*, **12**, 641–653.

Becker, K. M. (1971). Measurement of burnout conditions for flow of boiling water in horizontal round tubes. Rep. No. AERL-1262, Atomenergia-Aktieb, Sweden.

Bell, K. J., and Owhadi, A. (1969). Local heat transfer measurements during forced convection boiling in a helically coiled tube. *IMECHE*, **184**(3C), 52–58.

Benedek, S. (1976). Heat transfer at the condensation of steam on turbulent water jet. *Int. J. Heat Mass Transfer*, **19**, 448–450.

Benjamin, T. B. (1957). Wave formation in laminar flow down an inclined plane. *J. Fluid Mech.*, **2**, 554–574.

Bennett, D. L., and Chen, J. C. (1980). Forced convective boiling in vertical tubes for saturated pure components and binary mixtures. *AIChE J.*, **24**, 223–230.

Bennett, M. K., and Rohani, S. (2001). Solution of polulation balance equations with a new combined Lax–Wendroff/Cranck–Nicholson method. *Chem. Eng. Sci.*, **56**, 6623–6633.

Bennett, J. A. R., Collier, J. G., Pratt, H. R. C., and Thornton, D. (1961). Heat transfer to two-phase gas–liquid systems. Steam–water mixtures in the liquid-dispersed region in an annulus. *Trans. Inst. Chem. Eng.*, **39**, 113–126.

Benson, C. (2009). *Physical Chemistry*. Global Media, Delhi, India.

Bercic, G., and Pintar, A. (1997). The role of gas bubbles and liquid slug lengths on mass transport in the Taylor flow through capillaries. *Chem. Eng. Sci.*, **52**, 3709–3719.

Berenson, P. J. (1961). Film-boiling heat transfer from a horizontal surface. *J. Heat Transfer*, **83**, 351–358.

Bergles, A. E. (1962). Subcooled burnout in tubes of small diameter. ASME Paper 63-WA-182.

Bergles, A. E. (1978). Instabilities in two-phase flow. In *Two-Phase Flow and Heat Transfer in Power and Process Industries*, Bergles, A. E., Collier, J. G., Delhaye, J. M., Hewitt, G. F., and Mayinger, F., Eds., Hemisphere, Washington, DC.

Bergles, A. E., and Kandlikar, S. G. (2005). On the nature of critical heat flux in microchannels. *J. Heat Transfer*, **127**, 101–107.

Bergles, A. E., and Rohsenow, W. M. (1964). The determination of forced convection surface boiling heat transfer. *Int. J. Heat Mass Transfer*, **86**, 365–372.

Bergles, A. E. and Scarola, L. S. (1966). Effect of a volatile additive on the critical heat flux for surface boiling of water in tubes. *Chem. Eng. Science*, **21**, 721–723.

Berman, L. D., and Fuks, S. N. (1958). Mass transfer in condensers with horizontal tubes when the steam contains air. *Teploenergetika*, **5**(8), 66–74.

Bertsch, S. S., Groll, E. A., and Garimella, S. V. (2008). Refrigerant flow boiling heat transfer in parallel microchannels as a function of local vapor quality. *Int. J. Heat Mass Transfer*, **51**, 4775–4787.

Bertsch, S. S., Groll, E. A., and Garimella, S. V. (2009a). Effect of heat flux, mass flux, vapor quality, and saturation temperature on flow boiling heat transfer in microchannels. *Int. J. Multiphase Flow*, **35**, 142–154.

Bertsch, S. S., Groll, E. A., and Garimella, S. V. (2009b). A composite heat transfer correlation for saturated flow boiling in small channels. *Int. J. Heat Mass Transfer*, **52**, 2110–2118.

Berthoud, G., and Jayanti, S. (1990). Characterization of dryout in helical coils. *Int. J. Heat Mass Transfer*, **33**, 1451–1463.

Bertoletti, S., Gaspari, G. P., Lombardi, C., Peterlongo, G., Silvestri, M., and Tacconi, F. A. (1965). Heat transfer crisis with steam water mixtures. *Energia Nucleare*, **12**, 121–172.

Bestion, D. (1990). The physical closure laws in the CATHARE code. *Nucl. Eng. Design* **124**, 229–245.

Bhaga, D. and Weber, M. E. (1981). Bubbles in viscous liquids: shapes, wakes and velocities. *J. Fluid Mech.*, **105**, 61–85.

Bhagwat, S. M., and Ghajar, A. (2014). A flow pattern independent drift flux model based void fraction correlation for a wide range of gas–liquid two phase flow. *Int. J. Multiphase Flow*, **59**, 186–205.

Bharathan, D., and Wallis, G. B. (1983). Air–water countercurrent annular flow. *Int. J. Multiphase Flow*, **9**, 349–366.

Bharathan, D., Wallis, G. B., and Richter, H. J. (1979). Air–water countercurrent annular flow. Electric Power Research Institute Rep. EPRI NP-1165, Palo Alto, CA.

Bohdal, T., Charun, H., and Sikora, M. (2011). Comparative investigations of the condensation of R134a and R404A refrigerants in pipe minichannels. *Int. J. Heat Mass Transfer*, **54**, 1963–1974.

Bibeau, E. L., and Salcudean, M. (1990). The effect of flow direction on void growth at very low velocities and low pressure. *Int. Comm. Heat Mass Transfer*, **17**, 19–25.

Bibeau, E. L., and Salcudean, M. (1994a). Subcooled voidage growth mechanisms and prediction at low pressure and low velocity. *Int. J. Multiphase Flow*, **20**, 837–863.

Bibeau, E. L., and Salcudean, M. (1994b). A study of bubble ebullition in forced-convective subcooled nucleate boiling at low pressure. *Int. J. Heat Mass Transfer*, **37**, 2245–2259.

Binnie, A. M. (1957). Experiments on the onset of wave formation on a film of water flowing down a vertical plate. *J. Fluid Mech.*, **2**, 551–553.

Bird, R. B., Stewart, W. E., and Lightfoot, E. N. (2002). *Transport Phenomena*, 2nd Edn, Wiley, New York.

Birkhoff, G., Margulies, R. S., and Horning, W. A. (1958). Spherical bubble growth. *Phys. Fluids*, **1**, 201.

Bjonard, T. A., and Griffith, P. (1977). PWR blowdown heat transfer. In *Symposium on Thermal and Hydraulic Aspects of Nuclear Reactor Safety*, Jones, O. C., and Bankoff, S. G., Eds., ASME, New York, Vol. 1.

Bjorge, R. W., Hall, G. R., and Rohsenow, W. M. (1982). Correlation of forced convection boiling heat transfer data. *Int. J. Heat Mass Transfer*, **25**, 753–757.

Blasick, A. M., Dowling, M. F., Abdel-Khalik, S. I., Ghiaasiaan, S. M., and Jeter, S. M. (2002). Onset of flow instability in uniformly-heated thin horizontal annuli. *Exp. Thermal Fluid Sci.*, **26**, 1–14.

Blasius, P. R. H. (1913). Das Ähnlichkeitsgesetz bei Reibungsvorgängen in Flüssigkeiten. *Forschungsheft*, **131**, 1–41.

Blinkov, V. N., Jones, O. C., Jr. and Nigmatulin, B. I. (1993). Nucleation and flashing in nozzles – 2 Comparison with experiments using a five-equation model for vapor void development. *Int. J. Multiphase Flow*, **19**, 965–986.

Block, J. A., and Crowley, C. J. (1976). Effect of steam upflow and superheated walls on ECC delivery in a simulated multiloop PWR geometry. CREARE Rep. TN-2110, Creare, NH.

Bolinder, C. J., and Sundén, B. (1995). Flow visualization and LDV measurement of laminar and flow in a helical square duct with finite pitch. *Exp. Thermal Fluid Sci.*, **11**, 348–363.

Bolstad, M. M., and Jordan, R. C. (1948). Theory and use of the capillary tube expansion device. *Refrig. Eng.*, **56**, 519.

Bonnecaze, R. H., and Erskine, W., Jr., Greskovich, E.J. (1971). Holdup and pressure drop for two phase slug flow in inclined pipelines. *AIChE*, **17**, 1109–1113.

Bonjour, J., and Lallemand, M. (1998). Flow patterns during boiling in a narrow space between two vertical surfaces. *Int. J. Multiphase Flow*, **24**, 947–960.

Borishanskiy, V. M., *et al.* (1977). Effect of uncondensable gas content on heat transfer in steam condensation in a vertical tube. *Heat Transfer Sov. Res.*, **9**(2), 35–42.

Borishanskiy, V. M., *et al.* (1978). Heat transfer from steam condensing inside vertical pipes and coils. *Heat Transfer Sov. Res.*, **10**(4), 44–58.

Bortolin, S. (2010). Two-Phase Heat Transfer Inside Minichannels, PhD thesis, University of Padua, Italy.

Bose, F., Ghiaasiaan, S. M., and Heindel, T. J. (1997). Hydrodynamics of dispersed liquid droplets in agitated synthetic fibrous slurries. *Ind. Eng. Chem. Res.*, **36**, 5028–5039.

Bouré, J. A. (1977). The critical flow phenomena with reference to two-phase flow and nuclear reactor systems. *Proc. ASME Symp. Thermal-Hydraulic Aspects of Nuclear Reactor Safety*, ASME, New York, pp. 195–216.

Bouré, J. A., and Delhaye, J. M. (1981). Chapter 12. In *Thermohydraulics of Two-Phase Systems for Industrial Design and Nuclear Engineering*, Delhaye, M., Giot, M., Riethermuller, L. M., Eds., Hemisphere, Washington, DC, pp. 353–403.

Bouré, J. A., Bergles, A. E., and Tong, L. S. (1973). Review of two-phase flow instability. *Nucl. Eng. Design*, **25**, 165–192.

Bousman, W. S., McQuillen, J. B., and Witte, L. C. (1996). Gas–liquid flow patterns in microgravity: Effects of tube diameter, liquid viscosity and surface tension. *Int. J. Multiphase Flow*, **22**, 1035–1053.

Bowers, M. B., and Mudawar, I. (1994). High flux boiling in low flow rate, low pressure drop mini-channel and micro-channel heat sinks. *Int. J. Heat Mass Transfer*, **37**, 321–332.

Bowring, R. W. (1972). A simple but accurate round tube, uniform heat flux, dryout correlation over the pressure range 0.7–17 MN/m² (100–2500 psia). UKAEA Rep. AEEW-R789, Winfrith, England.

Bowring, R. W. (1979). WSC-2: A subcooled dryout correlation for water-cooled clusters over the pressure range 3.4–15.9 MPa (500–2300 PSIA). UKAEA Rep. AEEW–R983,Winfrith, England.

Bowring, R. W. (1962). Physical model of bubble detachment and void volume in subcooled boiling. OECD Halden Reactor Project Rep. HPR-10.

Boyce, B. E., Collier, J. G. and Levy, J. (1969). Hold-up and pressure drop measurement in the two-phase flow of air–water mixing tubes in helical coils, *Proc. Int Symp. on Research in Co-current Gas and Liquid Flow*, Rhodes, E., and Scott, D. S., Eds., Plenum Press, New York.

Boyd, R. D. (1985a). Subcooled flow boiling critical heat flux (CHF) and its application to fusion energy components. Part I. A review of fundamentals of CHF and related data base. *Fusion Technol.*, **7**, 1–30.

Boyd, R. D. (1985b). Subcooled flow boiling critical heat flux (CHF) and its application to fusion energy components. Part II. A review of microconvective, experimental, and correlational aspects. *Fusion Technol.*, **7**, 31–52.

Boyd, R. D. (1988). Subcooled water flow boiling experiments under uniform high heat flux conditions. *Fusion Technol.*, **13**, 131–142.

Boyd, R. D. (1990). Subcooled water flow boiling transition and the L/D effect on CHF for a horizontal uniformly heated tube. *Fusion Technol.*, **18**, 317–324.

Brauer, H. (1956). *Strömung und Wärmeübertragung bei Rieselfilmen*, VDI-Verlag, Düsseldorf, p. 457.

Braum, B., Ikier, C., and Klein, H. (1993). Thermocapillary migration of droplets in a binary mixture with miscibility gap during liquid–liquid phase separation under reduced gravity. *J. Colloid Interface Sci.*, **159**, 515–516.

Brauner, N. (1989). Modeling of wavy flow in turbulent free falling films. *Int. J. Multiphase Flow*, **15**, 505–520.

Brauner, N., and Molem Maron, D. (1982). Characteristics of inclined thin films, waviness and the associated mass transfer. *Int. J. Heat Mass Transfer*, **25**, 99–110.

Brauner, N., and Molem Maron, D. (1983). Modeling of wavy flow in inclined thin films. *Chem. Eng. Sci.*, **38**, 775–788.

Brauner, N., and Moalem Maron, D. (1992). Identification of the range of 'small diameter' conduits, regarding two-phase flow pattern transitions. *Int. Comm. Heat Mass Transfer*, **19**, 29–39.

Breber, D., Palen, J. W., and Taborek, J. (1980). Prediction of horizontal tubeside condensation of pure components using flow regime criteria. *J. Heat Transfer*, **102**, 471–476.

Breen, B. P., and Westwater, J. W. (1962). Effect of diameter of horizontal tubes on film boiling heat transfer. *Chem. Eng. Prog.*, **58**(7), 67.

Bretherton, F. B. (1961). The motion of long bubbles in tubes. *J. Fluid Mech.*, **10**, Part 2, 166–188.

Brodkey, R. S. (1967). *The Phenomena of Fluid Motion*, Addison-Wesley, Reading, MA.

Bromley, L. A. (1950). Heat transfer in stable film boiling. *Chem. Eng. Prog. Symp. Ser.*, **46**, 221–227.

Brötz, W. (1954). Über die Vorausberechnung der Absorptions geschwindigkeit von Gasen in strömenden Flüssigkeiter. *Chem. Eng. Technol.*, **26**, 470–478.

Brouwers, H. J. H. (1991). An improved tangency condition for fog formation in coolercondensers. *Chem. Eng. Sci.*, **47**, 3023–3036.

Brouwers, H. J. H. (1992). A film model for heat and mass transfer with fog formation. *Int. J. Heat Mass Transfer*, **34**, 2387–2394.

Brouwers, H. J. H. (1996). Effect of fog formation on turbulent vapor condensation with non-condensable gases. *J. Heat Transfer*, **118**, 243–245.

Brutin, D., Topin, F., and Tadris, L. (2003). Experimental study of unsteady convective boiling in heated minichannels. *Int. J. Heat Mass Transfer*, **46**, 2957–2965.

Brutin, D., Topin, F., and Tadris, L. (2004). Pressure drop and heat transfer analysis of flow boiling in a minichannel: Influence of the inlet condition on two-phase flow stability. *Int. J. Heat Mass Transfer*, **47**, 2365–2377.

Bui, T. D., and Dhir, V. K. (1985). Transition boiling heat transfer on a vertical surface. *J. Heat Transfer*, **107**, 756–763.

Butterworth, D. (1975). A comparison of some void-fraction relationships for co-current gas–liquid flow. *Int. J. Multiphase Flow*, **1**, 845–850.

Butterworth, B. (1983). Condensation of vapor mixtures, in Schlunder, E. U., Ed., *Heat Exchanger Design Handbook*, Vol. **2**, pp. 2.6.3.1–10, Hemisphere Publishing Corp.

Butterworth, D. (1988). Condensers and their design. In *Two-Phase Flow Heat Exchangers*, Kakac, S., Bergles, A. E., and Oliveira Fernandes, E., Eds., Kluwer, Dordrecht, pp. 779–828.

Butuzov, A., Bezrodnyy, M. K., and Pustovit, M. M. (1975). Hydraulic resistance and heat transfer in forced flow in rectangular coiled tubes. *Heat Transfer–Soviet Res.*, **7**(4), 84–88.

Caira, M., Caruso, G., and Naviglio, A. (1995). A correlation to predict CHF in subcooled flow boiling. *Int. Comm. Heat Mass Transfer*, **22**, 35–45.

Calderbank, P. H., and Korchinski, I. J. O. (1956). Circulation in liquid drops. *Chem. Eng. Sci.*, **6**, 65–78.

Calus, W. F., and Leonidopoulos, D. J. (1974). Pool boiling – Binary liquid mixtures. *Int. J. Heat Mass Transfer*, **17**, 249–256.

Cammenga, H. K., Schulze, F. W., and Theuerl, W. (1977). Vapor pressure and evaporation coefficient of water. *J. Chem Eng. Data*, **22**, 131–134.

Camp, D. W., and Berg, J. C. (1987). The spreading of oil on water in surface-tension regime. *J. Fluid Mech.*, **184**, 445–462.

Carey, V. P. (2008). *Liquid–Vapor Phase-Change Phenomena*, 2nd Edn, CRC Press.

Carey, V. P (1999). *Statistical Thermodynamics and Microscale Thermophysics*, Cambridge University Press, Cambridge.

Carne, M. (1964). Studies of the critical heat flux for some binary mixtures and their components. *Can. Chem. Eng. J.*, **41**, 235–241.

Catton, I., Ghiaasiaan, S. M., and Duffey, R. B. (1988). Multi-dimensional thermal-hydraulics and two-phase phenomena during quenching of hot rod bundles. In *Transient Phenomena in Multiphase Flow*, Afgan, N. H., Ed., Hemisphere Publishing Corp., Washington DC, pp. 491–525.

Cavallini, A., Censi, G., Del Col, D., Doretti, L., Longo, G. A., and Rossetto, L. (2001). Experimental investigation on condensation of new HFC refrigerants (R134a, R125, R32, R410A, R236ea) in a horizontal smooth tube. *Int. J. Refrigeration*, **24**, 73–87.

Cavallini, A., Censi, G., Del Col, D., Doretti, L., Longo, G. A., and Rossetto, L. (2002). Condensation of halogenated refrigerants inside smooth tubes. *Heating, Ventilation, Air Conditioning and Refrigeration*, **8**, 429–451.

Cavallini, A., Censi, G., Del Col, D., Doretti, L., Longo, G. A., Rossetto, L., and Zilio, C. (2003). Condensation inside and outside smooth and enhanced tubes – A review of recent research. *Int. J. Refrigeration*, **26**, 373–392.

Cavallini, A., Del Col, D., Doretti, L., Matkovic, M., Rossetto, L., and Zilio, C. (2005). Two-phase frictional pressure gradient of R236ea, R134a and R410A inside multi-port minichannels. *Exp. Thermal-Fluid Sci.*, **29**, 861–870.

Cavallini, A., Doretti, L., Matkovic, M., and Rossetto, L. (2006). Update on condensation heat transfer and pressure drop inside minichannels. *Heat Transfer Eng.*, **27**, 74–87.

Cavallini, A., Del Col, D., Mancin, S., and Rossetto, L. (2009). Condensation of pure and near-azeotropic refrigerants in microfin tubes: A new computational procedure. *Int. J. Refrigeration*, **32**, 162–174.

Cavallini, A., Del Col, D., and Rossetto, L. (2013). Heat transfer and pressure drop of natural refrigerants in minichannels (low charge equipment). *Int. J. Refrigeration*, **36**, 287–300.

Celata, G. P., Cumo, M., Farello, G. E., and Incalcaterra, P. C. (1983). Critical flows of subcooled liquid and jet forces. Presented at ASME – JSME National Heat Transf. Conf., Seattle, Washington.

Celata, G. P., Cumo, M., D'Annibale, G. E., Farello, G. E., and Setaro, T. (1986). Flow transient experiments with refrigerant – 12. *Revue Générale de Thermique*, **25**, 513–519.

Celata, G. P. (1993). Recent achievements in the thermal-hydraulics of high heat flux components in fusion reactors. *Exp. Thermal. Fluid Sci.*, **7**, 263–278.

Celata, G. P., Cumo, M., D'Annibale, F., and Farello, G. E. (1988). The influence of noncondensible gas on two-phase critical flow. *Int. J. Multiphase Flow*, **14**, 175–187.

Celata, G. P., Cumo, M., and Setaro, T. (1992). Flooding in inclined pipes with obstructions. *Exp. Heat Transfer*, **5**, 18–25.

Celata, G. P., Cumo, M., Farello, G. E., and Focardi, G. (1989). A comprehensive analysis of direct contact condensation of saturated steam on subcooled liquid jets. *Int. J. Heat Mass Transfer*, **32**, 639–654.

Celata, G. P., Cumo, M., D'Annibale, F., and Farello, G. E. (1991). Direct contact condensation of steam on droplets. *Int. J. Multiphase Flow*, **17**, 191–211.

Celata, G. P., Cumo, M., and Mariani, A. (1993a). Burnout in highly subcooled water flow boiling in small diameter tubes. *Int. J. Heat Mass Transfer*, **36**, 1269–1285.

Celata, G. P., Cumo, M., and Setaro, T. (1993b). Flooding in inclined pipes with obstructions. *Exp. Thermal Fluid Sci.*, **5**, 18–25.

Celata, G. P., Cumo, M., Mariani, A., Nariai, H., and Inasaka, F. (1993c). Influence of channel diameter on subcooled flow boiling burnout at high heat fluxes. *Int. J. Heat Mass Transfer*, **36**, 3407–3409.

Celata, G. P., Cumo, M., and Mariani, A. (1994a). Assessment of correlations and models for the prediction of CHF in water subcooled flow boiling. *Int. J. Heat Mass Transfer*, **37**, 237–255.

Celata, G. P., Cumo, M., Mariani, A., Simoncini, M., and Zummo, G. (1994b). Rationalization of existing mechanistic models for the prediction of water subcooled flow boiling critical heat flux. *Int. J. Heat Mass Transfer*, **37**, Suppl. 1, 347–360.

Celata, G. P., Cumo, M., and Setaro, T. (1994c). A review of pool and forced convective boiling of binary mixtures. *Exp. Thermal Fluid Sci.*, **9**, 367–381.

Celata, G. P., Cumo, M., and Setaro, T. (1994d). Critical heat flux in upflow convective boiling of refrigerant binary mixtures. *Int. J. Heat Mass Transfer*, **37**, 1143–1153.

Celata, G. P., Mishima, K., and Zummo, G. (2001). Critical heat flux prediction for saturated flow boiling of water in vertical tubes. *Int. J. Heat Mass Transfer*, **44**, 4323–4331.

Chen, T., and Garimella, S. V. (2012). A study of critical heat flux during flow boiling in microchannel heat sinks. *J. Heat Transfer*, **134**, paper 011504.

Červenka, J., and Kolář, V. (1973). Hydrodynamics of plate columns. X. Analysis of operation of sieve plate without downcomers. *Coll. Czech. Chem. Commun.*, **38**, 3749–3761.

Chalfi, T.K., and Ghiaasiaan, S.M. (2008). Pressure drop caused by flow area changes in capillaries under low flow conditions. *Int. J. Multiphase Flow*, **34**, 2–12.

Chandrasekhar, S. (1961). *Hydrodynamic and Hydromagnetic Stability*, Cambridge University Press, Cambridge.

Chang, T.-H., and Chung, J. N. (1985). The effects of surfactants on the motion and transport mechanisms of a condensing droplet in a high Reynolds number flow. *AIChE J.*, **31**, 1149–1156.

Chang, J. Y., and You, S. M. (1996). Heater orientation effect on pool boiling of micro-porous enhanced surface in saturated FC-72. *J. Heat Transfer*, **118**, 937–943.

Chapman, S., and Cowling, T. G. (1970). *The Mathematical Theory of Non-uniform Gases*, 3rd Edn, Cambridge University Press, Cambridge.

Chedester, R. C., and Ghiaasiaan, S. M. (2002). Aproposed mechanism for hydrodynamically-controlled onset of significant void in microchannels. *Int. J. Heat Fluid Flow*, **23**, 769–775.

Chen, J. C. (1966). Correlation for boiling heat transfer to saturated fluids in convective flow. *Ind. Eng. Chem. Process Des. Dev.*, **5**, 322–329.

Chen, C.-N., Han, J.-T., Jen, T.-C., and Shao, L. (2011a). Thermo-chemical characteristics of R134a flow boiling in helically coiled tubes at low mass flux and low pressure. *Thermochimica Acta*, **512**, 163–169.

Chen, C.-N., Han, J.-T., Jen, T.-C., Shao, L., and Chen, C.-Wen. (2011b). Experimental study on critical heat flux characteristics of R134a flow boiling in horizontal helically-coiled tubes. *Int. J. Thermal Sci.*, **50**, 169–177.

Chen, I. Y., Wang, C.-C., and Lin, S. Y. (2004). Measurement and correlations of frictional single-phase and two-phase pressure drops of R-410A flow in small U-type return bends. *Int. J. Heat Mass Transfer*, **47**, 2241–2249.

Chen, I. Y., Yang, K. S., Chang, Y. J., and Wang, C. C. (2001). Two-phase pressure drop of air–water and R-410a in small horizontal tubes. *Int. J. Multiphase Flow*, **27**, 1293–1299.

Chen, J. C., Ozkaynak, F. T., and Sundaram, R. K. (1979). Vapor heat transfer in post-CHF region including the effect of thermodynamic non-equilibrium. *Nucl. Eng. Design*, **51**, 143–155.

Chen, S. L., Gerner, F. M., and Tien, C. L. (1987). General film condensation correlations. *Exp. Heat Transfer*, **1**, 93–107.

Chen, S. W., Liu, Y., Hibiki, T., Ishii, M., Yoshida, Y., Kinoshita, I., Murase, M., and Mishima, K. (2012). One-dimensional drift-flux model for two-phase flow in pool rod bundle systems. *Int. J. Multiphase Flow*, **40**, 166–177.

Chen, T., and Garimella, S. V. (2006). Measurements and high-speed visualizations of flow boiling of a dielectric fluid in a silicon microchannel heat sink. *Int. J. Multiphase Flow*, **32**, 957–971.

Chen, W. C., Klausner, J. F., and Mei, R. (1995). A simplified model for predicting vapor bubble growth rates in heterogeneous boiling. *J. Heat Transfer*, **117**, 976–980.

Chen, X.-J., and Zhang, M.-Y. (1984). An investigation of flow pattern transitions for gas–liquid two-phase flow in helical coils. Veziroglu, T. N., and Bergles, A. E., Eds., *Multiphase Flow and Heat Transfer III. Part III: Fundamentals*, pp. 185–200, Elsevier.

Chen, Y., and Cheng, P. (2005). Condensation of steam in silicon microchannels. *Int. Comm. Heat Mass Transfer*, **32**, 175–183.

Chen, Y., Kulenovic, R., and Mertz, R. (2009). Numerical study on the formation of Taylor bubbles in capillary tubes. *Int. J. Therm. Sci.*, **48**, 234–242.

Cheng, S. C., Wong, Y. L., and Groeneveld, D. C. (1988). CHF prediction for horizontal flow. *Proc. Int. Symp. on Phase Change Heat Transfer*, Chonqing, China, pp. 211–215.

Cheng, L., and Mews, D. (2006). Review of two-phase flow and boiling of mixtures in small and mini channels. *Int. J. Multiphase Flow*, **32**, 183–207.

Cheng, L., and Wu. H. Y. (2006). Mesoscale and microscale phase change heat transfer. *Int. J. Heat Mass Transfer*, **39**, 461–463.

Cheng, X. (2005). Experimental studies on critical heat flux in vertical tight 37-rod bundles using Freon-12. *Int. J. Multiphase Flow*, **31**, 1198–1219.

Chexal, B., Abdollahian, D., and Norris, D. (1984). Analytical prediction of single-phase and two-phase flow through cracks in pipes and tubes. *AIChE Symp. Ser.*, **80**(236), 19–23.

Chexal, B., Lelluche, G., Horowitz, J., Healzer, J., and Oh, S. (1991). The Chexal–Lelluche void fraction correlation for generalized Applications. Electric Power Research Institute, Rep. NSAC-139, Palo Alto, CA.

Chexal, B., Merilo, M., Maulbetsch, M., Horowitz, J., Harrison, J., Westacott, J., Peterson, C., Kastner, W., and Schmidt, H. (1997). Void fraction technology for design and analysis, Electric Power Research Institute, Palo Alto, CA.

Chisholm, D. (1967). A theoretical basis for the Lockhart–Martinelli correlation for two-phase flow. *Int. J. Heat Mass Transfer*, **10**, 1767–1778.

Chisholm, D. (1972). An equation for velocity ratio in two-phase flow. NEL Rep. 535.

Chisholm, D. (1973). Pressure gradients due to friction during the flow of evaporating two-phase mixture in smooth tubes and channels. *Int. J. Heat Mass Transfer*, **16**, 347–358.

Chisholm, D. (1980). Two-phase pressure drops in bends. *Int. J. Multiphase Flow*, **6**, 363–367.

Chisholm, D. (1981). Modern developments in marine condensers: Noncondensable gases: An overview. In *Power Condenser Heat Transfer Technology*, Marto, P. J., and Nunn, R. H., Eds., Hemisphere Publishing Corp., New York, pp. 95–142.

Chisholm, D. (1983). *Two Phase Flow in Pipelines and Heat Exchangers*, Longman, New York.

Chisholm, D. (1985). Two-phase flow in heat exchangers and pipelines. *Heat Transfer Eng.*, **6**, 48–57.

Chisholm, D., and Laird, A. D. K. (1958). Two-phase flow in rough tubes. *Trans. ASME*, **80**, 276–283.

Cho, C., Irvine, T. F., Jr., and Karni, J. (1992). Measurement of the diffusion coefficient of naphthalene into air. *Int. J. Heat Mass Transfer*, **35**, 957–966.

Choi, C. W., Yu, D. I., and Kim, M. K. (2011). Adiabatic two-phase flow in rectangular microchannels with different aspect ratios: Part I – Flow pattern, pressure drop and void fraction. *Int. J. Heat Mass Transfer*, **54**, 616–624.

Choi, J., Pereyra, E., Sarica, C., Park, C., and Kang, J. M. (2012). An efficient drift flux closure relationship to estimate liquid holdups of gas–liquid two-phase flow in pipes. *Energies*, **5**, 5294–5306.

Choi, K.-I., Pamitran, A. S., Oh, C.-Y., and Oh, J.-T. (2008). Two-phase pressure drop of R-410A in horizontal smooth minichannels. *Int. J. Refrigeration*, **31**, 119–129.

Choi, S. B., Barron, R. F., and Warrington, R. O. (1991). Fluid flow and heat transfer in micro-tubes. *Proc. ASME 1991 Winter Annual Meeting*, DSC-Vol. 32, pp. 123–134, ASME, New York.

Chun, K. R., and Seban, R. A. (1971). Heat transfer to evaporating liquid films. *J. Heat Transfer*, **93**, 391–396.

Chung, P. M.-Y., and Kawaji, M. (2004). The effect of channel diameter on adiabatic two-phase flow characteristics in microchannels. *Int. J. Multiphase Flow*, **30**, 735–761.

Chung, P. M.-Y., Kawaji, M., Kawahara, A., and Shibata, Y. (2004). Two-phase flow through square and circular microchannels – Effects of channel geometry. *J. Fluids Eng.*, **126**, 546–552.

Churchill, S. W. (1977). Frictional equation spans all fluid flow regimes. *Chem. Eng.*, **84**, 91–92.

Churchill, S. W., and Bernstein, M. (1977). Correlating equation for forced convection from gases and liquids to a circular cylinder in crossflow. *J. Heat Transfer*, **99**, 300–306.

Cicchitti, A., Lombardi, C., Silvestri, M., Solddaini, G., and Zavalluilli, R., (1960). Two-phase cooling experiments – Pressure drop, heat transfer and burnout measurement. *Energia Nucleare*, **7**, 407–425.

Cioncolini, A., and Santini, L. (2006). An experimental investigation regarding the laminar to turbulent flow transition in helically coiled pipes. *Exp. Thermal Fluid Sci.*, **30**, 367–380.

Cioncolini, A., and Thome, J. R. (2012). Void fraction prediction in annular two phase flow. *Int. J. Multiphase Flow*, **43**, 72–84.

Clark, N. N., and Flemmer, R. L. (1985). Predicting the holdup in two phase bubble upflow and downflow using the Zuber and Findlay drift flux model. *AIChE J.*, **31**, 500–503.

Clark, C., Griffiths, M., Chen, S.-W., Hibiki, T., Ishii, M., Ozaki, T., Kinoshita, I., and Yoshida, Y. (2014). Drift-flux correlation for rod bundle geometries. *Int. J. Heat Fluid Flow*, **48**, 1–14.

Clift, R., Grace, J. R., and Weber, M. E. (1978). *Bubbles, Drops and Particles*, Academic Press, New York.

Coddington, P., and Macian, R. (2002). A study of the performance of void fraction correlations used in the context of drift-flux two-phase flow models. *Nucl. Eng. Design*, **215**, 199–216.

Cole, R., and Shulman, H. L. (1966). Bubble departure diameters at subatmospheric pressures. *Chem. Eng. Symp. Ser.* **62**(64), 6–16.

Colebrook, C. R. (1939). Turbulent flow in pipes with particular reference to the transition region between the smooth and rough pipe laws. *J. Inst. Civil Eng.*, **11**, 133–156.

Coleman, J. W., and Garimella, S. (1999). Characteristics of two-phase flow patterns in small diameter round and rectangular tubes. *Int. J. Heat Mass Transfer*, **42**, 2869–2881.

Coleman, J. W., and Garimella, S. (2003). Two-phase flow regimes in round, square and rectangular tubes during condensation of refrigerant R134a. *Int. J. Refrigeration*, **26**, 117–128.

Collier, J. G. (1981). Forced convection boiling. In *Two-Phase Flow and Heat Transfer in Power and Process Industries*, Bergles, A. E., Collier, J. G., Delhaye, J. M., Hewitt, G. F., and Mayinger, F., Eds., Hemisphere, Washington, DC.

Collier, J. G., and Thome, J. R. (1994). *Convective Boiling and Condensation*, 3rd Edn, Clarendon Press, Oxford, England.

Collier, R. P., and Norris, D. M. (1983) Two-phase flow experiments through intergranular stress corrosion cracks. *Proc. CSNI Specialist Meeting on Leak-Before Break in Nuclear Reactor Piping*, U.S. Nuclear Regulatory Commission Rep. NUREG/CP-005, pp. 273–299.

Collier, R. P., Stuben, F. B., Mayfield, M. E., Pope, D. B., and Scott, P. M. (1984). Two-phase flow through intergranular stress corrosion cracks. Electric Power Research Institute Rep. EPRI-NP-3540-LD, Palo Alto, CA.

Colorado, D., Papini, D., Hernandez, J. A., Santini, L., and Ricotti, L. E. (2011). Development and experimental validation of a computational model for a helically coiled steam generator. *Int. J. Thermal Sci.*, **50**, 569–580.

Comish, R. J. (1928). Flow in a pipe of rectangular cross-sections. *Proc. R. Soc. Ser. A*, **120** (A786), 691–695.

Consolini, L., and Thome, J. R. (2010). A heat transfer model for evaporation of coalescing bubble in micro-channel flow, *Int. J. Heat Fluid Flow*, **31**, 115–125.

Cooper, M.G. (1984). Saturated nucleate pool boiling – A simple correlation. *First UK National Heat Transfer Conf., Inst. Chem. Eng. Symp. Ser. 86*, **2**, 785–793.

Cooper, M. G., and Lloyd, A. J. P. (1969). The microlayer in nucleate pool boiling. *Int. J. Heat Mass Transfer*, **12**, 895–913.

Cooper, M. G., Judd, A. M., and Pike, R. A. (1978). Shape and departure of single bubbles growing at a wall. *Proc. 6th Int. Heat Transfer Conf.*, Toronto, **1**, 115–120.

Cornish, R. J. (1928). Flow in a pipe of rectangular cross-section. *Proc. R. Soc. Ser. A*, **120**, 691–700.

Cornwell, K., and Kew, P. A. (1992). Boiling in small parallel channels. *Proc. CEC Conference on Energy Efficiency in Process Technology*, Athens, October, Paper 22, Elsevier Applied Science, pp. 624–638.

Costa-Patry, E., Olivier, J., Nichita, B. A., Michel, B., and Thome, J. R. (2011a). Two-phase flow of refrigerants in 85 mm-wide multi-microchannels: part I: pressure drop. *Int. J. Heat Fluid Flow*, **32**, 451–463.

Costa-Patry, E., Olivier, J., Michel, B., and Thome, J. R. (2011b). Two-phase flow of refrigerants in 85 mm-wide multi-microchannels: part II: heat transfer with 35 local heaters. *Int. J. Heat Fluid Flow*, **32**, 464–476.

Costa-Patry, E., Olivier, J., and Thome, J. R. (2012). Heat transfer characteristics in a copper microevaporator and flow pattern-based prediction method for flow boiling in microchannels. *Frontiers Heat Mass Transfer*, **3**, paper 013002.

Costa-Patry, E., and Thome, J. R., (2013). Flow pattern based flow boiling heat transfer model for microchannels. *Int. J. Refrigeration*, **36**, 414–420.

Coulaloglou, C. A., and Tavlarides, L. L. (1977). Description of interaction processes in agitated liquid–liquid dispersions. *Chem. Eng. Sci.*, **32**, 1289–1297.

Colburn, A. P., and Drew, T. B. (1937). The condensation of mixed vapors. *Trans. AIChE*, **33**, 0197–0215.

Courtaud, M., Deruaz, R., and D'Aillon, L. G. (1988). The French thermal-hydraulic program addressing the requirements of the future pressurized water reactors. *Nucl. Technol.*, **80**, 73–82.

Cubaud, T., and Ho, C. M. (2004). Transport of bubbles in square microchannels. *Phys. Fluids*, **16**, 4575–4585.

Cubaud, T., Ulmanella, U., and Ho, C. M. (2006). Two-phase flow in microchannels with surface modifications. *Fluid Dyn. Res.*, **38**, 772–786.

Cumo, M., *et al.* (1980). Experimental advanced on boiler heat transfer. *Proc. 2nd Int. Conf. Boiler Dynamics and Control in Nuclear Power Stations*, British Nuclear Society, p. 367.

Cussler, E. C. (2009). *Diffusion Mass Transfer in Fluid Systems*, 3rd Edn, Cambridge University Press, Cambridge.

Dagan, R., Elias, E., Wacholder, E., and Olek, S. (1993). A two-fluid model for critical flashing flows in pipes. *Int. J. Multiphase Flow*, **19**, 15–25.

Dagan, Z. (1984). Spreading of films of adsorption on a liquid surface. *PCH PhysicoChem. Hydrodyn.*, **5**, 43–51.

Da Silva Lima, R. J., and Thome, J. R. (2012). Two-phase flow patterns in U-bends and their contiguous straight tubes for different orientations, tube and bend diameters. *Int. J. Refrigeration*, **35**, 1439–1454.

Daleas, R. S., and Bergles, A. E. (1965). Effects of upstream compressibility on subcooled critical heat flux. Paper 65–HT–67, ASME, New York.

Daiguji, H., Hihara, E., and Saito, T. (1977). Mechanism of absorption enhancement by surfactants. *Int. J. Heat Mass Transfer*, **40**, 1743–1752.

Dalle Donne, M., and Hame, W. (1985). Critical heat flux for triangular arrays of rod bundles with tight lattices, including the spiral spacer effects. *Nucl. Technol.*, **71**, 111–124.

Damianides, C. A., and Westwater, J. W. (1988). Two-phase flow patterns in a compact heat exchanger and in small tubes. *Proc. 2nd UK National Heat Transfer Conf.*, pp. 1257–1268.

Das, P. Y., Kumar, R., and Ramkrishna, D. (1987). Coalescence of drops in stirred dispersions, a white noise model for coalescence. *Chem. Eng. Sci.*, **42**, 213–220.

Davidson, J. F., and Harrison, D. (1971). *Fluidization*, Academic Press, New York.

Davies, J. T., and Rideal, E. K. (1963). *Interfacial Phenomena*, Academic Press, New York.

Davis, E. J., and Anderson, G. H. (1966). The incipience of nucleate boiling in forced convection flow. *AIChE J.*, **12**, 774–780.

Davis, R. M., and Taylor, G. I. (1950). The mechanism of large bubbles rising through extended liquids and through liquids in tubes. *Proc. R. Soc. Ser. A*, **200**, 375–390.

Del Col, D., Torresin, D., and Cavallini, A. (2010). Heat transfer and pressure drop during condensation of the low GWP refrigerant R1234yf. *Int. J. Refrigeration*, **33**, 1307–1318.

Dean, W. R. (1927). Note on the motion of the fluid in a curved pile. *Phil. Magazine*, **4**, 208–223.

Ded, J. S., and Lienhard, J. H. (1972). The peak pool boiling from a sphere. *AIChE J.*, **18**, 337–342.

Deendarlianto, Höhne, T., Lucas, D., and Vierow, K. (2012). Gas–liquid countercurrent two-phase flow in a PWR hot leg: A comprehensive research review. *Nucl. Eng. Design*, **243**, 214–233.

Deissler, R. G. (1954). Analysis of turbulent heat transfer, mass transfer and friction in smooth tubes at high Prandtl and Schmidt numbers. NACA Tech Rep. 1210.

Delhaye, J. M. (1969). General equations for two-phase systems and their applications to air–water bubble flow and to steam–water flashing flow. ASME-69-HT-63

Delhaye, J. M., and Bricard, P. (1994). Interfacial area in bubbly flow: Experimental data and correlations. *Nucl. Eng. Design*, **151**, 65–77.

Dengler, C. E., and Addoms, J. N. (1956). Heat transfer mechanism for vaporization of water in a vertical tube. *Chem. Eng. Prog. Symp. Ser.*, **52** (18), 85–103.

Derby, M., Lee, H.J., Peles, Y., and Jensen, M.K. (2012). Condensation heat transfer in square, triangular, and semi-circular mini-channels, *Int. J. Heat Mass Transfer*, **55**, 187–197.

De Salve, M., Panella, B., and Scorta, G. (1986). Heat and mass transfer by direct condensation of steam on a subcooled turbulent water jet. *Proc. 8th Int. Heat Transfer Conf.*, San Francisco, **4**, 1653–1658.

Dey, D., Boulton-Stone, J. M., Emery, A. N., and Blake, J. R. (1997). Experimental comparisons with a numerical model of surfactant effects on the burst of a single bubble. *Chem. Eng. Sci.*, **52**, 2769–2783.

De Young, T. L. (1975). Homogeneous equilibrium critical flow model. Aerojet Nuclear Company Internal Rep. TLD-1-75.

Dhir, V. K. (1991). Nucleate and transition boiling heat transfer under pool and external flow conditions. *Int. J. Heat Fluid Flow*, **12**, 290–314.

Dhir, V. K. (1993). Boiling heat transfer. *Annu. Rev. Heat Transfer*, **5**, 303–350.

Dhir, V. K. (1994). Boiling and two-phase flow in porous media. *Annu. Rev. Fluid Mech.*, **30**, 365–401.

Dhir, V. K. (1998). Boiling heat transfer. *Annu. Rev. Fluid Mech.*, **30**, 265–401.

Dhir, V. K., and Liaw, S. P. (1989). Framework for a unified model for nucleate and transition pool boiling. *J. Heat Transfer*, **111**, 739–746.

Dhir, V. K., and Lienhard, J. H. (1974). Peak pool boiling heat flux in viscous liquids. *J. Heat Transfer*, **96**, 71–78.

Dhir, V. K., Castle, J. N., and Catton, I. (1977). Role of Taylor instability on sublimation of a horizontal slab of dry ice. *J. Heat Transfer*, **99**, 411–418.

Dhir, V. K., Warrier, G. R., and Aktinol, E. (2013). Numerical simulation of pool boiling: a review. *J. Heat Transfer*, **135**, paper no. 061502.

Dittus, F. W., and Boelter, L. M. K. (1930). Heat transfer in automobile radiators of the tubular type. *University of California Publication on Engineering*, Vol. 2, No. 13, Berkeley, California.

Dix, G. E. (1971). Vapor void fractions for forced convection with subcooled boiling at low flow rates. PhD thesis, University of California, Berkeley. Also, General Electric Rep. NEDO-10491.

Dobran, F. (1987). Nonequilibrium modeling of two-phase critical flow in tubes. *J. Heat Transfer*, **109**, 731–738.

Dobson, M. K., and Chato, J. C. (1998). Condensation in smooth horizontal tubes. *J. Heat Transfer*, **120**, 193–213.

Dobson, M. K., Chato, J. C., Hinde, D. K., and Wang, S. P. (1994). Experimental evaluation of internal condensation of refrigerants R-12 and R-134a. *ASHRAE Trans.*, **100**, 744–754.

Domanski, P. A., and Hermes, C. J. L. (2008). An improved correlation for two-phase pressure drop of R-22 and R-410A in 180° return bends. *Appl. Thermal Eng.*, **28**, 793–800.

Donaldson, A. A., Macchi, A., and Kirpalani, D. M. (2011). Predicting inter-phase mass transfer for idealized Taylor flow: A comparison of numerical frameworks. *Chem. Eng. Sci.*, **66**, 3339–3349.

Doretti, L., Zilio, C., Mancin, S., and Cavallini, A. (2013). Condensation flow patterns inside plain and microfin tubes: A review. *Int. J. Refrigeration*, **36**, 567–587.

Doroschuk, V. E., Levitan, L. L., and Lantzman, F. P. (1975). Investigation into burnout in uniformly heated tubes. ASME Paper 75-WA/HT-22.

Dougall, R. L., and Rohsenow, W. M. (1963). Film boiling on the inside of vertical tubes with upward flow of the fluid at low qualities. Rep. MIT-TR-9070–26, Massachusetts Institute of Technology, Cambridge, MA.

Downar-Zapolski, Z., Bilicki, Z., Bolle, L., and Franco, J. (1996). The non-equilibrium relaxation model for one-dimensional flashing liquid flow. *Int. J. Multiphase Flow*, **22**, 473–483.

Downing, R. S., and Kojasoy, G. (2002) Single and two-phase pressure drop characteristics in miniature helical channels. *Exp. Thermal Fluid Sci.*, **26**, 535–546.

Drew, D. A., and Lahey, R. T. (1987). The virtual mass and lift force on a sphere in rotating and straining inviscid flow. *Int. J. Multiphase Flow*, **13**, 113–121.

Drew, D. A., Cheng, L. Y., and Lahey, R. T. (1979). The analysis of virtual mass effect in two-phase flow. *Int. J. Multiphase Flow*, **5**, 233–242.

Droso, E. I. P., Paras, S. V., and Karabelas, A. J. (2006). Counter-current gas–liquid flow in a vertical narrow channel – Liquid film characteristics and flooding phenomena. *Int. J. Multiphase Flow*, **32**, 51–81.

Dukler, A. E., and Hubbard, M. G. (1975). A model for gas-liquid slug flow in horizontal and near-horizontal tubes. *Ind. Eng. Chem. Fundam.*, **14**, 337–347.

Dukler, A. E., and Smith, L. (1979). Two-phase interactions in countercurrent flow: Studies of the flooding mechanism. NUREG/CR-0617, US Nuclear Regulatory Commission, Washington, DC.

Dukler, A. E., and Taitel, Y. (1986). Flow pattern transitions in gas–liquid systems: Measurement and modeling. In *Multiphase Science and Technology*, Hewitt, G. F., Delhaye, J. M., and Zuber, N., Eds., **2**, 1–94.

Dukler, A. E., Wicks, M., III, and Cleveland, R. G. (1964). Pressure drop and hold-up in two-phase flow. *AIChE J.*, **10**, 38–51.

Dumitrescu, D. T. (1943). Strömung an einer Luftblase im senkrechten Rohr. *Z. Angew Math. Mech.*, **23**, 139–149.

Duncan, A. B., and Peterson, G. P. (1994). Review of microscale heat transfer. *Appl. Mech. Rev.*, **47**, 397–428.

Dupont, V., Thome. J. R., and Jacobi, A. M. (2004). Heat transfer model for evaporation in microchannels. Part II: comparison with the database. *Int. J. Heat and Mass Transfer*, **47**, 3387–3401.

Eames, I. W., Marr, N. J., and Sabir, H. (1997). The evaporation of water: A review. *Int. J. Heat Mass Transfer*, **40**, 2963–2973.

Ebadian, M. A., and Dong, Z. F. (1998). Forced convection, internal flow in ducts, in *Handbook of Heat Transfer*, Rohsenow, W. M., Hartnett, J. P., and Cho, Y. I., Eds., 3rd Edn, McGraw-Hill, New York.

Edvinsson, R., and Irandoust, S. (1996). Finite-element analysis of Taylor flow. *AIChE J.*, **42**, 1815–1823.

Edwards, D. K., Denny, V. E., and Mills, A. F. (1979). *Transfer Processes*, 2nd Edn, Hemisphere, Washington, DC.

Eichhorn, R. (1980). Dimensionless correlation of the hanging film phenomenon. *J. Fluids Eng.*, **102**, 372–375.

Ekberg, N. P., Ghiaasiaan, S. M., Abdel-Khalik, S. I., Yoda, M., and Jeter, S. M. (1999). Gas–liquid two-phase flow in narrow horizontal annuli. *Nucl. Eng. Design*, **192**, 59–80.

El-Genk, M. S., and Rao, D. V. (1991a). Critical heat flux in rod bundles at low flow and low pressure conditions. ASME Heat Transfer Division, **150**, *Thermal Hydraulics of Advanced Nuclear Reactors*, pp. 25–30.

El-Genk, M. S., and Rao, D. V. (1991b). On the predictions of critical heat flux in rod bundles at low flow and low pressure conditions. *Heat Transfer Eng.*, **12**, 48–57.

El-Genk, M. S., Haynes, S. J., and Kim, S. H. (1988). Critical heat flow of water in vertical annuli. *Int. J. Heat Mass Transfer*, **31**, 2291–2303.

El Hajal, J., Thome, J. R., and Cavallini, A. (2003). Condensation in horizontal tubes. Part I: Two-phase flow pattern map. *Int. J. Heat Mass Transfer*, **46**, 3349–3363.

Elias, E., and Chambre, P. L. (1984). A mechanistic nonequilibrium model for two-phase critical flow. *Int. J. Multiphase Flow*, **10**, 21–40.

Elias, E., and Lelluche, G. S. (1994). Two-phase critical flow. *Int. J. Multiphase Flow*, **20**, Suppl., 91–168.

Elkassabgi, Y., and Lienhard, J. H. (1988). The peak pool boiling heat fluxes from horizontal cylinders in subcooled liquids. *J. Heat Transfer*, **110**, 479–492.

El-Sayed Mosaad, M., Al-Hajeri, M., Al-Ajami, R., and Koliub, A. M. (2009). Heat transfer and pressure drop of R-134a condensation in a coiled, double tube. *Heat and Mass Transfer*, **45**, 1107–1115.

Elsayed, A. M., Al-Dadah, R. K., Mahmoud, S., and Rezk, A. (2012). Investigation of flow boiling heat transfer inside small diameter helically coiled tubes. *Refrigeration*, **35**, 2179–2187.

Emiliani, E. (1992). *Planet Earth*, Cambridge University Press, Cambridge.

Emmert, R. E., and Pigford, R. L. (1954). A study of gas absorption in falling liquid films. *Chem. Eng. Prog.*, **50**, 87–93.

Epstein, M., and Hauser, G. M. (1991). Simultaneous fog formation and thermophoretic droplet deposition in a turbulent pipe flow. *J. Heat Transfer*, **113**, 224–231.

Epstein, M., Henry, R. E., Midvidy, W., and Pauls, R. (1983). One-dimensional modeling of two-phase jet expansion and impingement, *Thermal-Hydraulics of Nuclear Reactors 11, 2nd Int. Topical Meet. Nuclear Reactor Thermal-Hydraulics*, Santa Barbara, CA (Jan., 1983).

Faghri, A., and Zhang, Y. (2006). *Transport Phenomena in Multiphase Systems*, Elsevier/Academic Press, Amsterdam.

Fairbrother, F., and Stubbs, A. E. (1935). Studies in electroendosmosis. Part VI. The bubbletube method of measurements. *J. Chem. Soc.*, **1**, 527–529.

Fang, X. D., Xu, Y., and Zhou, Z. R. (2011a). New correlations of single-phase friction factor for turbulent pipe flow and evaluation of existing single-phase friction factor correlations. *Nucl. Eng. Design*, **241**, 897–902.

Fang, X. D., Zhang, H. G., Xu, Y., and Su, X. H. (2011b). Evaluation of using two-phase frictional pressure drop correlations for normal gravity to microgravity and reduced gravity. *Adv. Space Res.*, **49**, 351–364.

Feburie, V., Giot, M., Granger, S., and Seynhaever, J. M. (1993). A model for choked flow through cracks with inlet subcooling. *Int. J. Multiphase Flow*, **19**, 541–562.

Feng, Z., and Serizawa, A. (2000). Two-phase flow patterns in ultra-small channels. *Japanese–European Two-Phase Flow Group Meeting*, Tsukuba, Japan, 2000.

Fiori, M. P., and Bergles, A. E. (1970). Model of CHF in subcooled boiling. *Proc. 4th Int. Heat Transfer Conf.*, Paris–Versailles, **6**, Paper b6.3, Elsevier, Dordrecht.

Fisher, S. A., Harrison, G. S., and Pearce, D.C. (1978). Premature dryout in conventional and nuclear power station evaporators. *Proc. 6th Int. Heat Transfer Conf.*, Toronto, **2**, 49–54.

Fletcher, D. F. (1991). An improved mathematical model of melt/water detonation – I. Model formulation and example results. *Int. J. Heat Mass Transfer*, **34**, 2435–2448.

Fluent Inc. (2005). *Fluent 6.2.16 User's Guide*.

Fluid Dynamics International, Inc., Evanston, IL 60210 (1991).

Foda, M., and Cox, R. G. (1980). The spreading of thin liquid films on a water–air interface. *J. Fluid Mech.*, **101**, 33–51.

Ford, J. D., and Lekic, A. (1973). Rate of growth of drops during condensation. *Int. J. Heat Mass Transfer*, **16**, 61–64.

Forster, H. K., and Greif, R. (1959). Heat transfer to boiling liquid, mechanism and correlations. *J. Heat Transfer*, **81**, 43–53.

Forster, H. K., and Zuber, N. (1954). Growth of a vapor bubble in a superheated liquid. *J. Appl. Phys.*, **25**, 474–478.

Forster, H. K., and Zuber, N. (1955). Dynamics of vapor bubbles and boiling heat transfer. *Chem. Eng. Prog.*, **1**(4), 531–535.

Fourar, M., and Bories, S. (1995). Experimental study of air–water two-phase flow through a fracture (narrow channel). *Int. J. Multiphase Flow*, **21**, 621–637.

Friedel, L. (1979). Improved pressure drop correlations for horizontal and vertical two-phase pipe flow. *3R Int.*, **18**, 485–492.

Friedlander, S. K. (2000). *Smoke, Dust, and Haze*, 2nd edn, Oxford University Press, London.

Fries, D. M., Trachsel, F., and Von Rohr, P. R. (2008). Segmented gas–liquid flow characterization in rectangular microchannels. *Int. J. Multiphase Flow*, **34**, 1108–1118.

Fritz, W. (1935). Maximum volume of vapor bubbles. *Phys. Z.*, **36**, 379–384.

Fronk, B. M. (2014). Coupled heat and mass transfer during condensation of high-temperature-glide zeotropic mixtures in small diameter channels. PhD thesis, Georgia Institute of Technology, Atlanta, Georgia.

Fronk, B. M., and Garimella, S. (2013). In-tube condensation of zeotropic fluid mixtures: A review. *Int. J. Refrigeration*, **36**, 534–561.

Fu, B. R., and Pan, C. (2005). Flow pattern transition instability in a microchannel with CO_2 bubbles produced by chemical reactions. *Int. J. Heat Mass Transfer*, **48**, 4397–4409.

Fu, X., Qi, S. L., Zhang, P., and Wang, R. Z. (2008). Visualization of flow boiling of liquid nitrogen in a vertical mini-tube. *Int. J. Multiphase Flow*, **34**, 333–351.

Fujii, T., Uehara, H., Hirata, K., and Oda, K. (1972). Heat transfer and flow resistance in condensation of low pressure steam flowing through tube banks. *Int. J. Heat Mass Transfer*, **15**, 247–260.

Fujita, T., and Ueda, T. (1978). Heat transfer to falling liquid films and film breakdown – I Subcooled liquid films. *Int. J. Heat Mass Transfer*, **21**, 97–108.

Fujita, Y., and Bai, Q. (1998). Critical heat flux of binary mixtures in pool boiling and its correlation in terms of Marangoni number. *Int. J. Refrigeration*, **8**, 616–622.

Fujita, Y., Bai, Q., and Tsutsui, M. (1995). Critical heat flux of binary mixtures in pool boiling. *Proc. ASME/JSME Therm. Eng. Joint Conf.*, Vol. 2, pp. 193–200.

Fujita, Y., and Tsutsui, M. (1997). Heat transfer in nucleate boiling of binary mixtures (Development of a heat transfer correlation). *JSME Int. J. – Series B: Fluid Mechanics and Thermal Engineering*, **40**, 134–141.

Fukano, T., and Kariyasaki, A. (1993). Characteristics of gas–liquid two-phase flow in a capillary. *Nucl. Eng. Design*, **141**, 59–68.

Fukano, T., Kariyasaki, A., and Kagawa, M. (1989). Flow patterns and pressure drop in isothermal gas–liquid concurrent flow in a horizontal capillary tube. *Proc. 1989 National Heat Transfer Conf.*, Tilmaz, S. B., Ed., American Nuclear Society, pp. 153–161.

Fulford, G. D. (1964). The flow of liquids in thin films. *Adv. Chem. Eng.*, **5**, 151–236.

Furno, J. S., Taylor, R., and Krishna, R. (1986). Condensation of vapor mixtures. 2. Comparison with experiment. *Ind. Eng. Chem. Process Des. Development*, **25**, 98–101.

Gadis, E. S. (1972). The effect of liquid motion induced by phase change and thermocapillary on the thermal equilibrium of a vapor bubble. *Int. J. Heat Mass Transfer*, **15**, 2241–2250.

Gaertner, R. F. (1965). Photographic study of nucleate pool boiling on a horizontal surface. *J. Heat Transfer*, **87**, 17–29.

Galloway, J. E., and Mudawar, I. (1993). CHF mechanism in flow boiling from a short heated wall – I. Examination of near-wall conditions with the aid of photomicrography and high-speed video imaging. *Int. J. Heat Mass Transfer*, **36**, 2511–2526.

Galvis, E., and Culham, R. (2012). Measurement and flow pattern visualization of two-phase flow boiling in single channel microevaporators. *Int. J. Multiphase Flow*, **42**, 52–61.

Ganapathy, H., Al-Hajri, E., and Ohadi, M. M. (2013a). Phase field modeling of Taylor flow in mini/microchannels, part I: Bubble formation mechanisms and phase field parameters. *Chem. Eng. Sci.*, **94**, 138–149.

Ganapathy, H., Al-Hajri, E., and Ohadi, M. M. (2013b). Phase field modeling of Taylor flow in mini/microchannels, part II: Hydrodynamics of Taylor flow. *Chem. Eng. Sci.*, **94**, 156–165.

Ganapathy, H., Al-Hajri, E., and Ohadi, M. M. (2013c). Mass transfer characteristics of gas–liquid absorption during Taylor flow in mini/microchannel reactors. *Chem. Eng. Sci.*, **101**, 69–80.

Ganchev, B., Zozlov, V., and Lozovetskiy, V. (1972). A study of heat transfer to a falling liquid film at a vertical surface. *Heat Transfer Sov. Res.*, **4**(2), 102–110.

Ganic, E. N., and Mastanaiah, K. (1983). Hydrodynamics and heat transfer in falling film flow. In *Low Reynolds Number Flow Heat Exchangers*, Kakac, S., Shah, R., and Bergles, A. E., Eds., Hemisphere, Washington, DC.

Garimella, S. V., Agarwal, A., and Killion, J. D. (2005). Condensation pressure drop in circular microchannels. *Heat Transfer Eng.*, **26**, 28–35.

Garimella, S., and Bandhauer, T. M. (2001). Measurement of condensation heat transfer coefficients in microchannel tubes. *Proc. 2001 IMECE*, ASME, New York.

Garimella, S. V., and Chen, T. (2012). A study of critical heat flux during flow boiling in microchannel heat sinks. *J. Heat Transfer*, **134**, paper 011504.

Garimella, S., Killion, J. D., and Coleman, J. W. (2002). An experimentally validated model for two-phase pressure drop in the intermittent flow regime for circular microchannels. *J. Fluids Eng.*, **124**, 205–214.

Garrels, R. M., and Christ, C. L. (1965). *Solutions, Minerals and Equilibria*, Harper and Rowe, New York.

Garstecki, P., Fuerstman, M. J., Stone, H. A., and Whitesides, G. M. (2006). Formation of droplets and bubbles in a microfluidic T-junction—scaling and mechanism of break-up. *The Royal Society of Chemistry, Lab Chip*, **6**, 437–446.

Geiger, G. E., and Rohrer, W. M. (1966). Sudden contraction losses in two-phase flow. *J. Heat Transfer*, **88**, 1–9.

Geng, H., and Ghiaasiaan, S. M. (1998). Mechanistic modeling of critical flow of initially sub-cooled liquid containing dissolved noncondensables through cracks and slits based on the homogeneous-equilibrium mixture method. *Nucl. Sci. Eng.*, **129**, 294–304.

Geng, H., and Ghiaasiaan, H. (2000). Mechanistic nonequilibrium modeling of critical flow of subcooled liquids containing dissolved noncondensables using the dynamic flow regime model. *Proc. 8th Int. Conf. on Nuclear Engineering (ICONE-8)*, Paper No. ICONE-8708, Baltimore, MD.

Germano, M. (1982). On the effect of torsion in a helical pipe flow. *J. Fluid Mech.*, **125**, 1–8.

Ghiaasiaan, S. M. (2011). *Convective Heat and Mass Transfer*. Cambridge University Press, Cambridge.

Ghiaasiaan, S. M. and Abdel-Khalik, S. I. (2001). *Two-Phase Flow in Microchannels, Advances in Heat Transfer*, J. P. Hartnett and T. F. Irvine, Jr., Eds., Vol. 34, pp. 145–254, Academic Press.

Ghiaasiaan, S. M., and Catton, I. (1983). Multi-dimensional and two-phase flow effects in PWR core reflooding. Electric Power Research Institute Rep. EPRI NP-3437, Palo Alto, CA.

Ghiaasiaan, S. M., and Chedester, R. C. (2002). Boiling incipience in microchannels. *Int. J. Heat Mass Transfer*, **45**, 4599–4606.

Ghiaasiaan, S. M., and Eghbali, D. A. (1997). On modeling of turbulent vapor condensation with noncondensables. *J. Heat Transfer*, **119**, 373–376.

Ghiaasiaan, S. M., and Geng, H. (1997). Mechanistic non-equilibrium modeling of critical flashing flow of subcooled liquids containing dissolved noncondensables. *Num. Heat Transfer B*, **32**, 435–457.

Ghiaasiaan, S. M., and Laker, T. S. (2001). Turbulent forced convection in microtubes. *Int. J. Heat Mass Transfer*, **44**, 2777–2782.

Ghiaasiaan, S. M., Catton, I., and Duffey, R. B. (1985). Thermal-hydraulics and two-phase phenomena during reflooding of nuclear reactor cores. *J. Fluids Eng.*, **85**, 89–96.

Ghiaasiaan, S. M., Wassel, A. T., and Lin, C. S. (1991). Direct contact condensation in the presence of noncondensables in OC-OTEC condensers. *J. Solar Energy Eng.*, **113**, 228–235.

Ghiaasiaan, S. M., Kamboj, B. K., and Abdel-Khalik, S. I. (1995). Two-fluid modeling of condensation in the presence of noncondensables in two-phase channel flow. *Nucl. Sci. Eng.*, **119**, 1–17.

Ghiaasiaan, S. M., Wu, X., Sadaowski, D. L., and Abdel-Khalik, S. I. (1997a). Hydrodynamic characteristics of counter-current two-phase flow in vertical and inclined channels: Effects of liquid properties. *Int. J. Multiphase Flow*, **23**, 1063–1083.

Ghiaasiaan, S. M., Muller, J. R., Sadowski, D. L., and Abdel-Khalik, S. I. (1997b). Critical flow of initially highly subcooled water through a short capillary. *Nucl. Sci. Eng.*, **126**, 229–238.

Giavedoni, M. D., and Saita, F. A. (1997). The axisymmetric and plane cases of a gas phase steadily displacing a Newtonian liquid – A simultaneous solution of the governing equations. *Phys. Fluids*, **9**, 2420–2428.

Giavedoni, M. D., and Saita, F. A. (1999). The rear meniscus of a long bubble steadily displacing a Newtonian liquid in a capillary tube. *Phys. Fluids*, **11**, 786–794.

Giot, M., and Fritz, A. (1972). Two-phase two- and one-component critical flow with the variable slip model. *Prog. Heat Transfer*, **6**, 651–670.

Gnielinski, V. (1976). New equations for heat and mass transfer in Turbulent pipe and channel flow. *Int. Chem. Eng.*, **16**, 359–368.

Golden, S. (1964) *Elements of the Theory of Gases*, Addison-Wesley, Reading, MA.

Gombosi, T. I. (1994). *Gaskinetic Theory*, Cambridge University Press, Cambridge.

Gomez, L., Shoham, O., Schmidt, Z., Choshki, R., and Northug, T. (2000). Unified mechanistic model for steady state two phase flow: horizontal to upward vertical flow. *Soc. Petrol. Eng. J.*, **5**, 339–350.

Gong, M., Ma, J., Wu, J., Zhang, Y., Sun, Z., and Zhou, Y. (2009). Nucleate pool boiling of liquid methane and its natural gas mixtures. *Int. J. Heat Mass Transfer*, **52**, 2733–2739.

Gorenflo, D. (1993). Pool boiling. In *VDI-Heat Atlas* (English Version), Schlünder, E. U., Ed., pp. Ha 1–13, VDI-Verlag, Düsseldorf, Germany.

Gorenflo, D., Chandra, U., Kotthoff, S., and Luke, A. (2004). Influence of thermophysical properties on pool boiling heat transfer of refrigerants. *Int. J. Refrigeration*, **27**, 392–502.

Gorenflo, D., Knabe, V., and Bieling, V. (1986). Bubble density on surfaces with nucleate boiling – Its influence on heat transfer and burnout heat flux at elevated saturation pressures. *Proc. 8th Int. Heat Transfer Conf.*, San Francisco, **4**, 1995–2000.

Govier, F. W., and Aziz, K. (1972). *The Flow of Complex Mixtures in Pipes*, Robert E. Krieger, Malabar, FL.

Greskovich, E. J., and Cooper, W. T. (1975). Correlation and prediction of gas–liquid holdups in inclined upflows. *AIChE J.*, **21**, 1189–1192.

Griffith, P., and Wallis, J. D. (1960). The role of surface conditions in nucleate boiling. *Chem. Eng. Symp. Ser. 56*, **30**, 49–63.

Griffith, P., Clark, J. A., and Rohsenow, W. W. (1958). Void volumes in subcooled boiling systems. ASME Paper 58-HT-19.

Grober, H., Erk, S., and Grigull, U. (1961). *Fundamentals of Heat Transfer*, McGraw-Hill, New York.

Groeneveld, D. C. (1973). Post-dryout heat transfer at reactor operating conditions. *American Nuclear Society Topical Meeting on Water Reactor Safety*, Salt Lake City.

Groeneveld, D. C., and Delorme, G. G. J. (1976). Prediction of thermal non-equilibrium in the post-dryout regime. *Nucl. Eng. Design*, **36**, 17–26.

Groeneveld, D. C., and Snoek, C. W. (1986). A comprehensive examination of heat transfer correlations suitable for reactor safety analysis. In *Multiphase Science and Technology*, Hewitt, G. F., Delhaye, J. M., and Zuber, N., Eds., Hemisphere, Washington, DC, **2**, 181–274.

Groeneveld, D. C., Cheng, S. C., and Doan, T. (1986). AECL-UO critical heat flux look-up table. *Heat Transfer Eng.*, **7**, 46–62.

Groeneveld, D. C., Leung, L. K. H., Kirillov, P. I., Bobkov, V. P., Smogalev, I. P., Vinogradov, V. N., Huang, X. C., and Royer, E. (1996). The 1995 look-up table for critical heat flux in tubes, *Nucl. Eng. Design*, **163**, 1–23.

Groeneveld, D. C., Leung, L. K. H., Vasic, A. Z., Guo, Y. J., and Cheng, S. C. (2003). A look-up table for fully developed film boiling heat transfer. *Nucl. Eng. Design*, **225**, 83–97.

Grohmann, S. (2005). Measurement and modeling of single-phase and flow boiling heat transfer in microtubes. *Int. J. Heat Mass Transfer*, **48**, 4073–4089.

Grolmes, M. A., and Yue, M. H. (1993). Relief vent sizing and location for long tubular reactors. *J. Hazardous Materials*, **33**, 261–273.

Guermit, T. (2011). Simulation study of the condensation of mixed refrigerant R407d (R32/R125/R134a). *J. Appl. Sci. Environmental Sanitation*, **6**, 105–113.

Guglielmini, G., Lorenzi, A., Muzzio, A., and Sotgia, G. (1986). Two-phase flow pressure drops across sudden area contractions. *Proc. 8th Int. Heat Transfer Conf.*, Vol. 5, pp. 2361–2366, ASME, New York.

Gungor, K. E., and Winterton, R. H. S. (1986). A general correlation for flow boiling in tubes and annuli. *Int. J. Heat Mass Transfer*, **29**, 351–358.

Gungor, K. E., and Winterton, R. H. S (1987). Simplified general correlation for saturated flow boiling and comparison of correlations with data. *Chem. Eng. Res. Des.*, **65**, 148–156.

Guo, L., Feng, Z., and Chen, X. (2001). An experimental investigation of the frictional pressure drop of steam–water two-phase flow in helical coils. *Int. J. Heat Mass Transfer*, **44**, 2601–2610.

Gupta, R., Fletcher, D. F., and Haynes, B. S. (2009). On the CFD modeling of Taylor flow in microchannels. *Chem. Eng. Sci.*, **64**, 2941–2950.

Gupta, R., Fletcher, D. F., and Haynes, B. S. (2010). Taylor flow in microchannels: a review of experimental and computational work. *J. Comput. Multiphase Flow*, **2**, 1–31.

Gupta, A., Kumar, R., and Gupta. A. (2014). Condensation of R-134a inside a helically coiled tube-in-shell heat exchanger. *Exp. Thermal Fluid Sci.*, **54**, 279–289.

Habib, I. S., and Na, T. Y. (1974). Prediction of heat transfer in turbulent pipe flow with constant wall temperature. *J. Heat Transfer*, **96**, 253–254.

Hadamard, J. (1911). Mouvement permanant lent d'une sphère liquide et visqueuse dans un liquide visqueux. *J. Comp. Rend*, **152**, 1735.

Hainoun, A., Hicken, E., and Wolters, J. (1996). Modeling of void formation in the subcooled boiling regime in the ATHLET code to simulate flow instability for research reactors. *Nucl. Eng. Design*, **167**, 175–191.

Hall, D. D., and Mudawar, I. (1997). Evaluation of subcooled CHF correlations using the PU-BTPFL CHF database for vertical upflow of water in a uniformly heated round tube. *Nucl. Technol.*, **117**, 234–246.

Hall, D. D., and Mudawar, I. (2000a). Critical heat flux (CHF) for water flow in tubes – I. Compilation and assessment of world CHF data. *Int. J. Heat Mass Transfer*, **43**, 2573–2604.

Hall, D. D., and Mudawar, I. (2000b). Critical heat flux (CHF) for water flow in tubes – II. Subcooled CHF correlations. *Int. J. Heat Mass Transfer*, **43**, 2605–2640.

Hampton, H. (1951). The condensation of steam on a metal surface. *Proc. General Discussion of Heat Transfer*, Inst. Mech. Eng. and ASME, **84**, 58–64.

Han, C. Y., and Griffith, P. (1965). The mechanism of heat transfer in nucleate pool boiling, Part I. Bubble initiation, growth and departure. *Int. J. Heat Mass Transfer*, **8**, 887–904.

Han, J. T., Lin, C. X., and Ebadian, M. A. (2005). Condensation heat transfer and pressure drop characteristics of R-134a in an annular helical pipe. *Int. Commun. Heat Mass Transfer*, **32**, 1307–1316.

Han, Y., and Shikazono, N. (2009). Measurement of the liquid film thickness in micro tube slug flow. *Int. J. Heat Fluid Flow*, **30**, 842–853.

Hanratty, T. J., and Hershman, A. (1961). Initiation of roll waves. *AIChE J.*, **7**, 488–497.

Hapke, J., Boye, H., and Schmidt, J. (2000). Onset of nucleate boiling in microchannels. *Int. J. Thermal Sci.*, **39**, 505–513.

Haraguchi, H., Koyama, S., and Fujii, T. (1994). Condensation of refrigerants HCFC22, HFC134a and HCFC123 in a horizontal smooth tube (2nd report, Proposal of empirical expressions for the local heat transfer coefficient), *Trans. JSME (B)*, **60**(574), 245–252.

Haramura, Y. (1999). Critical heat flux in pool boiling. In *Handbook of Phase Change*, Kandlikar, S. G., Shoji, M., and Dhir, V. K., Eds., Taylor and Francis, London, Chapter 6.

Haramura, Y., and Katto, Y. (1983). A new hydrodynamic model of critical heat flux, applicable widely to both pool and forced convection boiling on submerged bodies in saturated liquids. *Int. J. Heat Mass Transfer*, **26**, 389–399.

Hardy, P., and Mali, P. (1983). Validation and development of a model describing subcooled critical flow through long tubes. *Energie Primaire*, **18**, 5–23.

Hari, S., and Hassan, Y. A. (2002). Improvement of the subcooled boiling model for low pressure conditions in thermal-hydraulic codes. *Nucl. Eng. Design*, **216**, 139–152.

Harirchian, T., and Garimella, S. V. (2008). Microchannel size effects on local flow boiling heat transfer to a dielectric fluid. *Int. J. Heat Mass Transfer*, **51**, 3724–3735.

Harirchian, T., and Garimella, S. V. (2009). Effects of channel dimension, heat flux, and mass flux on flow boiling regimes in microchannels. *Int. J. Multiphase Flow*, **35**, 349–362.

Harirchian, T., and Garimella, S. V. (2010). A comprehensive flow regime map for microchannel flow boiling with quantitative transition criteria. *Int. J. Heat Mass Transfer*, **53**, 2694–2702.

Hart, J., Ellenberger, I., and Hamersma, P. J. (1987). Single and two-phase flow through helically coiled tubes. *Chem. Eng. Science*, **43**, 775–783.

Harmathy, T. Z. (1960). Velocity of large drops and bubbles in media of infinite and restricted extent. *AIChE. J.*, **6**, 281–288.

Harper, M. J., and Rich, J. C. (1993). Radiation-induced nucleation in superheated liquid droplet neutron detectors. *Nucl. Instrum. Methods Phys. Res. A*, **336**, 220–225.

Hassan, I., Vaillancourt, M., and Pehlivan, K. (2005). Two-phase flow regime transitions in microchannels: a comparative experimental study. *Microscale Thermophysical Engineering*, **9**, 165–182.

Hassanvand, A., and Hashemabadi, S. H. (2012). Direct numerical simulation of mass transfer from Taylor bubble flow through a circular capillary. *Int. J. Heat Mass Transfer*, **55**, 5959–5971.

Hasson, D., Luss, D., and Peck, R. (1964). Theoretical analysis of vapor condensation on laminar liquid jets. *Int. J. Heat Mass Transfer*, **7**, 969–981.

Hatton, A. P., and Hall, I. S. (1966). Photographic study of boiling on prepared surfaces. *Proc. 3rd Int. Heat Transfer Conf.*, Chicago, **4**, 24–37.

Hausen, H. (1983). *Heat Transfer in Counter Flow, Parallel Flow, and Cross-Flow*, McGraw-Hill, New York.

Hayashi, K., and Tomiyama, A. (2012). Effects of surfactant on terminal velocity of a Taylor bubble in a vertical pipe. *Int. J. Multiphase Flow*, **39**, 78–87.

Haynes, B. S., and Fletcher, D. F. (2003). Subcooled flow boiling heat transfer in narrow passages. *Int. J. Heat Mass Transfer*, **46**, 3673–3682.

Hegab, H. E., Bari, A., and Ameel, T. (2002). Friction and convection studies of R-134a in microchannels within transition and turbulent flow regimes. *Exp. Thermal Fluid Sci.*, **15**, 245–259.

Heil, M. (2001). Finite Reynolds number effects in the Bretherton problem. *Phys. Fluids*, **13**, 2517–2531.

Heiszwolf, J. J., Engelvaart, L. B., Van den Eijnden, M. G., Kreutzer, M. T., Kapteijn, F., and Moulijn, J. A. (2001). Hydrodynamic aspects of the monolith loop reactor. *Chem. Eng. Sci.*, **56**, 805–812.

Henry, R. E. (1970). Two-phase critical flow at low qualities. *Nucl. Sci. Eng.*, **41**, 79–98.

Henry, R. E. (1974). A correlation for the minimum film boiling temperature. *Chem. Eng. Prog. Symp. Series*, **70**, 81–90.

Henry, R. E., and Fauske, H. K. (1971). The two-phase critical flow of one-component mixtures in nozzles, orifices, and short tubes. *J. Heat Transfer*, **93**, 179–187.

Henstock, W. H., and Hanratty, T. J. (1976). The interfacial drag and height of the wall layer in annular flows. *AIChE J.*, **22**, 990–1000.

Heo, J., Park, H., and Yun, R. (2013). Condensation heat transfer and pressure drop characteristics of CO_2 in rectangular microchannels. *Int. J. Refrigeration*, **36**, 1657–1668.

Herwig, H., and Hausner, O. (2003). Critical view on 'New results in micro-fluid mechanics: An example.' *Int. J. Heat Mass Transfer*, **46**, 935–937.

Hetsroni, G., Mosyak, A., Segal, Z., and Pogrebnyak, E. (2003). Two-phase flow patterns in parallel micro-channels. *Int. J. Multiphase Flow*, **29**, 341–360.

Hetsroni, G., Mosyak, A., Pogrebnyak, E., and Segal, Z. (2005). Explosive boiling of water in parallel micro-channels. *Int. J. Multiphase Flow*, **31**, 371–392.

Hetsroni, G., Mosyak, A., Pogrebnyak, E., and Segal, S. (2006). Periodic boiling in parallel micro-channels at low vapor quality. *Int. J. Multiphase Flow*, **32**, 1141–1159.

Heun, M. K. (1995). Performance and optimization of microchannel condensers. PhD thesis, University of Illinois, Urbana-Champagne, IL.

Hew, J., Park, H., and Yun, R. (2013). Condensation heat transfer and pressure drop characteristics of CO_2 in a microchannel. *Int. J. Refrigeration*, **36**, 1657–1668.

Hewitt, G. F. (1977). Mechanism and prediction of burnout. In *Two-Phase Flows and Heat Transfer*, Kakac, S., and Veziroglu, T. N., Eds., Hemisphere, Washington, DC, **2**, 721–745.

Hewitt, G. F. (1983). Gas–liquid flow. In *Heat Exchanger Design Handbook*, Schlünder, E. U., Editor-in-chief, Hemisphere, Washington, DC, **2**, 229–238.

Hewitt, G. F., and Govan, A. H. (1990). Phenomena and prediction in annular two-phase flow. In *ASME Advances in Gas-Liquid Flows*, ASME, New York, FED-99, 41–56.

Hewitt, G. F. and Jayanti, S. (1992). Prediction of film inversion in two-phase flow in coiled tubes. *J. Fluid Mech.*, **236**, 497–511.

Hewitt, G. F., and Roberts, D. N. (1969). *Studies of Two-Phase Flow Patterns by Simultaneous X-Ray and Flash Photography*. AERE-M 2159.

Hewitt, G. F., and Wallis, G. B. (1963). Flooding and associated phenomena: Falling film flow in a vertical tube. AERE-R4022, UKAEA, Harwell, England.

Hibiki, T., and Ishii, M. (2001). Interfacial area concentration in steady fully-developed bubbly flow. *Int. J. Heat Mass Transfer*, **44**, 3443–3461.

Hibiki, T., and Ishii, M. (2002). One-dimensional drift flux model for two-phase flow in a large diameter pipe. *Int. J. Heat Mass Transfer*, **46**, 1773–1790.

Hibiki, T., and Ishii, M. (2003a). One-dimensional drift – flux model for two-phase flow in a large diameter pipe. *Int. J. Heat Mass Transfer*, **46**, 1773–1790.

Hibiki, T., and Ishii, M. (2003b). One-dimensional drift flux model and constitutive equations for relative motion between phases in various two-phase flow regimes. *Int. J. Heat Mass Transfer*, **46**, 4935–4948.

Hibiki, T., and Ishii, M. (2005). Erratum to "One-dimensional drift flux model and constitutive equations for relative motion between phases in various two-phase flow regimes." *Int. J. Heat Mass Transfer*, **48**, 1222–1223.

Hibiki, T., and Mishima, K. (2001). Flow regime transition criteria for upward two-phase flow in vertical narrow rectangular channels. *Nucl. Eng. Design*, **203**, 117–131.

Hickox, C. E. (1971). Instability due to viscosity and density stratification in axisymmetric pipe flow. *Phys. Fluids*, **14**, 251–262.

Hindmarsh, A. C. (1980). LSODE and LSODI, two new initial value ordinary differential equation solvers. *ACM Newsl.*, **15**(5), 10.

Hino, R., and Ueda, T. (1985). Studies on heat transfer and flow characteristics in subcooled flow boiling – Part I. Boiling characteristics. *Int. J. Heat Mass Transfer*, **11**, 269–281.

Hinze, J. O. (1955). Fundamentals of the hydrodynamic mechanism of splitting in dispersion processes. *AIChE J.*, **1**, 289–295.

Hinze, J.O. (1975). *Turbulence*, McGraw-Hill, New York.

Hirschfelder, J., Curtiss, C. F., and Bird, R. (1954). *Molecular Theory of Gases and Liquids*, Wiley, New York.

Hopkins, N. E. (1950). Rating the restrictor tube. *Refrig. Eng.*, **58**, 1087–1095.

Hosaka, S., Hirata, M., and Kasagi, N. (1990). Forced convective subcooled boiling heat transfer and CHF in small diameter tubes. *Proc. 9th Int. Heat Transfer Conf.*, **2**, 129–134.

Hovestreijdt, J. (1963). The influence of the surface tension difference on the boiling of mixtures. *Chem. Eng. Sci.*, **18**, 631–639.

Howard, J. A., and Walsh, P. A. (2013). Review and extensions to film thickness and relative bubble drift velocity prediction methods in laminar Taylor or slug flows. *Int. J. Multiphase Flow*, **55**, 32–42.

Howell, J. R., and Siegel, R. (1967). Activation, growth and detachment of boiling bubbles in water from artificial nucleation sites of known geometry and size. NASA TN-0–4201.

Hsu, Y. Y. (1962). On the size range of active nucleation sites on a heating surface. *J. Heat Transfer*, **84C**, 207–216.

Hsu, Y. Y. (1981). Boiling heat transfer equations. In *Thermohydraulics of Two-Phase Systems for Industrial Design and Nuclear Engineering*, Delhaye, J. M., Giot, M., and Riethermuller, M. L., Eds., McGraw-Hill, New York, 255–296.

Hsu, Y. Y., and Graham, R. W. (1961). An analytical and experimental study of the thermal boundary layer and ebullition cycle in nucleate boiling. NASA TND-594.

Hsu, Y. Y., and Graham, R. W. (1986). *Transport Processes in Boiling and Two-Phase Systems*, American Nuclear Society, La Grange Park, IL.

Hsu, Y. Y., and Westwater, J. W. (1960). Approximate theory for film boiling on vertical surfaces. *Chem. Eng. Prog. Symp. Ser.* 56, **30**, 15–24.

Hu, L.-W., and Pan, C. (1995). Prediction of void fraction in convective subcooled boiling channels using a one-dimensional two-fluid model. *J. Heat Transfer*, **117**, 799–803.

Huang, W. S., and Kintner, R. C. (1968). Effects of surfactants on mass transfer inside drops. *AIChE J.* **15**, 735–744.

Huang, X., Ding, G., Hu, H., Zhu, Y., Peng, H., Gao, Y., and Deng, B. (2010). Influence of oil on flow condensation heat transfer of R410A inside 4.18 mm and 1.6 mm inner diameter horizontal smooth tubes. *Int. J. Refrigeration*, **33**, 158–169.

Huang, X. G., Yang, Y. H., Hu, P., and Bao, K. (2014). Experimental study of water–air countercurrent flow characteristics in large scale rectangular channel. *Annals Nucl. Energy*, **69**, 125–133.

Hubbard, G. L., Mills, A. F., and Chung, D. K. (1976). Heat transfer across a turbulent falling film with cocurrent vapor flow. *J. Heat Transfer*, **98**, 319–320.

Huber, J. B., Rewerts, L. E., and Pale, M. B. (1994a). Shell-side condensation heat transfer of R-134a – Part I: Finned-tube performance. *ASHRAE Trans.*, **100**(2), 239–247.

Huber, J. B., Rewerts, L. E., and Pale, M. B. (1994b). Shell-side condensation heat transfer of R-134a – Part II: Enhanced tube performance. *ASHRAE Trans.*, **100**(2), 248–256.

Huber, J. B., Rewerts, L. E., and Pale, M. B. (1994c). Shell-side condensation heat transfer of R-134a – Part III: Comparison with R-12. *ASHRAE Trans.*, **100**(2), 257–264.

Huebsch, W. W., and Pale, M. B. (2004). A comprehensive study of shell-side condensation on integral-fin tubes with R-114 and R-236ea. *ASHRAE Trans.*, **110**, Part 1, 40–52.

Hulburt, H. M., and Katz, S. (1964). Some problems in particle technology. A statistical mechanical formulation. *Chem. Eng. Sci.*, **19**, 555–574.

Hughmark, G. A. (1962). Hold-up in gas–liquid flow. *Chem. Eng. Prog.*, **58**, 62–65.

Hwang, J. J., Tseng, F. G., and Pan, C. (2005). Ethanol–CO_2 two-phase flow in diverging and converging microchannels. *Int. J. Multiphase Flow*, **31**, 548–570.

Hwang, Y. W., and Kim, M. S. (2006). The pressure drop in microtubes and the correlation development, *Int. J. Heat Mass Transfer*, **49**, 1804–1812.

Ide, H., Kimura, R., Hashiguchi, H., and Kawaji, M. (2012) Effect of channel length on the gas–liquid two-phase flow phenomena in a microchannel. *Heat Transfer Eng.*, **33**, 225–233.

Idelchik, I. E. (1994). *Handbook of Hydraulic Resistances*, 3rd Edn, CRC Press, London.

Iguchi, T., Okubo, T., and Murao, Y. (1988). Effect of loop seal on reflood phenomena in PWR. *Nucl. Sci. Technol.*, **25**, 520–527.

Ilyukhin, Yu. N., Balunov, B. F., Smirnov, E. L., and Gotovskii, M. A. (1989). Hydrodynamic characteristics of two-phase annular countercurrent flows in vertical channels. *High Temperature Physics*, **26**, 717–725.

Ilyukhin, Yu. N., Svetlov, S. V., Alekseev, S. B., Kukhtevich, V. O., and Sidorov, V. G. (1999). *High Temperature Physics*, **37** (3), 463.

Inasaka, F., and Nariai, H. (1993). Critical heat flux of subcooled flow boiling with water for high heat flux application. *Proc. High Heat Flux Engineering II*, SPIE, Vol. 1997, 328–339.

Inasaka, F., Nariai, H., and Shimura, T. (1989). Pressure drops in subcooled boiling in narrow tubes. *Heat Transfer Jpn. Res.*, **18**, 70–82.

Incropera, F. P., Dewitt, D. P., Bergman, T. L., and Lavine, A. S. (2007). *Fundamentals of Heat and Mass Transfer*, 6th Edn, Wiley, New York.

International Association for the Properties of Water and Steam (1994). IAPWS release on surface tension of ordinary water substance. Available at http://www.iapws.org/.

Isachenko, V. P., *et al.* (1971). Investigation of heat transfer with steam condensation on turbulent liquid jets. *Teploenergetika*, **18**(2), 7–10.

Ishigai, S., Nakanisi, S., Koizumi, T., and Oyabi, Z. (1972). Hydrodynamics and heat transfer of vertical falling films. *Bull. JSME*, **15**(83), 594–602.

Ishii, M. (1971). Thermally-induced flow instabilities in two-phase mixtures in thermal equilibrium. PhD thesis, Georgia Institute of Technology, Atlanta, GA.

Ishii, M. (1975). *Thermo-Fluid Dynamic Theory of Two-Phase Flow*, Eyrolles, Paris.

Ishii, M. (1976). Study of flow instabilities in two-phase mixtures. ANL-76–23, Argonne National Laboratory, IL.

Ishii, M. (1977). One-dimensional drift flux model and constitutive equations for relative motion between phases in various two-phase flow regimes. ANL Rep. ANL-77–47.

Ishii, M., and Hibiki, T. (2011). *Thermo-Fluid Dynamics of Two-Phase Flow*, 2nd Edn, Springer.

Ishii, M., and Mishima, K. (1984). Two-fluid model and hydrodynamic constitutive relations. *Nucl. Eng. Design*, **82**, 107–126.

Ishii, M., Kim, S., and Uhle, J. (2002). Interfacial area transport equation: Model development and benchmark experiments. *Int. J. Heat Mass Transfer*, **45**, 3111–3123.

Ishii, M., Paranjape, S. S., Kim, S., and Sun, X. (2004). Interfacial structures and interfacial area transport in downward two-phase bubbly flow. *Int. J. Heat Mass Transfer*, **30**, 779–801.

Ito, H. (1959). Friction factor for turbulent flow in a curved pipe. *J. Basic Engineering*, **81**, 123–134.

Ivey, H. J., and Morris, D. J. (1962). On the relevance of the vapour liquid exchange mechanism for sub-cooled boiling heat transfer at high pressure. UKAEA Rep. AEEW-R 137.

Iwamura, T., Watanabe, H., and Murao, Y. (1994). Critical heat flux experiments under steady-state and transient conditions and visualization of CHF phenomena with neutron radiography. *Nucl. Eng. Design*, **149**, 195–206.

Ishii, M., and Zuber, N. (1979). Drag coefficient and relative velocity in bubbly, droplet and particulate flow. *AIChE J.*, **25**, 843–855.

Jasper, J. J. (1972). The surface tension of pure liquid compounds. *J. Phys. Chem. Ref. Data*, **1**, 841–1010.

Jassim, E. W., and Newell, T. A. (2006). Prediction of two-phase pressure drop and void fraction in microchannels using probabilistic flow regime mapping. *Int. J. Heat Mass Transfer*, **49**, 2446–2457.

Jassim, E. W., Newell, T. A., and Chato, J. C. (2007). Probabilistic determination of two-phase flow regimes in horizontal tubes utilizing an automated image recognition technique. *Exp. Fluids*, **42**, 563–573.

Jassim, E. W., Newell, T. A., and Chato, J. C. (2008). Prediction of two-phase condensation in horizontal tubes using probabilistic flow regime maps. *Int. J. Heat Mass Transfer*, **51**, 485–496.

Jaster, H., and Kosky, P. G. (1976). Condensation heat transfer in a mixed flow regime. *Int. J. Heat Mass Transfer*, **19**, 95–99.

Jayakumar, S. J., Mahajani, S. M., Mandal, J. C., Vijayan, P. K., and Bhoi, R. (2008). Experimental and CFD estimation of heat transfer in helically coiled heat exchangers. *Chem. Eng. Res. Design*, **86**, 221–232.

Jayanti, S., and Hewitt, G. F. (1992). Prediction of the slug-to-churn flow transition in vertical two-phase flow. *Int. J. Multiphase Flow*, **18**, 847–860.

Jayanti, S., and Hewitt, G. F. (1997). Hydrodynamics and heat transfer in wavy annular gas–liquid flow: a computational fluid dynamic study. *Int. J. Heat Mass Transfer*, **40**, 2445–2460.

Jayawardena, S. S., Balakotaiah, V., and Witte, L. C. (1997). Flow pattern transition maps for microgravity two-phase flows. *AIChE J.*, **43**, 1637–1640.

Jens, W. H., and Lottes, P. A. (1951). Analysis of heat transfer, burnout, pressure drop and density data for high pressure water. Rep. ANL-4627, US Argonne National Laboratory, Argonne, IL.

Jense, M. K., and Bergles, A. E. (1981). Critical heat flux in helically coiled tubes. *J. Heat Transfer*, **103**, 660–666.

Jepsen, D. M., Azzopardi, B. J., and Whalley, P. B. (1989). The effect of gas properties on drops in annular flow. *Int. J. Multiphase Flow*, **15**, 327–339.

Jiang, L., Wong, M., and Zohar, Y. (1999). Phase change in microchannel heat sinks with integrated temperature sensors. *J. Microelectromech. Syst.*, **8**, 358–365.

Jiang, L., Wong, M., and Zohar, Y. (2001). Forced convection boiling in a microchannel heat sink. *J. Microelectromech. Syst.*, **10**, 80–87.

Jin, D. X., Kwon, J. T., and Kim, M. H. (2003). Prediction of in-tube condensation heat transfer characteristics of binary refrigerant mixtures. *Int. J. Refrigeration*, **26**, 593–600.

John, H., Reimann, J., Westphal, F., and Friedel, L. (1988). Critical two-phase flow through rough slits. *Int. J. Multiphase Flow*, **14**, 155–174.

Jones, D. D. (1977). Subcooled counter-current flow limiting characteristics of the upper region of a BWR fuel bundle. General Electric Company Nuclear Systems Products Division, NEDG-NUREG-23549.

Jones, O. C., Jr. (1976). An improvement in the calculation of turbulent friction in rectangular ducts. *J. Fluid Eng.*, **98**, 173–181.

Jones, O. C., Jr., and Zuber, N. (1979). Slug-annular transition with particular reference to narrow rectangular ducts. In *Two-Phase Momentum, Heat and Mass Transfer in Chemical, Process and Energy Engineering Systems*, Durst, F., Tsiklauri, G. V., and Afgan, N., Eds., Hemisphere, Washington, DC, **1**, 345–355.

Joos, P., and Pinters, J, (1977). Spreading kinetics of liquids on liquids. *J. Colloid Interface Sci.*, **60**, 507–513.

Judd, R. L., and Chopra, A. (1993). Interaction of the nucleation process occurring at adjacent nucleation sites. *J. Heat Transfer*, **115**, 955–962.

Julia, J. E., Hibiki, T., Ishii, M., Yun, B.-J., and Park, G.-C. (2009). Drift-flux model in a subchannel of rod bundle geometry. *Int. J. Heat Mass Transfer*, **52**, 3032–3041.

Jung, D. S., McLinden, M., Radermacher, R., and Didion, D. (1989a). Horizontal flow boiling heat transfer experiments with a mixture of R22/R114. *Int. J. Heat Mass Transfer*, **32**, 131–145.

Jung, D. S., McLinden, M., Radermacher, R., and Didion, D. (1989b). A study of flow boiling heat transfer with refrigerant mixtures. *Int. J. Heat Mass Transfer*, **32**, 1751–1764.

Kadambi, V. (1983). Heat transfer and pressure drop in a helically coiled rectangular duct. ASME Paper No. 83-WA/HT-1.

Kagayama, T., Peterson, P. F., and Schrock, V. E. (1993). Diffusion layer modeling for condensation in vertical tubes with noncondensable gases. *Nucl. Eng. Design*, **141**, 289–302.

Kaji, M., Mori, K., Nakanishi, S., and Ishigai, S. (1984). Flow transitions in air-water flow in helically coiled tubes. Veziroglu, T. N., and Bergles, A. E., Eds., *Multiphase Flow and Heat Transfer. Part III: Fundamentals*, Elsevier Amsterdam, pp. 201–214.

Kandlikar, S. G. (1990). A general correlation for two-phase flow boiling heat transfer coefficients inside horizontal and vertical tubes. *J. Heat Transfer*, **112**, 219–228.

Kandlikar, S. G. (1991). Development of a flow boiling map for subcooled and saturated flow boiling of different fluids in circular tubes. *J. Heat Transfer*, **113**, 190–200.

Kandlikar, S. G. (1997). Further development in subcooled flow boiling heat transfer. *Engineering Foundation Conf. on Convective and Pool Boiling*, May 18–25, Irsee, Germany.

Kandlikar, S. G. (1998a). Heat transfer and flow characteristics in partial boiling, fully developed boiling, and significant void flow regions of subcooled flow boiling. *J. Heat Transfer*, **120**, 395–401.

Kandlikar, S. G. (1998b). Boiling heat transfer with binary mixtures: part 1 – a theoretical model for pool boiling. *J. Heat Transfer*, **120**, 380–387.

Kandlikar, S. G. (1998c). Boiling heat transfer with binary mixtures: part 2 – flow boiling in plane tubes. *J. Heat Transfer*, **120**, 388–394.

Kandlikar, S. G. (2001). A theoretical model to predict pool boiling CHF incorporating effects of contact angle and orientation. *J. Heat Transfer*, **123**, 1071–1079.

Kandlikar, S. G. (2002). Fundamental issues related to flow boiling in minichannels and microchannels. *Exp. Thermal Fluid Sci.*, **26**, 389–407.

Kandlikar, S. G. (2004). Heat transfer mechanisms during flow boiling in microchannels. *J. Heat Transfer*, **126**, 8–16.

Kandlikar, S. G. (2005). High flux heat removal with microchannels – A roadmap of challenges and opportunities. *Heat Transfer Eng.*, **26**(8), 5–14.

Kandlikar, S. G. (2010). A scale analysis based theoretical force balance model for critical heat flux (CHF) during saturated flow boiling in microchannels and minichannels. *J. Heat Transfer*, **132**, paper 081501.

Kandlikar, S. G., and Balasubramanian, P. (2004). An extension of the flow boiling correlation to transition, laminar and deep laminar flows in mini channels and microchannels. *Heat Transfer Eng.*, **25**(3), 86–93.

Kandlikar, S. G., and Balasubramanian, P. (2005). An experimental study on the effect of gravitational orientation on flow boiling of water in 1054 × 197 µm parallel minichannels. *J. Heat Transfer*, **127**, 820–829.

Kandlikar, S. G., and Nariai, H. (1999). Flow boiling in circular tubes. In *Handbook of Phase Change*, Kandlikar, S. G., Shoji, M., and Dhir, V. K., Eds., Taylor & Francis, London, pp. 367–402.

Kandlikar, S. G., and Spiesman, P. H. (1997). Effect of surface characteristics on flow boiling heat transfer. *Engineering Foundation Conf. on Convective and Pool Boiling*, May 18–25, Irsee, Germany.

Kandlikar, S. G., and Steinke, M. E. (2003). Predicting heat transfer during flow boiling in minichannels and microchannels. *ASHRAE Trans.*, **109**, Part 1, 667–676.

Kandlikar, S. G., Mizo, V., Cartwright, M., and Ikenze, E. (1997). Bubble nucleation and growth characteristics in subcooled flow boiling of water. *National Heat Transfer Conf.*, ASME, HTD-342, pp. 11–18.

Kandlikar, S. G., Willistein, D. A., and Borelli, J. (2005). Experimental evaluation of pressure drop elements and fabricated nucleation sites for stabilizing flow boiling in minichannels and microchannels. *Proc. 3rd Int. Conf. on Microchannels and Minichannels*, Part B, pp. 115–124.

Kang, H. J., Lin, C. X., and Ebadian, M. A. (2000). Condensation of R134a flowing inside helicoidal pipe. *Int. J. Heat Mass Transfer*, **43**, 2553–2564.

Kaniowski, R., and Poniewski, M. (2013). Measurements of two-phase flow patterns and local void fraction in vertical rectangular minichannel. *Archives Thermodynamics*, **34**, 2, 3–21.

Kao, Y. S., and Kenning, D. B. R. (1972). Thermocapillary flow near a hemispherical bubble on a heated wall. *J. Fluid Mech.*, **53**, 715–735.

Kapitza, P. L. (1948). Wave flow of thin layers of a viscous fluid, I. The free flow. *Zh. Eksperim. Teor. Fiz.*, **18**, 3.

Karimi, G., and Kawaji, M. (1998). An experimental study of freely falling films in a vertical tube. *Chem. Eng. Sci.* **53**, 3501–3512.

Karimi, G., and Kawaji, M. (1999). Flow characteristics and circulatory motion in wavy falling films with and without counter-current gas flow. *Int. J. Multiphase Flow.*, **25**, 1305–1319.

Kariyasaki, A., Fukano, T., Ousaka, A., and Kagawa, M. (1992). Isothermal air–water two-phase up- and downward flows in vertical capillary tube (1st report, Flow pattern and void fraction). *Trans. JSME Ser. B.*, **58**, 2684–2690.

Karl, J. (2000). Spontaneous condensation in boundary layers. *Heat Mass Transfer*, **36**, 37–44.

Kasturi, G., and Stepanek, J. B. (1972a). Two-phase flow I: Pressure drop and void fraction measurement in co-current gas–liquid flow in a coil. *Chem. Eng. Sci.*, **27**, 1871–1880.

Kasturi, G., and Stepanek, J. B. (1972b). Two-phase flow II: Parameters for pressure drop and void fraction correlations. *Chem. Eng. Sci.*, **27**, 1881–1891.

Kataoka, I., and Ishii, M. (1983). Entrainment of and deposition rates of droplets in annular two-phase flow. *Proc. ASME/JSMG Thermal Eng. Joint Conf.*, Vol. 1.

Kataoka, I., Ishii, M., and Mishima, K. (1983). Generation and size distribution of droplets in annular two-phase flow. *J. Fluids Eng.*, **105**, 230–238.

Kataoka, I., and Ishii, M. (1987). Drift flux model for large diameter pipe and new correlation for pool void fraction. *Int. J. Heat Mass Transf.*, **30**, 1927–1939.

Kattan, N., Thome, J. R., and Favrat, D. (1998a). Flow boiling in horizontal tubes. Part I: Development of a diabatic two-phase flow pattern map. *J. Heat Transfer*, **120**, 140–147.

Kattan, N., Thome, J. R., and Favrat, D. (1998b). Flow boiling in horizontal tubes. Part II: New heat transfer data for five refrigerants. *J. Heat Transfer*, **120**, 148–155.

Kattan, N., Thome, J. R., and Favrat, D. (1998c). Flow boiling in horizontal tubes. Part III: Development of a new heat transfer model based on flow pattern. *J. Heat Transfer*, **120**, 156–165.

Katto, Y. (1990). Prediction of critical heat flux of subcooled flow boiling in round tubes. *Int. J. Heat Mass Transfer*, **33**, 1921–1928.

Katto, Y. (1992). A prediction model of subcooled flow boiling CHF for pressure in the range 0.1–20.0 MPa. *Int. J. Heat Mass Transfer*, **35**, 1115–1123.

Katto, Y. (1994). Critical heat flux. *Int. J. Multiphase Flow*, **20**, Suppl., 53–90.

Katto, Y. and Ohno, H. (1984). An improved version of the generalized correlation of critical heat flux for the forced convective boiling in uniformly heated vertical tubes. *Int. J. Heat Mass Transfer*, **27**, 1641–1648.

Katto, Y., and Yokoya, S. (1984). Critical heat flux of liquid helium (I) in forced convective boiling. *Int. J. Multiphase Flow*, **10**, 401–413.

Kawahara, A., Chung, P. M.-Y., and Kawaji, M. (2002). Investigation of two-phase flow pattern, void fraction and pressure drop in a microchannel. *Int. J. Multiphase Flow*, **28**, 1411–1435.

Kawahara, A., Sadatomi, M., Okayama, K., Kawaji, M., and Chung, P. M.-Y. (2005). Effects of channel diameter and liquid properties on void fraction in adiabatic two phase flow through microchannels. *Heat Transfer Engineering*, **26**, 13–19.

Kawahara, A., Sadotomi, M., Nei, K., and Matsuo, H. (2009). Experimental study on bubble velocity, void fraction and pressure drop for gas-liquid two-phase flow in a circular microchannel. *Int. J. Heat Fluid Flow*, **30**, 831–841.

Kawaji, M., Mori, K., and Bolintineanu, D., 2009. The effects of inlet geometry and gas–liquid mixing on two-phase flow in microchannels. *J. Fluids Eng.*, **131**, paper 041302.

Kays, W., Crawford, M., and Weigand, B. (2005). *Convective Heat and Mass Transfer*, 4th Edn, McGraw-Hill.

Kays, W. M., and London, A. L. (1984). *Compact Heat Exchangers*, 3rd Edn, McGraw-Hill.

Kefer, V., Kastner, W. and Krätzer, W. (1986). Leckraten bei unterkritischen Rohrleitungsrissen. *Jahrestagung Kerntechnik*, Aachen, Germany.

Kefer, V., Kohler, W., and Kastner, W. (1989). Critical heat flux (CHF) and post-CHF heat transfer in horizontal and inclined evaporator tubes. *Int. J. Multiphase Flow*, **15**, 385–392.

Kelessidis, V. C., and Dukler, A. E. (1989). Modeling flow pattern transitions for upward gas–liquid flow in vertical concentric and eccentric annuli. *Int. J. Multiphase Flow*, **15**, 173–191.

Kelly, J. (1994). VIPRE-02 – A two-fluid thermal-hydraulic code for reactor core and vessel analysis: Mathematical modeling and solution methods. *Nucl. Technol.*, **100**, 246–259.

Kendall, G. E., Griffith, P., Bergles, A. E., and Lienhard, J. V. (2001). Small diameter effects on internal flow boiling. *Proc. IMECE-2001*, Nov. 11–16, New York.

Kendzierski, M. A., Chato, J. C., and Rabas, T. J. (2003). Condensation. In Bejan, A., and Kraus, A. D., Eds., *Heat Transfer Handbook*, Wiley, New York, Chapter 10.

Kennedy, J. E., Roach, G. M., Jr., Dowling, M. F., Abdel-Khalik, S. I., Ghiaasiaan, S. M., Jeter, S. M., and Qureshi, Z. H. (2000). The onset of flow instability in uniformly heated horizontal microchannels. *J. Heat Transfer*, **122**, 118–125.

Kenning, D. B. R. (1989). Wall temperature in nucleate boiling. *Proc. 8th Eurotherm Seminar on Advances in Pool Boiling Heat Transfer*, Paderborn, Germany, pp. 1–9.

Kew, P. A., and Cornwell, K. (1997). Correlations for the prediction of boiling heat transfer in small-diameter channels. *Appl. Therm. Eng.*, **17**, 705–715.

Killian, J. D., and Garimellas, S. (2003). A critical review of models of coupled heat and mass transfer in falling-film absorption. *Int. J. Refrigeration*, **24**, 755–797.

Kim, D., Ghajar, A. J., and Dougherty, R. L. (2000). Robust heat transfer correlation for turbulent gas–liquid flow in vertical pipes. *J. Thermophys. Heat Transfer*, **14**, 574–578.

Kim, J., and Ghajar, A. J. (2006). A general heat transfer correlation for non-boiling gas–liquid flow with different flow patterns in horizontal pipes. *Int. J. Multiphase Flow*, **32**, 447–465.

Kim, J.-W., Im, Y.-B., and Kim, J.-S. (2006). A study on performance analysis of helically coiled evaporator with circular minichannels. *J. Mechanical Science and Technology (KSME Int. Journal)*, **20**, 1059–1067.

Kim, M. H., Shin, J. S., Kim, T. J., and Seo, K.W. (2003b). A study of condensation heat transfer in a single mini-tube and review of Korean micro- and mini-channel studies. *Proc. 1st Int. Conf. on Microchannesl and Minichannels*, Rochester, New York, pp. 47–58.

Kim, N. H., Cho, J. P., Kim, J. O., and Youn, B. (2003a). Condensation heat transfer of R-22 and R-410A in flat aluminum multi-channel tubes with or without micro-fins. *Int. J. Refrigeration*, **26**, 830–839.

Kim, S., and Mills, A. F. (1989). Condensation on coherent turbulent liquid jets: Part I – Experimental study. *J. Heat Transfer*, **111**, 1068–1082.

Kim, S., Sun, X., Ishii, M., Beus, S. G., and Lincoln, F. (2002). Interfacial area transport and evaluation of source and sink terms for confined air–water bubbly flow. *Nucl. Eng. Design*, **219**, 61–75.

Kim, S.-M., and Mudawar, I. (2012a). Flow condensation in parallel micro-channels – Part 1: experimental results and assessment of pressure drop correlations. *Int. J. Heat Mass Transfer*, **55**, 971–983.

Kim, S.-M., and Mudawar, I. (2012b). Flow condensation in parallel micro-channels – Part 2: Heat transfer results and correlation technique. *Int. J. Heat Mass Transfer*, **55**, 984–994.

Kim, S.-M., and Mudawar, I. (2012c). Universal approach to predicting two-phase frictional pressure drop for adiabatic and condensing mini/micro-channel flows. *Int. J. Heat Mass Transfer*, **55**, 3246–3261.

Kim, S.-M., and Mudawar, I. (2013). Universal approach to predicting heat transfer coefficient for condensing mini/micro-channel flow. *Int. J. Heat Mass Transfer*, **56**, 238–250.

Kim, Y. J., Jang, J., Hrnjak, P. S., and Kim, M. S. (2009). Condensation heat transfer of carbon dioxide inside horizontal smooth and microfin tubes at low temperatures. *ASME J. Heat Transfer*, **131**, paper # 021501.

Kirillov, P. L., Bobkov, V. P., Boltanko, E. A., Katan, I. B., Smogalev, I. P., and Vinogradov, V. N. (1991a). New CHF table for water in round tubes. Rep. IPPE-2225, Obninsk, Russia.

Kirillov, P. L., Bobkov, V. P., Boltanko, E. A., Katan, I. B., Smogalev, I. P., and Vinogradov, V. N. (1991b). Lookup tables of critical heat flux. *Atomnaya Energiya*, **71**, 18–28.

Kirillov, P. L., Smogalev, I. P., Ivacshkevitch, A. A., Vinogradov, V. N., Sudnitsina, M. O., and Mitrofanova, T. V. (1996). The look-up table for heat transfer coefficient in post-dryout region for water flowing in tubes (the 1996 version). Reprint FEI-2525 Institute of Physics and Power Engineering, Obninsk, Russia.

Kistler, S. F., and Scriven, L. E. (1984). Coating flows. In *Computational Analysis of Polymer Processing*, Pearson, J. K. A. and Richardson, S. M., Eds., Applied Science, London, pp. 243–299.

Klausner, J. F., Chao, B. T., and Soo, S. L. (1990). An improved method for simultaneous determination of frictional pressure drop and vapor volume fraction in vertical flow boiling. *Exp. Therm. Fluid Sci.*, **3**, 404–415.

Klausner, J. F., Mei, R., and Zeng, L. Z. (1997). Predicting stochastic features of vapor bubble detachment in flow boiling. *Int. J. Heat Mass Transfer*, **40**, 3547–3552.

Klimenko, V. V. (1988). A generalized correlation for two-phase forced flow heat transfer. *Int. J. Heat Mass Transfer*, **31**, 541–552.

Klimenko, V. V. (1990). A generalized correlation for two-phase forced flow heat transfer – Second assessment. *Int. J. Heat Mass Transfer*, **33**, 2073–2088.

Kocamustafaogullari, G., and Ishii, M. (1983). Interfacial area and nucleation site density in boiling systems. *Int. J. Heat Mass Transfer*, **26**, 1377–1387.

Kocamustafaogullari, G., and Ishii, M. (1995). Foundations of the interfacial area transport equation and its closure relations. *Int. J. Heat Mass Transfer*, **38**, 481–493.

Kocamustafaogullari, G., Smits, S. R., and Razi, J. (1994). Maximum and mean droplet sizes in annular two-phase flow. *Int. J. Heat Mass Transfer*, **37**, 955–965.

Kohl, M. J., Abdel-Khalik, S. I., Jeter, S. M., and Sadowski, D. L. (2005). An experimental investigation of microchannel flow with internal pressure measurements. *Int. J. Heat Mass Transfer*, **48**, 1518–1533.

Koizumi, H., and Yokohama, K. (1980). Characteristics of refrigerant flow in a capillary tube. *ASHRAE Trans.*, Part 2, **86**, 19–27.

Kokkonen, I., and Tuomisto, H. (1990). Air/water countercurrent flow limitation experiments with full-scale fuel bundle structures. *Exp. Therm. Fluid. Sci.*, **3**, 581–587.

Kolb, W. B., and Cerro, R. L. (1993a). The motion of long bubbles in tubes of square cross section. *Phys. Fluids*, **A5**, 1549–1557.

Kolb, W. B., and Cerro, R. L. (1993b). Film flow in the space between a circular bubble and a square tube. *J. Colloid Interphase Sci.*, **159**, 302–311.

Komaya, S., and Yu, J. (1999). Heat transfer and pressure drop in internal flow condensation. In *Handbook of Phase Change*, Kandlikar, S. G., Shoji, M., and Dhir, V. K., Eds., Taylor & Francis, London, pp. 621–637.

Konno, M., Aoki, M., and Saito, S. (1983). Scale effect on breakup process in liquid–liquid agitated tanks. *J. Chem. Eng. Jpn.*, **16**, 312–319.

Konno, M., Aoki, M., and Saito, S. (1988). Coalescence of dispersed drops in an agitated tank. *J. Chem. Eng. Jpn.*, **21**, 335–338.

Kordyban, E., and Okleh, A. H. (1995). The effect of surfactants on the wave growth and transition to slug flow. *J. Fluids Eng.*, **117**, 389–393.

Kordyban, E. S., and Ranov, T. (1970). Mechanism of slug formation in horizontal two-phase flow. *J. Basic Eng.*, **92**, 857–864.

Kosar, A. (2009). A model to predict saturated critical heat flux in minichannels and microchannels. *Int. J. Thermal Sci.*, **48**, 261–270.

Kosar, A., Kuo, C.-J., and Peles, Y. (2006). Suppression of boiling oscillations in parallel microchannels with inlet restrictors. *J. Heat Transfer*, **128**, 251–260.

Kosar, A., Kuo, C.-J., and Peles, Y. (2005). Boiling heat transfer in rectangular microchannels with reentrant cavities. *Int. J. Heat Mass Transfer*, **48**, 4867–4886.

Kosar, A. and Peles, Y. (2007). Critical heat flux of R-123 in silicon-based microchannels. *J. Heat Transfer*, **129**, 844–851.

Kosar, A., Peles, Y., Bergles, A. E., and Cole, G. S. (2009). Experimental investigation of critical heat flux in microchannels for flow-field probes. ASME Paper ICNMM2009-82214

Kosky, P. G., and Staub, F. W. (1971). Local condensation heat transfer coefficients in the annular flow regime. *AIChE J.*, **17**, 1037–1043.

Koyama, S., and Yu, J. (1999). Heat transfer and pressure drop in internal flow condensation. In *Handbook of Phase Change*, Kandlikar, S. G., Shoji, M., and Dhir, V. K., Eds., Taylor & Francis, London, pp. 621–678.

Koyama, S., Kuwahara, K., Nakashita, K., Kudo, S., and Yamamoto, K. (2001). An experimental study on pressure drop and local heat transfer characteristics of refrigerant R-134a condensing in a multi-port extruded tube. IIF-IIR Commission B1, Paderborn, Germany.

Koyama, S., Kuwara, K., and Nakashita, K. (2003). Condensation of refrigerant in a multiport channel. *Proc. 1st Int. Conf. on Micro- and Mini-channels*, Rochester, New York, pp. 193–205.

Kotake, S. (1978). Film condensation of binary mixture flow in a vertical channel. *Int. J. Heat Mass Transfer*, **21**, 875–884.

Koyama, S., Kuwahara, K., Nakashita, K., and Yamamoto, K. (2003). An experimental study on condensation of refrigerant R134a in a multi-port extruded tube. *Int. J. Refrigeration*, **24**, 425–432.

Koyama, S., Miata, A., Takamatsu, H., and Fujii, T. (1990). Condensation heat transfer of binary refrigerant mixtures of R22 and R114 inside a horizontal tube with internal spiral grooves. *Int. J. Refrigeration*, **13**, 256–263.

Kozeki, M., Nariai, H., Furukawa, T., and Kurosu, K. (1970). A study of helically coiled tube once-through steam generator. *Bull. JSME*, **13**, 1485–1494.

Kreutzer, M. T., Kapteijn, F., Moulijn, J. A., Kleijn, C. R., and Heiszwolf, J. J. (2005). Inertial and interfacial effects on the pressure drop of Taylor flow in capillaries. *AIChE J.*, **51**, 2428–2440.

Krishna, R. (1981). Ternary mass transfer in a wetted-wall column significance of diffusional interactions Part I. Stefan diffusion. *Chem. Eng. Res. Design*, **59**, 35–43.

Krishna, R., Panchal, C. B., Webb, D. R., and Coward, I. (1976). An Ackermann–Colburn and Drew type analysis for condensation of multicomponent mixtures. *Letters Heat Mass Transfer*, **3**, 163–172.

Krishna, R., and Standard, G. L. (1976). A multicomponent film model, incorporating a general matrix method of solution to the Maxwell–Stefan equations. *AIChE J.*, **22**, 383–389.

Krishna, V. S., and Kowalski, J. E. (1984). Stratified-slug flow transition in a horizontal pipe containing a rod bundle. *AIChE Symp. Ser.*, 236, **80**, 282–289.

Kroeger, P. G. (1978). Application of a non-equilibrium drift-flux model to two-phase blowdown experiments. Paper presented at OECD/NEA Specialists' Meeting on Transient Two-Phase Flow, Toronto, Canada, August 1998.

Kronig, R., and Brink, J. C. (1950). On the theory of extraction from falling droplets. *Appl. Sci. Res.*, **A2**, 142–154.

Kuan, W. K., and Kandlikar, S. G. (2008). Critical heat flux measurement and model for Refrigerant-123 under stabilized flow conditions in microchannels. *J. Heat Transfer*, **130**, paper 034503.

Kubair, V., and Kuloor, N. B. (1965). Non-isothermal pressure drop data for liquid flow in helical coils. *Indian J. Technol.*, **3**, 5–7.

Kukita, Y., Katayama, J., Nakamura, H., and Tasaka, K. (1990). Loop seal clearing and refilling during a PWR small-break LOCA. *Nucl. Eng. Design*, **121**, 431–440.

Kumar, V., Saini, S., Sharma, M., and Nigam, K. D. P. (2006). Pressure drop and heat transfer study in tube-in-tube helical heat exchanger. *Chem. Eng. Sci.*, **61**, 4403–4416.

Kumar, S., Cherlo, R., Kariveti, S., and Pushpavanam, S., 2010. Experimental and numerical investigation of two-phase (liquid–liquid) flow behavior in rectangular microchannels. *Ind. Eng. Chem. Res.*, **49**, 893–899.

Kuo, C. J., and Peles, Y. (2008). Critical heat flux of water at sub atmospheric pressures in microchannels. *J. Heat Transfer*, **130**, paper 072403.

Kutateladze, S. S. (1952). *Heat Transfer in Condensation and Boiling*, Moscow. English translation in US Atomic Energy Commission AEC-tr-3770, 2nd Edn.

Kutateladze, S. S. (1961). Boiling heat transfer. *Int. J. Heat Mass Transfer*, **4**, 31–45.

Kutateladze, S. S. (1972). Elements of hydrodynamics of gas–liquid systems. *Fluid Mech. Sov. Res.*, **1**, 29–50.

Kuwahara, A., Chung, P. M.-Y., and Kawaji, M. (2002). Investigation of two-phase flow patterns, void fraction and pressure drop in a microchannel. *Int. J. Multiphase Flow*, **28**, 1411–1435.

Kuwahara, K., Koyama, S., and Hashimoto, Y. (2000). Characteristics of evaporation heat transfer and flow pattern of pure refrigerant HFC134a in a horizontal capillary tube. *Proc. 4th JSME-KSME Conf.*, pp. 385–390.

Laborie, S., Cabassud, C., Durand-Bourlier, L., and Laine, J. M. (1999). Characterization of gas–liquid two-phase flow inside capillaries. *Chem. Eng. Sci.*, **54**, 5723–5835.

Lackme, C. (1979). Incompleteness of the flashing of supersaturated liquid and sonic ejection of the produced phases. *Int. J. Multiphase Flow*, **5**, 131–141.

Lahey, R. T., Jr., and Drew, D. A. (1988). The three-dimensional time and volume averaged conservation equations of two-phase flow. In *Advances in Nuclear Science and Technology.*, J. Lewis and M. Becker, Eds., Plenum Press, New York, **20**, 1–69.

Lahey, T. R., Jr., and Moody, F. J. (1993). *The Thermal-Hydraulics of Boiling Water Nuclear Reactors*, 2nd Edn, American Nuclear Society, LaGrange Park, IL.

Lakehal, D., Larrignon, G., and Narayanan, C. (2008). Computational heat transfer and two-phase flow topology in miniature tubes. *Microfluidics and Nanofluidics*, **4**, 261–271.

Lamb, Sir H. (1932). *Hydrodynamics*, 6th Edn, Cambridge University Press, Cambridge.

Lazarek, G. M., and Black, H. S. (1982). Evaporative heat transfer, pressure drop and critical heat flux in a small vertical tube with R-113. *Int. J. Heat Mass Transfer*, **25**, 945–960.

Ledinegg, M. (1938). Instabilität der Strömung bei natürlichem und Zwangsumlauf. *Wärme*, **61**, 891–898.

Lee, C. Y., and Lee, S. Y. (2008). Influence of surface wettability on transition of two-phase flow pattern in round mini-channels. *Int. J. Multiphase Flow*, **34**, 706–711.

Lee, C. H., and Mudawar, I. (1988). A mechanistic critical heat flux model for subcooled flow boiling based on local bulk flow conditions. *Int. J. Multiphase Flow*, **14**, 711–728.

Lee, H. J., and Lee, S. Y. (2001a). Pressure drop correlations for two-phase flow within horizontal rectangular channels with small heights. *Int. J. Multiphase Flow*, **27**, 783–796.

Lee, H. J., and Lee, S.Y. (2001b). Heat transfer correlation for boiling flows in small rectangular horizontal channels with low aspect ratios. *Int. J. Multiphase Flow*, **27**, 2043–2062.

Lee, J., and Mudawar, I. (2005a). Two-phase flow in high-heat-flux micro-channel heat sink for refrigeration cooling applications: Part I – Pressure drop characteristics. *Int. J. Heat Mass Transfer*, **48**, 928–940.

Lee, J., and Mudawar, I. (2005b). Two-phase flow in high-heat-flux micro-channel heat sink for refrigeration cooling applications: Part II – Heat transfer characteristics. *Int. J. Heat Mass Transfer*, **48**, 941–955.

Lee, H. J., Liu, D. Y., Alyousef, Y., and Yao, S.-C. (2010). Generalized two-phase pressure drop and heat transfer correlations in evaporative micro/minichannels. *J. Heat Transfer*, **132**, paper 041004.

Lee, K.-W., No, H. C., and Song, C.-H. (2007). Onset of water accumulation in the upper plenum with a perforated plate. *Nucl. Eng. Design*, **237**, 1088–1095.

Lee, M., Cheung, L. S. L., Lee, Y.-K., and Zohar, Y. (2005). Height effect on nucleation-site activity and size-dependent bubble dynamics in microchannel convective boiling. *J. Micromech. Microeng.*, **15**, 2121–2129.

Lee, R. C., and Nydahl, J. E. (1989). Numerical calculation of bubble growth in nucleate boiling from inception through departure. *J. Heat Transfer*, **111**, 474–479.

Lee, S. C., and Bankoff, S. G. (1983). Stability of steady-water countercurrent flow in an inclined channel: Flooding. *J. Heat Transfer*, **105**, 713–718.

Lee, S., and Kim, H.-J. (1992). Prediction of loop seal formation and clearing during small break loss of coolant accident. *J. Korean Nuclear Society*, **24**, 243–251.

Lee, J. Y., Kim, M. H., Kaviany, M., and Son, S. Y. (2011). Bubble nucleation in micro-channel flow boiling using single artificial cavity. *Int. J. Heat Mass Transfer*, **54**, 5139–5148.

Lee, K.-W., No, H. C., and Song, C.-H. (2007). Onset of water accumulation in the upper plenum with a perforated plate. *Nucl. Eng. Design*, **237**, 1088–1095.

Lee, S. Y., and Schrock, V. E. (1988). Homogeneous non-equilibrium critical flow model for liquid stagnation states. *Proc. 7th National Heat Transfer Conf.*, ASME, New York, HTD Vol. 96, pp. 507–513.

Lee, W. C., and Rose, J. W. (1984). Forced convection film condensation on a horizontal tube with and without noncondensing gases. *Int. J. Heat Mass Transfer*, **27**, 519–528.

Lenzing, T., Friedel, L., and Alhusein, M. (1998). Critical mass flow rate in accordance with the omega-method of DIERS and the Homogeneous Equilibrium Model. *J. Loss Prevention in the Process Industries*, **11**, 391–395.

Leonard, J. E., Sun, K. H., Anderson, J. G. M., Dix, G. E., and Yuoh, T. (1978). Calculation of low flow boiling heat transfer for BWR LOCA analysis. Rep. NEDO-20566–1 Rev. 1, General Electric Company, San Jose, CA.

Leung, J. C. (1986). A generalized correlation for one-component homogeneous equilibrium flashing flow. *AIChE J.*, **32**, 1743–1746.

Leung, J. C., and Epstein, M. (1990a). The discharge of two-phase flashing flow from an inclined duct. *J. Heat Transfer*, **112**, 524–528.

Leung, J. C., and Epstein, M. (1990b). A generalized correlation for two-phase nonflashing homogeneous choked flow. *J. Heat Transfer*, **112**, 528–530.

Leung, J. C., and Grolmes, M. A. (1988). A generalized correlation for flashing choked flow of initially subcooled liquid. *AIChE J.*, **34**, 688–691.

Leung, L. K. H., Hammouda, N., and Groeneveld, D. C. (1997). A look-up table for film boiling heat transfer coefficients in tubes with vertical upward flow. *Proc. 8th Int. Topical Meeting on Nuclear Reactor Thermal Hydraulics (NURETH-8)*, Kyoto, Japan, Sept. 30–Oct.4.

Leung, R. S. S. Y., Gupta, R., Fletcher, D. F., and Haynes, B. S. (2012). Effect of flow characteristics on Taylor flow heat transfer. *Ind. Eng. Chem. Res.*, **51**, 2010–2020.

Levich, V. G. (1962). *Physiochemical Hydrodynamics*, Prentice Hall, Englewood Cliffs, NJ.

Levy, S. (1967). Forced convection subcooled boiling: Prediction of vapor volumetric fraction. *Int. J. Heat Mass Transfer*, **10**, 951–965.

Levy, S. (1999). *Two-Phase Flow in Complex Systems*, Wiley, New York.

Levy, S., Healzer, J. M., and Abdollahian, D. (1981). Prediction of critical heat flux in vertical pipe flow. *Nucl. Eng. Design*, **65**, 131–140.

Lezzi, A. M., Niro, A., and Beretta, G. P. (1994). Experimental data of CHF for forced convection water boiling in long horizontal capillary tubes. In *Heat Transfer 1994: Proceedings of the Tenth International Heat Transfer Conference*, Hewitt, G. W., Ed., Rugby, UK, **7**, 491–496.

Li, J., and Peterson, G. P. (2005a). Microscale heterogeneous boiling on smooth surfaces – From bubble nucleation to bubble dynamics. *Int. J. Heat Mass Transfer*, **48**, 4316–4332.

Li, J., and Peterson, G. P. (2005b). Boiling nucleation and two-phase flow patterns in forced liquid flow in microchannels. *Int. J. Heat Mass Transfer*, **48**, 4797–4810.

Li, J., and Wang, B. (2003). Size effect on two-phase flow regime for condensarion in micro/mini tubes. *Heat Transfer – Asian Research*, **32**, 65–71.

Li, W., and Wu, Z. (2010a). A general correlation for adiabatic two-phase pressure drop in micro/mini-channels. *Int. J. Heat Mass Transfer*, **53**, 2732–2739.

Li, W., and Wu, Z. (2010b). A general criterion for evaporative heat transfer in micro/mini-channels. *Int. J. Heat Mass Transfer*, **53**, 1967–1976.

Li, H. J., Liu, D. Y., Alyousef, Y., and Yao, S.-C. (2010). Generalized two-phase pressure drop and heat transfer correlations in evaporative micro/minichannels. *J. Heat Transfer*, **132**, paper 041004.

Li, S., Han, J.-T., Su, G.-P., and Pan, J.-H. (2007). Condensation heat transfer of R-134a in horizontal straight and helically coiled tube-in-tube heat exchangers. *J. Hydrodynamics, Ser. B.*, **19**, 677–682.

Liaw, S. P., and Dhir, V. K. (1986). Effect of surface wettability on transition boiling heat transfer from a vertical surface. *Proc. 8th Int. Heat Transfer Conf.*, San Francisco, **4**, 2031–2036.

Lie, Y. M., Su, F. Q., Lai, R. L., and Lin, T. F. (2006). Experimental study of evaporation heat transfer characteristics of refrigerants R-134a and R-407C in horizontal small tubes. *Int. J. Heat Mass Transfer*, **49**, 207–218.

Lienhard, J. H. (1976). Correlation of the limiting liquid superheat. *Chem. Eng. Sci.*, **31**, 847–849.

Lienhard, J. H., and Dhir (1973). Hydrodynamic prediction of peak pool-boiling heat fluxes from finite bodies. *J. Heat Transfer*, **95**, 152–158.

Lienhard, J. H., and Karimi, A. (1981). Homogeneous nucleation and the spinodal line. *J. Heat Transfer*, **103**, 61–64.

Lienhard, J. H., and Witte, L. C. (1985). An historical review of the hydrodynamic theory of boiling. *Rev. Chem. Eng.*, **2**, 197–280.

Lienhard, J. H., IV, and Lienhard, J. H., V. (2005). *A Heat Transfer Textbook*, 3rd Edn, Phlogiston Press, Cambridge, MA.

Lin, C. X., and Ebadian, M. A. (1997). Developing turbulent convective heat transfer in helical pipes. *Int. J. Heat Mass Transfer*, **40**, 3861–3873.

Lin, C. X., and Ebadian, M. A. (2007). Condensation heat transfer and pressure drop of R134a in annular helicoidal pipe at different orientations. *In. J. Heat and Mass Transfer*, **50**, 4256–4264.

Lin, L., Udell, K. S., and Pisano, A. P. (1993). Vapor bubble formation on a micro heater in confined and unconfined micro channels. *Heat Transfer on the Microscale*, ASME HTD Vol. **253**, 85–93.

Lin, S., Kew, P. A., and Cornwell, K. (2001a). Two-phase heat transfer to a refrigerant in a 1 mm diameter tube. *Int. J. Refrigeration*, **24**, 51–56.

Lin, S., Kew, P. A., and Cornwell, K. (2001b). Flow boiling of refrigerant R141b in small tubes. *Trans. Inst. Chem. Eng.*, **79-A**, 417–424.

Lin, S., Kwok, C. C. K., Li, R.-Y., Chen, Z.-H., and Chen, Z.-Y. (1991). Local frictional pressure drop during vaporization of R-12 through capillary tubes. *Int. J. Multiphase Flow*, **17**, 95–102.

Li, J.-M., and Wang, B.-X. (2003). Size effect on two-phase flow regime for condensarion in micro/mini tubes. *Heat Transfer – Asian Research*, **32**, 65–71.

Li, W., and Wu, Z. (2010). A general criterion for evaporative heat transfer in micro/mini-channels. *Int. J. Heat Mass Transfer*, **53**, 1967–1976.

Lin, W.-C., Ferng, Y.-M., and Chieng, C.-C. (2013). Numerical computations on flow and heat transfer characteristics of a helically coiled heat exchanger using different turbulence models. *Nucl. Eng. Design*, **263**, 77–86.

Liu, C. P., McCarthy, G. E., and Tien, C. L. (1982). Flooding in vertical gas–liquid countercurrent flow through multiple short paths. *Int. J. Heat and Mass Transfer*, **25**, 1301–1312.

Liu, D., Lee, P.-S., and Garimella, S. V. (2005). Prediction of the onset of nucleate boiling in microchannel flow. *Int. J. Heat Mass Transfer*, **48**, 5234–5149.

Liu, D., and Wang, S. (2008). Hydrodynamics of Taylor flow in noncircular capillaries. *Chem. Eng. Process.*, **47**, 2098–2106.

Liu, D., and Wang, S. (2011). Gas–liquid mass transfer in Taylor flow through circular capillaries. *Ind. Eng. Chem. Res.*, **50**, 2323–2330

Liu, H., Vandu, C. O., and Krishna, R. (2005). Hydrodynamics of Taylor flow in vertical capillaries: Flow regimes, bubble rise velocity, liquid slug length, and pressure drop. *Ind. Eng. Chem. Res.*, **44**, 4884–4897.

Liu, Z., and Winterton, R. H. S. (1991). A general correlation for saturated and subcooled flow boiling in tubes and annuli, based on a nucleate pool boiling equation. *Int. J. Heat Mass Transfer*, **34**, 2759–2766.

Lockhart, R. W., and Martinelli, R. C. (1949). Proposed correlations of data for isothermal two-phase, two-component flow in a pipe. *Chem. Eng. Prog.*, **45**, 39–48.

Loomsmore, C. S., and Skinner, B. C. (1965). Subcooled critical heat flux for water in round tubes. MS thesis, MIT, Cambridge, MA.

Lowdermilk, W. H., Lanzo, C. D., and Siegel, B. L. (1958). Investigation of boiling burnout and flow stability for water flowing in tubes. NACA TN 4382.

Lowry, B., and Kawaji, M. (1988). Adiabatic vertical two-phase flow in narrow flow channels. *AIChE Symp. Ser.*, **48**, 133–139.

Maa, J. R. (1967). Evaporation coefficient of liquids. *Ind. Eng. Chem. Fundam.*, **6**, 504–518.

McEligot, D. M. (1964). Generalized peak heat flux for dilute binary mixtures. *AIChE J.*, **10**, 130–131.

Mahafy, J. H. (1982). A stability enhancing two-step method for fluid flow calculations. *J. Comput. Phys.*, **46**, 329–341.

Mahoney, A. W., and Ramkrishna, D. (2002). Efficient solution of population balance equations with discontinuities by finite elements. *Chem. Eng. Sci.*, **57**, 1107–1119.

Mala, G. M., and Li, D. (1999). Flow characteristics of water in microtubes. *Int. J. Heat Fluid Flow*, **20**, 142–148.

Mandal, S. N., and Das, S. K. (2003). Gas–liquid flow through coils. *Korean J. Chem. Eng.*, **20**, 624–630.

Mandhane, J. M., Gregory, G. A., and Aziz, K. (1974). A flow pattern map for gas–liquid flow in horizontal pipes. *Int. J. Multiphase Flow*, **1**, 537–553.

Manlapaz, R. L., and Churchill, S. W. (1980). Fully developed laminar flow in a helically coiled tube of finite pitch. *Chem. Eng. Comm.*, **7**, 57–78.

Manlapaz, R. L., and Churchill, S. W. (1981). Fully developed laminar convection from a helical coil. *Chem. Eng. Comm.*, **9**, 185–200.

Maracy, M., and Winterton, R. H. S. (1988). Hysteresis and contact angle effects in transition pool boiling of water. *Int. J. Heat Mass Transfer*, **31**, 1443–1449.

Marak, K. A. (2009). Condensation heat transfer and pressure drop for methane and binary methane fluids in small channels, PhD thesis, Norwegian University of Science and Technology, Trondheim.

Marchessault, R. N., and Mason, S. G. (1960). Flow of entrapped bubbles through a capillary. *Ind. Eng. Chem.* **52**, 79–84.

Marcy, G. P. (1949). Pressure drop with change of phase in a capillary tube. *Refrig. Eng.*, **57**, 53–57.

Marek, R., and Straub, J. (2001). The origin of thermocapillary convection in subcooled nucleate pool boiling. *Int. J. Heat Mass Transfer*, **44**, 619–632.

Marsh, W. J., and Mudawar, I. (1989). Predicting the onset of nucleate boiling in wavy freefalling turbulent liquid films. *Int. J. Heat Mass Transfer*, **32**, 361–378.

Martin-Callizo, C., Ali, R., and Palm, B. (2008). Dryout incipience and critical heat flux in saturated flow boiling of refrigerants in a vertical uniformly heated microchannel. *Proc. ASME Sixth International Conference on Nanochannels, Microchannels and Minichannels*, pp. 708–712.

Martinelli, R. C. (1947). Heat transfer to molten metals. *Trans. ASME*, **69**, 947–951.

Martinelli, R. C., and Nelson, D. B. (1948). Prediction of pressure drop during forced circulation boiling of water. *Trans. ASME*, **70**, 695–702.

Marto, P. J. (1984). Heat transfer and two-phase flow during shell-side condensation. *Heat Transfer Eng.*, **5**, 31–60.

Marto, P. J. (1988). Fundamentals of condensation. In *Two-Phase Flow Heat Exchangers*, Kakac, S., Bergles, A. E., and Oliveira, Fernandes, E., Eds., Kluwer Academic, Dordrecht, pp. 221–291.

Matkovic, M., Cavallini, A., Del Col, D. D., and Rossetto, L. (2009). Experimental study on condensation heat transfer inside a single circular minichannel. *Int. J. Heat Mass Transfer*, **52**, 2311–2323.

Matsumoto, K., Nakamura, S., Gotoh, N., Nabarayashi, T., Tanaka, Y., and Horimizu, Y. (1989). Study on coolant leak rates through pipe cracks: Part 2 – Pipe test. *Proc. ASME Pressure Vessels and Piping Conf., JSME Co-sponsorship*, ASME, New York, ASME PVP-Vol. 165, pp. 113–120.

Mauro, A. W, Thome, J. R, Toto, D., and Vanoli, G. P. (2010). Saturated critical heat flux in a multi-microchannel heat sink fed by a split flow system. *Exp. Thermal Fluid Sci.*, **34**, 81–92.

McAdams, W. H. (1954). *Heat Transmission*, 3rd Edn, McGraw-Hill, New York.

McAdams, W. H., Minden, C. S., Carl, R., Picornell, D. M., and Dew, J. E. (1949). Heat transfer at high rates to water with surface boiling. *Ind. Eng. Chem.*, **41**, 1945–1963.

McAdams, W. H., Woods, W. K., and Heroman, L. C., Jr. (1942). Vaporization inside horizontal tubes – II – Benzene–oil mixtures. *Trans. ASME*, **64**, 193–200.

McBeth, R. V. (1965–66). An appraisal of forced convection burnout data. *Proc. Inst. Mech. Eng.*, **180**, 47–48.

McBeth, R. V., and Thompson, B. (1964). Boiling water heat transfer burnout in uniformly heated round tubes: A compilation of world data with accurate correlations. UKAEA Rep. AEEW-R356, Winfrith, England.

McFadden, J. H., et al. (1992). RETRAN-03. A program for transient thermal-hydraulic analysis of complex fluid systems. Electric Power Research Institute Rep. EPRI NP-7450, Vol. 1, Palo Alto, CA.

McGillis, W. R., and Carey, V. P. (1996). On the role of Marangoni effect on the critical heat flux for pool boiling of binary mixtures. *J. Heat Transfer*, **118**, 103–109.

McQuillan, K. W., and Whalley, P. B. (1985a). Flow patterns in vertical two-phase flow. *Int. J. Multiphase Flow*, **11**, 161–175.

McQuillan, K. W., and Whalley, P. B. (1985b). A comparison between flooding correlations and experimental flooding data for gas–liquid flow in vertical circular tubes. *Chem. Eng. Sci.*, **40**, 1425–1440.

Meisenburg, S. J., Boarts, R. M., and Badger, W. L. (1935). The influence of small concentrations of air in steam on the steam film coefficient of heat transfer. *Trans. Am. Inst. Chem. Eng.*, **31**, 622–631.

Merilo, M. (1977). Critical heat flux experiments in a vertical and horizontal tube with Freon-12 and water as coolant. *Nucl. Eng. Design*, **44**, 1–16.

Merilo, M. (1979). Fluid-to-fluid modeling and correlation of flow boiling crisis in horizontal tubes. *Int. J. Multiphase Flow*, **5**, 313–325.

Merte, J., Jr., and Clark, J. A. (1964). Boiling heat transfer with cryogenic fluids at standard and near-zero gravity. *J. Heat Transfer*, **86**, 351–359.

Michaelides, E. E. (1997). Review – The transient equation of motion for particles, bubbles, and droplets. *J. Fluids Eng.*, **119**, 233–247.

Michiyoshi, I. (1978). Two-phase two-component heat transfer. *Proc. 6th Int. Heat Transfer Conf.*, **6**, 219–233.

Mikic, B. B., Rohsenow, W. M., and Griffith, P. (1970). On bubble growth rates, *Int. J. Heat Mass Transfer*, **13**, 647–666.

Mikol, E. P. (1963). Adiabatic single and two-phase flow in small bore tubes. *ASHRAE J.*, **5**, 75–86.

Miller, W. A., and Keyhani, M. (2001). The effect of roll waves on the hydrodynamics of falling films observed in vertical column absorbers. *American Society of Mechanical Engineers, Advanced Energy Systems Division (Publication) AES*, **41**, 45–56.

Millies, M., Drew, D. A., and Lahey, R. T., Jr. (1996). A first order relaxation model for the prediction of the local interfacial area density in two-phase flows. *Int. J. Multiphase Flow*, **22**, 1073–1104.

Mills, A. F. (2001). *Mass Transfer*, Prentice Hall, Upper Saddle River, NJ.

Mills, A. F., and Chung, D. K. (1973). Heat transfer across turbulent falling films. *Int. J. Heat Mass Transfer*, **16**, 694–696.

Mills, A. F., and Seban, R. A. (1967). The condensation coefficient of water. *Int. J. Heat Mass Transfer*, **10**, 1815–1827.

Minami, N., Murase, M., and Tomiyama, A. (2010). Countercurrent gas–liquid flow in a PWR hot leg under reflux cooling: II. Numerical simulation of 1/15-scale air–water tests. *J. Nucl. Sci. Technol.* **47**, 149–155.

Mishima, K. (1984). Boiling burnout at low flow rate and low pressure conditions. PhD thesis, Research Reactor Institute, Kyoto University, Japan.

Mishima, K., and Hibiki, T. (1996). Some characteristics of air–water two-phase flow in small diameter vertical tubes. *Int. J. Multiphase Flow*, **22**, 703–712.

Mishima, K., and Ishii, M. (1980). Theoretical prediction of onset of horizontal slug flow. *J. Fluids Eng.*, **102**, 441–445.

Mishima, K., and Ishii, M. (1984). Flow regime transition criteria for two-phase flow in vertical tubes. *Int. J. Heat Mass Transfer*, **27**, 723–737.

Mishima, K., Hibiki, T., and Nishihara, H. (1993). Some characteristics of gas–liquid flow in narrow rectangular ducts. *Int. J. Multiphase Flow*, **19**, 115–124.

Mizukami, K. (1977). Entrapment of vapor in re-entrant cavities. *Lett. Heat Mass Transfer*, **2**, 279–284.

Modine, A. D. (1963). Ternary Mass Transfer, Department of Chemical Engineering. Carnegie Institute of Technology, Pittsburgh.

Moin, P. (2001). *Fundamentals of Engineering Numerical Analysis*, Cambridge University Press, Cambridge.

Moissis, R., and Berenson, P. J. (1963). On the hydrodynamic transitions in nucleate boiling. *J. Heat Transfer*, **85**, 221–229.

Montes, F. J., Galan, M. A., and Cerro, R. L. (1999). Mass transfer from oscillating bubbles in bioreactors. *Chem. Eng. Sci.*, **54**, 3127–3136.

Montes, F. J., Galan, M. A., and Cerro, R. L. (2002). Comparison of theoretical and experimental characteristics of oscillating bubbles. *Ind. Eng. Chem. Res.*, **41**, 6235–6245.

Moncalvo, D., and Friedel, L. (2007). Explicit critical pressure ratio for choked two-phase homogeneous nozzle flow. *Chem. Eng. Technol.*, **30**, 530–533.

Moncalvo, D., and Friedel, L. (2010). A simple, explicit formula for the critical pressure of homogenous two-phase nozzle flows. *J. Loss Prevention in the Process Industries*, **23**, 178–182.

Moody, F. J. (1965). Maximum flow rate of a single-component two-phase mixture. *J. Heat Transfer*, **87**, 134–142.

Moody, F. J. (1966). Maximum two-phase vessel blowdown from pipes. *J. Heat Transfer*, **88**, 285–295.

Moody, F. J. (1975). Maximum discharge rate of liquid/vapor mixtures from vessels. In *Non-Equilibrium Two-Phase Flow*, ASME Special Publication, ASME, New York.

Moody, F. J. (1979). Maximum discharge rate of liquid–vapor mixtures from vessels. General Electric Co. Rep. NEDO-21052-A, San Jose, CA.

Moody, F. J. (1990). *Introduction to Unsteady Thermofluid Mechanics*, Wiley Interscience, New York.

Moore, F. D., and Mesler, R. B. (1961). The measurement of rapid surface temperature fluctuations during nucleate boiling of water. *AICHE J.*, **7**, 620–624.

Morel, C., Goreaud, N., and Delhaye, J.-M. (1999). The local volumetric interfacial area transport equation: Derivation and physical significance. *Int. J. Multiphase Flow*, **25**, 1099–1128.

Morgan, C. D., and Rush, G. C. (1983). Experimental measurements of condensation heat transfer with noncondensible gases present in a vertical tube at high pressure. Presented in *21st National Heat Transfer Conf., Seattle, Washington, July 24–28*, ASME-HTD-Vol. 27.

Morris, S. D. (1985). Two-phase pressure drop across valves and orifice plates, European Two-phase flow group meeting, Southampton, UK, 4–7 June.

Moser, K. W., Webb, R. L., and Na, B. (1998). A new equivalent Reynolds number model for condensation in smooth tubes. *J. Heat Transfer*, **120**, 410–417.

Mosyak, A., Rodes, L., and Hetsroni, G. (2012). Boiling incipience in parallel micro-channels with low mass flux subcooled water flow. *Int. J. Multiphase Flow*, **47**, 150–159.

Mudawar, I., and El-Masri, M. A. (1986). Momentum and heat transfer across freely-falling turbulent liquid films. *Int. J. Heat Mass Transfer*, **12**, 771–790.

Muller-Steinhagen, H., and Heck, K. (1986). A simple friction pressure drop correlation for two-phase flow in pipes. *Chem. Eng. Prog.*, **20**, 297–308.

Munis, A. A., Ghiaasiaan, S. M., and Wassel, A. T. (1987). *Water Retention in Primary Coolant System of Pressurized Water Reactors during Severe Accidents*. EPRI NP-5144, Electric Power Research Institute, Palo Alto, California.

Munoz-Cobo, J. L., Herranz, L., Sancho, J., Tkachenko, I., and Verdu, G. (1996). Turbulent vapor condensation with noncondensable gases in vertical tubes. *Int. J. Heat Mass Transfer*, **39**, 3249–3260.

Munson, B. R., Young, D. F., and Okishii, T. H. (1998). *Fundamentals of Fluid Mechanics*, 3rd Edn, Wiley, New York.

Murai, Y., Yoshikawab, S., Todac, S., Ishikawad, M., and Yamamoto, M. (2006). Structure of air–water two-phase flow in helically coiled tubes. *Nucl. Eng. Design*, **236**, 94–106.

Muralidhar, R., Ramkrishna, D., and Kumar, R. (1988). Coalescence of rigid droplets in a stirred dispersion – II. Band-limited force fluctuations. *Chem. Eng. Sci.*, **43**, 1559–1568.

Murase, M., Tomiyama, A., Lucas, D., Kinoshita, I., Utanohara, Y., and Yanagi, C. (2012). Correlation of countercurrent flow limitation in a PWR hot leg. *J. Nucl. Sci. Technol.*, **49**, 398–407.

Muzychka, Y. S., Walsh, E. J., and Walsh, P. (2011). Heat transfer enhancement using laminar gas–liquid segmented plug flows. *J. Heat Transfer*, **133**(4), paper 041902.

Nabarayashi, T., Ishiyama, T., Fujii, M., Matsumoto, K., Harimizu, Y., and Tanaka, Y. (1989). Study on coolant leak rates through pipe cracks: Part 1 – Fundamental tests. *Proc. ASME Pressure Vessels and Piping Conf., JSME Co-sponsorship*, ASME, New York, ASME PVP, Vol. 165, pp. 121–127.

Naitoh, H., Chino, K., and Kawabe, R. (1978). Restrictive effect of ascending steam on falling water during top spray emergency core cooling. *J. Nucl. Sci. Technol.*, **15**, 806–815.

Naphon, P., and Wongwises, S. (2002). An experimental study of the in-tube convective heat transfer coefficients in a spiral coil heat exchanger. *Int. Comm. Heat Mass Transfer*, **29**, 797–809.

Nariai, H., and Inasaka, F. (1992). Critical heat flux and flow characteristics of subcooled flow boiling with water in narrow tubes. In *Dynamics of Two-Phase Flow*, Jones, O. C., and Michiyoshi, I., Eds., CRC Press, Boca Raton, FL, pp. 689–708.

Nariai, H., Inasaka, F., and Shimuara, T. (1987). Critical heat flux of subcooled flow boiling in narrow tube. *Proc. ASME/JSME Thermal Energy Joint Conf., 1987*, **5**, 455–462.

Nariai, H., Inasaka, F., and Uehara, K. (1989). Critical heat flux in narrow tubes with uniform heating. *Heating Transfer Jpn. Res.*, **18**, 21–30.

Nariai, H., Kobayashi, M., and Matsuoka, T. (1982). Friction pressure drop and heat transfer coefficient oftwo phase flow in helically coiled tube once-through steam generator for inte-grted type marine water reactor. *J. Nucl. Sci. Technol.*, **19**, 936–947.

Narrow, T. L., Ghiaasiaan, S. M., Abdel-Khalik, S. I., and Sadowski, D. L. (2000). Gas–liquid two-phase flow patterns and pressure drop in a horizontal micro-rod bundle. *Int. J. Multi-phase Flow*, **26**, 1281–1294.

Narsimhan, G., Gupta, J. P., and Ramkrishna, D. (1979). A model for transitional breakage probability of droplets in agitated lean liquid-liquid dispersions. *Chem. Eng. Sci.*, **34**, 257–265.

Narsimhan, G., Ramkrishna, D., and Gupta, J. P. (1980). Analysis of drop size distributions in lean liquid–liquid dispersions. *AIChE J.*, **26**, 991.

Narsimhan, G., Nejfelt, G., and Ramkrishna, D. (1984). Breakage functions of droplets in agi-tated liquid–liquid dispersions. *AIChE J.*, **30**, 457.

Nema, G., Garimella, S., and Fronk, M.B. (2014). Flow regime transitions during condensation in microchannels. *Int. J. Refrigeration*, **40**, 227–240.

Neufeld, P. N., Janzen, A. R., and Aziz, R. A. (1972). Empirical equations to calculate 16 of the transport collision integrals Ω for the Lennard-Jones (12–6) potential. *J. Chem. Phys.*, **57**, 1100–1102.

Nichols, B. D., Hirt, C. W., and Hotchkiss, R. S. (1980). SOLA-VOF: A solution algorithm for transient fluid flow with multiple free boundaries. Rep. LA-8355, Los Alamos National Lab-oratory, Los Alamos, NM.

Nicklin, D. J., Wilkes, J. O., and Wilkes, F. F. (1962). Two-phase flow in vertical tubes, *Trans. Inst. Chem. Engrs.*, **40**, 61–68.

Nigmatulin, R. I. (1979). Spatial averaging in the mechanics of heterogeneous and dispersed systems. *Int. J. Multiphase Flow*, **5**, 353–385.

Nijhawan, S., Chen, J. C., Sundaram, R. K., and London, E. J. (1980). Measurement of vapor superheat in post-critical-heat-flux boiling. *J. Heat Transfer*, **102**, 465–470.

Nijhuis, T. A., Kreutzer, M. T., Romijn, A. C. J., Kapteijn, F., and Moulijn, J. A. (2001). Mono-lithic catalysts as efficient three-phase reactors. *Chem. Eng. Sci.*, **56**, 823–829.

Nishikawa, K., Fujita, Y., Yuchida, S., and Ohta, H. (1983). Effect of heating surface orienta-tion on nucleate boiling heat transfer. *Proc. ASME/JSME Thermal Eng. Joint Conf.*, Vol. **1**, ASME, New York, pp. 129–136.

Nishikawa, K., Fujita, Y., Yuchida, S., and Ohta, H. (1984). Effect of surface configuration on nucleate boiling heat transfer. *Int. J. Heat Mass Transfer*, **27**, 1559–1571.

Nukiyama, S. (1934). The maximum and minimum values of heat Q transmitted from metal to boiling water under atmospheric pressure. *J. Jpn. Soc. Mech. Eng.*, **37**, 367–374.

Nusselt, W. (1916). Die Oberflächenkondensation des Wasserdampfes. *Z. Ver. Dtsch. Ininuere*, **60**, 541–575.

Ogg, D. G. (1991). Vertical downflow condensation heat Transfer in gas–steam mixtures. MS thesis, University of California, Berkeley.

Oh, C. H., and Englert, S. B. (1993). Critical heat flux for low flow boiling in vertical uniformly heated thin rectangular channels. *Int. J. Heat Mass Transfer*, **36**, 325–335.

Oh, H. K., Katsuta, M., and Shibata, K. (1998). Heat transfer characteristics of R-134a in a capillary tube heat exchanger. *Proc. 11th Int. Heat Transfer Conf.*, **6**, 131–136.

Oh, H. K., and Son, C. H. (2011). Condensation heat transfer characteristics of R-22, R-134a, R-410A in a single circular microtube. *Exp. Therm. Fluid Sci.*, **35**, 706–716.

Ohnuki, A. (1986). Experimental study of counter-current two-phase flow in horizontal tube connected to inclined riser. *J. Nucl. Sci. Technol.*, **23**(3), 219–232.

Olson, C. O., and Sunden, B. (1994). Pressure drop characteristics of small-sized tubes. ASME Paper 94-WA/HT-1.

Ong, C. L., and Thome, J. R. (2011a). Macro-to-microchannel transition in two-phase flow: Part 1 – Two-phase flow patterns and film thickness measurements. *Exp. Thermal Fluid Sci.*, **35**, 37–47.

Ong, C. L., and Thome, J. R. (2011b). Macro-to-microchannel transition in two-phase flow. Part 2 – Flow boiling heat transfer and critical heat flux. *Exp. Thermal Fluid Sci.*, **35**, 873–886.

Orlov, V. K., and Tselishchev, P. A. (1964). Heat exchange in a spiral coil with turbulent flow of water. *Teploenergetika*, **11**(12), 97–99.

Ormiston, S. J., Raithby, G. D., and Carlucci, L. N. (1995a). Numerical modeling of power station steam condensers – Part 1: Convergence behavior of finite-volume model. *Num. Heat Transfer*, **27B**, 81–102.

Ormiston, S. J., Raithby, G. D., and Carlucci, L. N. (1995b). Numerical modeling of power station steam condensers – Part 2: Improvement of solution behavior. *Num. Heat Transfer*, **27B**, 103–125.

Ornatskiy, A. P. (1960). The influence of length and tube diameter on critical heat flux for water with forced convection and subcooling. *Teploenergetika*, **4**, 67–69.

Ornatskiy, A. P., and Kichigan, A. M. (1962). Critical thermal loads during the boiling of subcooled water in small diameter tubes. *Teploenergetika*, **6**, 75–79.

Ornatskiy, A. P., and Vinyarskiy, L. S. (1964). Heat transfer crisis in a forced flow of under heated water in small bore tubes. *Teplofizika Vysokikh Temperatur*, **3**, 444–451.

Osakabe, M., and Kawasaki, K. (1989). Top flooding in thin rectangular and annular passages. *Int. J. Multiphase Flow*, **15**, 747–754.

Osamusali, S. E., Groeneveld, D. C., and Cheng, S. C. (1992). Two-phase flow regimes and onset of flow instability in horizontal 37-rod bundles. *Heat Technol.*, **10**, 46–74.

Osamusali, S. I., and Chang, J. S. (1988). Two-phase flow regime transition in a horizontal pipe and annulus flow under gas-liquid two-phase flow. ASME, New York, ASME FED, Vol. 72, pp. 63–69.

Osher, S., and Fedkiw, R. P. (2003). *Level Set Methods and Dynamic Implicit Surfaces*, Springer, New York.

Osmachkin, V. S., and Borisov, V. (1970). Pressure drop and heat transfer for flow of boiling water in vertical rod bundles. In *IVth Int. Heat Transfer Conference*, Paris-Versailles, France.

Othmer, D. F. (1929). The condensation of steam. *Ind. Eng. Chem.*, **21**, 576–583.

Ould Didi, M. B., Kattan, N., and Thome, J. R. (2002). Prediction of two-phase pressure gradients of refrigerants in horizontal tubes. *Int. J. Refrigeration*, **25**, 935–947.

Owens, W. L., and Schrock, V. E. (1960). Local pressure gradients for subcooled boiling of water in vertical tubes. Paper 60-WA-249, ASME, New York.

Owhadi, A., Bell, K. J., and Berry, C., Jr. (1968). Forced convection boiling inside helically-coiled tubes. *Int. J. Heat Mass Transfer*, **11**, 1779–1793.

Oya, T. (1971). Upward liquid flow in small tube into which air streams (Second report, Pressure drop at the confluence). *Bull. JSME*, **14**, 1330–1339.

Padilla, M., Revellin, R., and Bonjour, J. (2009). Prediction and simulation of two-phase pressure drop in return bends. *Int. J. Refrigeration*, **32**, 1776–1783.

Padilla, M., Revellin, R., and Bonjour, J. (2012). Two-phase flow visualization and pressure drop measurement of HFO-1234yf and R-134a in horizontal return bends. *Exp. Thermal Fluid Sci.*, **39**, 98–111.

Padilla, M., Revellin, R., Wallet, J., and Bonjour, J. (2013). Flow regime visualization and pressure drops of HFO-1234yf, R-134a and R-410A during downward two-phase flow in vertical return bends. *Int. J. Heat Mass Transfer*, **40**, 116–134.

Palen, J. W., Breber, D., and Taborek, J. (1979). Prediction of flow regimes in horizontal tubeside condensation. *Heat Transfer Eng.*, **1**(2), 47–57.

Pamitran, A. S., Choi, K. I., Oh, J. T., and Hrnjak, P., 2010. Characteristics of two-phase flow pattern transitions and pressure drop of five refrigerants in horizontal circular small tubes. *Int. J. Refrigeration*, **33**, 578–588.

Panday, P. K. (2003). Two-dimensional turbulent film condensation of vapours flowing inside a vertical tube and between parallel plates: A numerical approach. *Int. J. Refrigeration*, **26**, 492–503.

Paranjape, S., Chen, S.-W., Hibiki, T., and Ishii, M. (2011). Flow regime identification under adiabatic upward two-phase flow in a vertical rod bundle geometry. *J. Fluids Eng.*, **133**, paper 091302.

Park, C. Y., and Hrnjak, P. S. (2009). CO_2 flow condensation heat transfer and pressure drop in multiport microchannels at low temperatures. *Int. J. Refrigeration*, **32**, 1129–1139.

Park, J.-E., and Thome, J. R. (2010). Critical heat flux in multi-microchannel copper elements with low pressure refrigerants. *Int. J. Heat Mass Transfer*, **53**, 110–122.

Park, J. E., Vakili-Farahani, F., Consolini, L., and Thome, J. R. (2011). Experimental study on condensation heat transfer in vertical minichannels for new refrigerant R1234ze(E) versus R134a and R236fa. *Exp. Thermal Fluid Sci.*, **35**, 442–454.

Park, K., and Lee, K. S. (2003). Flow and heat transfer characteristics of the evaporating extended meniscus in capillary tubes. *Int. J. Heat Mass Transfer*, **46**, 4587–4594.

Pasamehmetoglu, K. O., and Nelson, R. A. (1987). Transient direct contact condensation on liquid droplets. Presented at the *ASME-ANE-AIChE National Heat Transfer Conf.*, Pittsburgh, PA.

Pasch, J., and Anghaie, S. (2008). An improved model for two-phase hydrogen flow dynamics. *Int. J. Heat Mass Transfer*, **51**, 2784–2800.

Pauken, M.T., and Abdel-Khalik, S. I. (1995). Evaporation suppression from spent-fuel storage basins with monolayer films. *Trans. ANS*, **72**, 308–309.

Pearson, K. G., Cooper, C. A., and Jowitt, D. (1984). *The THETIS 80% Blocked Cluster Experiment, Part 5*: Level Swell Experiments, AEEW-R 1767, AEEE Winfrith Safety and Engineering Science Division, London, UK.

Peles, Y. P., Yarin, L. P., and Hetsroni, G. (2000). Thermodynamic characteristics of two-phase flow in a heated capillary. *Int. J. Multiphase Flow*, **26**, 1063–1093.

Peng, X. F., and Wang, B.-X. (1993). Forced convection and flow boiling heat transfer for liquid flowing through microchannels. *Int. J. Heat Mass Transfer*, **36**, 3421–3427.

Peng, X. F., and Wang, B. X. (1994). Liquid flow and heat transfer in microchannels with/without phase change. *Heat Transfer 1994, Proc. 10th Int. Heat Transfer Conf.*, **5**, 159–177.

Peng, X. F., and Wang, B. X. (1998). Forced-convection and boiling characteristics in microchannels. *Proc. 11th Int. Heat Transfer Conf.*, pp. 371–390.

Peng, X. F., Hu, H. Y., and Wang, B.-X. (1994a). Boiling nucleation during liquid flow in microchannels. *Int. J. Heat Mass Transfer*, **41**, 101–106.

Peng, X. F., and Peterson, G. P. (1995). The effect of thermofluid and geometrical parameters on convection of liquids through rectangular microchannels. *Int. J. Heat Mass Transfer*, **38**, 755–758.

Peng, X. F., Peterson, G. P., and Wang, B. X. (1994b). Frictional flow characteristics of water flowing through rectangular microchannels. *Exp. Thermal-Fluid Sci.*, **7**, 249–264.

Peng, X. F., Wang, B.-X., Peterson, G. P., and Ma, H. P. (1995). Experimental investigation of heat transfer in flat plates with rectangular microchannels. *Int. J. Heat Mass Transfer*, **38**, 127–137.

Peterson, P. F. (1996). Theoretical basis for the Uchida correlation for condensation in reactor containments. *Nucl. Eng. Design*, **162**, 301–306.

Peterson, P. F., Schrock, V. E., and Kagayama, T. (1993). Diffusion layer theory for turbulent vapor condensation with noncondensable gases. *J. Heat Transfer*, **115**, 998–1003.

Pettersen, J. (2004). Flow vaporization of CO_2 in microchannel tubes. *Exp. Thermal Fluid Sci.*, **28**, 111–121.

Petukhov, B. S. (1970). Heat transfer and friction in turbulent pipe flow with variable physical properties. *Adv. Heat Transfer*, **6**, 503–565.

Petukhov, B. S., and Popov, V. N. (1963). Theoretical calculation of heat exchange in turbulent flow in tubes of an incompressible fluid with variable physical properties. *High Temp.*, **1**, 69–83.

Piazza, I. D., and Ciofalo, M. (2010). Numerical prediction of turbulent flow and heat transfer in helically coiled pipes. *Int. J. Thermal Sci.*, **49**, 653–663.

Plesko, C., and Leutheusser, H. J. (1982). Dynamic effects of bubble motion. *Chem. Eng. Commun.*, **17**, 195–218.

Plesset, M. S., and Zwick, S. A. (1954). Growth of vapor bubbles in superheated liquids. *J. Appl. Phys.*, **25**, 493–500.

Pokhalov, Y. E., Kronin, G. H., and Kurganova, I. V. (1966). Correlation of experimental data on heat transfer with nucleate boiling of subcooled liquids in tubes. *Teploenergetika*, **13**, 63–68.

Pratt, N. H. (1947). The heat transfer in a reaction tank cooled by means of a coil. *Trans. Inst. Chem. Eng.*, **25**, 163–180.

Premoli, A., Francesco, D., and Prina, A. (1970). An empirical correlation for evaluating two-phase mixture density under adiabatic conditions. European Two-Phase Flow Group Meeting, Milan.

Premoli, A., Francesco, D., and Prina, A. (1971). A dimensionless correlation for determining the density of two-phase mixtures. *Lo Termotecnica*, **25**, 17–26.

Press, H.W., Teukolsky, S. A., Vetterling, W. T., and Flannery, B. P. (1992). *Numerical Recipes for FORTRAN 77*, Vol. **1**, Cambridge University Press, Cambridge.

Probstein, R. F. (2003). *Physicochemical Hydrodynamics*, 2nd Edn, Wiley, New York.

Prodanovic, V., Fraser, D., and Salcudean, M. (2002a). Bubble behavior in subcooled flow boiling of water at low pressures and low flow rates. *Int. J. Multiphase Flow*, **28**, 1–19.

Prodanovic, V., Fraser, D., and Salcudean, M. (2002b). On transition from partial to fully developed subcooled flow boiling. *Int. J. Heat Mass Transfer*, **45**, 4727–4738.

Pushkina, O. L., and Sorokin, Y. L. (1969). Breakdown of liquid film motion in vertical tubes. *Heat Transfer Sov. Res.*, **1**, 56–64.

Qi, S., Zhang, P., Wang, R., and Xu, L. (2007). Flow boiling of liquid nitrogen in microtubes: Part II – Heat transfer characteristics and critical heat flux. *Int. J. Heat Mass Transfer*, **50**, 5017–5030.

Qian, D., and Lawal, A. (2006). Numerical study on gas and liquid slugs for Taylor flow in a T-junction microchannel. *Chem. Eng. Sci.*, **61**, 7609–7625.

Qu, W., and Mudawar, I. (2002). Prediction and measurement of incipient boiling heat flux in micro channel heat sinks. *Int. J. Heat Mass Transfer*, **45**, 3933–3945.

Qu, W., and Mudawar, I. (2003a). Measurement and prediction of pressure drop in two-phase micro channel heat sinks. *Int. J. Heat Mass Transfer*, **46**, 2737–2753.

Qu, W., and Mudawar, I. (2003b). Flow boiling heat transfer in two-phase micro-channel heat sinks – I. Experimental investigation and assessment of correlation method. *Int. J. Heat Mass Transfer*, **46**, 2755–2771.

Qu, W., and Mudawar, I. (2004a). Transport phenomena in two-phase micro-channel heat sinks. *J. Electronic Packaging*, **126**, 213–224.

Qu, W., and Mudawar, I. (2004b). Measurement and correlation critical heat flux in microchannel heat sinks. *Int. J. Heat Mass Transfer*, **47**, 2045–2059.

Qu, W., Mala, G. M., and Li, D. (2000). Heat transfer for water in trapezoidal silicon microchannels. *Int. J. Heat Mass Transfer*, **43**, 3925–3936.

Ramu, K., and Weisman, J. (1974). A method for the correlation of transition boiling heat transfer data. *Proc. Fifth Int. Heat Transfer Conf.*, Tokyo, Vol. **IV**, B4.4.

Ranz, W. E., and Marshall, W. R., Jr. (1952). Evaporation from drops. Parts I and II. *Chem. Eng. Prog.*, **48**, 141–146 and 173–180.

Rattner, A. S., and Garimella, S. (2015). Vertical upward intermediate scale Taylor flow: Experiments and kinematic closure. *Int. J. Multiphase Flow*, **75**, 107–123.

Rebrev, E. V. (2010). Two-phase flow regimes in microchannels. *Theor. Found. Chem. Eng.*, **44**, 355–367.

Reddy, D. G., and Fighetti, C. F. (1983). A generalized subchannel CHF correlation for PWR and BWR fuel assemblies. Electric Power Research Institute, Rep. EPRI NP-2609-Vol. 2, Palo Alto, CA.

Reddy, R. P., and Lienhard, J. H. (1989). The peak heat flux in saturated ethanol–water mixtures. *J. Heat Transfer*, **111**, 480–486.

Reichardt, H. (1951a). Vollständige Darstellung der turbulenten Geschwindigkeitsverteilung in glatten leitungen. *Z. Angew. Math. Mech.*, **31**, 208–219.

Reichardt, H. (1951b). Die Grundlagen des turbulenten Wärmeüberganges. *Arch. Ges.Wärmetech.*, **2**, 129–142.

Reid, R. C., Prausnitz, J. M., and Sherwood, T. K. (1977). *The Properties of Gases and Liquids*, 3rd Edn, McGraw-Hill, New York.

RELAP5–3D Code Development Team (2012). *RELAP5–3D Code Manuals*, Version 2.3, Vols. 1–5, INEEL-EXT-98–00834.

Ren, W. M., Ghiaasiaan, S. M., and Abdel-Khalik, S. I. (1994a). GT3F: An implicit finite difference computer code for transient three-dimensional three-phase flow. Part I: Governing equations and solution scheme. *Num. Heat Transfer B: Fundam.*, **25**, 1–20.

Ren, W. M., Ghiaasiaan, S. M., and Abdel-Khalik, S. I. (1994b). GT3F: An implicit finite difference computer code for transient three-dimensional three-phase flow. Part II: Applications. *Num. Heat Transfer B: Fundam.*, **25**, 21–38.

Revankar, S. T., and Ishii, M. (1992). Local interfacial area measurement in bubbly flow. *Int. J. Heat Mass Transfer*, **35**, 913–925.

Revellin, R., Dupont, V., Ursenbacher, T., Thome, J. R., and Zun, I. (2006). Characterization of diabatic two-phase flows in microchannels: Flow parameter results for R-134a in a 0.5 mm channel. *Int. J. Multiphase Flow*, **32**, 755–774.

Revellin, R., Mishima, K., and Thome, J. R. (2009). Status of prediction methods for critical heat fluxes in mini and microchannels. *Int. J. Heat and Fluid Flow*, **30**, 983–992.

Revellin, R., and Thome, J. R. (2007a). New type of diabatic flow pattern map for boiling heat transfer in microchannels. *J. Micromech. Microeng.*, **17**, 788–796.

Revellin, R., and Thome, J. R. (2007b). Experimental investigation of R-134a and R-245fa twophase flow in microchannels for different flow conditions, *Int. J. Heat Fluid Flow*, **28**, 63–71.

Revellin, R., and Thome, J. R. (2008). A theoretical model for the prediction of the critical heat flux in heated microchannels. *Int. J. Heat Mass Transfer.* **51**, 1216–1225.

Rezkallah, K. S. (1996). Weber number based flow-pattern maps for liquid–gas flows at microgravity. *Int. J. Multiphase Flow*, **22**, 1265–1270.

Richter, H. J. (1981). Flooding in tubes and annuli. *Int. J. Multiphase Flow*, **7**, 647–658.

Richter, H. J. (1983). Separated two-phase flow model: Application to critical two-phase flow. *Int. J. Multiphase Flow*, **9**, 511–530.

Richter, H. J., Wallis, G. B., and Speers, M. S. (1979). Effect of scale on two-phase countercurrent flow flooding. NUREG/CR-0312, U.S. Nuclear Regulatory Commission, Washington, DC.

Rippell, G. R., Eidt, C. M., and Jornan, H. B. (1966). Two-phase flow in a coiled tube. *Ind. Eng. Chem. Fundamentals*, **5**, 32–39.

Rivard, W. C., and Travis, J. R. (1980). A nonequilibrium vapor production model for critical flow. *Nucl. Sci. Eng.*, **74**, 40–48.

Riznic, J., Kojasoy, G., and Zuber, N. (1999). On the spherically symmetric phase change problem. *Int. J. Fluid Mech. Res.*, **26**, 110–145.

Roach, G. M., Jr., Abdel-Khalik, S. I., Ghiaasiaan, S. M., and Jeter, S. M. (1999a). Low-flow onset of flow instability in heated microchannels. *Nucl. Sci. Eng.*, **133**, 106–117.

Roach, G. M., Jr., Abdel-Khalik, S. I., Ghiaasiaan, S. M., and Jeter, S. M. (1999b). Low-flow critical heat flux in heated microchannels. *Nucl. Sci. Eng.*, **131**, 411–425.

Roday, A. P., and Jensen, M. K. (2009). A review of the critical heat flux condition in mini- and microchannels. *J. Mechanical Sci. Technol.*, **23**, 2529–2547.

Rogers, J. T., and Li, J.-H. (1992). Prediction of the onset of significant void in flow boiling of water. *Fundamentals of Subcooled Flow Boiling*, ASME, HTD, Vol. 217, pp. 41–52.

Rogers, T. J., Salcudean, M., Abdullah, Z., McLeond, D., and Poirier, D. (1987). The onset of significant void in up-flow boiling of water at low pressure and velocities. *Int. J. Heat Mass Transfer*, **30**, 2247–2260.

Rohsenow, W. H. (1973). Boiling, in *Handbook of Heat Transfer*, Rohsenow, W. H., and Hartnett, J. P., Eds., McGraw-Hill, New York, Chapter 13.

Rohsenow, W. M. (1952). A method of correlating heat transfer data for surface boiling of liquids. *Trans. ASME*, **74**, 969–975.

Rohsenow, W. M. (1956). Heat transfer and temperature distribution in laminar film condensation. *Trans. ASME*, **78**, 1645–1648.

Rohsenow, W. M., Webber, J. H., and Ling, T. (1956). Effect of vapor velocity on laminar and turbulent film condensation. *Trans. ASME*, **78**, 1637–1643.

Rose, J., Uehara, H., Koyama, S., and Fujii, T. (1999). Film condensation. In *Handbook of Phase Change*, Kandlikar, S. G., Shoji, M., and Dhir, V. K., Eds., Taylor & Francis, London, pp. 523–580.

Rose, J. W. (1984). Effect of pressure gradient in forced convection film condensation on a horizontal tube. *Int. J. Heat Mass Transfer*, **27**, 39–47.

Rouhani, Z., and Axelsson, E. (1970). Calculation of void volume fraction in the subcooled and quality boiling regions. *Int. J. Heat Mass Transfer*, **13**, 383–393.

Rowley, R. L. (1994). *Statistical Mechanics for Thermophysical Property Calculations*, Prentice Hall, Englewood Cliffs, NJ.

Rudman, M. (1997). Volume-tracking methods for interfacial flow calculations. *Int. J. Num. Methods Fluids*, **24**, 671–691.

Ruffell, A. E. (1974). The application of heat transfer and pressure drop data to the design of helical coil once-through boiler. *Symp. Multi-Phase Flow Systems*, Inst. Chem. Eng. Symp. Ser., 37, Paper 15.

Sadatomi, Y., Sato, Y., and Saruwatari, S. (1982). Two-phase flow in vertical noncircular channels. *Int. J. Multiphase Flow*, **8**, 641–655.

Saffman, P. G., and Turner, J. J. (1956). On the collision of drops in turbulent clouds. *J. Fluid Mech.*, **1**, 16–30.

Saha, P., and Zuber, N. (1974). Point of net vapor generation and vapor void fraction in subcooled boiling. *Proc. 5th Int. Heat Transfer Conf.*, **4**, 175–179.

Saha, S. K, and Celata, G. P. (2011). Thermofluid dynamics of boiling in microchannels. Part I. *Advances in Heat Transfer*, **43**, 77–159.

Saito, T., Hughes, E. D., and Carbon, M. W. (1978). Multi-fluid modeling of annular two-phase flow. *Nucl. Eng. Design*, **50**, 225–271.

Saitoh, S., Daiguji, H., and Hihara, E. (2005). Effect of tube diameter on boiling heat transfer of R-134a in horizontal small-diameter tubes. *Int. J. Heat Mass Transfer*, **48**, 4973–4984.

Saitoh, S., Daiguji, H., and Hihara, E. (2007). Correlation for boiling heat transfer of R-134a in horizontal tubes including effect of tube diameter. *Int. J. Heat Mass Transfer*, **50**, 5215–5225.

Sakashita, H., Ono, A., and Nakabayashi, Y. (2010). Measurements of critical heat flux and liquid-vapor structure near the heating surface in pool boiling of 2-propanol/water mixtures. *Int. J. Heat Mass Transfer*, **53**, 1554–1562.

Sakata, E. K. (1969). Surface diffusion in monolayers. *Ind. Eng. Chem. Res.*, **8**, 570–575.

Sam, R. G., and Patel, B. R. (1984). An experimental investigation of OC-OTEC direct-contact condensation and evaporation processes. *J. Solar Energy*, **106**, 120–127.

Samokhin, G. I., and Yagov, V. V. (1988). Heat transfer and critical heat flux with liquids boiling in the region of low reduced pressures. *Thermal Eng.*, **35**, 115–118.

Sandal, O. C. (1974). Gas absorption into turbulent liquids at intermediate contact times. *Int. J. Heat Mass Transfer*, **17**, 459–461.

Sander, R. (2015). Compilation of Henry's law constants (version 4.0) for water as solvent. *Atmos. Chem. Phys.*, **15**, 4399–4981 (www.atmos-chem-phys.net/15/4399/2015/).

Santini, A., Cioncolini, A., Lombardi, C., and Ricotti, M. (2008). Two-phase pressure drops in a helically coiled steam generator. *Int. J. Heat Mass Transfer*, **51**, 4926–4939.

Santos, R. M., and Kawaji, M. (2010). Numerical modeling and experimental investigation of gas–liquid slug formation in a microchannel T-junction. *Int. J. Multiphase Flow*, **36**, 314–323.

Sato, T., and Matsumura, H. (1963). On the conditions of incipient subcooled boiling and forced-convection. *Bull. JSME*, **7**, 392–398.

Saxena, A. K., Schumpe, A., Nigam, K. D. P., and Decker, W. D. (1990). Flow regime, hold-up and pressure drop for two-phase flow in helical coils. *Can. J. Chem. Eng.*, **68**, 553–559.

Schlichting, H. (1968). *Boundary Layer Theory*, 6th Edn, McGraw-Hill, New York.

Schlünder, E. U. (1982). Heat transfer in nucleate pool boiling of mixtures. *Proc. 7th Int. Heat Transfer Conf.*, Vol. 4, pp. 2073–2079.

Schmidt, E. F. (1967). Wärmeübergang und Druckverlust in Rohrschlangen. *Chem. Ing. Tech.*, **39**, 781–789.

Schmidt, J., and Friedel, L. (1997). Two-phase pressure drop across sudden contractions in duct areas. *Int. J. Multiphase Flow*, **23**, 283–299.

Schrage, R. W. (1953). *A Theoretical Study of Interphase Mass Transfer*, Columbia University Press, New York.

Schrock, V. E., and Grossman, L. M. (1962). Forced convection boiling in tubes. *Nucl. Sci. Eng.*, **12**, 474–481.

Schrock, V. E., Wang, C.-H., Revankar, S., Wei, L.-H., Lee, S. Y., and Squarer, D. (1984). Flooding in particle beds and its role in dryout heat fluxes. *Proc. 6th Information Exchange Meeting on Debris Coolability*, UCLA, Los Angeles, CA.

Schrock, V. E., Revankar, S. T., and Lee, S. Y. (1988). Critical flow through pipe cracks. In *Particulate Phenomena and Multiphase Transport*, N. Veziroglu, Ed., Hemisphere, Washington, DC, **1**, 3–17.

Schultze, H. D. (1984). *Physico-Chemical Elementary Processes in Flotation*, Elsevier, Amsterdam, pp. 123–129.

Schwartz, A. M., and Tejada, S. B. (1972). Studies of dynamic contact angle on solids, *J. Colloid Interface Sci.*, **38**, 359–375.

Schwellnus, C. F., and Shoukri, M. (1991). A two-fluid model for non-equilibrium two-phase critical discharge. *Can. J. Chem. Eng.*, **69**, 187–197.

Scriven, L. E. (1959). On the dynamics of phase growth. *Chem. Eng. Sci.*, **10**, 1–13.

Scriven, L. E., and Sterling, C. V. (1964). On cellular convection driven by surface tension gradients – Effect of mean surface tension and surface viscosity. *J. Fluid Mech.*, **19**, 321–340.

Seban, R. A. (1954). Remarks on film condensation with turbulent flow. *Trans. ASME*, **76**, 299–302.

Seban, R. A., and Faghri, A. (1976). Evaporation and heating with turbulent falling liquid films. *J. Heat Transfer*, **98**, 315–318.

Seban, R. A., and Hodgson, J. A. (1982). Laminar film condensation in a tube with upward vapor flow. *Int. J. Heat Mass Transfer*, **25**, 1291–1300.

Seban, R. A. and McLaughlin, E. F. (1963). Heat transfer in tube coils with laminar and turbulent flow. *Int. J. Heat Mass Transfer*, **6**, 387–395.

Sepold, L., Hofmann, P., Leiling, W., Miassoedov, A., Piel, D., Schmidt, L., and Steinbruck, M. (2001). Reflooding experiments with LWR-type fuel rod simulators in the QUENCH facility. *Nucl. Eng. Design*, **204**, 205–220.

Serizawa, A., Feng, Z., and Kawara, Z. (2002). Two-phase flow in microchannels. *Exp. Thermal Fluid Sci.*, **26**, 703–714.

Shah, M. (1977). A general correlation for heat transfer during subcooled boiling in pipes and annuli. *ASHRAE Trans.*, **83**, Part 1, 205–215.

Shah, M. M. (1979). A general correlation for heat transfer during film condensation inside pipes. *Int. J. Heat Mass Transfer*, **22**, 547–556.

Shah, M. M. (1987). Improved general correlation for critical heat flux during upflow in uniformly heated vertical tubes. *Int. J. Heat Fluid Flow*, **8**, 326–335.

Shah, M. M. (2009). An improved and extended general correlation for heat transfer during condensation in plain tubes, *HVAC&R Res.*, **15**, 889–913.

Shah, R. K., and Joshi, S. D. (1987). Convective Heat Transfer in Curved Ducts. In Kakac, S., Shah, R. K., and Aung, W., Eds., *Handbook of Single-Phase Convective Heat Transfer*, John Wiley & Sons.

Shah, R. K., and London, A. L. (1978). *Laminar Flow Forced Convection in Ducts: A Source Book for Compact Heat Exchanger Analytical Data*, Academic Press, New York (Suppl. 1).

Shao, N., Gavriilidis, A., and Angeli, P. (2009). Flow regimes for adiabatic gas–liquid flow in micro channels. *Chem. Eng. Sci.*, **64**, 2749–2761.

Shao, N., Gavriilidis, A., and Angeli, P. (2011). Effect of inlet conditions on Taylor bubble length in microchannels. *Heat Transfer Eng.*, **32**, 1117–1125.

Sharp, K. V., and Adrian, R. J. (2004). Transition from laminar flow to turbulent flow in liquid filled microtubes. *Exp. Fluids*, **36**, 741–747.

Shatto, D. P., and Peterson, G. P. (1999). Pool boiling critical heat flux in reduced gravity. *J. Heat Transfer*, **121**, 865–873.

Shekriladze, I. G., and Gomelauri, V. I. (1966). Theoretical study of laminar film condensation of flowing vapor. *Int. J. Heat Mass Transfer*, **9**, 581–591.

Shima, A. (1970). The natural frequency of a bubble oscillating in a viscous compressible liquid. *J. Basic Eng.*, **92**, 555–562.

Simpson, H. C., Rooney, H. D., and Grattan, E. (1983). Two-phase flow through gate valves and orifice plates. *Int. Conf. on Physical Modeling of Multiphase Flow*, Coventry, UK, paper A2.

Shin, T. S., and Jones, O. C. (1993). Nucleation and flashing in nozzles – 1: A distributed nucleation model. *Int. J. Multiphase Flow*, **19**, 943–964.

Shin, J. S., and Kim, M. H. (2005). An experimental study of flow condensation heat transfer inside circular and rectangular mini-channels. *Heat Transfer Eng.*, **26**, 36–44.

Shinnar, R. (1961). On the behavior of liquid dispersions in mixing vessels. *J. Fluid Mech.*, **10**, 259–275.

Shipley, D. G. (1982). Two phase flow in large diameter pipes. *Chem. Eng. Sci.*, **39**, 163–165.

Shmerler, J. A., and Mudawar, I. (1988). Local evaporative heat transfer coefficient in turbulent free-falling liquid films. *Int. J. Heat Mass Transfer*, **31**, 731–742.

Shoji, M. (2004). Studies of boiling chaos: A review. *Int. J. Heat Mass Transfer*, **47**, 1105–1128.

Siddique, M. (1992). The effects of noncondensable gases on steam condensation under forced convection conditions. PhD thesis, Massachusetts Institute of Technology, Cambridge, MA.

Siddique, M., Golay, M. W., and Kazimi, M. S. (1994). Theoretical modeling of forced convection of steam in the presence of a noncondensable gas. *Nucl. Technol.*, **106**, 202–215.

Sieder, E. N., and Tate, G. E. (1936). Heat transfer and pressure drop of liquids in tubes. *Ind. Eng. Chem.*, **28**, 1429.

Singh, S. G., Jain, A., Sridharan, A., Duttagupta, S. P., and Agrawal, A. (2009). Flow map and measurement of void fraction and heat transfer coefficient using an image analysis technique for flow boiling of water in a silicon microchannel. *J. Micromech. Microeng.*, **19**, paper 075004.

Skelland, A. H. P. (1974). *Diffusional Mass Transfer*, Krieger, Malabar, FL.

Skinner, L. A., and Bankoff, S. G. (1964). Dynamics of vapor bubbles in binary liquids with spherically symmetric initial conditions. *Phys. Fluids*, **7**, 643–648.

Sklover, G. G., and Rodivilin, M. D. (1975). Heat and mass transfer with condensation of steam on water jet. *Teploenergetika*, **22**(11), 65–68.

Sleiti, A. K. (2011). Heat transfer and pressure drop through rectangular helical ducts. *J. Renewable and Sustainable Energy*, **3**, 043119.

Smedley, G. (1990). Preliminary drop-tower experiments on liquid-interface geometry in partially filled containers at zero gravity. *Exp. Fluids*, **8**, 312–318.

Smith, S. L. (1969–70). Void fraction in two-phase flow: A correlation based upon an equal velocity head model. *Inst. Mech. Eng.*, **184**, 647–657.

Smith, T. R., Schlegel, J. P., Hibiki, T., and Ishii, M. (2012). Mechanistic modeling of interfacial area transport in large diameter pipes. *Int. J. Multiphase Flow*, **47**, 1–16.

Snyder, N. R., and Edwards, D. K. (1956). *Summary of Conference on Bubble Dynamics and Boiling Heat Transfer. Memo 20–137*, Jet Propulsion Laboratory, Pasadena, CA, pp. 14–15.

Sobajima, M. (1985). Experimental modeling of steam–water countercurrent flow limit for perforated plates. *J. Nucl. Sci. Technol.*, **22**, 723–732.

Sohn, H. Y., Johnson, S. H., and Hindmarsh, A. C. (1985). Application of the method of lines to the analysis of single fluid–solid reactions in porous media. *Chem. Eng. Sci.*, **40**, 2185–2190.

Soliman, H. M. (1982). On the annular-to-wavy flow pattern transition during condensation inside horizontal tubes. *Can. J. Chem. Eng.*, **60**, 475–481.

Soliman, H. M. (1986). The mist-annular transition during condensation and its influence on heat transfer mechanism. *Int. J. Multiphase Flow*, **12**, 277–288.

Soliman, H. M., Schuster, J. R., and Berenson, P. J. (1968). A general heat transfer correlation for annular flow condensation, *J. Heat Transfer*, **90**, 267–276.

Song, A., Steiff, A., and Weinspach, P.-M. (1997). Very efficient new method to solve the population balance equation with particle-size growth. *Chem. Eng. Sci.*, **52**, 3493–3498.

Souza, A. L., and Pimenta, M. M. (1995). Prediction of pressure drop during horizontal two phase flow of pure and mixed refrigerants. *Proc. ASME Conf.* FED-210, pp. 161–171.

Sovova, H., and Prochazka, J. (1981). Breakage and coalescence of drops in a batch stirred vessel – I Comparison of continuous and discrete models. *Chem. Eng. Sci.*, **36**, 163–171.

Spalding, D. B. (1980). Numerical calculation of multiphase fluid flow and heat transfer. In *Recent Advances in Numerical Methods in Fluids*, Taylor, C., and Morgan, K., Eds., Pineridge Press, Swansea, UK.

Spalding, D. B. (1983). Development of the IPSA procedure for numerical computation of multiphase flow phenomena with interphase slip, unequal temperatures, etc. In *Numerical Methods and Methodologies in Heat Transfer*, Shih, T. M., Ed., Hemisphere, Washington, DC.

Spedding, P. L., and Spence, D. R. (1993). Flow regimes in two-phase gas–liquid flow. *Int. J. Multiphase Flow*, **19**, 245–280.

Springer, T. G., and Pigford, R. L. (1970). Influence of surface turbulence and surfactants on gas transport through liquid surfaces. *Ind. Eng. Chem. Fundam.*, **9**, 458–465.

Srinivasan, S., Nadapurkar, S., and Holland, F. A. (1970). Friction factors for coils. *Trans. Inst. Chem. Eng.*, **48**, T156–T161.

Srivastava, R. P. S. (1973). Liquid film thickness in annular flows. *Chem. Eng. Sci.*, **28**, 819–824.

Staniszewski, B. E. (1959). Nucleate boiling bubble growth and departure. MIT Technical Rep. No. 16, Div. Sponsored Research, Cambridge, MA.

Stanley, R. S., Barron, R. F., and Ameel, T. A. (1997). Two-phase flow in microchannels. In *ASME Microelectromechanical Systems*, ASME, New York, DSC, Vol. 62/HTD, Vol. 354, pp. 143–152.

Staub, F. W. (1968). The void fraction in subcooled boiling: Prediction of the initial point of net vapor generation. *J. Heat Transfer*, **90**, 151–157.

Steinbruck, M. (2001). Reflooding experiments with LWR-type fuel rod simulators in the QUENCH facility. *Nucl. Eng. Design*, **204**, 205–220.

Steiner, D. (1993). Heat transfer to boiling saturated liquids. In *VDI-Wärmeatlas (VDI Heat Atlas)*, Verein Deutscher Ingenieure (Eds.), VDI-Gesellschaft Verfahrenstechnik und Chemieingenieurwesen (GCV) Düsseldorf, Germany, (J.W. Fullarton, translator).

Steiner, D., and Taborek, J. (1992). Flow boiling heat transfer in vertical tubes correlated by an asymptotic model. *Heat Transfer Eng.*, **13**, 43–69.

Stephan, K., and Abdelsalam, M. (1980). Heat-transfer correlations for natural convection boiling. *Int. J. Heat Mass Transfer*, **23**, 73–87.

Stephan, K., and Körner, M. (1969). Calculation of heat transfer in evaporating binary mixtures. *Chem.-Ing. Tech.*, **41**, 409–417.

Sterman, L., Abramov, A., and Checheta, G. (1968). Investigation of boiling crisis at forced motion of high temperature organic heat carriers and mixtures. *Int. Symp. Research into Cocurrent Gas–Liquid Flow*, University of Waterloo, Canada, Paper E2.

Sternling C. V., and Tichacek, L. J. (1961). Heat transfer coefficients for boiling mixtures: Experimental data for binary mixtures of large relative volatility. *Chem. Eng. Sci.*, **16**, 297–337.

Straub, J., Betz, J., and Marek, R. (1994). Enhancement of heat transfer by thermocapillary convection around bubbles – A numerical study. *Num. Heat Transfer A*, **25**, 501–518.

Stuhmiller, J. H. (1986). A dynamic flow regime model of two-phase flow. EPRI Rep. RP888–1, Electric Power Research Institute, Palo Alto, CA.

Stuhmiller, J. H. (1987). Implementation of the dynamic flowregime model in thermal-hydraulic codes. EPRI Report RP2806–1, Electric Power Research Institute, Palo Alto, CA.

Subbotin, V. I., Deev, V. I., and Arkhipov, V. V. (1982). Critical heat flux in flow boiling of helium. *Proc. 7th Int. Heat Transfer Conf.*, **4**, 357–361.

Subramanian, R. S. (1975). Gas absorption into a turbulent liquid film. *Int. J. Heat Mass Transfer*, **18**, 334–336.

Sudo, Y., Usui, T., and Kaminaga, M. (1991). Experimental study of falling water limitation under a counter-current flow in a vertical rectangular channel (First report, Effect of flow channel configuration and introduction of CCFL correlation). *JSME Int. J.*, **34**, 169–174.

Sugawara, S. (1990a). Analytical prediction of CHF by FIDAS code based on three-fluid and film dryout model. *J. Nucl. Sci. Technol.*, **27**, 12–29.

Sugawara, S. (1990b). Droplet deposition and entrainment modeling based on the three-fluid model. *Nucl. Eng. Design*, **122**, 67–84.

Sumith, B., Kaminaga, F., and Matsumura, K. (2003). Saturated flow boiling of water in a vertical small diameter tube. *Exp. Thermal Fluid Sci.*, **27**, 789–801.

Sun, K. H., Duffey, R. B., and Peng, C. M. (1980). A thermal-hydraulic analysis of core uncover. in: *Proceedings of the 19th National Heat Transfer Conference, Experimental and Analytical Modeling of LWR Safety Experiments*, 1980, pp. 1–10. Orlando, Florida, USA.

Sun, K., Duffey, R., and Peng, C. (1981). The prediction of two phase mixture level and hydrodynamically controlled dryout under low flow conditions. *Int. J. Multiphase Flow*, **7**, 521–543.

Sun, K. H., and Lienhard, J. H. (1970). The peak pool boiling heat fluxes on horizontal cylinders. *Int. J. Heat Mass Transfer*, **13**, 1425–1439.

Sun, L., and Mishima, K. (2009). Evaluation analysis of prediction methods for two phase flow pressure drop in mini-channels. *Int. J. Multiphase Flow*, **35**, 47–54.

Sun, X., Kim, S., Ishii, M., and Beus, S. G. (2004a). Modeling of bubble coalescence and disintegration in confined upward two-phase flow. *Nucl. Eng. Design*, **230**, 3–26.

Sun, X., Kim, S., Ishii, M., and Beus, S. G. (2004b). Model evaluation of two-group interfacial area transport equation for confined upward flow. *Nucl. Eng. Design*, **230**, 27–47.

Suo, M., and Griffith, P. (1964). Two-phase flow in capillary tubes. *J. Basic Eng.*, **86**, 576–582.

Sussman, M., and Fatemi, E. (2003). A second-order coupled level set and volume-of-fluid method for computing growth and collapse of vapor bubbles. *J. Comput. Phys.*, **187**, 110–136.

Svehla, R. A. (1962). Estimated viscosities and thermal conductivities of gases at high temperatures. NASA Technical Rep. R-132.

Svetlov, S. V., Ilyukhin, Y. N., Alekseev, S. B., Sidorov, V. G., Kukhtevich, V. O., and Paramonova, I. L. (1999). True void fraction in rod bundle under conditions of low circulation rate and bubbling. *High Temp.*, **37**, 302–308.

Szczukiewicz, S., Borhani, N., and Thome, J. R. (2013a). Two-phase flow operational maps for multi-microchannel evaporators. *Int. J. Heat Fluid Flow*, **42**, 176–189.

Szczukiewicz, S., Borhani, N., and Thome, J. R. (2013b). Two-phase heat transfer and high speed visualization of refrigerant flows in 100×100 μm^2 silicon multi microchannels. *Int. J. Refrigeration*, **36**, 402–413.

Szczukiewicz, S., Mangini, M., and Thome, J. R. (2014). Proposed models, ongoing experiments, and latest numerical simulations of microchannel two-phase flow boiling. *Int. J. Multiphase Flow*, **59**, 84–101.

Taboas, F., Valles, M., Bourouis, M., and Coronas, A. (2007). Pool boiling of ammonia/water and its pure components: Comparison of experimental data in the literature with the predictions of standard correlations. *Int. J. Refrigeration*, **30**, 778–788.

Taghavi-Tafreshi, K., Dhir, V. K., and Catton, I. (1979). Thermal and hydrodynamic phenomena associated with melting of a horizontal substrate placed beneath a heavier immiscible liquid. *J. Heat Transfer*, **101**, 318–325.

Taha, T., and Cui, Z. F. (2004). Hydrodynamics of slug flow inside capillaries. *Chem. Eng. Sci.*, **59**, 1181–1190.

Taha T., and Cui. Z. F. (2006a). CFD modelling of slug flow in vertical tubes, *Chem. Eng. Sci.*, **61**, 676–687.

Taha, T., and Cui, Z. F. (2006b). CFD modelling of slug flow inside square capillaries. *Chem. Eng. Sci.*, **61**, 665–675.

Taitel, Y. (1990). Flow pattern transition in two-phase flow. *Proc. 9th Int. Heat Transfer Cont.*, Hemisphere, New York, pp. 237–254.

Taitel, Y., and Dukler, A. E. (1976a). A theoretical approach to the Lockhart–Martinelli correlation for stratified flow. *Int. J. Multiphase Flow*, **2**, 591–595.

Taitel, Y., and Dukler, A. E. (1976b). A model for predicting flow regime transitions in horizontal and near horizontal gas–liquid flow. *AIChE J.*, **22**, 47–55.

Taitel, Y., Bornea, D., and Dukler, A. E. (1980). Modeling flow pattern transitions for steady upward gas–liquid flow in vertical tubes. *AIChE J.*, **26**, 345–354.

Taitel, Y., Lee, N., and Dukler, A. E. (1978). Transient gas–liquid flow in horizontal pipes: Modeling the flow pattern transitions. *AIChE J.*, **24**, 920–924.

Takeuchi, K., Young, M. Y., and Hochreiter, L. M. (1992). Generalized drift flux correlation for vertical flow. *Nucl. Eng. Design*, **112**, 170–180.

Takahama, H., and Kato, S. (1980). Longitudinal flow characteristics of vertically falling liquid films without concurrent gas flow. *Int. J. Multiphase Flow*, **6**, 203–215.

Takahashi, T., Akagi, Y., and Kisimoto, T. (1979). Gas and liquid velocities at incipient liquid stagnant on a sieve tray. *Int. Chem. Eng.*, **19**, 113–118.

Talimi, V., Muzychka, Y. S., and Kocabiyik, S. (2012). A review on numerical studies of slug flow hydrodynamics and heat transfer in microtubes and microchannels. *Int. J. Multiphase Flow*, **39**, 88–104.

Tammaro, M., Mauro, W. A., Bonjour, J., Matrullo, R. M. A., and Revellin, R. (2013). Curvature ratio effect on two-phase pressure drops in horizontal return beds: Experimental data for R-134a. *Heat Transfer Engineering*, **34**, 1124–1132.

Tandon, T. N., Varma, H. K., and Gupta, C. P. (1982). A new flow regimes map for condensation inside horizontal tubes. *J. Heat Transfer*, **104**, 763–768.

Tandon, T. N., Varma, H. K., and Gupta, C. P. (1985). A void fraction model for annular two-phase flow. *Int. J. Heat Mass Transfer*, **28**, 191–198.

Tandon, T. N., Varma, H. K., and Gupta, C. P. (1995). Heat transfer during forced convection condensation inside horizontal tube. *Int. J. Refrig.*, **18**, 210–214.

Tangren, R. F., Dodge, C. H., and Seifert, H. S. (1949). Compressibility effects in two-phase flow, *J. Appl. Phys.*, **20**, 736.

Tarasova, N. Y., Leontiev, A. I., Hlopushin, V. I., and Orlov, V. M. (1966). Pressure drop of boiling subcooled water and steam–water mixture flowing in heated channels. *Proc. 3rd Int. Heat Transfer Conf., Chicago*, **4**, 178–183.

Taylor, D. D., *et al.* (1984). TRAC/BD1-MOD1: An advanced best estimate computer program for boiling water reactor transients. NUREG/CR-3633, US Nuclear Regulatory Commission, Washington, DC.

Taylor, G. I. (1961). Deposition of a viscous fluid on the wall of a tube. *J. Fluid Mech*, **10**, 161–165.

Taylor, R., Krishna, R., and Furno, J. S. (1986). Condensation of vapor mixtures. 1. Nonequilibrium models and design procedures. *Ind. Eng. Chem. Process Des. Development*, **25**, 83–97.

Theofanous, T. G., Tu, J. P., Dinh, A. T., and Dinh, T. N. (2002a). The boiling crisis phenomenon. Part I: Nucleation and nucleate boiling heat transfer. *Exp. Thermal Fluid Sci.*, **26**, 775–792.

Theofanous, T. G., Dinh, T. N., Tu, J. P., and Dinh, A. T. (2002b). The boiling crisis phenomenon. Part II: Dryout dynamics and burnout. *Exp. Thermal Fluid Sci.*, **26**, 793–810.

Thom, J. R. S. (1964). Prediction of pressure drop during forced circulation boiling water. *Int. J. Heat Mass Transfer*, **7**, 709–724.

Thom, J. R. S., Walker, W. M., Fallon, T. A., and Reising, G. F. S. (1965). Boiling in subcooled water during flow in tubes and annuli. Paper 6, *Symp. Boiling Heat Transfer in Steam Generating Units and Heat Exchangers*, Sept. 15–16, Manchester, UK.

Thome, J. R. (2003). Boiling. In *Heat Transfer Handbook*, Bejan, A., and Kraus, A. D., Eds., Wiley, New York, Chapter 12.

Thome, J. R. (2007). *Engineering Data Book III*, Wolverine Tube, Inc., Chapter 10.

Thome, J. R., Bar-Cohen, A., Revellin, R., and Zun, I. (2013). Unified mechanistic multiscale mapping of two-phase flow patterns in microchannels. *Int. J. Heat Mass Transfer*, **44**, 1–22.

Thome, J. R., and Consolini, L. (2009). Mechanisms of boiling in microchannels: Critical assessment. *Heat Transfer Eng.*, **31**, 288–297.

Thome. J. R., Dupont, V., and Jacobi, A. M. (2004). Heat transfer model for evaporation in microchannels. Part I: Presentation of the model. *Int. J. Heat and Mass Transfer*, **47**, 3375–3385.

Thome, J. R., El Hajal, J., and Cavallini, A. (2003). Condensation in horizontal tubes. Part II: New heat transfer model based on flow regimes. *Int. J. Heat Mass Transfer*, **46**, 3365–3387.

Thome, J. R., and Shakir, S. (1987). A new correlation for nucleate boiling of binary mixtures. *AIChE Symp. Series*, **83**, 46–51.

Thome, J. R., and Shock, A. W. (1984). Boiling of multicomponent liquid mixtures. *Adv. Heat Transfer*, **16**, 59–155.

Thulasidas, M. A., Abraham, M. A., and Cerro, R. L. (1995). Bubble-train flow in capillaries of circular and square cross-section. *Chem. Eng. Sci.*, **50**, 183–199.

Thulasidas, T. C., Abraham, M. A., and Cerro, R. L. (1997). Flow patterns in liquid slugs during bubble-train flow inside capillaries. *Chem. Eng. Sci.*, **52**, 2947–2962.

Thulasidas, T. C., Abraham, M. A., and Cerro, R. I. (1999). Dispersion during bubble train flow in capillaries. *Chem. Eng. Sci.*, **54**, 61–76.

Tian, Y., Liu, J.-T., and Peng, X.-F. (2005). Characteristics of nucleation and bubble growth during microscale boiling. *ASME 2005 Summer Heat Transfer Conf.,* July 17–22, San Francisco.

Tien, C. L. (1977). A simple analytical model for countercurrent flow limiting phenomena with condensation. *Lett. Heat Mass Transfer*, **4**, 231–237.

Tien, C. L., and Liu, C. P. (1979). Survey of vertical two-phase countercurrent flooding. Electric Power Research Institute Rep. EPRI NP-984, Hillview, CA.

Tien, C. L., Qin, T. Q., and Norris, P. M. (1994). Microscale thermal phenomena in contemporary technology. *Thermal Sci. Technol.*, **2**, 1–11.

Tiselj, I., Hetsroni, G., Mavko, B., Mosyak, A., Pogrebnyak, E., and Segal, Z. (2004). Effect of axial conduction on the heat transfer in micro-channels. *Int. J. Heat Mass Transfer*, **47**, 2551–2565.

Tobin, T., Muralidhar, R., Wright, H., and Ramkrishna, D. (1990). Determination of coalescence frequencies in liquid–liquid dispersions: Effect of drop size dependence. *Chem. Eng. Sci.*, **45**, 3491–3504.

Todreas, N. E., and Kazimi, M. S. (2011). *Nuclear Systems I: Thermal-Hydraulic Fundamentals*, 2nd edn, Taylor & Francis, Boca Raton, Florida.

Tolubinsky, V. I., and Matorin, P. S. (1973). Forced convection boiling heat transfer crisis with binary mixtures. *Heat Transfer – Soviet Res.*, **5**(2), 98–101.

Tong, L. S. (1967). Heat transfer in water cooled reactors. *Nucl. Eng. Design*, **6**, 301.

Tong, L. S. (1969). Boundary layer analysis of the flow boiling crisis. *Int. J. Heat Mass Transfer*, **11**, 1208–1211.

Tong, L. S. (1972). *Boiling Crisis and Critical Heat Flux*, AEC Critical Review Series, USAEC, Washington, DC.

Tong, L. S. (1975). A phenomenological study of critical heat flux. ASME Paper 75-HT-68, *National Heat Transfer Conference*, San Francisco.

Tong, W., Bar-Cohen, A., Simon, T. W., and You, S. M. (1990). Contact angle effects on boiling incipience of highly-wetting liquids. *Int. J. Heat Mass Transfer*, **33**, 91–103.

Tong, L. S., Currin, H. B., Larsen, P. S., and Smith, O. G. (1965). Influence of axially nonuniform heat flux on DNB. *Chem. Eng. Symp. Series*, **62**(64), 35–40.

Tong, L. S., and Tang, Y. S. (1997). *Boiling Heat Transfer and Two-Phase Flow*, Taylor & Francis, London.

Tong, L. S., and Young, J. D. (1974). A phenomenological transition boiling and film boiling correlation. *Proc. Fifth Int. Heat Transfer Conf., Tokyo*, Vol. IV, B3.9.

Tong, W., Bergles, A. E., and Jensen, M. K. (1997). Pressure drop with highly subcooled flow boiling in small-diameter tubes. *Exp. Thermal Fluid Sci.*, **15**, 202–212.

Tran, T. N., Wambsganss, M. W., France, D. M., and Jendrzejczyk, J. A. (1993). Boiling heat transfer in small, horizontal, rectangular channels. *AIChE Symp. Ser.*, **89**, 253–261.

Tran, T. N., Wambsganss, M. W., and France, D. M. (1996). Small circular and rectangular channel boiling with two refrigerants. *Int. J. Multiphase Flow*, **22**, 485–498.

Tran, T. N., Chyu, M.-C., Wamsganss, M. W., and France, D. M. (2000). Two-phase pressure drop of refrigerants during flow boiling in small channels: An experimental investigation and correlation development. *Int. J. Multiphase Flow*, **26**, 1739–1754.

Traviss, D. P., Rohsenow, W. M., and Baron, A. B. (1973). Forced convective condensation in tubes: A heat transfer correlation for condenser design. *ASHRAE Trans.*, **79**(1), 157–165.

Tribbe, C., and Muller-Steinhagen, H. M. (2000). An evaluation of the performance of phenomenological models for predicting pressure gradient during gas–liquid flow in horizontal pipelines. *Int. J. Multiphase Flow*, **26**, 1019–1036.

Triplett, K. A., Ghiaasiaan, S. M., Abdel-Khalik, S. I., and Sadowski, D. L. (1999a). Gas–liquid two-phase flow in microchannels. Part I: Two-phase flow patterns. *Int. J. Multiphase Flow*, **25**, 377–394.

Triplett, K. A., Ghiaasiaan, S. M., Abdel-Khalik, S. I., LeMouel, A., and McCord, B. N. (1999b). Gas–liquid two-phase flow in microchannels. Part II: Void fraction and pressure drop. *Int. J. Multiphase Flow*, **25**, 395–410.

Troniewski, L., and Ulbrich, R. (1984). Two-phase gas–liquid flow in rectangular channels. *Chem. Eng. Sci.*, **39**, 751–765.

Truesdell, C. L., and Adler, R. J. (1970). Numerical treatment of fully developed laminar flow in helically coiled tubes. *AIChE J.*, **16**, 1010–1015.

Tryggvason, G., Bunner, B., Esmaeeli, A., Juric, D., Al-Rawahi, A., Tauber, W., Han, J., Nas, S., and Jan, Y. J. (2001). A front-tracking method for the computations of multiphase flow. *J. Comput. Phys.*, **169**, 708–759.

Tsouris, C., and Tavlarides, L. L. (1994). Breakage and coalescence models for drops in turbulent dispersions. *AIChE J.*, **40**, 395–406.

Tu, J. Y., and Yeoh, G. H. (2002). On numerical modeling of low-pressure subcooled boiling flows. *Int. J. Heat Mass Transfer*, **45**, 1197–1209.

Tuckermann, D. B., and Peasa, R. F. (1981). High performance heat sinking for VLSI. *IEEE Electron Device Lett.*, EDL-2, 126–129.

Turner, J. M., and Wallis, G. B. (1965). The separate-cylinders model for two-phase flow. Paper NYO-3114-6, Thayer School of Eng., Darthmouth College, NH.

Uehara, H., and Kinoshita, E. (1994). Wave and turbulent film condensation on a vertical surface (correlation for local heat transfer coefficient). *Trans. JSME*, **60**, 3109–3116.

Uehara, H., and Kinoshita, E. (1997). Wave and turbulent film condensation on a vertical surface (correlation for average heat transfer coefficient). *Trans. JSME*, **63**, 4013–4020.

Ullmann, A., and Brauner, N. (2007). The prediction of flow pattern maps in mini channels. *Multiphase Sci. Technol.*, **19**, 49–73.

Unal, H. C. (1975). Determination of the initial point of net vapor generation in flow boiling systems. *Int. J. Heat Mass Transfer*, **18**, 1095–1099.

Unal, H. C. (1976). Maximum bubble diameter, maximum bubble-growth time and bubble-growth rate during the subcooled nucleate flow boiling of water up to 17.7 MN/m^2. *Int. J. Heat Mass Transfer*, **19**, 643–649.

Unal, H. C. (1978). Determination of void fraction, incipient point of boiling, and initial point of net vapor generation in sodium-heated helically-coiled steam generator tubes. *J. Heat Transfer*, **100**, 268–274.

Unal, H. C. (1986). Prediction of nucleate boiling heat transfer coefficient for binary mixtures. *Int. J. Heat Mass Transfer*, **29**, 637–640.

Unal, H. C., van Gasselt, M. L., and van 't Veerlat, P. M. (1981). Dryout and two-phase flow pressure drop in sodium heated helically coiled steam generator tubes at elevated pressures. *Int. J. Heat Mass Transfer*, **24**, 285–298.

Ungar, K. E., and Cornwell, J. D. (1992). Two-phase pressure drop of ammonia in small diameter horizontal tubes. Paper presented at AIAA 17th Aerospace Ground Testing Conf., Nashville, TN, July 6–8, 1992.

Usui, K., Aoki, S., and Inouie, A. (1980). Flow behavior and pressure drop of two-phase flow through C-shaped bend in vertical plane: (I) Upflow. *J. Nucl. Sci. Technol.*, **17**, 875–887.

Usui, K., Aoki, S., and Inouie, A. (1981). Flow behavior and pressure drop of two-phase flow through C-shaped bend in vertical plane: (II) Downward flow. *J. Nucl. Sci. Technol.*, **18**, 179–190.

Vachon, R. I., Nix, G. H., and Tanger, G. E. (1967). Evaluation of constants for the Rohsenow pool boiling correlation. *XX National Heat Transfer Conference*, ASME Paper 67-HT-33.

Vafai, K., and Sözen, M. (1990). A comparative analysis of multiphase transport in porous media. *Annu. Rev. Heat Transfer*, **3**, 145–162.

Vakili-Farahani, F., Agostini, B., and Thome, J. R. (2012). Experimental study on flow boiling heat transfer of multiport tubes with R245fa and R1234ze(E). *ECI 8th International Conference on Boiling and Condensation Heat Transfer*.

Vakili-Farahani, F., Agostini, B., and Thome, J. R. (2013). Experimental study on boiling heat transfer on flow boiling heat transfer of multiport tubes with R236fa and R123ze(E). *Int. J. Refrigeration*, **36**, 335–352.

van Baten, J. M., and Krishna, R. (2004). CFD Simulations of mass transfer from Taylor bubbles rising in circular capillaries. *Chem. Eng. Sci.*, **59**, 2535–2545.

van Baten, J. M., and Krishna, R. (2005). CFD Simulations of wall mass transfer for Taylor flow in circular capillaries. *Chem. Eng. Sci.*, **60**, 1117–1126.

Van Driest, E. R. (1956). On turbulent flow near a wall. *J. Aeronautical Sci.*, **23**, 1007–1011.

Van Stralen, S. J. D. (1956). Heat transfer to boiling binary liquid mixtures at atmospheric and subatmospheric pressures. *Chem. Eng. Sci.*, **5**, 290–296.

Van Stralen, S. J. D. (1962). Heat transfer to boiling binary liquid mixtures. *Br. Chem. Eng.*, **7**, 90–97.

Van Stralen, S. J. D. (1966). The mechanism of nucleate boiling in pure liquids and binary mixtures, Parts I and II. *Int. J. Heat Mass Transfer*, **9**, 995–1046.

Van Stralen, S. J. D. (1967). The mechanism of nucleate boiling in pure liquids and binary mixtures, Parts III and IV. *Int. J. Heat Mass Transfer*, **10**, 1469–1498.

Van Stralen, S., and Cole, R. (1979). *Boiling Phenomena: Physicochemical and Engineering Fundamentals and Applications* (2 volumes), Hemisphere, Washington, DC.

Van Wijk, W. R., Vos, A. S., and van Stralen, S. J. D. (1956). Heat transfer to boiling binary liquid mixtures. *Chem. Eng. Sci.*, **5**, 68–80.

Van Wijngaarden, L. (1976). Hydrodynamic interaction between gas bubbles in liquid. *J. Fluid Mech.*, **77**, 27–44.

Vandervort, C. L., Bergles, A. E., and Jensen, M. K. (1992). Heat transfer mechanisms in very high heat flux subcooled boiling. *Fundamentals of Subcooled Flow Boiling*, ASME HTD Vol. **217**, pp. 1–9.

Vandervort, C. L., Bergles, A. E., and Jensen, M. K. (1994). An experimental study of critical heat flux in very high heat flux subcooled boiling. *Int. J. Heat Mass Transfer*, **37**, Suppl. 1, 161–173.

Vandu, C. O., Liu, H., and Krishna, R. (2005). Mass transfer from Taylor bubbles rising in single capillaries. *Chem. Eng. Sci.*, **60**, 6430–6437.

Vardhan, A., and Dunn, E. E. (1997). Heat transfer and pressure drop characteristics of R-22, R-134a, and R-407C in microchannel tubes. ACRC TR-133, University of Illinois at Urbana-Champaign.

Venkateswararao, P., Semiat, R., and Dukler, A. E. (1982). Flow pattern transition for gas–liquid flow in a vertical rod bundle. *Int. J. Multiphase Flow*, **8**, 509–524.

Vierow, K. M. (1990). Behavior of steam–water condensing in cocurrent vertical downflow. MS thesis, University of California, Berkeley.

Vijaykumar, R., and Dhir, V. K. (1992a). An experimental study of subcooled film boiling on a vertical surface – Hydrodynamic aspects. *J. Heat Transfer*, **114**, 161–168.

Vijaykumar, R., and Dhir, V. K. (1992b). An experimental study of subcooled film boiling on a vertical surface – Thermal aspects. *J. Heat Transfer*, **114**, 169–178.

Waelchli, S., and von Rohr, P. R. (2006). Two-phase flow characteristics in gas–liquid microreactors. *Int. J. Multiphase Flow*, **32**, 791–806.

Wagner, E., and Stephan, P. (2008). Experimental study of local temperature distribution and heat transfer mechanisms during nucleate boiling of binary mixtures. *5th European Thermal Science Conf.*, The Netherlands.

Wallis, G. B. (1961). Flooding velocities for air and water in vertical tubes. AAEW-R123, UKAEA, Harwell, England.

Wallis, G. B. (1969). *One-Dimensional Two-Phase Flow*, McGraw-Hill, New York.

Wallis, G. B. (1990). Inertial coupling in two-phase flow: Macroscopic properties of suspension in an inviscid fluid. In *Multiphase Science and Technology*, Hewitt, G. F., Delhaye, J. M., and Zuber, N., Eds., Hemisphere, New York, Vol. 5, Chapter 4.

Wallis, G. B., and Makkenchery, S. (1974). The hanging film phenomenon in vertical annular two-phase flow. *J. Heat Transfer*, **96**, 297–298.

Wallis, G. B., Steen, D. A., and Brenner, S. N. (1963). AEC Rep. NYO-10487, EURAEC 890.

Walsh, E. J., Muzychka, Y. S., Walsh, P. A., Egan, V., and Punch, J. (2009). Pressure drop in two phase slug/bubble flows in mini scale capillaries. *Int. J. Multiphase Flow*, **35**, 879–884.

Wambsganss, M. W., Jendrzejczyk, J. A., and France, D. M. (1991). Two-phase flow patterns and transitions in a small, horizontal, rectangular channel. *Int. J. Multiphase Flow*, **7**, 327–342.

Wambsganss, M. W., France, D. M., Jendrzejczyk, J. A., and Tran, T. N. (1993). Boiling heat transfer in a horizontal small-diameter tube. *J. Heat Transfer*, **115**, 963–972.

Wang, B.-X, and Peng, X. F. (1994). Experimental investigation of liquid forced-convection heat transfer through microchannels. *Int. J. Heat Mass Transfer*, **37**, 73–82.

Wang, C. C., Chiang, C. S., and Lu, D. C. (1997). Visual observation of two-phase flow patter of R-22, R-134a, and R-407C in a 6.5-mm smooth tube. *Exp. Thermal Fluid Sci.*, **15**, 395–405.

Wang, C.-C., Chen, I. Y., Yang, Y.-W., and Chang, Y.-J. (2003). Two-phase flow pattern in small diameter tubes with the presence of horizontal return bend. *Int. J. Heat Mass Transfer*, **46**, 2975–2981.

Wang, C.-C., Tseng, C.-Y., and Chen, I. Y. (2010). A new correlation and the review of two-phase flow pressure change across sudden expansion in small channels. *Int. J. Heat Mass Transfer*, **53**, 4287–4295.

Wang, C. H. (1992). Experimental and analytical study of the effects of wettability on nucleation site density during pool boiling. PhD thesis. University of California, Los Angeles.

Wang, C. H., and Dhir, V. K. (1993). Effect of surface wettability on active nucleation site density during pool boiling of water on a vertical surface. *J. Heat Transfer*, **115**, 659–669.

Wang, C. Y., and Cheng, P. (1997). Multiphase flow and heat transfer in porous media. *Adv. Heat Transfer*, **30**, 93–196.

Wang, S. K., Lee, S. J., Jones, O. C., Jr., and Lahey, R. T., Jr. (1987). 3-D turbulence structure and phase distribution measurements in bubbly two-phase flows. *Int. J. Multiphase Flow*, **13**, 327–343.

Wang, W. C., Ma, X. H., Wei, Z. D., and Yu, P. (1998). Two-phase flow patterns and transition characteristics for in-tube condensation with different surface inclinations. *Int. J. Heat Mass Transfer*, **41**, 4341–4349.

Wang, W. W. (1999). Condensation and single-phase heat transfer coefficient and flow regime visualization in microchannel tubes for HCFC-134a. PhD thesis, Ohio State University.

Wang, W.-W., Radcliff, T. D., and Christensen, R. N. (2002). A condensation heat transfer correlation for millimeter-scale tubing with flow regime transition. *Exp. Thermal Fluid Sci.*, **26**, 473–485.

Warnier, M. J. F., de Croon, M. H. J. M., Rebrob, E. V., and Schouten, J. C. (2010). Pressure drop of gas–liquid Taylor flow in round micro-capillaries for low to intermediate Reynolds numbers. *Microfluid Nanofluid*, **8**, 33–45.

Warnier, M. J. F., Rebrob, E. V., de Croon, M. H. J. M., Hessel, V., and Schouten, J. C. (2008). Gas hold-up and liquid film thickness in Taylor flow in rectangular microchannels. *Chem. Eng. J.*, **135S**, S153–S158.

Warrier, G. R., Dhir, V. K., and Momoda, L. A. (2002). Heat transfer and pressure drop in narrow rectangular channels. *Exp. Therm. Fluid Sci.*, **26**, 53–64.

Watanabe, T., Hirano, M., Tanabe, F., and Kamo, H. (1990). The effect of virtual mass on the numerical stability and efficiency of system calculations. *Nucl. Eng. Design*, **120**, 181–192.

Watel, B. (2003). Review of saturated flow boiling in small passages of compact heat exchangers. *Int. J. Thermal Sci.*, **42**, 107–140.

Weatherhead, R. J. (1963). Heat transfer, flow instability, and critical heat flux for water in a small tube at 200 psia. Rep. ANL-6715, Argonne National Laboratory, Argonne, IL.

Webb, D. R., and Sardesai, R. G. (1981). Verification of multicomponent mass transfer models for condensation inside a vertical tube. *Int. J. Multiphase Flow*, **7**, 507–520.

Webb, R. L., and Ermis, K. (2001). Effect of hydraulic diameter on condensation of R-134a in flat, extruded aluminum tubes. *Enhanced Heat Transfer*, **8**, 77–90.

Webb, R. L., and Zhang, M. (1998). Heat transfer and friction in small diameter channels. *Microscale Thermophys. Eng.*, **2**, 189–202.

Weisman, J. (1992). The current status of theoretically based approaches to the prediction of the critical heat flux in flow boiling. *Nucl. Technol.*, **99**, 1–121.

Weisman, J., and Ileslamlou, S. (1988). A phenomenological model for prediction of critical heat flux under highly subcooled conditions. *Fusion Technol.*, **13**, 654–659.

Weisman, J., and Kang, S. Y. (1981). Flow pattern transitions in vertical and upwardly inclined lines. *Int. J. Multiphase Flow*, **7**, 271–291.

Weisman, J., and Pei, B. S. (1983). Prediction of critical heat flux in flow boiling at low qualities. *Int. J. Heat Mass Transfer*, **26**, 1463–1477.

Weisman, J., Duncan, D., Gibson, J., and Crawford, T. (1979). Effects of fluid properties and pipe diameter on two-phase flow patterns in horizontal lines. *Int. J. Multiphase Flow*, **5**, 437–462.

Welsh, S. A., Ghiaasiaan, S. M., and Abdel-Khalik, S. I. (1999). Countercurrent gas–pseudoplastic liquid two-phase flow, *Ind. Eng. Chem. Res.*, **38**, 1083–1093.

Whalley, P. B. (1977). The calculation of dryout in a rod bundle. *Int. J. Multiphase Flow*, **3**, 501–515.

Whalley, P. B. (1980). Air–water two-phase flow in a helically coiled tube. *Int. J. Multiphase Flow*, **6**, 345–356.

Whalley, P. B. (1987). *Boiling, Condensation and Gas–Liquid Flow*, Oxford Scientific Publications, Clarendon Press, Oxford, UK.

Whalley, P. B. (1996). *Two-Phase Flow and Heat Transfer*, Oxford University Press, Oxford.

Whalley, P. B., Hutchinson, P., and Hewitt, G. F. (1974). The calculation of critical heat flux in forced convection boiling. *Proc. 5th Int. Heat Transfer Conf., Tokyo*, Paper B.6, pp. 290–2904.

White, F. M. (1999). *Fluid Mechanics*, 3rd Edn, McGraw-Hill, New York.

Wilke, C. R. (1950). A viscosity equation for gas mixtures, *J. Chem. Phys.*, **18**, 517–519.

Wilke, C. R., and Chang, P. (1954). Correlation of diffusion coefficients in dilute solutions, *AIChE J.*, **1**, 264–270.

Wilmarth, T., and Ishii, M. (1994). Two-phase flow regimes in narrow rectangular vertical and horizontal channels. *Int. J. Heat and Mass Transfer*, **37**, 1749–1758.

Wilmarth, T., and Ishii, M. (1997). Interfacial area concentration and void fraction of two-phase flow in narrow rectangular vertical channels. *J. Fluids Eng.*, **119**, 916–922.

Wilson, M. J., Newell, T. A., Chato, J. C., and Infante Ferreira, C. A. (2003). Refrigerant charge, pressure drop, and condensation heat transfer in flattened tubes. *Int. J. Refrigeration*, **26**, 442–451.

Winkler, J., Killion, J., Garimella, S., and Fronk, B. M. (2012). Void fraction for condensing refrigerant flow in small channels: Part I literature review. *Int. J. Refrigeration.*, **35**, 219–245.

Woldesemayat, M. A., and Ghajar, A. J. (2007). Comparison of void fraction correlations for different flow patterns in horizontal and upward inclined pipes. *Int. J. Multiphase Flow*, **33**, 347–370.

Wojtan, L., Revellin, R., and Thome, J. R. (2006). Investigation of critical heat flux in single, uniformly heated microchannels. *Exp. Thermal Fluid Sci.*, **30**, 765–774

Wojtan, L., Ursenbacher, T., and Thome, J. R. (2005a). Investigation of flow boiling in horizontal tubes: Part I – A new diabatic two-phase flow pattern map. *Int. J. Heat Mass Transfer*, **48**, 2955–2969.

Wojtan, L., Ursenbacher, T., and Thome, J. R. (2005b). Investigation of flow boiling in horizontal tubes: Part II – Development of a new heat transfer model for stratified-wavy, dryout, and mist flow regimes. *Int. J. Heat Mass Transfer*, **48**, 2970–2985.

Won, Y. S., and Mills, A. F. (1982). Correlation of the effects of viscosity and surface tension on gas absorption rates into freely falling turbulent liquid films. *Int. J. Heat Mass Transfer*, **25**, 223–229.

Wong, S., and Hochreiter, L. E. (1981). Analysis of the FLECHT SEASET unblocked bundle steam cooling and boiloff tests. NUREG/CR-1533, U.S. Nuclear Regulatory Commission, Washington, DC.

Wong, Y. L., Groeneveld, D.C., and Cheng, S. C. (1990). CHF prediction for horizontal tubes. *Int. J. Multiphase Flow*, **16**, 123–138.

Wongwises, S., and Polsongkram, M. (2006a). Evaporation heat transfer and pressure drop of HFC-134a in a helically coiled concentric tube-in-tube heat exchanger. *Int. J. Heat Mass Transfer*, **49**, 658–670.

Wongwises, S., and Polsongkram, M. (2006b). Condensation heat transfer and pressure drop of HFC-134a in a helically coiled concentric tube-in-tube heat exchanger. *Int. J. Heat Mass Transfer*, **49**, 4386–4398.

World Watch Institute (2006). *Vital Signs 2006–2007*, Norton, New York.

Wozniak, G. (1991). On the thermocapillary motion of droplets under reduced gravity. *J. Colloid Interface Sci.*, **141**, 245–254.

Wu, H. Y., and Cheng, P. (2003). Visualization and measurements of periodic boiling in silicon microchannels. *Int. J. Heat Mass Transfer*, **46**, 2603–2614.

Wu, H. Y., and Cheng, P. (2004). Boiling instability in parallel silicon microchannels at different heat flux. *Int. J. Heat Mass Transfer*, **47**, 3631–3641.

Wu, H. Y., and Cheng, P. (2005). Condensation flow patterns in silicon microchannels. *Int. J. Heat Mass Transfer*, **48**, 2186–2197.

Wu, P., and Little, W. A. (1983). Measurement of friction factors for the flow of gases in very fine channels used for microminiature Joule–Thompson refrigerators. *Cryogenics*, **23**, 273–277.

Wu, Q., Kim, S., Ishii, M., and Beus, S. G. (1998). One-group interfacial area transport in vertical bubbly flow. *Int. J. Heat Mass Transfer*, **31**, 1103–1112.

Wu, X. (1996). Hydrodynamic characteristics of countercurrent two-phase flows involving highly viscous liquids. M. S. thesis, Georgia Institute of Technology, Atlanta.

Wu, Z., Li, W., and Ye, S. (2011). Correlation for saturated critical heat flux in microchannels. *Int. J. Heat Mass Transfer*, **54**, 379–389.

Wu, Z., and Li, W. (2011). A new predictive tool for saturated critical heat flux in micro/mini-channels: Effect of the heated length-to-diameter ratio. *Int. J. Heat Mass Transfer*, **54**, 2880–2889.

Wu, Z., and Sundén, B. (2016). Heat transfer correlations for elongated bubbly flow in flow boiling micro/mini channels. *Heat Transfer Eng.*, **37**, pp. 985–993.

Wulff, W. (1990). Computational methods for multiphase flow. In *Multiphase Science and Technology*, Hewitt, G. F., Delhaye, J. M., and Zuber, N., Eds., Hemisphere, New York, **5**, 85–238.

Xie, J. C., Lin, H., Han, J. H., Dong, X. Q., and Hu, W. R. (1998). Experimental investigation on Marangoni drop migrations using drop shaft facility. *Int. J. Heat Mass Transfer*, **41**, 2077–2081.

Xie, T. (2004). Hydrodynamic characteristics of gas/liquid/fiber three-phase flows based on objective and minimally-intrusive pressure fluctuation measurements. PhD thesis, Georgia Institute of Technology, Atlanta.

Xin, R. C., Awwad, A., Dong, Z. F., and Ebadian, M. A. (1996). An investigation and comparative study of the pressure drop in air–water two-phase flow in vertical helicoidal pipes. *Int. J. Heat Mass Transfer*, **39**, 735–743.

Xin, R. C., Awwad, A., Dong, Z. F., and Ebadian, M. A. (1997). An experimental study of single-phase and two-phase flow pressure drop in annular helicoidal pipes. *Int. J. Heat Fluid Flow*, **18**, 482–488.

Xin, R. C., and Ebadian, M. A. (1997). The effect of Prandtl number on local and average convective heat transfer characteristics in helical pipes. *J. Heat Transfer*, **119**, 467–473.

Xiong, R., and Chung, J. N. (2007). An experimental study on the size effect of adiabatic gas–liquid two-phase flow patterns and void fraction in micro-channels. *Phys. Fluids*, **19**, 033301-1–033301-8.

Xu, J. L., Cheng, P., and Zhao, T. S. (1999). Gas–liquid two-phase flow regimes in rectangular channels with mini/micro gaps. *Int. J. Multiphase Flow*, **25**, 411–432.

Xu, J. L., Wong, T. N., and Huang, X. Y. (2006). Two-fluid modeling for low-pressure subcooled flow boiling. *Int. J. Heat Mass Transfer*, **49**, 377–386.

Xu, Y., and Fang, X. (2014). Correlation of void fraction for two-phase refrigerant flow in pipes. *Appl. Thermal Eng.*, **64**, 242–251.

Xu, Y., Fang, X., Su, X., Zhou, Z., and Chen, W. (2012). Evaluation of frictional pressure drop correlations for two-phase flow in pipes. *Nucl. Eng. Des.*, **253**, 86–97.

Yadigaroglu, G. (1981a). Regime transitions in boiling heat transfer. In *Thermohydraulics of Two-Phase Systems for Industrial Design and Nuclear Engineering*, Delhaye, M., Giot, M., and Riethermuller, L. M., Eds., Hemisphere, Washington, DC, pp. 353–403.

Yadigaroglu, G. (1981b). Two-phase flow instabilities and propagation phenomena. In *Thermohydraulics of Two-Phase Systems for Industrial Design and Nuclear Engineering*, Delhaye, M., Giot, M., and Riethermuller, L. M., Eds., Hemisphere, Washington, DC, pp. 307–351.

Yadigaroglu, G., and Lahey, R. T., Jr. (1976). On the various forms of the conservation equations in two-phase flow. *Int. J. Multiphase Flow*, **2**, 477–494.

Yagov, V. V. (2004). Critical heat flux prediction for pool boiling of binary mixtures. *Chem. Eng. Res. Design*, **82** (A4), 457–461.

Yamaguchi, K., and Yamazaki, Y. (1982). Characteristics of countercurrent gas–liquid two-phase flow in vertical tubes. *J. Nucl. Sci. Eng.*, **19**, 985–996.

Yan, Y.-Y., and Lin, T.-F. (1998). Evaporation heat transfer and pressure drop of refrigerant R-134a in a small pipe. *Int. J. Heat Mass Transfer*, **41**, 4183–4194.

Yan, Y.-Y., and Lin, T.-F. (1999). Condensation heat transfer and pressure drop of refrigerant R-134a in a small pipe. *Int. J. Heat Mass Transfer*, **42**, 697–708.

Yang, C., Li, D., and Masliyah, J. H. (1998). Modeling forced liquid convection in rectangular microchannels with electrokinetic effects. *Int. J. Heat Mass Transfer*, **41**, 4229–4249.

Yang, C.-Y., and Shieh, C.-C. (2001). Flow patterns of air–water and two-phase R-134a in small circular tubes. *Int. J. Multiphase Flow*, **27**, 1163–1177.

Yang, C.-Y., and Webb, R. L. (1996). Friction pressure drop of R-12 in small hydraulic diameter extruded aluminum tubes with and without micro-fins. *Int. J. Heat Mass Transfer*, **39**, 801–809.

Yang, G., Dong, Z. F., and Ebadian, M. A. (1993). The effect of torsion on convective heat transfer in a helicoidal pipe. *J. Heat Transfer*, **115**, 796–800.

Yang, S., Schlegel, J. P., Liu, Y., Paranjape, S., Hibiki, T., and Ishii, M. (2013). Experimental study of interfacial area transport in air–water two phase flow in a scaled 88 BWR rod bundle. *Int. J. Multiphase Flow*, **50**, 16–32.

Yang, S. R., and Kim, P. H. (1988). A mathematical model of the pool boiling nucleation site density in terms of the surface characteristics. *Int. J. Heat Mass Transfer*, **31**, 1127–1135.

Yang, Z. L., Palm, B., and Sehgal, B. R. (2002). Numerical simulation of bubbly two-phase flow in a narrow channel. *Int. J. Heat Mass Transfer*, **45**, 631–639.

Yao, G., and Ghiaasiaan, S. M. (1996a). Wall friction in annular-dispersed two-phase flow. *Nucl. Eng. Design*, **163**, 149–161.

Yao, G. F., and Ghiaasiaan, S. M. (1996b). Numerical modeling of condensing two-phase flows. *Num. Heat Transfer B: Fundam.*, **30**, 137–159.

Yao, G. F., Ghiaasiaan, S. M., and Eghbali, D. A. (1996). Semi-implicit modeling of condensation in the presence of noncondensables in the RELAP5/MOD3 computer code. *Nucl. Eng. Design*, **166**, 277–291.

Yashar, D. A., Wilson, M. J., Kopke, H. R., Graham, D. M., Chato, J. C., and Newell, T. A. (2001). An investigation of refrigerant void fraction in horizontal, microfin tubes. *HVAC&R Res.*, **7**, 67–82.

Yen, T.-H., Kasagi, N., and Suzuki, Y. (2003). Forced convective boiling heat transfer in microtubes at low mass and heat fluxes. *Int. J. Multiphase Flow*, **29**, 1771–1792.

Yin, S. T., and Abdelmessih, A. H. (1974). Prediction of incipient flow boiling from a uniformly heated surface. *AIChE Symp. Ser.*, **164**, 236–243.

Yoder, G. L., Morris, D. G., Mullins, C. B., Ott, L. J., and Reed, D. A. (1982). Dispersed flow film boiling in rod bundle geometry – Steady state heat transfer data and correlation comparisons. NUREC/CR-2435, US Nuclear Regulatory Commission, Washington, DC.

Yoshimoto, Y., Bessho, Y., Yamashita, J., Masuhara, Y., Yokomizo, O., Nishida, K., Isoda, K., and Yoshida, H. (1993). Critical power experiments of tight fuel rod lattice for light water reactors. *J. Nucl. Sci. Technol.*, **30**, 1120–1130.

You, S. M., Simon, T. W., Bar-Cohen, A., and Tong, W. (1990). Experimental investigation of nucleate boiling with a highly-wetting dielectric fluids (R-113). *Int. J. Heat Mass Transfer*, **33**, 105–117.

Young, N. D., Goldstein, J. S., and Block, M. J. (1959). The motion of bubbles in a vertical temperature gradient. *J. Fluid Mech.*, **6**, 350–356.

Yu, B., Han, J. T., Kang, H. J., Lin, C. X., Awwad, A., and Ebadian, M. A. (2003). Condensation heat transfer of HFC-134a flow inside helical pipes at different orientations. *Int. Commun. Heat Mass Transfer.* **30**, 745–754.

Yu, D., Warrington, R., Barron, R., and Ameal, T. (1995). An experimental and theoretical investigation of fluid flow and heat transfer in microtubes. *Proc. ASME/JSME Thermal Eng. Conf.*, **1**, 523–530.

Yu, W., France, D. M., Wambsganss, M. W., and Hull, J. R. (2002). Two-phase pressure drop, boiling heat transfer, and critical heat flux to water in a small-diameter horizontal tube. *Int. J. Multiphase Flow*, **28**, 927–941.

Yu, Z., Hemminger, O., and Fan, L. S. (2007). Experiment and lattice Boltzmann simulation of two-phase gas–liquid flows in microchannels. *Chem. Eng. Sci.*, **62**, 7172–7183.

Yue, J., Chen, G., Yuan, Q., Luo, L., and Gonthier, Y. (2007). Hydrodynamics and mass transfer characteristics in gas–liquid flow through a rectangular microchannel. *Chem. Eng. Sci.*, **62**, 2096–2108.

Yue, J., Luo, L., Gonthier, Y., Chen, G., and Yuan, Q. (2009). An experimental study of air–water Taylor flow and mass transfer inside square microchannels. *Chem. Eng. Sci*, **64**, 3697–3708.

Yun, B. J., Park, G. C., Julia, J. E., and Hibiki, T. (2008). Flow structure of subcooled boiling water flow in a subchannel of 3×3 rod bundles. *J. Nucl. Sci. Technol.*, **45**, 1–21.

Yun, R., Heo, H., and Kim, Y. (2006). Evaporative heat transfer and pressure drop of R410a in microchannels. *Int. J. Refrigeration*, **29**, 92–100.

Zaloha, P., Kristal, J., Jiricny, V., Völkel, N., Xuereb, C., and Aubin, J. (2012). Characteristics of liquid slugs in gas–liquid Taylor flow in microchannels. *Chem. Eng. Sci.*, **68**, 640–649.

Zeggel, W., Erbacher, F. J., Cheng, X., and Bethke, S. (1990). Critical heat flux in Freon-cooled tight 7-rod bundles (P/D = 1.15). ASME, New York, ASME HTD, Vol. 150, pp. 61–72.

Zeitoun, O., and Shoukri, M. (1996). Bubble behavior and mean diameter in subcooled flow boiling. *J. Heat Transfer*, **118**, 110–116.

Zeitoun, O., and Shoukri, M. (1997). Axial void fraction profile in low pressure subcooled flow boiling. *Int. J. Heat Mass Transfer*, **40**, 869–879.

Zeitoun, O., Shoukri, M., and Chatoorgoon, V. (1995). Interfacial heat transfer between steam bubbles and subcooled water in vertical upward flow. *J. Heat Transfer*, **117**, 402–407.

Zeng, L. Z., Klausner, J. F., and Mei, R. (1993a). A unified model for the prediction of bubble detachment diameter in boiling systems – I. Pooling boiling. *Int. J. Heat Mass Transfer*, **36**, 2261–2270.

Zeng, L. Z., Klausner, J. F., Bernhard, D. M., and Mei, R. (1993b). A unified model for the prediction of bubble detachment diameter in boiling systems – II. Flow boiling. *Int. J. Heat Mass Transfer*, **36**, 2271–2279.

Zhang, C. (1994). Numerical modeling using a quasi-three-dimensional procedure for large power plant condensers. *J. Heat Transfer*, **116**, 180–188.

Zhang, H.-Y., Li, J.-M., Liu, N., and Wang, B.-X. (2012). Experimental investigation of condensation heat transfer and pressure drop of R22, R410A and R407C in mini-tubes. *Int. J. Heat Mass Transfer*, **55**, 3522–3532.

Zhang, M. (1998). A new equivalent Reynolds number model for vapor shear-controlled condensation inside smooth and micro-fin tubes. PhD thesis, Pennsylvania State University, University Park, PA.

Zhang, L., Wang, E. N., Goodson, K. E., and Kenny, T. W. (2005). Phase change phenomena in silicon microchannels. *Int. J. Heat Mass Transfer*, **48**, 1572–1582.

Zhang, M., and Webb, R. L. (1998). Condensation heat transfer in small diameter tubes. *Proc. 11th Heat Transfer Conf.*, Seoul, South Korea, August 1998.

Zhang, M., and Webb, R. L. (2001). Correlation of two-phase friction for refrigerants in small-diameter tubes. *Exp. Therm. Fluid Sci.*, **25**, 131–139.

Zhang, P., and Fu, X. (2009). Two-phase flow characteristics of liquid nitrogen in vertically upward 0.5 and 1.0 mm micro-tubes: Visualization studies. *Cryogenics*, **49**, 565–575.

Zhang, W., Hibiki, T., and Mishima, K. (2005). Correlation for flow boiling heat transfer at low liquid Reynolds number in small diameter channels. *J. Heat Transfer*, **127**, 1214–1221.

Zhang, W., Hibiki, T., Mishima, K., and Mi, Y., 2006. Correlation of critical heat flux for flow boiling of water in mini-channels. *Int. J. Heat Mass Transfer*, **49**, 1058–1072.

Zhang, W., Hibiki, T., and Mishima, K., 2010. Correlations of two-phase frictional pressure drop and void fraction in mini-channel. *Int. J. Heat Mass Transfer*, **53**, 453–465.

Zhao, L., Guo, L., Bai, B., Hou, Y., and Zhang, X. (2003). Convective boiling heat transfer and two-phase flow characteristics inside a small horizontal helically coiled tubing once-through steam generator. *Int. J. Heat Mass Transfer*, **46**, 4779–4788.

Zhao, L., and Rezkallah, K. S. (1993). Gas–liquid flow patterns at microgravity conditions. *Int. J. Multiphase Flow*, **19**, 751–763.

Zhao, T. S., and Bi, Q. C. (2001). Co-current air–water two-phase flow patterns in vertical triangular microchannels. *Int. J. Multiphase Flow*, **27**, 765–782.

Zijl, W., Ramakers, F. J. M., and Van Stralen, S. J. D. (1979). Global numerical solutions of growth and departure of a vapor bubble at a horizontal superheated wall in a pure liquid and a binary mixture. *Int. J. Heat Mass Transfer*, **22**, 401–420.

Zivi, S. M. (1964). Estimation of steady-state steam void-fraction by means of the principle of minimum entropy production. *J. Heat Transfer*, **68**, 247–252.

Zuber, N. (1959). Hydrodynamic aspects of boiling heat transfer. USAEC Rep. AECU-4439.

Zuber, N. (1964). On the dispersed two-phase flow in laminar flow regime. *Chem. Eng. Sci.*, **19**, 897.

Zuber, N., and Findlay, J. (1965). Average volumetric concentration in two-phase flow systems. *J. Heat Transfer*, **87**, 453–468.

Zuber, N., Tribus, M., and Westwater, J. W. (1963). The hydrodynamic crisis in pool boiling of saturated and subcooled liquids. In *International Developments in Heat Transfer*, ASME, New York, Part II, pp. 230–236.

Zuber, N., Staub, F. W., Bijwaard, G., and Kroeger, P. G. (1967). Steady state and transient void fraction in two-phase flow systems. DEAP-5417, General Electric Company.

Zurcher, O., Thome, R. J., and Favrat, D. (1999). Evaporation of ammonia in a smooth horizontal tube: Heat transfer measurements and predictions. *J. Heat Transfer*, **121**, 89–101.

Index